Oceanic Micropalaeontology

Oceanic Micropalaeontology

Edited by

A. T. S. RAMSAY

Department of Geology and Oceanography
University College of Swansea, Wales

Volume 2

ACADEMIC PRESS · 1977
LONDON · NEW YORK · SAN FRANCISCO
A Subsidiary of Harcourt Brace Jovanovich, Publishers

ACADEMIC PRESS INC. (LONDON) LTD
24/28 Oval Road
London NW1

United States Edition published by
ACADEMIC PRESS INC.
111 Fifth Avenue
New York, New York 10003

Library of Congress Catalog Number: 76-1097
ISBN: 0-12-577302-1

PRINTED IN GREAT BRITAIN AT THE PITMAN PRESS BATH

Contributors to Volume 2

RICHARD E. CASEY: *Department of Geology, Rice University, Houston, Texas 77001, U.S.A.*

KURT R. GEITZENAUER: *Department of Geology and Oceanography, Herbert H. Lehman College of C.U.N.Y., Bronx, N.Y., U.S.A.*

WILLIAM W. HAY: *Division of Marine Geology and Geophysics, Rosenstiel School of Marine and Atmospheric Sciences, University of Miami, Florida 33149, U.S.A.*

SUSUMU HONJO: *Woods Hole Oceanographic Institution, Woods Hole, Massachusetts 02543, U.S.A.*

MAURICE L. JORGENS: *Department of Oceanography, University of Washington, Seattle, Washington 98195, U.S.A.*

E. MARTINI: *Geologisch Paläontologisches Institut der Johann Wolfgang Goethe Universität, Frankfurt-am-Main, Germany*

ANDREW MCINTYRE: *CLIMAP, Lamont–Doherty Geological Observatory of Columbia University, Palisades, New York 10964, U.S.A.*

EMILE A. PESSAGNO, JR: *The University of Texas at Dallas, Box 3065, Dallas, Texas 75230, U.S.A.*

A. T. S. RAMSAY: *Department of Geology and Oceanography, University College of Swansea, Singleton Park, Swansea SA2 8PP, Wales*

W. R. REIDEL: *Scripps Institution of Oceanography, University of California, San Diego, La Jolla, California 92093, U.S.A.*

MICHAEL B. ROCHE: *Jersey Central Power and Light Co., Morristown, New Jersey 07960, U.S.A.*

ANNIKA SANFILIPPO: *Scripps Institution of Oceanography, University of California, San Diego, La Jolla, California 92093, U.S.A.*

NAHUM SCHNEIDERMANN: *Gulf Research and Development Co., Houston Technical Services Center, P.O. Box 36506, Houston, Texas 77036, U.S.A.*

JAN VAN DONK: *University of Botswana, Lesotho and Swaziland, Private Bag 0022, Gaborone, Botswana*

GRAHAM L. WILLIAMS: *Atlantic Geoscience Center, Geological Survey of Canada, Bedford Institute of Oceanography, Dartmouth, Nova Scotia, Canada*

THOMAS R. WORSLEY: *Department of Oceanography, University of Washington, Seattle, Washington 98195, U.S.A.*

Preface

The intensive investigation of the ocean since the Second World War and the increasing availability of pelagic sediments from dredge hauls, gravity and piston cores, and more recently from the continuous or nearly continuous sequences recovered by the Deep Sea Drilling Project has opened up new areas for research in micropalaeontology. The micropalaeontological investigation of oceanic sediments has as a result of this stimulus developed into an expanding and exciting field of palaeontological research. Micropalaeontologists are now involved in investigating and testing both the accepted and latest concepts in marine geology. They are also utilising the information gleaned from the ocean to improve existing schemes or develop new schemes of biostratigraphical zonation, and to investigate the climatic and oceanographic history of the Earth since the Mesozoic.

It was not possible to cover all fields of oceanic micropalaeontology in this work. The papers that follow present a review of knowledge and recent developments in several aspects of oceanic micropalaeontology.

I am grateful to all the authors who contributed to this work for their cooperation and to Academic Press for their assistance and enthusiastic support.

May 1977

A. T. S. Ramsay

Contents

Contents of Volume 1

7: The Ecology and Distribution of Recent Radiolaria

RICHARD E. CASEY

Department of Geology,
Rice University,
Houston, Texas
77001, U.S.A.

Polycystine radiolarians are considered to be herbivores and/or suspended organic feeders and a significant component of the pelagic ecosystem. Their life spans are probably less than a month and their densities may reach as high as 82 000 individuals per cubic meter, with the highest densities usually below the upper 25 m and above a few hundred meters depth. They exhibit greatest diversity in the equatorial waters with the diversity decreasing and the tendency towards dominancy increasing poleward.

Daily fluctuations in polycystine distributions are currently poorly understood; however, seasonal fluctuations are at least in some cases easily correlated with seasonal fluctuations in the physical oceanography of the area, such as the seasonal invasions of water masses and currents. The polycystine radiolarians do exhibit vertical and horizontal zonations which in some species appear to be related to such physical factors as water mass zonations, convergence zones, circulation patterns, and pycnoclines, and such biological factors as their association with symbiotic zooxanthellae and probably predator–prey and more subtle associations.

Enough is known about many of the living polycystine radiolarians so that they may be placed in their correct biogeographic zones. Knowledge of their present day distribution and ecology has resulted in their use as palaeoecologic indicators in determining palaeotemperatures and recent investigations using radiolarians have resulted in attempts to evaluate palaeodiversities, evolutions, mass mortalities, extinctions and selective extinctions in the geologic record as related to palaeo-oceanographic conditions.

Outline

Introduction

Radiolaria are marine planktonic protozoans. The radiolarians herein considered are the polycystine radiolarians which possess skeletons of almost pure opaline silica, and comprise the suborders Spumellaria and Nassellaria (Ehrenberg, 1838). The polycystines are commonly found in marine sediments, therefore they are used in biostratigraphic and palaeoecologic studies, and because of this their ecology and distribution is of great interest to marine micropalaeontologists. Some representative polycystine radiolarians may be seen on Plate 1.

Life spans of polycystine radiolarians

Huth (1913) suggested a life span of several months for radiolarians and Casey *et al.* (1971) determined that the average life span for two species of polycystins off the southern California coast to be about one month (37 days). The measurements by Casey *et al.* (1971) were made by determining the average densities of species in the water column and the rates of deposition of these species in a varved anaerobic basin off southern California: because of the assumptions made in this study the authors believed that their life spans of about one month may well be too long.

Probable niches for polycystine radiolarians

Radiolarians are considered to be herbivores and/or suspended organic feeders (Casey and Sloan, 1971). There is little direct evidence for this type of nutrition, however tests of phytoplankton are commonly found within the tests of living polycystins and their greatest densities (of polycystins) are found in, or not far below the photic zone. It is interesting that the greatest densities of living forms are not found in the pigment zone but usually around the base of the euphotic zone (Beers and Stewart, 1971). This suggests that many are suspended organic feeders. It is also intriguing that their density peak with depth is similar to that of the decomposer bacteria (Russell-Hunter, 1970) which might also suggest a suspended organic feeding existence or perhaps the use of these bacteria as a food source. However many of the species of polycystins are believed to be good nannoherbivores living off phytoplankton, indirect evidence of this also comes from fossil evidence where the extinction of many "shallow water" polycystins coincides with extinctions in phytoplankton groups, while many "deep water" forms (considered to be suspended organic feeders) do not become extinct during these phytoplankton extinctions (Casey and Sloan, 1971).

Some polycystins (especially members of the families possessing symbiotic zooxanthellae within their protoplasm) probably gain much or all of their nitrition from these symbionts. In fact Khmeleva (1967) found that radiolarians with symbiotic zooxanthellae in the surface waters of the Gulf of Aden showed a carbon production by zooxanthellae several times that of the surrounding free living phytoplankton.

Lipps and Valentine (1970) made some interesting observations on the niche of planktonic foraminifera in the pelagic community, these observations may well be appropriate for the niche of many of the radiolarians. They consider the planktonic foraminifera to be nannoherbivores or microherbivores in the open ocean. They state that the phytoplankton of coastal waters are usually larger than those of the open ocean and are consumed many times by metazoans, whereas in the open ocean the need for intermediaries such as radiolarians or foraminiferans may be necessary. This may also account for the reason why there are no good records of planktonic foraminifera (Lipps and Valentine, 1970) or radiolaria (Casey, 1971a) which are truly endemic to coastal waters.

The radiolarian fauna appears to be an important part of the planktonic ecosystem. Using the data of Beers and Stewart (1967 and 1971) the author finds that the polycystins in the pigment layer may account for the consumption of a few percent of the total daily phytoplankton carbon production. Using this same data it appears that planktonic foraminifera may consume about twice as much as the radiolarians do. Also, the majority of radiolarians live below the pigment layer and may be important to the plankton community as herbivores on bacteria or as scavengers on suspended and/or dropping organic matter, and in converting this to a usable size for larger plankton and nekton. Figure 1 shows the probable niches in the ecosystem for many of the Radiolaria.

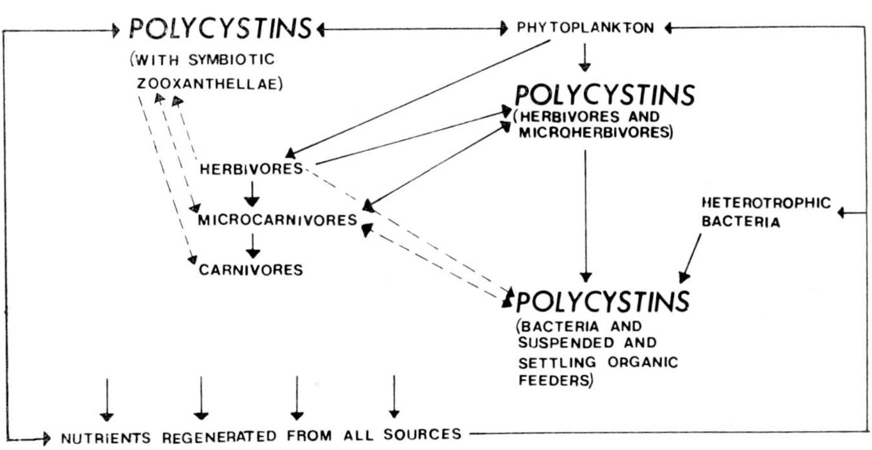

FIG. 1. Probable niches of polycystine radiolarians in the pelagic ecosystem. Solid lines represent probable direct interactions and broken lines represent probable partial or indirect interactions.

Diversity

Many authors have noted that the greatest radiolarian diversities are found in the equatorial regions of the world oceans. Petrushevskaya (1971b) recorded not less than one hundred frequently occurring species in tropical sediments and encountered several hundred rarer species. In Antarctic sediments she found a total of about 100 species of which 32 occurred frequently, their shells constituting over 69% and often up to 99% of the total number of shells in the sediment samples. She also noted (as has been mentioned by many authors) that cold water species generally have larger and more massive shells than warm water species. In the Pacific, Renz (1972), found over 110 radiolarian species in an area 11° North of the equator. She also recorded a decrease in diversity North and South of the equator with approximately 75 species occurring at 25° South and 80 species at 22° North. Renz (1972) also found a greater diversity in the plankton than in sediment and like Petrushevskaya (1966, 1971a) suggested that the selective dissolution of radiolarian tests was responsible.

The index of radiolarian species diversity for the Pacific illustrated on Fig. 2 indicates greatest diversity in the equatorial region, decreased diversity in the oligotrophic gyres and a slight increase in diversity in the high latitude eutrophic regions.

Casey and Sloan (1971) suggested that radiolarian diversities are related to (1) a reasonable amount of geological time for diversity to develop (2) a physical environment which is climatically and structurally stable so that many narrow, straightline community niches may develop and increase the diversity within each biogeographic zone, (3) a constant food source, which can be either eutrophic or oligotrophic but which contains a variety of potential nutriments (e.g. a diversity of phytoplankton types for herbivores), and (4) a physically "unstructured" environment which is homogeneous, at least on a small scale, so that evolutionary pressures such as competition will result in the necessity for a very efficient and highly structured biological system. Highly structured physical environments may result in large "isolated" habitats within which diversity arises due to competitive exclusion.

Density

The greatest densities of polycystins appear to occur in equatorial waters (Haeckel, 1887), Popofsky (1908, 1912, 1913), Petrushevskaya (1966, 1971a), Beers and Stewart (1971) (Renz, 1972). In these areas radiolarian densities of over 10 000 per cubic meter of sea water are commonly recorded with 82 000 per cubic meter being found at one station within the euphotic zone by Beers and Stewart (1971).

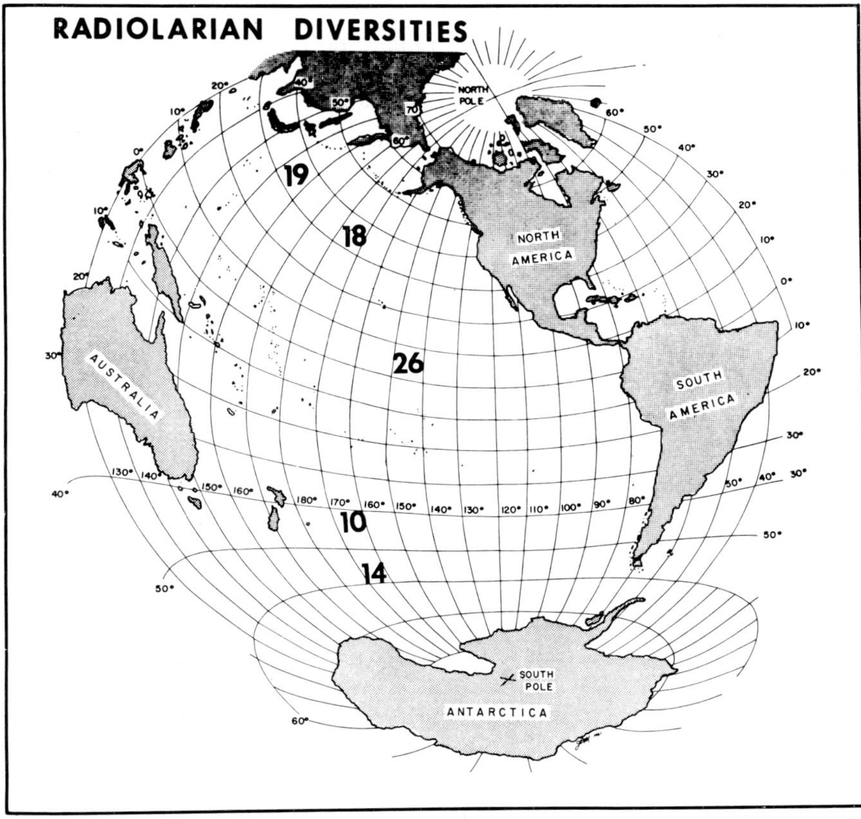

FIG. 2. Radiolarian diversities from Holocene Pacific sediments. The numbers (diversity index) represent the average number of species in one field of vision on a strewn slide. Eight fields of vision were averaged for each locality and at least 30 individuals were present in each field of vision.

Work by Beers and Stewart (1967, 1969a, b, 1971) suggests that the next highest densities may be found in upwelling, "near shore" temperate waters. During their investigations off southern California they commonly obtained densities of over 10 000 per cubic meter with maxima of about 30 000.

Radiolarians occur in the Antarctic in tens or hundreds of specimens per cubic meter (Petrushevskaya, 1971a). The lowest radiolarian densities probably occur in the oligotrophic regions at the centres of the major ocean gyres. Although no published data exist for these regions, studies by Renz (1972), Petrushevskaya (1971a) and Beers and Stewart (1971) all indicate that radiolarian densities decrease away from the equatorial region towards these areas.

In the Pacific the greatest densities in the tropics occur at depths of 25–75 m from about 27° North to 25° South (Renz, 1972). Off southern California radiolarians are more abundant from the bottom of the pigment layer (about 50 m on average) to 100 m, than within the pigment layer (Beers and Stewart, 1971). Beers and Stewart (1969b) in their study of waters from San Diego to Isla Guadalupe found in general, that the greatest abundance of radiolarians occurred between the 80 and 100 m samples while foraminifera were generally most abundant from the surface to 30 m. There are almost no radiolarians in the upper 25 m of the Antarctic waters to the south of the Antarctic Convergence they begin to appear at the 30–50 m horizons and attain their greatest densities at depths of 200–400 m and decline in density at greater depths (Petrushevskaya, 1966).

The densities recorded above were obtained by means of filtering techniques. Casey (1966, 1971a) calculated density values off southern California from tows made with a 62 micron mesh net. The density values obtained for April 1963 (280/m^3 from 0–100 m, 190/m^3 from 100–500 m, and 33/m^3 meter from 500–1000 m) and August 1973 (74/m^3 from 25–50 m, 380/m^3 from 50–123 m, 250/m^3 from 123–194 m and 29/m^3 from 480–960 m) are much lower than those of other workers. An examination of some of Beers and Stewarts' samples showed that the majority of radiolaria on their filters are the collar structures of nasellarians such as *Peridium spinipes* Haeckel this is by far the most abundant polycystine in the author's tows.

Moderate densities of radiolarians are recorded at depths below the optimum depth for the group by Casey (1966, 1971a) and Petrushevskya (1971a). Petrushevskaya reports densities which range from less than 50 to over 250 tests per cubic meter at depths of up to 5000 m. Her data for the vertical density of radiolarians along 154° West and for the vertical distribution of living and discarded tests at her equatorial station 5117 are shown on Fig. 3.

Petrushevskaya's (1971b) data show that the Nassellaria predominate over the Spumellaria in the water column, and that nassellarians constitute from 82–87% of the radiolarians sampled. Within the Nassellaria members of the family Cyrtoidea (suborder Cyrtellaria) predominate. Members of the suborder Larcoidea constitute the most numerous forms within the Spumellaria.

Distribution of polycystine radiolarians

The first notable work on the distribution of living polycystins was that of Haeckel (1887), in which he stated that radiolarians occur in all seas, at all climatic zones, and from the surface to great depths. The following is an attempt to review, comment on and add to the knowledge on radiolarian distribution by dealing with this distribution in the following terms: daily

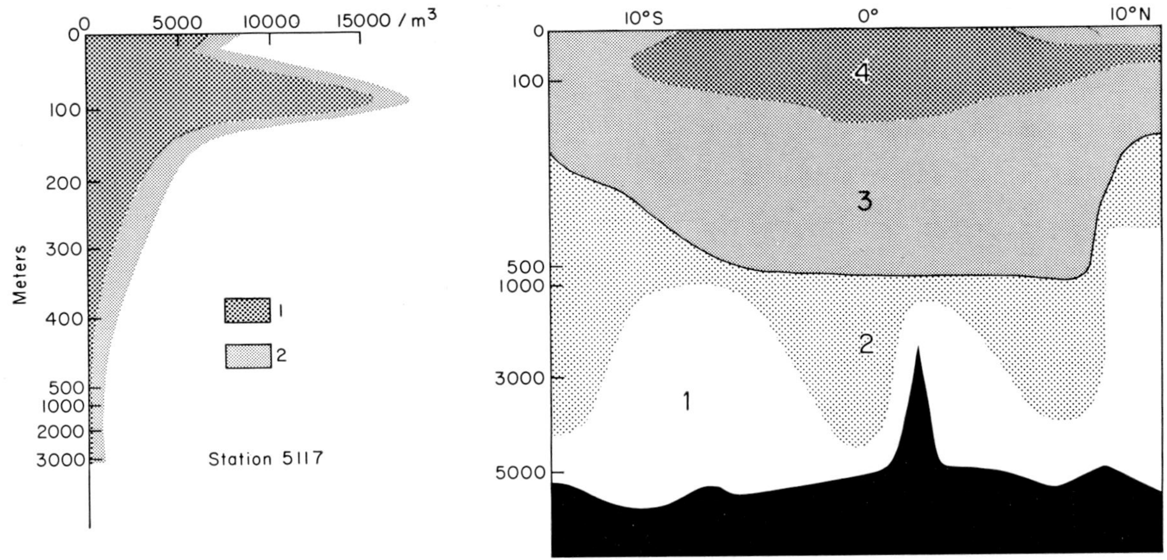

FIG. 3. (Right) Distribution and density (no./m^3 of water filtered) of living Spumellaria and Nassellaria in the plankton along 154° west in the Pacific. 1, <250; 2, 250–1250; 3, 1250–6000; 4, 6000–30 000. (From Petrushevskaya, 1971a.)
(Left) The quantity of living radiolarians (1) and discarded skeletons (2) in the water column from station 5117 at the equator in the Pacific (From Petrushevskaya, 1971a.)

fluctuations in distribution, seasonal distribution vertical and horizontal distributions. A review of polycystine distribution may be found in Casey (1971a), and much of the material presented in that publication is herein repeated and expanded upon.

DAILY FLUCTUATIONS IN DISTRIBUTION

There are few records of daily fluctuations in the distributions of radiolarians. Wolfenden (1905) records no vertical response of certain radiolarians to rain, although acantharians supposedly descend from the surface during rains or storms these radiolarians may possess a hydrostatic adaptation not found in polycystins. Haeckel (1887) states that, within his Pelagic Faunal Zone radiolarians swim at the surface but may descend to a depth of 20 to 30 fathoms.

Casey (1966, 1971a, b) noted a shallow daily fluctuation in polycystins off the southern California coast. During the late afternoon radiolarian density in the surface waters increased. This at first appeared to be a vertical migration; however, the timing of this "vertical migration" corresponded with afternoon onshore winds common to the area which probably caused a stirring of the mixed layer bringing radiolarians to the shallow waters from a deeper preferred depth. This shallow mixing was confirmed by bathythermograph records taken simultaneously with the plankton collections. Many polycystins do possess a locomotory sarcoflagellum, and although the evidence of active vertical migration of polycystins is at present inconclusive it is quite possible.

SEASONAL DISTRIBUTION

Cleve (1900) discussed the seasonal distribution of Atlantic plankton (including some polycystins) and concluded that some forms are introduced into regions seasonally during fluctuations of current systems. Even water masses which dive below water masses of lower density still retain their endemic forms.

Popofsky (1908) interpreted the appearance of some radiolarians in his Antarctic samples during the summer months in terms of seasonal fluctuations which he attributed to the southward displacement of warm currents.

Mielck (1913) noted that in the North Sea polycystins which were ordinarily restricted to depths greater than 100 m in the spring and summer, rise above 100 m in the autumn and winter. He considered these distributions to be related to the influx of waters from surrounding areas.

Schewiakoff (1926) noted that polycystins exhibited the same seasonal vertical pattern in the Mediterranean as had been observed in the North Sea by Mielck (1913). Perhaps both Mielck and Schewiakoff's results may be explained by the passive migration of polycystins due to the stirring of the mixed layer.

Bernstein (1934) found that some species were introduced seasonally into the Kara Sea from the Barents Sea with the inflow of Atlantic Water.

Casey (1966, 1971a, b) recorded distinct seasonal changes off southern California. These changes were apparently related to changes in the local circulation pattern which resulted in the introduction of radiolarians from the surrounding water masses during different seasons of the year. The seasonal fluctuations were concerned mainly with radiolarian faunas of the upper 200 m, the appearance however of radiolarians from depths of up to 400 m at the surface during one winter fluctuation was probably associated with local upwelling. Figure 4 shows the different radiolarian faunas of Casey (1966) and the caption to that figure explains their seasonally distributions in the waters off southern California.

It is apparent from the above discussion that seasonal changes in radiolarian populations are marked in regions where the oceanographic conditions undergo seasonal fluctuations. The studies of Mielck (1913), Schewiakoff (1926) and Casey (1966, 1971a) record the probably passive introduction of radiolarians to the surface via winter upwelling. Horizontal movements probably have greater palaeo-oceanographic significance. The effects of horizontal seasonal movements are recorded in the studies of Cleve (1900), Popofsky (1908), Mielck (1913), Schewiakoff (1926) and Casey (1966, 1971a).

VERTICAL DISTRIBUTION

Haeckel (1887) recognised three bathymetric faunal zones: Pelagic, Zonarial and Abyssal. The fauna of his Pelagic Zone extended from the surface to approximately 46 m and consisted primarily of radiolarians of the "order" Porulosida (suborder Spumellaria, plus the Acantharia) together with a few members of the "order" Osculosida (suborder Nassellaria plus the Phaeodaria). Haeckel noted that many of the radiolarians in the Pelagic Faunal Zone possess incomplete skeletons or none at all, that skeletons in general were thinner than those of abyssal forms, and that within the same family the dimensions of pelagic species seem to be smaller than those of related abyssal forms.

Haeckel's Zonarial Faunal Zone was dominated by members of the "order" Porulosida between 46 and 3656 m, at greater depths this group was gradually replaced by members of the "order" Osculosida. The Abyssal fauna was composed mainly of members of the Osculosida which floated above the sea floor.

Lo-Bianco (1903) distinguished four depth zones in the Mediterranean Sea which he correlated with the degree of illumination. These zones are as follows:

1. 0–50 m–Phaeoplankton (illuminated zone).

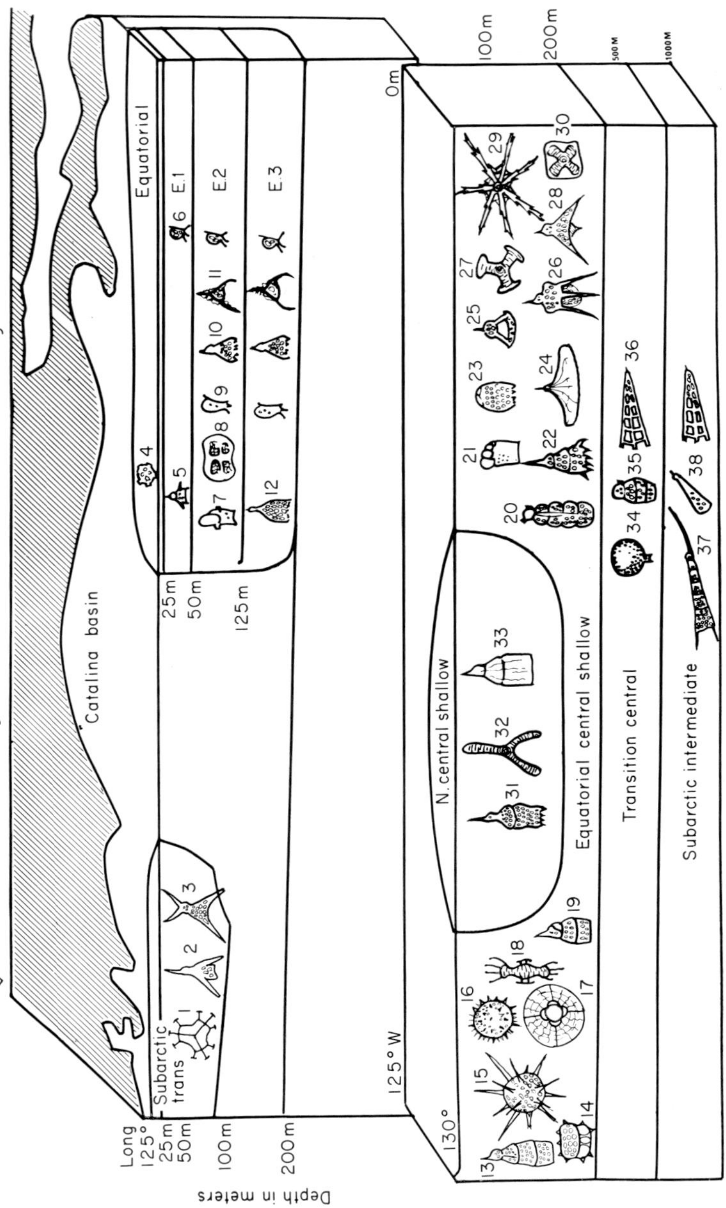
Lat
35°
30°
Catalina basin
Equatorial
E.1
E.2
E.3
25m
50m
125m
Subarctic trans
Long
125°
25m
50m
100m
200m
Depth in meters
125°W
0m
100m
200m
500M
1000M
130°
N. central shallow
Equatorial central shallow
Transition central
Subarctic intermediate

FIG. 4. Radiolarian seasonal fluctuation and zonation off southern California. The radiolarian faunas off southern California are depicted. In the shallow water (upper 200 m) the faunas fluctuate seasonally with invasions of Subarctic, Equatorial and North Central Shallow water faunas during different oceanographic seasons. Beneath these faunas from 200 to about 500 m the radiolarians may be described as a Transition Central Fauna which dives at the Subtropical Convergence and from about 500 to 1000 m a Subarctic Intermediate Fauna which dives at the Polar Convergence. The species illustrated are listed below and a few of these species are now placed in different zones by the author (such as *Spongotrochus glacialis* to the Subarctic-Transition Zone and *Spongosphaera streptacantha* to the Transitional Zone), (modified from Casey, 1966).

1. *Sethophormis rotula* Haeckel; 2. *Pterocorys hirundo* Haeckel; 3. *Pterocanium* sp.; 4. *Acrosphaera murrayana* (Haeckel); 5. *Lipmanella vichowii* (Haeckel); 6. *Peridium spinipes* Haeckel; 7. *Acrobotrissa cribrosa* Popofsky; 8. *Amphispyris costata—thorax* group Haeckel; 9. *Lithomelissa monoceras* Popofsky; 10. *Anthocyrtidium cineraria* Haeckel; 11. (?)*Clathrocorys murrayi* Haeckel; 12. *Eucecruphalus* sp. Haeckel; 13. *Eucyrtidium hexagonatum* Haeckel; 14. *Larcospira quadrangula* Haeckel; 15. *Spongotrochus glacialis* Popofsky; 16. *Spongotrochus brevispinus* Haeckel; 17. *Theophormis callipilium* Haeckel; 18. *Spongocore puella* Haeckel; 19. *Pterocorys zancleus* (Muller); 20. *Lithamphora furcaspiculata* Popofsky; 21. *Botryocyrtis* sp.; 22. *Lamprocyclas maritalis* Haeckel; 23. *Carpocanium* sp.; 24. *Litharachnium tentorium* Haeckel; 25. *Clathrocanium ornatum* Popofsky; 26. *Pterocanium praetextum* (Ehrenberg); 27. *Hymeniastrum profundum* Ehrenberg; 28. *Pterocanium trilobum* Haeckel; 29. *Spongosphaera streptacantha* Haeckel; 30. *Spongaster tetras* Ehrenberg; 31. *Calocyclas amicae* Haeckel; 32. *Euchitonia furcata-elegans* group; 33. *Eucyrtidium hertwigii* Haeckel; 34. *Spongopyle osculosa* Dreyer; 35. *Siphocampe erucosa* Haeckel; 36. *Peripyramis circumtexta* Haeckel; 37. *Cyrtopera laguncula* Haeckel; 38. *Corntuella profounda* Ehrenberg.

2. 50–400 m–Knephoplankton (partially illuminated zone).
3. 400–1500 m–Skotoplankton (barely illuminated zone).
4. 1500–5000 m–Nyctoplankton (non-illuminated zone).

Haecker (1907) found that some Sphaerellaria and Cyrtellaria are restricted to 400–5000 m. These deep water forms have massive skeletons, reduced radial spines and a laterally compressed form. Haecker (1908a) distinguished a Colloid Zone, (0–50 m), a Challengerid Zone (50–350 or 400 m), a Tuscarorid Zone (350 or 400–1000 or 1500 m) and a Pharyngellid Zone (1000 or 1500–5000 m). Although these are basically phaeodarian zones, the polycystins also show a similar zonation. Haecker (1908b) stated that the morphologies of some species are apparently related to environmental conditions. Surface-living forms are in general small and spherical and possess delicate skeletons, whereas, deep living forms are generally bilaterally symmetrical with larger and heavier skeletons.

Popofsky (1913) found that almost all nassellarians obtained from the Atlantic and southwest Indian Oceans occurred at depths from 0–400 m.

Reshetnjak (1955) established the following depth zonation for radiolarians of the Kuril-Kamchatka Deep:

1. Surface Radiolaria 0–50 m.
2. Subsurface forms, 50–200 m.
3. Moderately deep forms 200–1000 m.
4. Bathypelagic forms, 1000–2000 m.
5. A transitional fauna, from 50–1000 m.

Although these are mainly phaeodarian zones some polycystins also occur within her zones. Reshetnjak also designated forms which occur at one or several horizons as stenobathic and forms which are distributed through a wide depth range as eurybathic. Of the species she considered 45% were referred to an stenobathic forms.

Hulsemann (1963) found two distinct depth zones in the Arctic, a shallow zone of endemic forms, and a deeper zone of cosmopolitan forms.

Kling (1966) found Castanellid and Circoporid radiolarians in the eastern North Pacific to be zoned with depth, water masses and hydrographic conditions.

Petrushevskaya's (1966) studies of the Radiolaria have contributed a series of significant findings, those which pertain to depth are as follows:

1. In the central Pacific Ocean the species composition is nearly homogenous from the surface to a depth of 100–300 m. However, at greater depths in the Pacific a general impoverishment of the fauna is observed and new species appear which are not encountered at the surface.
2. Polycystins are found in all layers from 0–5000 m.

3. Almost no radiolarians occur in the surface layers of the Antarctic waters south of the Antarctic Convergence.
4. In the tropical Pacific the surface zone abounds with Nassellaria and Spumellaria with the greatest number of radiolarians and the greatest variety of species being found at the 150–500 m horizons. Down to 800 m no increase in the number of specimens and varieties of species was noticed.

Renz (1972) found in her studies that newly developed forms (forms without a complete skeleton) show their maximum abundance at the same depths as fully developed forms (usually from 25–75 m). Although her samples were, in general, taken from the upper 200 m of the water column she also found some evidence for the influence of depth on species distribution.

Casey (1966, 1971a, b) noticed a distinct vertical zonation of polycystins into three major depth zones from the surface to 1000 m. From the surface to 200 m there is a distinct zone the fauna of which is subject to seasonal variations. Another zone is recognised from 200 to 400 or 500 m, and a third major zone occurs between 400 or 500 m to 1000 m (the greatest depth sampled. During the summer the shallow zone (0–200 m) is subject to the incursion of different faunas into the area, and as a result can be subdivided into zones from 0–25, 25–50, 50–125 and 125–200 m. The faunal breaks at these depths were related to specific hydrological conditions. The 25 and 50 m breaks correlated with thermoclines, the 125 m break with a weak pycnocline and the 200 m break with the boundary between the Pacific Central Water and the overlying surface water. These physical and chemical barriers may also have acted as shear zones along which one "micro water mass" may be pushed over or under another.

From the above data it appears that a pronounced vertical zonation of polycystins occurs with thinner vertical zones in shallow water and thicker zones at depth. Also certain species are eurybathic, whereas others are stenobathic.

The results of all these studies indicate that vertical radiolarian faunal breaks occur at depths of approximately 50, 200, 400, and 1000 m in temperate and tropical waters. These breaks may be related to physical and chemical conditions in the following ways as has been suggested by Casey (1971a), and added to herein:

1. The 50 m and perhaps the 200 m faunal breaks may correlate with either thermoclines or pycnoclines as is seen in the waters off southern California (Casey, 1966, 1971a, b). These breaks may also be influenced by light penetration, particularly if some forms are truly dependent upon associations with symbiotic zooxanthellae. Species that are nanno or microhebivores may well be restricted to these waters due to preditor-prey relationships of a specific or general nature, or to some other biological association.

2. The 200 m division may correspond with the bottom of the mixed layer, the bottom of the surface water masses (water masses in the sense of Murometsey, 1963), or the beginning of the permanent thermocline.

3. The break at about 400 m may correspond with a water mass boundary such as Central Water on top of Intermediate Water (as occurs off southern California in Casey's work), or surface water on top of Central Water, as might occur in equatorial regions.

4. The breaks at greater depths than the above may also correlate with water mass boundaries, even though all waters approximate a single water type at depth. Therefore, deeper boundaries may be related in some instances to pressure, biological associations, the biologic history of the water, microconstituents of the waters, oxygen minimum zones, deep water circulation, or many other parameters, some of which may never be measured by our present sampling techniques.

Although the vertical and horizontal distributions are separated here (Casey, 1971a), these distributions are closely related and many forms which exist in shallow waters in high latitudes submerge to occupy deeper zones in the lower latitudes.

HORIZONTAL DISTRIBUTION

Haeckel (1887) made the following observations about the horizontal distribution of radiolarians: the richest development of forms and the greatest number of species occur in the tropics, the "frigid zones" possess few genera and species, from the tropics the abundance of species seems to diminish regularly towards the poles (more rapidly in the northern than in the southern hemisphere), the southern hemisphere seems to possess more species than the northern, and the Pacific Ocean radiolarian fauna seems to be quantitatively and qualitatively richer than that of any other ocean.

There appear to be no accounts of endemic, neritic polycystine species.

Popofsky (1907, 1908, 1912, 1913) distinguished warm water and cold water species of polycystins. He found several species to have a bipolar distribution, some to have different open-sea and coastal forms, and some that have morphologically distinct cold and warm water forms.

Riedel (1958) in comparing polycystins form the Antarctic sediments with sediments from other oceans concluded that: certain species are endemic to the Antarctic, some are bipolar, cold water types. certain species are cosmopolitan. He also states that the majority of species in shallow samples (to about 200 m) are restricted either to Antarctic waters or have a bipolar, cold water pattern of distribution, and that many of the apparently cosmopolitan species are found only at depths below 200 m.

Hulsemann (1963) found a shallow endemic and deeper cosmopolitan fauna in the Arctic.

Hays (1965) considered that the Antarctic Convergence formed an important break between radiolarian faunas in the Antarctic.

Nigrini (1967) distinguished warm, cool, and cold water assemblages and found a pronounced faunal break at the Antarctic Convergence, in her studies of the Indian and Atlantic Oceans.

Kling (1966) investigated castanellid and circoporid radiolarians from the eastern North Pacific and found that species distributions correlated with Sub-Arctic, Transition, Central and Equatorial-Central zones. He related these distributions to water masses of the region and noted that some forms dive with the Central and Intermediate Water Masses of the North Pacific Ocean.

Petrushevskaya (1966) considered that many species were related to more or less definite hydrological conditions. She stated that tropical and temperate radiolarian faunas do not transgress the Antarctic Convergence zone and that although the Antarctic Divergence may play a substantial role in the quantitative distribution of radiolarians it does not affect their distribution. She stated that the species characteristics of the Antarctic surface water layer are restricted by the Antarctic Convergence Zone.

Kruglikova (1969) divided the Northern Pacific into a boreal radiolarian fauna, restricted basically to sediments north of 40 degrees and a tropical fauna which is restricted to sediments south of this latitude. In the sediments of the boreal zone.

The prunoideans are represented by species related to the families Druppulidae and, to a lesser extent, the Ellipsidae. Members of the families Cyphinidae, Panartidae, Zygatidae, Cenodiscidae, Phacodiscidae and Coccodiscidae are lacking. The Larcoidea are represented primarily by forms assigned to the family Lithelidae. The greatest variety of Larcoidea is encountered in warm water areas, and forms assigned to the Spyroidae and Stephoidae sometimes account for as much as one third of the radiolarians in the sediments of the equatorial zone. Kruglikova (1969) stated that the panartids, zygartids, euchitonids and some groups of larcoids are exclusively tropical.

Between the tropical and boreal zones Kruglikova defines a wide mixed zone which contains tropical and boreal forms. The northern boundary of this zone is defined by the sparse occurrence of tropical species, its southern boundary is recognised by the reduction in boreal forms to below 5 to 7%.

Petrushevskaya (1971b) recognised similar restrictions in the distribution of radiolarian families in Antarctic waters south of the Antarctic Convergence to those described for the North Pacific by Kruglikova (1969). She also recorded the presence of a pronounced transitional zone between Antarctic and tropical water faunas in the Indian Ocean. Within this zone most Antarctic species are absent, and many tropical forms occur together with the characteristically abundant genera *Peromelissa* (?) and *Lithomelissa* and the endemic species *Androcycles gamphonycha*.

Nigrini (1970) applied a recurrent group analysis to radiolarian assemblages in the North Pacific. This approach enabled her to recognise six assemblages which reflect both the present day surface water circulation pattern and the distribution of the upper water masses. Nigrini recognised three major assemblages, with distribution breaks at approximately 40° and 20° North which she termed subarctic, transitional and tropical.

Casey (1966, 1971a, b) related the local distribution of radiolarians in the waters off southern California to radiolarian distribution in the North Pacific by comparing his plankton samples with sediment samples from the North Pacific.

Casey (1971a, b) divided the Pacific into radiolarian biogeographic faunal zones using sediment and plankton materials. This zonation (see Fig. 5) is a combination of vertical and horizontal distributional patterns and it shows a good correlation with the major water masses described for the area (Svedrup *et al.*, 1942; Murometzev, 1963). This suggests that water mass distribution is an important factor in radiolarian zonation. Casey (1971b) also mentioned a number of species which he considered as endemic to or nearly endemic to these zones, these species are shown on Plates 2–4. These species are not necessarily the dominant species in the Pacific, in fact they were chosen because they are relatively abundant in the waters off southern California where this study was initiated. Some species commonly used by other authors as indicative of various geographical regions in the Pacific, but which were not originally considered by Casey (1971b), are placed into Casey's zonation scheme and are shown on Plates 5 and 6. The inclusion of these species in Casey's zonation is based on published information and on the results of the author's own investigations.

A comparison between the biogeographical distributions of radiolarians proposed by Casey (1971a, b) and by other workers shows, that even though different workers have used different species the same general patterns of faunal distribution emerge. There appears to be an easily recognisable faunal break in the radiolarian assemblages at about the latitudes of the Subtropical and especially the Polar Convergences. Whereas some species are endemic to or characteristic of the Polar, Subpolar, Central and Equatorial shallow water faunal zones of Casey (1971b) other species such as *Lithelius nautiloides* Popofsky appear to have a bipolar distribution.

Many authors have mentioned that their warm water or tropical faunas do not enter the colder waters poleward of the Polar Convergence, but that there

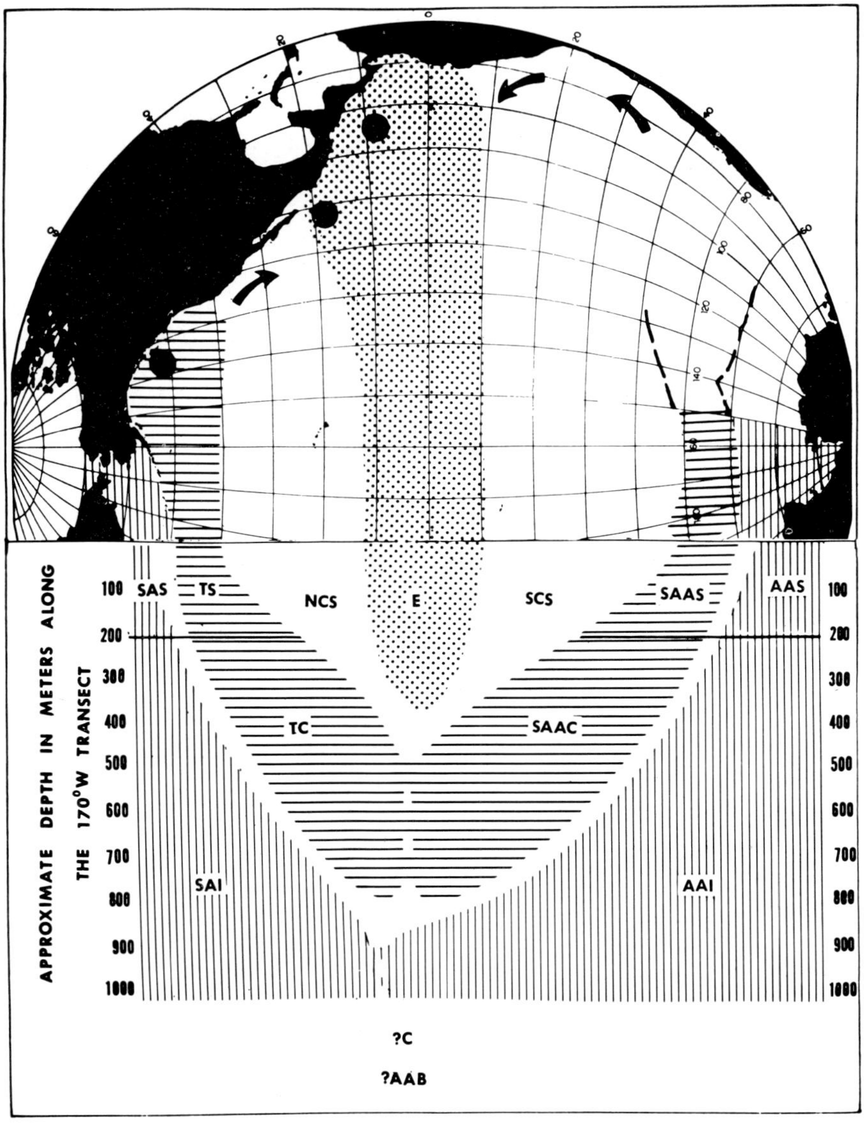

FIG. 5. Biogeographic zones for polycystine radiolarians in the Pacific Ocean (from Casey, 1971a). AAS, Antarctic Shallow Faunal Zone; AAI, Antarctic-Intermediate Faunal Zone; SAAS, Subantarctic Shallow Faunal Zone; SAAC, Subantarctic Central Faunal Zone; SCS, South Central Shallow Faunal Zone; E, Equatorial Faunal Zone; NCS, North Central Shallow Faunal Zone; TC, Transition-Central Faunal Zone; TS, Transition Shallow Faunal Zone; SAI, Subarctic Intermediate Faunal Zone; SAS, Subarctic Shallow Faunal Zone.

(Perhaps a C, Common Faunal Zone may exist below the Intermediate Zones, and an AAB, Antarctic Bottom Faunal Zone below the Common Zone. The large dots represent subzones which appear to be oceanographically stable areas which allow the development of slightly different faunas within the major zone.) SAS and AAS faunas are restricted to high latitudes whereas SAI and AAI faunas are tropical submergent faunas.

is a "leakage" of their polar faunas into their warm water faunas. This author believe that there are quite distinct, endemic, "nonleaking" members of polar faunas but they are masked many times by the "leaking" members which may for the most part be species which appear to be endemic to or mainly restricted to the water masses which dive at the Polar Convergence. *Cornutella profunda* Ehrenberg and *Cytropera laguncula* Haeckel appear to be species that show this distributional phenomenon and were distinguished as members of the Subarctic-Intermediate Fauna by Casey (1971b). There are probably many other forms which show this type of distribution and may include those polar species which leak into tropical waters.

Casey (1971b) suggested the possibility that some deep living forms may be members of a Deep Cosmopolitan Fauna. Riedel (1958) and Petrushevskaya (1966 and 1971b) have made similar suggestions. It is possible that further investigations of the depth distribution of the radiolarians may provide a means of palaeobathymetric determination.

Although the Pacific has been used here to demonstrate radiolarian biogeography areas of the Indian Ocean (Nigrini 1967; Petrushevskaya, 1966, 1969b, 1971b), Atlantic Ocean (Goll and Björklund, 1971; Petrushevskaya, 1969a; Nigrini, 1967), Arctic Ocean (Hulsemann, 1963; Tibbs, 1967) and Antarctic Ocean (Petrushevskaya, 1967) have also been investigated. Petrushevskaya (1971c) has compiled a large volume on radiolarian distribution of the world's oceans. From these studies it appears that the tropical zones of the Ocean have similar radiolarian faunas, particularly the Pacific and Indian Oceans. The radiolarian faunas of the Antarctic are similar around the Southern Ocean, and the diving deep water faunas appear to be cosmopolitan. The arctic radiolarian fauna apparently exhibits the greatest degree of endemism however endemism among the radiolarian species does not appear to be as pronounced as that exhibited by many other planktonic and especially benthonic groups.

Palaeoecology of polycystine radiolarians

PALAEOTEMPERATURES

There have been few attempts to use radiolarians for palaeotemperature analyses. Hays (1965) and Hays and Donahue (in press) drew some generalised palaeotemperature curves for Antarctic waters which, were based on the relative abundance of warm and cold water radiolarian species. Nigrini (1970) established palaeotemperature curves, based on radiolarians, for a North Pacific core. In this study she used a technique which was applied to the same core Kanaya and Koizumi (1966). These workers had used diatoms. Casey (1970) has used a technique based on ratios of radiolarian species as a means of determining palaeotemperatures. Essentially this method depends on establishing the relationship between the ratios of radiolarian species in Recent sediments and the mean summer surface temperature of the overlying waters. Once this relationship is established the ratios of the same radiolarians in deep sea or continental sections can be used to determine paleotemperatures. Ratios of species in Recent sediments are used instead of ratios derived from the plankton because these ratios are markedly different even for the same area (Casey *et al.*, 1971). Bandy *et al.* (1971) were the first workers to apply this technique. Casey (1972) correlated palaeotemperature curves which he established for the Antarctic, experimental Mohole and southern California, these results are shown on Fig. 6.

PALAEODIVERSITY AND SELECTIVE EXTINCTIONS

Casey and Sloan (1971) studied palaeodiversities in the time intervals of the end of the Mesozoic (a time of major diversity change and extinction) and at the end of the Miocene (a time of lesser diversity change and extinction). The end of the Miocene was studied in much greater detail and the following trends were noticed: (1) that diversity increased prior to extinction, (2) the severest declines in radiolarian diversities and the greatest numbers of extinctions occurred in tropical and temperate assemblages, whereas polar and cold temperate assemblages were less affected, (3) specialists (complex, ornamented, narrow endemic forms, which perhaps specialised on particular and/or specialised phytoplankters) appear to have been more affected than generalists (simple, unornamented, broad endemic forms, which perhaps eat a variety of phytoplankton types or suspended organic material), (4) low latitude shallow-water forms were more severely affected than polar, deep-water or tropical submergent forms.

These authors suggested that the increase in diversity results in a very efficient (greatest biomass for available nutrient supply) biological system with many narrow niches. This type of system would be most highly adapted for a specific environment but would be very unstable under conditions of either physical or biological change. Therefore, an environmental stress might result in differential extinctions, which eliminate the specialists (i.e. most of the species if enough geological time has lapsed prior to the stress) and allowing the generalists to survive if the change is not too severe. The suggestion was made that phytoplankton extinctions therefore may well result in differential radiolarian extinctions. For the periods considered this may have been accomplished by a decrease in the world heat budget and/or polar cooling (increased glaciation) disrupting the stable conditions that had existed for long periods and perhaps even causing eutrophism of the environment. Changes of this kind would cause great stress on the highly efficient, narrow niche, highly

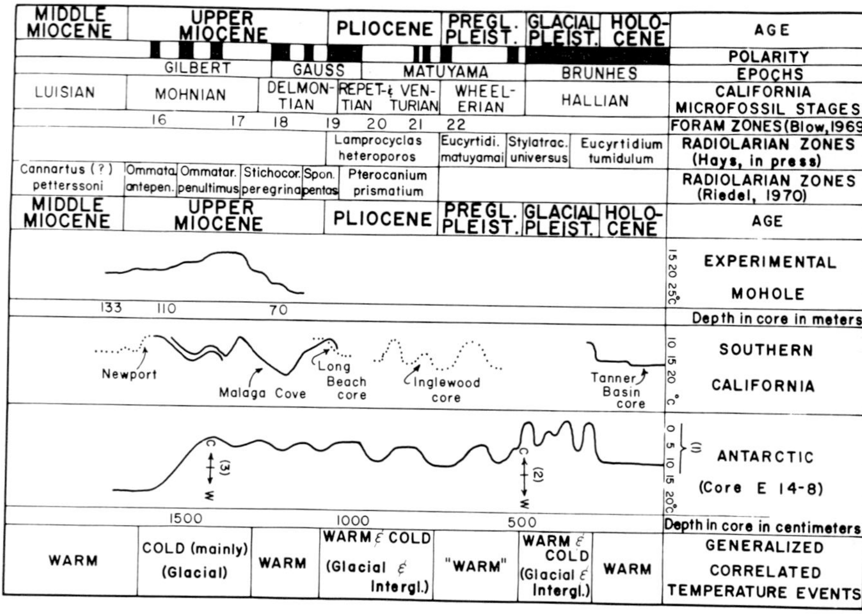

FIG. 6. Correlated palaeotemperature curves are correlated with each other biostratigraphically and temperature-wise. The palaeotemperature curves were derived in the manner described in the text, as an example the method for evaluating the Antarctic palaeotemperatures (Core E14-8) is given below.

Palaeotemperature variation, Core E14-8. There are three levels of temperature-related groups. (1) The first represents temperatures of less than 5°C and is based upon the ratio of the Antarctic *Spongotrochus glacialis* and *Lithelius nautiloides* to the Subantarctic *Lithamphora furcaspiculata* and *Theocalyptra bicornis* (these four species are illustrated on Plate 1). When the ratio of the Antarctic to Subantarctic group is about 9:1 or more, the surface temperature is 0°C or less than 0°C. When the ratio is about 1:1, the surface water temperature is about 5°C. (2) Added to the criteria used in (1) are fluctuations in the abundance of temperate radiolarians *Eucyrtidium acuminatum*, *Lamprocyclas maritalis*, and *Saturnulus planetes*; fluctuations in these temperate species suggest variations in surface water temperatures from about 5°C to about 15°C. Where they are very common the temperature approaches 15°C; where they are rare the temperatures are considered to be about 10°C; and where this group is very rare the temperatures are thought to be near the lower limit of about 5°C. (3) Subtropical temperatures of about 20°C are represented by the influx of the collosphaerid radiolarian group (from Casey, 1972). The polarity scale has been changed.

specialised ecosystems and allow the less "efficient" but more "opportunistic" generalists to not only survive but "out-compete" the specialists.

Casey *et al.* (1972) extended the work of Casey and Sloan (1971) in a detailed investigation of the palaeodiversities and selective extinctions of polycystins represented in an upper Miocene to Pliocene sequence in southern California. Figure 7 illustrates a comparison between a palaeotemperature curve for this core and palaeodiversities determined by the method described in the caption

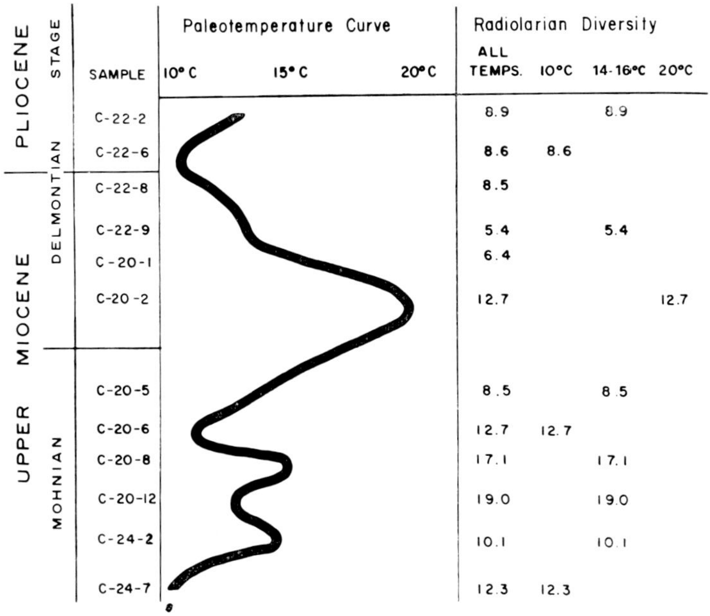

FIG. 7. Palaeotemperatures and radiolarian diversities within the Monetery Formation, Malaga Cove, Palos Verdes, California (from Casey, Price and Swift, 1972). The palaeotemperature curve was developed by the method mentioned in the text and the palaeodiversities were determined by the method as described in the caption to Fig. 2.

to Fig. 2. Temperatures of course appear to be related to diversity in that the diversity of radiolarians is higher in warm waters than it is in cold waters (Casey and Sloan, 1971); therefore in order to compare the densities for different times in the geological past it is best to compare assemblages which have accumulated at the same palaeotemperatures. The 14° to 16° diversities show a decline upsection with the lowest diversity at the Miocene-Pliocene boundary which is followed by an increase in diversity into the Pliocene. The decrease in radiolarian diversity prior to and at the Miocene-Pliocene boundary is not just a reflection of fewer species due to a cold interval but rather a real drop in radiolarian diversity due to the extinctions of species. Using the same line of reasoning it appears that the increase in radiolarian diversities in the Pliocene are due to the evolution of new radiolarian species and not just a temperature increase.

Casey *et al.* (1972) also grouped selected radiolarians, from the southern Californian section, into their respective biogeographical faunal zones (Fig. 8). These zones were established by Casey (1971a, b) from his work on living radiolarians from Holocene core and plankton samples. When these

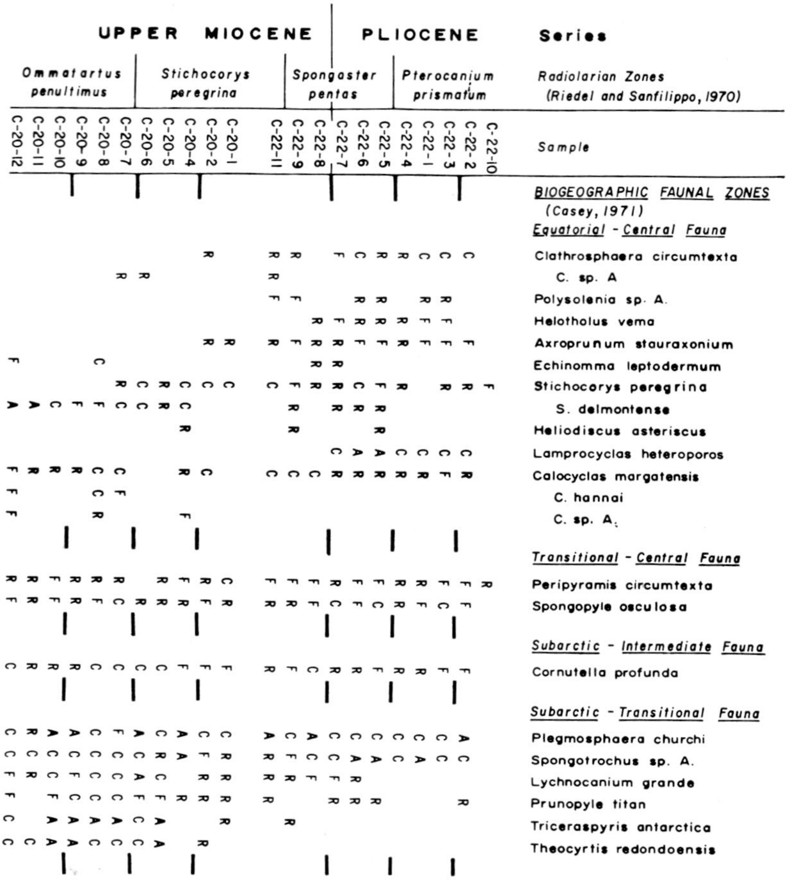

FIG. 8. Selected Neogene Radiolaria from Malaga Cove, Palos Verdes, California. The radiolarians are placed in their biogeographic faunal zones. A, abundant; C, common; F, few; R, rare; spaces, not observed. (From Casey *et al.*, 1972.)

radiolarians are placed in their respective or suggested faunal zones some interesting trends come to light as can be seen on Fig. 8. The Transition-Central and Subarctic-Intermediate faunas range throughout the section and do not appear to have been severely affected at the Miocene-Pliocene transition. Some members of the Subarctic Transition Fauna are strongly affected and some are not. However, members of the Equatorial-Central Fauna are severely affected during this time interval. Casey and Sloan (1971) suggested that the greatest numbers of extinctions occurred in tropical and temperate assemblages (the Equatorial-Central Fauna), while polar and cold temperate assemblages were less affected at the Miocene-Pliocene boundary It was also suggested by these authors that low latitude shallow water form were more severely affected than polar, deep water or tropical submergent forms. The findings of Casey *et al.* (1972) supports the suggestions of Casey and Sloan (1971). On Fig. 8 it can be seen that the Transition-Central and Subarctic-Intermediate faunas are essentially unaffected at the Miocene-Pliocene boundary; both of these faunas are tropical submergent faunas. The Subarctic-Transitional Fauna is affected and the species that drop out appear to be good cold water indicators of the Miocene-Pliocene boundary. However, the most severely affected species are members of the Equatorial-Central Fauna, only one may still be extant.

There are many possible reasons for these differential radiolarian extinctions and some of these have been discussed herein and by Casey and Sloan (1971). One direct or indirect cause may have been the temperature fluctuations before and across the Miocene-Pliocene transition. One temperature fluctuation, shown in Fig. 7, may have been as great as 10°C.

On a larger scale this author has noticed, in preliminary investigations that many of the indicator species used in Casey (1971b) which are believed to be endemic to surface water masses of the Pacific first evolve around the early or mid-Miocene. Since these forms may only become endemic to water masses once the water masses are present, these sparce data suggest that the major water masses of the Pacific originated or developed their present properties during this period.

MASS MORTALITY AND IT'S EFFECT ON THE FOSSIL RECORD

In their attempt to elucidate the life spans for radiolarians Casey *et al.* (1971) found an even more interesting phenomenon, namely in the turnover rates of radiolarians. In their study area in the southern Californian waters the turnover rates shown in Table 1 were found using the technique described in that paper. These rates range from less than one day to over 50 days. In order to understand the significance ot these turnover rates and, in order to differentiate between those which represent "true" life spans and those which represent mass mortalities, the turnover rates must be viewed in the light of the natural ecological and distributional information available for the living species given in Table 2.

Pterocorys zancleus (Muller) and *Eucyrtidium hexagonatum* Haeckel are considered to be members of the equatorial, central shallow fauna (Casey, 1971b). They are considered to be within their endemic reproductive area and the turnover rates are believed to represent "true" life spans for these radiolarians. The other nine species all have shorter "life spans" which are not considered to represent life spans in the sense of *T. zancleus* and *E. hexa-*

TABLE 1. Radiolarian turnover rates from southern California waters. (From Casey *et al.*, 1971.)

Species	Turnover rates (days)	
	core (230)	core (241)
Collosphaerids	4.2	4.0
Cornutella profunda Ehrenberg	0.10	0.7
Eucyrtidium hexagonatum Haeckel	26.0	37.5
Spongosphaera streptacantha (Haeckel)	7.5	7.1
Lithomelissa monoceras Popofsky	14.1	12.8
Peripyramis circumtexta Haeckel	8.7	19.4
Pterocorys hirundo Haeckel	1.17	7.6
Pterocorys zancleus (Muller)	33.8	51.4
Siphocampe erucosa Haeckel	1.4	2.0
Spongotrochus glacialis Popofsky	18.0	4.7
Spongopyle osculosa Dreyer	8.4	18.0

TABLE 2. Radiolarian turnover rates from southern California with the radiolarians placed in their respective biogeographical zones and the average determined for each zone. (From Casey *et al.*, 1971.)

Radiolarian biogeographical zone (Casey, 1971b)	Radiolarian species	Average rate	Average rate each rad. zone
Transition shallow	*Pterocorys hirundo*	4.4	
Transition shallow	*Spongosphaera streptacantha*	7.30	5.9
Subarctic-transition shallow	*Spongotrochus glacialis*	11.35	11.35
Equatorial shallow	Collosphaerids	4.1	
Equatorial shallow	*Lithomelissa monoceras*	13.3	8.7
Equatorial-central shallow	*Pterocorys zancleus*	42.6	
Equatorial-central shallow	*Eucyrtidium hexagonatum*	31.40	37.0
Transition-central	*Siphocampe erucosa*	1.7	
Transition-central	*Spongopyle osculosa*	13.2	9.7
Transition-central	*Peripyramis circumtexta*	14.1	
Subarctic intermediate	*Cornutella profunda*	0.44	0.44

gonatum but rather to represent high rates of mortality as these exotics enter water outside of their normal range or enter a less conclusive part of their range. *Spongotrochus glacialis* Popofsky is a member of the Subarctic-Transition Fauna, a shallow water fauna, which is introduced by the California Current. It is also believed to be subject to a fairly high rate of mortality as it is affected by change during the southward movement of the California Current. This species does not however, exhibit as high a mortality rate as the other eight non-endemic forms, this is probably due to its partial success as a tropical submergent form. All the other species show high mortality rates especially *Cornutella profunda* Ehrenberg which is a member of the Subarctic Intermediate Fauna and is probably endemic to the North Pacific Intermediate Water Mass. The average turnover rate for *C. profunda* is less than one day. In the waters over the Continental Borderland of southern California intermediate waters tend to be mixed with central waters from about 400–500 m. Below 500 m only intermediate water is found (Casey, 1971b). It is therefore not surprising that *C. profunda* is found only sparsely in the water column between 0 to 480 m. However, it is common in the sediments of the Santa Barbara Basin. This strongly suggests that this species may be invading the basin at or near sill depth and dies as it enters the anaerobic conditions of the basin.

Table 3 gives the relative abundances of radiolarians, used in this study, both in the plankton and in the sediments. As might be expected they do not correlate, and striking trends are evident in a comparison between the dominant species present in the plankton and in the sediments of the same area. The second and third most abundant species in the plankton, *Pterocorys zancleus* and *Eucyrtidium hexagonatum*, are only 5th and 6th in abundance in the sediment sample. This was explained by the high mortality rates of *Siphocampe erucosa* Haeckel, *Spongosphacra streptacantha* Haeckel, and *Spongotrochus glacialis*, and the possibility that *Spongopyle osculosa* Dreyer may also be within its endemic reproductive area. It was suggested by the authors that these anomalies are mainly due to the rates of mortality in a given area and that this mortality greatly complicates the oceanographic and pale-oceanographic interpretations that are to be derived from recent and fossil sediments.

Current research

Currently two major projects on radiolarian distribution and ecology are being carried out in the United States. Robert Goll of the Lamont-Doherty Geological Observatory of Columbia University and the author of this chapter are working on radiolarian distributions in the water column and sediments of the North Atlantic, and of the Gulf of Mexico and Caribbean respectively.

TABLE 3. The abundance rankings (1 most abundant and 11 least abundant) for radiolarians from southern California plankton and sediment samples. (From Casey *et al.*, 1971.)

Sediment abundance ranking (1 = most abundant)	Plankton abundance ranking (11 = most abundant)
1. *Siphocampe erucosa*	1. *Spongopyle osculosa*
2. *Spongosphaera streptacantha*	2. *Pterocorys zancleus*
3. *Spongopyle osculosa*	3. *Eucyrtidium hexagonatum*
4. *Spongotrochus glacialis*	4. *Siphocampe erucosa*
5. *Pterocorys zancleus*	5. *Spongotrochus glacialis*
6. *Eucyrtidium hexagonatum*	6. *Spongosphaera streptacantha*
7. *Cornutella profunda*	7. *Lithomelissa monoceras*
8. *Pterocorys hirundo*	8. *Pterocorys hirundo*
9. *Lithomelissa monoceras*	9. *Periphyramis circumtexta*
10. collosphaerids	10. Collosphaerids
11. *Peripyramis circumtexta*	11. *Cornutella profunda*

Although these are separate projects funded by the Oceanography Section of the National Science Foundation both parties are cooperating and using similar collecting techniques. The collecting techniques of the author consists of the use of Nansen closing nets, high speed plankton nets, Niskin bottles and pumping and deck filtration devices along with other types of water column and sediment sampling gear.

Preliminary results of the author's work shows that many radiolarians of the Gulf of Mexico and Caribbean are "endemic" to the water masses of that area and that many radiolarians suggested by the author to exhibit tropical submergence in the Pacific (Casey, 1971a, b) exhibit these same patterns in the Gulf of Mexico and Caribbean. Another interesting phenomenon is that a relict radiolarian fauna (in part) exists in the Gulf of Mexico and Caribbean containing living species that died out in the Pacific millions of years earlier. It has also been noted that radiolarians are preserved in the surface sediments of the Gulf of Mexico and Caribbean but are absent a few centimetres down core only to be found again when sediments as old or older than mid-Miocene are encountered.

Acknowledgements

Support of portions of this study came from the National Science Foundation under Grants GV-25749, GA-34145 and GB-8628, from the Foundation of California State University, Northridge and some research including taxonomic reversions and other additions were supported by the Oceanography Section, National Science Foundation, NSF Grant DES 74-21805 to which the author is most grateful.

Most of the scanning electron micrographs were taken by Dr. Ramil C. Wright of Beloit College, and typing assistance was provided by Mrs. Barbara Hawkins of Rice University.

The author acknowledges the invaluable assistance of the works of Yu. A. Orlov (1962) and W. R. Riedel (1967) in preparing sections dealing with some aspects of radiolarian distribution.

References

Bandy, O. L., Casey, R. E. and Wright, R. C. (1971). Late Neogene planktonic zonation, magnetic reversals, and radiometric dates, Antarctic to the Tropics *In* (Reid, J., ed.), *Antarctic Oceanology I, Antarctic Res. Ser.* **15,** Amer. Geoph. Un., Washington D.C., 1–26.

Beers, J. R. and Stewart, G. L. (1967). Micro-zooplankton in the euphotic zone at five localities across the California Current. *J. Fish. Res. Bd. Can.* **24,** 2053–2068.

Beers, J. R. and Stewart, G. L. (1969a). The vertical distribution of micro-zooplankton and some ecological observations. *J. cons. Int. Explor. Mer.* **33,** 30–44.

Beers, J. R. and Stewart, G. L. (1969b). Micro-zooplankton and its abundance relative to the larger zooplankton and other seston components. *Mar. Biol.* **4**(3), 182–189.

Beers, J. R. and Stewart, G. L. (1971). Micro-zooplankters in the plankton communities of the upper waters of the eastern tropical Pacific. *Deep-Sea Res.* **18**(9), 861–883.

Bernstein, T. (1934). Zooplankton des nordlichen Teiles des Kareschen Meeres. *Trudy Arktichesk. Inst., Leningrad* **9,** 3–58.

Casey, R. E. (1966). A seasonal study on the distribution of polycystine radiolarians from waters overlying the Catalina Basin, Southern California. Unpub. Ph.D. thesis, University of South California.

Casey, R. E. (1970). Paleotemperature analyses based on radiolarian ratios. *Geol. Soc. Am. Ann. Meeting*, 1970, *Abstracts* 736–738.

Casey, R. E. (1971a). Distribution of polycystine radiolarians in the oceans in relation to physical and chemical conditions. *In* (Funnell, B. M. and Riedel, W. R., eds), *The Micropalaeontology of Oceans* 151–159. Cambridge University Press, Cambridge.

Casey, R. E. (1971b). Radiolarians as indicators of past and present water masses. *In* (Funnell, B. M. and Riedel, W. R., eds), *The Micropalaeontology of Oceans*, 331–341. Cambridge University Press, Cambridge.

Casey, R. E. (1972). Neogene radiolarian biostratigraphy and paleotemperatures: southern California, the experimental Mohole, Antarctic core E 14–8. *Palaeogeography, Palaeoclimatology, Palaeoecology* **12,** 115–130. Elsevier, Amsterdam.

Casey, R. E., Partridge, T. M. and Sloan, J. R. (1971). Radiolarian life spans, mortality rates, and seasonality gained from Recent sediment and plankton samples. *In* (Farinacci, A., ed.), *Proc. II Plankt. Conf.* **2,** 159–165. Tecnoscienza, Roma.

Casey, R. E., Price, A. B. and Swift, C. A. (1972). Radiolarian definition and paleoecology of the late Miocene to early Pliocene in southern California. *In* (Stinemeyer, E. H. and Church, C. C., eds), *Proc. of the Pacific Coast Miocene Biostratigraphic Symposium*, 226–238. Pac. Sec. S.E.P.M., Bakersfield, California.

Casey, R. E. and Sloan, J. R. (1971). Possible causes of diversities and extinctions of radiolarians and other microplankton. *Geol. Soc. Am. Ann. Meeting*, 1971, *Abstracts* 763–765.

Cleve, P. T. (1900). The seasonal distribution of Atlantic plankton organisms. *Handl. Goteborgs Kgl. Vettenskoch Vitterh.-Samh.*, ser. 4, **3**, 1–369.

Ehrenberg, C. G. (1838). Über die Bildung der Kreidefelsen und des Kreidemergels durch unsichtbare Organismen. *Abh. Kgl. Akad. Wiss. Berlin Jahrg.* **1838**, 59–147.

Goll, R. M. and Björklund, K. R. (1971). Radiolaria in surface sediments of the North Atlantic Ocean. *Micropaleont.* **17**(4), 434–457.

Haeckel, E. (1887). Report on the Radiolaria collected by *H.M.S. Challenger* during the years 1873–76. *Rept. Voyage 'Challenger', Zool.* **18,** clxxxviii + 1803 pp.

Haecker, V. 1907. Altertümliche Spharellarien und Crytellarien aus grossen Meerestiefen. *Archiv. Protistenk*, **10,** 114–126.

Haecker, V. (1908a). Tiefsee-Radiolarien, Spezieller Teil, Lfg. 2. Die Tripyleen, Collodarien und Mikroradiolarien der Tiefsee. *Wissenschaftl. Ergebn. Deutschen Tiefsee Exped.* **14,** 337–476.

Haecker, V. (1908b). Tiefsee-Radiolarien. Allgemeiner Teil. Form und Formbildung bei den Radiolarien. *Wissenschaftl. Ergebn. Deutschen Tiefsee Exped.* **14,** 477–706.

Hays, J. D. (1965). Radiolaria and late Tertiary and Quaternary history of Antarctic seas. *Biol. Antartic Seas* (vol. 2) *Antarctic Research Ser.* **5,** 125–84.

Hays, J. D. and Donahue, J. G. (in press) Antarctic Quaternary climatic record and radiolarian and diatom extinctions. SCAR/IUGS Symposium.

Hulsemann, K. (1963). Radiolaria in plankton from the Arctic Drifting Station T-3, including the description of three new species. *Arctic Inst. Nth. Amer., Tech Paper* **13,** 1–52.

Huth, W. (1913). Zur Entwicklungsgeschichte der Thalassicollen. *Archiv Protistenk.* **30,** 1–124.

Kanaya, T. and Koizumi, I. (1966). Interpretation of diatom thanatocoenoses from the North Pacific applied to a study of core V20–130. *Tohoku Univ. Sci. Repts.* Sendai, Ser. 2 (geol.), **37,** 2, 89–130.

Khmeleva, N. N. (1967). Role of radiolarians in the estimation of the primary production in the Red Sea and the Gulf of Aden. *Trans. Doklady Akad. Nauk. U.S.S.R.* **172,** 1430–1433.

Kling, S. A. (1966). Castanellid and Circoporid Radiolarians: Systematics and zoogeography in the Eastern North Pacific. Unpub. Ph.D. thesis, University of California, San Diego.

Kruglikova, S. B. (1969). Radiolarians in the surface layer of the sediments of the northern half of the Pacific Ocean. *In* (Shirshov, P. P., ed.), *Microflora and Microfauna in the Recent Sediments of the Pacific Ocean*, 48–72. Nauka, Moskow (in Russian).

Lipps, J. H. and Valentine, J. W. (1970). The role of Foraminifera in the tropic structure of marine communities. *Lethaia* **3,** 279–286.

Lo-Bianco, S. (1903). Le pesche abissale eseguite da F. A. Krupp col Yacht *Puritan* nelle adiacenze di Capri ed in altre localita del Mediterraneo. *Mitt. Zool. Sta. Neapol.* **16**(1–2), 109–279.

Mielck, W. (1913). Résumé des observations sur le plankton des mers explorées par le *Conseil* pendant les années 1902–1908: Radiolaria. *Conseil Permanent Internat. pur l'Exploration de la Mer, Bull. Trimest.* **3,** 303–402.

Murometsey, A. M. (1963). *The principal hydrological features of the Pacific Ocean.* 417 pp. Israel Prog. for Sci. Transl.

Nigrini, C. (1967). Radiolaria in pelagic sediments from the Indian and Atlantic Oceans. *Bull. Scripps Instn. Oceanogr.* **11,** 1–106.

Nigrini, C. (1968). Radiolaria from eastern tropical Pacific sediments. *Micropaleont.* **14**(1), 51–63.

Nigrini, C. (1970). Radiolarian assemblages in the North Pacific and and their applications to a study of Quaternary sediments in core V 20–130. *In* (Hays, J. D., ed.), *Geol. Soc. Am. Mem.* **126,** 139–183.

Orlov, Yu. A. (1962). *Fundamentals of palaeontology, a manual for palaeontologists and geologists of the U.S.S.R., gen. pt. Protozoa.* 728 pp. Israel Prog. for Sci. Transl.

Petrushevskaya, M. G. (1966). Radiolyarii v planktone iv donnykh osadkakh. *In Geokhimiya Kremnesema,* 219–45. Nauka, Moscow.

Petrushevskaya, M. G. (1967). Nassellaria antarkticheskoe oblasti (Antarctic spumelline and nasseline radiolarians). *Issled. Fauny Morei* **4**(12) (*Rez. biol. Issled. sov. antarkt. Eksped.* 1955–58) 3, 5–186.

Petrushevskaya, M. G. (1969a). Distribution of radiolarian skeletons in sediments of the North Atlantic Ocean. *In Second all-Union seminar on radiolarians, Lvov, U.S.S.R.,* 123–132. Lvov University Press (in Russian).

Petrushevskaya, M. G. (1969b). Spumellarian and Nassellarian radiolarians in bottom sediments as indicators of hydrological conditions. *In Basic Problems of Micropaleontology and of the Accumulation of Organogenic Sediments in Oceans and Seas,* 127–150. Nauka, Moscow.

Petrushevskaya, M. G. (1971a). Spumellarian and nassellarian Radiolaria in the plankton and bottom sediments of the Central Pacific. *In* (Funnell, B. M. and Riedel, W. R., eds), *The Micropalaeontology of Oceans,* 309–317. Cambridge University Press, Cambridge.

Petrushevskaya, M. G. (1971b). Radiolaria in the plankton and Recent sediments from the Indian Ocean and Antarctic. *In* (Funnell, B. M. and Riedel, W. R., eds), *The Micropalaeontology of Oceans,* 319–329. Cambridge University Press, Cambridge.

Petrushevskaya, M. G. (1971c). Radiolarians of the Ocean, Reports on the Soviet Expeditions. *In Explorations of the Fauna of the Seas,* IX (XVII), 1–418. Academy of Sciences of the USSR, Zoological Institute (in Russian).

Popofsky, A. (1907). Neue Radiolarien der deutschen Sudpolar-Expedition. *Zool. Anz.* **31**(23), 697–705.

Popofsky, A. (1908). Die Radiolarien der Antarktis. *Dt. Sudpol. Exped.* 1901–1903, **10**(3), 183–305.

Popofsky, A. (1912). Die Sphaerellarien des Warmwassergebietes. *Dt. Sudpol. Exped.* 1901–1903, **13**(2), 73–159.

Popofsky, A. (1913). Die Nassellarien des Warmwassergebietes. *Dt. Sudpol. Exped.* 1901–1903, **14,** 217–416.

Renz, G. E. (1972). Unpublished Ph.D. thesis, Scripps Inst. of Oceanog., University of California, San Diego.

Reshetnjak, V. V. (1955). Vertikalnoe raspredelenie radiolyarii Kurilo-Kamchatskoi vpadiny. *Trudy Zool. Inst. Akad. Nauk. U.S.S.R.* **21,** 94–101.
Riedel, W. R. (1958). Radiolaria in Antarctic sediments. *Repts. B.A.N.Z. Antarctic Research Exped.*, ser. B, **6**(10), 217–55.
Riedel, W. R. (1967). *An annotated and indexed bibliography of polycystine radiolaria.* 220 pp. Privately printed by W. R. Riedel, Scripps Institution of Oceanography, La Jolla, California, U.S.A.
Russell-Hunter, W. D. (1970). *Aquatic Productivity—An Introduction to some aspects of biological oceanography and limnology.* 306 pp. Collier-Macmillan Limited, London.
Schewiakoff, W. (1926). Acantharia: *Fauna e Flora del Golfo di Napoli. Monogr.* **37,** 755 pp.
Sverdrup, H. W., Johnson, M. W. and Fleming, R. H. (1942). *The Oceans—their physics, chemistry, and general biology.* 1087 pp. Prentice-Hall Inc. New York.
Tibbs, J. F. (1967). On some planktonic protozoa taken from the track of drift station Arlis I, 1960–61. *Arctic* **20**(4), 247–254.
Wolfenden, R. N. (1905). Biscayan plankton collected during a cruise of H.M.S. *Research* 1900. Part VI, the colloid Radiolaria. *Trans. Linnean Soc., London* **10,** 131–5.

PLATES

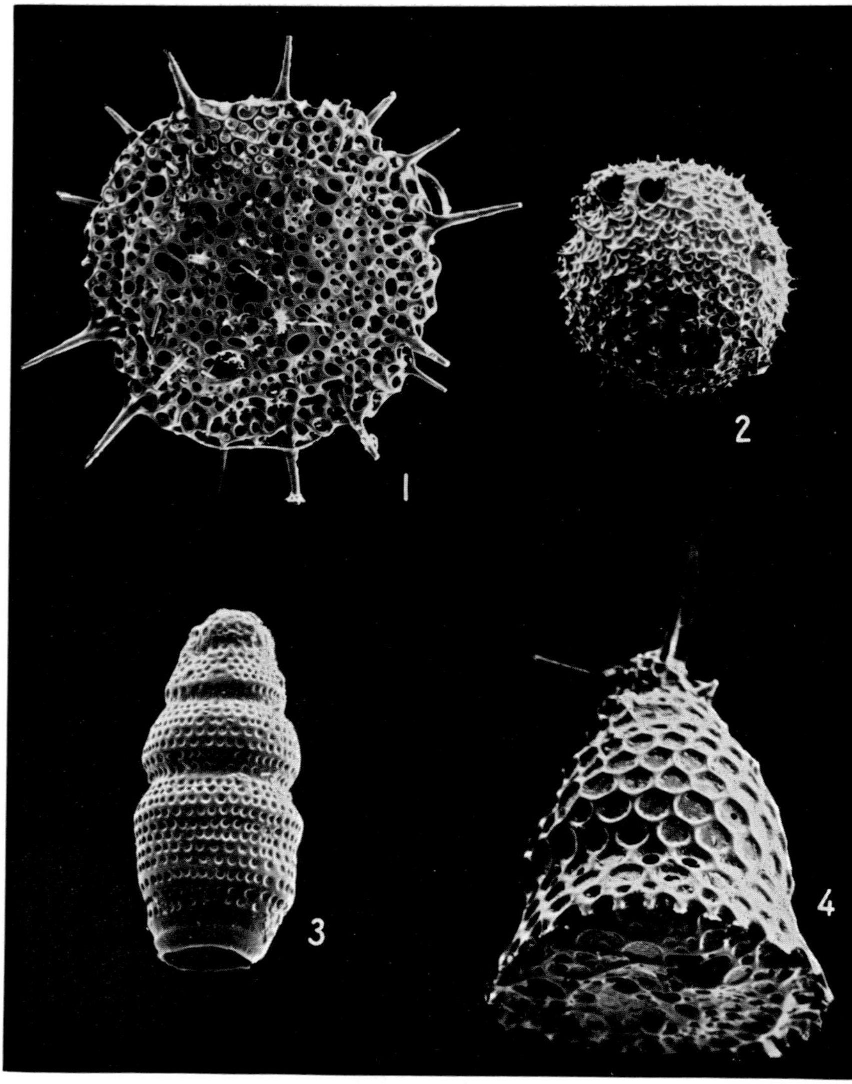

PLATE 1. Representative polycystine radiolarians.

Figures 1 and 2 are *Spongotrochus glacialis* and *Lithelius nautiloides* respectively and are members of the suborder Spumellaria. Figures 3 and 4 are *Lithamphora furcaspiculata* and *Theocalyptra bicornis* respectively and are members of the sub-order Nassellaria. All specimens figured were taken from core E14-8 from the Antarctic Ocean.

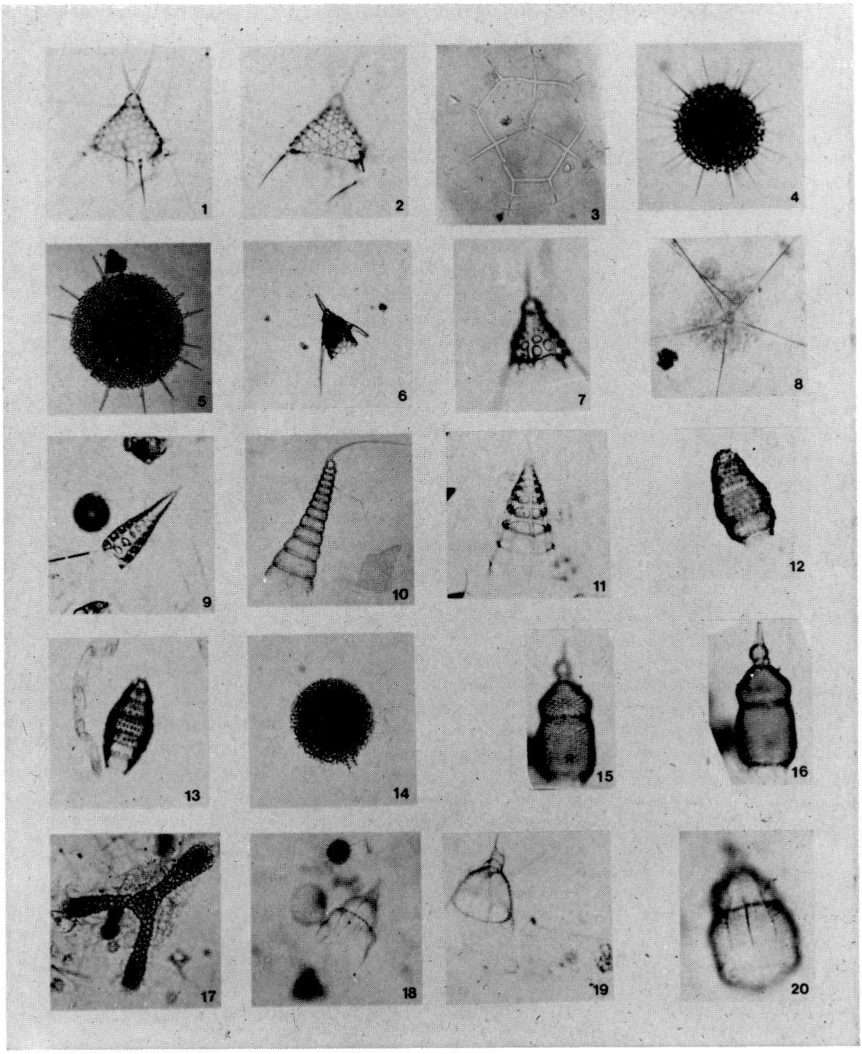

PLATE 2. Subarctic-Transition, Transitional, Subarctic-Intermediate, Transition-Central and Central "shallow" radiolarian faunas (from Casey, 1971b).

Subarctic-Transition fauna: 1–2. *Pterocanium* sp.; 3. *Sethophormis rotula* Haeckel; 4–5. *Spongotrochus glacialis* Popofsky.

Transitional fauna: 6–7. *Pterocorys hirundo* Haeckel; 8. *Spongosphaera streptacantha* (Haeckel).

Subarctic-Intermediate fauna: 9. *Cornutella profounda* Ehrenberg; 10. *Cyrtopera laguncula* Haeckel.

Transition-Central fauna: 11. *Peripyramis circumtexta* Haeckel; 12–13. *Siphocampe erucosa* Haeckel; 14. *Spongopyle osculosa* Dreyer.

Central "shallow" fauna: 15–16. *Calocyclas amicae* Haeckel; 17. *Euchitonia furcata* Ehrenberg; 18–20. *Eucyrtidium hertwigii* Haeckel.

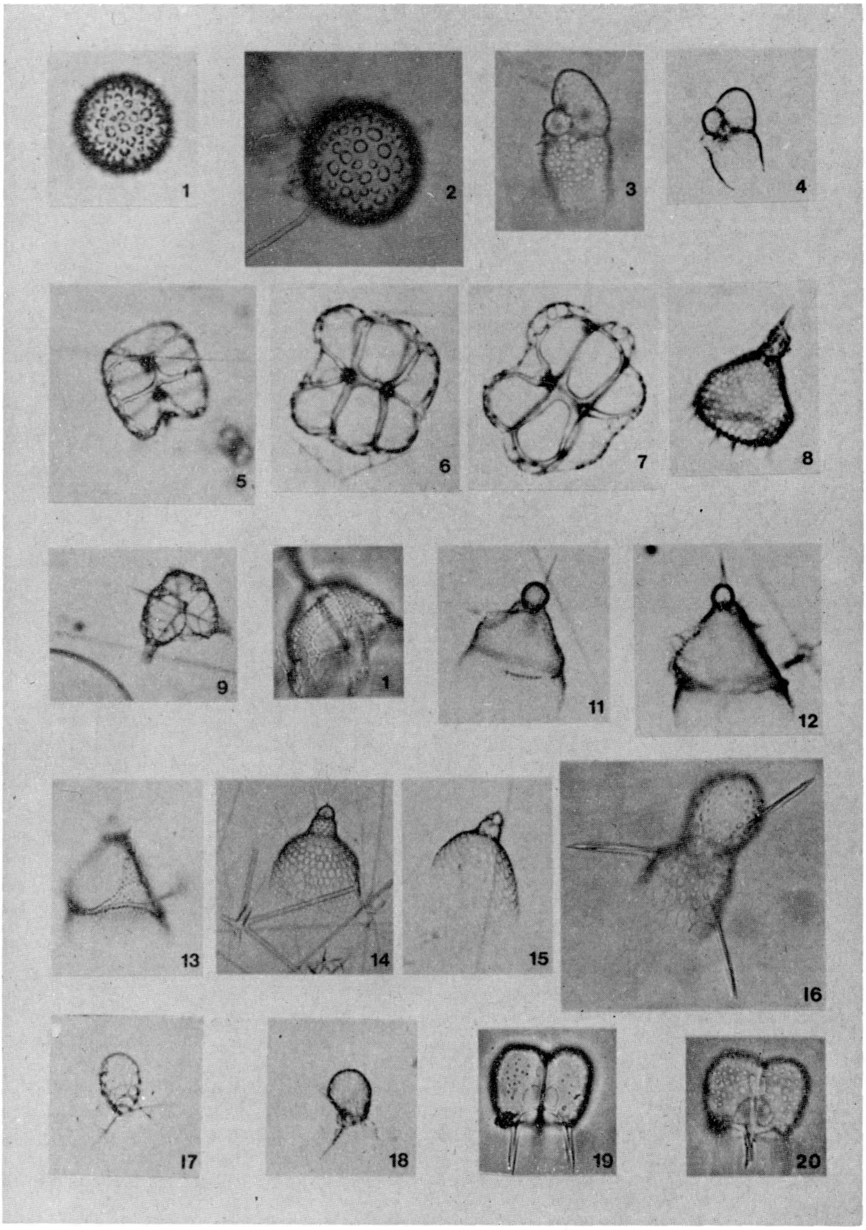

PLATE 3. Equatorial radiolarian fauna (from Casey, 1971b).
Equatorial fauna: 1–2. *Acrosphaera murrayana* (Haeckel); 3–4. *Acrobotrissa cribrosa* Popofsky; 5–7. *Amphispyris costata-thorax* gr.; 8. *Anthocyrtidium cineraria* Haeckel; 9–10. ?*Clathrocorys murrayi* Haeckel; 11–13. *Lipmanella vichowii* (Haeckel); 14–15. *Eucecryphalus* sp.; 16. *Lithomelissa monoceras* Popofsky; 17–18. *Peridium spinipes* Haeckel; 19–20. *Tristylospyris scaphipes* Haeckel.

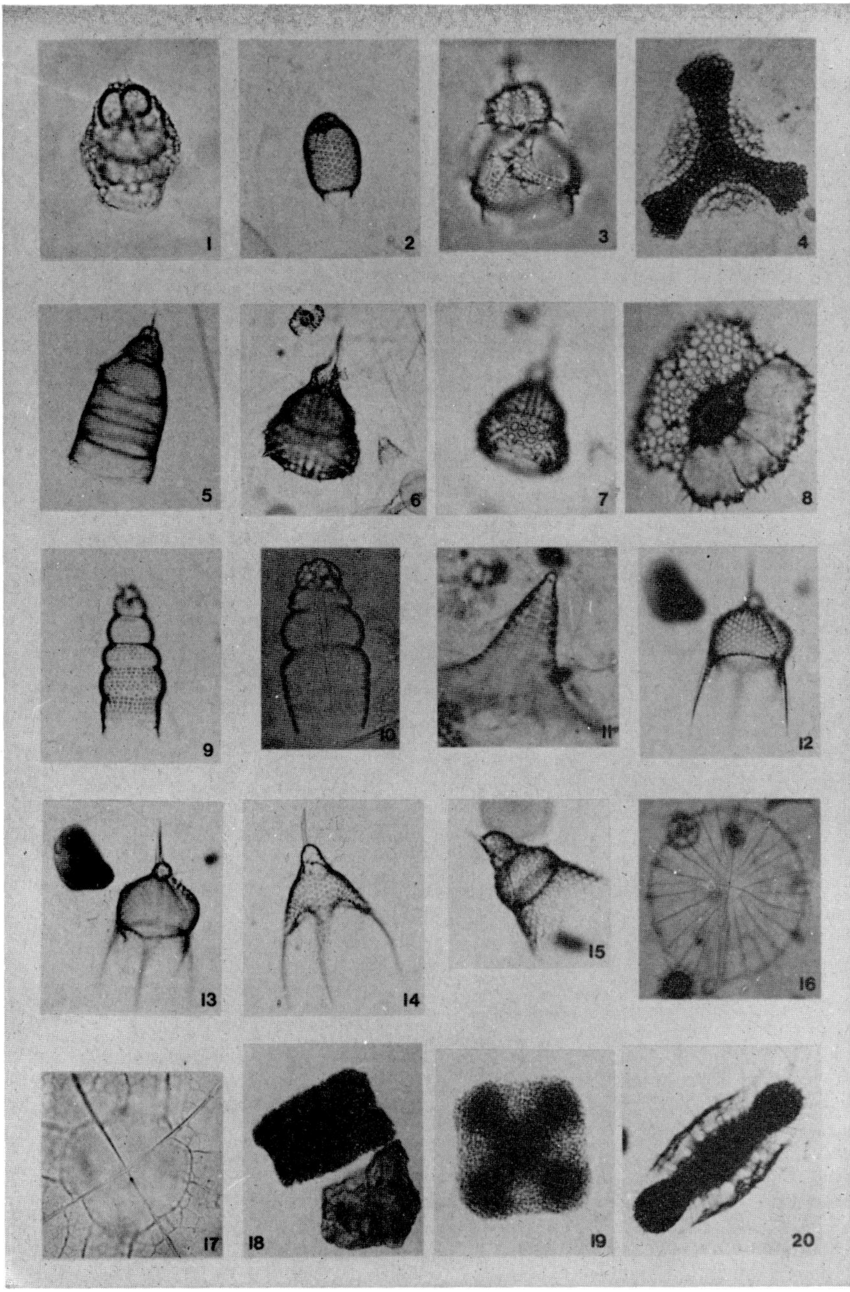

PLATE 4. Equatorial-Central "shallow" radiolarian fauna (from Casey, 197.1b).

Equatorial-Central "shallow" fauna: 1. *Botrycyrtis* sp.; 2. *Carpocanium* sp.; 3 *Clathrocanium ornatum* Popofsky; 4. *Hymeniastrum profundum* (Ehrenberg); 5. *Eucytridium hexagonatum* Haeckel; 6–7. *Lamprocyclas maritalis* Haeckel; 8. *Larcospira quadrangula* Haeckel; 9–10. *Lithamphora furcaspiculata* Popofsky; 11. *Litharachnium tentorium* Haeckel; 12–13. *Pterocanium praetextum* (Ehrenberg); 14. *Pterocanium trilobum* Haeckel; 15. *Pterocorys zancleus* (Muller); 16–17. *Theophormis callipilium* Haeckel; 18–19. *Spongaster tetras* Ehrenberg; 20. *Spongocore puella* Haeckel.

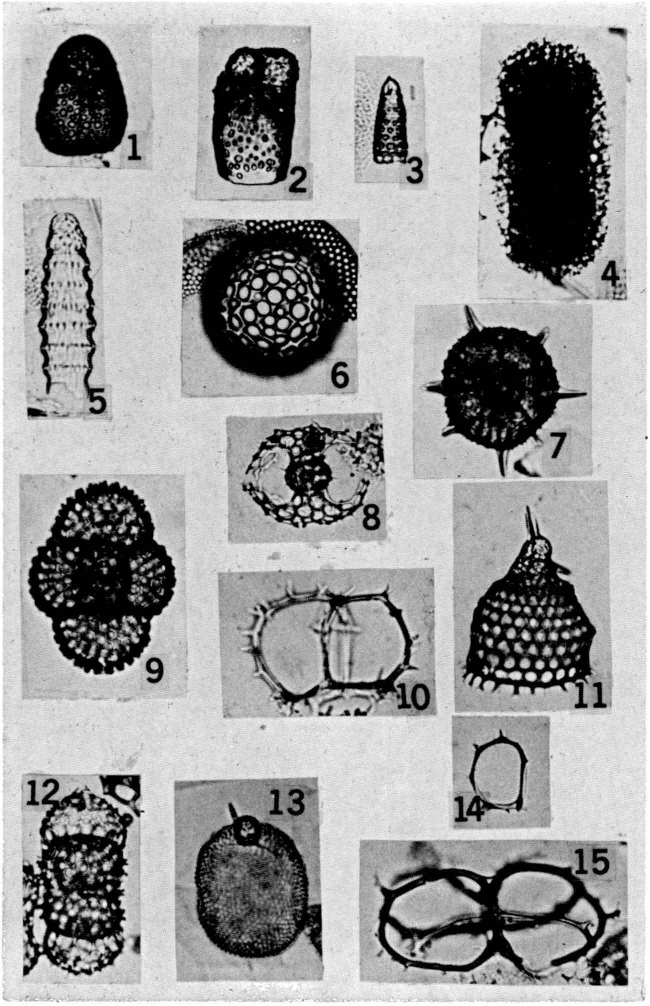

PLATE 5. Some polycystine radiolarians commonly used as oceanographic indicators of water masses or geographic regions. The species illustrated are placed in the biogeographic zonation of Casey (1971b) by attempting to interpret their distributional characteristics from published work and some original work by the author.

1. *Antarctissa denticulata* (Ehrenberg)—Antarctic and Subarctic Shallow Faunal Zone; 2. *Saccospyris*? *antarctica* Haecker—Antarctic Shallow Faunal Zone; 3. *Artostrobus annulatus* (Bailey)—Antarctic and Subarctic Intermediate Faunal Zones; 4. *Spongurus pylomaticus* Riedel—Antarctic and Subarctic Shallow Faunal Zones; 5. *Lithomitra arachnea* (Ehrenberg)—Antarctic and Subarctic Intermediate Faunal Zones; 6. *Cenosphaera nagatai* Nakaseko—Antarctic and Subarctic Shallow Faunal Zones; 7. *Cromyechinus borealis*—(Cleve)—Antarctic and Subarctic Shallow Faunal Zones; 8. *Tetrapyle* sp.—Central Shallow and Equatorial Faunal Zones; 9. *Tholonid* sp.—Central Shallow, Equatorial and subpolar-Intermediate Faunal Zones; 10. *Ceratospyris* sp.—Central Shallow and Equatorial Faunal Zones; 11. *Corocalyptra craspedota* (Jorgensen)—Central Shallow Faunal Zone; 12. *Ommatartus tetrathalamus* (Haeckel)—Central Shallow and Equatorial Faunal Zones. 13. *Lithopera bacca* Ehrenberg—Central Shallow and Equatorial Faunal Zones; 14. *Zygocircus* sp.—Central Shallow and Equatorial Faunal Zones; 15. ?*Toxarium* sp.—Central Shallow and Equatorial Faunal Zones.

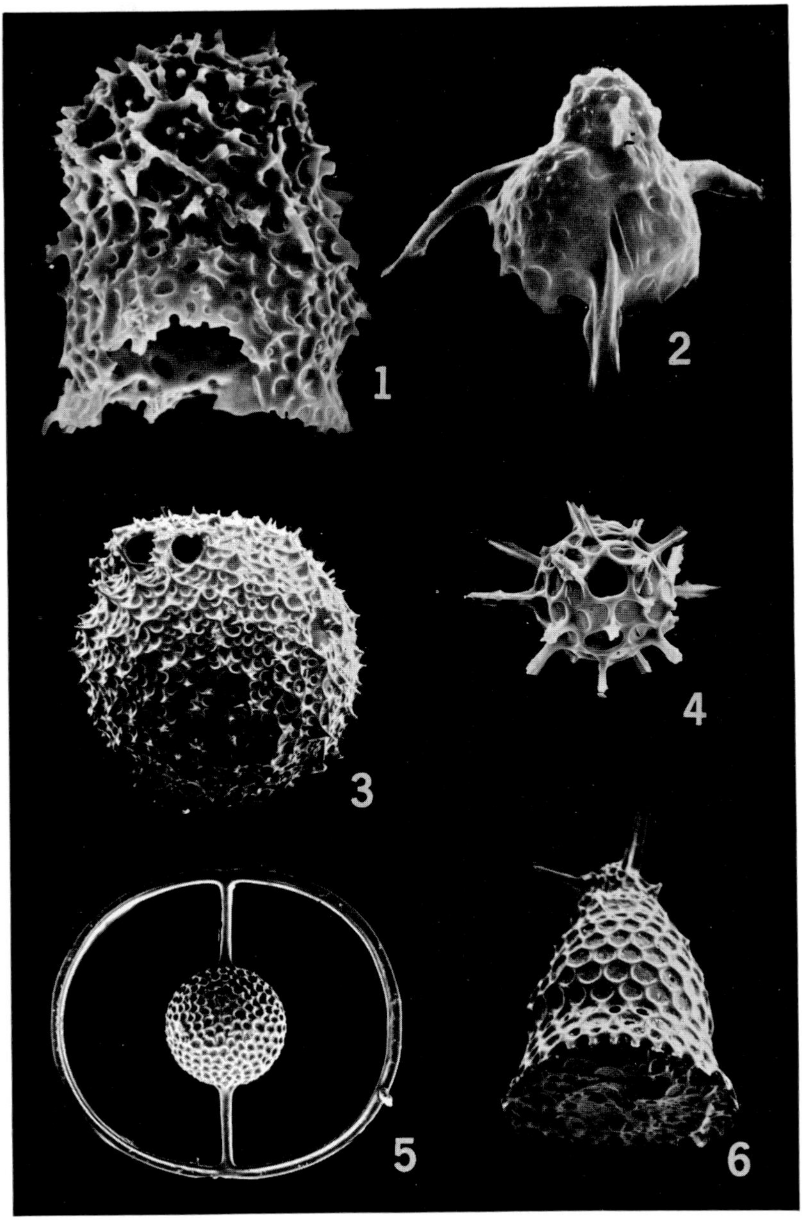

PLATE 6. Some polycystine radiolarians commonly used as oceanographic indicators of water masses or geographic regions. The species illustrated are placed in the biogeographical zonation of Casey (1971b) by attempting to interpret their distributional characteristics from published work and some original work by the author.

1. *Helotholus histricosa* Jorgensen—Antarctic and Subarctic Shallow Faunal Zones; 2. *Dictyophimus gracilipes* Bailey—Antarctic and Subarctic Shallow Faunal Zones; 3. *Lithelius nautiloides* Popofsky—Antarctic and Subarctic Shallow Faunal Zones; 4. *Cromyechinus borealis*—Antarctic and Subarctic Shallow Faunal Zones; 5. *Saturnulus planetes* Haeckel—Central Shallow and Equatorial Faunal Zones; 6. *Theocalyptra bicornis* (Popofsky) —Subantarctic Shallow Faunal Zones.

8: Cainozoic Radiolaria

W. R. RIEDEL and ANNIKA SANFILIPPO
Scripps Institution of Oceanography,
University of California, San Diego,
La Jolla, California 92093, U.S.A.

The siliceous skeletons of polycystine radiolarians form a substantial component of many Cainozoic pelagic sediments, and considerable advances in our understanding of the stratigraphy and taxonomy of this group are resulting particularly from investigation of sequences sampled by the Deep Sea Drilling Project. Cainozoic radiolarian faunas apparently comprise much larger numbers of species than do the faunas of most other planktonic microfossil groups, thus providing the potential for high-resolution biostratigraphy and for eventual detailed palaeoecological interpretations. The possibility of tracing phyletic lineages in some radiolarian groups assists in establishing a natural taxonomic system, and provides information on the process of evolution in planktonic protozoans. In this chapter, illustrations and stratigraphic ranges are given for many low-latitude Cainozoic polycystine radiolarians, and the taxonomic system, which is currently undergoing profound changes, is summarised.

Outline

Introduction

Soon after the discovery of radiolarians as a component of living plankton (Meyen, 1834), the extent of their contribution to Quaternary and Tertiary sediments began to be investigated (Ehrenberg, 1838, 1844a, b, 1847a, b). The 1850s to 1880s saw the appearance of extensive descriptions of living faunas (e.g. Müller, 1858; Haeckel, 1860a, b, 1862; Hertwig, 1879) and fossil assemblages (e.g. Ehrenberg, 1854, 1872a, b, 1873, 1875; Zittel, 1876; Stöhr, 1880), and successive attempts to develop a taxonomic system capable of accommodating the rapidly increasing number of forms discovered (Müller, 1858; Haeckel, 1862, 1881; Hertwig, 1879; Bütschli, 1882a, b). The *Challenger* Expedition of 1873–1876 collected a very large number of radiolarians in plankton and sediment samples, and their investigation by Haeckel (1887) resulted in a monographic treatment that has remained the most important single reference work on radiolarians to this day. Plankton samples including radiolarians continued to be collected by later oceanographic expeditions (Cleve, 1900a, b, 1901; Joergensen, 1905; Haecker, 1906, 1907, 1908; Popofsky, 1908, 1912, 1913, 1917, 1920; Dreyer, 1913), and by the early part of the twentieth century the general nature of these protozoans, and their occurrence in plankton and sediments, were known in broad outline.

In this chapter we deal only with polycystine radiolarians (with skeletons consisting of solid structures of opaline silica), since this is the only group with a substantial fossil record. Phaeodarian (tripylean) radiolarians occur in some Neogene sediments containing exceptionally well preserved siliceous microfossils (Riedel, 1963; Dumitrica, 1964; Reshetnyak, 1969; Stadum and Ling, 1969), but their skeletons consisting of hollow structures of silica with admixed organic material are much more readily dissolved during sedimentation and early diagenesis.

Polycystine radiolarians are exclusively marine protozoans, and are, except in one family (Collosphaeridae), solitary. Their protoplasm is divided into extracapsular and intracapsular parts by the central capsular membrane. The nucleus is within the central capsule, and pseudopods radiate from the extracapsular cytoplasm. Reproduction can occur by binary fission or by spore formation, but it is not known whether the spores represent gametes. Food particles are trapped by the pseudopods, and some nutrition may be derived also from the zooxanthellae that live symbiotically within many species. The radiolarians have no power of directional movement except through changing their specific gravity by the formation of fat globules and gas vacuoles. By far the greatest number of polycystins live in the upper few hundred meters of the oceanic water column, but a few occur even at abyssal depths. The skeletons (usually about 50–400 μm in length or diameter), which in the living organisms are included within the extracapsular cytoplasm, form a substantial component of some marine sediments, and provide a record of this group extending back at least as far as the Ordovician. For recent information on aspects of Quaternary radiolarians not treated in this chapter, the reader is referred to the following papers: Cytology; Hollande and Enjumet (1960), Hollande *et al.* (1970), Cachon *et al.* (1968), Cachon and Cachon (1969, 1971a, 1971b). Depth-distribution of species in the water column; Reshetnyak (1955), Casey (1971a, and this volume), Casey *et al.* (1971).

Distribution of radiolarians

AS A COMPONENT OF CAINOZOIC PELAGIC SEDIMENTS

Radiolarians occur in all marine waters of normal salinity, the numbers of large individuals being approximately equivalent to those of foraminifera (Cifelli and Sachs, 1966) and the numbers of small individuals being considerably greater (Beers and Stewart, 1971; of the organisms reported as radiolarians in that paper, no more than a few percent were phaeodarians—J. R. Beers, personal communication). Though data are scanty, their abundance in plankton seems to conform to the general pattern of total plankton abundance, being greatest in the regions of high biological productivity associated with the equatorial current system, eastern and western boundary currents, and the east-flowing currents in latitudes higher than about 40° (Petrushevskaya, 1971d, e).

The preservation of siliceous microfossils is not dependent upon water depth as is the preservation of calcareous skeletons, and therefore the occurrence of radiolarians in sediments is determined principally by their rate of production in the overlying water column and the rate at which they are buried and thus protected from dissolution in the bottom water (Riedel, 1959a; Lisitzin, 1971). The general pattern of radiolarian distribution in the sediments of the northern-hemisphere portion of a large ocean such as the Pacific is illustrated in Fig. 1. Radiolarians are abundant and well preserved beneath the biologically productive equatorial current system, though they are diluted by calcareous nannofossils and foraminifera at localities shallower

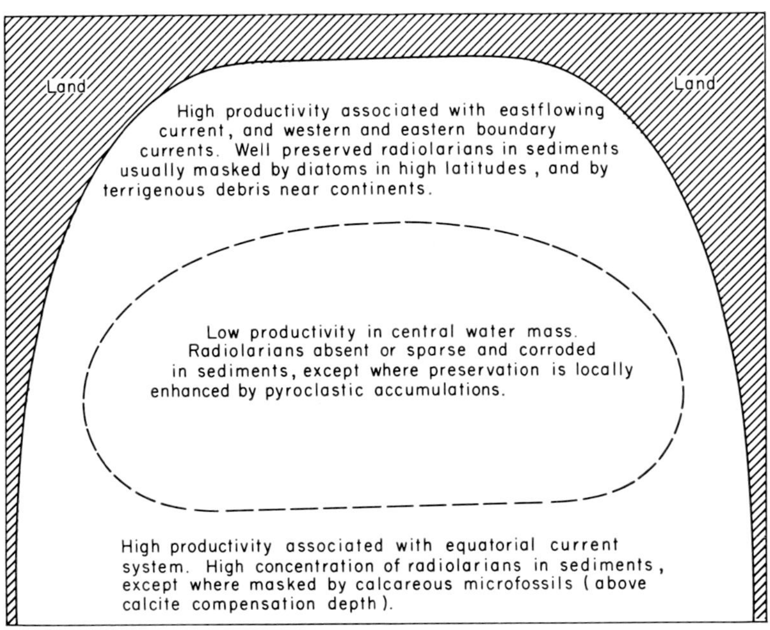

FIG. 1. Distribution of radiolarians in the northern half of a large ocean.

than the calcite compensation depth. They are also well preserved beneath the biologically productive eastward-flowing currents of high latitudes, and beneath the eastern and western boundary currents, but in these regions they do not generally constitute a large fraction of the sediments because of dilution by diatoms in high latitudes and by terrigenous debris near continents. The central water mass is stably stratified and less productive biologically, the sediments beneath it accumulating slowly (at approximately 1 m per million years as compared to about 5–30 m per million years in more productive regions), and for these reasons radiolarians are absent from the sediment, or present only as sparse, corroded specimens—except where large contributions of pyroclastic debris create local conditions favourable to their preservation by increasing the rate of sediment accumulation and the concentration of dissolved silica in the interstitial waters. The situation in the southern hemisphere is the mirror-image of that illustrated. The pattern of distribution described is of course highly generalised, and susceptible to modifications as a result of local abnormalities in biological productivity, rate of sediment accumulation, and physico-chemical conditions in interstitial waters. Nevertheless, the pattern seems to have prevailed in the major ocean basins through most of the Cainozoic, at least (Riedel and Funnell, 1964, Ramsay, 1971).

BIOGEOGRAPHY AND DIVERSITY

In recent years, studies of the distribution of polycystins in plankton have been made for the Antarctic, Indian Ocean and Central Pacific by Petrushevskaya (1966, 1967, 1971a, d, e) and for the northeastern Pacific by Casey (1971b). Species distributions in Recent sediments have been indicated by Petrushevskaya and Casey in the papers referred to, and in addition by Nigrini (1967, 1968, 1970, 1971) for the Indian Ocean and the equatorial and northern Pacific, by Kruglikova (1969) for the northern Pacific, and by Goll and Bjørklund (1971) for the northern Atlantic.

The combined results of these studies are not yet sufficient to permit the detailed delineation of radiolarian biogeographic provinces for most of the world's oceans, but they allow some generalisations to be made. The polycystine fauna of the tropical and equatorial regions of present-day oceans comprises perhaps 100–300 species, and is rather uniform throughout the tropics. A transitional assemblage in latitudes of about 30–40° is characterised by lacking many of the tropical species, and having a few endemic species and subspecies. The coldwater fauna has still fewer species and again includes endemic forms, some of which are different in the subarctic and subantarctic regions. This biogeographic regionality permits the radiolarians to be used in palaeoecological interpretations, as exemplified by Nigrini's (1970) study of a Quaternary core from the North Pacific.

The extent to which assemblages in sediments differ from the populations in the overlying water column is not yet clear, and is presently being actively investigated. Certainly, some of the more delicate polycystins (e.g. myelastrins) are rarely if ever preserved in sediments, and dissolution (most of which probably occurs while the skeletons are in the top few centimetres of the sediment) distorts the proportions of species originally present in the plankton (Petrushevskaya, 1966, 1967, 1971d; Killmar, personal communication). It has been reported, however, that no nassellarian species occurring in the plankton is completely absent from the sediments of the corresponding region (Petrushevskaya, 1971a).

Biostratigraphical applicability

For about sixty years following the publication of Haeckel's *Challenger* Report, it seemed that radiolarians could not be applied stratigraphically in Cainozoic deposits, since many of the species in (presumed Quaternary) sediments on the deep sea floor were identical with those in Paleogene assemblages from land-based localities. Also, an assemblage described by Tan (1927) from the Indonesian island of Roti and believed to be Pliocene in age, was strikingly similar to some Cretaceous assemblages. Thus radiolarian genera and species seemed to have very long time-ranges, until it was shown that sea-floor sediments at many localities are Tertiary in age, or have Tertiary microfossils admixed with Quaternary assemblages (Riedel, 1952, 1957), and that the assemblage described from Roti is in fact Cretaceous (Riedel, 1953; Bramlette *in* Foreman, 1966).

It is now apparent that Cainzoic radiolarians have evolved as rapidly as have other groups of planktonic microfossils, that species have wide geographical distributions, and that the total number of species living at any given time numbered one or two hundred or more—these factors together ensuring the usefulness of the group for stratigraphical purposes. Their stratigraphical application is at present limited by the unsatisfactory state of their taxonomy, by the limited number of Tertiary land-based localities containing radiolarians, and by the lack of thorough descriptions of many of the assemblages that are known.

The principal descriptions of land-based Tertiary radiolarian faunas are the following:

PLIOCENE

Europe: Seguenza (1880, pp. 230–234)

MIOCENE

Europe: Bachmann *et al.* (1963); Carnevale (1908); Dreyer (1890); Lucchese

(1927); Principi (1909); Sanfilippo *et al.* (1973); Stöhr (1880); Vinassa (1900)
Asia: Kozlova (1960); Nakaseko (1955, 1959, 1960, 1963, 1971)
America: Campbell and Clark (1944); Martin (1904)

EOCENE
Europe: Brandt (1935)
Asia: Borisenko (1960b); Kozlova and Gorbovets (1966); Krasheninnikov (1960); Lipman (1950, 1953, 1965, 1969); Mamedov (1969a, b, c, 1970)
America and Caribbean: Clark and Campbell (1942, 1945); Ehrenberg (1873, 1875)

PALAEOCENE
Asia: Borisenko (1958, 1960a)
America: Frizzell and Middour (1951)

Many additional Tertiary radiolarian localities have become available since the advent of intensive coring of the deep sea floor beginning with the Swedish Deep Sea Expedition (Riedel, 1957, 1959b; Hays *et al.*, 1969), and currently the Deep Sea Drilling Project is multiplying spectacularly the number and geographical coverage of sequences available for radiolarian stratigraphic studies (Riedel and Hays, 1969; Cita *et al.*, 1970; Riedel and Sanfilippo, 1970, 1971; Olsson and Goll, 1970; Kling, 1971; Moore, 1971; Goll, 1972a; Sanfilippo and Riedel, 1973; Foreman, 1973).

Largely as a result of investigations of Deep Sea Drilling Project cores, the Cainozoic radiolarian sequence in the tropics has been divided into about thirty zones, as indicated in Table 1. In addition to the radiolarian events used to define zonal boundaries, the table includes the earliest and latest occurrences of some other species mentioned in the systematic section, or illustrated. Over some parts of the column, the precise order of the events is uncertain because complete radiolarian sequences are not yet available. Higher-latitude species and zones are not included. Reasons for the placement of the boundaries of epochs in relation to the radiolarian zones, on the evidence of co-occurring calcareous microfossils, may be found in the papers by Riedel and Sanfilippo (1970), Brönnimann *et al.* (1971), Tracey *et al.* (1971), Riedel (1971b), Hays *et al.* (1972) and Foreman (1973). A slightly different zonation for the Oligocene has been proposed by Moore (1971).

Methods of Investigation

COLLECTION OF SAMPLES

In sampling sediments and sedimentary rocks for radiolarian investigations, procedures do not differ substantially from those employed in collecting

TABLE 1. Radiolarian zonation applicable to the tropical Pacific and Caribbean.

In the list of events, (B) indicates the earliest appearance of a taxon, (T) the latest occurrence, and → an evolutionary transition. (m) indicates that the limit of a taxon is defined morphotypically, and (e) that it is defined evolutionarily.

Epochs	Radiolarian Zones	Radiolarian events
Quaternary	No Zone	
		(eT) *Pterocanium prismatium*
Pliocene	*Pterocanium prismatium*	
		(eT) *Stichocorys peregrina*
	Spongaster pentas	*Spongaster pentas* → *S. tetras*
		Spongaster berminghami → *S. pentas*
		(mB) *Pterocanium prismatium*
	Stichocorys peregrina	(mT) *Solenosphaera omnitubus*
		(mT) *Acrobotrys tritubus*
		Stichocorys delmontensis → *S. peregrina*
Late Miocene	*Ommatartus penultimus*	
		Ommatartus antepenultimus → *O. penultimus*
	Ommatartus antepenultimus	(mB) *Solenosphaera omnitubus*
		(eT) *Ommatartus hughesi*
		(mB) *Acrobotrys tritubus*
		(mB) *Spongaster berminghami*
		Cannartus petterssoni → *Ommatartus hughesi*
Middle Miocene	*Cannartus petterssoni*	*Cannartus laticonus* → *Ommatartus antepenultimus*
		(eB) *Lithopera bacca*
		(eT) *Dorcadospyris alata*
		(mT) *Cyrtocapsella cornuta*
		(mB) *Cannartus petterssoni*
	Dorcadospyris alata	*Cannartus mammifer* → *C. laticonus*
		(mT) *Cyrtocapsella tetrapera*
		(mT) *Liriospyris parkerae*
		(mT) *Calocycletta virginis*
		(mT) *Calocycletta costata*
		Dorcadospyris dentata → *D. alata*
Early Miocene	*Calocycletta costata*	*Liriospyris stauropora* → *L. parkerae*
		(mT) *Cannartus prismaticus*
		Cannartus violina → *C. mammifer*
		(mT) *Lychnocanoma elongata*
		(mB) *Calocycletta costata*
		(mB) *Dorcadospyris dentata*
		(mB) *Liriospyris stauropora*
		(mT) *Dorcadospyris ateuchus*

Epochs	Radiolarian Zones	Radiolarian events
Early Miocene	*Calocycletta virginis*	*Cannartus tubarius* → *C. violina*
		(eT) *Cyclampterium pegetrum*
		(mT) *Theocyrtis annosa*
		(mB) *Cyrtocapsella cornuta*
		(mB) *Cyrtocapsella tetrapera*
		(mB) *Calocycletta virginis*
	Lychnocanoma elongata	
		(mB) *Lychnocanoma elongata*
Late Oligocene	*Dorcadospyris ateuchus*	(mB) *Cannartus tubarius*
		Lithocyclia angusta → *Cannartus prismaticus*
		Tristylospyris triceros → *Dorcadospyris ateuchus*
Early Oligocene	*Theocyrtis tuberosa*	*Centrobotrys petrushevskayae* → *C. thermophila*
		(mB) *Theocyrtis annosa*
		Cylcampterium milowi → *C. pegetrum*
		(mT) *Theocyrtis tuberosa*
		(mT) *Theocampe amphora group*
		Centrobotrys gravida → *C. petrushevskayae*
		(mB) *Centrobotrys gravida*
		Lithocyclia aristotelis → *L. angusta*
Late Eocene	*Thyrsocyrtis bromia*	(mT) *Lychnocanoma babylonis* group
		(mT) *Theocampe mongolfieri*
		(mT) *Lychnocanoma amphitrite*
		(mT) *Theocampe armadillo* group
		(mT) *Periphaena decora*
		(mB) *Theocyrtis tuberosa*
		(mT) *Thyrsocyrtis bromia*
		(mT) *Thyrsocyrtis tetracantha*
		(mT) *Calocyclas turris*
		(eT) *Lithapium mitra*
		(eT) *Podocyrtis geotheana*
		Thyrsocyrtis triacantha ± *T. tetracantha*
		(mT) *Eusyringium fistuligerum*
		(mT) *Artophormis dominasinensis*
		(mT) *Calocyclas hispida*
		Lychnocanoma bellum → *L. amphitrite*

Epochs	Radiolarian Zones	Radiolarian events
Late Eocene	*Thyrsocyrtis bromia*	(mT) *Podocyrtis papalis*
		Lithocyclia ocellus → *L. aristotelis*
		(mB) *Thyrsocyrtis bromia*
	Podocyrtis goetheana	(mB) *Podocyrtis goetheana*
Middle Eocene	*Podocyrtis chalara*	(mB) *Tristylospyris triceros*
		(mB) *Calocyclas turris*
		(mB) *Artophormis dominasinensis*
		(eT) *Lithochytris vespertilio*
		Podocytris mitra → *P. chalara*
	Podocyrtis mitra	*Lithapium anoectum* → *L. mitra*
		(mB) *Theocampe armadillo* group
		(mT) *Lophocyrtis biaurita*
		(mT) *Rhopalocanium ornatum*
		(mT) *Spongatractus pachystylus*
		(mT) *Phormocyrtis striata striata*
		(eT) *Podocyrtis ampla*
		(mT) *Lithapium plegmacantha*
		Podocyrtis sinuosa → *P. mitra*
	Podocyrtis ampla	(mT) *Theocorys anapographa*
		Eusyringium lagena → *E. fistuligerum*
		Podocyrtis phyxis → *P. ampla*
	Thyrsocyrtis triacantha	*Podocyrtis diamesa* → *P. phyxis*
		(mT) *Spongodiscus phrix*
		(mT) *Lamptonium fabaeforme constrictum*
		(mB) *Eusyringium lagena*
		(eB) *Thyrsocyrtis triacantha*
	Theocampe mongolfieri	(mT) *Periphaena delta*
		(mB) *Lithapium anoectum*
		(mT) *Theocorys anaclasta*
		Lithochytris archaea → *L. vespertilio*
		(mB) *Theocampe mongolfieri*
	Theocotyle cryptocephala cryptocephala	(mT) *Stylotrochus quadribrachiatus*
		Podocyrtis aphorma → *P. sinuosa*
		(mB) *Lithapium plegmacantha*
		Theocotyle cryptocephala nigriniae → *T. c. cryptocephala*
	Phormocyrtis striata striata	(eB) *Dictyophimus craticula*
		(mB) *Theocorys anaclasta*
		(mB) *Theocorys anapographa*
		(mB) *Periphaena delta*
		(mB) *Podocyrtis aphorma*
		(mB) *Stylotrochus quadribrachiatus*
		(mB) *Lamptonium fabaeforme constrictum*
		(mB) *Rhopalocanium ornatum*

Epochs	Radiolarian Zones	Radiolarian events
Middle Eocene	*Phormocyrtis striata striata*	(mB) *Spongodiscus phrix*
		(mT) *Spongodiscus quartus quartus*
		Phormocyrtis striata exquisita → *P. s. striata*
Early Eocene	*Buryella clinata*	(mB) *Lychnocanoma belluoma*
		(mB) *Theocampe amphora* group
		(mB) *Lithocyclia ocellus* group
		Spongatractus balbis → *S. pachystylus*
		(mB) *Spongatractus balbis*
		(mT) *Bekoma bidartensis*
		(mB) *Periphaena decora*
		(mB) *Caloycclas hispida*
		Lychnocanoma auxilla → *Lithochytris archaea*
		(mB) *Lophocyrtis biaurita*
		Buryella tetradica → *B. clinata*
	Bekoma bidartensis	(eT) *Thyrsocyrtis tarsipes*
		(mB) *Theocotyle cryptocephala nigriniae*
		(mT) *Stylosphaera goruna*
		(mB) *Bekoma bidartensis*
Late Palaeocene	No Zone	(mB) *Lychnocanoma babylonis* group
		(mB) *Podocyrtis papalis*
		(mT) *Buryella pentadica*
		(mB) *Thyrsocyrtis tarsipes*
		(mT) *Bekoma divaricata*
		(mB) *Spongodiscus quartus*
		(mB) *Stylosphaera goruna*

other types of microfossils. Usually, samples of about a kilogram will provide more than enough material for the planned investigation, so that some remains for later use and for exchange with colleagues. Collecting localities should be recorded on topographic maps at a scale of about 1:25 000, and sufficiently detailed field notes kept so that the locality can later be re-sampled precisely if necessary. Publications resulting from the investigation should include data sufficient to enable other workers to find and re-sample the locality.

It is usually impossible to determine, by use of a hand-lens in the field, whether or not radiolarians are present in sedimentary rocks being sampled. Though radiolarians are sufficiently large to be seen on smooth, fractured surfaces of radiolarites and cherts which occur more commonly in Mesozoic deposits, they can generally not be distinguished on the rougher surfaces of less indurated rocks. A field microscope with a magnification of about 100 may be useful during collecting, since a simple sediment smear on a glass slide, mounted in water and with a coverglass, will usually reveal radiolarian fragments if these microfossils are present in useful quantity.

ISOLATION OF RADIOLARIANS

A common method of disaggregating sediments for the isolation of radiolarians combines the mechanical effect of bubbling hydrogen peroxide with the wetting (detergent) effect of tetrasodium pyrophosphate. To about 50 ml of water in a 250 ml beaker are added about 25 ml of 35% hydrogen peroxide solution and 1–2 g of tetrasodium pyrophosphate. The mixture is heated until the pyrophosphate dissolves, and then approximately 5–10 cc of sediment (or of sedimentary rock broken into 5 mm pieces) is added, and the mixture is boiled for 2–20 min. The coarse fraction is obtained by passing through a sieve with mesh-openings of about 63 μm, and is washed back into the beaker. Calcareous components are dissolved by adding hydrochloric acid, and the excess acid and salts in solution are removed by successive dilutions and decantations. After removing any additional fine material by again using a 63 μm sieve, the coarse material (including radiolarians) may be ready for mounting. If lumps of sediment still remain, the sample may be dried, and treated again with peroxide-pyrophosphate solution—by successively drying and boiling two or three times, the radiolarian skeletons can usually be cleaned effectively.

Some shales and marls that do not respond satisfactorily to the above treatment may be disaggregated by the use of kerosene or paint solvent. The sample is broken into 5 mm pieces, dried thoroughly in an oven, and while still warm is covered with kerosene or paint solvent. When the sediment is saturated (usually within a few minutes), the excess liquid is decanted and the sample immediately covered with water. This is allowed to stand, with occasional stirring, until it is judged that no further disaggregation will occur—approximately 5–30 min. The fine material is removed by the use of a 63 μm sieve, and if lumps remain the sample is dried and treated again. When disaggregation is complete, the coarse residue is treated with hydrochloric acid, the acid and salts are removed by decantations, and the sample sieved once more.

If the microfossils are not sufficiently cleaned by these treatments, use of an ultrasonic vibrator for a few seconds is sometimes effective, or if silica is cementing the clay particles a brief treatment with a hot solution of sodium hydroxide may help. Care must be exercised in using these procedures, however since the ultrasonic vibrator may break the microfossils, and the sodium hydroxide will begin to dissolve them.

The result of these separation procedures is a concentration of the non-calcareous fraction of the sediment coarser than 63 μm, in water.

PREPARATION AND EXAMINATION OF MICROSLIDES

For some investigations, it is useful to pass the residue coarser than 63 μm through a sieve with a mesh-openings of 150 μm, in order that separate preparations may be made of the larger radiolarians and of the smaller ones. Glass slides of approximately 75 × 25 mm are labelled, and saliva smeared very thinly over the area to be covered by the microfossils. The radiolarians are transferred (with some water) from the beaker to the slide by means of a small pipette with a rubber bulb, in such a way as to cover the required area with the appropriate density. The preparation is then dried, a few drops of xylene added to expel air from the skeletons, and the preparation completed by the addition of Canada balsam and a cover glass (usually 22 × 40 mm). Most of the procedure of separating and mounting the radiolarians is carried out without drying the residue, thus minimising the possibility of later preparations being contaminated by radiolarians in the laboratory dust.

In addition to the course-fraction strewn-slides for investigation of the radiolarians, it is desirable to prepare a smear-slide from a minute portion of the untreated sediment sample, in order that the general lithology may be determined. If the sediment contains calcareous nannofossils, the smear-slide may also provide useful information on that group.

Practically all radiolarian investigations are performed with a compound microscope at magnifications of 50–400 and using transmitted light, because important internal structures cannot be seen if reflected light is used. Canada balsam has the ideal refractive index for examination of most Cainozoic radiolarians, but if skeletons are unusually delicate or have an abnormal refractive index other media may be necessary. There is no convincing evidence that the scanning electron microscope reveals structures of Cainozoic radiolarians additional to those discernible with an optical microscope—and though electron micrographs may be useful for illustrating some details of structure, they provide images not easily comparable with those seen in the usual optical microscopy.

Positions of individual specimens on radiolarian strewn-slides may conveniently be recorded by using an England Finder (Riedel and Foreman, 1961)—a glass slide marked with a grid of numbered squares.

RECORDING AND PUBLICATION OF RESULTS

In order that research results may be evaluated critically by other workers, the basic observations should be recorded and published as objectively as possible, and clearly distinguished from subjective interpretations. Generally, the fundamental data should be presented in the form of tabulations of the abundances of clearly defined morphotypes in each of the samples examined. Some groups of radiolarian species are known to constitute evolutionary lineages showing a continuum of morphological change through time, and in these cases the limits between the species may be defined either morphotypically or evolutionarily (Riedel and Sanfilippo, 1971; Sanfilippo and Riedel, 1973). It is important that each author clearly specifies which of these types of limits are being used, and since evolutionary limits involve a certain amount of subjective judgement they should be employed not in the basic tabulation of abundances of morphotypes, but in a subsequent part of the published report.

Because the siliceous skeletons of radiolarians are resistant to solution, they frequently survive erosion and transportation to become incorporated in younger sediments. Therefore, in the interpretation of fossil assemblages, care should be taken to distinguish the reworked species, if any, from the autochthonous fauna. In some assemblages, this distinction is aided by differences in the preservation of the two components.

Only a small proportion of the total number of Cainozoic radiolarian species are known sufficiently well for information on them to be effectively transmitted by Linnean binomen alone, and therefore illustrations form an essential part of most papers on this group. Photomicrography using transmitted light is generally an effective means for obtaining suitable illustrations, and the use of Polaroid film rather than conventional wet-process photography has advantages of speed and simplicity. Several photographs of each species are often desirable to illustrate intraspecific variability. Morphological details are best illustrated by drawings made with the aid of a camera lucida, or by scanning electron micrographs.

Taxonomy

The taxonomic system erected by Haeckel (1887) for the *Challenger* radiolarians (summarised conveniently, but not completely accurately, by Campbell, 1954) has provided the framework within which most radiolarian assemblages have since been described, despite the fact that it has long been recognised as not reflecting true relationships adequately (Haeckel, 1887, p. ci; Bütschli, 1889; Guppy, 1909; Tan Sin Hok, 1927; Khabakov, 1937; Deflandre, 1953). Hilmers' (1906) reclassification of the collosphaerids remained, until very recently, the only large-scale modification of a part of Haeckel's taxonomic system for the polycystins.

In recent years, however, profound changes in the taxonomic system have been proposed on the basis of two types of studies—comparative morphology and phylogeny. Comparative-morphological studies have involved a re-evaluation of the significance of certain elements of skeletal structure, without any particular regard to geologic time, and have resulted in revisions of the classification at the level of families, subfamilies and genera, as for example the genera of botryoids by Petrushevskaya (1965), genera of orosphaerids by Friend and Riedel (1967), families of polycystins by Riedel (1971a), families, subfamilies and genera of nassellarians by Petrushevskaya (1967, 1971b, c) and Petrushevskaya and Kozlova (1972), genera of collosphaerids by Strelkov and Reshetnyak (1971), and families and genera of discoidal spumellarians by Petrushevskaya and Kozlova (1972). Many of these proposed taxonomic revisions are so recently published that their applicability has not yet been confirmed, and neither has any large amount of stratigraphical information been accumulated for the categories defined in them. Therefore it has not been practicable to incorporate some of the most recent suggestions in the systematic section below.

Phylogenetic studies are resulting in taxonomic revisions mostly at the level of species, subgenera and genera, and the resulting changes are generally more firmly substantiated than the higher-level revisions based on comparative-morphological studies. The general outlines of evolution within the genera *Cannartus*, *Ommatartus* and *Dorcadospyris* were described by Riedel (1959b). In 1968 and 1969, Goll published the first results of his broad revision of a number of spyrid genera, parts of which have been confirmed and elaborated by later, more detailed investigations (Goll, 1972a, b). Many other evolutionary lineages in Cainozoic polycystins have been traced by Riedel and Sanfilippo (1970, 1971), Sanfilippo and Riedel (1970, 1973), Hays (1970), Moore (1971, 1972) and Foreman (1973). Some of the phylogenetic relationships thus determined could not have been deduced on the basis of morphological studies alone (e.g. the species of the genus *Lithapium*, and the origin of *Cannartus* from *Lithocyclia*), and it therefore seems prudent at present to investigate Cainozoic radiolarian evolution in sequences such as those being sampled by the Deep Sea Drilling Project, rather than to base taxonomic revisions on comparative-morphological studies without supporting stratigraphic evidence.

Because some parts of the taxonomic system have remained unchanged since the time of Haeckel, and other parts have been modified on the basis of comparative-morphological studies or phylogenetic investigations, it is not possible at present to provide a uniform summary of radiolarian taxonomy. On the following pages we describe and illustrate a selection of Cainozoic radiolarians, emphasising those with well understood relationships and stratigraphic significance and treating others in less detail. The brief descriptions given for genera and higher-level taxa are not definitive diagnoses, but only indications of the most obvious features characterising the species included. It is not possible to define briefly many of the natural genera representing phyletic lineages, since the course of evolution often resulted in great changes in skeletal morphology. The species illustrated are chosen to overcome this deficiency in the written descriptions, and further details can be found in the literature cited.

To complement the systematic arrangement of the taxonomic text, the illustrations are arranged stratigraphically—the oldest forms on the earliest plates and younger ones on later plates. If we were to illustrate only those forms which can be confidently identified to genus or species in accordance with recent, critical taxonomic studies, the plates would give a distorted impression of the radiolarian assemblages of different ages, and therefore we have included illustrations of a large number of unidentified forms to reduce this potential distortion.

Systematic section

Subclass **Radiolaria** Müller, 1858

Superorder Polycystina Ehrenberg 1838, emend. Riedel 1967b.

Radiolaria with skeleton of opaline silica without admixed organic compounds. The bars forming the skeleton are generally solid, rather than hollow.

Order SPUMELLARIA Ehrenberg, 1875

Skeleton generally in the form of a sphere, or derived from a sphere (e.g. ellipsoidal, discoidal, lenticular, spiral). Many species have several concentric shells, the robust outer one being termed "cortical" and the usually much smaller, more delicate inner one(s) termed "medullary"—see, for example, pl. 4, fig. 16; pl. 21, fig. 10.

Family OROSPHAERIDAE Haeckel, 1887. Pl. 9, fig. 1; pl. 11, fig. 10; pl. 15, figs 12–14; pl. 18, fig. 1.

Large robust skeletons consisting of a lattice shell, or two approximately concentric shells, and radial spines generally circular in cross section. In some forms, a spicule of usually 6–8 rays is present at or near the shell apex. Shell in some members subspherical, but actually bipolar (and allopolar) in all well-known forms.

The family ranges from Eocene to Recent, having its greatest diversity in Oligocene and Early Miocene assemblages. Of the species illustrated, *Orostaurus simplex* Friend and Riedel (1967) (pl. 9, fig. 1) is known only from the *Podocrytis mitra* Zone (Middle Eocene), *Oropelex pagoda* Friend and Riedel (1967) (pl. 11, fig. 10) ranges from the *Podocyrtis mitra* Zone to the *Dorcadospyris ateuchus* Zone (Middle Eocene to Late Oligocene), *Orodapis spongiosa* Friend and Riedel (1967) (pl. 15, fig. 12) ranges from the *Dorcadospyris ateuchus* Zone to the *Calocycletta virginis* Zone (Late Oligocene to Early Miocene), *Oropagis dolium* Friend and Riedel (1967) (pl. 15, fig. 13) ranges from the *Lychnocanoma elongata* Zone to the *Calocycletta virginis* Zone (early Early Miocene), *Oroscena carolae* Friend and Riedel (1967) (pl. 15, fig. 14) ranges from about the *Lychnocanoma elongata* Zone to the *Dorcadospyris alata* Zone (Late Oligocene or Early Miocene to early Middle Miocene), and the genus *Orodendrum* Haeckel (1887) (pl. 18, fig. 1) ranges from the *Calocycletta virginis* Zone (Early Miocene) to the present.

Family COLLOSPHAERIDAE Müller, 1858. Pl. 10, fig 4(?); pl. 17, fig. 4; pl. 18, figs 2–7; pl. 20, figs 3, 4, 12–15; pl. 21, figs 6, 7.

Colonial forms, the spherical or subspherical shells of which, though dissociated in sediments, are usually recognisable by having thin shell walls with relatively wide flat areas between pores, commonly conical or tubular protuberances, spines (if present) circular in section, and never possessing medullary shells.

The family is known to range from the *Calocycletta virginis* Zone (Early Miocene) to the present, and may extend back as far as the Late Eocene *Thyrsocyrtis bromia* Zone (see pl. 10, fig. 4). The only species for which the stratigraphic range is known is *Solenosphaera omnitubus* Riedel and Sanfilippo (1971) (pl. 20, fig. 12) which occurs from the *Ommatartus penultimus* Zone to the *Spongaster pentas* Zone (Late Miocene to Pliocene).

Family ACTINOMMIDAE Haeckel, 1862, *sensu* Riedel, 1967b. Pl. 1, figs 1–7; pl. 4, fig. 16; pl. 5, figs 1, 4, 6; pl. 6, figs 7, 8; pl. 8, figs 10, 15; pl. 12, figs 13, 14; pl. 13, figs 1, 2, 4; pl. 15, fig. 19; pl. 17, figs 5, 6; pl. 18, figs 8–13; pl. 20, figs 5–7, 16, 17; pl. 21, figs 8–13.

Solitary spumellarians with shells spherical or ellipsoidal (or modifications of those shapes), usually without internal spicule, and generally much smaller than orosphaerids.

This polyphyletic family ranges through the entire Cainozoic and includes a great number of species, the relationships of most of which have not yet been worked out. We list below some low-latitude species of which the stratigraphic ranges have been established, followed by two groups sufficiently well understood to be treated as subfamilies.

Diploplegma somphum Sanfilippo and Riedel (1973) (pl. 1, fig. 2) is known only from the Late Palaeocene, but may range earlier.

Entapium chaenapium Sanfilippo and Riedel (1973) (pl. 5, fig. 4) ranges from the *Phormocyrtis striata striata* Zone to the *Thyrsocyrtis triacantha* Zone (early to middle Middle Eocene).

The genus *Lithapium* Haeckel (1887) appears to have evolved from *Entapium chaenapium* by outward migration of the medullary shell, its fusion with the cortical shell, and eventual transformation of the entire skeleton into an open cone. The species of this lineage are *Lithapium plegmacantha* Riedel and Sanfilippo (1970) (pl. 6, fig. 8), *L. anoectum* Riedel and Sanfilippo (1970) (pl. 8, fig. 10), and *L. mitra* (Ehrenberg, 1873) (pl. 8, fig. 15), and their combined range is from the *Theocotyle cryptocephala cryptocephala* Zone through the *Thrysocyrtis bromia* Zone (early Middle Eocene through Late Eocene).

Spongatractus pachystylus (Ehrenberg, 1873) (pl. 6, fig. 7) evidently developed from *S. balbis* Sanfilippo and Riedel (1973) (pl. 5, fig. 6), and their combined range is from the *Buryella clinata* Zone to approximately the *Thyrsocyrtis bromia* Zone (late Early Eocene to Late Eocene).

Stylosphaera goruna Sanfilippo and Riedel (1973) (pl. 1, fig. 5) ranges from Late Paleocene or earlier, to the *Bekoma bidartensis* Zone (early Early Eocene).

Thecosphaerella glebulenta Sanfililppo and Riedel (1973) (pl. 5, fig. 1) appears to be practically confined to the *Buryella clinata* Zone (late Early Eocene).

Subfamily SATURNALINAE Deflandre, 1953. Pl. 1, fig. 7; pl. 21, fig. 11.

Spherical shell with two opposite spines joined by a ring.

This subfamily had its principal development in the Mesozoic, but is represented throughout the Cainozoic by one or a few species. Forms with spines on the periphery of the ring are very rare in Palaeocene (and perhaps Eocene) assemblages, but are more common in Mesozoic faunas.

Subfamily Artiscinae Haeckel, 1881, *sensu* Riedel, 1967b. Pl. 13, fig. 4; pl. 15, fig. 19; pl. 17, figs 5, 6; pl. 18, figs 12, 13; pl. 20, figs 5–7, 16, 17; pl. 21, figs 12, 13.

Cortical shell ellipsoidal, usually equatorially constricted and enclosing a medullary shell; opposite poles of the shell generally bear spongy columns and/or single or multiple latticed caps.

Genus *CANNARTUS* Haeckel, 1881, *sensu* Riedel, 1971a. Pl. 13, fig. 4; pl. 15, fig. 19; pl. 17, figs 5, 6; pl. 18, figs 12, 13; pl. 20, fig. 5.

Ellipsoidal cortical shell, enclosing a double medullary shell and commonly with an equatorial constriction, and with two opposite spongy columns in the main axis. No polar caps between cortical shell and spongy columns.

The earliest member of this genus, *Cannartus prismaticus* (Haeckel, 1887) (pl. 13, fig. 4) evidently evolved from *Lithocyclia angusta* or *L. crux* by reduction of the number of spongy radii from three to two, and displacement of the axis of rotational symmetry through 90°. Most of the included species constitute an evolutionary lineage from *C. prismaticus* through *C. tubarius* (Haeckel, 1887) (pl. 15, fig. 19), *C. violina* Haeckel (1887) (pl. 17, fig. 5) and *C. mammifer* (Haeckel, 1887) (pl. 17, fig. 6) to *C. laticonus* Riedel (1959b) (pl. 20, fig. 5). *C. petterssoni* Riedel and Sanfilippo (1970) (pl. 18, figs 12, 13) and its apparent descendant *Ommatartus hughesi* (Campbell and Clark, 1944 (pl. 20, fig. 7) should probably be placed in a separate genus or subgenus, since they represent a branch from the main *Cannartus-Ommatartus* lineage. The genus (excluding *C. petterssoni*) ranges from the *Dorcadospyris ateuchus* Zone through the *Cannartus petterssoni* Zone (Late Oligocene through Middle Miocene).

Genus *OMMATARTUS* Haeckel, 1881, *sensu* Riedel, 1971a. Pl. 20, figs 6, 7, 16, 17; pl. 21, figs 12,13.

Equatorially constricted cortical shell, enclosing a double medullary shell, and bearing polar caps which in some species are surmounted by spongy columns.

With the exception of *Ommatartus hughesi* (see discussion of *Cannartus*), this genus comprises a single lineage arising from *Cannartus laticonus* and extending from *O. antepenultimus* Riedel and Sanfilippo (1970) (pl. 20, fig. 6) through *O. penultimus* (Riedel, 1957) (pl. 20, fig. 17) and *O. avitus* (Riedel, 1953) (pl. 20, fig. 16) to *O. tetrathalamus* (Haeckel, 1887) (pl. 21, figs 12, 13). In the course of the evolution of this lineage, the polar caps increased in size and the spongy columns decreased. The genus ranges from the *Ommatartus antepenultimus* Zone (early Late Miocene) to the present.

Family PHACODISCIDAE Haeckel, 1881. Pl. 3, fig. 14; pl. 5, figs 5, 6, 11–13; pl. figs 9, 10; pl. 11, fig. 11.

A lenticular cortical shell, not surrounded by spongy or chambered structure, encloses a much smaller, single or multiple medullary shell. The margin (and less commonly the surfaces) of the cortical shell may bear radial spines.

This family apparently ranges through the entire Cainozoic. The present generic taxonomy is not satisfactory, and the two genera treated below include only a small proportion of Cainozoic phacodiscids.

Genus *HELIOSTYLUS* Haeckel, 1881, *sensu* Sanfilippo and Riedel, 1973. Pl. 3, fig. 14; pl. 5, fig. 5; pl. 11, fig. 11.

Lenticular cortical shell joined to the medullary shell by two opposite bars in the equatorial plane (often continued as strong marginal spines) and by shorter bars in the central area.

Phyletic relationships of the species have not been investigated, and the genus ranges from Late Palaeocene or earlier to the *Dorcadospyris ateuchus* Zone (Late Oligocene).

Genus *PERIPHAENA* Ehrenberg, 1873 *sensu* Sanfilippo and Riedel, 1973. Pl. 5, figs 11–13; pl. 6, figs 9, 10.

Disc circular, triangular or elliptical in outline, with smooth surface and numerous small pores. Margin with spines, or a poreless girdle, or both. Without two opposite bars as *Heliostylus*.

This genus may be polyphyletic, and its total range is not yet known, but some of the included species are useful stratigraphically. *Periphaena decora* Ehrenberg (1873) (pl. 5, fig. 13; pl. 6, fig. 9) ranges from the *Buryella clinata* Zone to the *Thyrsocyrtis bromia* Zone (late Early Eocene to Late Eocene). *P. tripyramis* (Haeckel, 1887) (pl. 5, fig. 11; pl. 6, fig. 10) ranges from the *Phormocyrtis striata striata* Zone to at least the *Podocyrtis mitra* Zone (early Middle Eocene to Late Middle Eocene or higher). *P. delta* Sanfilippo and Riedel (1973) (pl. 5, fig. 12) ranges from the *Phormocyrtis striata striata* Zone to the *Theocampe mongolfieri* Zone (early to middle Middle Eocene).

Family COCCODISCIDAE Haeckel, 1862

Discoidal forms consisting of a lenticular cortical shell enclosing a small single or double medullary shell, and surrounded by an equatorial zone of spongy or concentrically-chambered structure.

As it is defined below, the genus *Lithocyclia* includes the great majority of forms assignable to this family. This is probably not a satisfactory long-term solution for the taxonomy of the family at the generic level, but phyletic studies are required to improve the classification.

Genus *LITHOCYCLIA* Ehrenberg, 1847a, *sensu* Riedel and Sanfilippo, 1970. Pl. 6, fig. 12; pl. 10, figs 2, 3; pl. 11, figs 12, 13.

Coccodiscids with the spongy or chambered equatorial girdle either entire or reduced to a number of separate arms (commonly 3 or 4). Radial spines are present in some species.

Members of this genus may be coarsely classified into the *Lithocyclia ocellus* group (*sensu* Riedel and Sanfilippo, 1970) (pl. 6, fig. 12), the *L. aristotelis* group (*sensu* Riedel and Sanfilippo, 1970) (pl. 10, figs 2, 3), *L. angusta* (Riedel, 1959b) (pl. 11, fig. 12) and *L. crux* Moore (1971) (pl. 11, fig. 13). The *L. ocellus* group evidently gave rise to the *L. aristotelis* group from which developed *L. angusta* and *L. crux*. *L. angusta* or *L. crux* apparently gave rise to the genus *Cannartus*. The genus ranges from the *Buryella clinata* Zone to the *Dorcadospyris ateuchus* Zone or *Lychnocanoma elongata* Zone (late Early Eocene to Late Oligocene).

Family PORODISCIDAE Haeckel, 1881, *sensu* Kozlova in Petrushevskaya and Kozlova, 1972. Pl. 1, figs 8–11; pl. 4, figs 1, 2; pl. 18, figs 14, 15; pl. 21, figs 14, 15.

Discoidal forms comprising a small central chamber surrounded by concentric, interrupted or spiral latticed bands. In some the disc is small and the major part of the shell is represented by two to four arms in the equatorial plane.

The family ranges through the entire Cainozoic, and includes a large number of species. The generic taxonomy is presently undergoing a profound change, largely as a result of the investigation of details of the central structure, and the present state of our understanding of this group is summarised by Kozlova in Petrushevskaya and Kozlova (1972). Because generic revisions are not yet tested and stabilised, the few stratigraphically significant species mentioned below are named according to older concepts of genera.

Amphymenium splendiarmatum Clark and Campbell (1942) (pl. 18, fig. 14) ranges from the *Phormocyrtis striata striata* Zone to at least the *Podocyrtis mitra* Zone (early Middle Eocene to late Middle Eocene and probably later).

Spongodiscus quartus (Borisenko, 1958, *sensu* Sanfilippo and Riedel, 1973) (pl. 1, fig. 9; pl. 4, fig. 1) ranges from Late Palaeocene or earlier to the lower part of the *Phormocyrtis striata striata* Zone (early Middle Eocene).

Stylotrochus alveatus Sanfilippo and Riedel (1973) (pl. 1, fig. 11) ranges from the Late Palaeocene or earlier to the lower part of the *Bekoma bidartensis* Zone (earliest Eocene).

S. nitidus Sanfilippo and Riedel (1973) (pl. 4, fig. 2) ranges from the Late Palaeocene or earlier through the *Phormocyrtis striata striata* Zone (early Middle Eocene).

Family SPONGODISCIDAE Haeckel, 1862, *sensu* Kozlova in Petrushevskaya and Kozlova, 1972. Pl. 1, fig. 12; pl. 4, figs 3, 4; pl. 5, figs 7–10; pl. 6, figs 11, 13; pl. 9, fig. 2; pl. 13, fig. 3; pl. 18, fig. 16; pl. 20, figs 8, 9, 18; pl. 21, figs 16–20.

Similar in general form to Porodiscidae, but the skeletal structure predominantly irregularly spongy, concentric chambers (when present) being confined to the central region.

As is the case with the Porodiscidae, this family ranges through the Cainozoic and is currently undergoing profound revision, and the species listed here are named according to older concepts of genera.

Amphicraspedum prolixum Sanfilippo and Riedel (1973) (pl. 6, fig. 13) ranges from Late Palaeocene to the *Thyrsocyrtis triacantha* Zone (middle Middle Eocene).

Spongodiscus americanus Kozlova (in Kozlova and Gorbovets, 1966) (pl. 5, fig 9) ranges from Late Palaeocene or earlier to the *Podocyrtis ampla* Zone (Middle Eocene). *S. cruciferus* (Clark and Campbell, 1942) (pl. 4, fig 3) ranges from Late Palaeocene to the *Thyrsocyrtis triacantha* Zone (middle Middle Eocene). *S. phrix* Sanfilippo and Riedel (1973) (pl. 5, fig. 10) ranges from the *Phormocyrtis striata striata* Zone through the *Thyrsocyrtis triacantha* Zone (early through middle Middle Eocene). *S. rhabdostylus* (Ehrenberg, 1873) (pl. 6, fig. 11; pl. 9, fig. 2) ranges from the *Phormocyrtis striata striata* Zone to the *Podocyrtis mitra* Zone (early to late Middle Eocene) and probably higher.

Stylotrochus quadribrachiatus Sanfilippo and Riedel (1973) (pl. 5, figs 7, 8) ranges from the *Phormocyrtis striata striata* Zone to the *Theocotyle crypto-*

cephala cryptocephala Zone (early Middle Eocene), evolving into a form with more than four arms toward the end of its range.

Genus *SPONGASTER* Ehrenberg, 1860, emend. Riedel and Sanfilippo, 1971. Pl. 20, figs 8, 18; pl. 21, fig. 19.

This genus comprises the evolutionary lineage extending from *Spongaster berminghami* (Campbell and Clark, 1944; *S. klingi* Riedel and Sanfilippo 1971 is probably a junior synonym) (pl. 20, fig. 8) through *S. pentas* Riedel and Sanfilippo (1970) (pl. 20, fig. 18) to *S. tetras* Ehrenberg (1860) (pl. 21, fig. 19). The range of the genus is from the *Ommatartus antepenultimus* Zone (early Late Miocene) to the present.

Family PYLONIIDAE Haeckel, 1881. Pl. 17, fig. 7; pl. 21, fig. 21.

Shell consisting of an ellipsoidal central chamber of complicated construction (Kozlova, 1967) surrounded by a series of successively larger elliptical latticed girdles in three mutually perpendicular planes, the major diameter of each girdle being the minor diameter of the next larger one.

This family ranges from the Eocene to the present, but is well developed and common only in Miocene and later assemblages.

Family THOLONIIDAE Haeckel, 1887. Pl. 18, fig. 17; pl. 21, fig. 22.

Shell completely latticed, without large openings, and with two or more annular constrictions or furrows separating dome-shaped protuberances.

Representatives of this family are rare, and range from the *Cannartus petterssoni* Zone (Middle Miocene) to the present.

Family LITHELIIDAE Haeckel, 1862. Pl. 1, fig.13; pl. 5, fig. 14; pl. 18, fig. 18; pl. 20, fig. 19.

Shell usually ellipsoidal to spherical in outer form, and of spiral structure (at least internally).

This probably polyphyletic family ranges throughout the Cainozoic, but the only species for which the phyletic relationship has been established are *Lithelius foremanae* Sanfilippo and Riedel (1973) (pl. 1, fig. 13) which evolved into *L. hexaxyphophorus* (Clark and Campbell, 1942) (pl. 5, fig. 14)—the cumulative range of these two species is from Late Palaeocene or earlier to at least the *Podocyrtis mitra* Zone (late Middle Eocene).

Order NASSELLARIA Ehrenberg, 1875

Skeleton heteropolar and generally bilaterally symmetrical, commonly in the form of a spicule with rays arising from a median bar, a ring (often D-shaped), or a cap-like construction often of several segments uniserially arranged.

A large number of special terms are used to describe nassellarian skeletons, and it is not possible to avoid using some of them here. The fundamental spicule is shown within a large, delicate, latticed cephalis in pl. 22, fig. 10—

from a short median bar arise three downwardly directed spines which form the "feet" (the one to the right broken), and two upwardly directed spines, the vertical one termed "apical" and the oblique one (to the left in the illustration) termed "vertical". In the spyrids, the median bar and apical and vertical spines are included in the sagittal ring (for example, forming the bottom, right and left, respectively, of the specimen illustrated in pl. 13, fig. 14). In the cyrtids, homologues of the elements of the fundamental spicule are included in the uppermost segment which is termed the "cephalis", the following two segments being termed the "thorax" and "abdomen". For a detailed discussion of nassellarian skeletal structure and terminology, see Petrushevskaya (1971a).

Suborder SPYRIDA Ehrenberg, 1847b, emend. Petrushevkaya, 1971c. Pl. 1, figs 14, 15; pl. 4, fig. 5; pl. 6, figs 14–17; pl. 9, fig. 3; pl. 10, figs 5, 6; pl. 11, figs 1, 2; pl. 12, figs 1–3; pl. 13, figs 5–15; pl. 15, fig. 15; pl. 16, figs 1, 12–16; pl. 17, figs 8–12; pl. 18, fig. 19; pl. 19, figs 1–5; pl. 22, figs 1–6.

Skeleton generally possessing a complete sagittal ring, and commonly also latticed lateral chambers forming a bilobed cephalis.

Families within this suborder, which ranges from Late Palaeocene or earlier to the present, are not yet satisfactorily defined.

Genus *DORCADOSPYRIS* Haeckel, 1881. Pl. 11, figs 1, 2; pl. 13, figs 5–8; pl. 16, fig. 12; pl. 17, fig. 8.

Shell thickwalled, small, bearing two (or in some species more) large cylindrical feet and usually a strong apical horn.

The two-footed members of this genus appear to constitute a lineage originating from *Tristylospyris triceros* (Ehrenberg, 1873) (pl. 10, fig. 5) and extending from *Dorcadospyris ateuchus* (Ehrenberg, 1873) (pl. 13, fig. 8) to *D. alata* (Riedel, 1959b) (pl. 17, fig. 8). The genus ranges from the *Dorcadospyris ateuchus* Zone to the *Cannartus petterssoni* Zone (Late Oligocene to Middle Miocene).

Genus *GORGOSPYRIS* Haeckel, 1881. Pl. 16, fig. 13; pl. 17, fig. 9.

Cephalis lacking horns, and thorax generally latticed, usually terminating in numerous lamellar feet.

This genus includes a lineage apparently originating from *Dendrospyris pannosa* Goll (1968) (pl. 16, fig. 1), comprising *Gorgospyris schizopodia* Haeckel (1887) (pl. 16, fig. 13) evolving to *G. perizostra* Sanfilippo and Riedel (in Sanfilippo *et al.*, 1973) (pl. 17, fig. 9). The cumulative range of the latter two species is from the *Calocycletta virginis* Zone through the *Dorcadospyris alata* Zone (Early Miocene to early Middle Miocene).

Genus *LIRIOSPYRIS* Haeckel, 1881, *sensu* Goll, 1968. Pl. 16, figs 14–16; pl. 17, fig. 10; pl. 19, fig. 1.

Species of this genus exhibit a variety of morphologies that cannot be summarised briefly.

Two evolutionary lineages are apparently well substantiated. *Liriospyris mutuaria* Goll (1968) (pl. 16, fig. 15) evolved to *L. reticulata* (Ehrenberg, 1872b) (pl. 19, fig. 1) by way of *L. globosa* Goll (1968) (pl. 16, fig. 14), this lineage extending from the *Lychnocanoma elongata* Zone (Early Miocene) to the Quaternary. *L. stauropora* (Haeckel, 1887) (pl. 16, fig. 16) evolved to *L. parkerae* Riedel and Sanfilippo (1971) (pl. 17, fig. 10), and this short lineage extends from the upper part of the *Calocycletta virginis* Zone to the lower part of the *Dorcadospyris alata* Zone (late Early Miocene to early Middle Miocene).

Genus *PSYCHOSPYRIS* Riedel and Sanfilippo, 1971. Pl. 15, fig. 15; pl. 17, fig. 11.

Large, discoidal forms with a degenerated cephalic wall and a peripheral spongy zone.

Psychospyris parva Riedel and Sanfilippo (1971) (pl. 15, fig. 15) evolved from *Dendrospyris anthocyrtoides* (Bütschli, 1882b) (pl. 13, figs 9, 10), thus initiating an evolutionary lineage that terminated in *P. grandis* (Riedel and Sanfilippo, 1971) (pl. 17, fig. 11). The genus ranges from the *Lychnocanoma elongata* Zone to the *Dorcadospyris alata* Zone (Early Miocene to early Middle Miocene).

Genus *RHABDOLITHIS* Ehrenberg, 1847b. Pl. 4, fig. 5; pl. 9, fig. 3.

Skeleton consisting mainly of a rod, circular in section, commonly thickened at one end and the other end in one species attached to the median bar. More delicate lateral branches commonly arise from the main rod.

The genus ranges from the Late Palaeocene or earlier to some undetermined level probably in the Late Eocene, the evolutionary transition from *Rhabdolithis ellida* Sanfilippo and Riedel (1973) (pl. 4, fig. 5) to *R. pipa* Ehrenberg (1854) (pl. 9, fig. 3) occurring in the *Buryella clinata* Zone (late Early Eocene).

Suborder CYRTIDA Haeckel, 1862, emend. Petrushevskaya, 1971c.

Skeleton generally conical or cap-shaped, consisting of segments uniserially and usually rectilinearly arranged.

Family PLAGONIIDAE Haeckel, 1881, emend. Riedel, 1967b. Pl. 2, figs 1–5; pl. 4, fig. 6; pl. 7, figs 1–5; pl. 8, fig. 11; pl. 13, fig. 16; pl. 14, figs 1–12; pl. 16, fig. 2; pl. 17, figs 13, 14; pl. 19, figs 6–11; pl. 20, fig. 20; pl. 22, figs 7–15.

Cephalis consisting entirely of the fundamental spicule, or having a latticed skeleton including a large cephalis within which this spicule is well developed.

This large, probably polyphyletic family ranges through the entire Cainozoic, and is subdivided in various ways by different authors. Petrushevskaya (1971a, c) treats it as a superfamily (Plagiacanthoidea). In addition to many forms whose relationships are not satisfactorily understood, it includes some apparently natural groups which may be treated as subfamilies—e.g. the flatly-expanded sethophormins (pl. 2, fig. 5; pl. 14, fig. 1; pl. 19, figs 8, 9; pl. 22, figs 12, 13) and the sethoperins, in which three latticed plates extend from the apical horn to the thoracic ribs, or the thorax is perforated by three large gates (pl. 14, figs 8–12; pl. 17, figs 13, 14; pl. 19, fig. 11; pl. 20, fig. 20; pl. 22, figs 14, 15).

Genus *TEPKA* Sanfilippo and Riedel, in Sanfilippo *et al.*, 1973. Pl. 16, fig. 2.

A thin, approximately circular plate with large central pore, smaller pores over remainder of plate, and undifferentiated margin.

Only one species is known, *T. perforata* Sanfilippo and Riedel (in Sanfilippo *et al.*, 1973), ranging from the *Lychnocanoma elongata* Zone to the *Dorcadospyris alata* Zone (early Early Miocene to early Middle Miocene).

Family THEOPERIDAE Haeckel, 1881, emend. Riedel, 1967b. Pl. 2, figs 8–14; pl. 3, figs 1–12; pl. 4, figs 7–15; pl. 5, figs 2, 3; pl. 6, figs 1–4; pl. 7, figs 7–18; pl. 8, figs 1–3, 12, 13; pl. 9, figs 4–10, 15; pl. 10, figs 7–15; pl. 11, figs 1–4; pl. 12, figs 4–8; pl. 14, figs 13–21; pl. 15, figs 1–6; pl. 16, figs 3–7, 17; pl. 17, figs 15–17; pl. 19, figs 12–18; pl. 20, figs 10, 21; pl. 21, figs 1, 5; pl. 22, fig. 16; pl. 23, figs 1–7.

Cephalis relatively small, approximately spherical, often poreless or sparsely perforate. The internal spicule, homologous with that of the plagoniids, is reduced to a less conspicuous structural element than in that group.

This diverse family ranges through the entire Cainozoic, and includes a considerable number of species of which the stratigraphic ranges and phyletic relationships are understood.

Genus *ARTOPHORMIS* Haeckel, 1881, *sensu* Riedel and Sanfilippo, 1970. Pl. 10, figs 11, 12; pl. 15, fig. 5.

A partly artificial grouping of three stratigraphically useful forms in which the abdomen bears numerous short teeth, or a post-abdominal segment tending to be more coarsely latticed proximally than distally.

Artophormis dominasinensis (Ehrenberg, 1873) (pl. 10, fig. 11) appears to be restricted to the *Thyrsocyrtis bromia* Zone (Late Eocene). *A. barbadensis* (Ehrenberg, 1873) (pl. 10, fig. 12) evolved into *A. gracilis*, Riedel (1959b) (pl. 15, fig. 5), and this short lineage ranges from approximately the *Podocyrtis goetheana* Zone through the *Lychnocanoma elongata* Zone (early Late Eocene to early Early Miocene).

Genus *BEKOMA* Riedel and Sanfilippo, 1971. Pl. 3, figs 3, 4; pl. 4, fig. 9.

Cephalis bears apical and vertical horns, and is not separated from the upper part of the campanulate thorax by an external stricture. Three feet usually subcylindrical, with branches in the distal portion which may join to form a ring.

Bekoma bidartensis Riedel and Sanfilippo (1971) (pl. 3, fig. 9) evidently evolved from *B. campechensis* Foreman (1973) (pl. 3, fig. 4) but the phylogenetic relationships of the other species are not understood. The genus ranges from Late Palaeocene or earlier to Late Early Eocene (*Buryella clinata* Zone).

Genus *BURYELLA* Foreman, 1973. Pl. 3, figs 6, 7; pl. 7, fig. 16.

Four or more segments, without appendages, not flared terminally, and with pores aligned transversely on at least one segment.

The three species comprising this genus represent an evolutionary lineage from *Buryella pentadica* Foreman (1973) (pl. 3, fig. 6) to *B. clinata* Foreman (1973) (pl. 7, fig. 16). The genus ranges from Late Palaeocene or earlier to the *Dorcadospyris ateuchus* Zone (Late Oligocene), but is very rare above the *Phormocyrtis striata striata* Zone (early Middle Eocene).

Genus *CALOCYCLAS* Ehrenberg, 1847b. Pl. 9, fig. 10; pl. 11, fig. 1.

Shell with strong apical horn, subspherical cephalis, hemispherical or inflated thorax, and numerous lamellar feet sometimes joined proximally to form an abdomen.

The genus ranges from the *Buryella clinata* Zone to the *Thyrsocyrtis bromia* Zone (late Early Eocene to late Late Eocene), and perhaps comprises only the two species *Calocyclas hispida* (Ehrenberg, 1873) (pl. 9, fig. 10) and *C. turris* Ehrenberg (1873) (pl. 11, fig. 1), the latter having evolved from the former by the feet becoming joined to form an abdomen.

Genus *CYCLAMPTERIUM* Haeckel, 1887, *sensu* Sanfilippo and Riedel, 1970. Pl. 12, figs 5, 6; pl. 16, fig. 3; pl. 19, fig. 16.

There is some question regarding the relationship between the type species and others assigned to this genus, but the best-known forms currently included comprise an evolutionary lineage especially useful for determination of ages of poorly preserved assemblages in which smaller, less robust skeletons are dissolved. The genus ranges from the *Theocyrtis tuberosa* Zone (Early Oligocene) to the present, and, of the species illustrated, *Cyclampterium milowi* Riedel and Sanfilippo (1971) (pl. 12, fig. 6) ranges from the *Theocyrtis tuberosa* Zone to the *Dorcadospyris ateuchus* Zone (Early Oligocene to Late Oligocene), *C. pegetrum* Sanfilippo and Riedel (1970) (pl. 12, fig. 5) ranges from the *Lychnocanoma elongata* Zone to the *Calocycletta virginis* Zone (Late Oligocene to Early Miocene), *C. leptetrum* Sanfilippo and Riedel (1970) (pl. 16, fig. 3) ranges from the *Calocycletta virginis* Zone to the *Dorcadospyris alata* Zone (Early Miocene to early Middle Miocene) and *C. brachythorax* Sanfilippo and Riedel (1970) (pl. 19, fig. 16) ranges from the *Cannartus petterssoni* Zone to the *Ommatartus antepenultimus* Zone (late Middle Miocene to early Late Miocene).

Genus *CYRTOCAPSELLA* Haeckel, 1887, emend. Sanfilippo and Riedel, 1970. Pl. 16, figs 4, 17.

Shell generally drop-shaped or pyriform, consisting of three or four segments of which the terminal one has a very constricted aperture. Some specimens have an additional, usually more delicate, generally closed conical segment below the constricted one.

The genus ranges from the *Lychnocanoma elongata* Zone to the *Cannartus petterssoni* Zone, two distinctive species being *Cyrtocapsella cornuta* Haeckel (1887) (pl. 16, fig. 4) and *C. tetrapera* Haeckel (1887) (pl. 16, fig. 17).

Genus *EUSYRINGIUM* Haeckel, 1881. Pl. 9, figs 4–6.

Cephalis subspherical, with a robust cylindro-conical horn. Thorax pyriform, with a constricted aperture which may be prolonged as a subcylindrical tube.

The best-known species are *Eusyringium lagena* (Ehrenberg, 1873) (pl. 9, fig. 4), which ranges from the *Thyrsocyrtis triacantha* Zone to approximately the *Podocyrtis mitra* Zone (Middle Eocene), and its direct descendant *E. fistuligerum* Ehrenberg (1873) (pl. 9, fig. 5) which ranges up to the *Thyrsocyrtis bromia* Zone (Late Eocene).

Genus *LAMPTONIUM* Haeckel, 1887, *sensu* Riedel and Sanfilippo, 1970. Pl. 3, fig. 12; pl. 6, fig. 2.

Shell generally robust, pyriform with inflated thorax, and usually a strong apical horn. Three spines extend laterally from the thorax in one species, and some have three indistinct feet terminally, or the thoracic aperture closed by a lattice-plate, or a short, narrow, subcylindrical abdomen.

In addition to some species whose relationships are not clearly understood, the genus includes an evolutionary lineage extending from *Lamptonium pennatum* Foreman (1973) (pl. 3, fig. 12) to *L. fabaeforme constrictum* Riedel and Sanfilippo (1970) (pl. 6, fig. 2), which ranges from Late Palaeocene or earlier to the lower part of the *Thyrsocyrtis triacantha* Zone (middle Middle Eocene).

Genus *LITHOCHYTRIS* Ehrenberg, 1847a. Pl. 5, fig. 2; pl. 7, fig. 15.

Three-segmented, closed distally, tending toward tetrahedral in general form. Short, robust apical horn and three robust feet terminally.

Lithochytris vespertilio Ehrenberg (1873) (pl. 7, fig. 15) evolved from *L. archaea* Riedel and Sanfilippo (1970) (pl. 5, fig. 2), which evidently developed from *Lychnocanoma auxilla*. The genus ranges from the *Buryella clinata* Zone to the *Podocyrtis goetheana* Zone (late Early Eocene to earliest Late Eocene).

Genus *LITHOPERA* Ehrenberg, 1847a. Pl. 17, fig. 16; pl. 19, fig. 17; pl. 20, fig. 21.

Terminally closed, two- or three-segmented forms in which the thorax forms the major part of the shell. This genus comprises the phyletic lineage leading to the development of *Lithopera bacca* Ehrenberg (1872a) on the one hand and *L. baueri* Sanfilippo and Riedel (1970) and *L. thornburgi* Sanfilippo and Riedel (1970) on the other, beginning with the earliest form in which the second segment constitutes the major part of the shell.

The genus originated in the *Calocycletta costata* Zone (late Early Miocene), and is still extant. Of the species illustrated, *Lithopera renzae* Sanfilippo and Riedel (1970) (pl. 17, fig. 16) ranges from the *Calocycletta costata* Zone to the *Dorcadospyris alata* Zone (late Early Miocene to early Middle Miocene), *L. thornburgi* (pl. 19, fig. 17) ranges from the *Dorcadospyris alata* Zone to the *Cannartus petterssoni* Zone (Middle Miocene), and *L. bacca* (pl. 20, fig. 21) ranges from the *Cannartus petterssoni* Zone to the present.

Genus *LYCHNOCANOMA* Haeckel, 1887, *sensu* Foreman (1973). Pl. 2, figs 10, 11; pl. 3, fig. 5; pl. 4, fig. 12; pl. 7, fig. 12; pl. 10, figs 8–10; pl. 16, fig. 5.

An apparently polyphyletic genus, ranging through the entire Cainozoic and comprising two-segmented forms with two or three feet and without corresponding ribs in the thorax. In some, abdominal lattice (usually delicate) joins the feet.

Among the many unrelated forms included under this name are a few whose evolutionary relationships are understood. *Lychnocanoma auxilla* Foreman (1973) (pl. 4, fig. 12) evidently evolved into *Lithochytris archaea*, and *Lychnocanoma amphitrite* Foreman (1973) (pl. 10, fig. 9) is the descendant of *L. bellum* (Clark and Campbell, 1942) (pl. 10, fig. 10). The *L. babylonis* (Clark and Campbell, 1942) group (pl. 7, fig. 17), known also as the *Sethochytris babylonis* group, ranges from Late Palaeocene to the *Thyrsocyrtis bromia* Zone (Late Eocene), and give rise to the late Middle Eocene *Sethochytris triconiscus* Haeckel (1887) (pl. 9, fig. 7).

Genus *ORBULA* Foreman, 1973. Pl. 3, figs 8–10.

Shell of two segments with rather large cephalis, and thorax constricted distally, with ribs terminating in short, subterminal feet.

Only three species have been described in this genus, which is restricted to Late Palaeocene (and perhaps older) assemblages.

Genus *PHORMOCYRTIS* Haeckel, 1887. Pl. 3, fig. 11; pl. 4, fig. 11; pl. 8, fig. 13.

This generic name is used rather loosely for a stratigraphically useful form with two subspecies. *Phormocyrtis striata exquisita* Foreman (1973) (pl. 4, fig. 11) evolved to *P. striata striata* Brandt (1935) (pl. 3, fig. 11), and this lineage ranges from Late Palaeocene or earlier to approximately the *Podocyrtis mitra* Zone (Middle Eocene).

Genus *PTEROCANIUM* Ehrenberg, 1847a. Pl. 12, figs 7, 8; pl. 19, fig. 18; pl. 21, fig. 5.

Cephalis bearing an apical horn; thorax including three ribs continued as well developed, usually proximally latticed feet. Most species have delicate abdominal lattice-work connecting the proximal parts of the three feet.

The only species of which the stratigraphic range is known is *Pterocanium prismatium* Riedel (1957) (pl. 21, fig. 5)—*Spongaster pentas* Zone through *Pterocanium prismatium* Zone (late Pliocene). The genus is extant, and ranges back at least as far as the Early Oligocene (*Theocyrtis tuberosa* Zone).

Genus *STICHOCORYS* Haeckel, 1881. Pl. 16, figs 6, 7; pl. 17, fig. 17; pl. 21, fig. 1.

Many-segmented forms in which the first three or four segments constitute a conical upper portion of the shell and the narrower subsequent segments constitute a cylindrical lower portion.

The genus ranges from the *Calocycletta virginis* Zone to the *Spongaster pentas* Zone (Early Miocene to Pliocene). Of the species illustrated, *Stichocorys delmontensis* (Campbell and Clark, 1944) (pl. 17, fig. 17) ranges from the *Calocycletta virginis* Zone to the *Ommatartus penultimus* Zone (Early Miocene to Late Miocene), *S. wolffii* Haeckel (1887) (pl. 16, fig. 6) and *S. armata* (Haeckel, 1887) (pl. 16, fig. 7) range from the *Calocycletta virginis* Zone to the *Cannartus petterssoni* Zone (Early Miocene to late Middle Miocene), and *S. peregrina* (Riedel, 1953) (pl. 21, fig. 1) ranges from the *Stichocorys peregrina* Zone to the *Spongaster pentas* Zone (late Late Miocene to Pliocene).

Genus *THEOCORYS* Haeckel, 1881. Pl. 4, fig. 13; pl. 6, fig. 1; pl. 8, fig. 12; pl. 15, fig. 6.

Three-segmented forms with an apical horn, distal margin undifferentiated or with a simple peristome, and pores irregularly or quincuncially arranged.

A polyphyletic genus including many species, the evolutionary relationships of most of which are not understood. *Theocorys acroria* Foreman (1973) (pl. 4, fig. 13) may be ancestral to both *T. anaclasta* Riedel and Sanfilippo (1970) (pl. 6, fig. 1) and *T. anapographa* Riedel and Sanfilippo (1970) (pl. 8, fig. 12). Members of the genus as presently defined occur throughout the Cainozoic.

Genus *THEOCOTYLE* Riedel and Sanfilippo, 1970. Pl. 4, fig. 14; pl. 5, fig. 3; pl. 6, fig. 4; pl. 7, fig. 18; pl. 9, fig. 8.

Three-segmented forms with an apical horn, pores quincuncially arranged, and a distinctly differentiated peristome bearing short, broad feet in early representatives.

This genus comprises two evolutionary lineages, separated as subgenera. The subgenus *Theocotyle* extends from *T. cryptocephala nigriniae* Riedel and

Sanfilippo (1970) (pl. 5, fig. 3) to *T. cryptocephala conica* Foreman (1973) (pl. 6, fig. 4) or *T. venezuelensis* Riedel and Sanfilippo (1970) (pl. 7, fig. 18), and the subgenus *Theocotylissa* extends from *T. auctor* Foreman (1973) (pl. 4, fig. 14) to *T. ficus* (Ehrenberg, 1873) (pl. 9, fig. 8). The genus ranges from the *Bekoma bidartensis* Zone (early Early Eocene) to the lower part of the *Thyrsocyrtis bromia* Zone (Late Eocene).

Genus *THYRSOCYRTIS* Ehrenberg, 1847b, *sensu* Riedel and Sanfilippo, 1970. Pl. 4, fig. 15; pl. 6, fig. 3; pl. 8, figs 2, 3; pl. 9, fig. 9; pl. 11, figs 2–4.

Phyletically related forms with three segments, robust apical horn, pores quincuncially or irregularly arranged, and usually three feet arising from a differentiated peristome—though in the terminal member of the principal lineage there is no differentiated peristome and the feet number four or more.

Most species of this genus comprise an evolutionary lineage extending from *Thyrsocyrtis tarsipes* Foreman (1973) (pl. 4, fig. 15) to *T. tetracantha* (Ehrenberg, 1873) (pl. 11, figs 3, 4). *T. rhizodon* Ehrenberg (1873) (pl. 9, fig. 9) and *T. hirsuta robusta* Riedel and Sanfilippo (1970) (pl. 6, fig. 3) are evidently related but not in the direct line. The genus ranges from Late Palaeocene into the *Thyrsocyrtis bromia* Zone (Late Eocene).

Family CARPOCANIIDAE Haeckel, 1881, emend. Riedel, 1967b. Pl. 11, fig. 5; pl. 12, fig. 9; pl. 16, figs 8, 9, 18; pl. 17, fig. 1; pl. 23, fig. 8.

Shell consisting usually of cephalis and thorax, only a few species having considerable abdominal structure. Cephalis small, not sharply distinguished in contour from thorax, and tending to be reduced to a few bars within top of thorax.

The genus extends through most of the Cainozoic, and is extant. Its evolutionary origin is not yet clear, but the Eocene species may have originated from a form close to *Cryptoprora ornata* Ehrenberg (1873).

Genus *CARPOCANISTRUM* Haeckel, 1887, *sensu* Riedel and Sanfilippo, 1971. Pl. 11, fig. 5; pl. 12, fig. 9; pl. 16, fig. 8; pl. 23, fig. 8.

Cephalis merged with ovate thorax, pores often longitudinally aligned, and somewhat constricted peristome often bearing numerous teeth.

This genus is probably polyphyletic, and ranges from approximately the *Thyrsocyrtis bromia* Zone (Late Eocene) to Recent.

Genus *CARPOCANOPSIS* Riedel and Sanfilippo, 1971. Pl. 16, figs 9, 18; pl. 17, fig. 1.

Shell robust, with abdomen separated from thorax by a well-developed internal lumbar stricture.

The phylogenetic relationships of the species included here have not been determined, and the group may be unnatural. The genus ranges from the

Lychnocanoma elongata Zone to the *Cannartus petterssoni* Zone (early Early Miocene through Middle Miocene).

Family PTEROCORYTHIDAE Haeckel, 1881, emend. Riedel, 1967b. Pl. 6, fig. 5; pl. 8, figs 4–6, 14; pl. 9, figs 11, 12; pl. 10, fig. 1; pl. 11, fig. 6; pl. 12, fig. 10; pl. 15, figs 7, 8, 16; pl. 16, fig. 10; pl. 17, figs 2, 3; pl. 19, figs 19–23; pl. 23, figs 9, 10.

In the cephalis, two lateral lobes are separated from the main cephalic lobe by two obliquely downwardly directed lateral furrows originating from the apical bar.

This family is still extant, and its earliest known representatives are forms evidently closely related to *Podocyrtis papalis*, in the earliest Eocene or latest Palaeocene.

Genus *ANTHOCYRTIDIUM* Haeckel, 1881, *sensu* Petrushevskaya *in* Petrushevskaya and Kozlova (1972). Pl. 15, fig. 16; pl. 19, figs 19, 20.

Delicate, two-segmented forms with subcylindrical cephalis bearing a bladed apical horn. A distinct peristome is usually present, and commonly also terminal and/or subterminal teeth.

The origin of the genus is obscure, and it ranges from the *Lychnocanoma elongata* Zone (Late Oligocene or Early Miocene) or earlier, to the present.

Genus *CALOCYCLETTA* Haeckel, 1887, *sensu* Moore (1972). P. 16, fig. 10; pl. 17, figs 2, 3; pl. 19, fig. 21.

Three-segmented shell with robust apical horn, inflated-hemispherical thorax, subcylindrical or distally tapering abdomen, and commonly lamellar or triangular feet.

The genus comprises an evolutionary lineage extending from *Calocycletta parva* Moore (1972) through *C. robusta* Moore (1971) (pl. 16, fig. 10) and *C. virginis* Haeckel (1887) (pl. 17, fig. 3) to terminate with *C. caepa* Moore (1972) (pl. 19, fig. 21), with side-branches to *C. serrata* Moore (1972) and *C. costata* (Riedel, 1959b) (pl. 17, fig. 2). The lineage from *C. robusta* through *C. caepa* ranges from the *Dorcadospyris ateuchus* Zone to the *Ommatartus antepenultimus* Zone or the *O. penultimus* Zone (Late Oligocene to Late Miocene).

Genus *LAMPROCYCLAS* Haeckel, 1881, *sensu* Petrushevskaya and Kozlova, 1972. Pl. 15, fig. 7; pl. 19, fig. 22; pl. 23, fig. 9.

Three-segmented forms with subcylindrical cephalis usually bearing a stout apical horn. Thorax campanulate, and abdomen commonly longer and wider than thorax, in many species with a peristome and with terminal and subterminal teeth.

The origin of this genus has not been determined. It ranges from the *Theocyrtis tuberosa* Zone (Early Oligocene), or earlier, to the present.

Genus *PODOCYRTIS* Ehrenberg, 1847a, *sensu* Riedel and Sanfilippo, 1970. Pl. 6, fig. 5; pl. 8, figs 4–6, 14; pl. 9, figs 11, 12; pl. 10, fig. 1; pl. 11, fig. 6.

Comprises two evolutionary series. One (subgenus *Podocyrtis*) extends from *P. papalis* Ehrenberg (1847b) (pl. 9, fig. 12) through *P. diamesa* Riedel and Sanfilippo (1970) (pl. 8, fig. 5) and *P. phyxis* Sanfilippo and Riedel (1973) (pl. 8, fig. 6) to terminate in *P. ampla* Ehrenberg (1873) (pl. 8, fig. 14), with branches represented by *P. dorus* Sanfilippo and Riedel (1973) and *P. platypus* Sanfilippo and Riedel (1973). The other (subgenus *Lampterium* Haeckel, 1881) extends from *Podocyrtis aphorma* Riedel and Sanfilippo (1970) (pl. 6, fig. 5) through *P. sinuosa* Ehrenberg (1873) (pl. 8, fig. 4), *P. mitra* Ehrenberg (1854) (pl. 9, fig. 11) and *P. chalara* Riedel and Sanfilippo (1970) (pl. 10, fig. 1) to terminate in *P. goetheana* (Haeckel, 1887) (pl. 11, fig. 6), with *P. trachodes* Riedel and Sanfilippo (1970) evidently related but not in the direct line. Most members of the genus have a robust apical horn and three broad lamellar feet, but the other skeletal features undergo marked evolutionary change—in the subgenus *Podocyrtis* the abdomen becomes broader and the pores lose their pronounced longitudinal alignment, in *Lampterium* the thorax decreases in size and the abdomen becomes subcylindrical with few large pores, and in both subgenera the feet are ultimately lost.

The subgenus *Podocyrtis* ranges from the *Bekoma bidartensis* Zone into the *Podocyrtis mitra* Zone (earliest Eocene to Middle Eocene), and *Lampterium* ranges from the *Phormocyrtis striata striata* Zone through the *Podocyrtis goetheana* Zone (early Middle Eocene to early Late Eocene).

Genus *THEOCYRTIS* Haeckel, 1887. Pl. 12, fig. 10; pl. 15, fig. 8.

Three-segmented, generally campanulate forms, in which the termination of the shell is ragged (without differentiated peristome).

Two species of stratigraphic significance are *Theocyrtis tuberosa* Riedel (1959b) (pl. 12, fig. 10), which ranges from the *Thyrsocyrtis bromia* Zone to the *Theocyrtis tuberosa* Zone (Late Eocene to Early Oligocene), and *T. annosa* (Riedel ,1959b) (pl. 15, fig. 8), which ranges from the *Dorcadospyris ateuchus* Zone to the *Calocycletta virginis* Zone (early Late Oligocene to Early Miocene).

Family AMPHIPYNDACIDAE Riedel, 1967a. Pl. 6, fig. 6.

Many-segmented forms in which the cephalis, generally with a poreless wall, is divided into two chambers by a transverse internal ledge.

Genus *AMPHITERNIS* Foreman, 1973. Pl. 6, fig. 6.

Amphipyndacids in which the collar structures extend downward into the thoracic cavity.

Amphipyndacids had their principal development in the Late Cretaceous, and the only known Tertiary representatives are rare members of this genus which range up to the *Podocyrtis ampla* Zone (Middle Eocene).

Family ARTOSTROBIIDAE Riedel, 1967a, *sensu* Foreman, 1973. Pl. 3, fig. 13; pl. 8, fig. 7; pl. 9, figs 13, 14, 16; pl. 11, figs 7–9; pl. 12, fig. 11; pl. 16, fig. 11; pl. 17, figs 18, 19; pl. 19, fig. 24; pl. 20, fig. 11; pl. 21, figs 2, 3; pl. 23, figs 11–14.

Shell with six collar pores, a well developed vertical tube, no lateral or terminal appendages, the last segment not flared, and the pores of at least one segment arranged in transverse rows.

The family ranges through the entire Cainozoic. No satisfactory generic classification is yet available (for the most recent attempt, see Petrushevskaya and Kozlova, 1972), but some species have been grouped according to their similarities with one another and with the type species of certain genera, as listed below.

Artostrobium auritum (Ehrenberg, 1844a) group, *sensu* Riedel and Sanfilippo, 1971. Pl. 23, fig. 11.

More than four segments separated by rounded constrictions, and with numerous rows of pores on each segment.

This species-group ranges from approximately the *Dorcadospyris alata* Zone (early Middle Miocene) to Recent.

Artostrobium doliolum Riedel and Sanfilippo, 1971. Pl. 21, fig. 2.

This species ranges from the *Cannartus petterssoni* Zone to approximately the *Stichocorys peregrina* Zone (late Middle Miocene to near the Miocene/Pliocene boundary).

Lithomitra lineata (Ehrenberg 1838) group, *sensu* Riedel and Sanfilippo, 1971. Pl. 17, fig. 18; pl. 21, fig. 3; pl. 23, fig. 12.

Three-segmented forms with subcylindrical abdomen perforated by rather widely spaced transverse rows of pores, and forms in which the abdomen has the appearance of being segmented as a result of rounded constrictions alternating with the rows of pores.

This species-group ranges from the *Bekoma bidartensis* Zone (Early Eocene) or earlier, to the present.

Phormostichoartus corona Haeckel, 1887. Pl. 20, fig. 11.

This species ranges from the *Calocycletta costata* Zone to approximately the *Stichocorys peregrina* Zone (late Early Miocene to Early Pliocene). An apparently related form, *Theocamptra formaster* Petrushevskaya (in Petrushevskaya and Kozlova, 1972) (pl. 12, fig. 11) occurs in the *Theocyrtis tuberosa* Zone and *Lychnocanoma elongata* Zone (early Oligocene to early Early Miocene), and may range further.

Siphocampe corbula Harting, 1863. Pl. 23, fig. 13.

This species ranges from the *Dorcadospyris alata* Zone (early Middle Miocene) to the present.

Spirocyrtis scalaris Haeckel, 1887. Pl. 23, fig. 14.

This species ranges from the *Stichocorys peregrina* Zone (Early Pliocene) to Recent, and evidently evolved from *Spirocyrtis subtilis* Petrushevskaya (in Petrushevskaya and Kozlova, 1972) (pl. 17, fig. 19).

Theocampe amphora (Haeckel, 1887) group, *sensu* Foreman, 1973. Pl. 9, figs 13, 14.

Three-segmented, spindle-shaped forms with constricted peristome, and pores largest where abdomen is widest.

This species-group ranges from the *Phormocyrtis striata striata* Zone to the *Theocyrtis tuberosa* Zone (early Middle Eocene to Early Oligocene), and an early member may have evolved to *Theocampe mongolfieri*.

Theocampe armadillo (Ehrenberg, 1873) group, *sensu* Riedel and Sanfilippo, 1971. Pl. 11, figs 8, 9.

Relatively large, three-segmented artostrobiids with a distinct peristome and commonly with a large apical horn and prominent vertical tube.

This species-group ranges from the *Podocyrtis mitra* Zone to the *Thyrsocyrtis bromia* Zone (late Middle Eocene to Late Eocene).

Theocampe mongolfieri (Ehrenberg, 1854). Pl. 8, fig. 7.

Abdominal pores are aligned both transversely and longitudinally.

Ranges from the *Theocampe mongolfieri* Zone through the *Thyrsocyrtis bromia* Zone (Middle Eocene to latest Eocene).

Theocampe pirum (Ehrenberg, 1873). Pl. 11, fig. 7.

Third segment wide in normal view, but compressed in the sagittal plane.

Restricted to the *Thyrsocyrtis bromia* Zone (Late Eocene).

Family CANNOBOTRYTHIDAE Haeckel, 1881, *sensu* Riedel, 1967b. Pl. 8, fig. 8; pl. 12, fig. 12; pl. 15, figs 9, 10, 17; pl. 19, figs 25, 26; pl 20, fig 1; pl 21, fig 4; pl 23, figs 15–17.

Skeleton consisting of cephalis, thorax and occasionally an abdomen, the cephalis consisting of several lobes of complex structure.

The family evidently arose in the Late Cretaceous or early Paleogene, but the origins of most of the genera remain obscure.

Genus *ACROBOTRYS* Haeckel, 1881, *sensu* Riedel and Sanfilippo, 1971. Pl. 19, fig. 25; pl. 21, fig. 4.

Two or more prominent tubes arise from the cephalis, their origins being above the level of the collar pores.

The genus ranges from the *Calocycletta virginis* Zone (Early Miocene) to the present.

Genus *BOTRYOCELLA* Haeckel, 1881, *sensu* Petrushevskaya *in* Petrushevskaya and Kozlova, 1972. Pl. 15, figs 9, 17; pl. 19, fig. 26; pl. 23, fig. 15.

On one side of the eucephalic lobe (the darkly-outlined spherical chamber in pl. 15, figs 9, 17) is a large antecephalic lobe, and on the other side a post-cephalic lobe. Tubes, if present, are at approximately the level of the collar pores.

The genus ranges from the *Thyrsocyrtis bromia* Zone (Late Eocene) or earlier, to the present.

Genus *BOTRYOCYRTIS* Ehrenberg, 1860, *sensu* Riedel and Sanfilippo, 1971. Pl. 20, fig. 1; pl. 23, fig. 16.

Cephalis consisting of three or four obvious lobes not markedly differing in size, often covered by spongy meshwork.

The genus ranges from the *Calocycletta virginis* Zone (early Miocene) to the present.

Genus *CENTROBOTRYS* Petrushevskaya, 1965. Pl. 12, fig. 12; pl. 15, fig. 10; pl. 23, fig. 17.

The eucephalic lobe is surrounded by a large chamber not sub-divided into ante- and post-cephalic parts.

This is the only cannobotrythid genus in which an evolutionary lineage has been described. The lineage extends from the *Theocyrtis tuberosa* Zone (Early Oligocene) to the present and comprises *Centrobotrys gravida* Moore (1971) (pl. 12, fig. 12), *C. petrushevskayae* Sanfilippo and Riedel (1973) (pl. 15, fig. 10) and *C. thermophila* Petrushevskaya (1965) (pl. 23, fig. 17).

Acknowledgements

Most of the information summarised in this chapter has resulted from investigations of samples from the Deep Sea Drilling Project, made available by the National Science Foundation, and its compilation has been supported by NSF grant GA-31284X. We are grateful to David Clegg for preparing many of the microslides, to Phyllis Flechsig for assisting in the preparation of the plates, and to M. A. Hanger, M. A. Robertson and P. B. Helms for typing and proof reading the manuscript.

References

Bachmann, A., Papp, A. and Stradner, H. (1963). Mikropaläontologische Studien im "Badener Tegel" von Frättingsdorf, N. Ö. *Mitt. geol. Ges. Wien* **56**(1), 117–210. Plates 1–24.

Beers, J. R. and Stewart, G. L. (1971). Micro-zooplankters in the plankton communities of the upper waters of the eastern tropical Pacific. *Deep-Sea Res.* **18,** 861–883.

Borisenko, N. N. (1958). Radiolyarii paleotsena zapadnoi Kubani. *Trudy vses. neftegaz. nauchno-issled. Inst.* Krasnodarskii Filial **17**, (Voprosy geologii, bureniya i expluatatsii skvazhin), 81–100. Plates 1–6.

Borisenko, N. N. (1960a). Novye radiolyarii iz paleotsenovykh otlozhenii Kubani. *Trudy vses. neftegaz. nauchno-issled. Inst.* Krasnodarskii Filial **4**, (Geologicheskii Sbornik), 199–207. Plates 1, 2.

Borisenko, N. N. (1960b). Radiolyarii nizhnego i srednego eotsena zapadnoi Kubani. *Trudy vses. neftegaz. nauchno-issled. Inst.* Krasnodarskii Filial **4**, (Geologicheskii Sbornik), 219–232. Plates 1–4.

Brandt, R. (1935). Radiolarien. *In* Wetzel, O. Die Mikropalaeontologie des Heiligenhafener Kieseltones (Ober-Eozän). *Jber. niedersächs. geol. Ver.* **27**, 41–81. Plates 8–10.

Brönnimann, P., Martini, E., Resig, J., Riedel, W. R., Sanfilippo, A. and Worsley, T. (1971). Biostratigraphic synthesis: Late Oligocene and Neogene of the western tropical Pacific. *In* (Winterer, E. L. *et al.*) *Initial Rep. Deep Sea Drilling Project* **7**, 1723–1745. U.S. Govt. Printing Office, Washington.

Bütschli, O. (1882a). Radiolaria. *In* (Bronn, H. G. *et al.*), Klassen und Ordnungen des Thier-Reichs, **1**, pt. 1, 332–478. Plates 17–32.

Bütschli, O. (1882b). Beiträge zur Kenntnis der Radiolarienskelette, insbesondere der der Cyrtida. *Z. wiss. Zool.* **36**, 485–540. Plates 31–33.

Bütschli, O. (1889). Kurze Uebersicht des Systems der Radiolaria. *In* (Bronn, H. G. *et al.*), *Klassen und Ordnungen des Thier-Reichs*, **1**, pt. 3, 1946–2004.

Cachon, J. and Cachon, M. (1969). Révision systématique des Nassellaires *Plectoidea* à propos de la description d'un nouveau représentant, *Plectagonidium deflandrei* nov. gen. nov. sp. *Arch. Protistenk.* **111**, 236–251. Plates 39–40.

Cachon, J. and Cachon, M. (1971a). Recherches sur le métabolisme de la silice chez les Radiolaires. Absorption et excrétion. *C. r. hebd. Séanc. Acad. Sci., Paris* **272**, ser D, 1652–1654. Plate 1.

Cachon, J. and Cachon, M. (1971b). Le Système axopodial des Radiolaires Nassellaires. *Arch. Protistenk.* **113**, 80–97. Plates 1–9.

Cachon, J., Cachon, M. and Ferru, G. (1968). Rapports du squelette et du système axopodial chez les Radiolaires Nassellaires. *C. r. hebd. Séanc. Acad. Sci., Paris* **267**, ser. D, 1602–1604.

Campbell, A. S. (1954). Radiolaria. *In* (Moore, R. C. ed.), *Treatise on Invertebrate Paleontology*, Part. D, Protista **3**, 11–163. Geol. Soc. Amer. and Univ. Kansas Press.

Campbell, A. S. and Clark, B. L. (1944). Miocene radiolarian faunas from Southern California. *Spec. Pap. geol. Soc. Am.* **51**, i–vii, 1–76. Plates 1–7.

Carnevale, P. (1908). Radiolarie e Silicoflagellati di Bergonzano (Reggio Emilia). *Memorie R. Ist. veneto Sci.* **28**(3), 1–46. Plates 1–4.

Casey, R. E. (1971a). Distribution of polycystine Radiolaria in the oceans in relation to physical and chemical conditions. *In* (Funnell, B. M. and Riedel, W. R. eds), *The Micropalaeontology of Oceans*, 151–159, Cambridge University Press.

Casey, R. E. (1971b). Radiolarians as indicators of past and present water-masses. *In* (Funnell, B. M. and Riedel, W. R. eds), *The Micropalaeontology of Oceans*, 331–341. Cambridge University Press.

Casey, R. E., Partridge, T. M. and Sloan, J. R. (1971). Radiolarian life spans, mortality rates, and seasonality gained from Recent sediment and plankton samples. *In* (Farinacci, A. ed.), *Proc. II Plankt. Conf.* 159–165. Tecnoscienza, Roma.

Cifelli, R. and Sachs, K. N., Jr. (1966). Abundance relationships of planktonic Foraminifera and Radiolaria. *Deep-Sea Res.* **13**, 751–753.

Cita, M. B., Nigrini, C. and Gartner, S. (1970). Biostratigraphy. *In* (Peterson, M. N. A. *et al.*), *Initial Rep. Deep Sea Drilling Project* **2,** 391–411. U.S. Govt. Printing Office, Washington.

Clark, B. L. and Campbell, A. S. (1942). Eocene radiolarian faunas from the Mount Diablo area, California. *Spec. Pap. geol. Soc. Am.* **39,** i–vii, 1–112. Plates 1–9.

Clark, B. L. and Campbell, A. S. (1945). Radiolaria from the Kreyenhagen Formation near Los Banos, California. *Mem. geol. Soc. Am.* **10,** i-vii, 1–66. Plates 1–7.

Cleve, P. T. (1900a). Notes on some Atlantic plankton-organisms. *K. svenska Vetensk-Akad. Handl.* **34**(1), 1–22. Plates 1–8.

Cleve, P. T. (1900b). The seasonal distribution of Atlantic plankton organisms. *Göteborgs K. Vetensk.—o VitterhSamh. Handl.* ser. 4, **3**,, 1–369.

Cleve, P. T. (1901). Plankton from the Indian Ocean and the Malay Archipelago. *K. svenska Vetensk-Akad. Handl.* **35**(5), 1–58. Plates 1–8.

Deflandre, G. 1953. Radiolaries fossiles. *In* (Grassé, P.-P. ed.), *Traité de Zoologie* **1,** pt. 2, 389–436. Masson, Paris.

Dreyer, F. (1890). Morphologische Radiolairenstudien. 2. Die Tripoli von Caltanisetta (Steinbruch Gessolungo) auf Sizilien. *Jena. Z. Naturw.* **24** (new ser. 7), 1–79. Plates 1–6.

Dreyer, F. (1913). Die Polycystinen der Plankton-Expedition. *Ergebn. Plankton-Exped. Humboldt-Stift.* **3** (L.d.e.), 1–104. Plates 1–3.

Dumitrica, P. (1964). Asupra prezentei unor radiolari din famioia *Challengeridae* (ord. *Phaeodaria*) in tortonianul din Subcarpati. *Studii Cerc. Geogr. Geofiz. Geol., ser. geol.* **9**(1), 217–222.

Ehrenberg, C. G. (1838). Über die Bildung der Kreidefelsen und des Kreidemergels durch unsichtbare Organismen. *Abh. preuss. Akad. Wiss.* Jahrg. 1838, 59–147. Plates 1–4.

Ehrenberg, C. G. (1844a). Über 2 neue Lager von Gebirgsmassen aus Infusorien als Meeres-Absatz in Nord-Amerika und eine Vergleichung derselben mit den organischen Kreide-Gebilden in Europa und Africa. *Mber. preuss. Akad. Wiss.* Jahrg. 1844, 57–97.

Ehrenberg, C. G. (1844b). Einige vorläufige Resultate seiner Untersuchungen der ihm von der Südpolreise des Capitain Ross, so wie von den Herren Schayer und Darwin zugekommenen Materialien über das Verhalten des kleinsten Lebens in den Oceanen und den grössten bisher zugänglichen Tiefen des Weltmeers. *Mber. preuss. Akad. Wiss.* Jahrg. 1944, 182–207.

Ehrenberg, C. G. (1847a). Über eine halibiolithische, von Herrn R. Schomburgk entdeckte, vorherrschend aus mikroskopischen Polycystinen gebildete, Gebirgsmasse von Barbados. *Mber. preuss. Akad. Wiss.* Jahrg. 1846, 382–385.

Ehrenberg, C. G. (1847b). Über die mikroskopischen kieselschaligen Polycystinen als mächtige Gebirgsmasse von Barbados und über das Verhältnis der aus mehr als 300 neuen Arten bestehenden ganz eigenthümlichen Formengruppe jener Felsmasse zu den lebenden Thieren und zur Kreidebildung. Eine neue Anregung zur Erforschung des Erdlebens. *Mber. preuss. Akad. Wiss.* Jahrg. 1847, 40–60, 1 pl.

Ehrenberg, C. G. (1854). *Mikrogeologie.* xxviii + 374 pp. Atlas, 31 pp. + 41 pls. Fortsetzung (1856), 89 pp. Leipzig.

Ehrenberg, C. G. (1860). Über den Tiefgrund des stillen Oceans zwischen Californien und den Sandwich-Inseln aus bis 15,600' Tiefe nach Lieut. Brooke. *Mber. preuss. Akad. Wiss.* Jahrg. 1860, 819–833.

Ehrenberg, C. G. (1872a). Mikrogeologischen Studien als Zusammenfassung seiner

Beobachtungen des kleinsten Lebens der Meeres-Tiefgrunde aller Zonen und dessen geologischen Einfluss. *Mber. preuss. Akad. Wiss.* Jahrg. 1872, 265–322.

Ehrenberg, C. G. (1872b). Mikrogeologischen Studien über das kleinste Leben der Meeres-Tiefgrunde aller Zonen und dessen geologischen Einfluss. *Abh. preuss. Akad. Wiss.* Jahrg. 1872, 131–399, Plates 1–12, 1 chart.

Ehrenberg, C. G. (1873). Grössere Felsproben des Polycystinen-Mergels von Barbados mit weiteren Erläuterungen. *Mber. preuss. Akad. Wiss.* Jahrg. 1873, 213–263.

Ehrenberg, C. G. (1875). Fortsetzung der mikrogeologischen Studien als Gesammt-Uebersicht der mikroskopischen Paläontologie gleichartig analysierter Gebirgsarten der Erde, mit specieller Rücksicht auf den Polycystinen-Mergel von Barbados. *Abh. preuss. Akad. Wiss.* Jahrg. 1875, 1–226. Plates 1–30.

Foreman, H. P. (1966). Two Cretaceous radiolarian genera. *Micropaleont.* **12**(3), 355–359.

Foreman, H. P. (1973). Radiolaria of Leg X, with systematics and ranges for the families Amphipyndacidae, Artostrobiidae and Theoperidae. *In* (Worzel, J. L. *et al.*) *Initial Rep. Deep Sea Drilling Project* **10**. U.S. Govt. Printing Office, Washington.

Friend, J. K. and Riedel, W. R. (1967). Cenozoic orosphaerid radiolarians from tropical Pacific sediments. *Micropaleont.* **13,** 217–237.

Frizzell, D. L. and Middour, E. S. (1951). Paleocene Radiolaria from southeastern Missouri. *Bull. Univ. Mo. Sch. Min. Metall.* tech. ser., **7,** 1–41. Plates 1–3.

Goll, R. M. (1968). Classification and phylogency of Cenozoic Trissocyclidae (Radiolaria) in the Pacific and Caribbean Basins, Part I. *J. Paleont.* **42,** 1409–1432. Plates 173–176.

Goll, R. M. (1969). Classification and phylogeny of Cenozoic Trissocyclidae (Radiolaria) in the Pacific and Caribbean Basins, Part II. *J. Paleont.* **43,** 322–339. Plates 55–60.

Goll, R. M. (1972a). Leg 9 synthesis, Radiolaria. *In* (Hays, J. D. *et al.*) *Initial Rep. Deep Sea Drilling Project* **9,** 947–1058. U.S. Govt. Printing Office, Washington.

Goll, R. M. (1972b). Systematics of eight *Tholospyris* taxa (Trissocyclidae: Radiolaria). *Micropaleont.* **18,** 443–475. Plates 1–16, 5 figures.

Goll, R. M. and Bjørklund, K. R. (1971). Radiolaria in surface sediments of the North Atlantic Ocean. *Micropaleont.* **17,** 434–454.

Guppy, R. J. L. (1909). The geological connections of the Caribbean region. *Trans. R. Can. Inst.* **8**(18), 373–391. 1 chart.

Haeckel, E. (1860a). Über neue, lebende Radiolarien des Mittelmeeres. *Mber. preuss. Akad. Wiss.* Jahrg. 1860, 794–817.

Haeckel, E. (1860b). Abbildungen und Diagnosen neuer Gattungen und Arten von lebenden Radiolarien des Mittelmeeres. *Mber. preuss. Akad. Wiss.* Jahrg. 1860, 835–845.

Haeckel, E. (1862). *Die Radiolarien* (*Rhizopoda Radilaria*). Eine Monographie, xiv + 572 pp., 35 pls. Reimer, Berlin.

Haeckel, E. (1881). Entwurf eines Radiolarien-Systems auf Grund von Studien der Challenger—Radiolarien. *Jena. Z. Naturw.* **15** (new ser. 8), no. 3, 418–472.

Haeckel, E. (1887). Report on the Radiolaria collected by *H.M.S. Challenger* during the years 1873–76. *Rept. Voyage 'Challenger', Zool.* **18,** clxxxviii + 1803 pp., 140 pls., 1 map.

Haecker, V. (1906). Über einige grosse Tiefsee-Radiolarien. *Zool. Anz.* **30**(26), 878–895.
Haecker, V. (1907). Altertümliche Sphärellarien und Cyrtellarien aus grossen Meerestiefen. *Arch. Protistenk.* **10,** 114–126.
Haecker, V. (1908). Tiefsee-Radiolarien. Spezieller Teil, Lfg. 2. Die Tripyleen, Collodarien und Mikroradiolarien der Tiefsee. *Wiss. Ergebn. dt. Tiefsee-Exped. 'Valdivia'* **14,** 337–476. Plates 63–85.
Harting, P. (1863). Bijdrage tot de kennis der mikroskopische fauna en flora van de Banda-Zee. *Verh. K. Akad. Wet.* **10,** 1–34 Plates 1–3.
Hays, J. D. (1970). Stratigraphy and evolutionary trends of Radiolaria in North Pacific deep-sea sediments *Mem. geol. Soc. Am.* **126,** 185–218.
Hays, J. D., Cook, H. E. III, Jenkins, D. G., Cook, F. M., Fuller, J. T., Goll, R. M., Milow, E. D. and Orr, W. N. (1972). Introduction. *In* (Hays, J. D. *et al.*) *Initial Rep. Deep Sea Drilling Project* **9,** 3–20A. U.S. Govt. Printing Office, Washington.
Hays, J. D., Saito, T., Opdyke, N. D. and Burckle, L. H. (1969). Pliocene-Pleistocene sediments of the equatorial Pacific: their paleomagnetic, biostratigraphic and climatic record. *Bull. geol. Soc. Am.* **80,** 1481–1513. Plate 1.
Hertwig, R. (1879). *Der Organismus der Radiolarien*, iv + 149 pp., 10 pls. G. Fischer, Jena.
Hilmers, C. (1906). Zur Kenntnis der Collosphaeriden. Doctoral Dissertation, 95 pp., 1 pl. Kgl. Christian-Albrecht-Univ., Kiel.
Hollande, A., Cachon, J. and Cachon, M. (1970). La signification de la membrane capsulaire des radiolaires et ses rapports avec le plasmalemme et les membrans du réticulum endoplasmique. Affinités entre Radiolaires, Héliozoaires et Péridiniens. *Protistologica* **6**(3), 311–318.
Hollande, A. and Enjumet, M. (1960). Cytologie, évolution et systématique des Sphaeroidés (Radiolaires). *Archs Mus. natn. Hist. nat., Paris, ser.* 7, **7,** 1–134. Plates 1–64.
Joergensen, E. (1905). The protist plankton and the diatoms in bottom samples. *Bergens Mus. Skr.* 1905, 49–151, 195–225. Plates 6–18.
Khabakov, A. V. (1937). Fauna radiolyarii nizhemelovykh i verkhneyurskikh fosforitov basseina verkhnei vyatki i kamy. (Die Radiolarien-Fauna aus mesozoischen Phosphoriten des Kama- und Wiatka-Gebietes.) *Ezheg. vses. paleont. Obshch.* **11,** 90–120. Plates 11–14.
Kling, S. A. (1971). Radiolaria: Leg 6 of the Deep Sea Drilling Project. *In* (Fischer, A. G. *et al.*) *Initial Rep. Deep Sea Drilling Project* **6,** 1069–1117. U.S. Govt Printing Office, Washington.
Kozlova, G. E. (1960). Radiolyarii srednego i verkhnego miotsena Severnogo Sakhalina. *Trudy vses. neft. nauchno-issled. geol.-razv. Inst.* **153,** (Mikrofauna SSSR 11), 307–325. Plates 1–4.
Kozlova, G. E. (1967). O filogeneticheskoi svyazi podotryadov Discoidea i Larcoidea (Radiolaria, Spumellaria). *Zool. Zh.* **46,** 1311–1320.
Kozlova, G. E and Gorbovets, A. N. (1966). Radiolyarii verknemelovykh i verkhneeotsenovykh otlozhenii Zapadno-Sibirskoi Nizmennosti. *Trudy vses. neft. nauchno-issled. geol.-razv. Inst.* **248,** 1–159. Plates 1–17.
Krasheninnikov, V. A. (1960). Nekotorye radiolyarii nizhnego i srednogo eotsena zapadnogo predkavkazya. *Trudy vses. neft. nauchno-issled. geol.-razv. Inst.* **16** (Paleontologicheskii Sbornik 3), 271–308. Plates 1–3.
Kruglikova, S. B. (1969). Radiolyarii v poverkhnostnom sloe osadkov severnoi

poloviny Tikhogo okeana. *In* (Kort, V. G. ed.), *Mikroflora i Mikrofauna v Sovremennykh Osadkakh Tikhogo Okeana*, 48–72. Nauka, Moscow.

Lipman, R. Kh. (1950). Radiolyarii eotsena Kyzyl-Kumov. *Trudy vses. nauchno-issled. geol. Inst.* **1,** 51–65. Plates 1–3.

Lipman, R. Kh. (1953). Materialy k izucheniyu radiolyarii paleogena zapadnykh raionov Srednei Azii. *Ezheg. vses. paleont. Obshch.* **14,** 135–157. Plates 7, 8.

Lipman, R. Kh. (1965). Kompleksy radiolyarii v paleogenovykh otlozheniyakh Turgaiskogo progiba i Severnogo Priaralya. *Trudy vses. nauchno-issled. geol. Inst.* new ser. **115,** 78–89.

Lipman, R. Kh. (1969). Novyi rod i novye vidy eotsenovykh radiolyarii SSSR. *Trudy vses. nauchno-issled. geol. Inst.* new ser. **130,** 181–192. Plates 1–4.

Lisitzin, A. P. (1971). Distribution of siliceous microfossils in suspension and in bottom sediments. *In* (Funnell, B. M. and Riedel, W. R. eds), *The Micropalaeontology of Oceans*, 173–195. Cambridge Univeristy Press.

Lucchese, C. (1927). Radiolari miocenici di Salsomaggiore. *G. Geol.* ser. 2. **2,** 80–116. Plates 3–8.

Mamedov, N. A. (1969a). Novye predstaviteli eotsenovykh radiolyarii Azerbaidzhana. *In* (Vyalov, O. S. ed.), *Iskopaemye i Sovremennye Radiolyarii* (Materialy vtorogo vsesoyuznogo seminara po radiolyariyam), 94–101. Lvovskoe Geol. Obshch., Univ. Lvov.

Mamedov, N. A. (1969b). Novye vidy radiolyarii iz eotsenovykh otlozhenii Azerbaidzhana. *Izv. Akad. Nauk azerb. SSR*, Ser. nauk o Zemle **4,** 21–30.

Mamedov, N. A. (1969c). Nekotorye novye predstaviteli eotsenovykh radiolyarii Azerbaidzhana. *Izv. Akad. Nauk. azerb. SSR*, Ser. nauk o Zemle **6,** 23–41.

Mamedov, N. A. (1970). Novye vidy radiolyarii iz eotsenovykh otlozhenii severo-vostochnogo Azerbaidzhana. *Uchen. Zap. azerb. Inst. neft. zimi*, ser. **9,** 3, 61–69.

Martin, G. C. (1904). Radiolaria. *Md. geol. Surv.* (*Miocene*), 447–459. Plate 130.

Meyen, F. J. F. (1834). Über das Leuchten des Meeres und Beschreibung einiger Polypen und anderer niederer Tiere. *In* Beiträge zur Zoologie, gesammelt auf einer Reise um die Erde. *Nova Acta Acad. Caesar. Leop. Carol.* **16**(=8), suppl. 1, abh. 5, 125–216. Plates 27–36.

Moore, T. (1971). Radiolaria. *In* (Tracey, J. I., Jr. *et al.*), *Initial Rep. Deep Sea Drilling Project* **8,** 727–775. U.S. Govt. Printing Office, Washington.

Moore, T. C., Jr. (1972). Mid-Tertiary evolution of the radiolarian genus *Calocycletta*. *Micropaleont.* **18,** 144–152.

Müller, J. (1858). Über die Thalassicollen, Polycystinen und Acanthometren des Mittelmeeres. *Abh. preuss. Akad. Wiss.* Jahrg. 1858, 1–62. Plates 1–11.

Nakaseko, K. (1955). Miocene radiolarian fossil assemblage from the southern Toyama Prefecture in Japan. *Sci. Rep. Osaka Univ.* **4,** 65–127. Plates 1–11.

Nakaseko, K. (1959). Applied micropaleontological research by means of radiolarian fossil in the oil bearing Tertiary, Japan (mainly in Akita and Yamagata sedimentary basins): Part I. Method, geological note and radiolarian assemblage in Akita sedimentary basin. *Sci. Rep. Osaka Univ.* **8,** 113–193.

Nakaseko, K. (1960). Applied micropaleontological research by means of radiolarian fossils in the oil bearing Tertiary, Japan (mainly in Akita and Yamagata sedimentary basins): Part II. Radiolarian assemblage in Yamagata basin, discussion and conclusion. *Sci. Rep. Osaka Univ.* **9,** 149–185.

Nakaseko, K. (1963). Neogene Cyrtoidea (Radiolaria) from the Isozaki Formation in Ibaraki Prefecture, Japan. *Sci. Rep. Osaka Univ.* **12**(2), 165–198. Plates 1–4.

Nakaseko, K. (1971). On some species of the genus *Thecosphaera* from the Neogene formations, Japan. *Sci. Rep. Osaka Univ.* **20**(2), 59–66.

Nigrini, C. (1967). Radiolaria in pelagic sediments from the Indian and Atlantic Oceans. *Bull. Scripps Instn. Oceanogr.* **11,** 1–106, Plates 1–9.

Nigrini, C. (1968). Radiolaria from eastern tropical Pacific sediments. *Micropaleont.* **14,** 51–63.

Nigrini, C. (1970). Radiolarian assemblages in the North Pacific and their application to a study of Quaternary sediments in core V20–130. *Mem. geol. Soc. Am.* **126,** 139–183.

Nigrini, C. (1971). Radiolarian zones in the Quaternary of the equatorial Pacific Ocean. *In* (Funnell, B. M. and Riedel, W. R. eds), *The Micropalaeontology of Oceans*, 443–461. Cambridge University Press.

Olsson, R. K. and Goll, R. (1970). Biostratigraphy. *In* (McManus, D. A. *et al.*) *Initial Rep. Deep Sea Drilling Project* **5,** 557–567. U.S. Govt. Printing Office, Washington.

Petrushevskaya, M. G. (1965). Osobennosti konstruktsii skeleta radiolyarii Botryoidae (otr. Nassellaria). *Trudy zool. Inst. Leningr.* **35,** 79–118.

Petrushevskaya, M. G. (1966). Radiolyarii v planktone i v donnykh osadkakh 219–245. *In* (Strakhov, N. M. ed.), *Geokhimiya Kremnezema*, 423 pp. Nauka, Moscow.

Petrushevskaya, M. G. (1967). Radiolyarii otryadov Spumellaria i Nassellaria antarkticheskoi oblasti (po materialam Sovetskoi antarkticheskoi ekspeditsii). *In* (Pavlovskii, E. N. ed.), *Issledovaniya Fauny Morei* **3**(12), 3–186. Nauka, Leningrad.

Petrushevskaya, M. G. (1971a). Radiolyarii Nassellaria v planktone Mirovogo okeana. *In* (Bykhovskii, B. E., ed.), *Issledovaniya Fauny Morei* **9**(17), 5–294. *Zool. Inst. Akad. Nauk SSSR, Leningrad.*

Petrushevskaya, M. G. (1971b). Sootnoshenie nekotorykh rodov radiolyarii Nassellaria. *Zool. Zh.* **50,** 1133–1142

Petrushevskaya, M. G. (1971c). On the natural system of polycystine Radiolaria (Class Sarcodina). *In* (Farinacci, A. ed.), *Proc. II Plankt. Conf.* 981–992. Tecnoscienza, Roma.

Petrushevskaya, M. G. (1971d). Spumellarian and nassellarian Radiolaria in the plankton and bottom sediments of the central Pacific. 309–317. *In* (Funnell, B. M. and Riedel, W. R. eds), *The Micropalaeontology of Oceans*, 309–317. Cambridge University Press.

Petrushevskaya, M. G. (1971e). Radiolaria in the plankton and Recent sediments from the Indian Ocean and Antarctic. *In* (Funnell, B. M. and Riedel, W. R. eds), *The Micropalaeontology of Oceans*, 319–329. Cambridge University Press.

Petrushevskaya, M. G. and Kozlova, G. E. (1972). Radiolaria. *In* (Hayes, J. D. *et al.*) *Initial Rep. Deep Sea Drilling Project* **14.** U.S. Govt, Printing Office. Washington.

Popofsky, A. (1908). Die Radiolarien der Antarktis (mit Ausnahme der Tripyleen). *Dt. Südpol.-Exped. 1901–1903* **10** (*Zool.* vol. 2), no. 3, 183–305. Plates 20–36, 1 table.

Popofsky, A. (1912). Die Sphaerellarien des Warmwasser-gebietes. *Dt. Südpol.-Exped. 1901–1903* **13** (*Zool.* vol 5), no. 2, 73–159. Plates 1–8.

Popofsky, A. (1913). Die Nassellarien des Warmwasser-gebietes. *Dt. Südpol.-Exped. 1901–1903*, **14,** (*Zool.* vol. 6), 217–416. Plates 28–38.

Popofsky, A. (1917) Die Collosphaeriden, mit Nachtrag zu den Spumellarien und

Nassellarien. *Dt. Südpol.-Exped. 1901–1903*, **16** (*Zool.* vol. 8) no. 3, 235–278. Plates 13–17.

Popofsky, A. (1920). Die Sphaerozoiden. *Dt. Südpol.-Exped. 1901–1903*, **16** (*Zool.* vol. 8), 541–587. Plates 57–61.

Principi, P. (1909). Contributo allo studio dei Radiolari Miocenici Italiani. *Boll. Soc. geol. ital.* **28,** 1–22. Plate 1.

Ramsay, A. T. S. (1971). Occurrence of biogenic siliceous sediments in the Atlantic Ocean. *Nature* **233,** 115–117.

Reshetnyak, V. V. (1955). Vertikalnoe raspredelenie radiolyarii Kurilo-Kamchatskoi vpadiny. *Trudy zool. Inst., Leningr.* **21,** 94–101.

Reshetnyak, V. V. (1969), Radiolyarii otryada *Phaeodaria* v sovremennykh donnykh osadkakh i drevnikh otlozheniyakh. *In* (Vyalov, O. S. ed.), *Iskopaemye i Sovremennye Radiolyarii* (Materialy vtorogo vsesoyuznogo seminara po radiolyariyam), 133–140 Lvovskoe Geol. Obshch., Univ. Lvov.

Riedel, W. R. (1952). Tertiary Radiolaria in western Pacific sediments. *Göteborgs K. Vetensk.-o. VitterhSamh. Handl.*, följ. 7, ser. B, **6**(3), 1–18. Plates 1–2.

Riedel, W. R. (1953). Mesozoic and Late Tertiary Radiolaria of Rotti. *J. Paleont.* **27,** 805–813. Plates 84–85.

Riedel, W. R. (1957). Radiolaria: a preliminary stratigraphy. *Rep. Swed. deep Sea Exped.* **6**(3), 59–96, Plates 1–4.

Riedel, W. R. (1959a). Siliceous organic remains in pelagic sediments. *In* (Ireland, H. A. ed.), *Silica in Sediments*, spec. Publ. Soc. econ. Paleont. Miner. **7,** 80–91.

Riedel, W. R. (1959b). Oligocene and Lower Miocene Radiolaria in tropical Pacific sediments. *Micropaleont.* **5**(3), 285–302.

Riedel, W. R. (1963). Paleontology of pelagic sediments. *In* (Hill, M. N. ed.), *The Sea* **3,** 866–887. Interscience, New York and London.

Riedel, W. R. (1967a). Some new families of Radiolaria. *Proc. geol. Soc. Lond.*, **1640,** 148–149.

Riedel, W. R. 1967b. Subclass Radiolaria. *In* (Harland, W. B. *et al.* eds), *The Fossil Record*, 291–298. Geological Society, London.

Riedel, W. R. (1971a). Systematic classification of polycystone Radiolaria. *In* (Funnell, B. M. and Riedel, W. R. eds), *The Micropalaeontology of Oceans*, 649–661. Cambridge University Press.

Riedel, W. R. (1917b). Radiolarians from Atlantic deep-sea drilling. *In* (Farinacci, A. ed.), *Proc. II Plankt. Conf.* 1057–1967 and erratum sheet distributed subsequently by author Tecnoscienza, Roma.

Riedel, W. R. and Foreman, H. P. (1961). Type specimens of North American Paleozoic Radiolaria. *J. Paleont.* **35,** 628–632.

Riedel, W. R. and Funnell, B. M. (1964). Tertiary sediment cores and microfossils from the Pacific Ocean floor. *Q. J. geol. Soc. Lond.* **120**, 305–368. Plates 14–32.

Riedel, W. R. and Hays, J. D. (1969) Cenozoic Radiolaria from Leg 1. *In* (Ewing, M. *et al.*), *Initial Rep. Deep Sea Drilling Project* **1,** 400–402. U.S. Govt. Printing Office, Washington.

Riedel, W. R. and Sanfilippo, A. (1970). Radiolaria, Leg 4, Deep Sea Drilling Project. *In* (Bader, R. G. *et al.*), *Inital Rep. Deep Sea Drilling Project* **4,** 503–575. U.S. Govt. Printing Office, Washington.

Riedel, W. R. and Sanfilippo, A. (1971). Cenozoic Radiolaria from the western tropical Pacific, Leg 7. *In* (Winterer, E. L. *et al.*), *Initial Rep. Deep Sea Drilling Project* **7,** 1529–1672. U.S. Govt. Printing Office, Washington.

Sanfilippo, A., Burckle, L. H., Martini, E. and Riedel, W. R. (1973). Radiolarians, diatoms, silicoflagellates and calcareous nannofossils in the Mediterranean Neogene. *Micropaleont.* **19,** 209–234, Plates 1–6.

Sanfilippo, A. and Riedel, W. R. (1970). Post-Eocene "closed" theoperid radiolarians. *Micropaleont.* **16,** 446–462.

Sanfilippo, A. and Riedel, W. R. (1973). Cenozoic Radiolaria (exclusive of theoperids, artostrobiids and amphipyndacids) from the Gulf of Mexico, DSDP Leg X. *In* (Worzel, J. L. *et al.*). *Initial Rep. Deep Sea Drilling Project* **10.** U.S. Govt. Printing Office, Washington.

Seguenza, G. (1880). Le formazioni terziarie nella provincia di Reggio (Calabria). *Atti R. Accad. Lincei Memorie*, Classe di scienze fisiche, mathematiche e naturali, ser. 3, **6,** 1–446. Plates 1–17.

Stadum, C. J. and Ling H.-Y. (1969). Tripylean Radiolaria in deep-sea sediments of the Norwegian Sea. *Micropaleont.* **15,** 481–489.

Stöhr. E. (1880). Die Radiolarienfauna der Tripoli von Grotte, Provinz Girgenti in Sicilien. *Paleontographica* **26,** (ser. 3, vol. 2), 69–124 (and Corrigenda). Plates 17–23 (1–7).

Strelkov, A. A. and Reshetnyak, V. V. (1971). Kolonialnye radiolyarii Spumellaria Mirovogo okeana. *In* (Bykhovskii, B. E. ed.), *Issledovaniya Fauny Morei* **9**(17) 295–373. Zool. Inst. Akad. Nauk SSSR, Leningrad.

Sutton, H. J. (1896). Radiolaria: a new species from Barbados. *Am. mon. microsc. J.* **17,** 58–60.

Tan Sin Hok (1927). Over de samenstelling en het onstaan van krijt- en mergelgesteenten van de Molukken. *Jaarb. Mijnw. Ned.-Oost-Indië, jaarg.* 1926, *Verhand.* pt. 3, 5–165. Plates 1–16.

Tracey, J. I. Jr., Sutton, G. H., Nesteroff, W. D., Galehouse, J., Von der Borch, C. C., Moore, T., Lipps, J., Bilal ul Haq, U. Z. and Beckmann, J. P. (1971). Introduction. *In* (Tracey, J. I. Jr. *et al.*), *Initial Rep. Deep Sea Drilling Project* **8,** 5–15. U.S. Govt. Printing Office, Washington.

Vinassa de Regny, P. E. (1900). Radiolari Miocenici Italiani. *Memorie R. Accad. Sci. Ist. Bologna* ser. 5, **8,** 565–595. Plates 1–3.

Zittel, K. A. (1876). Ueber einige fossile Radiolarien aus der norddeutschen Kreide. *Z. dt. geol. Ges.* **28,** 75–86. Plate 2.

PLATES

All illustrations are of species occurring in the tropical Pacific and Caribbean regions, and are arranged approximately in stratigraphic order. Each species is included on a plate containing forms of a certain age, and it is necessary to consult the text of the systematic section or the original literature to determine whether it also ranges into younger and/or older assemblages.

Author and date are given for the names of species not included in the foregoing text.

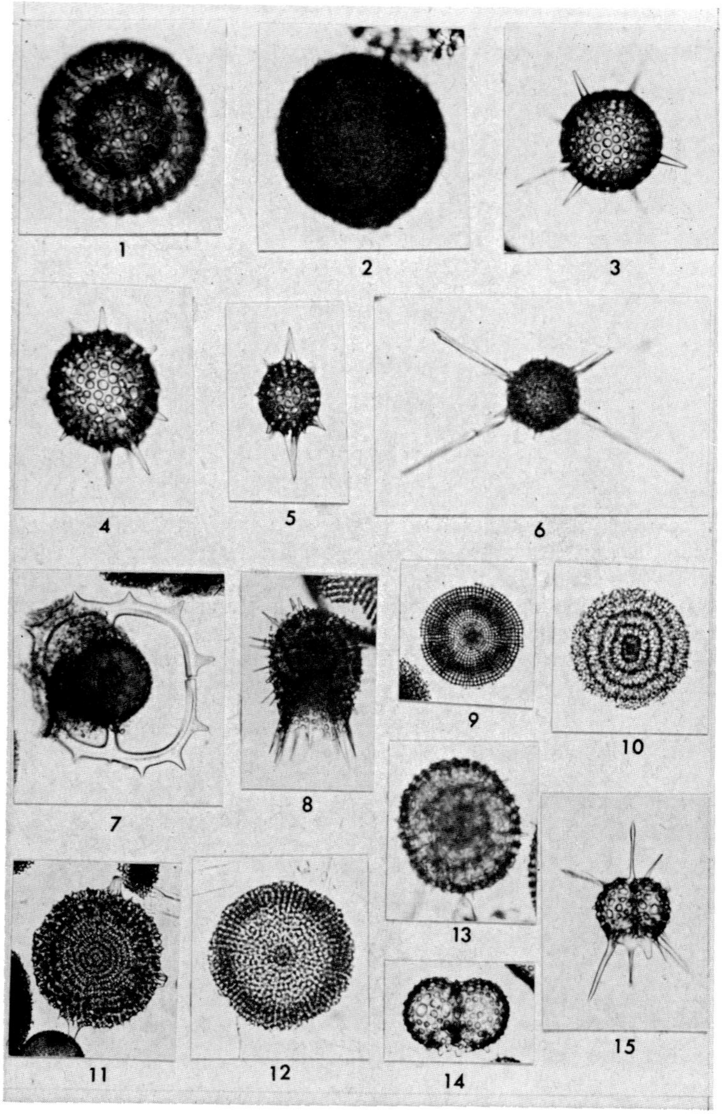

PLATE 1. Late Palaeocene radiolarians

ACTINOMMIDAE: 1. *Carposphaera subbotinae* (Borisenko, 1958) (×120), 2. *Diploplegma somphum* (×120), 3. *Hexalonche* (?) sp. A of Sanfilippo and Riedel (1973) (×120), 4. *Lithomespilus mendosa* (Krasheninnikov, 1960) (×120), 5. *Stylosphaera goruna* (×120), 6. Actinommid, gen. et sp. indet (×75), 7. Saturnalin, gen. et sp. indet (×120).
PORODISCIDAE: 8. *Spongopyle* sp. cf. *S. insolita* Kozlova (Kozlova and Gorboveerts, 1966) (×120), 9. *Spongodiscus quartus bosoculus* Sanfilippo and Riedel (1973) (×75), 10. *Xiphospira circularis* (Clark and Campbell, 1942) (×75), 11. *Stylotrochus alveatus* (×75).
SPONGODISCIDAE: 12. *Spongodiscus pulcher* Clark and Campbell (1945) (×75).
LITHELIIDAE: 13. *Lithelius foremanae* (×200).
SPYRIDA: 14, 15. Spyrids, gen. et sp. indet. (×120).

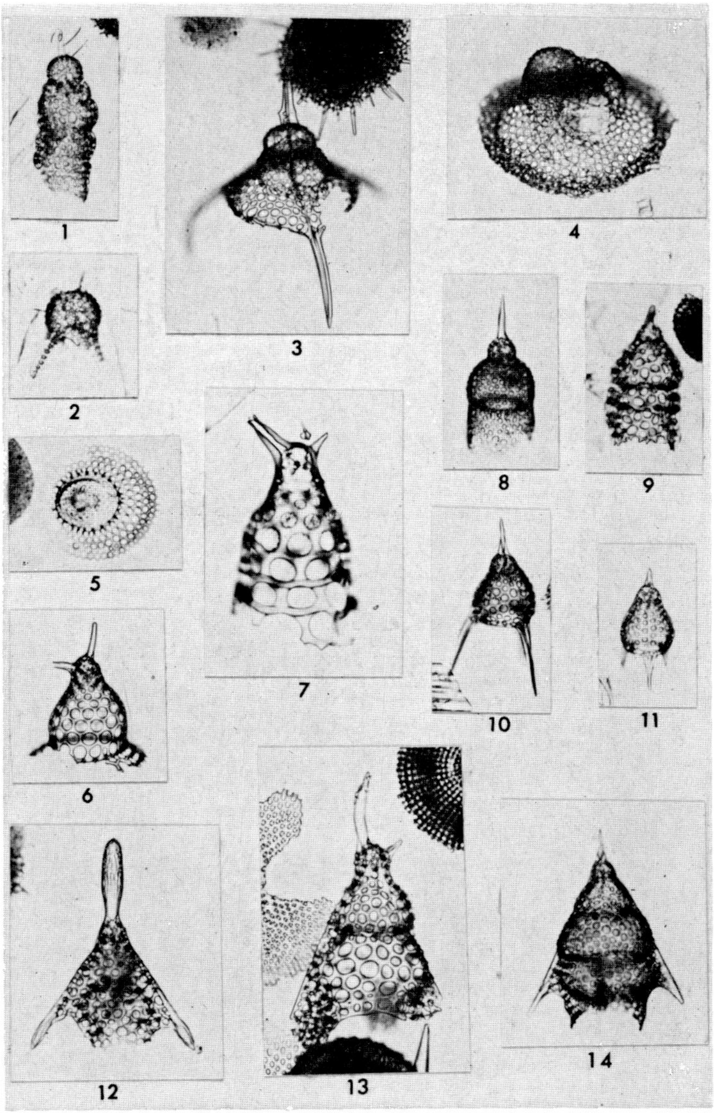

PLATE 2. Late Palaeocene radiolarians (×120)

PLAGONIIDAE: 1, 2. Plagoniids, gen. et sp. indet., 3. *Spongomelissa adunca* Sanfilippo and Riedel (1973), 4. *Velicucullus* sp. 5. *Theocalyptra* sp.

INCERTAE SEDIS (theoperids?): 6. *Clathrocycloma parcum* Foreman (1973), 7. *Clathrocycloma capitaneum* Foreman (1973).

THEOPERIDAE: 8. *Theocorys* sp., 9. Theoperid gen. et sp. indet., 10, 11. *Lychnocanoma* spp., 12. *Dictyoceras caia* Foreman (1973), 13, 14. Theoperids, gen. et sp. indet.

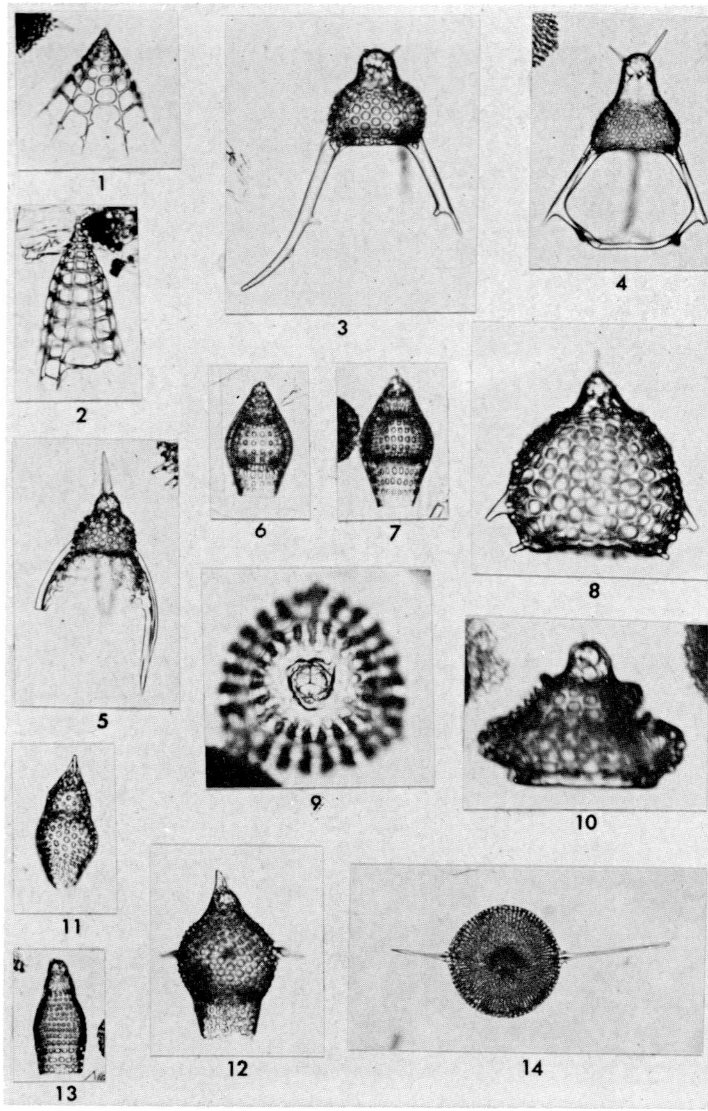

PLATE 3. Late Palaeocene and Early Eocene radiolarians. (1–13, from Late Palaeocene; 14, from the *Bekoma bidartensis* Zone.) (×120)

THEOPERIDAE: 1, 2. *Bathropyramis* sp.(p), 3. *Bekoma divaricata* Foreman (1973), 4. *Bekoma campechensis*, 5. *Lychnocanoma anacolum* Foreman (1973), 6. *Buryella pentadica*, 7. *Buryella tetradica* Foreman (1973), 8. *Orbula discipulus* Foreman (1973), 9. *Orbula ducalis* Foreman (1973), viewed from above, 10. *Orbula ducalis*, lateral view, 11. *Phormocyrtis striata striata* morphotype, 12. *Lamptonium pennatum*.

ARTOSTROBIIDAE: 13. *Lithomitra docilis* Foreman (1973).

PHACODISCIDAE: 14. *Heliostylus* sp. (×75).

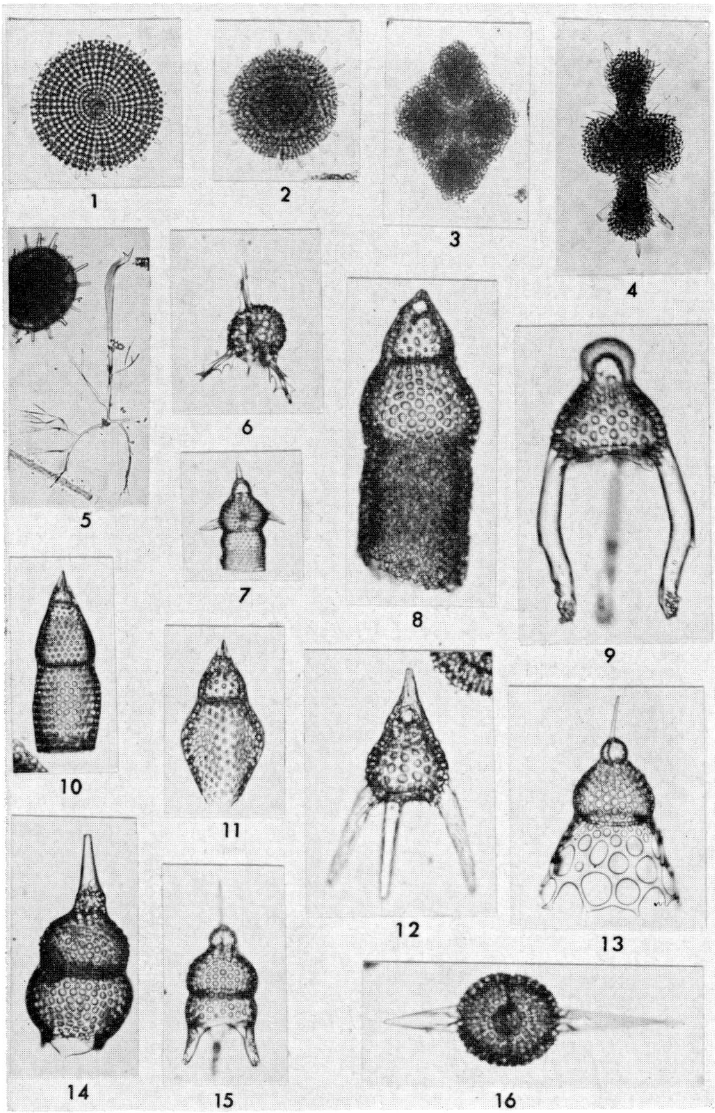

PLATE 4. Early Eocene radiolarians. (1–15, from the *Bekoma bidartensis* Zone; 16, from he *Buryella clinata* Zone.)
PORODISCIDAE (×75): 1. *Spongodiscus quartus quartus*, 2. *Stylotrochus nitidus*.
SPONGODISCIDAE (×75): 3. *Spongodiscus cruciferus*, 4. *Amphicraspedum murrayanum* Haeckel (1887).
SPYRIDA: 5. *Rhabdolithis ellida* (×75).
PLAGONIIDAE: 6. *Tripilidium* sp. (×120).
THEOPERIDAE (×120): 7. *Pterocanium contiguum* Ehrenberg (1873), 8. *Phormocyrtis cubensis* (Riedel and Sanfilippo, 1971), 9. *Bekoma bidartensis*, 10. *Pterocodon tenellus* Foreman (1973), 11. *Phormocyrtis striata exquisita* morphotype, 12. *Lychnocanoma auxilla*, 13. *Theocorys acroria*, 14. *Theocotyle auctor*, 15. *Thrysocyrtis tarsipes*.
ACTINOMMIDAE: 16. *Axoprunum pierinae* group (*sensu* Sanfilippo and Riedel, 1973) (×120).

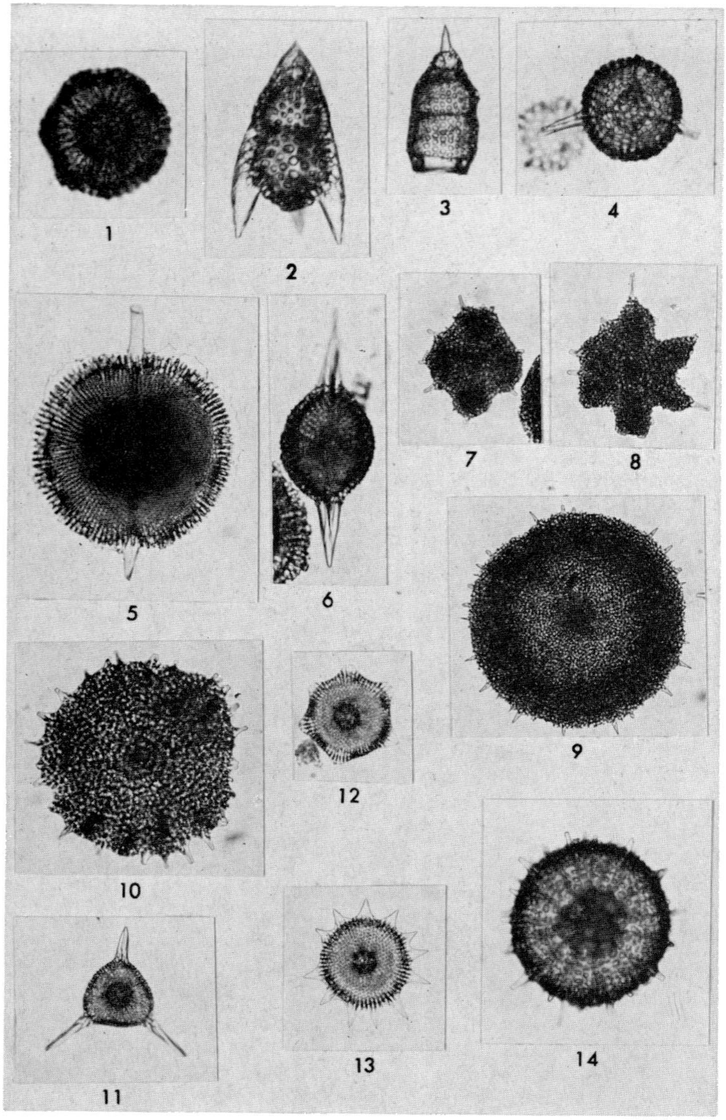

PLATE 5. Early and Middle Eocene radiolarians. (1–3, from the *Buryella clinata* Zone; 4, 5, from the *Phormocyrtis striata striata* Zone; 6–10, from the *Theocotyle cryptocephala cryptocephala* Zone; 11–14, from the *Theocampe mongolfieri* Zone.)
ACTINOMMIDAE: 1. *Thecosphaerella glebulenta* (×120).
THEOPERIDAE (×120): 2. *Lithochytris archaea*, 3. *Theocotyle cryptocephala nigriniae*.
ACTINOMMIDAE: 4. *Entapium chaenapium* (×120).
PHACODISCIDAE: 5. *Heliostylus* sp. (×75).
ACTINOMMIDAE: 6. *Spongatractus balbis*. (×120).
SPONGODISCIDAE (×75): 7. *Stylotrochus quadribrachiatus quadribrachiatus*, 8. *Stylotrochus quadribrachiatus multibrachiatus*, 9. *Spongodiscus americanus*, 10. *Spongodiscus phrix*.
PHACODISCIDAE (×75): 11. *Periphaena tripyramis tripyramis*, 12. *Periphaena delta*, 13. *Periphaena decora*.
LITHELIIDAE: 14. *Lithelius hexaxyphophorus* (×120).

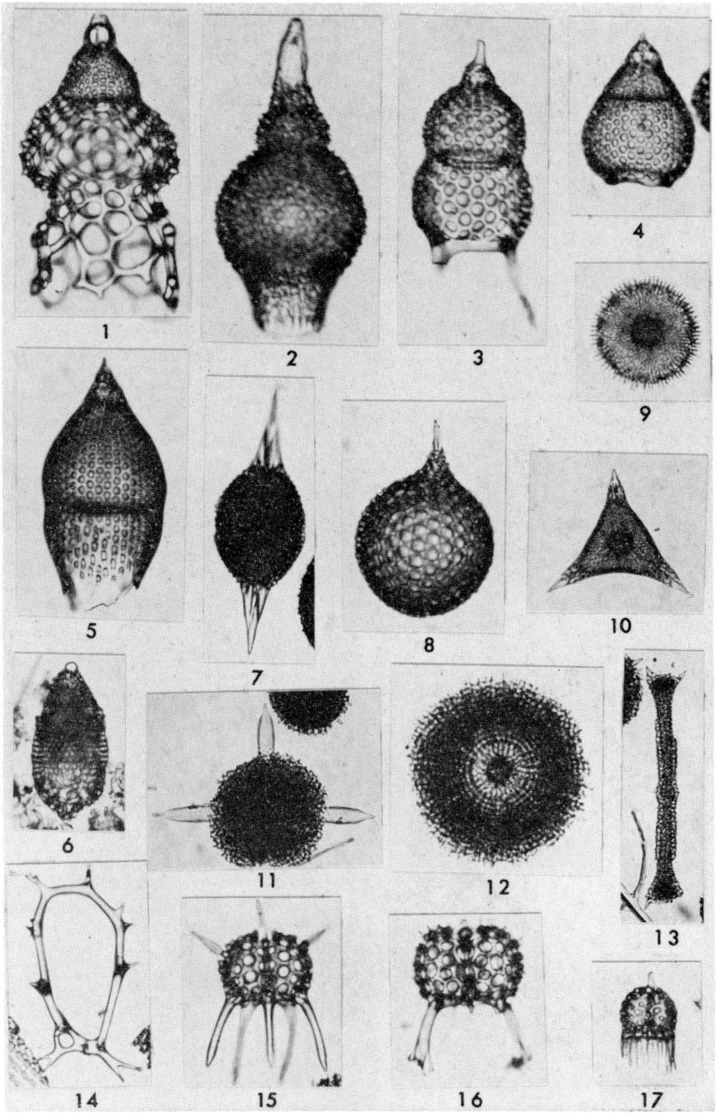

PLATE 6. Middle Eocene radiolarians. (1–6, from the *Theocampe mongolfieri* Zone; 7–17, from the *Thyrsocyrtis triacantha* Zone.)
THEOPERIDAE (×120): 1. *Theocorys anaclasta*, 2. *Lamptonium fabaeforme constrictum*, 3. *Thyrsocyrtis hirsuta robusta*, 4. *Theocotyle cryptocephala conica*.
PTEROCORYTHIDAE: 5. *Podocyrtis aphorma* (×120).
AMPHIPYNDACIDAE: 6. *Amphiternis clava* Foreman (1973) (×120).
ACTINOMMIDAE (×120): 7. *Spongatractus pachystylus*, 8. *Lithapium plegmacantha*.
PHACODISCIDAE (×75): 9. *Periphaena decora*, 10. *Periphaena tripyramis triangula*.
SPONGODISCIDAE: 11. *Spongodiscus rhabdostylus* (×75).
COCCODISCIDAE: 12. *Lithocyclia ocellus* group (×75).
SPONGODISCIDAE: 13. *Amphicraspedum prolixum* (×75).
SPYRIDA: 14–17. Spyrids, gen. et sp. indet. (×120).

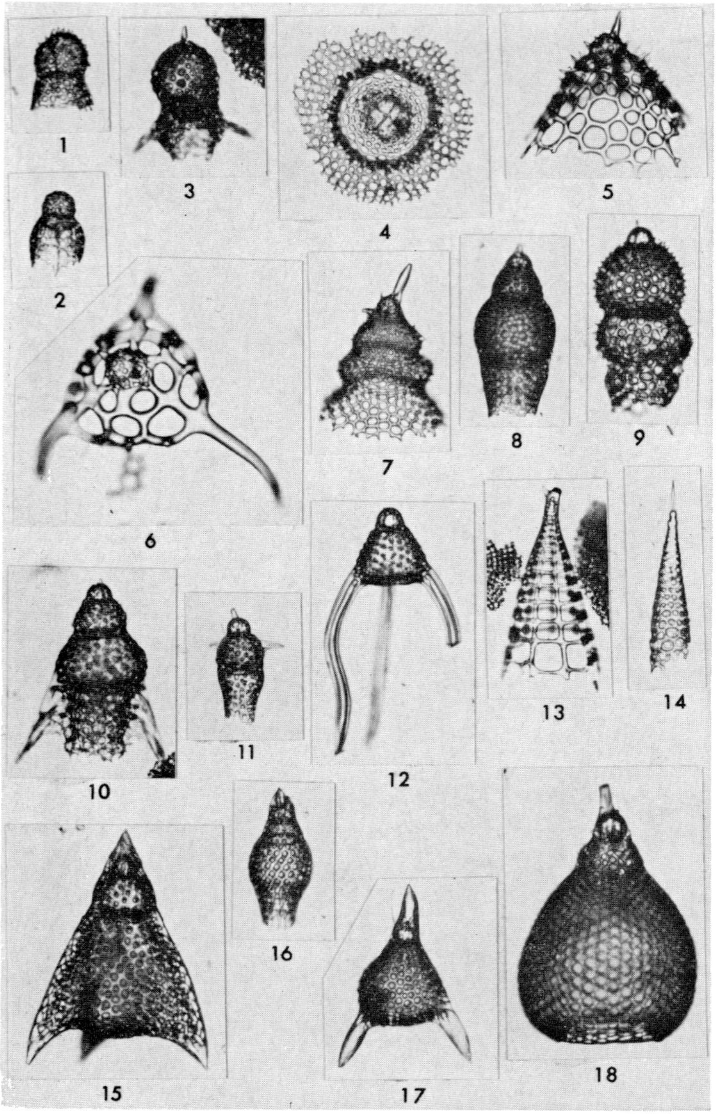

PLATE 7. Middle Eocene radiolarians (from the *Thyrsocyrtis triacantha* Zone) (×120)
PLAGONIIDAE: 1–3. Plagoniids, gen. et sp. indet., 4. *Velicucullus* sp., 5. *Lamprotripus* ? sp.
INCERTAE SEDIS (theoperid): 6. *Dictyophimus craticula* Ehrenberg (1873).
THEOPERIDAE: 7. Theoperid, gen. et sp. indet., 8. cf. *Tricolocampe vitrea* Krasheninnikov (1960), 9. *Theocorys* (?) sp., 10. *Pteropilium sphinx* (Ehrenberg, 1873), 11. *Lipmanella* ? sp., 12. *Lychnocanoma* sp., 13. *Bathropyramis* sp. (×75), 14. *Cornutella* sp., 15. *Lithochytris vespertilio*, 16. *Buryella clinata*, 17. *Lychnocanoma babylonis* group, 18. *Theocotyle venezuelensis*.

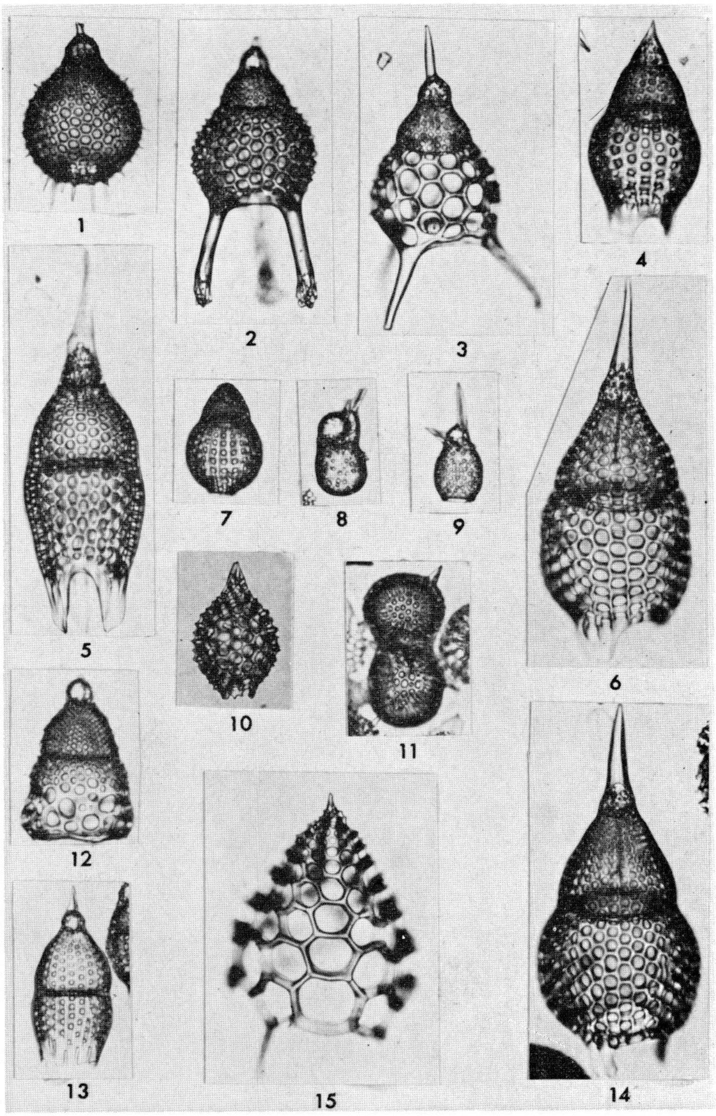

PLATE 8. Middle Eocene radiolarians. (1–9, from the *Thyrsocyrtis triacantha* Zone; 10–14, from the *Podocyrtis ampla* Zone; 15, from the *Podocyrtis mitra* Zone.) (×120)
THEOPERIDAE: 1. *Calocycloma ampulla* Ehrenberg (1854), 2, 3. *Thyrsocyrtis triacantha* Ehrenberg (1873).
PTEROCORYTHIDAE: 4. *Podocyrtis sinuosa*, 5. *Podocyrtis diamesa*, 6. *Podocyrtis phyxis*.
ARTOSTROBIIDAE: 7. *Theocampe mongolfieri*.
CANNOBOTRYTHIDAE: 8. Cannobotrythid, gen. et sp. indet.
INCERTAE SEDIS: 9. *Lophocyrtis biaurita* (Ehrenberg, 1873).
ACTINOMMIDAE: 10. *Lithapium anoectum*.
PLAGONIIDAE: 11. Plagoniid, gen. et sp. indet. (×130).
THEOPERIDAE: 12. *Theocorys anapographa*, 13. *Phormocyrtis embolum* (Ehrenberg, 1873)?
PTEROCORYTHIDAE: 14. *Podocyrtis ampla*.
ACTINOMMIDAE: 15. *Lithapium mitra*.

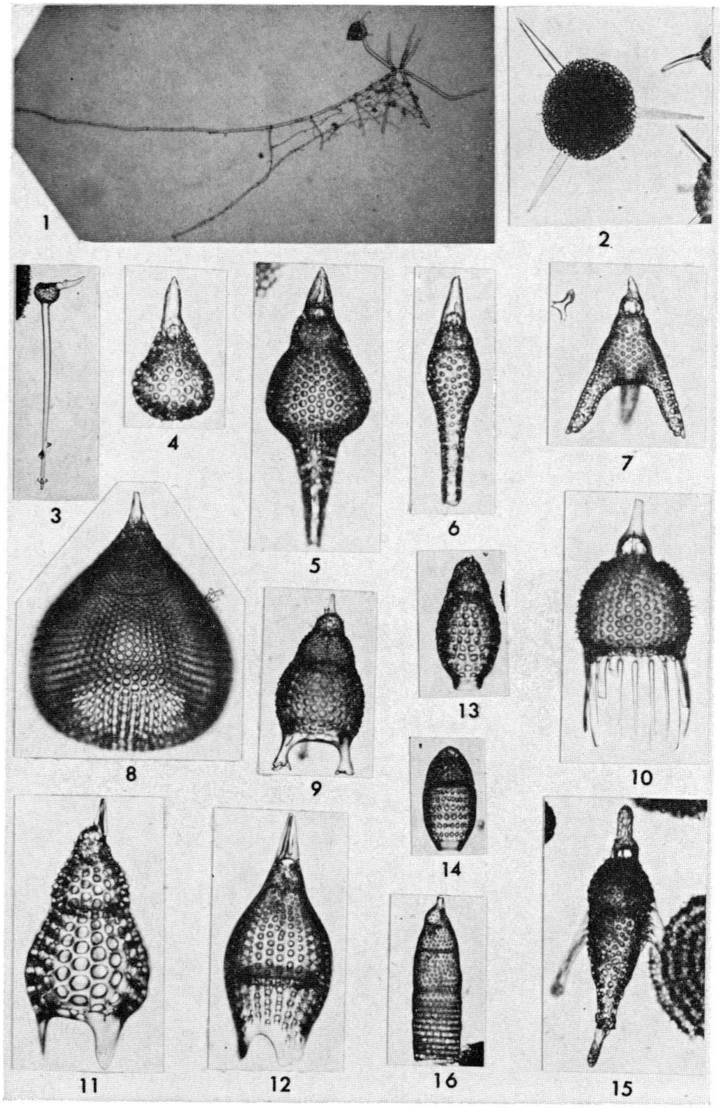

PLATE 9. Middle Eocene radiolarians. (1–14, from the *Podocyrtis mitra* Zone; 15, 16, from the *Podocyrtis chalara* Zone.)
OROSPHAERIDAE: 1. *Orostaurus simplex* (×15).
SPONGODISCIDAE: 2. *Spongodiscus rhabdostylus* (×75).
SPYRIDA: 3. *Rhabdolithis pipa* (×75).
THEOPERIDAE (×120): 4. *Eusyringium lagena*, 5. *Eusyringium fistuligerum*, 6. *Eusyringium fistuligerum*?, 7. *Sethochytris triconiscus*, 8. *Theocotyle ficus*, 9. *Thyrsocyrtis rhizodon*, 10. *Calocyclas hispida*.
PTEROCORYTHIDAE (×120): 11. *Podocyrtis mitra*, 12. *Podocyrtis papalis*.
ARTOSTROBIIDAE: 13, 14. *Theocampe amphora* group (×120).
THEOPERIDAE: 15. *Rhopalocanium ornatum* Ehrenberg (1847b) (×120).
ARTOSTROBIIDAE: 16. Artostrobiid, gen. et sp. indet. (×120).

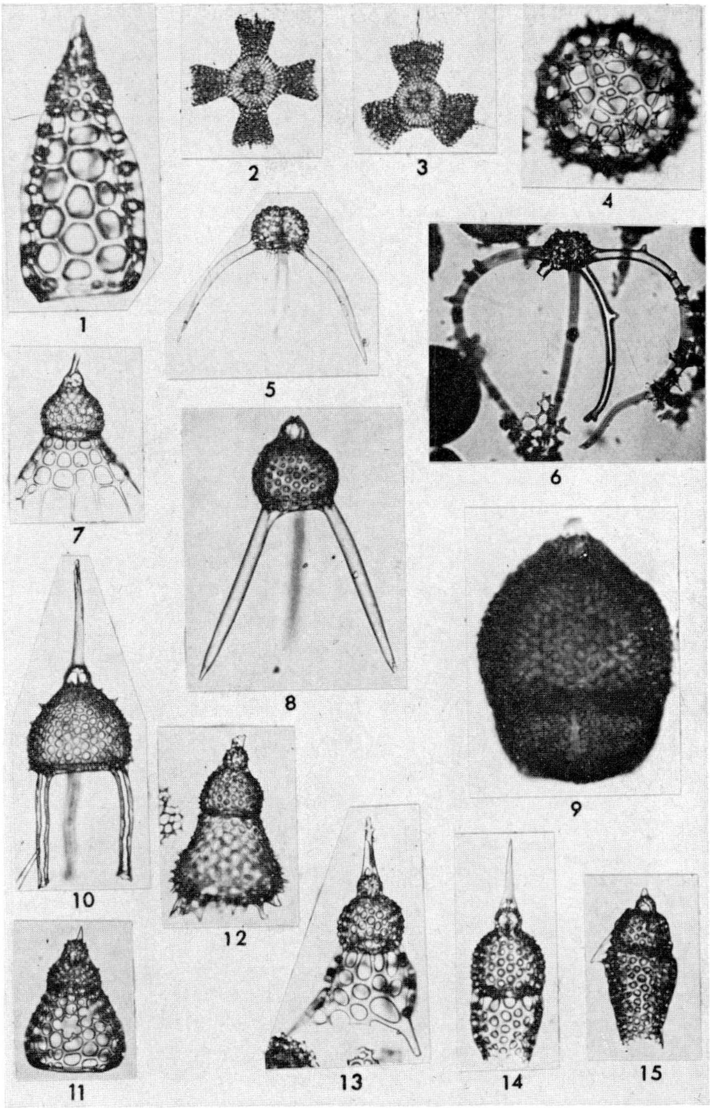

PLATE 10. Middle and Late Eocene radiolarians. (1, from the *Podocyrtis chalara* Zone; 2–15, from the *Thyrsocyrtis bromia* Zone.)
PTEROCORYTHIDAE: 1. *Podocyrtis chalara* (×120).
COCCODISCIDAE: 2, 3. *Lithocyclia aristotelis* group (×75).
COLLOSPHAERIDAE: 4. Collosphaerid (?), gen. et sp. indet. (×120).
SPYRIDA (×75): 5. *Tristylospyris triceros*, 6. *Dorcadospyris spinosa* Moore (1971).
THEOPERIDAE (×120): 7. *Anthocyrtella* sp., 8. *Lychnocanoma* sp., 9. *Lychnocanoma amphitrite*, 10. *Lychnocanoma bellum*, 11. *Artophormis dominasinensis*, 12. *Artophormis barbadensis*, 13, 14. *Lophocyrtis jacchia* (Ehrenberg, 1873), 15. *Lipmanella*? sp.

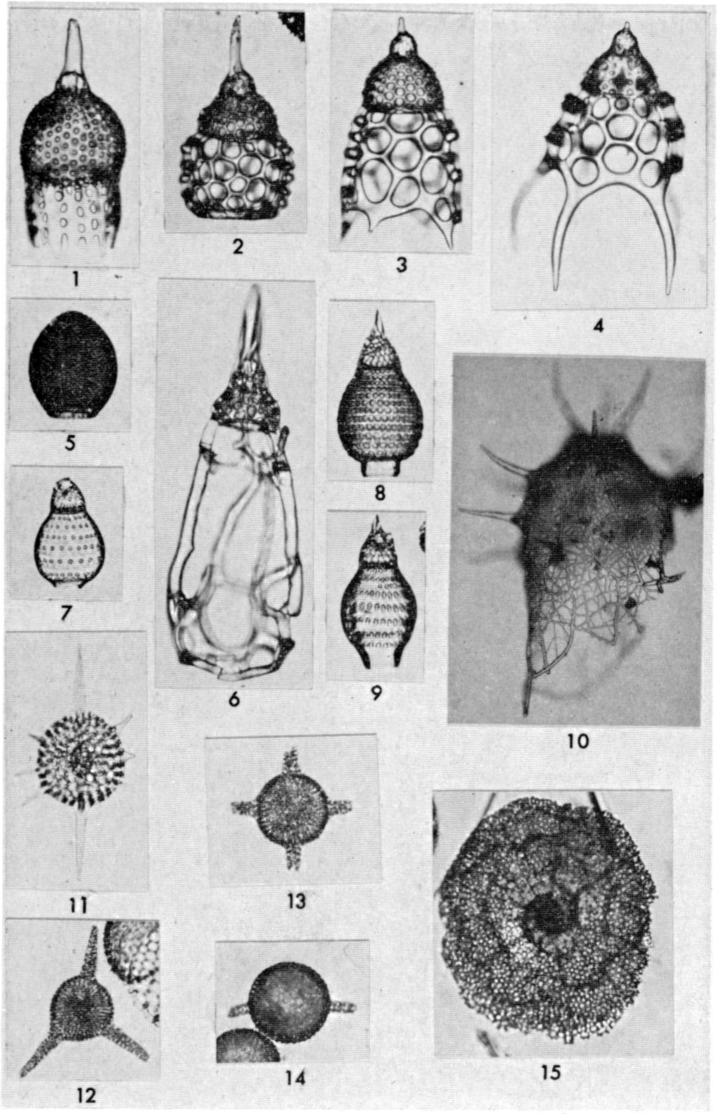

PLATE 11. Late Eocene and Early Oligocene radiolarians. (1–9, from the *Thyrsocyrtis bromia* Zone; 10–15, from the *Theocyrtis tuberosa* Zone.)

THEOPERIDAE (×120): 1. *Calocyclas turris*, 2. *Thyrsocyrtis bromia* Ehrenberg (1873), 3, 4. *Thyrsocyrtis tetracantha*.

CARPOCANIIDAE: 5. *Carpocanistrum azyx* Sanfilippo and Riedel (1973) (×120).

PTEROCORYTHIDAE: 6. *Podocyrtis goetheana* (×120).

ARTOSTROBIIDAE (×120) 7. *Theocampe pirum*, 8, 9. *Theocampe armadillo* group.

OROSPHAERIDAE: 10. *Oropelex pagoda* (×30).

PHACODISCIDAE: 11. *Heliostylus* sp. (×75).

COCCODISCIDAE (×75): 12. *Lithocyclia angusta*, 13. *Lithocyclia crux*, 14. Transitional form between *Lithocyclia angusta* (or *L. crux*) and *Cannartus prismaticus*.

LITHELIIDAE: 15. Litheliid (?), gen. et sp. indet. (×130).

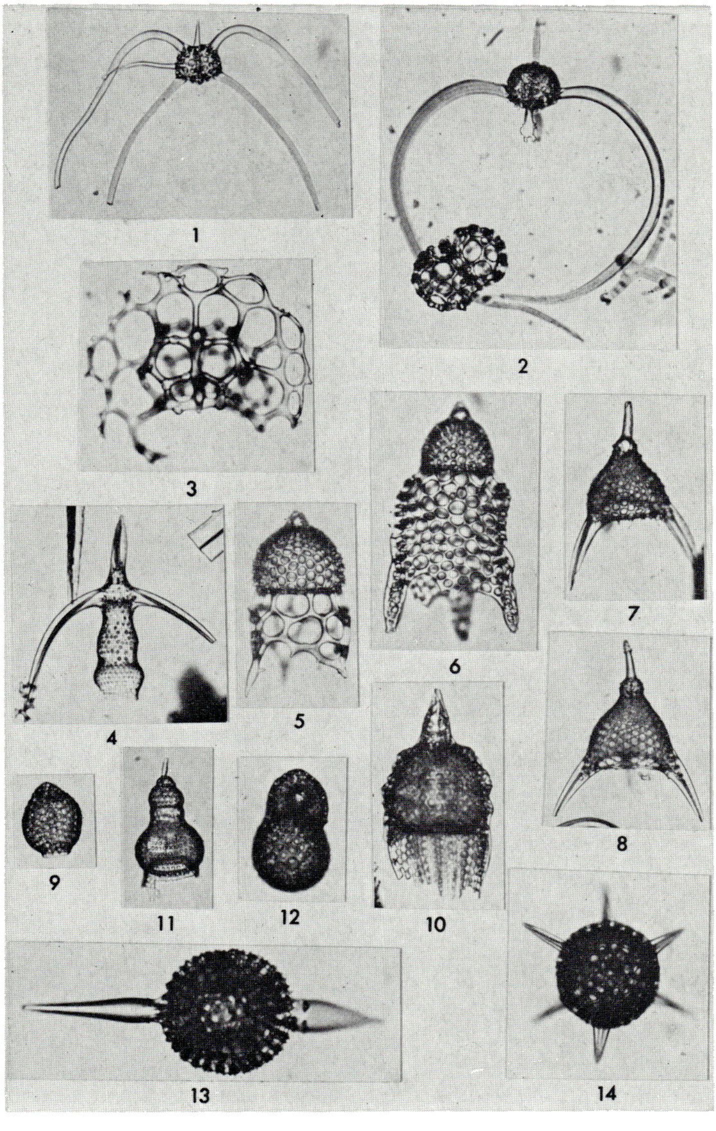

PLATE 12. Oligocene radiolarians. (1–12, from the *Theocyrtis tuberosa* Zone; 13, 14, from the *Dorcadospyris ateuchus* Zone.)

SPYRIDA (×75): 1. *Dorcadospyris quadripes* Moore (1971), 2. *Dorcadospyris pseudopapilio* Moore (1971), 3. Spyrid, gen. et sp. indet.

THEOPERIDAE: 4. Theoperid, gen. et sp. indet. (×120), 5. *Cyclampterium pegetrum* (×75), 6. *Cyclampterium milowi* (×75), 7. *Lychnodictyum* (?) sp. (120), 8. *Lychnodictyum audax* Riedel (1953) (×120).

CARPOCANIIDAE: 9. *Carpocanistrum* sp. (×120).

PTEROCORYTHIDAE: 10. *Theocyrtis tuberosa* (×120).

ARTOSTROBIIDAE: 11. *Theocamptra formaster* (×120).

CANNOBOTRYTHIDAE: 12. *Centrobotrys gravida* (×120).

ACTINOMMIDAE: 13, 14. Actinommids, gen. et sp. indet. (×120).

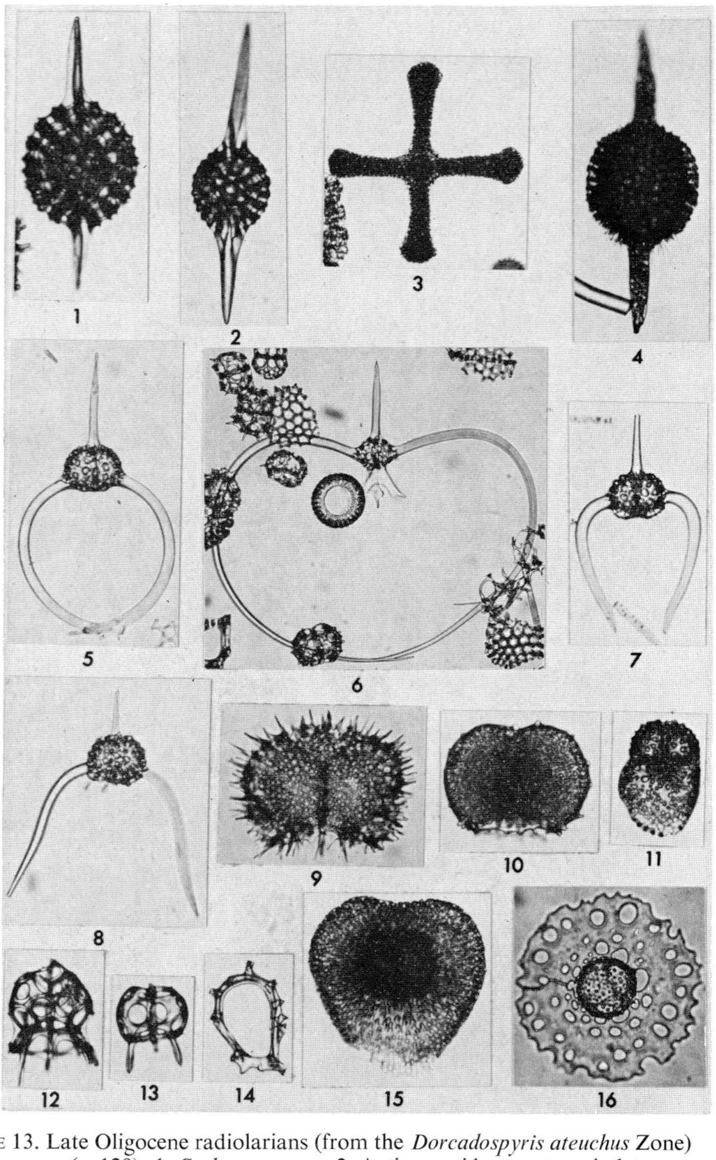

PLATE 13. Late Oligocene radiolarians (from the *Dorcadospyris ateuchus* Zone)
ACTINOMMIDAE (×120): 1. *Stylatractus* sp., 2. Actinommid, gen. et sp. indet.
SPONGODISCIDAE: 3. *Histiastrum* sp. (×75).
ACTINOMMIDAE: 4. *Cannartus prismaticus* (×120).
SPYRIDA: 5. *Dorcadospyris circulus* (Haeckel, 1887) (×75), 6. *Dorcadospyris papilio* (Riedel, 1959b) (×75), 7. *Dorcadospyris forcipata* (Haeckel, 1887) (×75), 8. *Dorcadospyris ateuchus* (×75), 9, 10. *Dendrospyris anthocyrtoides* (×120), 11–15. Spyrids, gen. et sp. indet. (×120)
PLAGONIIDAE: 16. Possible ancestor of *Tepka perforata* (×120).

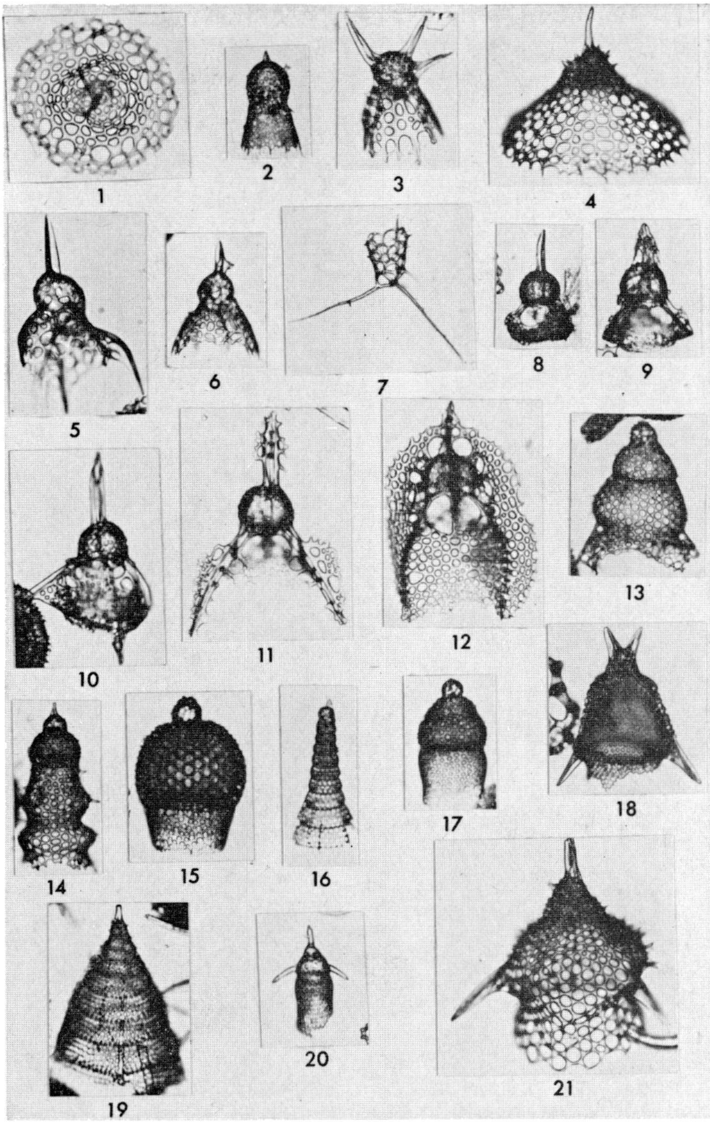

PLATE 14. Late Oligocene radiolarians (from the *Dorcadospyris ateuchus* Zone) (×120) PLAGONIIDAE: 1. Sethophormin, gen. et sp. indet., 2, 3. Plagoniids, gen. et sp. indet., 4. *Lamprotripus* sp., 5. *Pseudodictyophimus* sp. (×200), 6, 7. Plagoniids, gen. et sp. indet., 8, 9. *Clathrocanium* sp., 10. Sethoperin, gen. et sp. indet., 11, 12. *Clathrocorys* ? sp. THEOPERIDAE: 13. *Artophormis* sp. cf. *A. gracilis*, 14. *Theocorys* ? sp., 15. *Theocorys* sp., 16. *Stichophormis* (?) sp., 17. *Theocorys* sp., 18. Theoperid, gen. et sp. indet., cf. *Artopilium elegans* Haeckel (1887), 19. Theoperid, gen. et sp. indet. (×75), 20, 21. Theoperids, gen. et sp. indet.

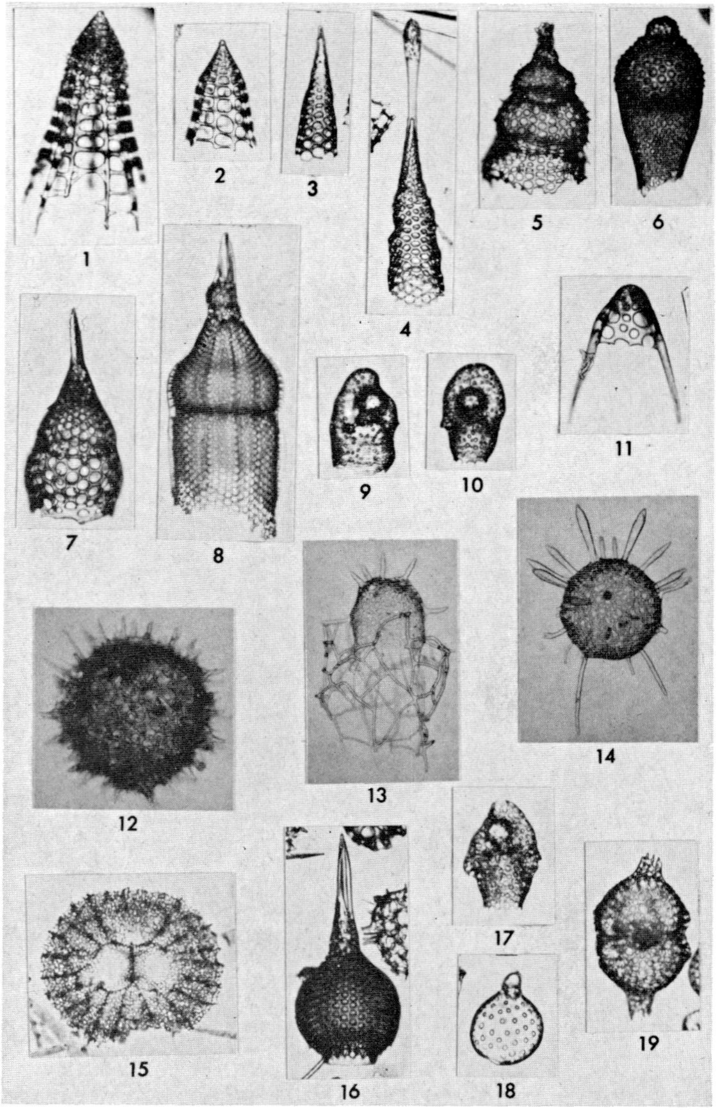

PLATE 15. Late Oligocene and Early Miocene radiolarians. (1–11, from the *Dorcadospyris ateuchus* Zone; 12–18, from the *Lychnocanoma elongata* Zone; 19, from the *Calocycletta virginis* Zone.)

THEOPERIDAE: 1. *Peripyramis* sp. (×75), 2. *Bathropyramis* sp. (×75), 3. *Cornutella* sp. (×120), 4. *Cornutella* sp. (×75), 5. *Artophormis gracilis* (×120), 6. *Theocorys spongoconum* Kling (1971) (×120).

PTEROCORYTHIDAE (×120): 7. *Lamprocyclas* sp., 8. *Theocyrtis annosa*.

CANNOBOTRYTHIDAE (×120): 9. *Botryocella* sp., 10. *Centrobotrys petrushevskayae*.

INCERTAE SEDIS: 11. *Sethopilium macropus* Haeckel (1887)? (×120).

OROSPHAERIDAE (×15): 12. *Orodapis spongiosa*, 13. *Oropagis dolium*, 14. *Oroscena carolae*.

SPYRIDA: 15. *Psychospyris parva* (×75).

PTEROCORYTHIDAE: 16. *Anthocyrtidium* sp. (×120).

CANNOBOTRYTHIDAE: 17. *Botryocella* sp. (×120).

INCERTAE SEDIS: 18. *Dicolocapsa microcephala* Haeckel (1887) (×120).

ACTINOMMIDAE: 19. *Cannartus tubarius* (×120).

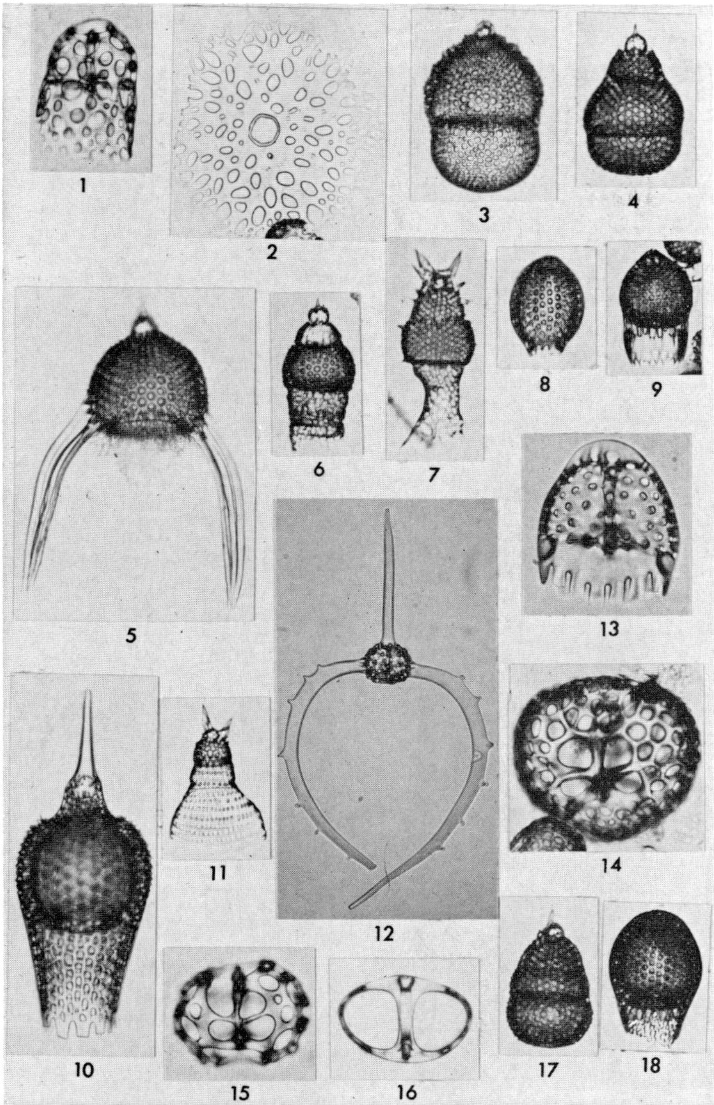

PLATE 16. Early Miocene radiolarians. (1–11 from the *Calocycletta virginis* Zone; 12–18, from the *Calocycletta costata* Zone.)
SPYRIDA: 1. *Dendrospyris pannosa* (×200).
PLAGONIIDAE: 2. *Tepka perforata* (×120).
THEOPERIDAE: 3. *Cyclampterium leptetrum* (×75), 4. *Cyrtocapsella cornuta* (×120), 5. *Lychnocanoma elongata* (Vinassa, 1900) (=*Lychnocanium bipes* Riedel, 1959b) (×120), 6. *Stichocorys wolffii* (×120), 7. *Stichocorys armata* (×120).
CARPOCANIIDAE (×120): 8. *Carpocanistrum* sp., 9. *Carpocanopsis bramlettei* Riedel and Sanfilippo (1971).
PTEROCORYTHIDAE: 10. *Calocycletta robusta* (×120).
ARTOSTROBIIDAE: 11. *Phormostichoartus* sp. aff. *P. corona* (×120).
SPYRIDA: 12. *Dorcadospyris dentata* Haeckel (1887) (×75), 13. *Gorgospyris schizopodia* (×202), 14. *Liriospyris globosa* (×202), 15. *Liriospyris mutuaria* (×200), 16. *Liriospyris stauropora* (×200).
THEOPERIDAE: 17. *Cyrtocapsella tetrapera* (×120).
CARPOCANIIDAE: 18. *Carpocanopsis cingulata* Riedel and Sanfilippo (1971) (×120).

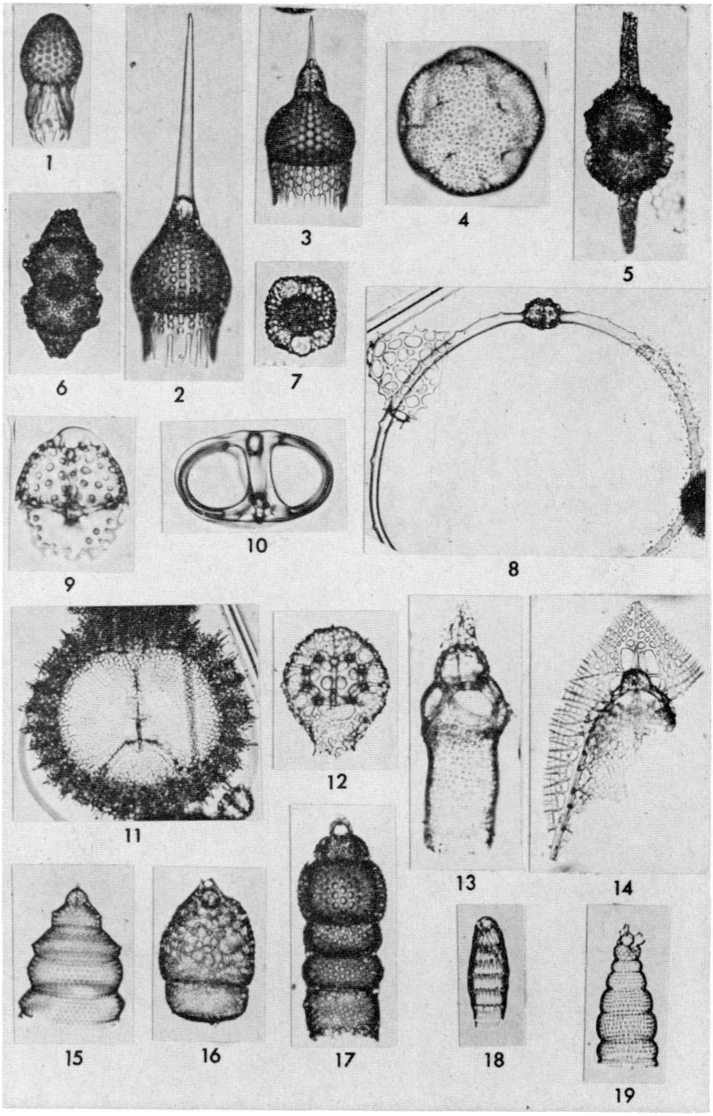

PLATE 17. Early and Middle Miocene radiolarians. (1–3, from the *Calocycletta costata* Zone; 4–19 from the *Dorcadospyris alata* Zone.)
CARPOCANIIDAE: 1. *Carpocanopsis favosa* (Haeckel, 1887) (×120).
PTEROCORYTHIDAE (×120): 2. *Calocycletta costata*, 3. *Calocycletta virginis*.
COLLOSPHAERIDAE: 4. *Tribonosphaera* sp. (×120).
ACTINOMMIDAE (×120): 5. *Cannartus violina*, 6. *Cannartus mammifer*.
PYLONIIDAE: 7. Pyloniid, gen. et sp. indet. (×120).
SPYRIDA: 8. *Dorcadospyris alata* (×75), 9. *Gorgospyris perizostra* (×202), 10. *Liriospyris parkerae* (×200), 11. *Psychospyris grandis* (×75), 12. Spyrid, gen. et sp. indet. (×120).
PLAGNOIIDAE: 13. *Clathrocanium* sp. (×200), 14. *Clathrocorys* ? sp. (×120).
THEOPERIDAE (×120): 15. *Lithostrobus* sp., 16. *Lithopera renzae*, 17. *Stichocorys delmontensis*.
ARTOSTROBIIDAE (×120): 18. *Lithomitra lineata* group, 19. *Spirocyrtis subtilis*.

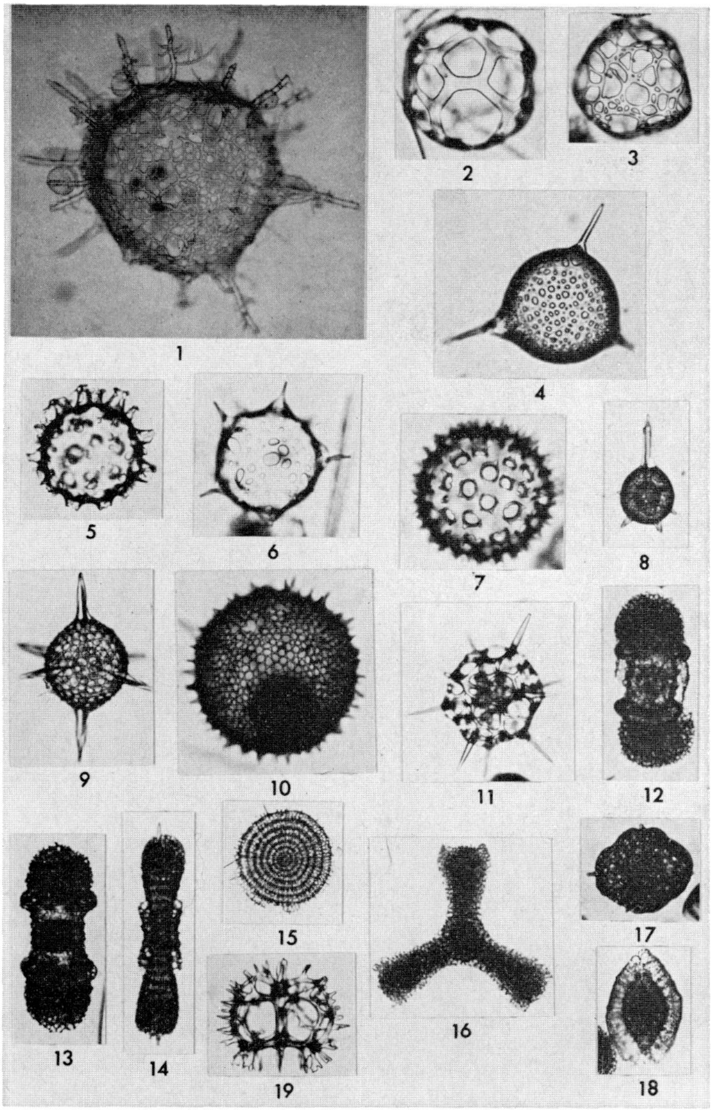

PLATE 18. Middle Miocene radiolarians (from the *Cannartus petterssoni* Zone)
OROSPHAERIDAE: 1. *Orodendrum* sp. (×15).
COLLOSPHAERIDAE (×120): 2, 3. *Collosphaera* sp., 4. *Solenosphaera* sp., 5. *Polysolenia* (?) sp., 6, 7. *Polysolenia* sp.
ACTINOMMIDAE: 8. *Stylosphaera* sp. (×120), 9, 10. Actinommids, gen. et sp. indet. (×120), 11. Actinommid, gen. et sp. indet. (×75), 12, 13. *Cannartus petterssoni* (×120).
PORODISCIDAE: 14. *Amphymenium* sp. cf. *A. splendiarmatum* (×120), 15. Porodiscid, gen. et sp. indet. (×75).
SPONGODISCIDAE: 16. *Euchitonia* ? sp. (×75).
THOLONIIDAE: 17. Tholoniid, gen. et sp. indet. (×120).
LITHELIIDAE: 18. *Larcospira* sp. (×75).
SPYRIDA: 19. Spyrid, gen. et sp. indet. (×120).

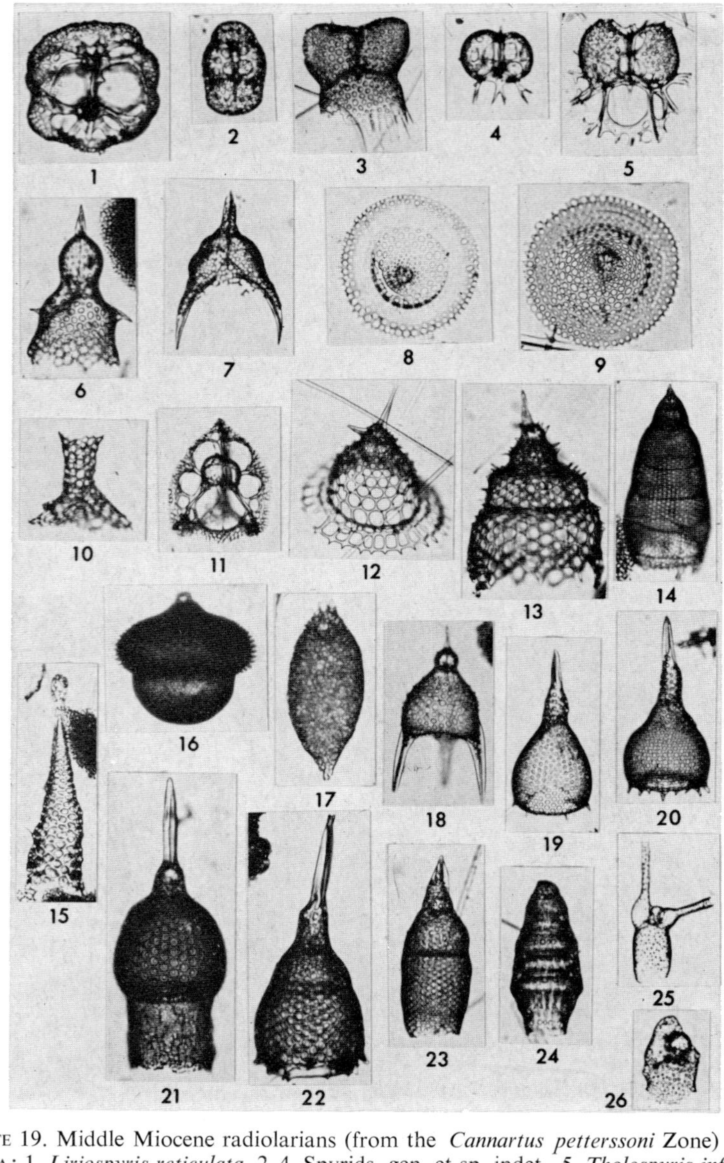

PLATE 19. Middle Miocene radiolarians (from the *Cannartus petterssoni* Zone) (×120)
SPYRIDA: 1. *Liriospyris reticulata*, 2–4. Spyrids, gen. et sp. indet., 5. *Tholospyris infericosta* Goll (1969).
PLAGONIIDAE: 6. Plagoniid, gen. et sp. indet., 7. *Verticillata* ? sp., 8, 9. *Theocalyptra* sp(p), 10. *Amphiplecta* sp., 11. *Clathrocorys* sp.
THEOPERIDAE: 12. *Clathrocycloma* ? sp., 13. *Calocyclas* ? sp., 14. *Eucyrtidium cienkowskii* Haeckel (1887), 15. *Cornutella* sp. (×75), 16. *Cyclampterium brachythorax* (×60), 17. *Lithopera thornburgi*, 18. *Pterocanium* sp.
PTEROCORYTHIDAE: 19, 20. *Anthocyrtidium* sp., 21. *Calocycletta caepa*, 22. *Lamprocyclas* sp., 23. *Theoconus* sp.
ARTOSTROBIIDAE: 24. *Artostrobium miralestense* (Campbell and Clark, 1944).
CANNOBOTRYTHIDAE: 25. *Acrobotrys* sp., 26. *Botryocella* sp.

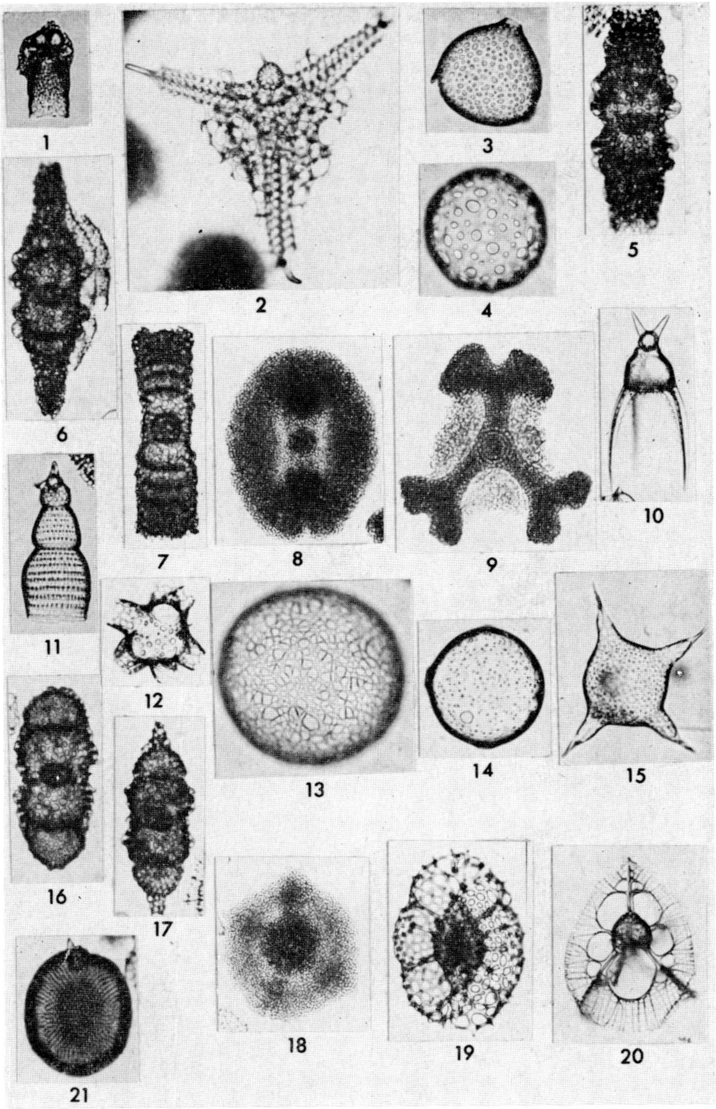

PLATE 20. Middle and Late Miocene radiolarians. (1, 2, from the *Cannartus petterssoni* Zone; 3–11, from the *Ommatartus antepenultimus* Zone; 12, from the *Ommatartus penultimus* Zone; 13–21, from the *Stichocorys peregrina* Zone.) (×120)

CANNOBOTRYTHIDAE: 1. *Botryocyrtis* sp.

INCERTAE SEDIS: 2. Gen. et sp. indet. cf. *Rhopalastrum* (?) *anomalum* Sutton (1896).

COLLOSPHAERIDAE: 3. *Solenosphaera* sp., 4. *Collosphaera* sp.

ACTINOMMIDAE: 5. *Cannartus laticonus* morphotype, 6. *Ommatartus antepenultimus*, 7. *Ommatartus hughesi*.

SPONGODISCIDAE (×75): 8. *Spongaster berminghami*, 9. *Dictyocoryne ontongensis* Riedel and Sanfilippo (1971).

THEOPERIDAE: 10. *Lychnocanoma* ? sp.

ARTOSTROBIIDAE: 11. *Phormostichoartus corona*.

COLLOSPHAERIDAE: 12. *Solenosphaera omnitubus*, 13. *Collosphaera* sp., 14. *Siphonosphaera* sp., 15. *Solenosphaera* sp.

ACTINOMMIDAE: 16. *Ommatartus avitus*, 17. *Ommatartus penultimus*.

SPONGODISCIDAE: 18. *Spongaster pentas* (×75).

LITHELIIDAE: 19. *Larcospira* sp.

PLAGONIIDAE: 20. *Clathrocorys* sp.

THEOPERIDAE: 21. *Lithopera bacca*.

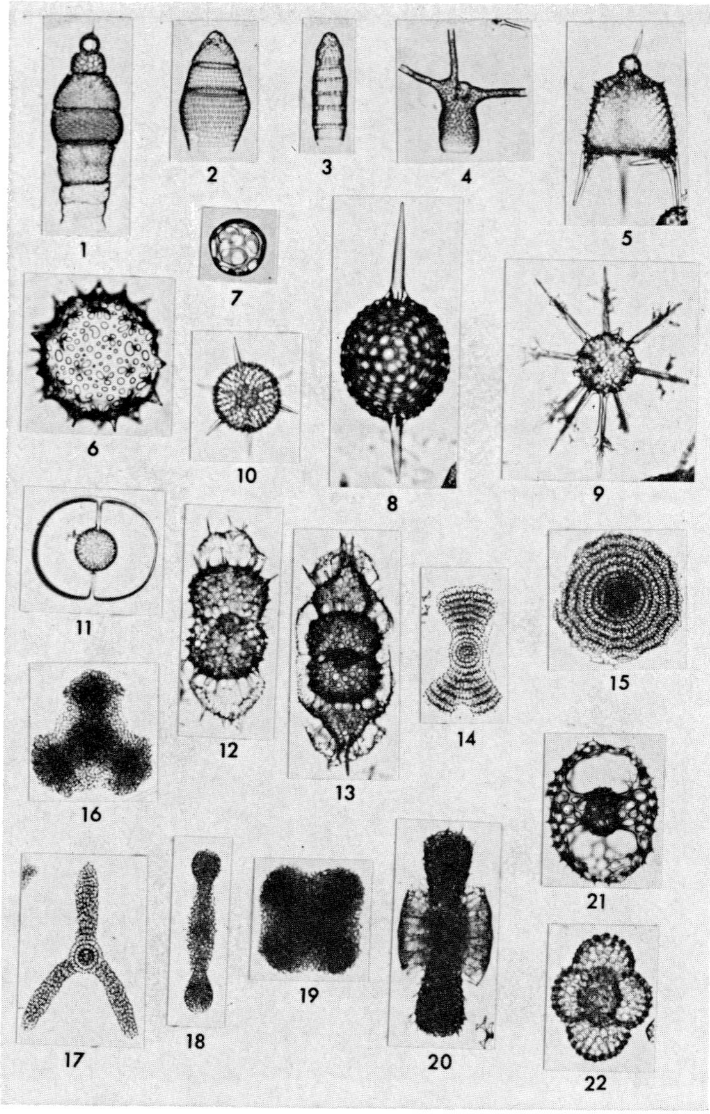

PLATE 21. Late Miocene to Quaternary radiolarians. (1–4, from the *Stichocorys peregrina* Zone; 5, from the *Pterocanium prismatium* Zone; 6–22 from the Quaternary.)
THEOPERIDAE: 1. *Stichocorys peregrina* (×120).
ARTOSTROBIIDAE (×120): 2. *Artostrobium doliolum*, 3. *Lithomitra lineata* group.
CANNOBOTRYTHIDAE: 4. *Acrobotrys tritubus* Riedel (1957) (×120).
THEOPERIDAE: 5. *Pterocanium prismatium* (×120).
COLLOSPHAERIDAE (×150): 6. *Polysolenia* sp., 7. *Collosphaera* sp.
ACTINOMMIDAE: 8. Actinommid, gen. et sp. indet. (×120), 9. *Cladococcus* sp. (×75), 10. Actinommid, gen. et sp. indet. (×75), 11. *Saturnalis circularis* Haeckel (1887) (×75), 12, 13. *Ommatartus tetrathalamus* (×120).
PORODISCIDAE (×95): 14. *Amphirhopalum ypsilon* Haeckel (1887), 15. *Stylotrochus* ? sp.
SPONGODISCIDAE: 16. *Dictyocoryne* sp. (×76), 17. *Euchitonia* sp. (×75), 18. *Spongocore* sp. (×75), 19. *Spongaster tetras* (×75), 20. *Spongocore* sp. (×120).
PYLONIIDAE: 21. Pyloniid, gen. et sp. indet. (×120).
THOLONIIDAE: 22. Tholoniid, gen. et sp. indet. (×120)

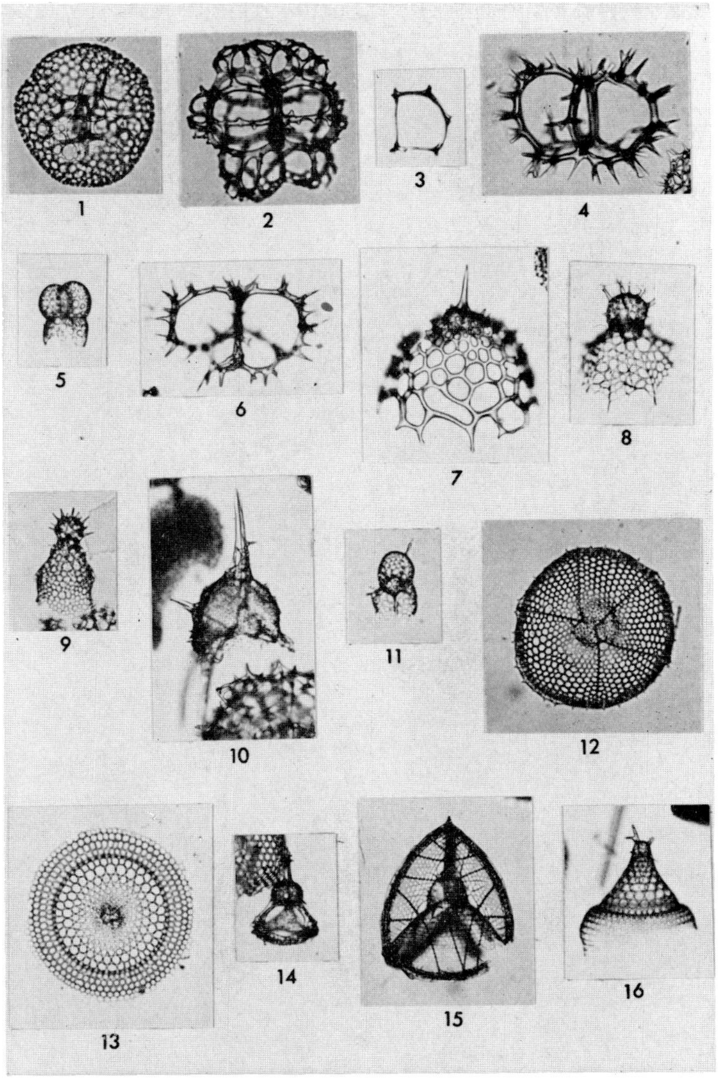

PLATE 22. Quaternary radiolarians (× 120)

SPYRIDA: 1–6. Spyrids, gen. et sp. indet.

PLAGONIIDAE: 7. *Lamprotripus* ? sp., 8. *Lophophaena* ? sp., 9. *Lophophaena* sp., 10. *Verticillata hexacantha* Popofsky (1913), 11. Plagoniid, gen. et sp. indet., 12. *Lampromitra* sp., 13. *Theocalyptra* sp., 14. *Clathrocanium* sp., 15. *Callimitra* sp.

THEOPERIDAE: 16. *Clathrocycloma* ? sp.

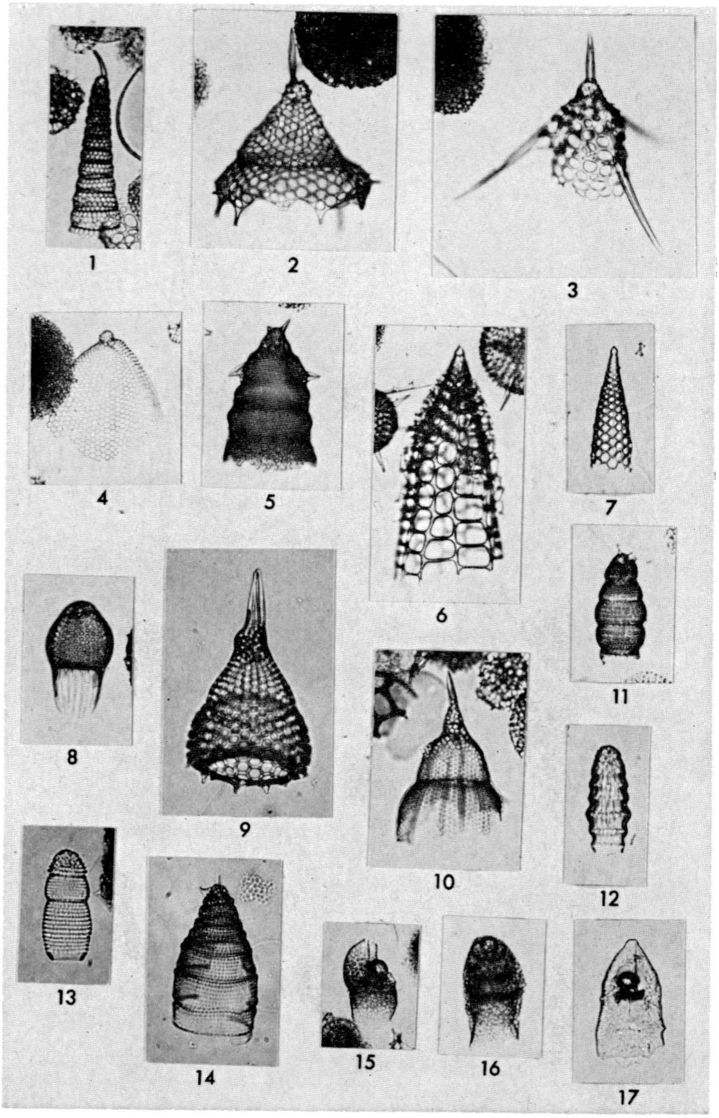

PLATE 23. Quaternary radiolarians (×120)

THEOPERIDAE: 1. *Cyrtopera* (?) sp., 2. *Calocyclas monumentum* Haeckel (1887), 3. *Dictyophimus crisiae* Ehrenberg (1854), 4. *Lampromitra parabolica* Popofsky (1913) (×75), 5. *Stichopilium* ? sp., 6. *Peripyramis* sp. (×75), 7. *Cornutella* sp.
CARPOCANIIDAE: 8. *Carpocanistrum* sp.
PTEROCORYTHIDAE: 9. *Lamprocyclas* sp.
ARTOSTROBIIDAE: 10. *Pterocorys hertwigii* (Haeckel, 1887), 11. *Artostrobium auritum* group, 12. *Lithomitra lineata* group, 13. *Siphocampe corbula*, 14. *Spirocyrtis scalaris*.
CANNOBOTRYTHIDAE: 15. *Botryocella* sp., 16. *Botryocyrtis* sp., 17. *Centrobotrys thermophila*.

9: Radiolaria in Mesozoic Stratigraphy

EMILE A. PESSAGNO, Jr.
The University of Texas at Dallas,
Box 3065, Dallas,
Texas 75230, U.S.A.

Until recently Radiolaria have been overlooked by Mesozoic stratigraphers and palaeontologists. To a large degree this has been due to a continual repetition of pessimistic statements in the literature which state (1) that Radiolaria are universally long ranging and (2) that Radiolaria are too minute and complex to study. Such statements are unwarranted and decidedly premature in that few workers have taken the time to study Radiolaria.

The present report deals with (1) the mode of occurrence of Radiolaria in Mesozoic marine strata; (2) methods of sample preparation and illustration of specimens; (3) the biostratigraphic potential of Mesozoic Radiolaria; and (4) the development of a meaningful classification; it is designed to familiarise the reader in a general way with Mesozoic Radiolaria.

Recent technological advances in sample processing and method of illustration have made Radiolaria much easier to study. The development of a technique for extracting Radiolaria from indurated chert samples has opened new avenues of research, many of which can lead to a deciphering of complex geological problems. The use of the scanning electron microscope in the study and illustration of Radiolaria has facilitated a better understanding of radiolarian morphology and phylogeny and has aided in the construction of a more meaningful classification. Furthermore, the use of this instrument has provided accurate illustrations to all workers involved in biostratigraphic investigations.

Outline

Introduction

Approximately twenty years ago a new era began in micropalaeontology and biostratigraphy—an era that might be called "the age of microplankton". This era began with a few workers such as Bolli, Brönnimann, Subbotina, Cita, Blow, Loeblich and Tappan, Reichel, and Sigal who recognised the importance of planktonic foraminifera in biostratigraphy. It began at a time in which most of their colleagues considered planktonic foraminifera to be universally long ranging and too small and difficult to work with. Now, after over two decades of intensive study, the planktonic foraminifera are regarded as one of the most important groups of fossils for interpreting the stratigraphy of Cretaceous, Tertiary, and Quaternary marine strata of the continental and oceanic crust. The momentum gained through the intensive study of planktonic foraminifera has spread to the study of other groups of planktonic microfossils such as the coccoliths, the Radiolaria, and the dinoflagellates. It is now further reinforced by the important role being played by planktonic microfossils in the current Deep Sea Drilling Project.

A number of negative statements have been made concerning the use of Radiolaria in biostratigraphy. Campbell (1954, p. D18), for example, stated:

> The value of fossil Radiolaria for stratigraphy is diminished by the long ranges of many genera. Some seem to extend throughout the whole post-pre-Cambrian geological column and others through a number of systems. Only a few genera are limited sharply to part of a single system. Furthermore, the number of species is tremendous, so that, despite general limitation to very short geological ranges (Riedel, 1952), identifications and establishment of the ranges of individual species are very difficult indeed. Many species remain to be described. The selection of index forms and correlation of faunas in widely separated geographic provinces cannot be undertaken satisfactorily under conditions of present knowledge.

Campbell hardly stood alone in making such a statement. Most textbooks dealing with invertebrate palaeontology voice the same air of pessimism. Indeed, it has become a tradition to state that Radiolaria have no biostratigraphic value because they are universally long ranging. Statements of this sort are difficult to understand. It is evident from even a casual examination of the literature that Radiolaria have not been sufficiently studied to prove their true worth.

The advent of the study of Mesozoic Radiolaria corresponds closely to the time of the famous *Challenger* expedition before the turn of the century. Perhaps stimulated by Haeckel's (1887a, b) comprehensive monographs dealing with the Radiolaria recovered by the *Challenger* expedition, a number of workers initiated investigations of Mesozoic radiolarina assemblages. Among these workers Hinde (1908) and Rüst (1892b) published reports dealing with Triassic faunas. Rüst (1885, 1898), Vinassa de Regny (1899) and Parona (1890) published monographic studies of Jurassic and Lower Cretaceous Radiolaria; Squinabol (1903, 1904, 1913, 1914), Zittel (1876), Holmes (1900), and Rüst (1892a) presented reports dealing with Upper Cretaceous Radiolaria. From 1914 with the last report of Squinabol until the mid 1940s few significant reports dealing with Mesozoic Radiolaria were published. Of these reports, that of Tan Sin Hok (1927) and that of Khabakov (1937) are the only ones worth noting.

As indicated by the references dealing with Mesozoic Radiolaria in the bibliography, the study of Mesozoic Radiolaria has gradually increased from the mid 1940s until the present day. It should be noted, however, that most of this research was undertaken by Lipman (1952), Khabakov, Aliev, (1961, 1965), Zhamoida (1972) and various other Russian workers. Even with this resurgence in activity, Mesozoic Radiolaria still remain poorly studied. One need only compare the number of publications dealing with the Foraminiferida with that dealing with Radiolaria during the period from 1945 to 1973 to demonstrate this.

Mode of occurrence of Mesozoic Radiolaria

Because Radiolaria are dominantly oceanopelagic rather than neritopelagic in character, one would expect to find them preserved in deposits which were originally laid down in ocean basins. It is not surprising, then, to find that the Mesozoic strata which are richest in Radiolaria occur within the ocean or in the "eugeosynclinal" facies of orogenic belts. In the continental United States, for example, Mesozoic Radiolaria are most abundant in the California Coast Ranges. Here, Late Mesozoic (Tithonian to Maestrichtian) rocks west of the Nevadian orogenic belt include two distinctive assemblages: (1) an eastern assemblage termed the Great Valley sequence (Bailey *et al.*, 1964) and a western assemblage termed the Franciscan complex (Lawson, 1895) (see Fig. 1). Whereas the Great Valley sequence consists for the most part of a monotonous flysch succession of sandstones and shales, the Franciscan complex includes a heterogeneous assemblage of tectonically emplaced rocks (melanges) typified by ophiolites and metamorphics. The flysch sediments of the Great Valley sequence were (except on the eastern edge of the Great Valley deposited at bathyal to abyssal depths below the compensation depth for calcium carbonate. The presence of abundant Radiolaria throughout the Mesozoic portion of the Great Valley sequence probably indicates that deposition occurred on the continental slope or rise.

As already noted polycystine Radiolaria today are predominantly oceanopelagic in character and are not as abundant or as diversified over the continental shelf areas. In most Upper Cretaceous strata within the Great Valley flysch succession, however, it is not uncommon to find abundant Radiolaria associated with abundant arenaceous and calcareous benthonic foraminifera At most horizons planktonic foraminifera are sparse. It is likely that this anomalous association was caused by the introduction of a shelf foraminiferal assemblage into waters of bathyal and abyssal depths via turbidity currents from a narrow continental shelf area bordering the Nevadian front. The fact that planktonic foraminifera are rare in most of the Great Valley sequence suggests that deposition took place below the compensation depth for calcium carbonate. It is likely that such calcareous planktonic and benthonic foraminifera

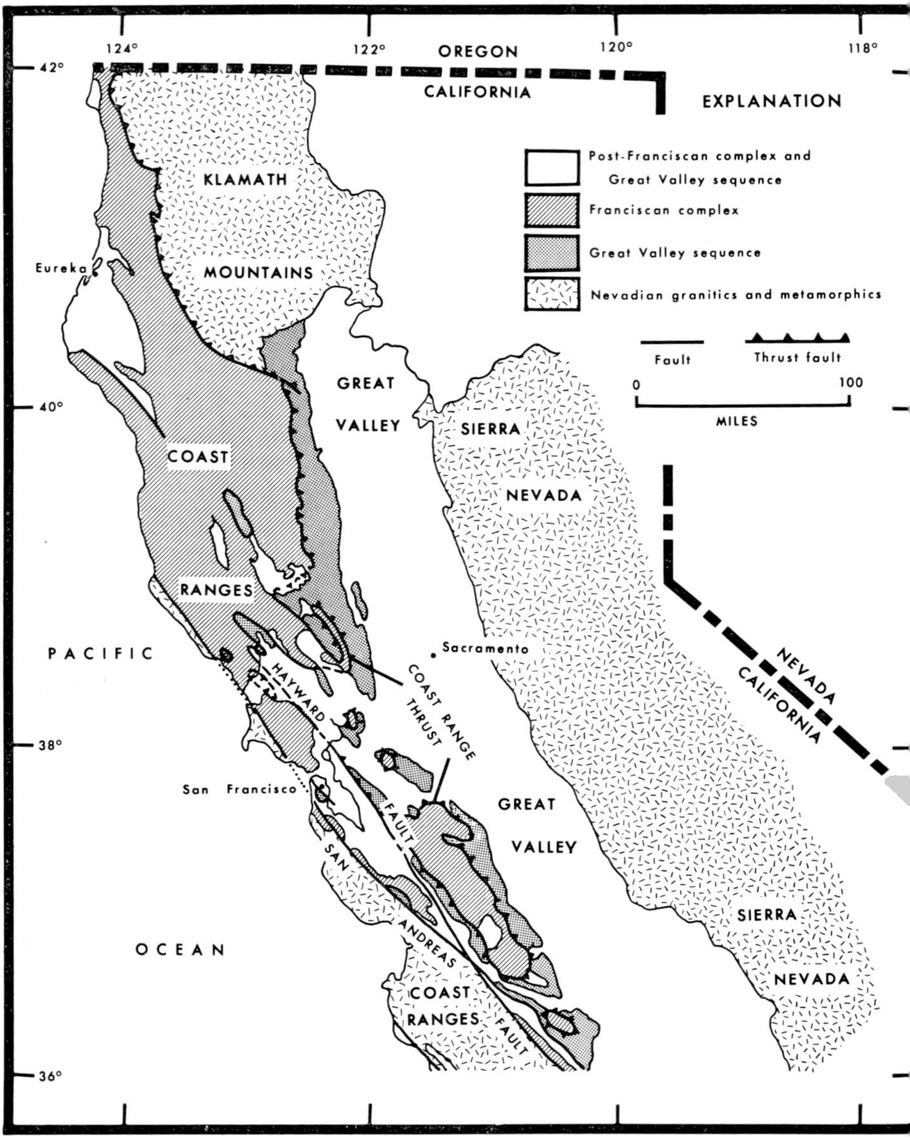

FIG. 1. Geologic map of northern and central California. Modified from Bailey *et al.*, 1970.

that do occur were preserved as a result of their rapid deposition and burial via turbidity currents.

In Leg I of the Deep Sea Drilling Project, Ewing *et al.* (1969) found that the only calcareous foraminifera preserved in the sediments of the Blake-Bahama Basin occurred in turbidites. Late Jurassic to Recent sedimentation in the Blake-Bahama Basin occurred at depths of 17 000 ft. Within the Great Valley sequence the rapidity of sedimentation (e.g., 3800 ft; 1176 m in the Coniacian) suggests a high frequency of turbidity currents. True pelagic sediments are rare. Apart from the radiolarian cherts in ophiolite successions at the base of the Great Valley sequence, the only other unquestioned pelagic sediments noted by the writer in this succession are pure radiolarites which occur in the upper part of the Forbes Formation (see Pessagno, 1976).

West of the Sacramento and San Joaquin Valleys the Franciscan complex has been thrust eastward beneath the Great Valley sequence probably as a result of subduction and sea-floor spreading (see Fig. 1). Although Radiolaria are common in many of the strata comprising the Franciscan complex, the most radiolarian-rich strata are the famed Franciscan cherts, which in many instances overlie pillow basalts. It is probable that these cherts represent sediments which accumulated on oceanic crust which was carried eastward via sea floor spreading and was subducted beneath the Great Valley sequence and the continental edge.

Mesozoic Radiolaria are also well-known from most major island arc systems (e.g. the East Indies and West Indies) where oceanic crust was dragged down into ancient subduction zones. They are not overly common or abundant in shallow Mesozoic deposits. Cretaceous Radiolaria occur sporadically in the deposits of the Western Interior of North America. This assemblage is however, not as diversified as the faunas of the California Coast Ranges. Cretaceous and Jurassic(?) Radiolaria are well known from the Western Siberian Lowland of the USSR.

From the writer's experience of continental geology Mesozoic Radiolaria occur in five major rock types: (1) mudstones and shales, (2) limestone (calcilutites and calcisilites), (3) marls, (4) cherts, and (5) radiolarites. In the deep sea the results of the Deep Sea Drilling Project show that Mesozoic Radiolaria occur in chalks, limestones, various sorts of clays, silts, oozes and cherts.

In the more consolidated strata it is usually possible to see Radiolaria with the aid of a ×20 hand lens. In many limestone samples, however, Radiolaria are replaced by calcite and, hence, are difficult if not impossible to extract from the rock. It is particularly important to find limestone samples where Radiolaria have been replaced by quartz, chalcedony or some sort of silicate mineral. This can be done by etching the sample with dilute HCl and determining (with a hand lens) whether the Radiolaria etch out in relief. In many cases it may be necessary to examine numerous samples at a given locality before one can find a sample with silicified Radiolaria.

In general, the best preserved Radiolaria can be obtained from limestones, radiolarites and cherts. In shales, mudstones, marls and siltstones the more fragile Radiolaria are frequently destroyed as a result of sedementary compaction. As a result, only sturdy specimens of Spumellariina and Nassellariina remain (e.g. pseudoaulophacids and archaeodictyomitrids); even these forms often show flattening and other forms of distortion due to sedimentary compaction.

Methods of sample preparation

LIMESTONES

The following procedure can be utilised in processing limestone samples:

(1) Crush sample in a rock crusher or mortar and pestle to fragments not exceeding 3 or 4 cm in size.

(2) Place crushed sample in a one quart jar.

(3) Slowly drip concentrated HCl on sample utilising a separatory funnel. This allows the reaction to continue for as long as 12 h depending on the speed in which HCl is released from the separatory funnel. Should the sample react too vigorously, control the reaction by adding methyl alcohol.

(4) When the reaction ceases and all calcium carbonate has been removed from the sample, pour off the spent acid.

(5) Wet sieve the sample through a 230 mesh sieve with a gentle spray of water. This removes most of the acid, alcohol and unneeded fine sediment.

(6) Place the sieved and washed residue in a pyrex beaker and cover it with hydrogen peroxide (30%). The hydrogen peroxide tends to free the matrix remaining on Radiolarian tests and to disaggregate any clay left in the sample.

In some samples the reaction of the residue with the hydrogen peroxide is vigorous and generates a great deal of heat. In others the reaction is extremely slow. In the latter case, it is necessary to bring the hydrogen peroxide to a boil on a hot plate. If the Radiolaria are still coated with minor amounts of matrix, the process should be repeated one or two more times.

(7) Wet sieve the sample with a gentle spray of water using 230 mesh, 80 mesh and 40 mesh sieves. Retain the residue on the 80 mesh (40–80 mesh fraction) and the 230 mesh (80–230 mesh fraction) sieves.

(8) Dry the washed residues in a oven.

(9) In many residues the Radiolaria may be nearly masked by abundant quartz sand and silt grains. To concentrate the Radiolaria, place the sample in carbon tetrachloride. The Radiolaria being lighter than the remainder of the residue will remain in suspension and can be poured off through filter

paper in a funnel. The filter paper dries in a matter of minutes and the Radiolaria can be placed in vials for storage.

SHALES AND MUDSTONES

Radiolaria occurring in shales and mudstones can be extracted from samples using the following procedure:

1. *Disintegration of samples.* Crushed sample (2 mm sized particles) is placed with water in a Waring Blender. Mudstones or shales are reduced to mud in 30–60 s. The disintegrated material is then wet sieved in a 230 mesh sieve to wash away the very fine clay fraction.

2. *Cleaning of microfossils.* The sieved material from the blender is then placed in an ultrasonic cleaner for two or more hours. Most or all of the matrix is cleaned from the fossils during this time. The more indurated the sample the longer it takes to clean matrix from the fossils. Samples are placed in 1 quart jars filled with water and a small amount of liquid detergent. The jars are placed in the cleaner tank and the water level of the tank is adjusted so that three quarters of a given jar is submerged.

3. *Concentration of microfossils.* The sample after being treated in the ultrasonic cleaner is wet sieved, broken down into 230–80 mesh and 80–40 mesh size fractions and dried in an oven.

If the sample contains a sparse microfauna, the 230–80 mesh and 80–40 mesh size fractions are placed in a Franz Isodynamic Separator. Foraminifera, radiolaria, and ostracods are concentrated in the nonmagnetic fraction whereas clay particles are drawn via the magnetic properties of the clay minerals into the magnetic fraction. Often a pure separate of radiolaria and foraminifera can be obtained utilising this last technique. (See Eckert *et al.*, 1961, pp. 876–877.) Furthermore, samples which would otherwise be too sparsely fossiliferous to yield biostratigraphic data, yield valuable information utilising this technique.

CHERTS

Recently, Dumitrica (1970, p. 46) and Pessagno and Newport (1972, pp. 231–234) independently discovered that Radiolaria could be extracted from cherts utilising hydrofluoric acid. A detailed discussion of the hydrofluoric acid technique was presented by Pessagno and Newport.

As noted by Pessagno and Newport, a basic prerequisite for the success of the hydrofluoric acid technique is that a chert sample must contain Radiolaria that are well preserved and not badly recrystallised. Hence, in thin-section radiolarian tests should appear to be sharply defined with respect to the matrix, and shell structures, such as fragile meshwork and spines, should be clearly visible. Pessagno and Newport summarised this technique in the following steps:

(1) A chert fragment is etched from three to five minutes with concentrated hydrofluoric acid (52–55%), removed from the acid, and briefly but gently washed in water. The result of this step is the best indicator of the ultimate success or lack of success of the method. If one can observe well-preserved Radiolaria standing out in relief on the etched surface of the sample fragment it is likely that numerous Radiolaria can be extracted from the chert through further solution.

(2) The sample is then crushed with a rock crusher or steel mortar and pestle into 1–5 cm fragments.

(3) Crushed chert is placed in a plastic beaker with dilute hydrofluoric acid (9 parts water to 1 part acid) for a period of 24 h. In some cases it may be necessary to vary the degree of dilution of the acid or to vary the length of the reaction.

(4) The acid is discarded and the sample is washed, sieved and dried with the use of standard micropalaeontological techniques.

MARLS, DEEP-SEA CLAYS, SILTS, AND OOZES

Riedel and Sanfilippo (Chapter 8) give a detailed account of the technique used for the disaggregation of these sediments.

METHOD OF ILLUSTRATION

The majority of Radiolaria illustrated in this report were extracted from limestone samples. Most illustrations are scanning electron micrographs taken with JEOL JSM-1 scanning electron microscope equipped with a goniometer stage.

In scanning electron microscopy it is critical that all cover slips and electron microscope plugs be free of organic material such as oil. Even oil derived from the touch of a covership or a plug with human hands can prevent uniform coating of specimens with gold or gold palladium in the vacuum evaporator.

To remove organic matter from coverslips and scanning electron microscope plugs, the following procedure developed by Urban and Padovani (1970, pp. 131–139) was utilised:

(1) SEM plugs and coverslips are cleaned by boiling in trichlorethyline for 10 min, decanting, covering with spectrographic grade acetone; decanting and covering with anhydrous ethyl alcohol. The ethyl alcohol is decanted once.

(2) The plugs and coverships are placed in jars, covered with anhydrous ethyl alcohol.

(3) Whenever plugs and coverslips are needed, they can be removed by means of forceps from their respective containers. They are ready to use once the alcohol has evaporated.

To assure an even background free of machine markings, a glass coverslip is glued to the surface of an electron microscope plug. A thin coating of Elmer's Glue (greatly diluted with water) is placed on the surface of the coverslip.

Radiolarian specimens are placed in rows on the coverslips by means of a slightly dampened picking brush.

The electron microscope plug is coated with gold or gold palladium in a vacuum evaporator. With Radiolaria the amount of coating needed to prevent "charge-ups" varies greatly with the type of Radiolaria being studied and with their state of preservation. Spongodiscacea, for example, often require a heavier coating than Spumellariina having latticed meshwork.

After the specimens have been photographed they can be removed from the coverslip on the plug surface with a slightly wet picking brush. Gold or gold palladium can be removed in seconds from the Radiolarian specimens with a drop of Aqua Regia. At this point, specimens can be mounted on a glass slide using Caedax or any other suitable mounting medium.

Biostratigraphical potential of Mesozoic Radiolaria

Because of their planktonic, cosmopolitan nature, great diversity in form, and rapid evolution, Radiolaria are ideally suited for developing detailed systems of zonation for Mesozoic oceanic strata. Their common occurrence in the otherwise poorly fossiliferous strata of major orogenic belts—particularly in the chert component of ophiolite successions—makes them important in deciphering the complex geological history of such regions.

The first comprehensive zonation using Mesozoic Radiolaria has recently been completed by the writer in his studies of the Upper Cretaceous portion of the Great Valley sequence, California Coast Ranges (see Pessagno, 1976) The zonal units defined in this scheme are for the most part what are termed *Oppel Zones* by Hedberg *et al.* (1972, p. 21). According to Hedberg *et al.*: "the Oppel Zone may be defined briefly as a zone characterised by a unique association or aggregation of selected taxons of restricted and largely concurrent range, chosen because they are considered to be indicative of approximate contemporaneity. The lower portion of an Oppel Zone is marked broadly by the first appearances and its upper portion by the last apperances of certain taxons. The body of the zone is marked largely by concurrence of diagnostic taxons." In addition to being defined by the concurrence of certain diagnostic taxa, an Oppel Zone may be based on a combination of other criteria having time significance such as lineage zones, abundance zones, and so forth.

The Upper Cretaceous portion of the Great Valley sequence has been divided into 8 zones and 9 subzones utilising some 115 species. Many of the diagnostic species used for this zonal scheme belong to the Suborder Spumellariina and to the Superfamily Spongodiscacea. Members of this superfamily possess sturdy tests which are preserved in a number of facies (e.g. mudstones) where sedimentary compaction tends to destroy more fragile Radiolaria.

Figure 2 shows the interrelation between the radiolarian zonation for northern California and planktonic foraminiferal zonation. It is impossible here to give a full explanation of these zonal units. However, a definition for two zonal units, the *Alievium praegallowayi* Zone (Coniacian) and the *Alievium gallowayi* Zone (Santonian), is presented below in order to demonstrate potential of Radiolaria in Mesozoic biostratigraphy.

Alievium praegallowayi Zone

The range zones and abundance zones of the diagnostic species of the *Alievium praegallowayi* Zone are shown in Fig. 3; illustrations of many of the species are presented in plates 1–7. The lower limit of the *Alievium praegallowayi* Zone is marked by the first appearance of a number of species. Foremost among these is the first evolutionary appearance of *A. praegallowayi* Pessagno (derived from *A. superbum* (Squinabol)). In addition, *Archaeospongoprunum bipartitum* Pessagno, *Pseudoaulophacus lenticulatus* (White), *Pseudoaulophacus praefloresensis* Pessagno, *P. venadoensis* Pessagno, *Phaseliforma concentrica* (Lipman), *Squinabolella putahensis* Pessagno, *Microsciadiocapsa radiata* Pessagno, *M. madisonae* Pessagno, *Crucella plana* Pessagno, *Patulibracchium ruesti* Pessagno, *P. torvitatis* Pessagno, *P. ungulae* Pessagno, *Paronaella solanoensis* Pessagno, *P. napaensis* Pessagno, *Orbiculifroma quadrata* Pessagno, *O. vacaensis* Pessagno, *Praecanacaryomma universa*, Pessagno, *Dictyomitra crassicostata*, Pessagno, and various other species make their first appearance at the base of this zone. Particularly noticeable at this horizon is the sudden increase in abundance and diversity of the Pseudoaulophacidae.

The top of the *A. praegallowayi* Zone is delimited by the final occurrence of *Microsciadiocapsa radiata* Pessagno, *M. madisonae* Pessagno, *Crucella irwini* Pessagno, *Praeconocaryomma lipmanae* Pessagno, *Orbiculiforma vacaensis* Pessagno, *Orbiculoforma persenex* Pessagno, *Patulibracchium ruesti* Pessagno, *Paronaella venadoensis* Pessagno, *Pyramispongia glascockensis* Pessagno, *Archaeospongoprunum triplum* Pessagno, and *Dictyomitra napaensis* Pessagno. It lies below the first evolutionary appearance of *Alievium gallowayi* (White).

The *A. praegallowayi* Zone is divided into a lower *Archaeospongoprunum triplum* Subzone and an upper *Archaeospongoprunum bipartitum* Subzone.

Archaeospongoprunum triplum Subzone

The lower limit of the *A. triplum* Subzone is defined by the same species that

European Stages	Northern California Radiolaria (Pessagno, 1976)		Northern California Planktonic Foraminifera (after Douglas, 1969)		Gulf Coast - Caribbean Planktonic Foraminifera (after Pessagno, 1969)	
MAESTRICHTIAN	*Orbiculiforma renillaeformis* Zone				*Globotruncana contusa-stuartiformis* Zone	*Abathomphalus mayaroensis* Subzone
						Globotruncana gansseri Subzone
			Globotruncanella havanensis Zone		*Globotruncana fornicata-stuartiformis* Zone	*Rugotruncana subcircumnodifer* Subzone
CAMPANIAN	*Patulibracchium dickinsoni* Zone		*Globotruncana arca* Zone	*Globotruncana churchi* Subzone		*Globotruncana elevata* Subzone
	Crucella espartoensis Zone	*Phaseliforma carinata* Subzone				
		Patulibracchium lawsoni Subzone		*Globotruncana stuartiformis* Subzone		*Archaeoglobigerina blowi* Subzone
		Protoxiphotractus perplexus Subzone				
SANTONIAN	*Alievium gallowayi* Zone		*Globotruncana coronata* Zone		*Globotruncana bulloides* Zone	*Globotruncana fornicata* Subzone
						Marginotruncana concavata Subzone
CONIACIAN	*Alievium praegallowayi* Zone	*Orbiculiforma vacaensis* Subzone	*Globotruncana cachensis* Zone		*Marginotruncana renzi* Zone	
		Archaeospongoprunum triplum Subzone				
TURONIAN	*Alievium Superbum* Zone	*Archaeospongoprunum venadoensis* Subzone	*Praeglobotruncana helvetica* Zone	*Globotruncana imbricata* Subzone	*Marginotruncana helvetica* Zone	*Whiteinella archaeocretacea* Subzone
		Halesium sexangulum Subzone	*Praeglobotruncana stephani* Zone	*Globotruncana sigali* Subzone		*Marginotruncana Sigali* Subzone
CENOMANIAN	*Rotaforma hessi* Zone	*Q. spinosa* Subzone		*Rotalipora cushmani* Subzone	*Rotalipora s.s.* Zone	*Rotalipora cushmani-greenhornensis* Subzone
		Cassideus riedeli Subzone	*Rotalipora appenninica* Zone			*Rotalipora evoluta* Subzone
	Archaeospongoprunum tehamensis Zone					

FIG. 2. Interrelation of proposed radiolarian zonation to planktonic foraminferal zonation.

define the base of the *Alievium praegallowayi* Zone. Its upper limit is defined by the last occurrence of *Microsciadiocapsa berryessaensis* Pessagno, *M. monticelloeniss* Pessagno, *Squinabolella putahensis* Pessagno, *Saturniforma abastrum* Pessagno, *S. caelestium* Pessagno, and *Paronaella solanoensis* Pessagno (see Fig. 3). *A. triplum* Pessagno makes its first appearance in the lower part of this subzone.

Orbiculiforma vacaensis Subzone

The upper limit of the *O. vacaensis* Subzone is defined by the same species that define the top of the *Alievium praegallowayi* Zone. The base of this zonal unit is defined by the first occurrence of *Archaeospongoprunum rumseyensis* Pessagno and *Orbiculiforma persenex*, Pessagno.

Alievium gallowayi Zone

The base of the *A. gallowayi* Zone is defined by the first evolutionary appearance of *A. gallowayi. Alievium praegallowayi* Pessagno becomes extinct in the lower part of this zone shortly after giving rise to *A. gallowayi* (White). *Crucella espartoensis* Pessagno, *Pseudoaulophacus floresensis* Pessagno, *P. venadoensis* and *P. colburni* Pessagno also appear above the base of this unit. The top of the *A. gallowayi* Zone is defined primarily by the last appearance of *Archaeospongoprunum bipartitum* Pessagno and *Orbiculiforma quadrata* Pessagno. Range zones and abundance zones of species characterising the *A. gallowayi* Zone are shown in Fig. 3, illustrations of various species are presented in plates 1–7.

It is evident from the writer's investigations of Upper Cretaceous Radiolaria that the faunal change is extremely rapid from Cenomanian to Maestrichtian times. This probably suggests an equally rapid rate of evolutionary change. Because of the rapidity of sedimentation (e.g. 3800 ft in the Coniacian) within the Great Valley sequence, rapid evolutionary changes between species are difficult to monitor; bed to bed control is lacking in most cases. However, it is possible to monitor some of the slower evolutionary changes. For example, Pessagno (1972b, pp. 287–288) demonstrated that a bioseries existed between three species of *Alievium* Pessagno: (1) *Alievium superbum* (Squinabol); (2) *Alievium praegallowayi* Pessagno; and (3) *Alievium gallowayi* (White).

Alievium superbum (Squinabol), the earliest member of this bioseries, makes its first appearance in the early Turonian. This form is characterised by having a triangular, somewhat inflated test with three spines which are triradiate in axial section (pl. 6, fig. 3). Each spine consists of three prominent grooves each separated by three prominent ridges. By early Coniacian times *Alievium superbum* gave rise to *Alievium praegallowayi* Pessagno (pl. 6, fig. 4). This species likewise has a triangular, somewhat inflated test. However, the distal half of each spine is circular in axial section and only the proximal half remains triradiate, having narrow grooves separating narrow ridges. By early Santonian times *Alievium praegallowayi* gave rise to *Alievium gallowayi* (White) (pl. 6, fig. 5), a species characterised by its more flattened triangular test and by spines which are completely circular in axial section. It should be noted that *Alievium superbum* became extinct shortly after giving rise to *A. praegallowayi* in the early Coniacian; *A. praegallowayi* became extinct shortly after giving rise to *A. gallowayi* in the early Santonian; and finally, *A. gallowayi* became extinct before the end of the Campanian.

Insufficient data exists at present to make a comprehensive comparison of the Boreal Upper Cretaceous assemblage of California with those of the Tethyan and Austral Faunal Provinces. Yet, the California radiolarian assemblage is quite diversified and is in fact, far more diversified than the foraminiferal or molluscan assemblages. Such studies of Tethyan Upper Cretaceous Radiolarian assemblages that have been made thus far (e.g. early Campanian of Puerto Rico, Mexico, and Cyprus; Santonian, DSDP Leg 15; Turonian of Iran) indicate that there is a large body of species in common between the Tethyan and Boreal Faunal Provinces. Furthermore, there is some indication that no marked diversity gradient existed between radiolarian assemblages of the Tethys and Boreal Faunal Provinces during Late Cretaceous times. Recent investigations of latest Campanian assemblages from the Antarctic Region (DSDP Leg 29, Site 275: southeastern portion of the Campbell Plateau) demonstrate (1) that there is likewise a large body of species in common between the Boreal and Austral Faunal Provinces and (2) that the Austral assemblage is nearly as diverse as the Boreal assemblage. Because of the large number of species in common between all three faunal provinces, it has been possible to utilise the system of zonation developed in the California Coast Ranges with no great difficulty in each of these faunal provinces. Careful cross-checking of the radiolarian and planktonic foraminiferal data for California and the Caribbean indicates that the Radiolarian zones are not time transgressive from north to south. The Antarctic data is still too sparse at the moment to determine whether this same observation is valid between California and the Antarctic. It is possible that many of the species in common between different faunal provinces are eurybathic forms. Such species would not be affected by temperature and salinity changes and would, hence, tend to be more cosmopolitan.

Development of a meaningful classification

Most classifications dealing with Mesozoic Radiolaria have been outgrowths of that proposed by Ernst Haeckel just before the turn of the century. Haeckel's classification was extremely detailed. However, it was based to a large degree

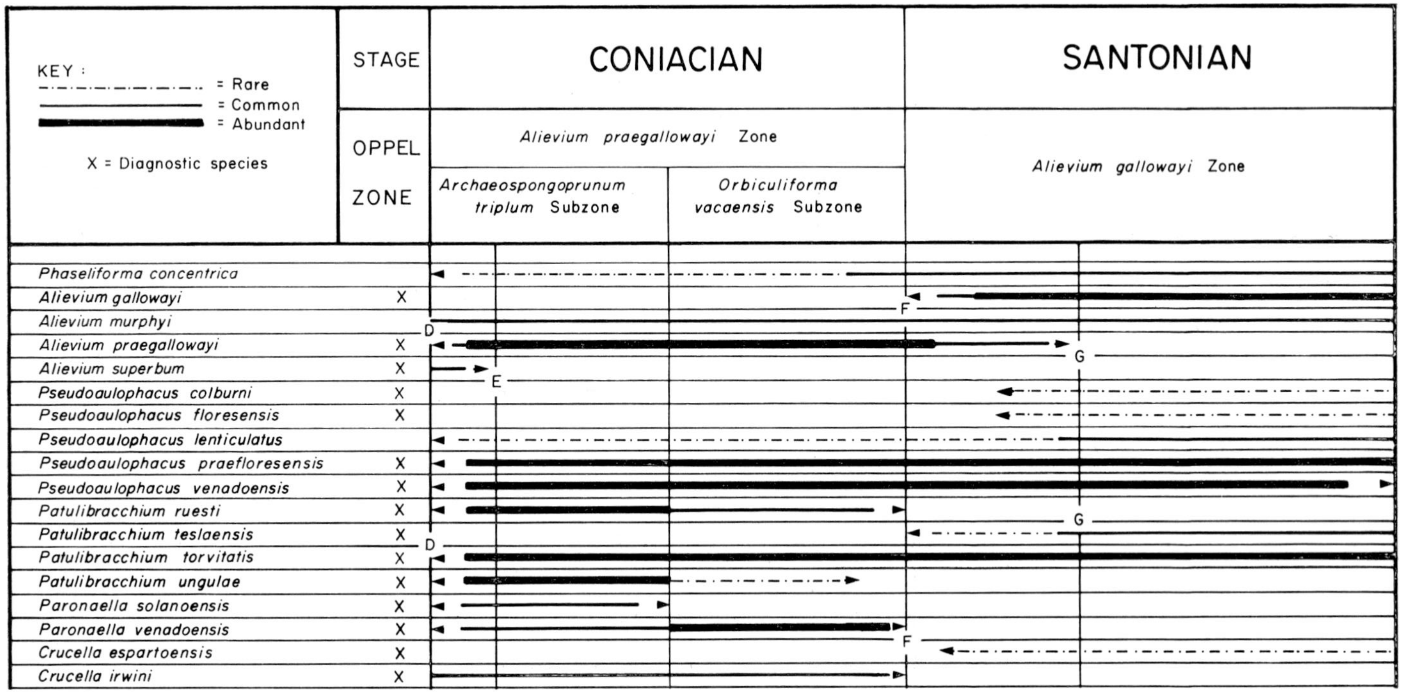
KEY :
= Rare
= Common
= Abundant
X = Diagnostic species
STAGE
CONIACIAN
SANTONIAN
OPPEL
ZONE
Alievium praegallowayi Zone
Archaeospongoprunum triplum Subzone
Orbiculiforma vacaensis Subzone
Alievium gallowayi Zone
Phaseliforma concentrica
Alievium gallowayi X
Alievium murphyi
Alievium praegallowayi X
Alievium superbum X
Pseudoaulophacus colburni X
Pseudoaulophacus floresensis X
Pseudoaulophacus lenticulatus
Pseudoaulophacus praefloresensis X
Pseudoaulophacus venadoensis X
Patulibracchium ruesti X
Patulibracchium teslaensis X
Patulibracchium torvitatis X
Patulibracchium ungulae X
Paronaella solanoensis X
Paronaella venadoensis X
Crucella espartoensis X
Crucella irwini X
D
E
F
G

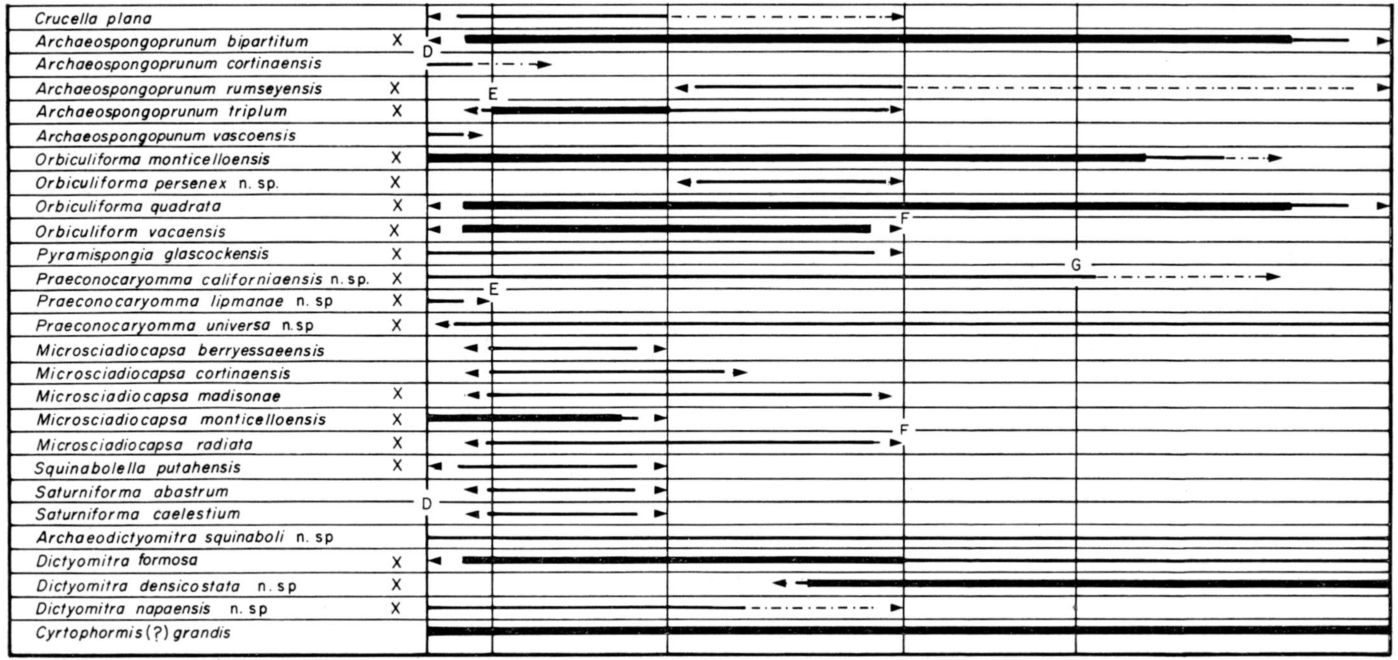

FIG. 3. Range, abundance and Oppel Zones of Upper Cretaceous Radiolaria (after Pessagno, 1976). Biohorizons: D, first evolutionary appearance of *Alievium praegallowayi* Pessagno; E, final evolutionary appearance of *Alievium superbum* (Squinabol); F, first evolutionary appearance of *Alievium gallowayi* (White); G, final evolutionary appearance of *Alievium praegallowayi* Pessagno.

on the geometry or shape of the radiolarian test. Riedel (1971), was the first to suggest that Haeckel's classification was artificial and should be abandoned by future workers. He proposed that a new classification be formulated and that this classification be more phylogenetic in character.

Previous classifications (e.g. Haeckel, 1887a, b; Campbell, 1954) were based on studies of Cainozoic Radiolaria. As a result, it was common practice to find Cainozoic generic and family names being applied to Mesozoic taxa which were at best only grossly similar to their Cainozoic successors.

It is clear that if one hopes to construct a more phylogentic classification of Mesozoic Radiolaria or for that matter Cainozoic Radiolaria, one must consider *how* a radiolarian test is constructed: This is particularly true at the supergeneric level. The following example illustrates the difference between Haeckelian classification stressing shape and form and more recent classifications based on the manner in which a Radiolarian test is constructed. Haeckel (1887a, pp. 284–286) placed *Spongoprunum* Haeckel in the "Prunoidea" largely on the basis of the ellipsoidal or cylindrical character of its test. Whereas *Spongoprunum* has (exclusive of polar spines) a completely spongy test, it was placed with other genera such as *Ellipsostylus* Haeckel and *Xiphotractus* Haeckel which possess one or more latticed shells. Pessagno (1971a, 1972a, b, 1973) placed all Spumellariina with predominantly spongy tests in the Superfamily Spongodiscacea without regard to test shape or symmetry. He further placed *Spongoprunum* within the Sub-superfamily Pseudoaulophacilae Riedel because its spongy meshwork is arranged in concentric layers.

Taxonomy: with emphasis on important Mesozoic families and groups

Order POLYCYSTIDA Ehrenberg, emend. Riedel, 1967

Remarks. All Radiolaria preserved in Mesozoic strata belong to this order. Riedel (1967, p. 291) emended the definition of the Polycystida to include only those Radiolaria having a skeleton comprised of opaline silica lacking organic compounds. The polycystida are divided in turn into two suborders: (1) the Spumellariina and (2) the Nassellariina.

Suborder SPUMELLARIINA

Remarks. This group has not been as extensively studied in recent years as the Nassellariina. This is particularly true in the case of Mesozoic Spumellariina. As a result, a phylogenetic classification is poorly developed at the superfamily level.

It is suggested here that the Spumellariina can be divided into three superfamilies on the basis of the construction of their tests.

The majority of Mesozoic Spumellariina belong to the Superfamily Sphaerallacea and the Superfamily Spongodiscacea. Of these two superfamilies the

Spongodiscacea are far more important and constitute the largest component of the Spumellariina assemblage. In general, the Sphaerallacea are not as abundant in the Jurassic and Early Cretaceous as they are in the Late Cretaceous. The Spongodiscacea becomes progressively more abundant and diversified from Jurassic until Late Cretaceous times. From late Campanian times to late Maestriction times they begin to decline both in numbers and diversity. Although the Spongodiscacea are well represented in the Cainozoic, they never attain the abundance and diversity that they displayed in the Mesozoic.

Superfamily COLLODARACEA Haeckel, 1882

Definition. Spumellariina lacking skeletal material or with isolated spicules in protoplasm or with spicules fused together to form discrete shells.

Remarks. It is further suggested that the Collodacea be divided into three subsuperfamilies:

(*a*) Subsuperfamily 1. Includes forms lacking skeletal matter.
(*b*) Subsuperfamily 2. Includes forms with spicules in the calymma.
(*c*) Subsuperfamily 3. Includes forms with discrete shells formed by the fusing of spicules (see pl. 11, fig. 3). No attempt to formalise these subsuperfamilies is made here.

Range. Palaeozoic to Recent.
Occurrence. Worldwide.

Superfamily SPHAERALLACEA Haeckel, 1882

Definition. Spumellariina with discrete lattice shells of variable shape.

Remarks. This is a large diverse group which is still poorly studied particularly in the Mesozoic. It is likely that the Sphaerallacea should be divided into subsuperfamilies on the basis of the following criteria:

(1) Structure of the cortical shell.
(2) Number and structure of medullary shells.
(3) Disposition and structure of spines and radial beams.

At present, the author regards the structure of the outermost shell or cortical shell as being the most important of the criteria listed above. This criterion is closely coupled with the disposition and structure of primary and secondary spines. The structure of the cortical shell and the placement and structure of primary and secondary spines may be directly related to development of pseudopodia and, hence, to food gathering.

Range. Palaeozoic to Recent.
Occurrence. Worldwide.

Family PRAECONOCARYOMMIDAE Pessagno

The Praeconocaryommidae Pessagno include Sphaerallacea with a test consisting of a latticed spherical cortical shell and three latticed medullary shells (pl. 3, figs 2–6; pl. 11, fig. 4). The cortical shell is particularly important in the recognition of this group. It is characterised by possessing numerous equally spaced cone-like tubercles called mammae which are connected to the first medullary shell by long, massive radial beams. The cortical shell may possess radial spines protruding from the centre of each mamma; where present, these are continuous with the radial beams. The first medullary shell and the two remaining medullary shells lack mammae and are connected to each other by means of short radial beams.

The Praeconocaryommidae are known from strata as old as Jurassic (Tithonian). "*Acanthosphaera*" *magnimamma* Rüst and "*Heliosphaera*" *mammallaria* Rüst from the Kieselkalk of Cittiglio appear to be early representatives of praeconocaryommids. These species are now recorded from Tithonian strata in the California Coast Ranges. Investigations recently completed on the Upper Cretaceous radiolarian assemblage of the California Coast Ranges indicate that the Praeconocaryommmidae became extinct in the late Campanian.

Range. Jurassic to Upper Cretaceous (Upper Campanian).

Superfamily SPONGODISCACEA Haeckel, emend. Pessagno, 1971a, 1973

Remarks. Pessagno (1971a, p. 16) defined the Spongodiscacea to include Spumellariina with spongy tests of variable shape lacking sieve plates, lattice shells or chambered rays. The pore frames comprising the dense, spongy meshwork may be arranged with or without symmetry. In later reports he emended this definition to include forms having a spongy cortical shell and a latticed medullary shell.

The Spongodiscacea are divided by Pessagno (1973, 1974) into 3 subsuperfamilies on the basis of test construction (1) the Spongodiscilae Haeckel, (2) the Pseudoaulophacilae Riedel, and (3) the Spongodruppilae Haeckel.

Range. Palaeozoic to Recent.

Occurrence. Worldwide.

Subsuperfamily SPONGODISCILAE Haeckel

Remarks. The Spongodiscilae are characterised by having tests comprised of irregular, dense, spongy meshwork with pore frames arranged unsymmetrically. Overall test shape varies with family or subfamily (pl. 10, fig. 2).

Range. Palaeozoic to Recent.

Occurrence. Worldwide.

Subsuperfamily PSEUDOAULOPHACILAE Riedel

Remarks. The Pseudoaulophacilae are characterised by having spongy tests comprised of meshwork arranged symmetrically in concentric rings, spirals, parallel layers, and so forth. Overall test shape varies with family or subfamily (pl. 10, figs 3–6).

Range. Palaeozoic (?); Mesozoic to Recent.

Occurrence. Worldwide.

Family HAGIASTRIDAE Riedel

Pessagno (1971a) divided this family into three subfamilies: (1) the Amphibrachinae Pessagno, (2) the Patulibracchinae Pessagno, and (3) the Hagiastrinae Riedel. Basically, these three subfamilies include respectively the two, three, and four rayed Hagiastridae (pl. 4, figs 1–6; pl. 11, figs 5–6.

The Amphibrachinae do not seem to occur above the Jurassic, but are figured by Rüst (1892b) from the Palaeozoic. Both the Patulibracchinae and the Hagiastrinae range from the Palaeozoic to Upper Cretaceous (Cainozoic (?)) and comprise a large, biostratigraphically important, segment of the entire Mesozoic assemblage. Pessagno, in his investigations of the Great Valley sequence noted twenty-four species and four genera of Hagiastridae from the Upper Cretaceous alone. Many of these species are short ranging and were found to be useful in developing a detailed system of zonation for the Upper Cretaceous.

Range. Palaeozoic to Mesozoic; Cainozoic(?).

Family SPONGURIDAE Haeckel

The Sponguridae Haeckel include Pseudoaulophacilae with ellipsoidal tests characterised by a meshwork arranged in a series of concentric layers.

The Sponguridae are divided into two subfamilies: (1) the Spongurinae Haeckel and (2) the Archaeospongopruninae Pessagno. The Spongurinae differ from the Archeospongopruninae by lacking polar spines (pl. 7, figs 3–4; pl. 10, figs 4–5). Of these two subfamilies the Archaeospongopruninae are the most important in the Mesozoic.

The Archaeospongopruninae include two genera: *Spongoprunum* Haeckel (Cainozoic) and *Archaeospongoprunum* Pessagno (Mesozoic). *Archaeospongoprunum* differs from *Spongoprunum* by having polar spines displaying alternating ridges and grooves (Pessagno, 1973).

Recent investigations indicate that *Archaeospongoprunum* occurs in strata as old as Late Jurassic (Tithonian) and that species assignable to this genus are common until Late Cretacous (latest Campanian) times. In California *Archaeospongoprunum* includes only one species (undescribed) in the Maestrichtian.

Range. Palaeozoic to Recent.

Family ORBICULIFORMIDAE Pessagno

The Orbiculiformidae are characterised by having disc-shaped tests with irregular polygonal pore frames arranged in concentric rings and with or without equatorially placed spines.

This is a large family group which includes a number of species. It is likely that a number of genera included by Campbell (1954, pp. D93–D94) under the Spongodiscidae will probably have to be reassigned to the Orbiculiformidae. Pessagno (1973, p. 71) restricted the Spongodiscidae to forms displaying spongy, disc-shaped tests lacking pore frames arranged in a symmetrical fashion.

Within the Mesozoic succession *Orbiculiforma* Pessagno includes a number of biostratigraphically important species. Morphologially, it is characterised by having a test which possesses a well-developed central cavity on opposing sides and short, peripheral spines (pl. 5, figs 1–6). The outline of the test and the size and shape of the central cavity vary with the species under consideration. This genus is now known to range from Early Cretaceous (Albian) to Late Cretaceous (Maestrichtian). It may well range into the Cainozoic.

Range. Early Cretaceous to Recent.

Family PSEUDOAULOPHACIDAE Riedel

The Pseudoaulophacidae are characterised by their spongy, lenticular tests composed of concentric layers of triangular pore frames. The family presently includes two genera: *Pseudoaulophacus* Pessagno and *Alievium* Pessagno.

Pseudoaulophacus differs from *Alievium* by possessing dome-like structures known as tholi, which are centrally positioned on opposing sides of the test. The meshwork of the tholi is comprised of considerably larger equilateral triangular pore frames than those comprising the remainder of the test (pl. 6, figs 3–6; pl. 7, figs 1–2).

Recently completed investigations (Pessagno, 1972b) indicate that *Alievium* first appeared in the Early Cretaceous and is definitely known from strata as old as Valenginian. During the Late Cretaceous *Alievium* gave rise to *Pseudoaulophacus* in early Turonian times. Before the early Turonian the Pseudoaulophacidae formed a minor proportion of the Mesozoic radiolarian assemblage. However, from early Turonian times until late Campanian times the Pseudoaulophacidae became increasingly more abundant and diversified. In latest Campanian times the Pseudoaulophacidae suffered a sudden decline and nearly became extinct. At this time *Pseudoaulophacus* s.s. became extinct and *Alievium* was survived by only one species in the latest Campanian and Maestrichtian.

Range. Late Valenginian to Maestrichtian.

Subsuperfamily SPONGODRUPPILAE Haeckel

Remarks. The Spongodruppilae are characterised by having spongy, concentrically layered cortical shells enclosing latticed medullary shells (pl. 11, figs 1–2).

Range. Mesozoic to Cainozoic.

Suborder NASSELLARIINA

Remarks. Most present-day students of the Nassellariina e.g. Petrushevskaya (1964, 1965, 1971a, b); Goll (1968), 1969, 1972); Foreman (1968); Pessagno (1969b, 1970, 1974) have stressed the importance of the arrangement of the cephalic skeletal elements in classification. The most comprehensive classification of the Nassellariina in recent years was made by Petrushevskaya (1971a, b.) In general, Petrushevskaya's classification seems to be a good one. However, the writer disagrees with her somewhat over her ranking of taxa. Petrushevskaya (1971b, p. 982) divides the Nassellariina into two groups: (1) the Order Cyrtida and (2) the Order Spyrida. In this chapter these groups are treated as superfamilies rather than orders.

Mesozoic Nassellariina are for the most part still poorly studied. It is likely that only 10% of the Mesozoic Nassellariinid assemblage has been described.

Superfamily CYRTOIDEA

Remarks. The Cyrtoidea include Nassellariina having cephalic skeletal elements consisting of spines, spines and bars, or a combination of spines, bars, and rings. The overall shape of the test is extremely variable. A variety of different types of Cyrtoid cephalic skeletal elements are illustrated in most treatises (see Campbell, 1954, pp. D26–D27) the simplest form being the tripod and the most advanced form being a combination of spines, bars and rings. In general, the phylogeny of the cyrtoidea can be deciphered by examining the cephalic skeletal elements. Scanning electron micrographs of cyrtoid cephalic skeletal elements are shown in pl. 9, fig. 5.

This superfamily includes a number of family groups, and most Mesozoic Cyrtoid families are still in need of a great deal of study.

Range Palaeozoic to Recent.

Occurrence. Worldwide.

Superfamily CYRTOIDEA (Group 1)

This group includes Cyrtoidea with multisegmented, turret-shaped tests displaying prominent circumferential ridges at closely spaced shell joints. Two or three rows of polygonal (usually hexagonal) pore frames occur between massive circumferential ridges (pl. 9, figs 2–4).

Range. This group is particularly prevalent in Late Jurassic (Tithonian) strata in both the Tethyan and Boreal Faunal Provinces. It persists into the Neocomian, but is not known from younger strata. Because of a similarity in test construction, it is likely that these distinctive Nassellariina will have to be placed in the same subsuperfamily under the Cyrtoidea.

Superfamily CYRTOIDEA (Group 2)

A second prominent group of Cyrtoid Nassellariina includes cryptocephalic and cryptothoracic forms; that is dicyrtid (two segments/chambers) forms which have either a cephalis encased by the thorax or tricyrtid (three segments/chambers) forms which have both cephalis and thorax encased by the abdomen (pl. 10, fig. 1).

This rather large group of Nassellariina is common in all Mesozoic deposits and includes a number of genera. Among these are *Gongylothorax* Foreman (Jurassic to Cretaceous; *Heliocryptocapsa* Dumitrica Upper Cretaceous); *Dicanthocapsa* Squinabol (Upper Cretaceous); and so forth. (See Dumitrica, 1970, pp. 45–124.)

Range. Palaeozoic to Recent.

Family ARCHAEODICTYOMITRIDAE Pessagno.

This group of Cyrtoidea includes the Archaeodictyomitridae of Pessagno (1976, in press). The Archaeodictyomitridae differ from all other multi-segmented Cyrtoidea (1) by possessing tests with aligned subparallel costae and one or more (usually one) subparallel rows of aligned pores in the inter-coastal areas and (2) by possessing primary and relict pores. The relict pores are usually covered by an accretion of later shell material; the primary pores occur in strictures at segment (chamber) joints and always remain open.

Scanning electron microscope analyses of a number of species assignable to the Archaeodictyomitridae indicate that a new segment or chamber is formed by the secretion of costal projections (pl. 7, figs 5–6; pl. 8, figs 1–4). This in turn is followed by the secretion of pore frames costal projections.

The Family Archaeodictyomitridae include at least two genera: (1) *Archaeodictyomitra* Pessagno and (2) *Dictyomitra* Zittel. It is probable that *Diplostrobus* Squinabol should also be assigned to this family. Pessagno (1976, in press) emended and restricted the definition of *Dictyomitra* to include forms having abdominal and post-abdominal segments (1) with well-developed strictures at joints between segments and (2) with pores arranged in single (rarely double) rows between costae, becoming relictic except in the position of strictures.

Archaeodictyomitra Pessagno differs from *Dictyomitra* Zittel s.s. (1) by having a test which lacks well-developed strictures and thus, is non-lobate in outline and (2) by possessing relict pores and lacking primary pores.

Studies by Pessagno (1976, in press) indicate that *Archaeodictyomitra* made its first occurrence in strata which are at least as old as Late Jurassic (Tithonian) and its final appearance in the Campanian. In Late Cretaceous (middle Turonian) times *Archaeodictyomitra* gave rise to *Dictyomitra* s.s. *Dictyomitra* s.s. became extinct at the end of the Late Cretaceous (Maestrichtian).

Range. Triassic; Jurassic to Cretaceous.

Family NEOSCIADIOCAPSIDAE Pessagno

The Neosciadiocapsidae Pessagno (1969b) include dicyrtid Cyrtoidea with (1) helmet shaped tests; (2) nine cephalic skeletal elements (pl. 9, fig. 5); and (3) a thoracic velum (=cover plate over thoracic mouth). A tubular structure, termed the cephalopyle, is usually present on the distal portion of the cephalis near the collar stricture (or in its absence, a vertical spine). (See pl. 9, fig. 6).

Range. It is apparent from investigations now underway that the Neosciadiocapsidae occur in strata at least as old as Late Jurassic (Tithonian). From Late Jurassic until Late Cretaceous times the Neosciadiocapsidae become progressively more abundant and diversified. The available evidence suggests that the Neosciadiocapsidae became extinct in the Eocene.

Family ROTAFORMIDAE Pessagno

The Rotaformidae Pessagno are a bizzare group of dicyrtid Cyrtoidea with cart-wheel shaped tests and nine cephalic skeletal elements which are identical to those of the Neosciadiocapsidae Pessagno. Transitional specimens suggest that the Neosciadiocapsidae gave rise to the Rotaformidae through (1) the rotation of the cephalis and the upper thorax into the plane of the thoracic skirt and (2) the loss of the thoracic skirt and the subsequent development of a thoracic ring.

Two genera are included in the Rotaformidae: *Rotaforma* Pessagno and *Saturniforma* Pessagno. *Rotaforma* (Pl. 9, fig. 1) differs from *Saturniforma* by possessing a cephalopyle, an apical horn, a large distinct thoracic mouth, a thoracic fringe, and by lacking portae and central canals (see pl. 2, figs 4–5; pl. 8, figs 5–6).

Range. Investigations in progress indicate that the Rotaformidae first appeared in the Early Cretaceous (Albian) and became extinct in the Late Cretaceous (latest Campanian).

Superfamily SPYROIDEA

Remarks. The Spyroidea includes Nassellariina whose cephalic skeletal elements consist strictly of rings (see Campbell, 1954; Goll, 1972).

Range. Cainozoic.

Occurrence. Worldwide.

Acknowledgements

This work has been supported by grants from the National Science Foundation, GA-4043 to the University of California at Davis, GA-1224 to the S.W. Center for Advanced Studies, Dallas, Texas, GA-15998, GA-25712 to the University of Texas at Dallas and by the general NASA grant (NGL-44-004-001) to the S.W. Center for Advanced Studies.

The writer wishes to thank Mr. Walter Brown, Mr. Allen White, and W. T. Rothwell for their care in taking the scanning electron micrographs.

References

Aliev, Kh. Sh. (1961). Novye vidy radiolyarii nizhnemelovykh ottzhenu severo-vostochnozo Azerbaidzhana. *Izvetsiya Akad. Nauk. Azerbaidzhanskoe SSR. Ser. Geol. Geog. Nauk i Nafti* **1,** 51–65.

Aliev, Kh. Sh. (1965). Radiolyarii nizhnemelolvykh otozhenii severo-vastochnogo Aberbaidzhana i ikh stratigraficheskoe znachenie. *Izdafelstvo Akad. Nauk. Azerbaidzhanskoi SSR, Baku* 124 pp. 15 pls.

Bailey, E. H., Irwin, W. P. and Jones, D. L. (1964). Franciscan and related rocks, and their significance in the geology of Western California. *Calif. Div. Mines and Geol.* **183,** 1–177.

Campbell, A. S. (1954). Radiolaria. *In* (Moore, R. C., ed.), *Treatise on Invertebrate Paleontology. Kansas Press and Geol. Soc. Amer., Pt. D, Protista* **3,** 1–163.

Campbell, A. S. and Clark, B. L. (1944). Radiolaria from Upper Cretaceous of middle California. *Geol. Soc. Amer., Spec. Paper* **57,** 1–61, pls 1–8.

Cita, M. B. (1964). Richerche micropaleontologiche e stratigrafiche sui seimenti del Giurassico superiore e del Cretaceo inferiore nella catena del Monte Baldo. *Rivista Ital. di Paleon. e Strat.* **10,** 1–182, pls 1–12.

Dumitrica, P. (1970). Cryptocephalic and cryptothoracic Nassellaria in some Mesozoic deposits of Romania. *Rev. Roum. Geol. Geophys. Geogr., Ser. Geol.* **14,** 1, 45–124, pls 1–21,

Eckert, R. Hay, W. W., Lorenz, G. and Vogt, P. (1961). The magnetic separator as a tool in micropaleontology. *J. Paleont.* **35,** 876–877.

Ewing, M. *et al.* (1969). *Initial Reports of the Deep Sea Drilling Project* **1,** 3–672. U.S. Government Printing Office, Washington.

Foreman, H. P. (1968). Upper Maestrichtian Radiolaria of California. *Special Papers in Paleontology* **3,** 1–82, pls 1–8.

Foreman, H. P. (1971). Cretaceous Radiolaria, Leg 7, Deep Sea Drilling Project., *In* (Winterer E. L. *et al.*), *Initial Reports of the Deep Sea Drilling Project* **7,** 1673–93. U.S. Government Printing Office, Washington.

Goll, R. M. (1968). Classification and phylogeny of Cenozoic Trissocylidae (Radiolaria) in the Pacific and Caribbean Basins. Part 1. *J. Paleont.* **42,** 1409–1432, pls 173–176.

Goll, R. M. (1969). Classification and phylogeny of Cenozoic Trissocylidae (Radiolaria) in the Pacific and Caribbean Basins, Part II. *J. Paleont.* **43,** 322–339, pls 55–60.

Goll, R. M. (1972). Systematics of eight Tholospyris taxa (Trissocyclidae, Radiolaria). *Micropaleont.* **18,** 443–475, pls 1–16.

Haeckel, E. (1887a). Report on the Radiolaria collected by *H.M.S. Challenger* during the years 1873–1876. *Rept. Voyage 'Challenger', Zool.* **18,** *i–xxxviii*, 1–1802, pls 1–140.

Haeckel, E. (1887b) *Die Radiolarien* (*Rhizopoda Radiolaria*). *Eine Monographie. Pt. 2. Grundriss einer allegemeinen Naturgeschichte der Radiolarien*, 248 pp, 64 pls.

Hedberg, H. D. *et al.* (1972). Preliminary report on biostratigraphic units. International Subcommission on stratigraphic Classification, Report No. 5, 1–50. *International Geological Congress, Montreal, Canada*, 1971.

Hinde, G. J. (1908). Radiolaria from Triassic and other rocks of the Dutch East Indian Archipelago. *Jaarb. Mijnw. Ned. Oost-Indie* **37,** 709–751, pls 5–10.
Holmes, W. M. (1900). On Radiolaria from the Upper Chalk at Coulsdon (Surrey). *Q. J. Geol. Soc. London* **56,** 694–704, pls 37–38.
Ichikawa, Koichiro (1950). A study on the radiolarian fauna of the Mt. Mitake in the southeastern part of the Kwanto Mountainland. Japan. *J. Fac. Sci. Univ. Tokyo sec. 2,* **7,** 281–315, pls 1–3.
Khabakov, A. V. (1961). Fauna radiolyarii nizhemelovykh i verkhneyurskikh fosforitov basseina verkhnei vyatki i kamy. *Ezhegadnik Vserossiiskogo Paleontologicheskogo Obshchestva* **11,** 90–120, pls 11–14.
Lawson, A. C. (1895). Sketch of the geology of the San Francisco Peninsula. *U.S. Geol. Survey 15th Ann. Rept.* 399–476.
Lipman, R. Kh. (1952). Materialy morphographicheskomu izucheniyu radiolyarii verkhnemelovykh otlozhenii russkoi platformy. *Paleontologiya i Stratigrafiya (Trudy VSEGEI)*, 24–51, pls 1–3.
Parona, C. F. (1890). Radiolarie nei noduli selciosi del calcare giurese di Cittiglio presso Laveno. *Boll. Soc. Geol. Ital.* **9,** 1–46, pls 1–6.
Pessagno, E. A., Jr. (1963). Upper Cretaceous Radiolaria from Puerto Rico. *Micropaleont.* **9,** 197–214, pls 1–7.
Pessagno, E. A., Jr. (1969a). Cenomanian-Turonian (Eaglefordian) Stratigraphy in the Western Gulf Coastal Plain Area. *In* (Brönnimann, P. and Renz, H. H., eds), *Proceedings, First International Conference on Planktonic Microfossils* **2,** 509–524.
Pessagno, E. A., Jr. (1969b). The Neosciadiocapsidae, a new family of Upper Cretaceous Radiolaria. *Bull. Amer. Paleont.* **58,** 377–439, pls 23–38.
Pessagno, E. A., Jr. (1970). The Rotaformidae, a new family of Upper Cretaceous Nassellariina (Radiolaria) from the Great Valley Sequence, California Coast Ranges. *Bull. Amer. Paleont.* **58,** 5–32.
Pessagno, E. A., Jr. (1971a). Jurassic and Cretaceous Hagiastridae from the Blake-Bahama Basin (Site 5A, JOIDES Leg 1) and the Great Valley Sequence, California Coast Ranges. *Bull. Amer. Paleont.* **60,** 1–80, pls 1–19.
Pessagno, E. A., Jr. (1971b). A new radiolarian from the Upper Cretaceous of the California Coast Ranges. *Micropaleont.* **17,** 361–364, pl. 1.
Pessagno, E. A., Jr. (1972a). The Phaseliformidae, new family, and other Spongodiscacea from the Upper Cretaceous portion of the Great Valley Sequence. *In* Cretaceous Radiolaria, part I, *Bull. Amer. Paleont.* **61,** 269–280, pls 22–23.
Pessagno, E. A., Jr. (1972b). Pseudoaulophacidae Riedel from the Cretaceous of California and the Blake-Bahama Basin (JOIDES Leg I). *In* Cretaceous Radiolaria part I, *Bull. Amer. Paleont.* **61,** 281–314, pls 24–31.
Pessagno, E. A., Jr. (1973). Upper Cretaceous Spumellariina from the Great Valley sequence, California Coast Ranges. *Bull. Amer. Paleont.* **63,** 49–102, pls 9–21.
Pessagno, E. A., Jr. (1976). Radiolarian zonation and stratigraphy of the Upper Cretaceous portion of the Great Valley Sequence, California Coast Ranges. *Micropaleont. Special Paper, No. 2,* 234 pp. (ms), 14 pls (in press).
Pessagno, E. A., Jr. and Newport, R. L. (1972). A technique for extracting Radiolaria from radiolarian cherts. *Micropaleont.* **18,** 231–234, pl. 1.
Petrushevskaya, M. G. (1964). O gomologiyakh elementov vnutrennego skeleta nekototykh radiolyarii nassellaria. (On homologies in the elements of the inner skeleton of some Nassellaria.) *Zoolog. Zhurnal* **43,** 1121–1128.

Petrushevskaya, M. G. (1965). Osobennosti konstruksii skeleta radiolyarii Botryoidae (ostr. Nassellaria). *Trudy Zoolog. Inst.* (Akad, Nauk SSSR) **35**, pp. 79–118.
Petrushevskaya, M. G. (1971a). Radiolyarii Nassellaria i v planktone mikosvogo okeana (Radiolarians Nasselaria in the ocean's plankton), pp 3–288. *In* Radiolarii mikovogo okeana po materialam Sovetskoy Ekspedtsi (Radiolaria of the Oceans; reports on the Soviet expeditions). *Akademiia Nauk SSSR, Zoologisheskii Institute, Issedovaniia flaoonyi* **9,** 3–418.
Petrushevskaya, M. G. (1971b). On the natural system of polycystine Radiolaria (Class Sarcodina). *In* (Farinacci, A., ed.), *Proc. II Plankt. Conf.* 981–992. Tecnoscienza, Roma.
Riedel, W. R. (1967). *In* (Harland, W. B. *et al.* eds), *The fossil record*, 291–332. London (Geological Society).
Riedel, W. R. (1971). Systematic classification of polycystine Radiolaria. *In* (Funnell, B. M. and Riedel, W. R., eds), *The Micropaleontology of the Oceans*, 649–661. Cambridge University Press.
Rüst, D. (1885). Beitrage zur Kenntniss der fossilen Radiolarien aus Gesteinen des Jura. *Palaeontographica* **31**, 269–321, pls 26–45.
Rüst, D. (1888). Beitrage zur Kenntniss der fossilen Radiolarien aus Gesteinen der Kreide. *Palaeontographica* **34,** 181–213, pls 22–29.
Rüst, D. (1892a). Radiolaria from the Pierre Formation of North-Western Manitoba. *Geol. Nat. Hist. Surv. Canada, Contr. to Canadian Micropaleontology*, pt. 4, pp. 101–110, pls 14–16.
Rüst, D. (1892b). Beitrage zur Kenntniss der fossilen Radiolarien aus Gesteinen der Trias und der palaezoischen Schicten. *Palaeontographica* **38,** 107–192, pls 6–30.
Rüst, D. (1898). Neue Beitrage zur Kenntniss der fossilen Radiolarien aus Gesteinen des Jura und der Kreide. *Palaeontographica* **45,** 1–67, pls 1–19.
Squinabol, S. (1903). Le Radiolarie dei noduoi selciosi nella Scaglia degli Euganei. *Riv. Italiana Paleon.* **9,** 105–150, pls 8–10.
Squinabol, S. (1904). Radiolarie cretacee degli Euganei. *Atti Mem. R. Accad. Sci. Lett. Arti Padova, new ser.* **20,** 171–244, pls 1–10.
Squinabol, S. (1913). Radiolarie della strada nazionale al Monginevro. *Boll. Reale Ufficio Geol. Italia, 43, fasc.* **2/3,** 281–289, 1 plate.
Squinabol, S. (1914). Contributo alla conoscenza dei Radiolarii fossili del Veneto. Appendice-Di un genere di Radiolari caratteristico del Secondario. *Mem. Inst. R., Univ. Padova* **2,** 249–306, pls 20–24.
Tan Sin Hok (1927). Over de samenstalling en het onstaan van krijten mergel-gesteenten von de Molukken. *Jaarb. Mijnw. Ned.-Indie, Verhand.* pt. 3, pp. 5–165, pls 1–16.
Vinassa de Regny, P. E. (1899). I Radiolari delle ftaniti titoniane de Carpena (Spezia). *Palaeontographia Italica* **4,** 217–238, pls 17–18.
Urban, J. and Padovani, E. R. (1970). A new scanning electron microscope holder for palynology. Pollen and Spores. *Mus. Nat. d'histoire Natur.* **12,** 131–139.
Zhamoida, A. I. (1972). Biostratigrafii Mesozoiskich Kremnistich tolch vostoka SSSR. (Na osnove izocheniya radiolarii). Ministerstvo Geologii SSSR. *Trudi Vsesovznogo Ordena Lenina Naukno—Issledovateleskogo Geologischeskogo Instituta (VSEGEI), N. S.* **183,** 3–242, pls 1–19.
Zittel, K. A. (1876). Ueber einige fossile Radiolarien aus der norddeutschen Kreide. *Zeitschr. Deutschen Geol. Gessell.* **28,** 75–86, pl. 2.

PLATES

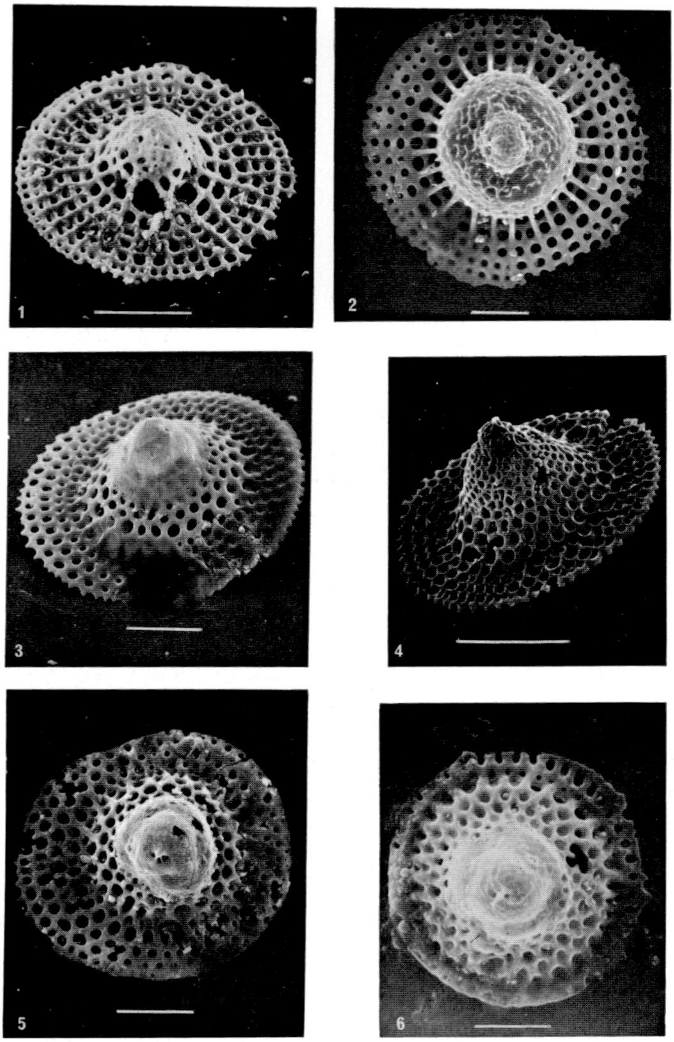

PLATE 1. All figures are scanning electron micrographs of Neosciadiocapsidae from the Coniacian portion of the Great Valley sequence. (*Alievium praegallowayi* Zone), see Fig. 3. Fig. 1. *Microsciadiocapsa madisonae* Pessagno. Yolo Formation (marker, 40 μm). fig. 2. *Microsciadiocapsa radiata* Pessagno. Sites Formation, (marker, 100 μm). fig. 3. *Microsciadiocapsa monticelloensis* Pessagno. Yolo Formation, (marker, 40 μm). fig. 4. *Squinabollella putahensis* Pessagno. Yolo Formation, (marker, 100 μm). fig. 5. *Microsciadiocapsa berryessaensis* Pessagno. Yolo Formation. Note triad of apical pores near horn, (see Plate 2, fig. 1), (marker, 40 μm). fig. 6. *Microsciadiocapsa cortinaensis* Pessagno. Sites Formation. Note four apical pores near horn, (see Plate 2, fig. 2), (marker, 40 μm).

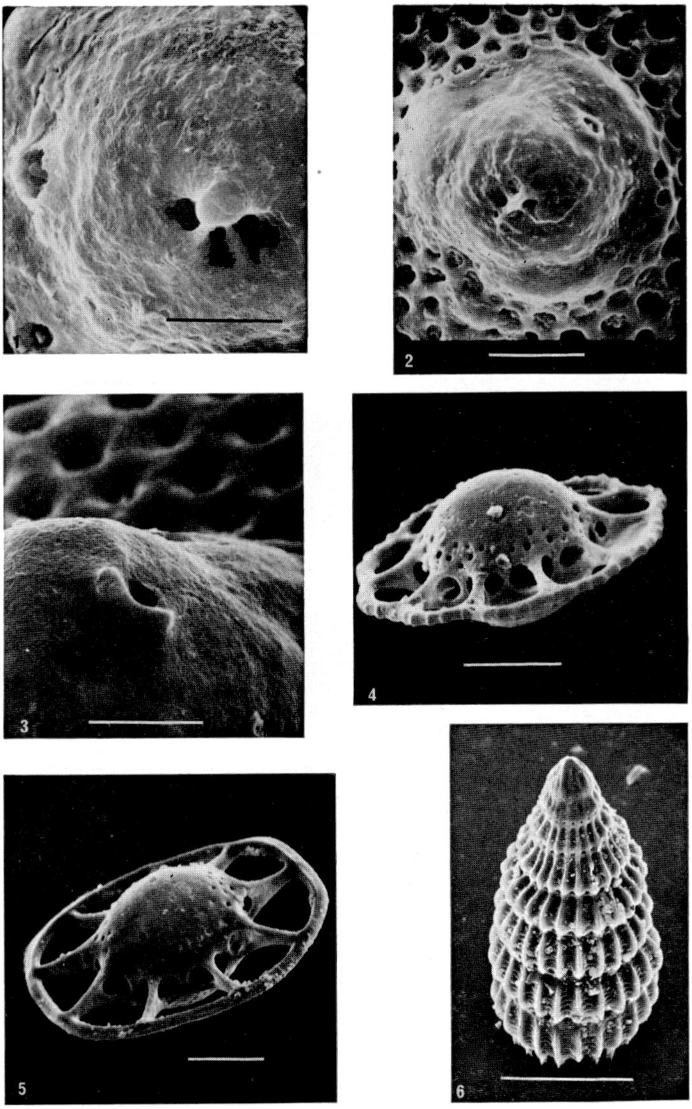

PLATE 2. All figures are scanning electron micrographs of Nassellariina from the Coniacian portion of the Great Valley sequence (*Aliveium praegallowayi* Zone), see Fig. 3. fig. 1. *Microciadiocapsa berryessaensis* Pessagno. Yolo Formation. Same specimen as that shown in Pl. 1, fig. 5. Cephalis and triad of apical pores at greater magnification, (marker, 10 μm). fig. 2. *Microsciadiocapsa cortinaensis* Pessagno. Sites Formation. Same specimen as that shown in Pl. 1, fig. 6. Cephalis and four apical pores at greater magnification, (marker, 20 μm). fig. 3. *Microsciadiocapsa monticelloensis* Pessagno. Yolo Formation. Same specimen as Pl. 1, fig. 3. Cephalis and two apical pores at greater magnification, (marker, 10 μm). fig. 4. *Saturniforma abastrum* Pessagno. Yolo Formation, (marker, 40 μm). fig. 5. *Saturniforma caelestium* Pessagno. Yolo Formation, (marker, 40 μm). fig. 6. *Dictyomitra formosa* Squinabol. Yolo Formation. Note primary pores at strictures between shell joints and relict pores between costae and stricture, (marker, 100 μm).

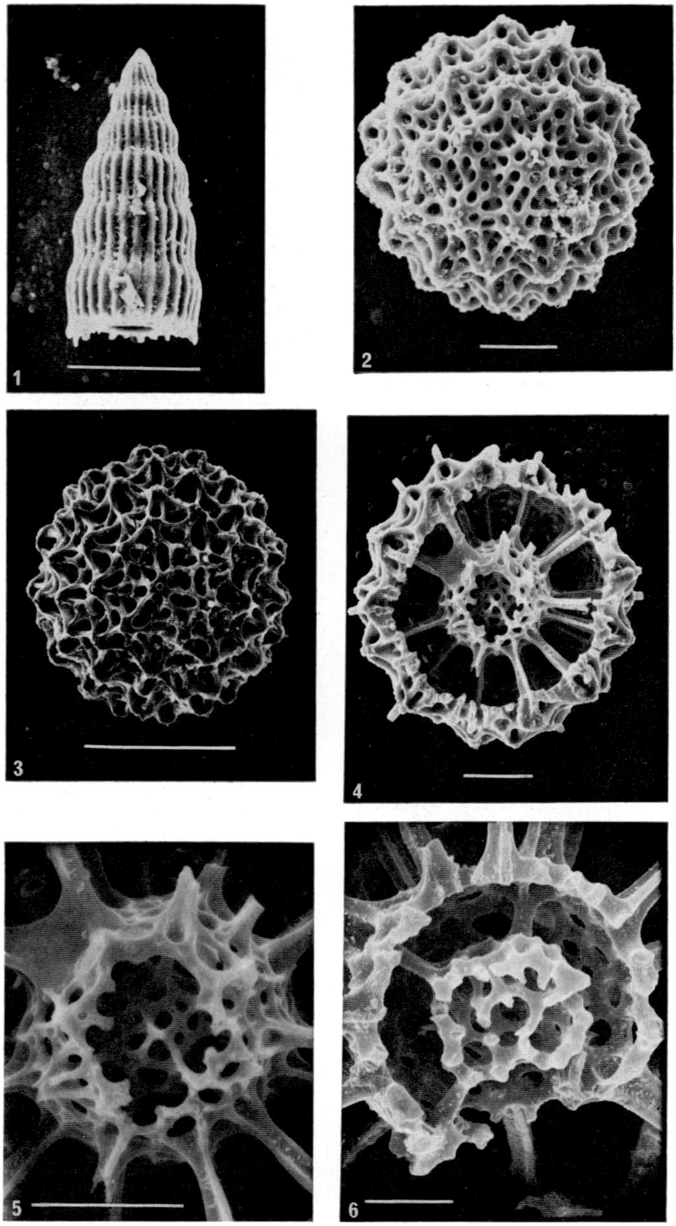

PLATE 3. All figures are scanning electron micrographs of Radiolaria from the Great Valley sequence, see Fig. 3.
fig. 1. *Dictyonitra napaensis* Pessagno. Venado Formation. *Alievium superbum* Zone (Turonian), (marker, 100 μm). fig. 2. *Praeconocaryomma universa* Pessagno. Yolo Formation. *Alievium praegallowayi* Zone (Coniacian). Note tubercles (mammae) on cortical shell, (marker, 40 μm). fig. 3. *Praeconocaryomma californiaensis* Pessagno. Sites Formation. *Alievium praegallowayi* Zone (Coniacian) (marker, 100 μm). figs 4, 5. *Praeconocaryomma californiaensis* Pessagno. Yolo Formation. *Alievium praegallowayi* Zone (Coniacian). Note cortical shell (outermost), first medullary shell, and second medullary shell. Medullary shells lack mammae, (marker in fig. 4, 40 μm; in fig. 5, 40 μm). fig. 6. *Praeconocaryomma californiaensis* Pessagno. Yolo Formation. *Alievium praegallowayi* Zone (Coniacian) Natural section showing first, second, and third medullary shells, (marker, 100 μm).

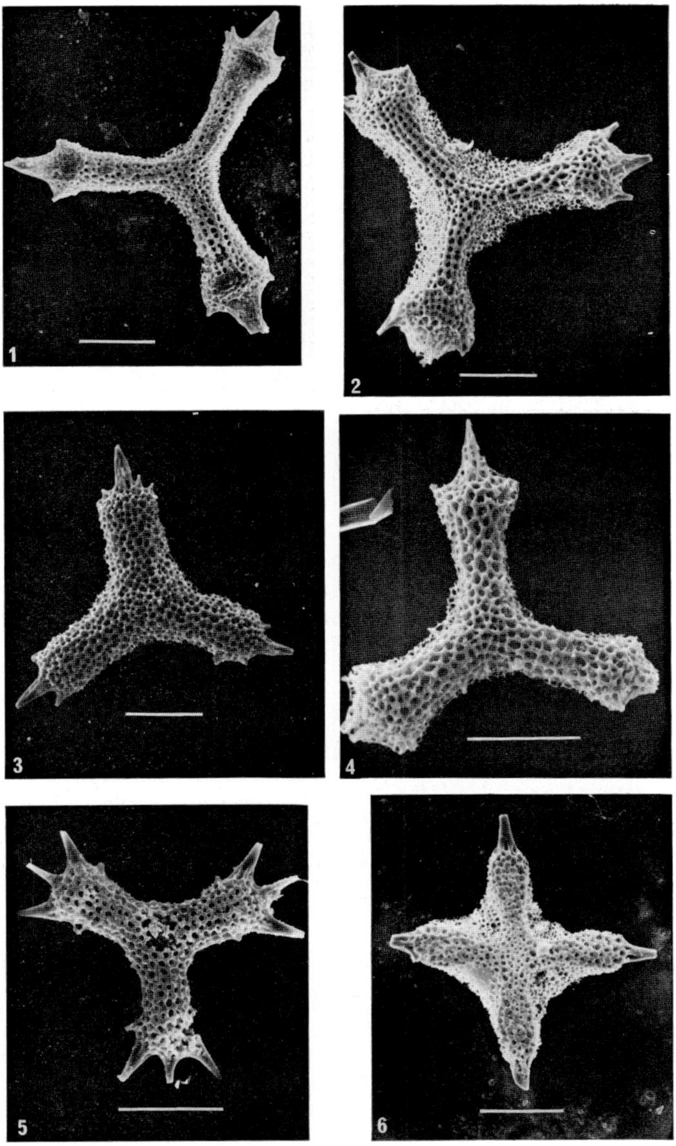

PLATE 4. All figures are scanning electron micrographs of Hagiastridae from the *Alievium praegallowayi* Zone (Coniacian), Great Valley sequence, (markers, 100 μm).
fig. 1. *Patulibracchium ruesti* Pessagno. Yolo Formation. fig. 2. *Patulibracchium torvitatis* Pessagno. Yolo Formation. fig. 3. *Paronaella venadoensis* Pessagno. Yolo Formation. fig. 4. *Paronaella solanoensis* Pessagno. Yolo Formation. fig. 5. *Patulibracchium ungulae* Pessagno. Yolo Formation. fig. 6. *Crucella plana* Pessagno. Yolo Formation.

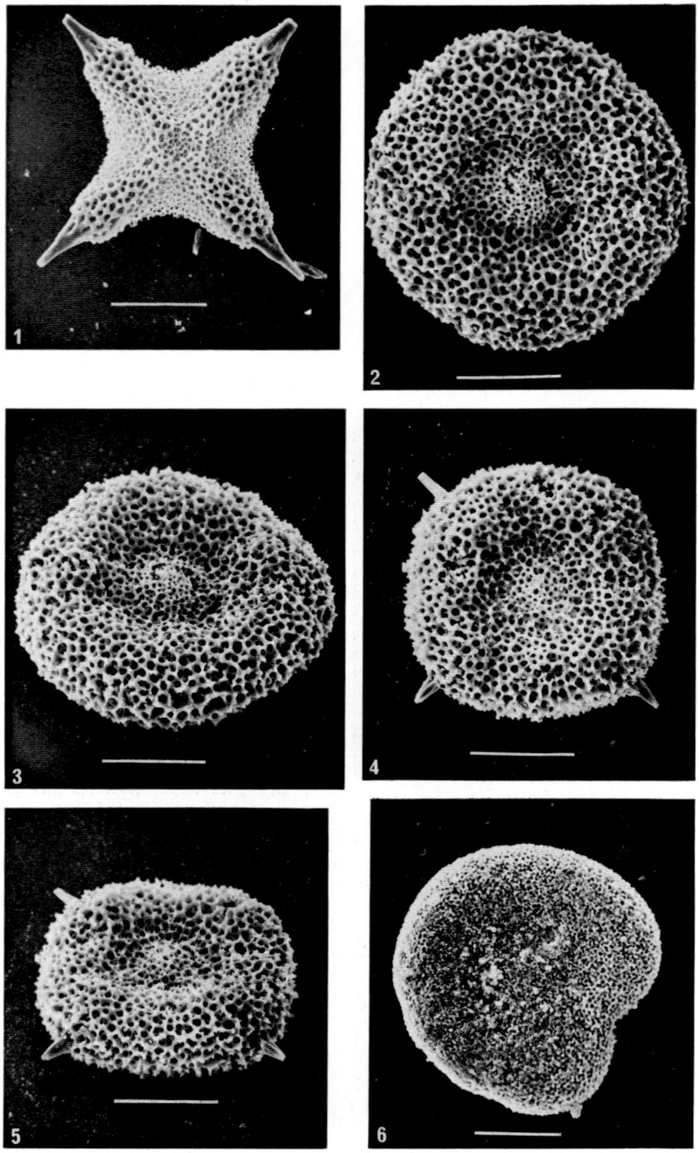

PLATE 5. All figures are scanning electron micrographs of Hagiastridae and Orbiculiformidae from the Great Valley sequence, (markers, 100 μm).
fig. 1. *Crucella espartoensis* Pessagno. Forbes Formation. This species makes its first appearance in the *Alievium gallowayi* Zone (Santonian). figs 2, 3. *Orbiculiforma monticelloensis* Pessagno. Yolo Formation. *Alievium praegallowayi* Zone (Coniacian). figs 4, 5. *Orbiculiforma quadrata* Pessagno. Yolo Formation. *Alievium praegallowayi* Zone (Coniacian). fig. 6. *Orbiculiforma persenex* Pessagno. Sites Formation. *Alievium praegallowayi* Zone (Coniacian).

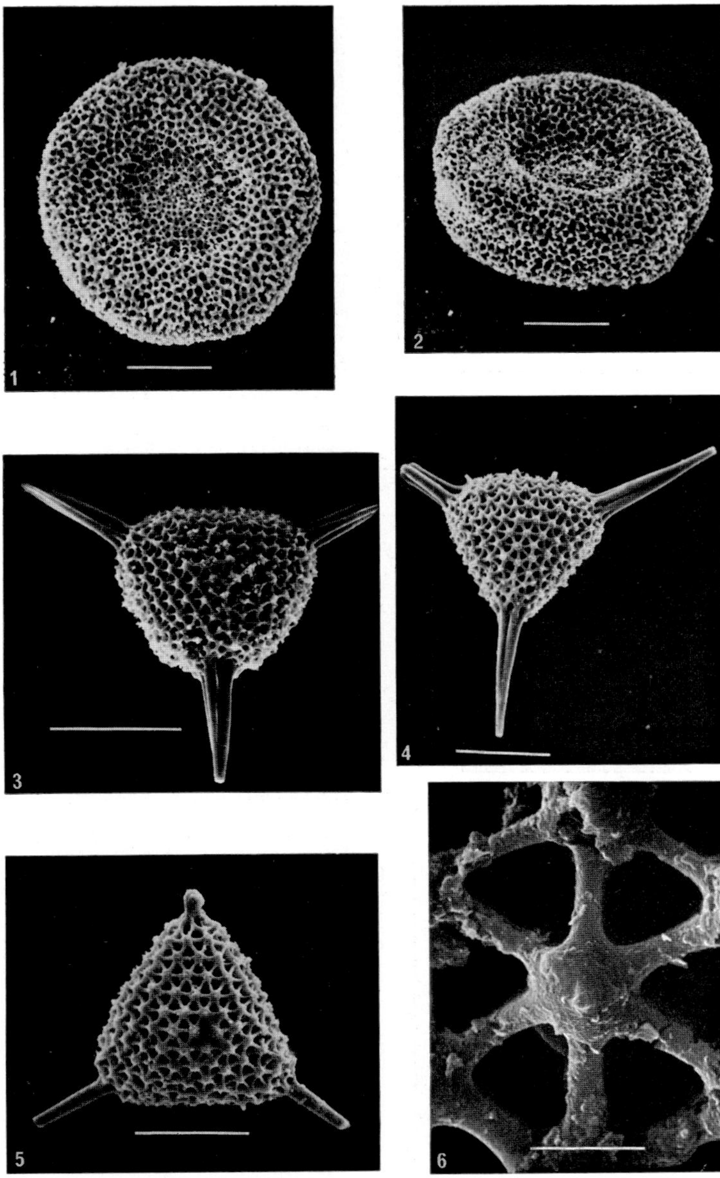

PLATE 6. All figures are scanning electron micrographs of Orbiculiformidae and Pseudoaulophacidae from the Great Valley sequence, (markers, 100 μm, except for fig. 6 which is 10 μm).
figs 1, 2. *Orbiculiforma vacaensis* Pessagno. Yolo Formation. *Alievium praegallowayi* Zone (Coniacian). fig. 3. *Alievium superbum* (Squinabol). Yolo Formation. *Alievium praegallowayi* Zone (Coniacian). Note triradiate primary spine. fig. 4. *Alievium praegallowayi* Pessagno. Yolo Formation. *Alievium praegallowayi* Zone (Coniacian). Note that distal half of primary spines are circular in axial section; proximal half of primary spines remain triradiate. figs 5, 6. *Alievium gallowayi* (White). Forbes Formation. This species makes it's first appearance at the base of the *Alievium gallowayi* Zone (Santonian), fig. 5: Note circular nature of primary spine in axial section, fig. 6: Triangular pore frames at higher magnification.

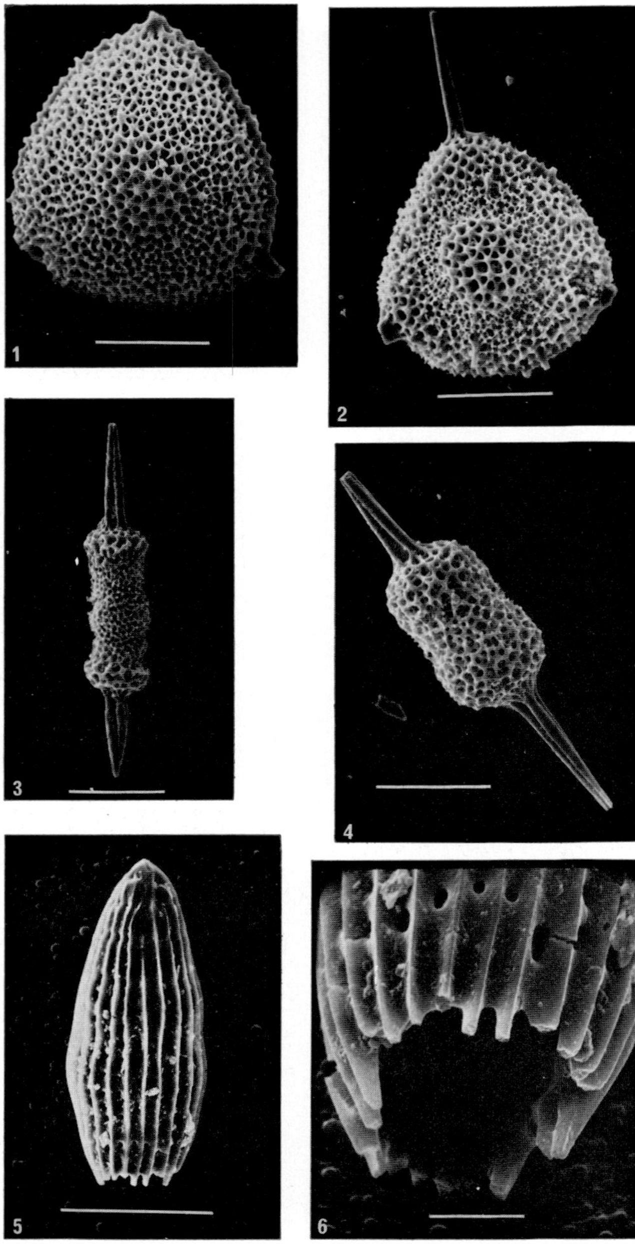

PLATE 7. All figures are scanning electron micrographs of Upper Cretaceous Radiolaria from the Great Valley sequence, (markers, 100 μm, except for fig. 6 which is 20 μm). fig. 1. *Pseudoaulophacus venadoensis* Pessagno. Paratype (USNM 165511). Sites Formation. *Alievium praegallowayi* Zone (Coniacian). fig. 2. *Pseudoaulophacus praefloresensis* Pessagno. Holotype. Coniacian portion of Yolo Formation. *Alievium praegallowayi* Zone. fig. 3. *Archaeospongoprunum triplum* Pessagno. Paratype (USNM 165606). Coniacian portion of Yolo Formation. *Alievium praegallowayi* Zone. fig. 4. *Archaeospongoprunum bipatitum* Pessagno. Holotype (USNM 165608). Coniacian portion of Yolo Formation. *Alievium praegallowayi* Zone. figs 5, 6. *Archaeodictypmitra squinaboli* Pessagno. Paratype (USNM 186361). Venado Formation (Turonian). Note lack of strictures and presence of relict pores. Coastal projections largely broken on final post-abdominal chamber (segment). *Alievium superbum* Zone.

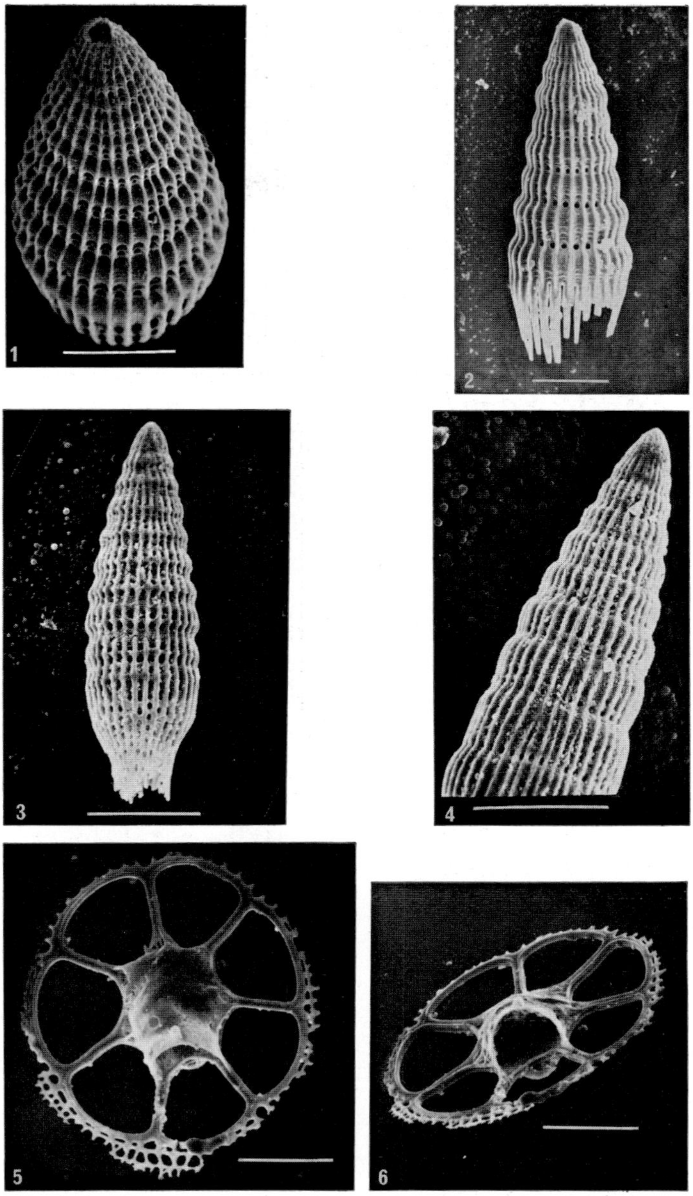

PLATE 8. All photographs are scanning electron micrographs of Upper Cretaceous Nassellariina from the Great Valley sequence, (markers, 100 μm).
fig. 1. *Dictyomitra multicostata* Zittel. Latest Campanian portion of Forbes Formation. Note primary pores occur only at strictures and that relict pores (between strictures) are completely obscured by accretion of shell material. Distal portion of final post-abdominal chamber shows formation of new pores between costal projections. The roof of the cephalis of this specimen is broken away, thus revealing the cephalic skeletal elements and collar pores. fig. 2. *Dictyomitra multicostata* Zittel. Latest Campanian portion of Forbes Formation. Note well developed coastal projections on final post abdominal chamber. figs 3, 4. *Dictyomitra densicostata* Pessagno. Latest Campanian portion of Forbes Formation. Note presence of a few relict pores between strictures. figs 5, 6. *Rotaforma mirabilis* Pessagno. Holotype (USNM 165481). "Fiske Creek Formation".

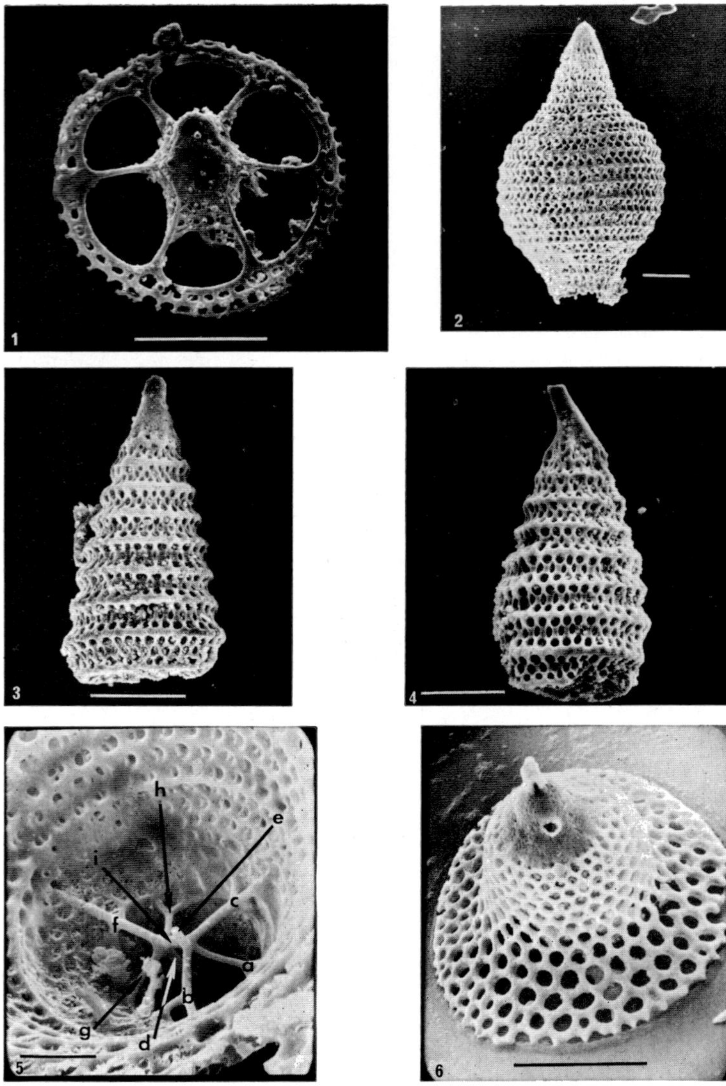

PLATE 9. Markers on all scanning electron micrographs 100 μm, except for fig. 5 which is 40 μm.
fig. 1. *Rotaforma hessi* Pessagno. Holotype (USNM 165478). Cenomanian portion of "Fiske Creek Formation". fig. 2. "*Lithocampe*" sp. aff. "*L.*" *mediodilatata* Rust. Unnamed Jurassic (Tithonian) cherts at Pt. Sal, California. Note concentric ridges with two rows of massive hexagonal pore frames between any two given ridges. figs 3, 4. "*Eucyrtidium*" *khabokovi* Zhamoida. Note concentric ribs with three rows of massive hexagonal pore frames between any two given ridges. From Tithonian cherts exposed east of Morro Bay, California. fig. 5. *Petasiforma glascockensis* Pessagno. Paratype (USNM 164211). Cenomanian portion of "Fiske Creek Formation". Interior of cephalis at high magnification showing cephalic skeletal elements and collar pores. a, vertical bar; b, primary left lateral bar; c, primary right lateral (lateral of some workers) bar; d, apical bar; e, axial spine; f, dorsal bar; g, secondary left lateral bar; h, secondary right lateral bar; i median bar. fig. 6. *Neosciadiocapsa diabloensis* Pessagno. Holotype (USNM 164257). Late Campanian portion of Del Valle Formation. Tubular opening in cephalis is the *cephalopyle*.

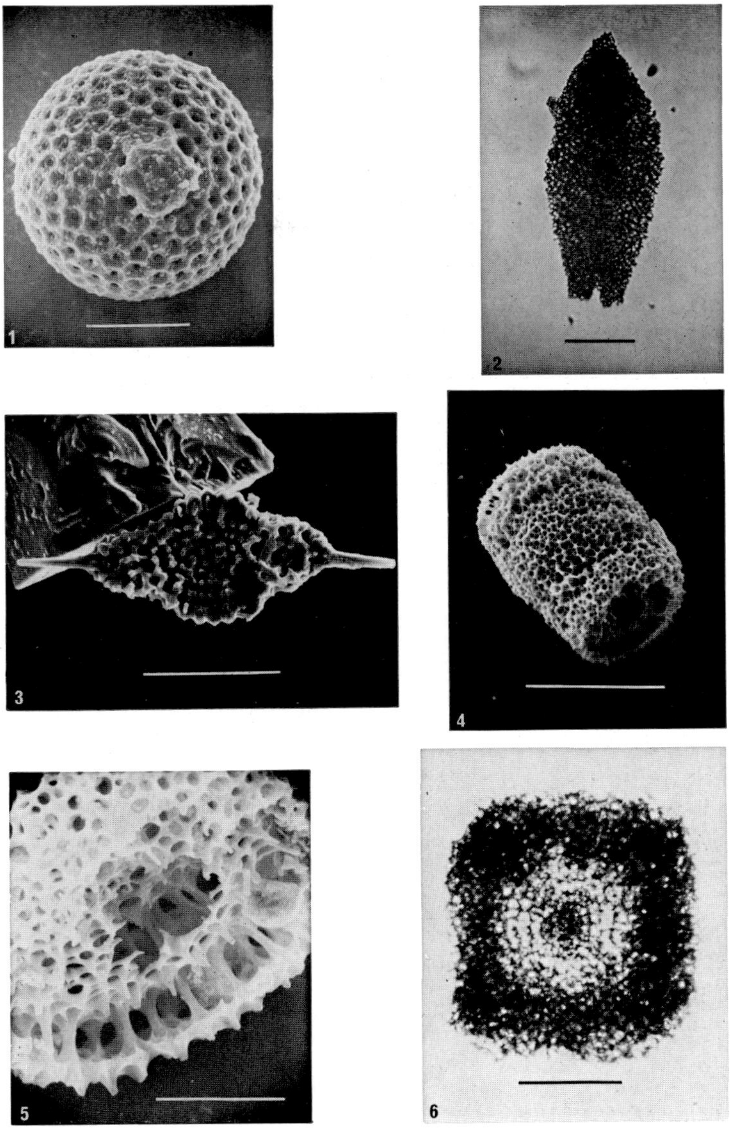

PLATE 10. figures 1–5 are scanning electron micrographs. figures 2 and 6 are light photo micrographs.
fig. 1. Dicyrtid (two chambered) Nassellariinid with thorax enclosing most of cephalis; button-like structure represents apical portion of cephalis. Specimen from Franciscan chert of probable Jurassic (Tithonian) age; Wilbur Springs, Cailfornia, (marker, 40 μm). fig. 2. *Parvicuspis shastaensis* Pessagno. Paratype (USNM 165583). Early Campanian portion of Forbes Formation, "Dobbins Shale Member". Lack of symmetry (e.g. concentric arrangement of pore frames typifies members of Subsuperfamily Spongodiscilae), (marker, 100 μm). fig. 3. *Pseudoaulophacus floresensis* Pessagno. Early Campanian portion of Forbes Formation. Natural section showing concentric layers of meshwork characteristic of Subsuperfamily Pseudoaulophacilae, (marker, 100 μm). figs 4, 5. *Spongurus* (?) sp. Early Campanian portion of Forbes Formation. Note concentric layers of spongy meshwork characteristic of Subsuperfamily Pseudoaulophacilae, (marker, in fig. 4, 100 μm; in fig. 5, 20 μm). fig. 6. *Orbiculiforma quadrata* Pessagno. Coniacian portion of Yolo Formation. Note concentric nature of meshwork, (marker, 100 μm).

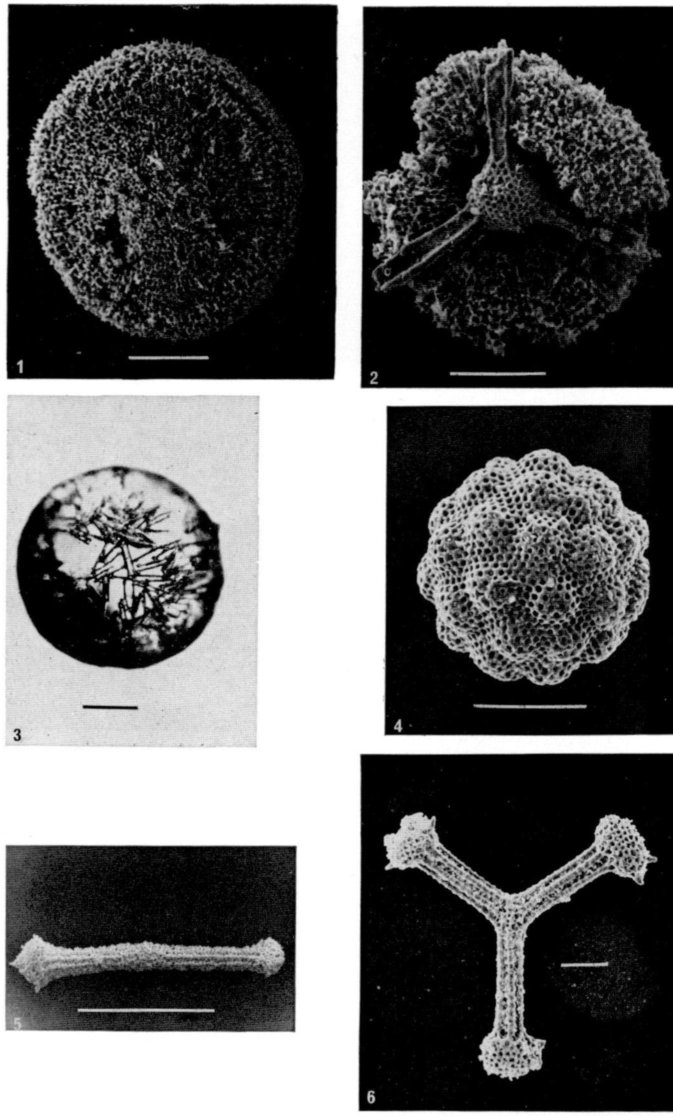

PLATE 11. All figures except fig. 3 are scanning electron micrographs, (markers, 100 μm). figs 1, 2. *Cavidiscus fiskensis* Pessagno. Paratypes (USNM 165661). Early Turonian portion of "Fiske Creek Formation", fig. 1: Specimen showing spongy cortical shell, fig. 2: Natural section showing relation between latticed first medullary shell (inner shell) and spongy cortical shell (outer shell). Note geometric arrangement of pore frames comprising spongy meshwork of cortical shell. Subsuperfamily Spongodruppilae. fig. 3. Unnamed species belonging to Superfamily Collodaracea (subsuperfamily 3 herein). Early Campanian portion of Forbes Formation. Note that the spherical test of this form consists of rod-like spicules that have been fused together at their ends. These forms are common throughout the Upper cretaceous. fig. 4. *Praeconocaryomma* (?) *sphaeroconus* Rust. Deep Sea Drilling Project, Leg 1, Site 5A, Core 4, Section 1. Hauterivian of Blake-Bahama Basin. The figured specimen is completely replaced by pyrite. fig. 5. *Amphibracchium petersoni* Pessagno. Paratype (USNM 165579). Deep Sea Drilling Project, Leg 1, Site 5A, Core 7, Section 1 (top). Late Jurassic (Tithonian) of Blake-Bahama Basin. fig. 6. *Paronaella* (?) *ewingi* Pessagno. From limestone lens in Jurrassic (Tithonian) cherts at Pt. Sal, California. This species was originally described from the Tithonian strata of the Blake-Bahama Basin.

10: Biogeography and Provincialism of Living Coccolithophorids in the Pacific Ocean

SUSUMU HONJO

Woods Hole Oceanographic Institution,
Woods Hole, Massachusetts 02543 U.S.A.

The main patterns of coccolithophorid species distribution and abundance in the North, West and Equatorial Pacific Ocean are outlined. The three-dimensional structure of the coccolithophorid community during the late summer of 1968 has been determined for the North and Equatorial Pacific, based on surface and subsurface samples collected along five traverses.

Ninety-four coccolithophorid species and varieties were identified. Statistical analyses of species assemblages revealed six geographical coccolithophorid zones: Subarctic Zone, Transitional Zone, Central Zone, Equatorial Zone North, Equatorial Zone South and Central Zone South. These zones are developed subparallel to latitude and their distribution pattern appears to be influenced by the current system of the North Pacific. The restriction of some species to specific depths produces a stratified community and species stratification becomes marked beneath the water masses at the southern margin of the Transitional Zone and within the Central Zone. Species diversity was found to increase at all depths within the photic zone from high to low latitudes, reaching its maximum value in a narrow, probably metastable subsurface zone of mixing between the Transitional and Central Zones. Coccolithophorid cell density varies with depth, and the density distribution patterns vary for different zones. The relationship between species diversity, cell density, hydrographic parameters and nutrient distribution is investigated.

Outline

Introduction

Coccolithophorids, autotrophic, unicellular, planktonic organisms generally smaller than 30 μm, are accepted as plants and referred to as Haptophyceae (Christensen, 1962; Parke and Dixon, 1964). Despite their small size, they are probably quantitatively one of the most abundant members of the phytoplankton in the ocean, serving as one of the more important sources of oceanic primary production. Living individuals disintegrate leaving greatly diversified calcitic shields (coccoliths) several microns in diameter, which have been generally well preserved in great quantity as nannofossils since the Jurassic period. It is known that large areas of the deep sea-floor are covered by "coccolith ooze" (Bramlette, 1958).

The distribution of pelagic coccolithophorids in the surface water* of oceans has been studied by many scientists (for bibliography refer to Gaarder 1971, pp. 101–102). The species distribution data presented by McIntyre and Bé (1967) for the Atlantic, and McIntyre *et al.* (1970) for the Pacific are particular examples of investigations on a regional scale. With the exception of one report (Honjo and Okada, 1974, in press) there are no available data based on systematic regional sampling concerning the vertical distribution of nannoplankton species within the photic zone.

The horizontal and vertical distribution of coccolithophorids has been investigated in a large number of filter residue samples from the Pacific Ocean and its marginal seas (Okada and Honjo, 1970, 1973; Honjo and Okada, 1974, in press). Ninety-four coccolithophorid species and varieties were found. The assemblages of these species was distributed in several floral zones which were developed sub-parallel to the latitude. Some species were found at consistent depths, resulting in a stratified community.

In the following chapter, the main patterns of species distribution and abundance in the North, West, and Equatorial Pacific Ocean will be outlined. The species distribution along meridional profiles throughout the photic layer of the North-Central Pacific Ocean will be discussed in order to describe both the three-dimensional development of the community and the provincialism of coccolithophorids. The results of the study presented here represent only the late summer phase of the development of the coccolithophorid community. Significant seasonal change can be expected in the community particularly with increasing distance from the equator.

Samples and methods

Surface and subsurface water samples were collected along five traverses

* The term "surface water" is used in this report to indicate a depth from 0–0.5 m. The "subsurface" refers to any depth greater than this.

in the North and Equatorial Pacific (Fig. 1). Traverses 1 and 2 (Fig. 1) were made in 1968 as part of Leg 30 of the R.V. *Oshoro Maru* Hokkaido University (Okada and Honjo, 1970); traverses 3, 4 and 5 were made in 1969 on Leg 69–4 of the R.V. *Hakuho Maru*, Tokyo University (Okada and Honjo, 1973; Honjo and Okada, 1974, in press). The results of various chemical analyses

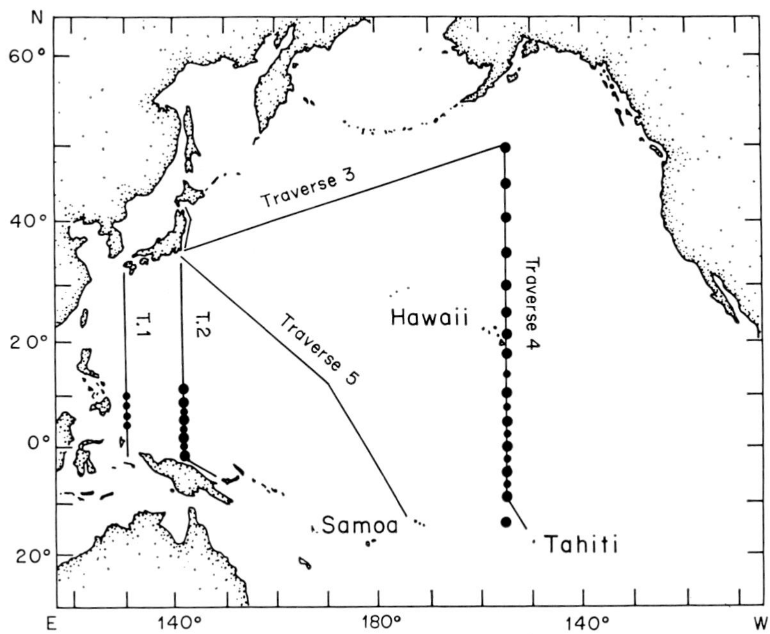

FIG. 1. Track chart of the transects and the location of stations. The surface water sampling was conducted between these major stations.

and hydrographic data are available in Tsujita (1970) and Ocean Research Institute (Marumo, (ed.) 1970). Part of traverse 4 coincided with the track of the *Ursa Major* expedition (1964) and the *Zetes* expedition (1966) of the Scripps Institution of Oceanography (Scripps Institution of Oceanography, 1967 and 1970).

A total of 231 surface samples were collected from the five traverses using a plastic bucket. Along traverses 1, 2 and 3, samples were normally collected at six hour intervals. Along traverses 4 and 5, surface samples were collected at approximately each degree of latitude, but in the tropical region they were taken at half degree intervals.

Three hundred and seventy-seven subsurface samples were also obtained along traverses 1, 2 and 4 with van Dorn-type water samplers and were usually collected from several levels within the upper 200 m of the water column.

The water samples were filtered with membrane filters on board ship, rinsed and dried. Cell numbers of coccolithophorids in a unit areas of the filter were counted under high power magnification in order to determine cell numbers in a litre of sample water. Another small piece of the same filter was processed for production of a carbon replica and observed under a transmission electron microscope (Hitachi HU 11 and HU 12) for species identification. The number of cells per species was counted by listing species identifications for 200 individuals in a sample. A computer was extensively used for listing, and retrieval and community indices computation. Methods of microscopic observation and computer software have been described in detail elsewhere (Okada and Honjo, 1973; Honjo *et al.*, 1973).

Zonal distribution of coccolithophorids

Ninety-four coccolithophorid species and varieties were identified from the North and Equatorial Pacific Ocean. Statistical analyses of species assemblages, and the distribution of cell numbers from the surface or near-surface water of the North and Equatorial Pacific enabled us to distinguish six coccolithophorid zones (Okada and Honjo, 1973). These zones coincide roughly with Brinton's (1962) euphausiid zones in the North Pacific, thus we applied his terminology to our coccolithophorid zonation which is described below and is illustrated on Fig. 2. No samples were obtained from the Arctic region for this study.

SUBARCTIC ZONE

Except for the 100 m level at Sta. 1 where *Calciopappus caudatus* (Gaarder and Ramstjell) was common, the subarctic ecophenotype of *Emiliania huxleyi* (Lohmann) monopolised the euphotic water column. This variety is characterised by a deformed shield which is poorly developed and formed of irregularly oriented "T"-shaped elements.

TRANSITIONAL ZONE

The cold variety of *E. huxleyi* (McIntyre and Bé, 1967) was still predominant at the surface but was rapidly replaced by *Rhabdosphaera clavigera* (Murray and Blackman). At 35°N *Umbellosphaera tenuis* (Kamptner) Paasche, dominated the middle euphotic layer, and *Florisphaera profunda* (Okada and Honjo, 1973) *Thorosphaera flabellata* (Halldal and Markali), *Cyclococcolithina fragilis* (Lohmann) and *Anthosphaera quadricornu* (Schiller), the lower euphotic layer, especially at the southern station (Sta. 5).

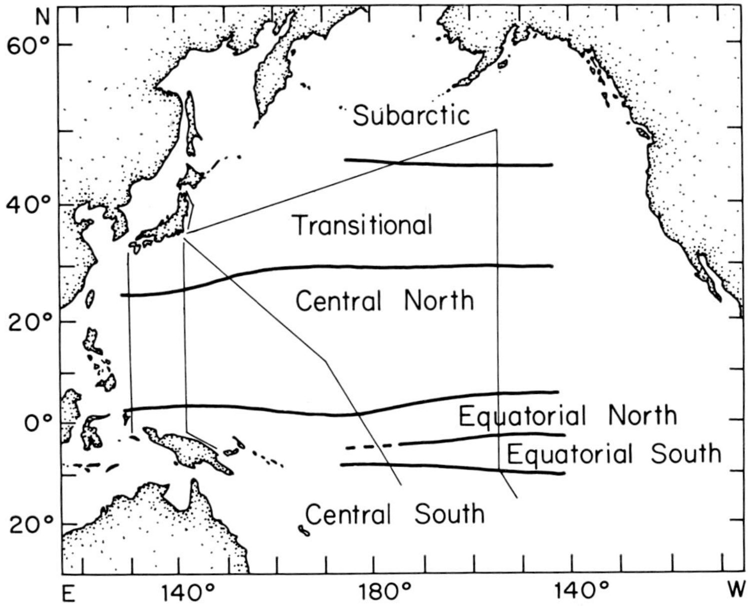

FIG. 2. Coccolithophorid zones in the northern Pacific (late summer) (Okada and Honjo, 1973).

CENTRAL ZONE, NORTH OR CENTRAL GYRE

Umbellosphaera irregularis (Paasche) was most abundant in the upper layer. The flora was similar in the middle and lower layers to the Transitional Zone except for Sta. 10 (10°N), where a cold undercurrent exists and consequently three species of the placolith group, *Gephyrocapsa ericsonii* (McIntyre and Bé), *Gephyrocapsa* sp. A and *Coccolithus* sp. A (Okada and Honjo, 1973, pl. 1) were predominant.

EQUATORIAL ZONE, NORTH

The species assemblage did not differ between the upper and middle photic zone. Five placolith species were common in this combined layer; *Emiliania huxleyi* and *Gephyrocapsa oceanica* (Kamptner) were abundant, and *Cyclococcolithina leptopora* (Murray and Blackmann) and *Umbilicosphaera hulburtiana* (Gaarder) were common. In the lower photic layer, the flora was similar to that in the northern zones.

EQUATORIAL ZONE, SOUTH

The species assemblage was close to that of the Equatorial Zone, North, but *Emiliania huxleyi* was replaced by *Umbellosphaera irregularis* in the upper layer.

CENTRAL ZONE, SOUTH

The southern boundary of this zone has not been determined. The species assemblage was quite similar to that of the Central Gyre, being its Southern Hemisphere counterpart, which suggests bipolar distribution of coccolithophorid zones.

Distribution of coccolithophorids in marginal seas

The distribution of coccolithophorid species in the Asian marginal seas,* such as the East, South China Sea and the Sunda Shelf area of the Java Sea, were greatly different from the pelagic floras at the equivalent latitude in the Pacific (Okada, 1973). Marginal seas are characterised by the following: (1) A lower diversity of coccolithophorid species than is observed in the open ocean. (2) Placolith-bearing coccolithophorids predominate and members of the species *Gephyrocapsa oceanica* dominate the floras. (3) Most coccolithophorids observed from the marginal seas bear malformed coccoliths. (4) The standing crop is considerably more aggregated in marginal sea facies than open sea facies. (5) A sharp biological boundary exists between the open ocean and the marginal seas and the percentage similarity of species was usually less than a few percent between two adjacent stations separated by the boundary. (6) The 1 mg/l sediment load contour coincided with such a boundary.

The relationship between cococlithophorid Zones and current systems

The current systems in the Northern Pacific appear to influence the zonal distribution pattern of the coccolithophorid species in the surface and near surface water in late summer (Table 1).

The boundary between the Subarctic and Transitional Zones coincides with that of the eastward flowing Pacific Subarctic current and the strong eastward flow of the Kuroshio current. The floral transition of approximately 45°N was defined by the northern limits of the Kuroshio extension. The difference in the floral contents of samples across this boundary was not sharply defined, and there was a percentage similarity of 76.5% between the species at surface stations at 45°N and 40°N along traverse 4. Malformed individuals of *Emiliania huxleyi* however, disappeared sharply at this boundary.

The boundary between the Central and Transitional Zones was established between 30°N and 29°N along the 155°W transect on the basis of the abundance of *Umbellosphaera irregularis*. At the 30°N station of traverse 4 this boundary

* "Marginal Sea" is defined here as an enclosed sea such as the South China Sea, separated from the ocean by archipelagos and/or continents. We have not studied the open neritic environment, such as the North American Pacific Coast.

TABLE 1. Coccolithophorid Zones along the surface water of Traverse 4 (155°W) compared with current, water mass and Euphausiid Zones (Okada and Honjo, 1973).

	Coccolithophorid Zones (Aug.–Oct., 1969)	Dominant species in surface waters	Current system (Marumo *et al.*, 1970) (Aug.–Oct., 1969)	Water mass (Marumo *et al.*, 1970) (Aug.–Oct., 1969)	Euphausiid Zones (Johnson and Brinton, 1963)
50°N	Subarctic Zone A (–45°N)	*Emiliania huxleyi* (subarctic variety)	Subarctic Current (–45°N)	Pacific Subarctic	Subarctic (ca. 57°N–41°N)
40°N 30°N	Transitional Zone B (45°N–30°N)	*Emiliania huxleyi* (cold variety) *Rhabdosphaera clavigera*	Kuroshio Extension (45°N–35°N) North Pacific Current (45°N–25°N)	Western North Pacific central	Transition (41°N–33°N)
20°N 10°N	Central–North Zone C (30°N–5°N)	*Umbellosphaera irregularis*	North Equatorial Current (25°N–10°N) Equatorial Countercurrent (10°N–5°N)		Central (33°N–12°N)
0°	Equatorial–North Zone D (5°N–3°N)	*Gephyrocapsa oceanica* *Emiliania huxleyi* (warm variety)	South Equatorial Current (5°N– °N)	Equatorial	Equatorial (12°N–15°S)
10°S	Equatorial–South Zone E (3°S–10°S)	*Gephyrocapsa oceanica*	A part of the South Pacific Circulation (5°S– °S)		
	Central–South Zone F (10°S– °S)	*Umbellosphaera irregularis*			

apparently corresponds with a metastable boundary between currents. The percentage similarity between the surface stations at 35°N and 25°N (stations 4 and 6) was only 7.5%.

The Central-North Zone contains two current systems, the broad westward flowing North Equatorial Current, and the swift eastward flowing equatorial counter-current. Mixing of the coccolithophorid floras of the Central Zone and the Equatorial North Zone occurs within the Equatorial Countercurrent, and *U. irregularis* becomes scarcer and *Gephyrocapsa oceanica* more abundant as one approaches the South Equatorial Current.

The Equatorial-North Zone and Equatorial-South Zone, coincide with the westward flowing South Equatorial Current. The division between these zones was less significant on the basis of differences between their species assemblages (percentage similarity, 39.5%) than at other boundaries. The two Equatorial Zones may almost be termed sub-zones.

South of 10°S the Central-South Zone was clearly separated from the Equatorial-South Zone, in terms of their species assemblages. The percentage similarity between two adjacent stations which were separated by the boundary was only 17.5% along traverse 4. However, there was no clear current system boundary at this latitude.

The Kuroshio Current which flows northwestward along the southwestern margin of the North pacific and into the northeast Pacific, disturbs the latitudinal arrangement of the floral zones and influences the species assemblage. An as yet incomplete study of the floral content of the Kuroshio current suggests that it contains a mixture of central and transitional floras with abundant *Syracosphaera* and *Umbellosphaera*.

The density distribution of coccolithophorid cells in the photic layer

PHOTIC WATER LAYER

Each floral zone was characterised by a distinctive density distribution pattern of coccolithophorid cells (see Fig. 3). In the Subarctic Zone the cell density was large, in the upper 50 m (10^5/l), decreased rapidly with depth, and virtually disappeared at 100 m. This may be due to the poor light penetration and relatively stable thermocline which is developed in this area. A large cell number such as 10^5/l was maintained above this thermocline in water temperatures higher than 6°C.

Cell density increased rapidly with depth and reached its maximum at approximately 75 m to 125 m in the Transitional Zone. The cell density of the surface water was maintained throughout the 200 m water column in the northern part of the Central Zone. In the Equatorial Zones, the cell density

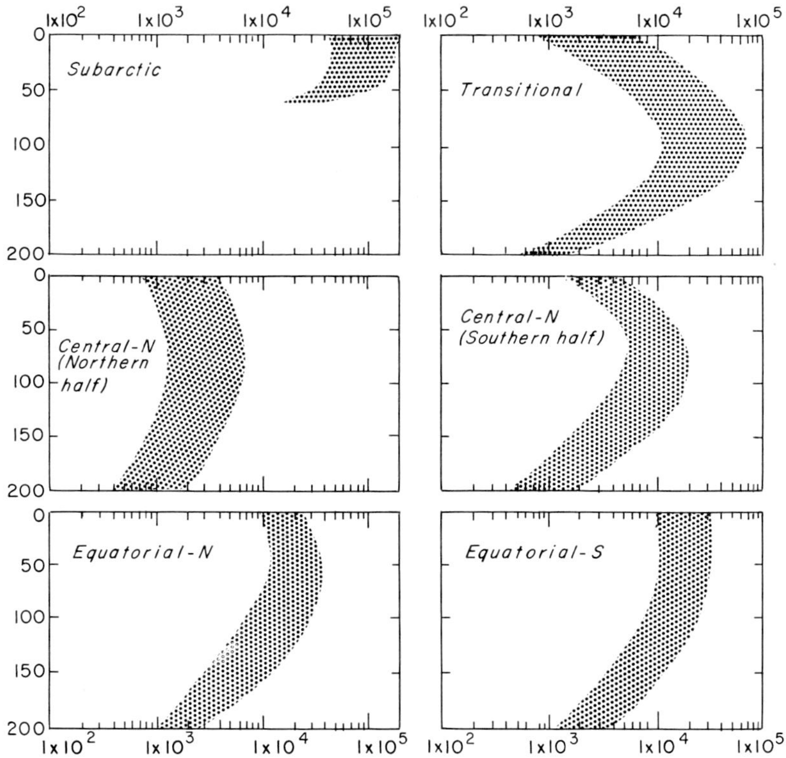

FIG. 3. Pattern of coccolithophorid cell density in the water column to 200 m in the North Equatorial Pacific Ocean (late summer).

was high, slowly increasing to 75 m, then decreasing down to 200 m. The cell density at 200 m was larger than at other zones.

THE SUBPHOTIC WATER MASS

A small but consistent number of apparently live coccospheres of a few species were found in the subphotic water masses beneath the Equatorial (100/L) and Transitional (10/L) to depths of 4000 m.

The vertical distribution of coccolithophorid species in the photic layer

The distribution of 10 major species of coccolithophorids in the entire photic zone at the 155°W meridian is illustrated in Fig. 4. *E. huxleyi* was ubiquitously distributed and the percentage of this species in the flora remained fairly constant within the profile of the photic zone of traverse 4. The vertical distribution of other coccolithophorid species however could not be predicted from their surface distribution. For example, *Umbellosphaera irregularis* was dominant in the surface water of the Central Zone but was replaced by *Umbellosphaera tenuis* at the 100 m level.

Gephyrocapsa oceanica and *Cycloccolithina leptopora* were distributed in the upper-middle section of the water column in the equatorial area. *Umbellosphaera irregularis, Umbilicosphaera hulburtiana, Discosphaera tuberifera*

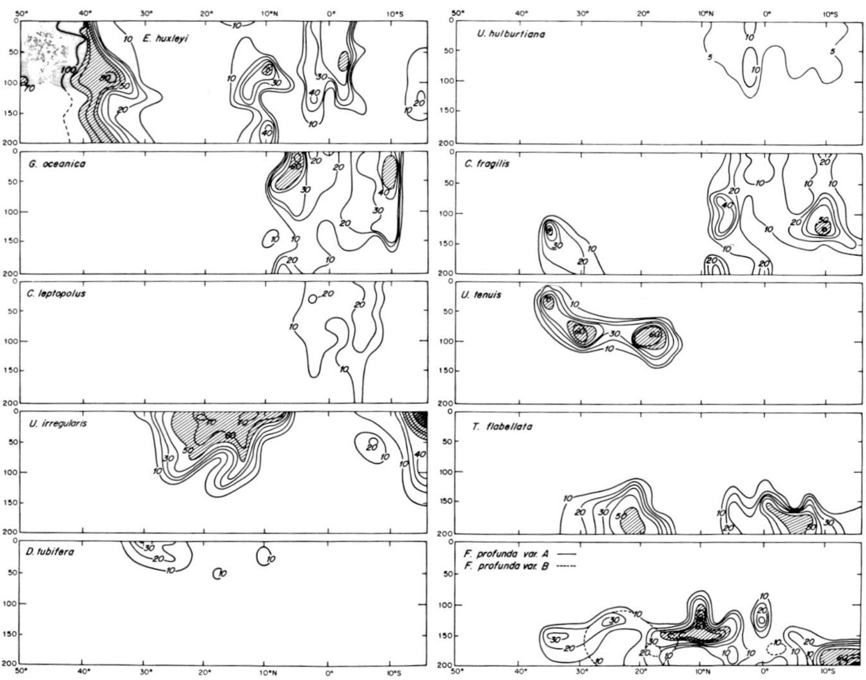

FIG. 4. Distribution of 10 representative coccolithophorid species through the photic layer profile along 155°W transect in percent for the total coccolithophorid population.

and *Rhabdosphaera clavigera* were usually distributed in surface or near-surface water and tended to decrease in abundance with depth. On the other hand, *Umbellosphaera tenuis* was rare in the surface water but common at depths from 50 to 100 m in the Central Zone. *Cyclococcolithina fragilis* was frequent in the surface water and abundant in the middle depths of the photic layer of the Equatorial Zone. This species was also pronounced at the 125 m level at 35°N.

The species assemblage up to a depth of 30 m was similar to that in the surface water along traverses 1, 2 and 4. At the 50 m level of traverse 4 more

Umbellosphaera tenuis appeared at the northern part of the Transitional Zone and replaced *Rhadosphaera clavigera*. At the Equatorial Zones, the vertical cell density of *Gephyrocapsa*, increased in relation to the total coccolithophorid population; *Geophyrocapsa ericsonii* and its variety, in particular, occupied as much as 50% at 10°N (Fig. 5).

There is a sizeable discrepancy between the species assemblage at 100 m and that of the surface water. In the Central Zone *Umbellosphaera irregularis* was replaced by *Umbellosphaera tenuis*. *Florisphaera profunda* (Okada and Honjo, 1973) first appeared in abundance at 10°N. The conspicuous difference in flora in the deep photic water at 10°N appears to be related to the strong upwelling of cold water in this area.

The species assemblage of coccolithophorids was almost entirely different from the surface assemblage at a depth of 125 m to 150 m, except in the Subarctic Zone. Two species of deep photic water coccolithophorids, *Florissphaera profunda* and *Thorosphaera flabellata* were outnumbered by other species throughout traverse 4. At the 200 m level, these two species were more dominant than at the 150 m level and almost monopolised the population in many stations. The high diversity of coccolithophorid species was maintained only in the neighbourhood of the Equator.

The diversity of coccolithophorid species in the Pacific

Using diversity index H′ (Shannon and Weaver, 1963)* as modified by Basharin (Basharin, 1959, cited from Pielou, 1966) we determined the nannoplankton species population of the Pacific to be "Type B" of Pielou (1966) during the months August to October.

DIVERSITY IN SURFACE WATER

Diversity of coccolithophorid species in the surface waters of the Subarctic Zone was low but increased rapidly southward (Fig. 6). The maximum diversity in the surface water, 2.4. nat. bel, was observed at 33°N, 155°W. The diversity

* The diversity index H′ and the percent similarity index *S* were computed by the following formula:

$$H' \text{ (nat. bel)} = -\sum_{i=1}^{S} Pi \ln Pi - (S-1)/2N$$

where *Pi* is the proportion of the *i*th species in the population and S is the number of species.

$$S = 100 \times (1.0 - 0.5 \sum_{i=1}^{n} [1Pij - Pik1])$$

where *Pij* and *Pik* are the proportion of the *i*th species in the *j*th and *k*th samples and n is the number of species.

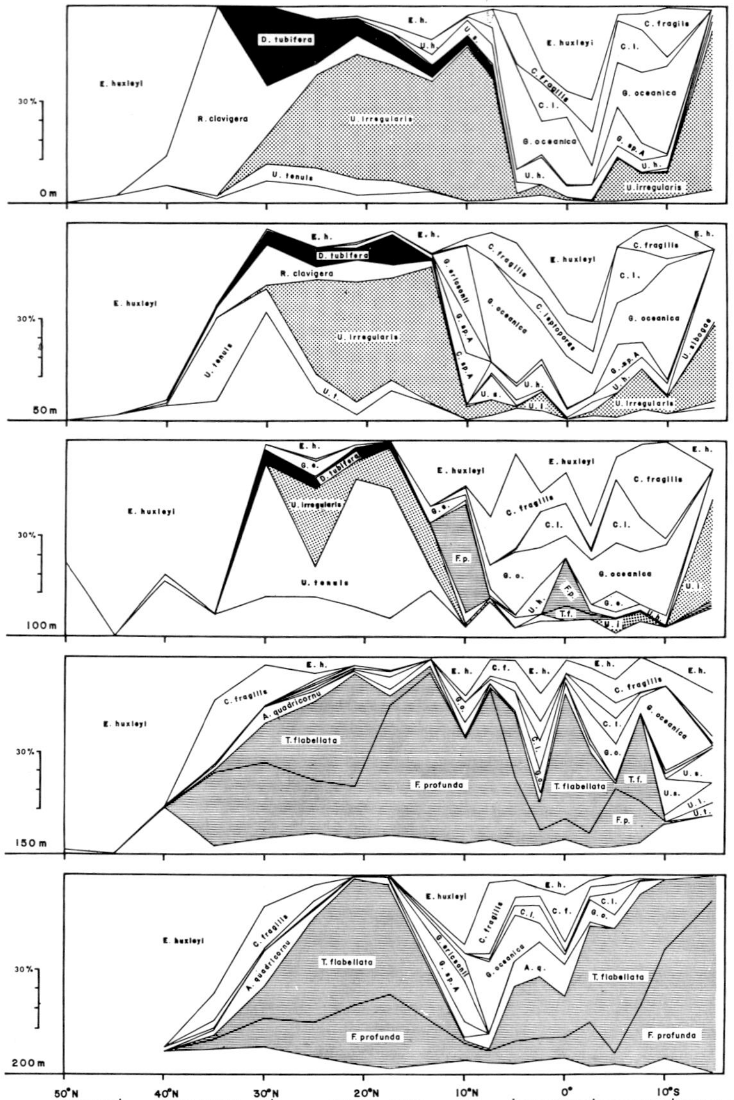

FIG. 5. Species assemblage of coccolithophorids at the four subsurface levels from 50 m to 200 m during the late summer, along the 155°W transect (Okada and Honjo, 1973; Honjo and Okada, 1974).

decreased at the northern edge of the Central Zone and maintained 1.0 to 1.6 nat. bel whereas it increased significantly at the boundary between the Central and Equatorial Zones of each traverse. In the Equatorial Zone diversity was usually more than 1.5 nat. bel. The second highest H′ value at any surface station (1.9 nat. bel) was recorded at 2°N, 155°W. Diversity of species in surface water was low near the equator, particularly in traverses 1 and 2 of the eastern equatorial Pacific. Along traverses 4 and 5 in the Equatorial South Zone a high species diversity was observed which decreased southward. It was noted that species diversity tended to decrease at the boundaries of coccolithophorid zones.

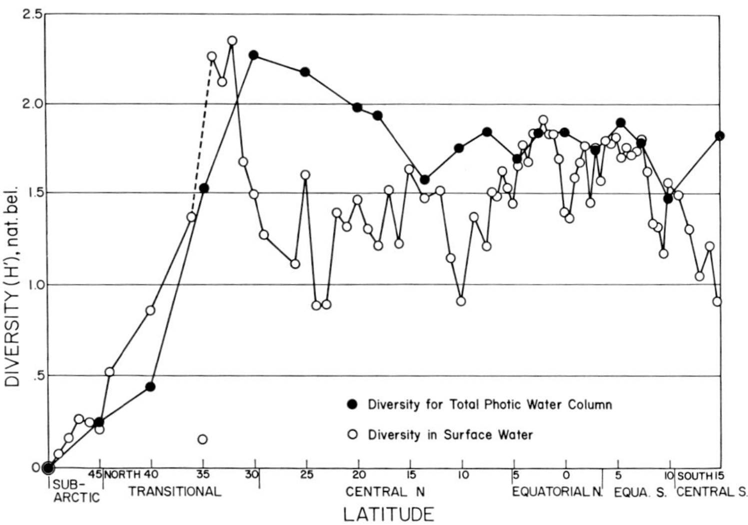

FIG. 6. The latitudinal change (155°W) of species diversity (H′) for total water column compared to the diversity in surface water community (Honjo and Okada, 1974).

DIVERSITY IN SUBSURFACE WATER

Utilising a linear interpolation computer program, contoured profiles were drawn for species diversity (H′)[3] in subsurface water (Fig. 7; Honjo *et al.*, 1973). The contouring of species diversity revealed the presence of a conspicuous community structure in the photic water profile. In particular, we observed the following: (1) Closely spaced, vertically elongated, contour lines were developed from 50°N to 30°N showing a rapid simultaneous increase of diversity at all depths towards the south, from a very low value in subarctic water to the highest value in the subsurface water of the Transitional Zone. (2) The subsurface of the Central Zone in general had a low H′ value and the diversity contours showed a horizontal pattern. The subsurface water of the Equatorial Zone was homogenous in terms of species diversity distribution, though a weak horizontal structure was observed between 150 m and 175 m, 0° and 15°S.

Subsurface samples at 30°N contained a high diversity of species, the highest species diversity (2.7 nat. bel) was recorded at the 50 m level. The high species diversity is attributed to a mixture of Transitional and Central Zone floras at the boundary between these zones. The zone of mixing is very narrow and

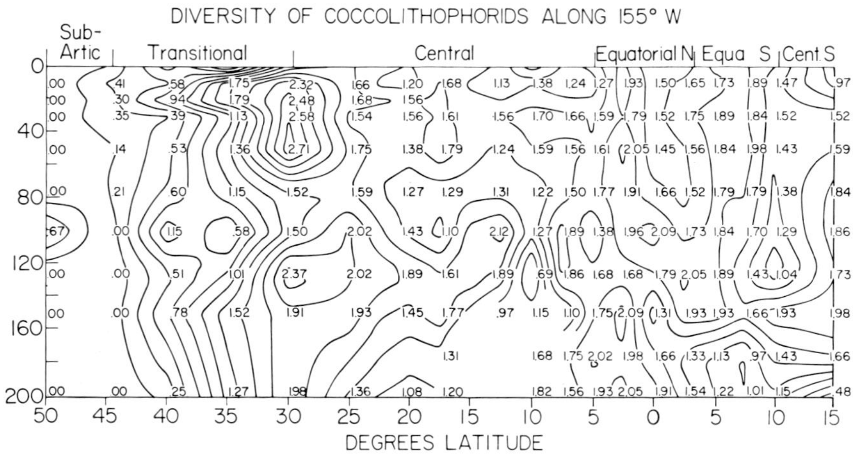

FIG. 7. The diversity contours (in nat. bels) of coccolithophorid species along the 155°W transect. The contours were produced by a linear interpolation program.

probably metastable. A similarly abrupt increase in species diversity has not been recorded for the boundary between the Central and Equatorial Zones.

Coccolithophorid provincialism in the photic zone

A quantitative value for the extent of floral isolation between floral zones was obtained by comparing species assemblages from the same depth at different localities (horizontal comparison; example shown in Fig. 8) with those found at various depths in each substation (vertical comparison). By computing the percentage similarity index (Whittaker and Fairbanks, 1958), see footnote on p. 961, values of indices (percents) were obtained and plotted by station and depth for visual comparison of "similarity curves" (Honjo and Okada, 1974). Analysis of curve characteristics and other statistical models such as species assemblage (Fig. 5) or diversity contours (Fig. 7) enabled us to observe the development of floral provinces (Fig. 9).

For example, surface water samples from stations 1, 2 and 3 were strikingly similar (Fig. 8, left row). The similarity between stations 1 and 2 was 96%; 76.5% for stations 2 and 3; and 76% between stations 1 and 3. This index however, decreases sharply to 0.5% between stations 1 and 4 and 15% between stations 3 and 4. The curves for stations 1, 2 and 3 are quite similar and show a similarity of more than 50% to the station 15 curve but very little

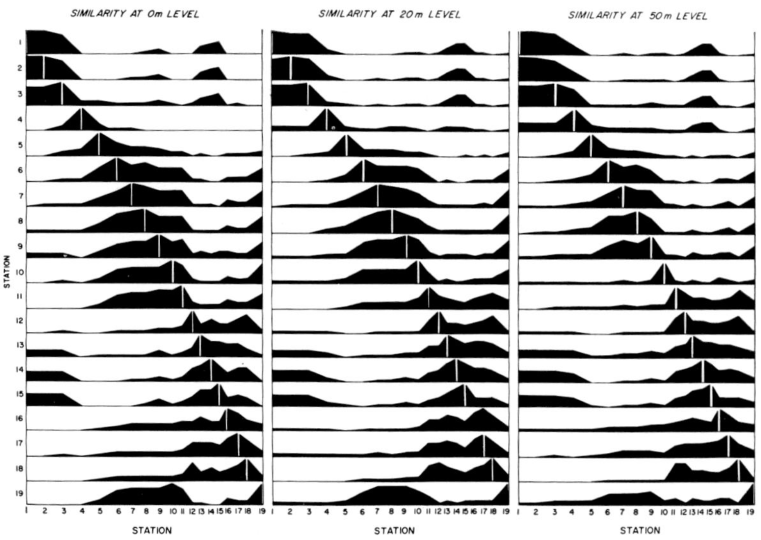

Fig. 8. Examples of similarity curves, horizontal comparison between the 19 stations along the 155°W meridian at the surface, 20 m and 50 m levels.

similarity to stations through 7 as well as stations 10, 11 and 12. The community at station 4 shows distinct independence from communities at other stations. Stations 5 to 11 are similar and show a marked similarity to station 19; station 10, in particular, had an 86.5% similarity to station 19. Station 19 would in terms of its species composition appear to be the Southern Hemisphere version of the Central Zone. Stations 12–15 were quite similar in composition but were not similar to other stations. Stations 14 and 15 were especially similar but did not resemble stations 16, 17 and 18 at all. Thus, the coccolithophorid community in the surface water can be divided into five groups: (1) stations 1, 2 and 3; (2) station 4; (3) stations 5–11 and station 19; (4) stations 12–15; and (5) stations 16, 17 and 18. Similar analyses made at eight other levels resulted in 19 vertical comparisons and in the recognition of floral provincialism (Honjo and Okada, 1974).

Surface assemblage vs. total assemblage

The hypothetical species assemblage of coccolithophorids in the photic water column (of 10 cm × 10 cm × 200 m, from the surface to 200 m) was computed by totalling the number of observed species per litre obtained from each substation. The values between stations were obtained by the averaging method, based upon the cell number curves.

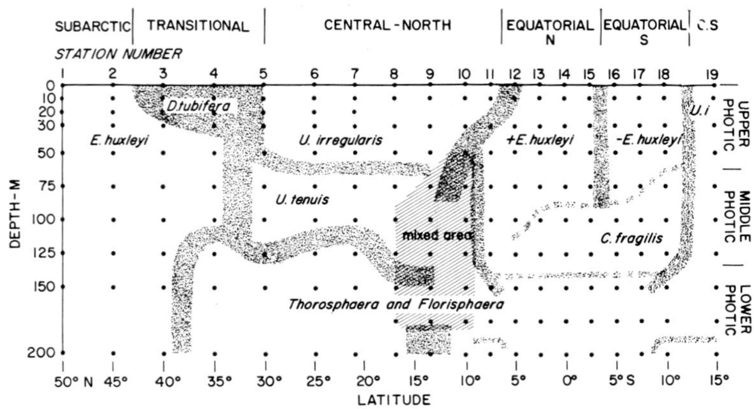

FIG. 9. Subsurface provinces of coccolithophorids developed in the photic water column during the late summer along 155°W meridian based upon available community values (Honjo and Okada, 1974).

In the Subarctic Zone and the northern portion of the Transitional Zone the surface water assemblage is generally representative of the total photic layer assemblage. On the southern edge of the Transitional Zone the total assemblage becomes significantly different from that of the surface water, this trend continues into the Central Zone. In the Central Zone the restriction of the maximum cell density distribution, and of *Thorosphaera flabellata* and *Florisphaera profunda*, to the subsurface water mass contributes to the marked discrepancy between the surface assemblage and total assemblage in the photic layer.

In the Equatorial Zone, coccolithophorid species were well mixed throughout the water column. The surface assemblage was consequently similar to the total species assemblage. A large number of *Gephyrocapsa* cells were concentrated at the boundary between the Central-North Zone and the Equatorial -North Zone. This resulted in a significant discrepancy between the total species assemblage and the species assemblage in a narrow zone between 5°N and 10°N.

The relationship between species diversity, cell density, hydrographic parameters and nutrient distribution

It is difficult to determine the factors which maintain the shape of community patterns of coccolithophorids. As was discussed before, the zonal, latitudinal development of coccolithophorid species can be related partly to the current

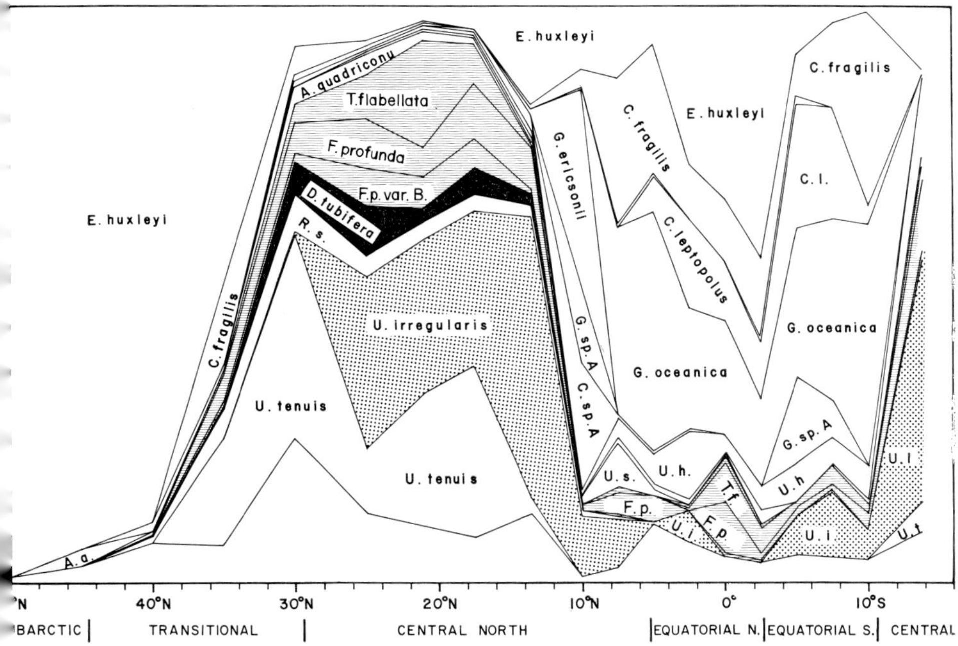

FIG. 10. The hypothetical species assemblage of coccolithophorids in the photic water column (200 m) along the 155°W meridian (Okada and Honjo, 1973).

pattern of the Pacific Ocean. The community pattern and provincialism of coccolithophorids in the meridional photic layer profile of traverse 4 has been compared with hydrographic and nutrient data obtained simultaneously (Marumo, 1970). The contour maps of these data were prepared by computing linear exploration values among the substations; the results were plotted on the photic profile by a line plotter and were visually compared.

SPECIES DIVERSITY, CELL DENSITY VS. HYDROGRAPHIC MEASUREMENTS

The diversity pattern was not well correlated with cell density distribution (Fig. 11). There was no common feature between the water temperature, species diversity and cell density pattern. The area of upwelling cold water at 10°N where the cell density, temperature and diversity values were all decrease is the only exception. The diversity pattern was somewhat similar to the salinity distribution pattern of stratified, low values in the Central Zone and homogeneous values in the Equatorial Zones. The oxgyen value distribution (O_2 ml/l and O_2—saturation percentage) was not related to the diversity and cell density pattern.

DIVERSITY, CELL DENSITY VS. NUTRIENT DISTRIBUTION

The distribution of NO_3 (as N μg/l) had a distinctively low value area (less than 2 μg/l) which coincided with the vertical profile of the Central Zone.

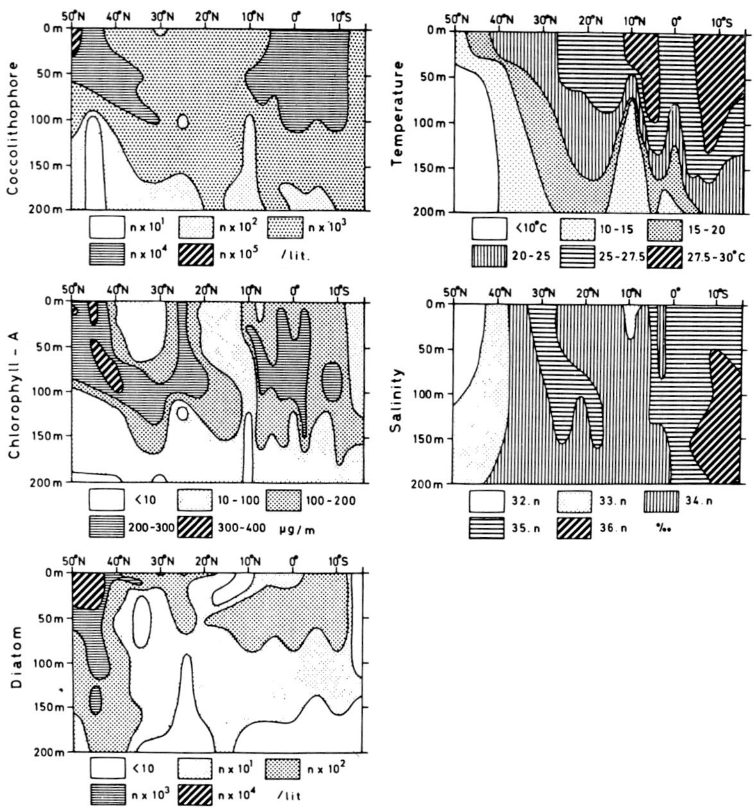

FIG. 11. The isotherms, salinity, chlorophyll-A distribution along the 155°W meridian were compared with the coccolithophorid cell density obtained simultaneously (late summer, 1969, R/V *Hakuho Maru*, KH-69, Ocean Research Institution, 1970). The diatom density profile was cited by permission from Dr. O. Asaoka, Nagasaki Marine Observatory. The profiles were compiled by H. Okada.

The upwelling area at 10°N brought NO_3 and PO_4 rich water as shallow as 100 m. The diversity and cell density was low in this area despite nutrient enrichment. This may be due to the low temperature which offset the advantageous conditions produced. The NH_4 (as N in μg/l) and NO_2 (as N in μg/l) by the enrichment in nutrients distribution patterns were not related to the species diversity or cell density pattern.

RELATION TO OTHER PHYTOPLANKTON

The chlorophyll-A density pattern ($\mu g/m^3$) resembled the species diversity pattern in the Transitional, Central and Equatorial Zones. Chlorophyll-A density was however, related to the cell density in the Subarctic Zone. The cell density pattern of diatoms and coccolithophorid appeared not to be related to one another. The shallow layer at 50°N to 45°N was characterised by the large cell density of coccolithophorids (monopolised by *Emiliana huxleyi*), as large as 10^5/L; diatoms were also very rich in this same layer (10^4 cells/L), and the highest densities were recorded along the profile of traverse 4.

PRELIMINARY RESULTS OF MULTIPLE REGRESSION ANALYSIS

A multiple regression analysis was conducted with fourteen variables simultaneously obtained from the same water sample. All the values were compared with each other, disregarding the latitude and depth of the samples. Seventeen most common coccolithophorid species were analysed against these variables. Preliminary results on regression coefficiency are described below.

H′ value. Species diversity (H′ value) appeared to be generally independent of other variables but correlated negatively with the dominance* of *Emiliana huxleyi* (−0.70). There appeared to be no pertinent relationship between the standing crop and the H′ value. The regression between the diversity index of cocolithophorids and salinity showed a moderate correlation (+0.66). The regression coefficient between H′ value and water temperature was −0.57. However, the salinity/temperature regression in this transect was also high (+0.65), so that a reasonable explanation of the relation between these three variables has not yet been provided.

Cell number density. The density of coccolithophorid cells was correlated with total chlorophyll content (+0.55). A weak negative correlation was observed with salinity.

Dominance of species. The dominance of *Emiliania huxleyi* (including all varieties, ecophenotypes) as compared with other variables showed negative correlation with salinity (−0.75) and temperature (−0.63) whereas other

* "Dominance" is the percentage of each species in the total coccolithophorid standing crop.

species (except for the genus *Umbellosphaera*) were independent of these variables. The dominance of *E. huxleyi* correlated positively with the concentration of O_2, SiO_2 and NH_4 while *Umbellosphaera* showed the reverse relationship.

Discussion

The coccolithophorid zonation established for the Atlantic Ocean generally agrees with that from the surface water of the Atlantic Ocean (McIntyre and Bé, 1967, Fig. 17, p. 593). The influence of the Kuroshio Current is analogous to the influence of the Gulf Stream in the Atlantic. The Equatorial Zone of the Atlantic is much broader, pushing the Subtropical (=Central Gyre or Sargasso Sea) northward; the Central Zone of the Pacific ranges from 30°N to 3 or 5°N in the Central Pacific during summer, however in the Central Atlantic it ranges from 40°N to 20°N. This may be caused by the seasonal fluctuation or discrepancy in definition of the zone.

One of the conspicuous differences between the two oceans in terms of coccolithophorid distribution is the scarcity of *Coccolithus pelagicus* (Wallich) in the Pacific Ocean. In the Atlantic Ocean, *C. pelagicus* was found in the north at the 14° isotherm (Transitional and Subarctic Zones; McIntyre and Bé, 1967, Fig. 12, p. 586). Contrarily, this species was not recorded throughout the photic layer of the Pacific, with the exception of a few specimens from 47°N, 155°W (surface water). McIntyre *et al.* (1970) in their coccolithophorids study of Pacific Ocean, described that this species distributes further polar direction such as the Bering Sea. At the 1000 m level, 50°N, 150°W (3.6×10^3/l), the coccolith density was high compared to the samples collected from neighbouring depths. The coccoliths of *Coccolithus pelagicus* were found in these samples. This may indicate that in some previous season *C. pelagicus* was dominant in this area. Assuming the rate of free descent of a coccolith of *C. pelagicus* is a few hundred meters a year, the above mentioned coccoliths could have originated during winter of the previous year.

The seasonal discontinuity of the samples used for this investigation is one of the built-in difficulties in obtaining the year-round community scheme of coccolithophorid. McIntyre and Bé (1967) reported the seasonal fluctuation of coccolith species from the Atlantic (surface subtropical water). The seasonal fluctuation in general may be most dominant in the Subarctic and Transitional Zones, where severe winter mixing will be disturbing. The Central and Equatorial Zones may be a more conservative environment and less influenced by seasonal fluctuation.

Another limitation of the study on the distribution of coccolithophorids is the influence of the aggregated distribution of cells. Evidence indicates that the distribution of diatoms in the open sea may be aggregated (Venrick, 1972).

According to an approximation of species and cell number distribution in the Central North Pacific, it is evident that aggregation does exist, particularly in the surface water of the Transitional Zone. The dimensions of the patch is not yet known; nevertheless we can discern an overall distribution pattern.

Acknowledgements

I thank Dr. Hisatake Okada, Lamont–Doherty Geological Observatory, Columbia University for his information and discussions. I also thank Virginia Goodrich-Mahoney, for her editorial assistance. This work was supported by the Oceanographic Section, National Science Foundation, Grant GA-30996. This paper is a contribution from the Woods Hole Oceanographic Institution, No. 3247.

References

Bramlette, M. N. (1958). Significance of coccolithophorids in calcium carbonate deposition. *Bull. Geol. Soc. Amer.* **69,** 121–126.

Brinton, E. (1962). The distribution of Pacific Euphausiids. *Bull. Scripps Inst. Oceanog., Univ. Calif.* **8,** 51–270.

Christensen, T. (1962). *Algen in Botanik.* **2,** p. 1–178. Kopenhagen.

Gaarder, K. R. (1971). Comments on the distribution of coccolithophorids in the oceans. *In* (Funnell, B. M. and Riedel, W. R., eds), *The Micropaleontology of Oceans*, 97–103. Cambridge University Press.

Honjo, S. and Okada, H. (1974), Community structure of coccolithophorids in the photic layer of the mid-Pacific Ocean. *Micropaleont.* **20,** 209–230.

Honjo, S., Schmidt, D. and Sass, W. (1973). Computer program package for the coccolithophorids and coccoliths distribution study in the ocean. Technical Report, Woods Hole Oceanographic Institution, Series 73. (In press.)

Johnson, M. W. and Brinton, E. (1963). Biological species water masses and currents. *In* (Hill, M. N., ed.), *The Sea*, **2,** 381–414, Interscience, New York and London.

McIntyre, A. and Bé, A. W. H. (1967). Modern coccolithophoridae of the Atlantic Ocean, I, placoliths and cyrtoliths. *Deep-Sea Res.* **14,** 561–597.

McIntyre, A., Bé, A. W. H. and Roche, M. B. (1970). Modern Pacific coccolithophoridae: a paleontological thermometer. *Trans. N.Y. Acad. Sci.* **32,** 720–731.

Marumo, R. (ed.) (1970). *Preliminary report of HAKUHO-MARU, Cruise HK-69-4*, p. 68. Ocean Research Institute, University of Tokyo.

Okada, H. (1973). Modern coccolithophroids in neritic environment of the western Pacific Ocean. *Geol. Soc. Amer.* 1973 Ann. Meeting, Abst. 5(7), 753–754.

Okada, H. and Honjo, S. (1970). Coccolithophoridae distributed in southwest Pacific. *Pacific Geology* **2,** 11–21.

Okada, H. and Honjo, S. (1973). The distribution of oceanic coccoliths in the Pacific *Deep-Sea Res.* **20,** 355–374.

Okada, H., and Honjo, S. (1973). *Distribution of coccolithophorids in the North and Equatorial Pacific Ocean: Quantitative data on samples collected during Leg 30, OSHORO MARU, 1968 and Leg HK69-4, HAKUHO MARU, 1969.* Technical Report WHOI-73-81. Woods Hole Oceanographic Institution, Woods Hole, Massachusetts.

Parke, M. and Dixon, P. S. (1964). A revised check-list of British marine algae. *J. Mar. Biol. Assoc. U.K.* **44,** 499–542.

Pielou, E. C. (1966). The measurement of diversity in different types of biological collections. *J. Theoret. Biol.* **13,** 131–144.

Scripps Institution of Oceanography (1967). *Physical, chemical and biological data, URSA MAJOR Expedition, 4 Aug.-4 Oct., 1964.* S.I.O. Ref. 67–5, 1–43.

Scrips Institution of Oceanography (1970). *Physical, chemical and biological data, ZETES Expedition, Leg 1, 11–24 Jan. 1966*, S.I.O. Ref. 70–5, 1–67.

Shannon, C. E. and Weaver, W. (1963). *The mathematical theory of communication*, p. 125. University of Illinois Press, Urbana.

Tsujita, T. (ed.) (1970). The OSHORO MARU Cruise 30 to the north and tropical Pacific Ocean and the Gulf of Carpentaria, November 1968–January 1969. *Data Rec. oceanogr. Obsns. Explor. Fishg.* **Z4,** 217–313.

Venrick, E. L. (1972). Small-scale distributions of oceanic diatoms. *Fishery Bull.* **70**(2) 363–373.

Whittaker, R. H. and Fairbanks, C. W. (1958) A study of plankton copepod community in the Columbian Basin, Southeastern Washington. *Ecology* **39,** 46–65.

11: Coccolith Biogeography from North Atlantic and Pacific Surface Sediments

A comparison of species distribution and abundances

KURT R. GEITZENAUER,* MICHAEL B. ROCHE† and ANDREW McINTYRE

CLIMAP, Lamont-Doherty Geological Observatory of Columbia University, Palisades, New York 10964, U.S.A.

The distribution and relative abundances of twenty-two species of coccolithophorids in surface sediments of the North Atlantic and Pacific Oceans has been determined. Inter-ocean comparison indicates that ten forms have analogous biogeographic ranges, with regard to surface water masses, in both the North Atlantic and North Pacific. Eleven species have roughly comparable distribution patterns in both Pacific hemispheres.

The degree of correlation between all species and latitude of sample, depth of sample, and summer and winter sea surface temperature was calculated by multiple regression. The strongest correlation exists between species and sea surface temperatures. For this sample set there was generally a low correlation between species abundance and water depth.

Outline

Introduction

Coccoliths are the calcite tests produced by the unicelular marine algae, Coccolithophorida. Coccolithophorida inhabit equatorial through subpolar waters, including some near-shore areas and marginal seas. Earlier papers have presented information on the biogeographic distribution of the Coccolithophorida (McIntyre and Bé, 1967; Gaarder, 1971; Okada and Honjo,

**Present address:* Department of Geology and Oceanography, Herbert H. Lehman College of C.U.N.Y., Bronx, N.Y., U.S.A.

† *Present address:* Jersey Central Power and Light Co., Morristown, New Jersey 07960, U.S.A.

1973; McIntyre *et al.*, 1970), and the coccoliths from surface sediments (McIntyre and Bé, 1967; McIntyre, 1967; McIntyre *et al.*, 1970; McIntyre and McIntyre, 1971). The intention of this paper is to: (1) present contoured relative abundance data for numerically important species from surface sediments in the North Atlantic and Pacific Oceans, (2) Analyse the data to elucidate the mechanisms responsible for their distribution.

Details concerning the sampling localities are given in the Appendices, pp. 1002–1007.

Methodology

In general, the surface samples examined were taken from the tops of trigger weight cores (gravity) to avoid the possible loss of surface sediments sometimes associated with piston coring. Sample preparation followed the routine established in McIntyre *et al.* (1967), with the omission of ultrasonic agitation. Instead, a laboratory shaker was used to disaggregate the samples. Approximately 350 individual coccoliths were counted per sample, using a transmission electron microscope.

Surface sediment samples

In the North Atlantic, 136 surface samples were examined for coccoliths (Appendix 1). Of these, eighteen were eliminated, nine being barren of coccoliths (Appendix 2), and nine others because of severe calcite dissolution (Appendix 3). The apparent lack of coccoliths in a sample may be caused by all or some combinations of the following factors: no coccolith production in the surface waters; calcite dissolution; dilution by large quantities of terrigenous detritus. The sample depth of the 118 samples used ranged from 187 to 6400 m, with a mean of 3231 m. The samples excluded because of obvious calcite dissolution have depths ranging from 3005 to 5159 m, with a mean of 4126 m.

Sixty-four samples from the Pacific were initially selected for examination (Appendices 4 and 5). Thirteen were eliminated because of significant calcite dissolution (Appendix 5). The remaining samples range in depth from 617 to 3867 m, with a mean of 2865 m. The dissolution samples range from 2033 to 3660 m, with a mean of 3099 m.

Loss of coccolith species with depth

Since the Coccolithophorida are one of the main constituents of the open ocean phytoplankton, their shells should, and often do, form a large percentage of the calcium carbonate found in deep sea sediments. Comparison of species abundances from studies of the plankton and surface sediments has revealed a major loss of species and a shift in relative abundance due to calcium carbonate solution during the process of deposition (McIntyre and McIntyre,

1971; Schneiderman, 1973 and this volume, p. 1009; Berger, 1973). Calcite dissolution may occur in the water column during post-mortem descent, while the shells are in the sediment and bottom-water interface, or through post-burial pore water exchange. Basic questions about the role of faecal pellets, herbivore gut conditions, and the oxygen minimum layer have yet to be answered.

New data (H. Okada, personal communication) on the relative abundance of free coccoliths found in Pacific water samples between 400 m and the bottom yield the following results when compared to abundance data from underlying surface sediments assemblages: (1) From 19 to 79% (mean of 47%) of the coccolith species found in the upper 200 m of the water column are generally missing at 400 m, and (2) only 4 to 6% of the species are lost subsequent to deposition. Thus, of the mean percentage of species lost between production and the coring operation (49%), roughly 90% are lost in the water column, and most of these above the 400 m level. (A few relatively unimportant species occur in higher concentrations and apparently live below 400 m.) More detailed analysis of this data and an exposition of species abundance modification will be presented in a future paper.

Data analysis

Analysis of the distribution of the coccolith species will be presented on two levels: interocean biogeographical comparison, and interocean statistical comparison. The purpose of the analysis is to: (1) determine and compare the distribution of each of eighteen species in the North Atlantic and Pacific Oceans, (2) relate any differences to causal mechanisms. An implicit assumption in this analysis is the equality of environmental tolerances for each species in the two oceans, an assumption warranted by a prior investigation (McIntyre *et al.*, 1970), and by research, presently underway, in which extensive data sets of living Coccolithophorida from the two oceans are compared (H. Okada, personal communication).

A property of biogeographical distributions which can be used for comparative analysis is the domain of high abundance for each species. Examination of Table 1 reveals eleven species out of the twenty-two studied have significantly different relative abundance maxima at a 95% confidence level. By mapping the species with a "percent of maximum transformation" (Imbrie and Kipp, 1971) interocean comparison is facilitated. The results of this kind of mapping exercise, coupled to a geographic comparison of distributions of each species, is presented below.

DESCRIPTION OF SPECIES' DISTRIBUTIONS

Twenty-two species from North Atlantic and Pacific surface sediments have abundances that reach one percent in any single sample (Table 1). Of the

TABLE 1. Maximum relative abundance of coccolith species.

Species name	North Atlantic maximum %	Pacific maximum %
Syracosphaera pulchra	8.1	5.4
Syracosphaera tuberculata	2.2	0.0
Syracosphaera variabilis	1.6	1.1
Umbellosphaera irregularis	4.0	11.7
Umbellosphaera tenuis	3.0	16.0
Helicopontosphaera kamptneri	17.0 (53.9 D)*	4.4
Calciosolenia sinuosa	6.6	7.7
Discosphaera tubifera	2.5	1.6
Rhabdosphaera clavigera	13.9	8.9
Cyclococcolithina fragilis	5.3	22.3
Cyclococcolithina leptopora var. B	17.7	7.2
Cyclococcolithina leptopora var. C	33.5 (42.8 D)*	43.6
Cyclococcolithina leptopora var. D	6.5	29 1
Coccolithus pelagicus	96.6	65.4
Emiliania huxleyi cold var.	82.6	
Emiliania huxleyi warm var.	75.3	66.6
Gephyrocapsa caribbeanica	57.9 (76.2 D)*	66.3
Gephyrocapsa ericsonii	16.0	6.2
Gephyrocapsa oceanica	43.2	47.6
Umbilicosphaera hulburtiana	2.2	3.2
Umbilicosphaera mirabilis cold var.	8.9	6.0
Umbilicosphaera mirabilis warm var.	25.0	32.4

* D indicates highest abundance found in samples affected by dissolution.

twenty-two species, abundance distribution maps of nineteen species are presented (Figs 1–18). The remaining three species, *Syracosphaera tuberculata* Kamptner, *Pontosphaera variabilis* Halldal and Markali, and *Gephyrocapsa ericsonii* McIntyre and Bé, all have very low abundances and do not show coherent patterns. Possibly, larger population counts per sample would enable interpretation of these species.

Many of the coccolith species have distributional ranges which extend throughout the entire North Atlantic and Pacific basins (subpolar through equatorial), with peak abundances occurring in much smaller domains. These species distributions will be presented first, and the smaller group of coccoliths, having more restricted ranges, will be discussed second.

Basin-wide species

Emiliania huxleyi (Lohmann) var. warm and cold. (Figs 1a–b)

North Atlantic. This species is found in most sediments of the North Atlantic and Pacific Oceans and is generally thought to be the most abundant modern coccolith. The species has two identifiable forms (McIntyre *et al.*, 1970) which have overlapping sea surface temperature response curves, but with discrete temperature-abundance optima. However, since morphologic intermediates do exist, and the architectural distinctiveness of the two forms is sometimes eliminated by calcite dissolution, definitive exposition of the forms' distribution and environmental relationships is not possible from these data. Generally, the warm form is found in the Gulf of Mexico, Caribbean, western Equatorial Atlantic and, in low numbers, as far north as 43°N. The cold form has high abundances north of 30°N, up to the edges of the Labrador and Norwegian Seas, and low abundances into the Equatorial Atlantic.

Pacific. Identification of the two forms of *E. huxleyi* was prevented by partial calcite dissolution. The species is found throughout the basin with abundance peaks in the central regions of the western South Pacific and North Pacific.

Gephyrocapsa caribbeanica Boudreaux and Hay. (Figs 2a–b)

North Atlantic. This species has a wide distributional range, slightly less extensive than *E. huxleyi*. It is found in low numbers in the Equatorial Atlantic Caribbean, Gulf of Mexico, throughout the North Atlantic up to the Labrador and Norwegian Seas. High abundances of this species occur in the northeast North Atlantic.

Pacific. This species is widely distributed in the Pacific. Peak abundances occur in the central regions of the western North Pacific and eastern South Pacific. It is found in very low numbers in the equatorial region.

Cyclococcolithina leptopora (Murray and Blackman) var. C. (Figs 3a–b)

North Atlantic. This species is distributed throughout the North Atlantic basin, Caribbean, and Gulf of Mexico. High abundances are found in the southern Equatorial Atlantic, and in an area from the Saragasso Sea up to the northeast North Atlantic.

Pacific. This species, var. C, is found basin-wide in low numbers; the highest abundances occur just south of the equatorial region in the eastern South Pacific and west to northeast in the central North Pacific.

Cyclococcolithina leptopora (Murray and Blackman) var. B and var. D. (Figs 4a–b, 5a–b)

North Atlantic. Var. B has a basin-wide range similar to *C. leptopora* var. C, with high abundances found in the Equatorial Atlantic and eastern Saragasso Sea. Var. D has a basin-wide range with no real trends. Its peak abundance occurs in the eastern equatorial region.

Pacific. Var. B is found in samples over much of the Pacific, but with peak abundances in the eastern equatorial region. Var. D is distributed across the south Equatorial Pacific and in the central region of the eastern South Pacific, with peak abundances in the latter region.

Syracosphaera pulchra Lohmann. (Figs 6a–b)

North Atlantic. This species is found in low abundances in the entire basin, with peak values beneath the southern Saragasso Sea.

Pacific. This species occurs in the central regions of the North and South Pacific, with abundance peaks in the western North Pacific and eastern South Pacific. It is not found in significant numbers in the Equatorial Pacific.

Helicopontosphaera kamptneri Hay and Mohler. (Figs 7a–b)

North Atlantic. This species' distribution extends across the Equatorial Atlantic where its peak abundances are found, into the Caribbean, the southern Sargasso Sea and, in very low abundances, in the northern central North Atlantic.

Pacific. This species occurs in most Pacific sediment samples, with peak abundances in the equatorial region.

Calciosolenia sinuosa Schlauder. (Figs 8a–b)

North Atlantic. This species is found throughout the basin, with highest abundances in the northern Equatorial Atlantic, Caribbean, and Saragasso Sea.

Pacific. This species occurs in much of the Pacific from the equator to the borders of the subpolar regions. High values are concentrated in the western equatorial and western central North Pacific regions.

Umbilicosphaera mirabilis (Lohmann) var. cold and warm. (Figs 9a–b, 10a–b)

North Atlantic. This species is found throughout much of the North Atlantic and in the adjacent basins. The "cold" variety has peak occurrences in the Saragasso Sea. The "warm" variety has the highest abundances in the Equatorial Atlantic, southeastern North Atlantic, and Caribbean Sea.

Pacific. This species is widely distributed with "cold" variety abundance peaks in the equatorial and western central North Pacific. The "warm" variety shows abundance peaks in the southern part of the central North Pacific and in the central South Pacific.

Geographically restricted species

Gephyrocapsa oceanica Kamptner. (Figs 11a–b)

North Atlantic. This species is found in the Equatorial Atlantic and the Caribbean and Saragasso Seas, with high values associated almost exclusively with the equatorial region. Extremely low abundances do occur in the central North Atlantic, particularly the northwest region.

Pacific. This species occurs across the Equatorial Pacific in relatively high abundances, with peak values in the eastern equatorial region. Lower values occur along the northwestern margin north of the equator and south of 40°N.

Rhabdosphaera clavigera (Murray and Blackman) Kamptner. (Figs 12a–b)

North Atlantic. This species is not found in significant abundances north of 45°N; its range extends from the central Equatorial Atlantic into the Caribbean, and across the entire North Atlantic south of 50°N. Peak abundances are found in the southern Saragasso Sea.

Pacific. This species is found in low abundances throughout most of the Equatorial Pacific. It reaches its greatest abundances in the central areas of the North Pacific and eastern South Pacific.

Umbellosphaera irregularis Paasche. (Figs 13a–b)

North Atlantic. This species occurs in sediments from the Equatorial Atlantic, Caribbean, and the central North Atlantic up to 51°N, with highest values from the Saragasso Sea west to the offshore southeastern United States.

Pacific. This species is found essentially west of 140°W and equatorward of 30° latitude in both hemispheres. Peak abundances are found in the western part of the North and South Central Pacific regions.

Umbellosphaera tenuis (Kamptner) Paasche. (Figs 14a–b)

North Atlantic. This species is found in sediments of the Equatorial Atlantic Caribbean Sea, Gulf of Mexico, and throughout the central North Atlantic up to 51°N. It has somewhat higher abundances than *U. irregularis*, but a similar range. Its highest sediment concentrations are found in the southern Saragasso Sea.

Pacific. This species occurs in sediments of the central regions of both North and South Pacific, with peak abundances in the western North Pacific and in samples near the Hawaiian Islands.

Discosphaera tubifera (Murray and Blackman). (Figs 15a–b)

North Atlantic. This species, having low abundances throughout its range, is found in the western Equatorial Atlantic, Caribbean Sea, and in the central North Atlantic up to 60°N. Peak abundances occur in the region east of the greater Antilles.

Pacific. This species occurred in only three samples with an abundance greater than 1%. The highest values are close to the Hawaiian Islands and in the western central region of the South Pacific.

Umbilicosphaera hulburtiana Gaarder. (Figs 16a–b)

North Atlantic. This species is found in low abundances in the Equatorial Atlantic, Caribbean Sea, Gulf of Mexico, and central North Atlantic up to

53°N. Peak abundances are found in the Caribbean, the Gulf, and in a semicircular band following the northern half of the North Atlantic gyre system.

Pacific. This species is distributed widely and occurred in most samples. Peak abundances are found in the central regions of the North and South Pacific.

Coccolithus pelagicus (Wallich) Schiller. (Figs 17a–b)

North Atlantic. This species' distribution is restricted to the southern Labrador Sea, across the entire southern border of the Norwegian Sea, up to the northern coast of Norway. The abundance gradient decreases abruptly both north (where samples are barren of coccoliths), and south of this band of very high abundances. Two samples south of this region contain significant abundances—one due to the effects of calcite dissolution, and one with a sample depth of 454 m, close to South America, which appears to have reworked older forms.

Pacific. This species is found only in the subarctic region. The peak abundance occurs in the northwest North Pacific.

Cyclococcolithus fragilis (Lohmann). (Figs 18a–b)

North Atlantic. This species, ranging from the equatorial to central North Atlantic, was found in five samples with significant abundances. The maximum value (one sample) was found in the northwestern North Atlantic.

Pacific. This species was found from the western equatorial to the central South Pacific region, with peak abundances occurring in the former area.

INTEROCEAN BIOGEOGRAPHICAL COMPARISON

Comparison of geographic ranges

Of the eighteen species mapped, nine species, *S. pulchra*, *C. sinuosa*, *R. clavigera*, *C. pelagicus*, *U. mirabilis* (warm and cold varieties), *G. oceanica*, *U. irregularis*, *U. tenuis*, and *D. tubifera* show similar biogeographic ranges in the North Atlantic and North Pacific. *C. sinuosa*, *G. caribbeanica*, *G. oceanica*, *S. pulchra*, *H. kamptneri*, *U. mirabilis* var. warm, *R. clavigera*, *U. irregularis*, *U. tenuis*, *C. fragilis*, and *U. hulburtiana* have roughly comparable distribution patterns in both Pacific hemispheres. Four forms with dissimilar ranges between the North Atlantic and North Pacific, have a more restricted distribution in the North Pacific (*H. kamptneri*, *C. fragilis*, *C. leptopora* var. B and D).

Sample-set dissimilarity however, is such that a definite difference can be established for only one of these five forms: *C. leptopora* var. D is found in samples up to 60°N in the North Atlantic, while in the North Pacific, it is not found in two samples at 48°N, but is present in samples south of 34°N.

Six species, *E. huxleyi*, *C. leptopora* varieties C, B and D, *D. tubifera*, *C. mirabilis* var. cold and *C. pelagicus*, have obviously dissimilar ranges between

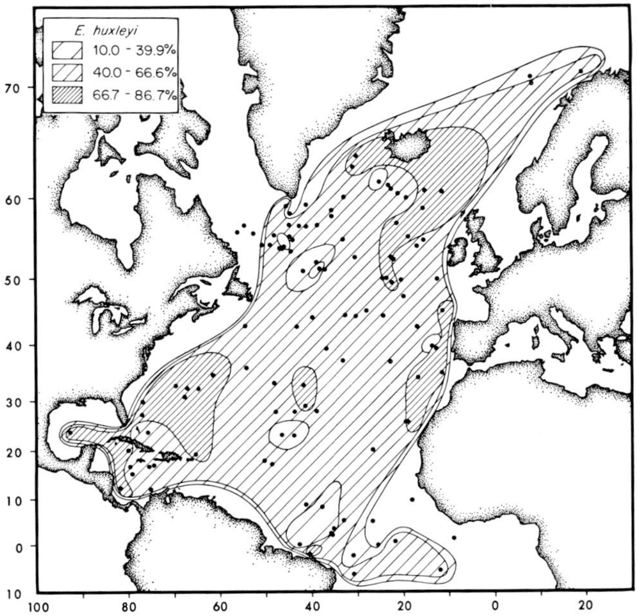

FIG. 1a. Relative abundance of *Emiliania huxleyi* in North Atlantic surface sediments.

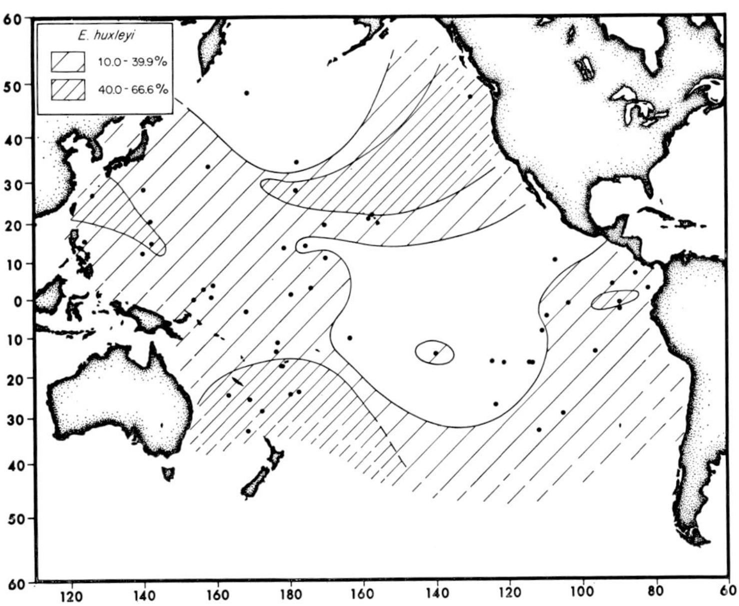

FIG. 1b. Relative abundance of *Emiliania huxleyi* in Pacific surface sediments.

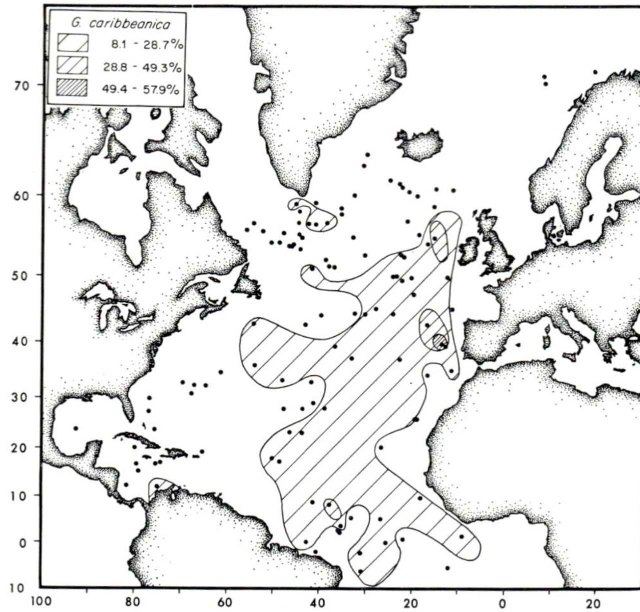

FIG. 2a. Relative abundance of *Gephyrocapsa caribbeanica* in North Atlantic surface sediments.

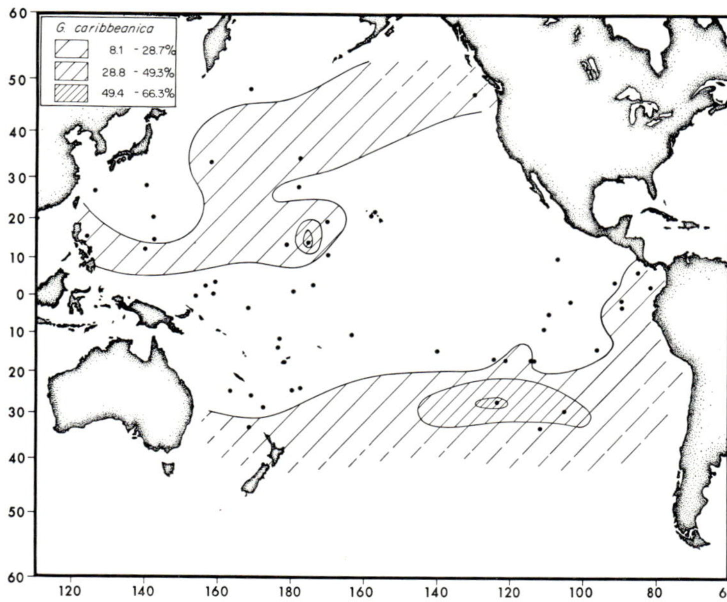

FIG. 2b. Relative abundance of *Gephyrocapsa caribbeanica* in Pacific surface sediments.

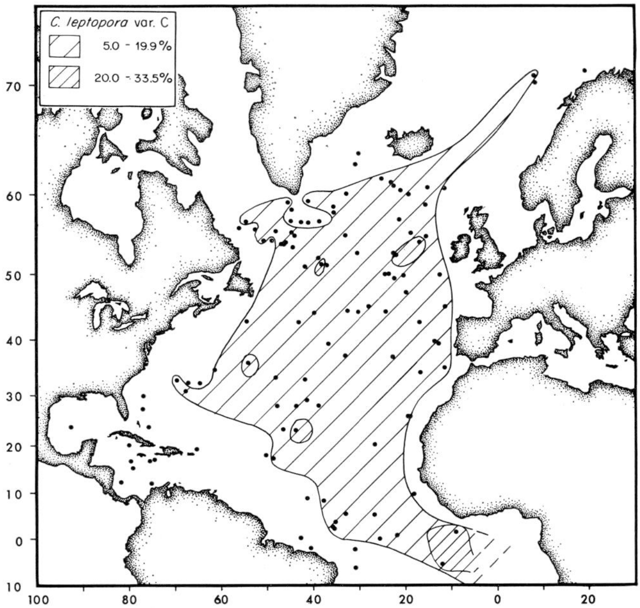

FIG. 3a. Relative abundance of *Cyclococcolithina leptopora* var. C in North Atlantic surface sediments.

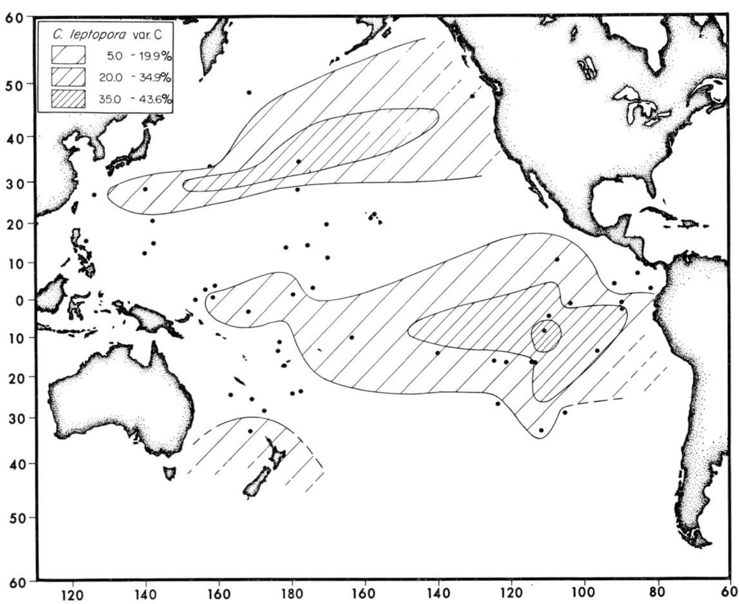

FIG. 3b. Relative abundance of *Cyclococcolithina leptopora* var. C in Pacific surface sediments.

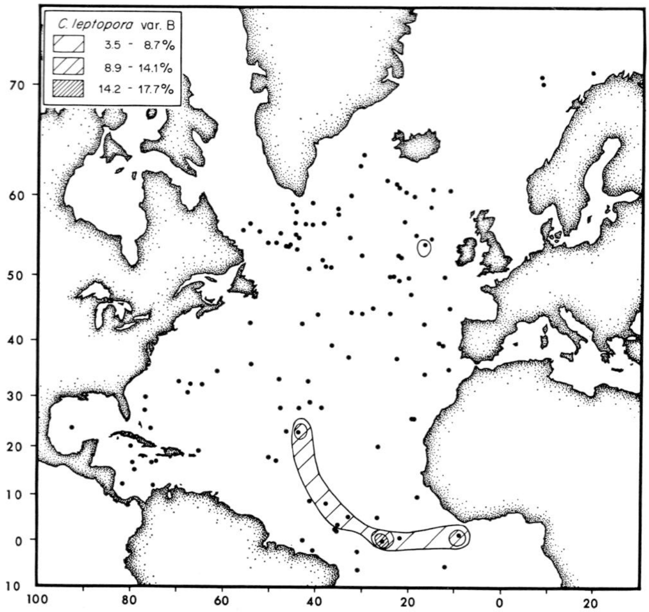

FIG. 4a. Relative abundance of *Cyclococcolithina leptopora* var. B in North Atlantic surface sediments.

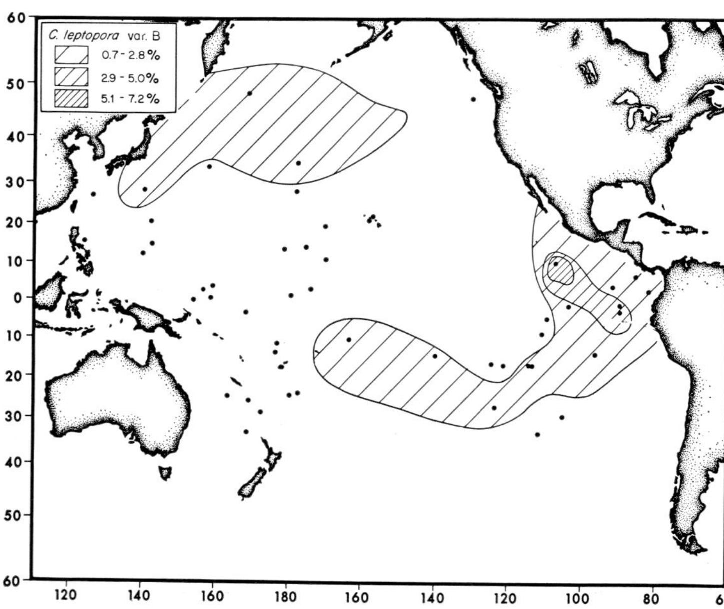

FIG. 4b. Relative abundance of *Cyclococcolithina leptopora* var. B. in Pacific surface sediments.

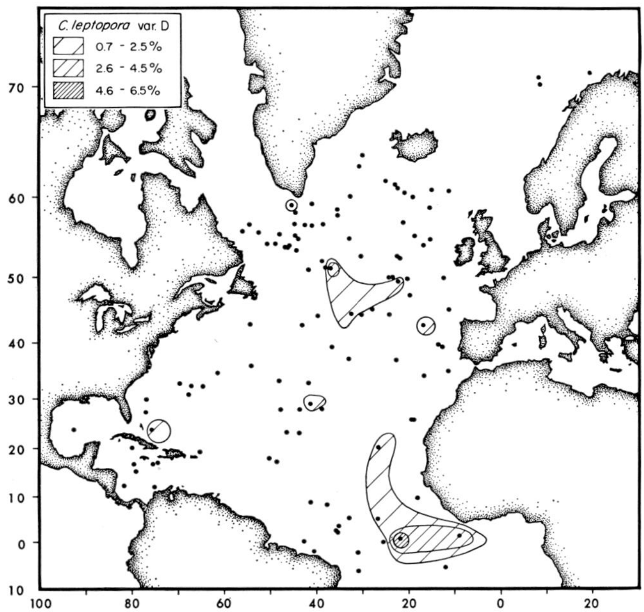

FIG. 5a. Relative abundance of *Cyclococcolithina leptopora* var. D. in North Atlantic surface sediments.

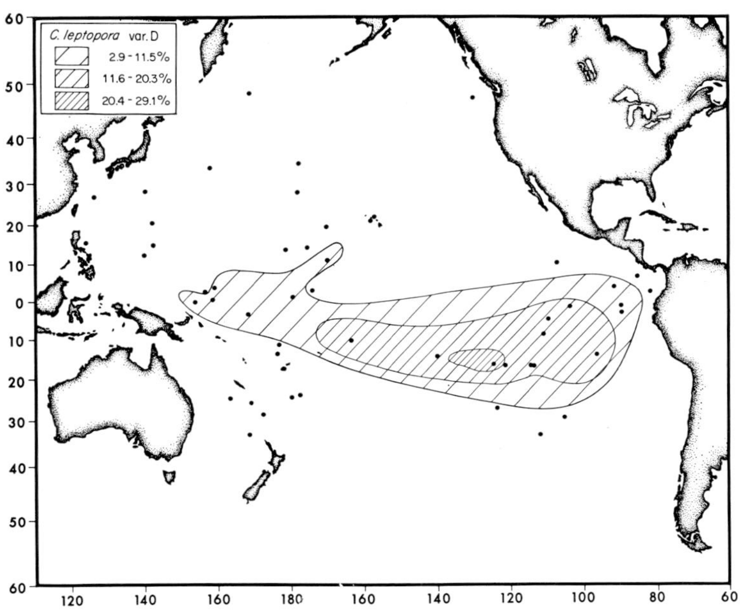

FIG. 5b. Relative abundance of *Cyclococcolithina leptopora* var. D. in Pacific surface sediments.

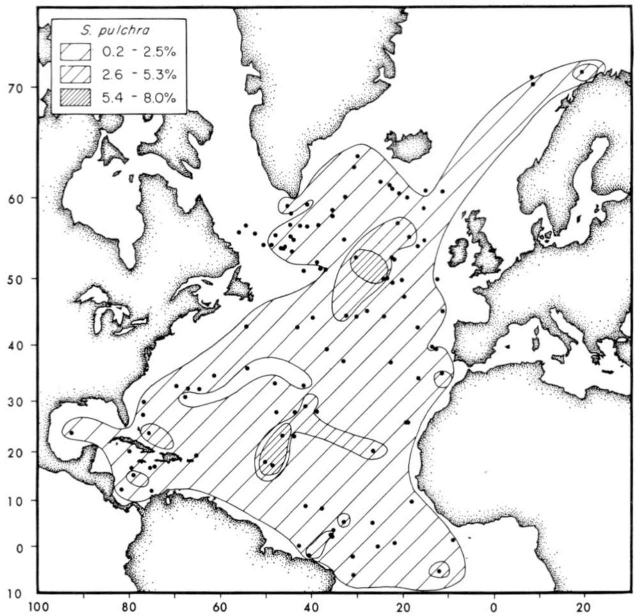

FIG. 6a. Relative abundance of *Syracosphaera pulchra* in North Atlantic surface sediments.

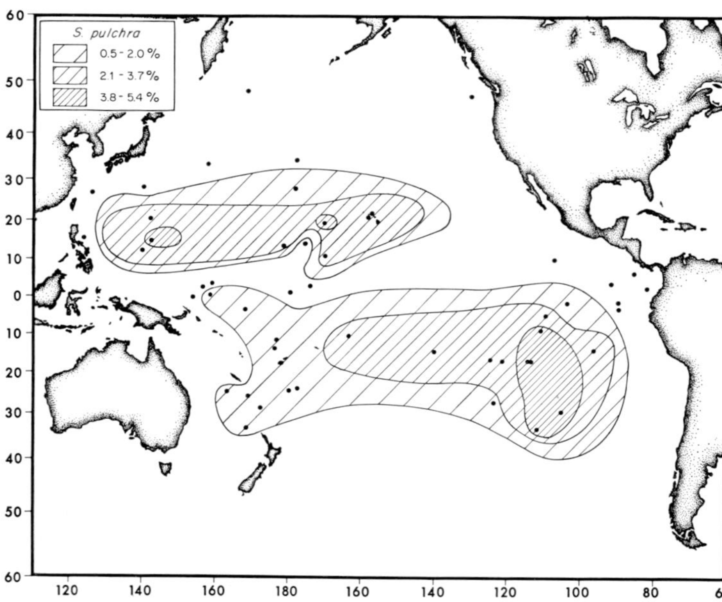

FIG. 6b. Relative abundance of *Syracosphaera pulchra* in Pacific surface sediments.

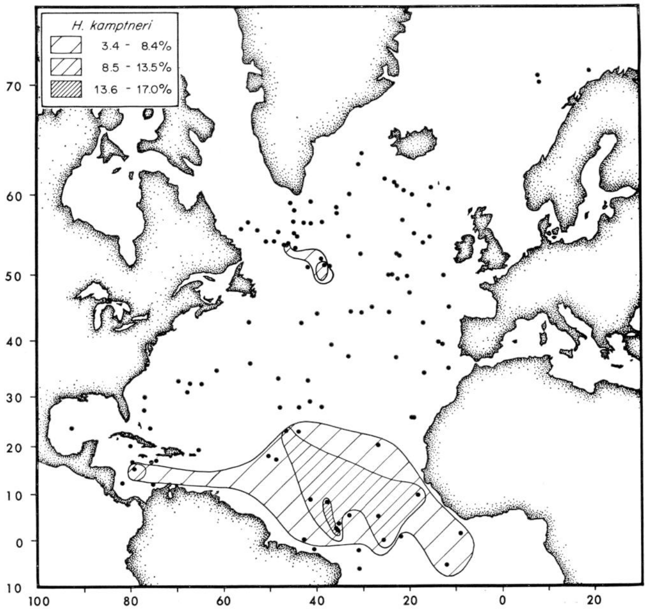

FIG. 7a. Relative abundance of *Helicopontosphaera kamptneri* in North Atlantic surface sediments.

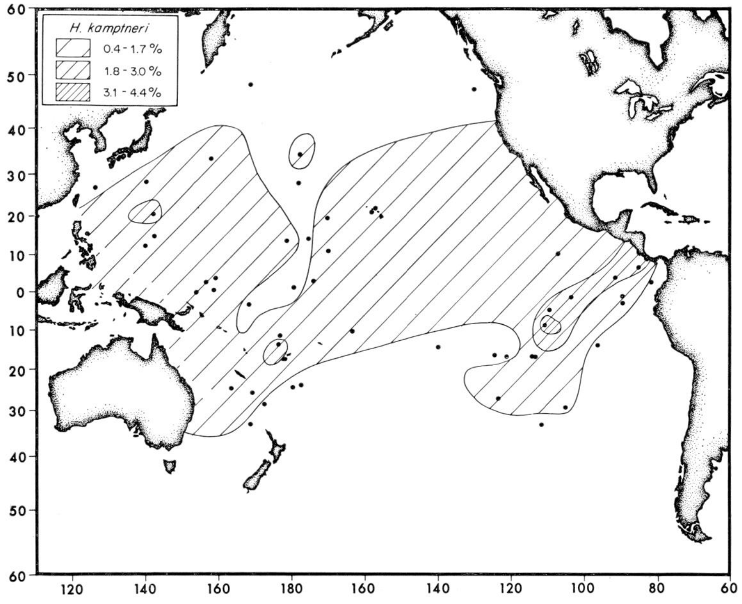

FIG. 7b. Relative abundance of *Helicopontosphaera kamptneri* in Pacific surface sediments.

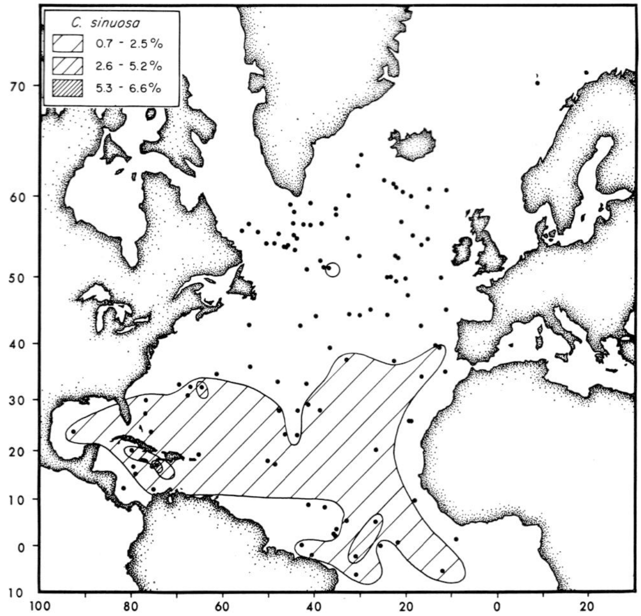

FIG. 8a. Relative abundance of *Calciosolenia sinuosa* in North Atlantic surface sediments.

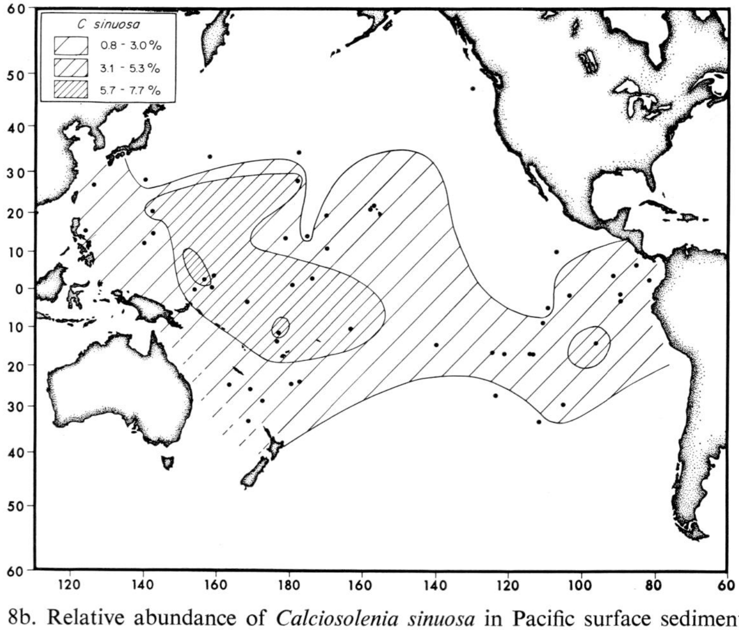

FIG. 8b. Relative abundance of *Calciosolenia sinuosa* in Pacific surface sediments.

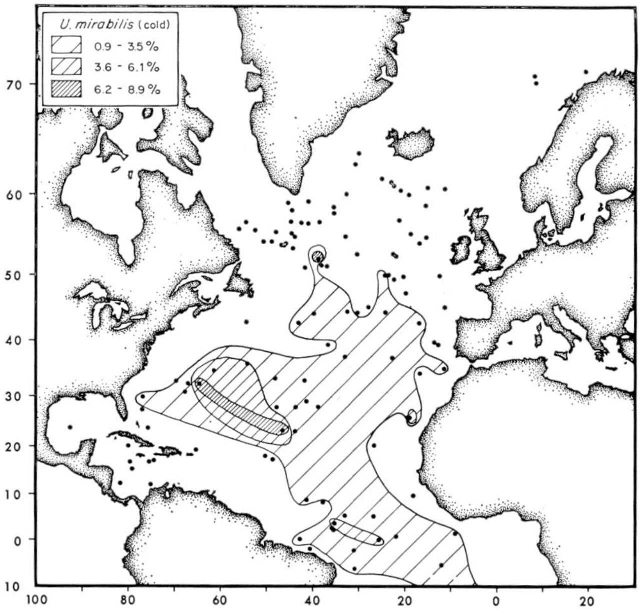

FIG. 9a. Relative abundance of *Umbilicosphaera mirabilis* (cold) in North Atlantic surface sediments.

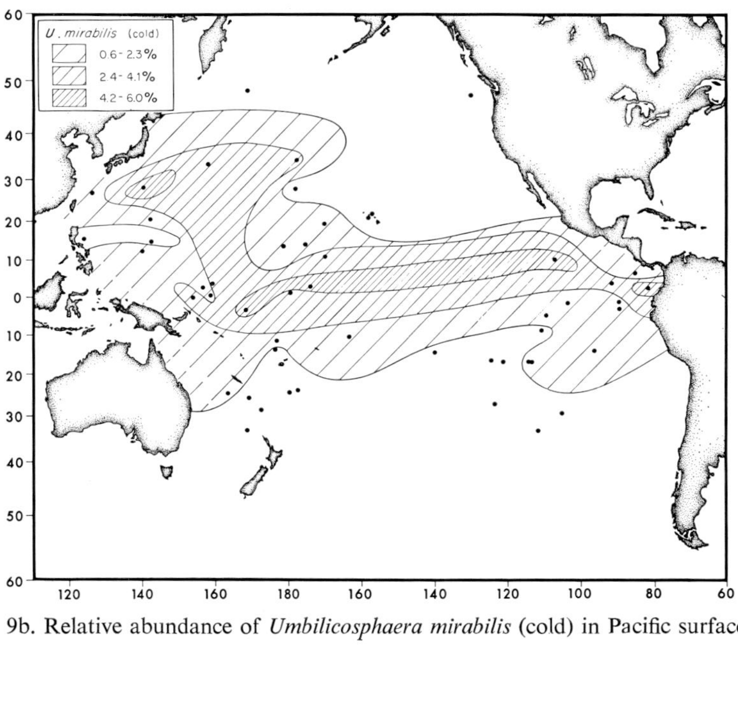

FIG. 9b. Relative abundance of *Umbilicosphaera mirabilis* (cold) in Pacific surface sediments.

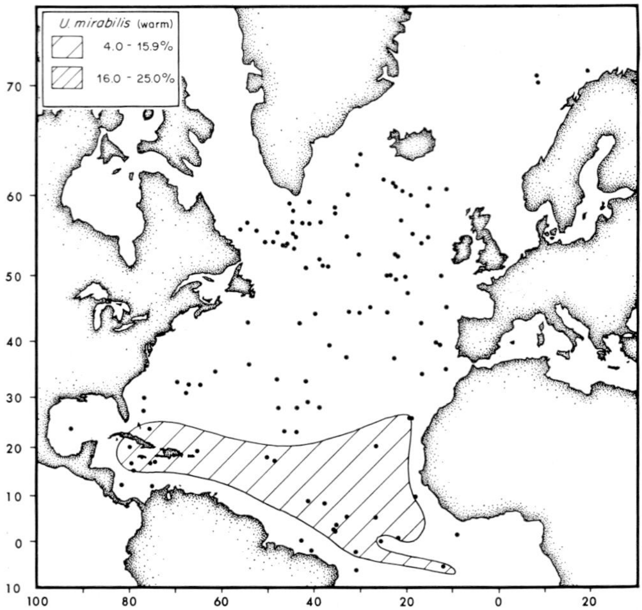

FIG. 10a. Relative abundance of *Umbilicosphaera mirabilis* (warm) in North Atlantic surface sediments.

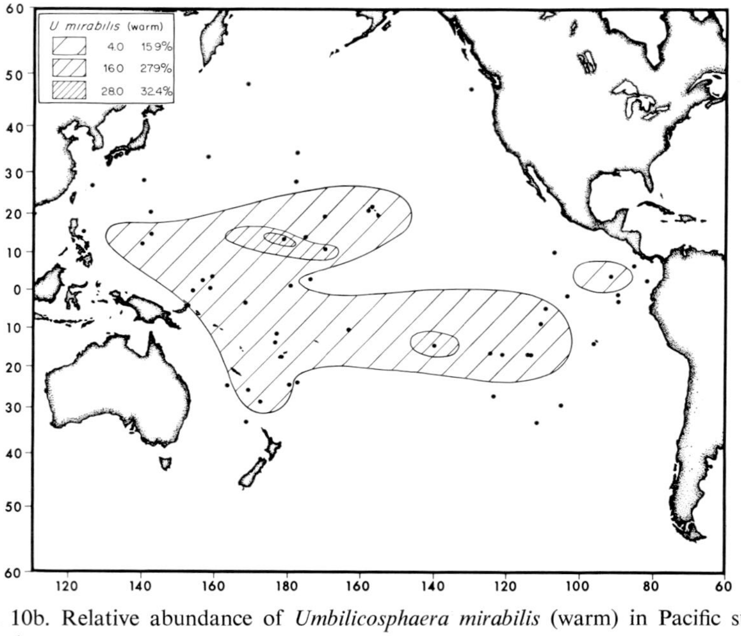

FIG. 10b. Relative abundance of *Umbilicosphaera mirabilis* (warm) in Pacific surface sediments.

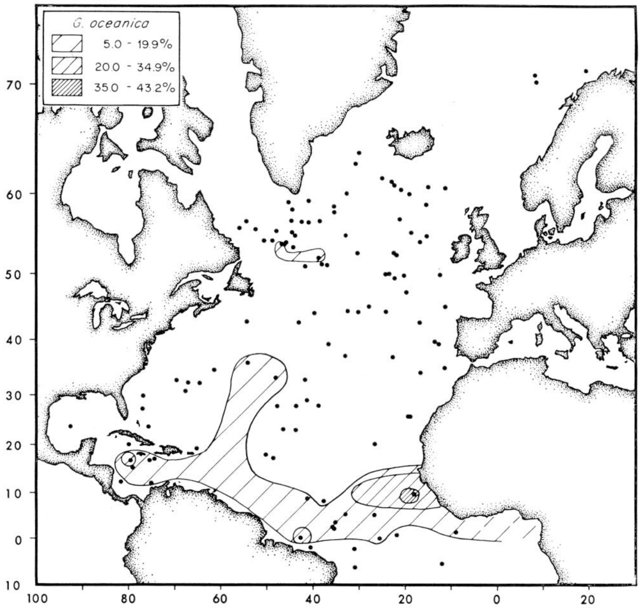

FIG. 11a. Relative abundance of *Gephyrocapsa oceanica* in North Atlantic surface sediments.

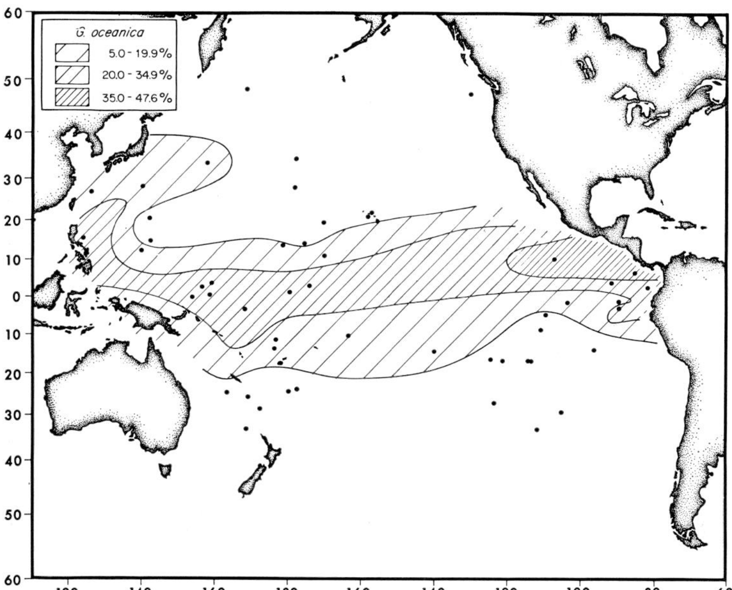

FIG. 11b. Relative abundance of *Gephyrocapsa oceanica* in Pacific surface sediments.

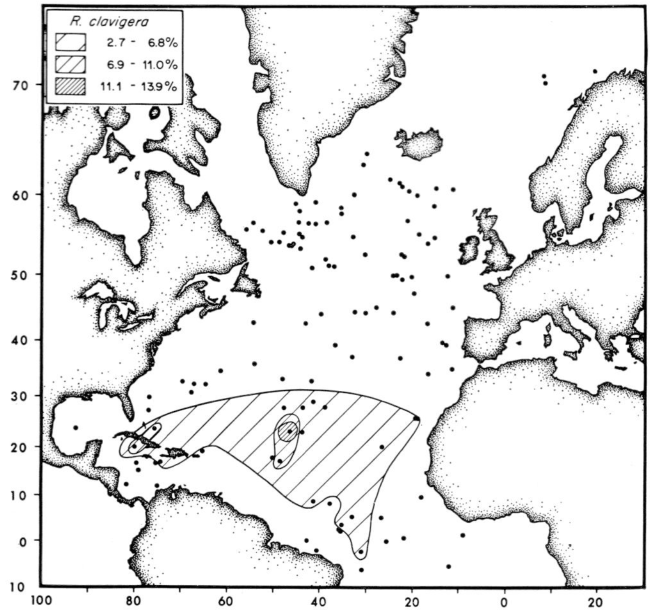

FIG. 12a. Relative abundance of *Rhabdosphaera clavigera* in North Atlantic surface sediments.

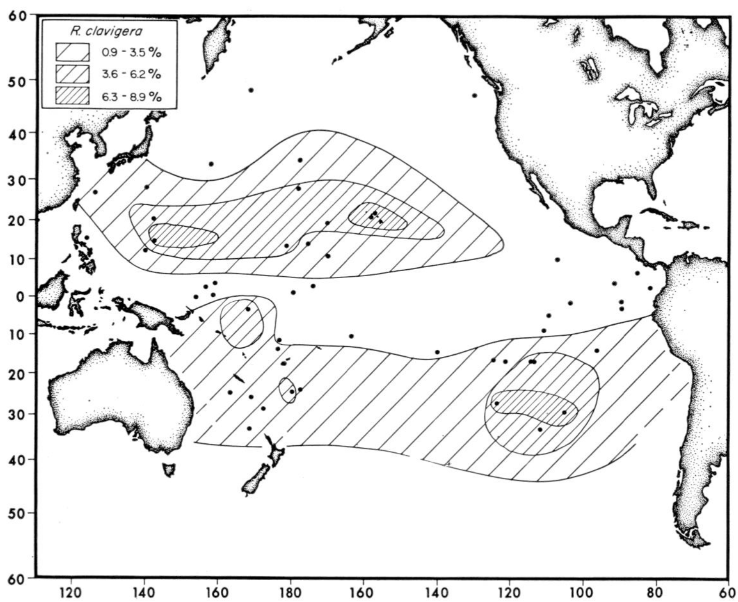

FIG. 12b. Relative abundance of *Rhabdosphaera clavigera* in Pacific surface sediments.

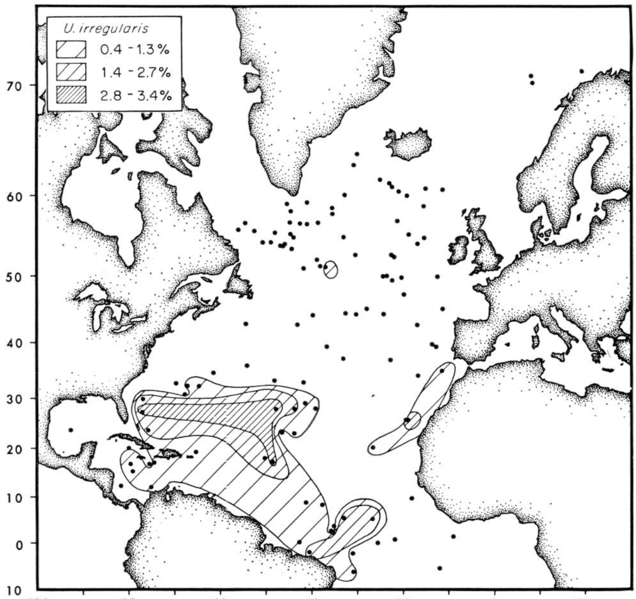

FIG. 13a. Relative abundance of *Umbellosphaera irregularis* in North Atlantic surface sediments.

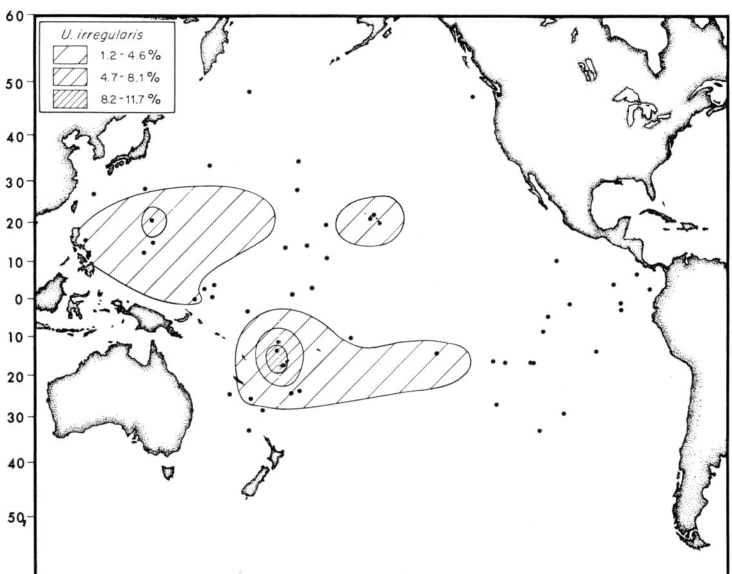

FIG. 13b. Relative abundance of *Umbellosphaera irregularis* in Pacific surface sediments.

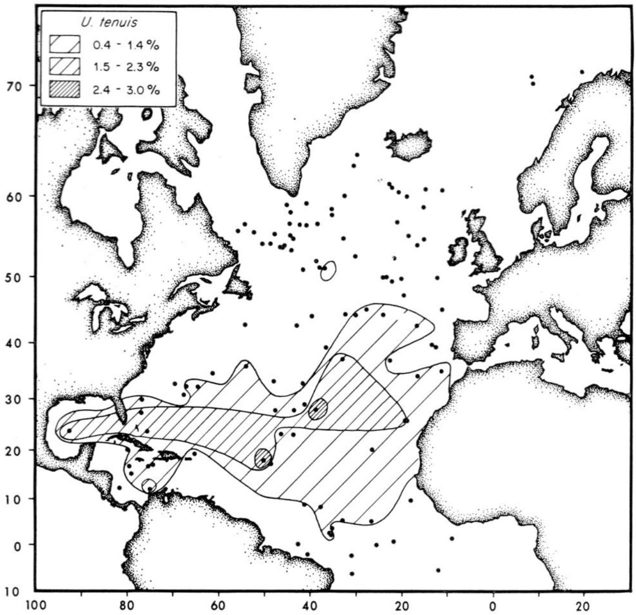

FIG. 14a. Relative abundance of *Umbellosphaera tenuis* in North Atlantic surface sediments.

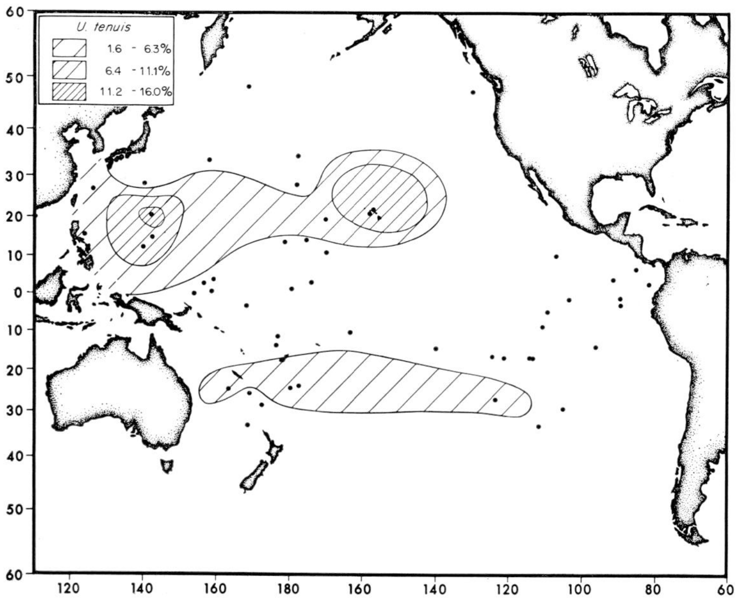

FIG. 14b. Relative abundance of *Umbellosphaera tenuis* in Pacific surface sediments.

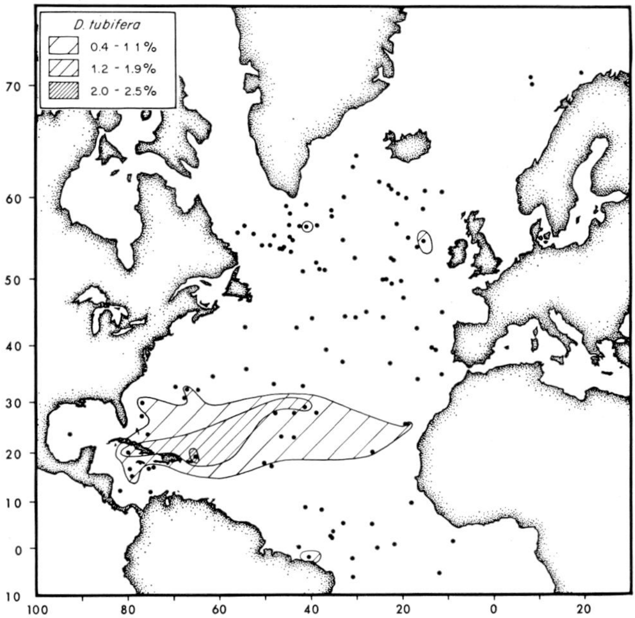

FIG. 15a. Relative abundance of *Discosphaera tubifera* in North Atlantic surface sediments.

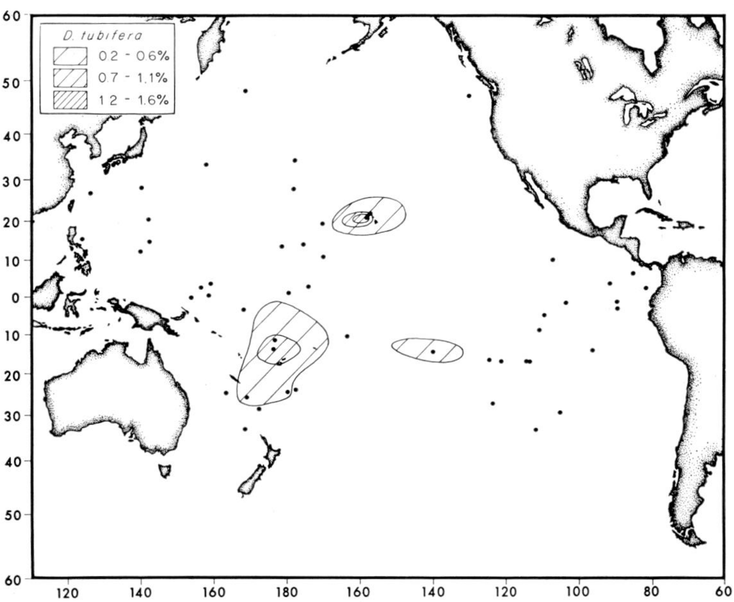

FIG. 15b. Relative abundance of *Discosphaera tubifera* in Pacific surface sediments.

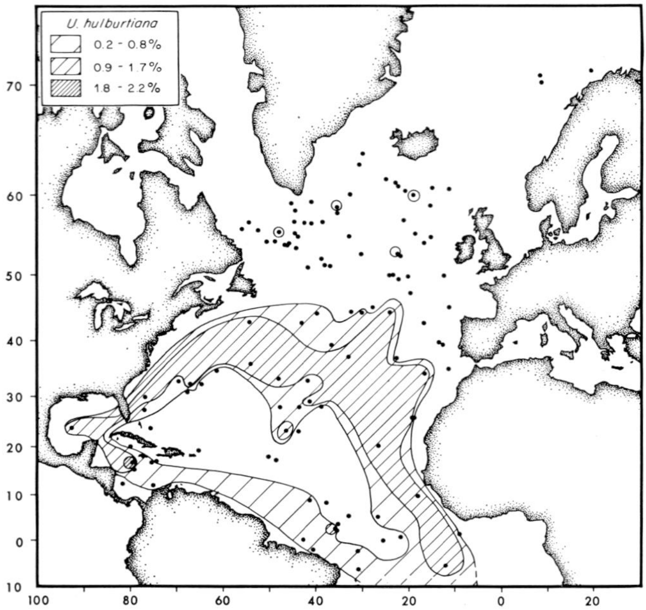

FIG. 16a. Relative abundance of *Umbilicosphaera hulburtiana* in North Atlantic surface sediments.

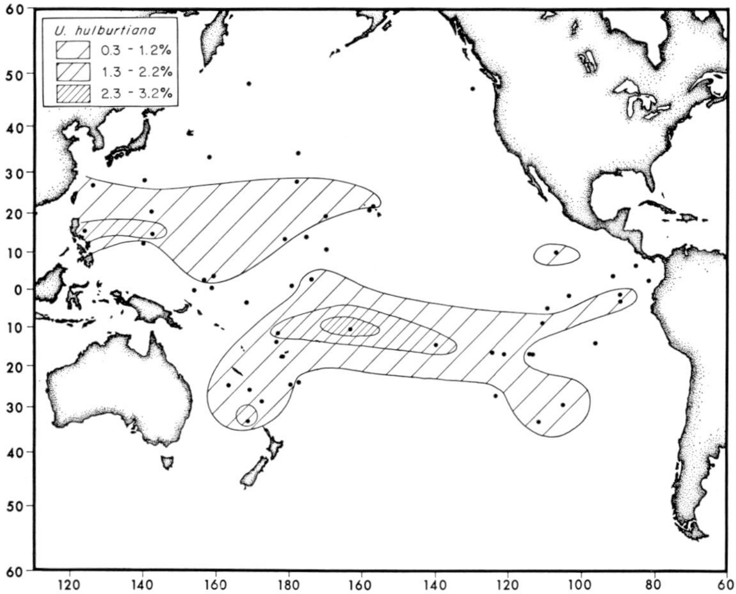

FIG. 16b. Relative abundance of *Umbilicosphaera hulburtiana* in Pacific surface sediments.

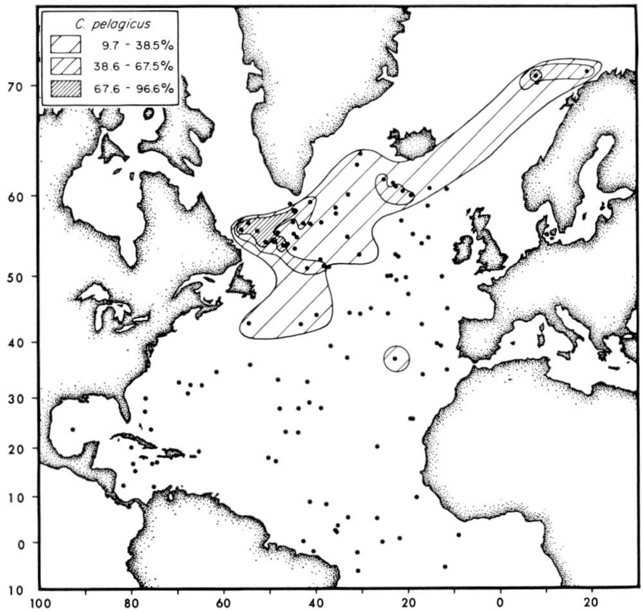

FIG. 17a. Relative abundance of *Coccolithus pelagicus* in North Atlantic surface sediments.

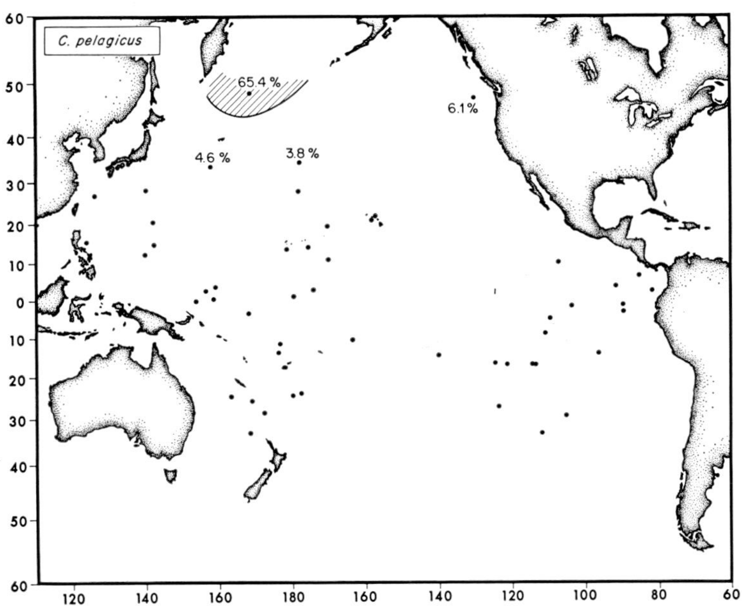

FIG. 17b. Relative abundance of *Coccolithus pelagicus* in Pacific surface sediments.

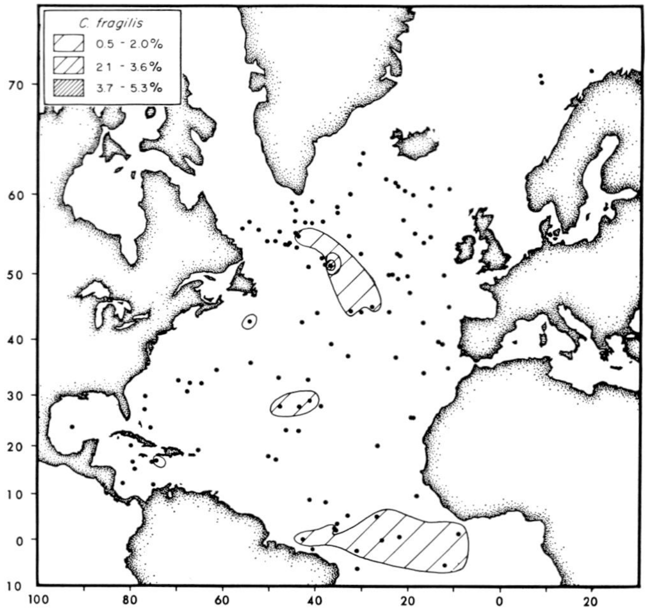

FIG. 18a. Relative abundance of *Cyclococcolithus fragilis* in North Atlantic surface sediments.

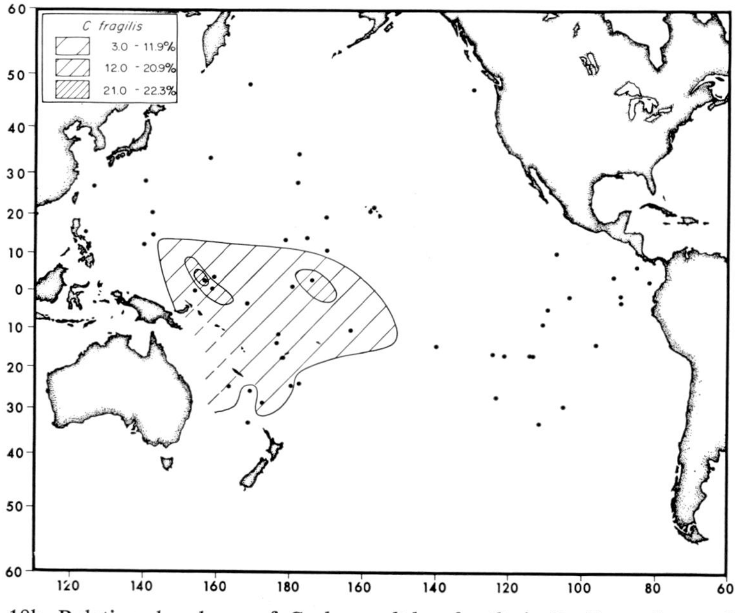

FIG. 18b. Relative abundance of *Cyclococcolithus fragilis* in Pacific surface sediments.

the North and South Pacific. The difference for *D. tubifera*, a species found in extremely low abundance, may be due to our choice of population size.

Comparison of peak abundance domains

Comparing the North Atlantic and North Pacific, ten species out of twenty-two have dissimilar high abundance regions. Of these ten, three species, *S. tuberculata*, *P. variabilis*, and *D. tubifera*, occur in concentrations which are too low to compare. One species, *C. leptopora* var. D., presents no trend in its peak abundances in the North Atlantic to warrant comparison. The six remaining species, *H. kamptneri*, *C. leptopora* var. B and C, *U. mirabilis* var. warm, and *C. fragilis*, have differences in peak abundance regions which may be due either to the Pacific's restricted sample distribution, or to true differences in the coccolith thanatacoenose. The discrepancy in *C. fragilis* distribution and abundance may also be due to arbitrary taxonomic differences and, therefore, count bias between the authors.

Comparing the North and South Pacific three species *S. tuberculata*, *P. variabilis*, and *D. tubifera*, have abundances which are too low to compare. Six species have different peak abundance domains which could be due to sample distribution, *S. pulchra*, *R. clavigera*, *C. leptopora* var. C and *D*, *G. caribbeanica*, and *U. mirabilis* var. warm, and one species, *C. pelagicus*, may have significantly different abundances. *C. pelagicus* which was previously unrecorded in surface sediments of the southern oceans, was found in one sample in the South Pacific in very low abundance. Reference to other studies of the subantarctic Pacific (Geitzenauer, 1972) confirms the non-match of high abundance for this species.

CORRELATION COEFFICIENT COMPARISON

Correlation coefficients provide a method by which numerical relationships between variables can be quantified. Causuality cannot be proven from such an analysis, but further hypothesis testing can be conducted to elucidate the basis for the observed correlations. Correlation coefficients were calculated by multiple regression between all species and latitude of sample, depth of sample, and summer and winter sea surface temperature separately for the North Atlantic and Pacific samples.

North Atlantic

Twenty out of twenty-two species from the North Atlantic show a significant correlation at the 95% confidence level (0.183) to both summer and winter sea surface temperature (Table 2). Fourteen species also show a significant correlation with salinity but, because a strong correlation (0.785) exists between salinity and temperature, the exact relationship between these parameters and the species distribution cannot be determined with certainty. However, for most of the species the correlation coefficient between species and temperature is higher than between species and salinity. Nineteen species show a significant correlation with latitude, but temperature is often a function of latitude and, although not calculated, a strong correlation probably exists between these two variables.

TABLE 2. Atlantic coccolith correlation coefficients. No. of samples, 117; significant correlation coefficient value, 0.183.

	Summer Temp.	Winter Temp.	Latitude	Depth	Salinity
S. pulchra	0.353	0.398	−0.253	0.015	0.321
S. tuberculata	0.204	0.197	−0.159	−0.090	0.153
S. variabilis	0.144	0.081	−0.088	−0.029	0.243
U. irregularis	0.487	0.520	−0.464	−0.099	0.409
U. tenuis	0.456	0.477	−0.433	0.002	0.420
H. kamptneri	0.396	0.483	−0.561	0.260	0.107
C. sinuosa	0.595	0.626	−0.594	0.056	0.423
D. tubifera	0.301	0.321	−0.278	0.080	0.286
R. clavigera	0.458	0.481	−0.492	0.130	0.565
C. fragilis	0.304	0.306	−0.245	0.071	0.072
C. leptopora (B)	0.256	0.288	−0.357	0.192	0.069
C. leptopora (C)	0.137	0.133	0.075	0.302	−0.025
C. leptopora (D)	0.238	0.220	−0.226	0.138	0.042
C. pelagicus	−0.743	−0.747	0.566	−0.168	−0.709
E. huxleyi (cold var.)	−0.296	−0.373	0.526	−0.142	0.003
E. huxleyi (warm var.)	0.706	0.758	−0.618	−0.059	0.420
G. caribbeanica	0.345	0.353	−0.235	0.321	0.297
G. ericsonii	0.247	0.223	−0.272	0.168	0.318
G. oceanica	0.425	0.505	−0.526	0.062	0.098
U. hulburtiana	0.451	0.439	−0.378	0.032	0.316
U. mirabilis (cold var.)	0.389	0.353	−0.336	0.187	0.369
U. mirabilis (warm var.)	0.494	0.564	−0.627	0.128	0.314

Significant correlation coefficients exist between depth and five species, *H. kamptneri*, *C. leptopora* var. C and B, *G. caribbeanica*, and *U. mirabilis* var. cold. All of these species are rather robust and are considered relatively resistant to dissolution (Berger, 1973).

Pacific

This smaller sample set yielded a significant correlation coefficient of 0.276. However, only seven out of nineteen species showed significant correlation

TABLE 3. Pacific coccolith correlation coefficients. No. of samples, 51; significant correlation coefficient value, 0.276. No data on salinity.

	Summer Temp.	Winter Temp.	Latitude	Depth
S. pulchra	0.044	0.099	0.223	0.105
S. tuberculata	—	—	—	—
S. variabilis	0.113	0.148	−0.042	−0.189
U. irregularis	0.291	0.312	−0.051	0.188
U. tenuis	0.190	0.142	−0.048	−0.074
H. kamptneri	0.338	0.323	−0.066	0.071
C. sinuosa	0.460	0.548	−0.171	0.135
D. tubifera	0.085	0.127	−0.066	−0.133
R. clavigera	0.079	0.065	0.106	−0.055
C. fragilis	0.400	0.464	−0.212	0.036
C. leptopora (B)	0.019	0.012	−0.198	−0.067
C. leptopora (C)	−0.149	−0.150	0.181	−0.030
C. leptopora (D)	0.104	0.198	0.034	0.065
C. pelagicus	−0.690	−0.641	−0.105	0.077
E. huxleyi (cold var.)	—	—	—	—
E. huxleyi (warm var.)	—	—	—	—
G. caribbeanica	−0.159	−0.263	0.079	0.115
G. ericsonii	0.101	0.069	0.188	0.156
G. oceanica	0.520	0.548	−0.348	0.147
U. hulburtiana	0.151	0.186	0.033	0.262
U. mirabilis (cold var.)	0.404	0.367	−0.279	0.136
U. mirabilis (warm var.)	0.348	0.474	−0.157	0.186

with summer and winter temperatures (Table 3). Two species correlate with latitude, while none of the species correlate with depth. A species to salinity correlation was not attempted.

In summary, although there is a significant correlation between five species and depth in the North Atlantic, the effect of depth on the samples has probably had little influence on the general distribution patterns and abundances of coccolithophorida in the sample sets discussed here. Thus, the coccolith species distributions and high abundance domains here noted for the North Atlantic and Pacific Oceans, are most closely related to sea surface temperatures.

Conclusions

(*a*) Ten coccolith species show approximately analogous distributions, with regard to surface water masses, in both the North Pacific and North Atlantic.

Coccolithus pelagicus is characteristic of the Sub-arctic region. *Syracosphaera pulchra, Rhabdosphaera clavigera, Umbellosphaera irregularis, Umbellosphaera tenuis* and *Discosphaera tubifera* show peak abundances in the central water mass regions. *Calciosolenia sinuosa*, and *Umbilicosphaera mirabilis* "cold" and "warm" varieties have peak abundances in equatorial and adjacent central water mass regions. *Gephyrocapsa oceanica* is predominantly an equatorial water mass species.

(*b*) North Atlantic coccolith abundances are strongly related to temperature with twenty of the twenty-two species showing a significant correlation. Only seven out of nineteen species in the Pacific show a significant correlation between abundance and temperature.

(*c*) Using a somewhat depth restricted sample set only five species in the North Atlantic, and none in the Pacific, show a significant correlation between distribution and depth.

APPENDIX 1. Atlantic trigger weight core data.

Core	Latitude	Longitude	Depth (m)
RC 9–14	27°23′N	76°57′W	1254
RC 9–35	20°03′N	79°51′W	2745
RC 9–210	39°12′N	12°46′W	4083
RC 9–225	54°58′N	15°23′W	2334
RC 10–49	16°34′N	79°32′W	1238
RC 11–14	02°04′S	40°36′W	1639
RC 12–23	15°16′N	79°07′W	2270
RC 13–146	11°55′N	75°02′W	3387
RC 13–152	16°43′N	75°27′W	2027
V 3–128	23°46′N	92°29′W	3495
V 4–8	37°14′N	33°08′W	1655
V 4–32	35°03′N	11°37′W	2230
V 5–1	32°36′N	69°30′W	5135
V 5–31	32°17′N	67°07′W	1554
V 5–40	32°13′N	64°43′W	1903
V 7–53	35°54′N	54°02′W	5258
V 7–67	34°40′N	61°28′W	4308
V 9–31	08°14′N	37°52′W	4204
V 10–89	23°02′N	43°48′W	3523
V 12–122	17°00′N	74°24′W	2800
V 16–20	17°56′N	50°21′W	4410
V 16–21	17°17′N	48°25′W	3975
V 16–206	23°20′N	46°29′W	3734
V 17–165	32°45′N	41°54′W	3924
V 17–198	55°49′N	52°27′W	3306

APPENDIX 1 (*contd.*)

Core	Latitude	Longitude	Depth (m)
V 18–24	02°25′N	35°50′W	3685
V 19–296	01°25′N	09°05′W	5017
V 19–308	29°01′N	41°24′W	3197
V 19–310	33°18′N	48°16′W	4807
V 20–233	02°00′N	35°36′W	3884
V 20–234	05°19′N	33°02′W	3133
V 22–26	08°43′N	41°15′W	3720
V 22–219	27°55′N	43°38′W	2582
V 23–22	54°12′N	45°58′W	3669
V 23–29	59°57′N	32°51′W	2186
V 23–60	70°03′N	08°19′E	2743
V 23–81	54°15′N	16°50′W	2393
V 23–82	52°35′N	21°56′W	3974
V 23–83	49°52′N	24°15′W	3871
V 25–50	00°12′N	42°46′W	3749
V 25–56	03°33′N	35°14′W	3512
V 26–31	27°56′N	47°46′W	4433
V 26–46	09°34′N	18°11′W	2898
V 26–54	05°28′S	11°56′W	2906
V 26–55	11°36′S	15°34′W	3457
V 26–96	06°17′S	30°53′W	5150
V 26–98	02°11′S	31°04′W	4667
V 27–7	42°47′N	54°20′W	4212
V 27–15	42°44′N	43°14′W	4856
V 27–16	44°10′N	39°52′W	4797
V 27–18	51°12′N	36°59′W	3580
V 27–19	52°06′N	38°48′W	3466
V 27–20	54°00′N	46°12′W	3510
V 27–21	54°03′N	46°52′W	3636
V 27–22	54°28′N	50°48′W	3188
V 27–24	55°59′N	56°04′W	2628
V 27–25	56°44′N	54°42′W	3146
V 27–28	58°11′N	44°37′W	2308
V 27–30	59°09′N	41°04′W	2759
V 27–34	63°01′N	30°57′W	2283
V 27–114	55°03′N	33°04′W	2532
V 27–116	52°50′N	30°20′W	3202
V 27–137	42°41′N	17°04′W	4883
V 27–142	36°59′N	22°54′W	4726
V 27–144	39°35′N	13°33′W	4894
V 27–162	34°11′N	16°51′W	4153
V 27–178	05°06′N	26°39′W	4327
V 27–181	00°04′N	25°30′W	3601

APPENDIX 1 (*contd.*)

Core	Latitude	Longitude	Depth (m)
V 27–254	20°11′N	26°37′W	4850
V 28–8	59°00′N	45°27′W	2325
V 28–65	60°33′N	11°27′W	1232
V 28–66	60°37′N	15°11′W	2043
V 28–82	49°27′N	22°16′W	3935
V 28–83	47°28′N	19°59′W	4559
V 28–88	45°15′N	27°54′W	2492
V 28–89	44°32′N	32°35′W	3643
V 28–124	12°16′N	81°42′W	452
V 29–193	55°24′N	18°44′W	1326
V 29–194	57°00′N	21°19′W	2085
V 29–198	58°44′N	15°34′W	1139
V 29–200	59°57′N	19°12′W	2650
V 29–202	60°23′N	20°58′W	2658
V 29–203	60°48′N	22°27′W	2085
V 29–204	61°11′N	23°01′W	1849
V 29–205	61°33′N	25°07′W	1450
V 30–115	54°24′N	48°54′W	3523
V 30–116	53°38′N	44°15′W	3609
V 30–117	56°56′N	44°52′W	3440
V 30–118	55°24′N	44°38′W	3344
V 30–120	55°01′N	43°53′W	3332
V 30–121	55°34′N	48°07′W	3398
V 30–122	56°48′N	38°33′W	2878
V 30–123	56°42′N	40°49′W	2822
V 30–124	56°44′N	42°28′W	3345
V 30–125	57°53′N	35°30′W	2597
V 30–126	58°34′N	35°30′W	2419
V 30–128	64°04′N	30°13′W	2277
A 152–84	44°21′N	30°16′W	2750
A 153–154	28°00′N	38°47′W	4020
A 157–3	50°56′N	41°45′W	4025
A 167–18	29°47′N	76°48′W	914
A 172–13	19°24′N	65°07′W	6401
A 179–13	23°56′N	75°45′W	1847
A 179–20	30°47′N	67°40′W	4993
A 180–15	39°16′N	36°42′N	4610
A 180–39	25°50′N	19°18′W	3470
A 180–72	00°36′N	21°47′W	3841
BS-1	71°00′N	19°54′E	187
D 2992	49°48′N	12°28′W	2085
D 3126	44°28′N	24°19′W	2979
D 4288	45°17′N	11°22′W	4042

APPENDIX 1 (*contd.*)

Core	Latitude	Longitude	Depth (m)
K 708–1	50°00′N	23°44′W	4053
K 708–7	49°46′N	20°20′W	3502
K 708–8	52°44′N	22°33′W	4009
R 5–54	25°52′N	19°03′W	3295
SP–10–1	51°23′N	38°04′W	3695
SP–11–6	70°29′N	07°57′E	2926

APPENDIX 2. Barren samples in the North Atlantic

Core	Latitude	Longitude	Depth (m)
R 9–7	59°39′N	22°46′W	2770
V 2–12	49°08′N	49°20′W	1267
V 14–4	15°29′N	40°31′W	4473
V 16–200	01°58′N	37°04′W	4093
V 17–196	60°44′N	57°50′W	2818
V 18–21	04°14′N	47°45′W	2374
V 20–7	11°33′N	60°31′W	1018
V 22–204	15°01′N	23°14′W	1723
V 27–104	62°52′N	17°56′W	1057

APPENDIX 3. Dissolution samples in the North Atlantic

Core	Latitude	Longitude	Depth (m)
V 20–242	23°22′N	43°39′W	4565
V 26–124	16°08′N	74°27′W	3005
V 26–165	30°27′N	67°59′W	5159
V 27–175	08°48′N	22°07′W	4548
V 27–183	03°33′S	30°33′W	4694
V 27–234	01°00′N	01°59′W	4618
V 28–7	55°32′N	47°06′W	3250
V 28–90	49°27′N	22°16′W	3935
V 30–119	56°50′N	43°42′W	3367

APPENDIX 4. Pacific trigger weight core data

Core	Latitude	Longitude	Depth (m)
RC 8–91	33°25′S	111°54′W	2723
RC 8–93	29°22′S	105°14′W	3157
RC 9–104	27°26′S	123°26′W	3632
RC 9–121	24°46′S	179°41′W	2153
RC 9–124	28°45′S	172°36′E	2540
RC 9–126	33°14′S	168°44′E	2060
RC 10–115	10°35′S	163°18′W	3135
RC 10–143	00°20′S	153°59′E	3074
RC 10–161	33°05′N	158°00′E	3587
RC 11–187	47°09′N	130°07′W	2670
RC 11–220	14°49′S	139°58′W	2950
RC 11–230	08°48′S	110°48′W	3259
RC 12–107	26°00′S	169°12′E	3115
RC 12–113	24°53′S	163°31′E	2454
RC 12–121	03°44′S	168°23′E	3519
RC 12–173	20°47′N	157°56′W	1476
RC 12–210	24°14′S	177°36′W	1529
RC 12–361	15°06′N	124°08′E	3528
RC 12–366	26°35′N	126°20′E	1644
RC 13–1	21°28′N	157°14′W	2114
RC 13–17	19°05′N	170°04′W	3374
RC 13–27	13°52′N	175°16′W	3555
RC 13–108	03°07′S	89°25′W	3308
RC 13–113	01°39′S	103°38′W	3195
RC 13–122A	10°03′N	107°00′W	3275
V 15–29	06°21′N	85°17′W	1889
V 19–25	02°28′N	081°42′W	2404
V 19–41	14°06′S	096°12′W	3243
V 19–53	17°01′S	113°31′W	3058
V 19–55	17°00′S	114°11′W	3177
V 19–64	16°56′S	121°12′W	3570
V 19–65	16°39′S	124°23′W	3867
V 19–110	11°51′N	140°03′E	3532
V 20–103	33°59′N	177°50′W	3442
V 20–119	47°57′N	168°47′E	2739
V 21–30	01°13′S	089°41′W	617
V 21–41	04°43′S	109°25′W	3435
V 21–85	27°58′N	140°30′E	1684
V 24–96	27°40′N	177°59′W	3305
V 24–101	13°10′N	178°53′E	3336
V 24–109	00°26′N	158°48′E	2367
V 24–110	02°21′N	156°42′E	2613

APPENDIX 4 (*contd.*)

Core	Latitude	Longitude	Depth (m)
V 28–148	03°42′N	091°32′W	2834
V 28–195	10°39′N	169°48′W	2439
V 28–201	02°46′N	173°56′W	3217
V 28–203	00°57′N	179°25′W	3248
V 28–211	11°46′S	177°00′E	2644
V 28–212	14°03′S	176°26′E	2972
V 28–239	03°15′N	159°11′E	3490
V 28–246	14°22′N	142°43′E	2745
V 28–255	20°06′N	142°27′E	3261

APPENDIX 5. Dissolution samples in the Pacific

Core	Latitude	Longitude	Depth (m)
RC 8–96A	19°45′S	095°37′W	3334
RC 9–76	16°38′S	078°08′W	3248
RC 9–77	18°00′S	079°09′W	2875 (approx.)
RC 9–95	22°18′S	109°43′W	3660
RC 9–97	22°30′S	112°46′W	3394
RC 10–230	40°28′N	128°25′W	3200
RC 11–188	46°44′N	131°35′W	3319
RC 14–156	58°46′N	147°23′W	2033
RC 14–172	50°23′N	142°34′W	2255
V 19–20	03°35′S	083°56′W	3157
V 19–30	03°23′S	083°31′W	3091
V 20–74	41°04′N	132°22′W	3749
V 20–78	47°15′N	131°02′W	2983

Acknowledgements

We thank the people who have given their time and effort, particularly those involved in the CLIMAP program of which this paper is in part a spin-off, and especially Ms. Barbara Molfino and Ms. Mary Perry. We also thank the reviewers for their helpful suggestions.

The manuscript was typed by Ms. Ursula Findikyan. This research was supported by National Science Foundation Grant and IDOE grant GX

28671 E. The core collection and processing was supported by the National Science Foundation grant DES72-01568-A03 and the Office of Naval Research grant N00014-75-C-0210. This paper is Lamont–Doherty Geological Observatory Contribution No. 2375.

References

Berger, W. H. (1973). Deep-sea carbonates: evidence for a coccolith lysocline. *Deep-Sea Res.* **20,** 917–921.

Gaarder, K. (1971). Comments on the distribution of Coccolithophorida in the oceans. *In* (Funnell, B. M., and Riedel, W. R., eds), *The Micropaleontology of Oceans*, 97–102. Cambridge University Press.

Geitzenauer, K. R. (1972). The Pleistocene calcareous nannoplankton of the subantarctic Pacific Ocean. *Deep-Sea Res.* **19,** 45–60.

Imbrie, J. and Kipp, N. G. (1971). A new micropaleontological method for quantitative paleoclimatology: application to a late Pleistocene Caribbean core. *In* (Turekian, K., ed.), *The late Cenozoic glacial ages*, 71–181, Yale University Press.

McIntyre, A. (1967). Coccoliths as paleoclimatic indicators of Pleistocene glaciations. *Science* **158,** 1314–1317.

McIntyre, A. and Bé, A. W. H. (1967). Modern Coccolithophoridae of the Atlantic Ocean. I. placoliths and cyrtoliths. *Deep-Sea Res.* **14,** 561–597.

McIntyre, A., Bé, A. W. H. and Preikstas, R. (1967). Coccoliths and the Plio-Pleistocene boundary. *In* (Sears, M., ed.), *Progress in Oceanography*, **4,** 3–24. Pergamon, Oxford.

McIntyre, A., Bé, A. W. H. and Roche, M. B. (1970). Modern Paciffc Coccolithophorida: a paleontological thermometer. *N.Y. Acad. Sci. Trans.* **32,** 720–731.

McIntyre, A. and McIntyre, R. (1971). Coccolith concentrations and differential solution in oceanic sediments. *In* (Funnell, B. M. and Reidel, W. R., eds), *The Micropaleontology of Oceans*, 253–261. Cambridge University Press.

Okada, H. and Honjo, S. (1973). The distribution of oceanic Coccolithophorida in in the Pacific. *Deep-Sea Res.* **20,** 355–374.

Schneiderman, N. (1973). Deposition of Coccoliths in the compensation zone of the Atlantic Ocean. *In* (Smith. L. A. and Hardenbol, J., eds), *Proceedings of Symposium on calcareous nannofossils*, pp. 140–151. Houston, Society of Economic Paleontologists and Mineralogists.

12: Selective Dissolution of Recent Coccoliths in the Atlantic Ocean

NAHUM SCHNEIDERMANN*

*University of Puerto Rico,
Department of Marine Sciences,
Mayaguez, Puerto Rico 00708*

The coccoliths produced by chrysomonad algae in the Atlantic Ocean undergo selective dissolution upon entering the $CaCO_3$ under-saturated waters below the surface. The level below which all coccoliths have been dissolved is termed the calcium carbonate compensation depth (CCCD). Coccoliths may settle through the water column as discrete particles so that their settling time approximates their residence time on the bottom before burial, or they may settle more rapidly as aggregates such as fecal pellets and thus be protected from corrosive deep ocean waters. The rate of dissolution affecting each species is a function of its skeletal ultrastructure, and within a single coccolith selective removal of skeletal elements proceeds in an orderly sequence. Holococcoliths are readily dissolved, and rarely occur in oceanic sediments but placoliths are highly resistant to solution and are concentrated near the CCCD; other heterococcoliths have intermediate levels of resistance. The variations in resistance to solution permits recognition of a series of dissolution zones in the ocean. These zones are deepest in the equatorial region, and ascend toward the poles.

The application of the data presented here to the recognition of ancient compensation zones is considered.

Outline

* *Present address:* Gulf Research and Development Company, Houston Technical Services Center, P.O. Box 36506, Houston, Texas 77036, U.S.A.

Introduction

Pelagic sediments are derived from particles which settle out of or precipitate from the water column with little terrigenous influx. The calcareous fraction of these deposits consists chiefly of tests and shells of planktonic animals from throughout the overlying water column (pteropods and foraminiferids) and the remains of coccolithophorids, which inhabit the photic zone of the ocean.

The relative abundance of planktonic foraminifera and coccoliths in pelagic sediments varies greatly. In some Tertiary sediments, calcareous nannofossils (coccoliths and related forms) may account for over 90% of the calcium carbonate fraction (Bramlette, 1958; Maxwell *et al.*, 1970). Turekian (1964) indicated that coccoliths may account for up to 30% of the total calcium carbonate of Recent pelagic sediments.

Lisitzin (1972) recognised three zones of pelagic carbonate sedimentation: (1) An upper zone of dilution by terrigenous influx in shallow margins of the ocean (see also Kuenen, 1950 and Turekian, 1965), (2) A middle zone of normal accumulation, and (3) A lower zone of dissolution. This third zone, in which solution of $CaCO_3$ occurs, is termed the "compensation zone".

The upper limit of the compensation zone is the level below which sea water is undersaturated with respect to $CaCO_3$–termed the "saturation depth" (Pytkowicz, 1970). The depth below which the rate of solution of $CaCO_3$ exceeds the rate of its supply to the ocean floor is termed the "calcium carbonate compensation depth". The true depth of $CaCO_3$ saturation in the ocean is a subject of controversy (Edmond, 1970; Edmond and Gieskes, 1970). A shallow saturation depth, separated from the compensation depth by thousands of metres, was postulated by Berner (1965), Howley and Pytkowicz (1969) and Pytkowicz (1970). On the other hand, Revelle and Fairbridge (1957), Turekian (1964, 1965) and Li *et al.* (1970) have favoured a proximity or even identity of the saturation depth with the compensation depth. Because the rapid solution of $CaCO_3$ occurs far above the compensation depth (Peterson, 1966; Berger, 1967, 1968), the concept of a shallower saturation depth is favoured here (Fig. 1).

Variations of the calcium carbonate compensation depth with latitude, surface productivity and local hydrography have been recognised since the "Challenger" expedition (Murray and Renard, 1891). Of the mineral phases of $CaCO_3$, aragonite (secreted by pteropods) dissolves at a much shallower depth than calcite, usually between 1000 to 2500 m (Murray and Chumley, 1924; Friedman, 1965; Chen, 1964; Li *et al.* 1969); the calcium carbonate compensation depth (CCCD) referred to in this study applies to the low magnesium calcite supplied by planktonic foraminifera and coccolithophores. It lies at depths of about 4500 m in the Pacific (Bramlette, 1961); in the Atlantic it is somewhat deeper, at about 5500 m (Turekian, 1965).

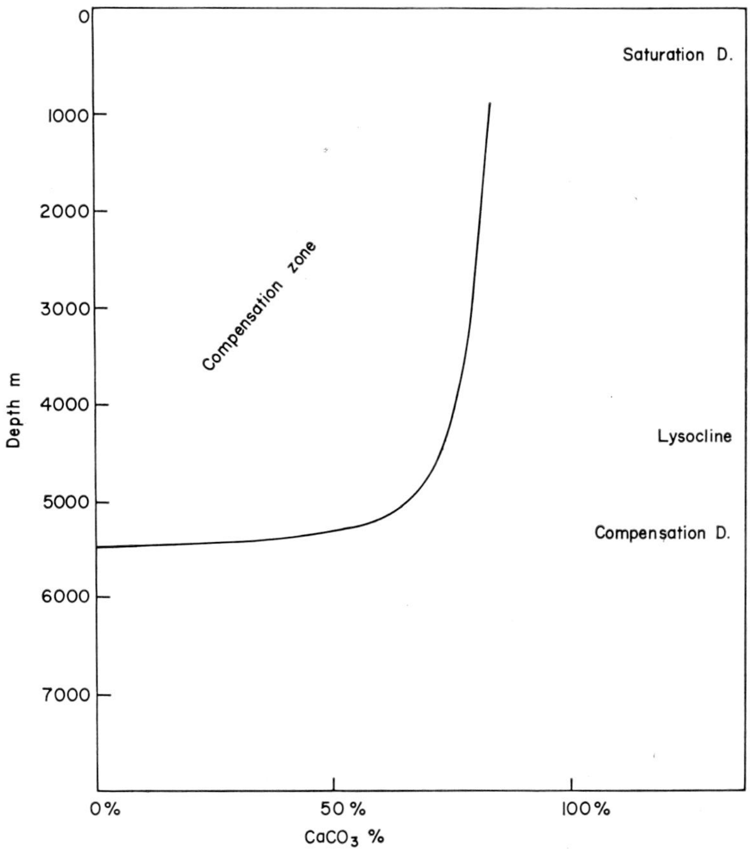

FIG. 1. Calcium carbonate compensation in the Atlantic Ocean. After Turekian (1964), Berner (1965) and Berger (1968).

The complex conditions influencing the position and thickness of the CCCD and the rate of accumulation of $CaCO_3$ within it have been studied by a number of workers (Heath and Culberson, 1970 and references therein). The CCCD separates "red" clays, devoid of $CaCO_3$ from calcareous pelagic sediments. The fraction of the ocean-floor presently covered by $CaCO_3$ appears to be a result of mass balance between the rate of removal of $CaCO_3$ from the ocean waters by planktonic organisms and the rate of supply of calcium to the ocean by rivers (Olausson, 1965, 1971; Li *et al.*, 1969; Broecker, 1971a). The rate of calcium fixation by organisms and thus the potential for removal from the ocean is five times the rate of supply of calcium from run-off. If the ocean system is to be in a steady state, about five sixths of the sinking biogenic carbonate must be dissolved, returning both calcium and carbon to the system (Li *et al.*, 1969).

The position of the CCCD in any given area of the ocean is determined by the following interelated factors: (1) Surface productivity; a high rate of production of $CaCO_3$ will depress the CCCD (Bramlette, 1961; Arrhenius, 1963), (2) Increase of $CaCO_3$ dissolution with depth (pressure), a major reason for the general variation in $CaCO_3$ content of sediments with depth (Pytkowicz, 1968 and references therein), (3) Amount of CO_2 dissolved in sea water, a function of the amount of organic debris oxidised in deep water (Li *et al.*, 1969) (4) Temperature; solubility of CO_2 increases with decreasing temperature, increasing the rate at which $CaCO_3$ will be dissolved (Sverdrup *et al.*, 1942), (5) Physical and chemical properties of the deposited $CaCO_3$ particles, and (6) Quantity of non-calcareous material supplied to the bottom; dilution by terrigenous sediments and siliceous biogenic particles (radiolaria, diatoms) will decrease the $CaCO_3$ content of a given sediment.

In investigations of the compensation zone, it has been found that inorganic and biogenic (planktonic foraminifera) calcite, suspended in deep ocean waters on a buoyed line for several months, underwent measurable dissolution more than 2500 m above the local compensation depth (Peterson, 1966; Berger, 1967), the rate of dissolution increasing with depth. Planktonic foraminiferal tests undergo selective dissolution with depth, so that certain species are removed from an assemblage before others are attacked (Schott, 1935; Ruddiman and Heezen, 1967; Berger, 1968, 1970). Samples from the vicinity of the CCCD contain remains of only the most resistant groups. From observations on the tests of planktonic foraminifera, Berger (1968) recognised a level of rapid increase in the rate of dissolution of the tests well above the CCCD, and termed it the "lysocline".

McIntyre and McIntyre (1971) studied dissolution of coccoliths in five closely spaced cores ranging from 4500 m to 5400 m depth in the Atlantic. Their results, compared with distribution of coccoliths in surface sediment samples, suggested selective dissolution of the coccoliths at deeper depositional sites, a phenomenon analogous to that which had been described for planktonic foraminiferal tests. Selectivity of dissolution of various nannofossils in older sediments has also been described by Hay (1970, 1971); Ramsay (1971, 1972) and Bukry (1971). Other parameters being equal, the solubility of any calcium carbonate particle will be influenced by the following: (1) Mineralogy—aragonite is more soluble than calcite under the conditions prevailing in the ocean and therefore will disappear from the sediment at shallower depth than calcite, (2) Composition—tests of high magnesium calcite will dissolve more rapidly than those made of low magnesium calcite, (3) Size–smaller particles tend to dissolve more rapidly than larger ones because they present a larger surface area to the corrosive medium, (4) Impurities–organic coatings will retard the solution of calcium carbonate particles in undersaturated waters, and (5) Ultrastructure–the selective dissolution observed for the tests of planktonic foraminifera (Berger, 1970) and for coccoliths (McIntyre and McIntyre, 1971) indicates that the preservation of each particle will also depend on its ultramicroscopic structure.

Selective dissolution in the vicinity of the CCCD might be a useful tool for determining relative palaeobathymetry of isochronous sediments recovered in various areas of the ocean. Hay, (1970) proposed that considerable vertical fluctuations in the CCCD had occurred during the Tertiary (see also Heath, 1969), as deduced from alternations of carbonate-poor and rich sediments, radiolarian oozes and red clays, and by variations in the state of preservation of the nannofossils. Detailed studies of carbonate compensation in modern oceans should provide a means of determining and describing ancient compensation levels and provide insight into long-term chemical changes in the oceans. The purposes of this study are: (1) To describe the selective dissolution of Recent coccoliths in the Atlantic Ocean, (2) To subdivide the compensation zone using selective dissolution of calcareous nannoplankton, and (3) To derive general rules for applying the Recent model to interpretation of fossil coccolith assemblages.

Sample source and preparation

Core samples from the Atlantic Ocean were obtained from the repositories at Woods Hole Oceanographic Institution, Lamont–Doherty Geological Observatory, and the Rosenstiel School of Marine and Atmospheric Sciences in Miami. The uppermost 2 to 5 cm^3 of trigger weight cores were collected whenever available, otherwise the tops of piston cores were sampled. Most of the samples are from the Equatorial Atlantic (Fig. 2), with several transects from Africa to South America. Scattered samples from 78°49.9′N to 50°52′S represent the biogeographical zones described by McIntyre and Bé (1967). The sample locations and depths are given in the Appendix (p. 1039) and in Fig. 2. Scattered samples from 1000 to 6700 m, with closely spaced samples from the vicinity of the compensation depth insure good coverage of the compensation zone.

Smear slides for optical studies and abundance counts were prepared from each of the 99 untreated samples. For scanning electron microscopy, the cleaning procedures described by Worsley (1970) were used, Between 200 and 300 specimens per sample were counted using the scanning electron microscope and the percentage abundance of the important species was calculated.

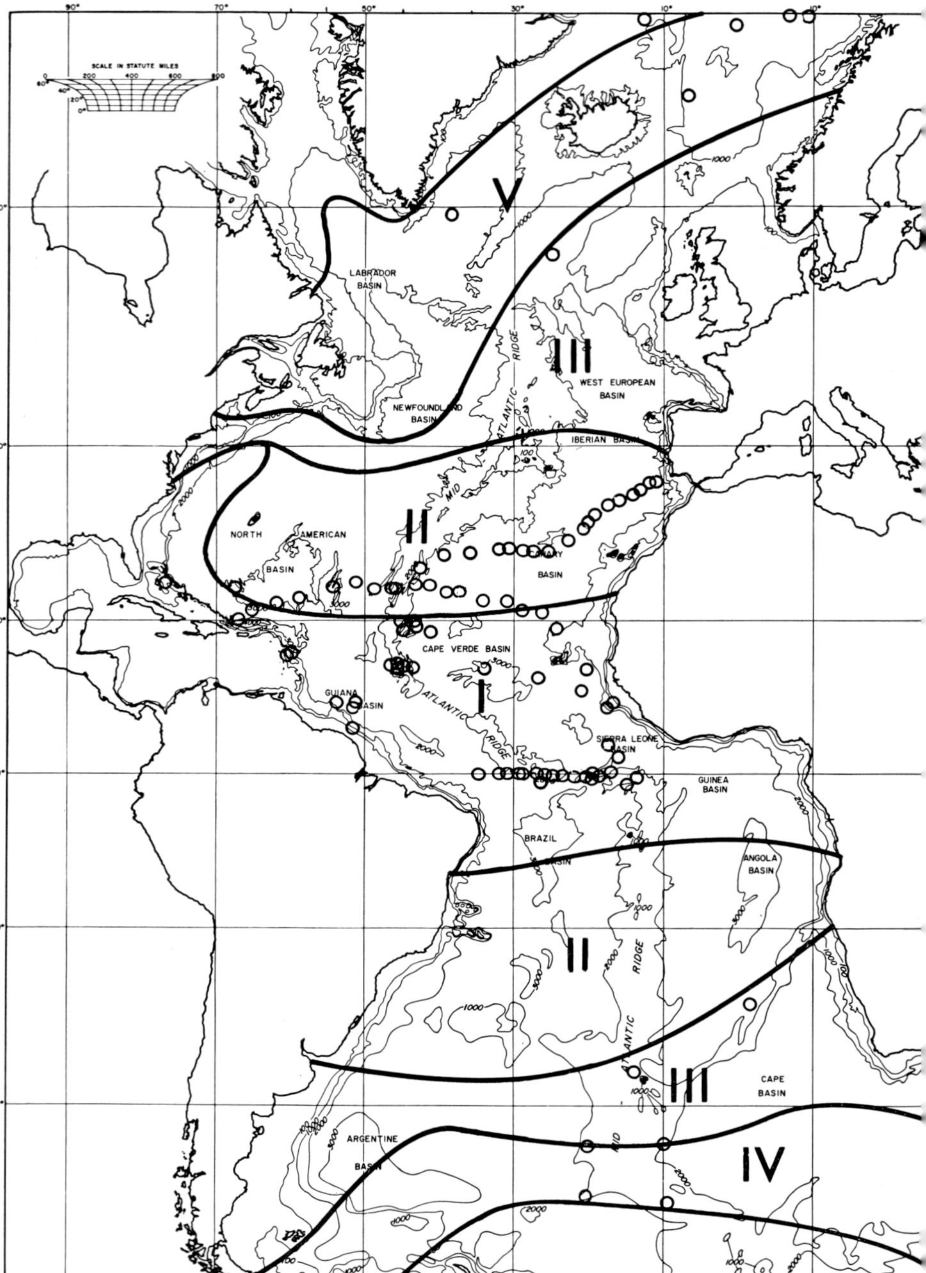

FIG. 2. Core location. The coccolithophorid floral zones after McIntyre and Bé (1967). I, tropical; II, subtropical; III, transitional; IV, subantarctic; V, subarctic.

All scanning electron microscopy was done using the Cambridge Stereoscan Mark II A at the Center for Electron Microscopy of the University of Illinois.

The distribution of coccolithophorids in the Atlantic Ocean: a review

To understand the effects of dissolution on coccolith assemblages deposited in the deep sea requires knowledge of the supply of coccoliths to the sediment, and distribution of the living coccolithophores which produce the coccoliths. The ecology and distribution of coccolithophorids in the Atlantic Ocean have been the subject of extensive qualitative studies but data suitable for quantitative treatment are still minimal. Coccolithophore abundances are given as cells per volume of filtered sea water (Hulbert, 1964) or as relative percentages. Often, however, abundance estimates are complicated by seasonal and diurnal variations in the phytoplankton population, as reflected in data by Hulbert (1966) and McIntyre and Bé (1967). Furthermore, the identification of the species by optical means may result errors in recognition of the smaller coccolithophores. No quantitative data are available for many of the species recognised in the early publications.

The most extensive studies of coccolith distribution in the water and in the sediment have been carried out at Lamont–Doherty Geological Observatory by McIntyre and Bé (1967). These authors were able to recognise definite water-mass and temperature preferences of various species, and to distinguish five floral zones, which they termed tropical, subtropical, transitional, subarctic and subantarctic. The coccolithophorid species which characterise these floral zones are indicated in Table 1. Boundaries of those zones (Fig. 2) parallel those of the living coccolithophore assemblages, although species characteristic of warm waters generally have a wider distribution in the water than in the sediment. One cold water form, *Coccolithus pelagicus* (Wallich), shows an opposite trend and has a wider distribution in the bottom sediments. McIntyre and Bé (1967) interpreted this in terms of a recent warming trend in the Atlantic Ocean, which has resulted in the migration of living tropical and subtropical species toward polar regions, a trend not yet reflected in the sedimentary assemblages.

The greatest species diversity is found in the tropical and subtropical zones. McIntyre and Bé (1967, Tables 9 and 10) list 19 dominant species, but other published data (Halldal and Markali, 1955; Markali and Paasche, 1955; Hulburt *et al.*, 1960; Gaarder, 1962, 1970; Black, 1968) indicate that more than 60 species of living coccolithophores may be present. Unfortunately, the relative abundances of these species in the population is poorly known, Species diversity decreases toward the polar regions, reaching a minimum of

TABLE 1. Atlantic coccolithophorid floral assemblages. Species arranged in descending order of abundance (after McIntyre and Bé, 1967).

I. TROPICAL
Umbilicosphaera irregularis
Cyclolithella annula
Cyclococcolithus fragilis
Umbilicosphaera tenuis
Discosphaera tubifera
Rhabdosphaera claviger
Helicopontosphaera kamptneri
Gephyrocapsa oceanica
Emiliania huxleyi
Cyclococcolithina leptopora
**Umbilicosphaera mirabilis*
**Syracosphaera pulchra*
**Coccolithus pelagicus*
**Thoracosphaera* spp.
**Scapholithus fossilis*
**Pontosphaera* spp.
**Ceratolithus cristatus*

II. SUBTROPICAL
Umbilicosphaera tenuis
Rhabdosphaera claviger
Discosphaera tubifera
Cyclolithella annula
Gephyrocapsa oceanica
Umbilicosphaera mirabilis
Helicopontosphaera kamptneri
Cyclococcolithina leptopora
Cyclococcolithus fragilis
Emiliania huxleyi
**Syracosphaera pulchra*
**Coccolithus pelagicus*
**Thoracosphaera* spp.
**Scapholithus fossilis*
**Pontosphaera* spp.

III. TRANSITIONAL
Emiliania huxleyi
Cycococcolithina leptopora
Gephyrocapsa ericsonii
Rhabdosphaera claviger
Gephyrocapsa oceanica
Umbilicosphaera tenuis
Coccolithus pelagicus
Umbilicosphaera mirabilis
**Helicopontosphaera kamptneri*
**Syracosphaera pulchra*

IV. SUBARCTIC
Coccolithus pelagicus
Emiliania huxleyi
Cyclococcolithina leptopora
**Gephyrocapsa oceanica*
**Helicopontosphaera kamptneri*
**Syracosphaera pulchra*

V. SUBANTARCTIC
Emiliania huxleyi
Cyclococcolithina leptopora
**Coccolithus pelagicus*
**Gephyrocapsa oceanica*
**Helicopontosphaera kamptneri*
**Syracosphaera pulchra*

* Indicate coccoliths found in the sediment assemblages studied, but not reported by McIntyre and Bé (1967).

four species, *Emiliania huxleyi* (Lohman), *Coccolithus pelagicus*, *Cyclococcolithina leptopora* (Murray and Blackman) and *Syracosphaera pulchra* Lohman, in the subarctic (Zone V of Fig. 2) and two species, *E. huxleyi* and *C. leptopora*, in the subantarctic (Zone IV of Fig. 2).

Emiliania huxleyi is the most ubiquitous and abundant coccolithophore in modern oceans. It constitutes over 50% of the flora in tropical regions and increases to almost 100% in subarctic and subantarctic waters.

Coccospheres of *Umbilicosphaera irregularis* Paasche, *Umbilicosphaera tenuis* (Kamptner), *Discosphaera tubifera* (Murray and Blackman) and *Rhabdosphaera claviger* Murray and Blackman (which McIntyre and Bé considered to be conspecific with *R. stylifer* (Lohman)) are abundant among tropical and subtropical phytoplankton, and often intrude into the transitional zone (see also Hulbert, 1962, 1964). The seasonal variations indicated by the data of McIntyre and Bé and of Hulbert prevent a detailed evaluation of their abundance, but an estimate of about 30% to 40% total for these species may be correct. The relative abundance of each coccolithophore varies with latitude, surface temperature and vertical distribution in the water column (McIntyre and Bé, 1967). *Umbilicosphaera irregularis* accounts for 56% and *U. tenuis* for 64% of the living population. Sediment assemblages studied by McIntyre and Bé (1967, Table 8) indicated a marked decrease in the abundance of these species relative to others.

Gephyrocapsa oceania Kamptner is not found in the subarctic and subantarctic floral zones. In the tropical and subtropical zones, this species accounts for 10% to 15% of the population. In the transitional zone, its abundance increases to 25%.

Cyclococcolithina leptopora (Murray and Blackman) has a wide geographic range, similar to that of *E. huxleyi*. It is reported to occur in most plankton samples, but usually in very small abundances (less than 5%). In the subarctic zone, McIntyre and Bé reported only one occurrence of *C. leptopora*, out of several plankton samples. In the Southern Atlantic the relative abundance of *C. leptopora* is also usually below 5% of the population, with only one occurrence of 24% reported.

Coccolithus pelagicus (Wallich) has been found in plankton samples only in the North Atlantic (Black, 1965; McIntyre and Bé, 1967). McIntyre and Bé indicated the presence of this species in several samples, accounting for less than 0.1% of the population, but in two samples *C. pelagicus* averaged 10%, and in one 50% of the living assemblage.

Among the less common species, *Umbililicosphaera mirabilis* Lohman and *Helicopontosphaera kamptneri* (Hay and Mohler) account for maximum of 6%, each, of the living population.

Variations in coccolith ultrastructure with temperature are recognised. Three varieties of *C. leptopora* have been described by McIntyre *et al.* (1967). Separation of these varieties is based on the number of elements that make up the shield and the diameter of the shield. The largest form, *C. leptopora* var. A, has an average element count of 40. It occurs rarely in post Pliocene sediments, and has been recently renamed *Cyclococcolithina macintyrei* by Bukry. *C. leptopora* var. B has an element count averaging 30, and is a warm water form. The number of elements of *C. leptopora* var. C averages 20. This coccolithophore has a wide geographic distribution, but is distinct in occupying a similar water mass in the Southern Hemisphere as *C. pelagicus* does in the Northern Hemisphere (McIntyre *et al.*, 1970). In this study no attempt was made to separate these varieties because in many assemblages a gradual transition between the different forms was observed. Similar phenotypic variations in morphology are recognised in the ultrastructure of *Emiliania huxleyi* (Lecal-Schlauder, 1966; McIntyre and Bé, 1967). The types of elements of the distal shield indicate the temperature at which they were formed, I-shaped elements being characteristic of warm water, and solid elements indicating cold water.

The formation, sedimentation and burial of coccoliths

In this section, we shall follow the path of the coccoliths from the living cell to their ultimate burial on the bottom of the ocean.

Coccoliths are formed by calcification or organic scales within or at the vicinity of the Golgi apparatus and are gradually expelled toward the periphery of the coccolithophore (Wilbur and Watable, 1963; Manton and Leedale, 1969; Outka and Williams, 1971). In motile, flagellate cells the coccoliths are dispersed loosely on the outside of the cell. In non-motile cells (Parke and Adams, 1960), the coccoliths interlock to form a solid shell—the coccosphere.

TABLE 2. Average number of coccolith per coccolithopore. The left column contains species abundant in the sediment assemblage, arranged in descending order of resistance to dissolution; after Lohman (1902), Schiller (1930), Markali and Paasche (1955), Halldal and Markali (1955), Gaardner (1962, 1970), McIntyre and Bé (1967) and Odkada and Honjo (1970).

Species	No.	Species	No.
Cyclococcolithina leptopora	20	*Syracosphaera nodosa*	60
Emiliania huxleyi	20	*Syracosphaera pulchra*	30
Gephyrocapsa oceanica	20	*Syracosphaera conifera*	110
Umbilicosphaera mirabilis	100	*Syracosphaera adriatica*	60
Helicopontosphaera kampfneri	20	*Syracosphaera molischi*	28
Coccolithus pelagicus	20	*Syracosphaera brandti*	200
Rhabdosphaera claviger	20	*Syracosphaera dalmatica*	100
Umbilicosphaera tenuis	16	*Pontosphaera* spp.	16–80
Cyclolithella annula	12	*Michaelsarsia* sp.	150
Umbilicosphaera tubifera	50	*Scyphosphaera lopadolith*	6
Umbilicosphaera irregularis	25	*Scyphosphaera coccolith*	11–40
Cyclococcolithus fragilis	60	*Acanthoica quattrospina*	80
Gephyrocapsa ericsonii	20	*Holococcoliths*	60

The number of coccoliths per cell varies among the coccolithophorid species (Table 2), motile cells generally having three to four times as many coccoliths as non-motile cells. Relative proportions of coccoliths in representative coccolithophore populations (McIntyre and Bé, 1967) were computed (Table 3). Abundances of coccolithophores of *Umbilicosphaera tenuis*, *Rhabdosphaera clavigera* and *Gephyrocapsa oceanica* were rounded to 1%. In a hypothetical assemblage composed of coccoliths released from the coccospheres of Table 3, the abundance of the coccoliths of *Acanthoica quattrospina* is expected to increase threefold, and those of *Pontosphaera nana* and *Syracosphaera* spp. twofold in the sediment assemblages as compared with the abundances of cells in the living population. The coccoliths of *Emiliania huxleyi* would account for only 56% of a coccolith assemblage, compared with 78% for the abundance of the species in the living population.

The relative abundance of coccoliths in a living assemblage was recalculated for a plankton sample, Vidal 1965—10 (from McIntyre and Bé, 1967, Table 2), which had a large species diversity. The coccolithophores recorded for this sample by McIntyre and Bé account for only 75% of the population. For the present study this was rounded to 100% by assuming, that *Cyclococclithus fragilis* and *Umbilicosphaera mirabilis* account for 1% each (by comparison with other samples) and that the remaining 23% of the population was composed of species of holococcoliths and syracosphaeriids (Table 4). According to this recalculation placoliths have decreased in abundance, while the proportion of cyrtoliths and caneoliths has increased relative to the proportion

TABLE 3. Distribution of coccoliths in a representative living population. (Coccolithopore proportions from McIntyre and Bé, 1967, Table 9.)

Species	Coccolithopore proportion (%)	No. of coccoliths/cell	Coccolith proportion (%)
Acanthoica quattrospina	1	80	2.9
Pontosphaera nana	1	50	1.8
Syracosphaera spp.	15	60	32.7
Discosphaera tubifera	1	50	1.8
Umbilicosphaera tenuis	1	16	0.6
Rhabdosphaera claviger	1	20	0.7
Emiliania huxleyi	78	20	56.6
Cyclococcolithina leptopora	2	20	1.5
Coccolithus pelagicus	1	20	0.7
Gephyrocapsa oceanica	1	20	0.7

of coccolithophores in the living population. These results indicate that in a sediment composed of well-preserved coccoliths, the relative abundance of cyrtoliths and caneoliths can be expected to be almost twice as high as that of their coccolithophores.

TABLE 4. Distribution of coccoliths in a living population. (Coccolithophore data modified from McIntyre and Bé, 1967; plankton sample Vidal–10; 21°N; 31°W.)

Species	Coccolithopore proportion (%)	No. of Coccoliths/cell	Coccolith proportion (%)
Discosphaera tubifera	11	50	17.1
Umbilicosphaera irregularis	1	25	0.8
Umbilicosphaera tenuis	40	16	20.0
Rhabdosphaera claviger	10	20	6.2
Cyclolithella annula	1	12	0.4
Emiliania huxleyi	8	20	5.0
Cyclococcolithina leptopora	1	20	0.6
Gephyrocapsa oceanica	2	20	1.2
Helicopontosphaera kamptneri	1	20	0.6
Cyclococcolithus fragilis	1	60	1.9
Umbilicosphaera mirabilis	1	100	3.1
Syracosphaera spp. + other	23	60	43.0

The means by which the coccoliths are detached from the cell and enter the environment are largely unknown. A high rate of coccolith production, resulting in multiple layers of coccoliths covering the cell was observed in natural populations of *E. huxleyi* by Lohman (1902) and in cultures by Braarud (1963) and Wilbur and Watabe (1963). Okada and Honjo (1970) have reported a coccosphere of *Syracosphaera nodosa* covered with two layers of coccoliths. Expulsion of coccoliths of *E. huxleyi* by the growing cell was noted in cultures by Mjaaland (1956), Braarud (1963) and Paasche (1968b). Braarud observed that the coccoliths are absent on cells during dark periods, suggesting that their presence was associated with levels of illumination. Paasche (1967, 1968a) was not able to confirm Braarud's hypothesis and suggested that coccoliths are shed due to excessive production.

The life span of coccolithophores is indefinite because they usually reproduce by binary fission (Paasche 1968a). The coccoliths are passed on to daughter cells (Lohman, 1902; Lecal-Schlauder, 1951; Paasche, 1968a), so that none are released into the environment. Braarud (1963) has reported reproduction of *E. huxleyi* involving shedding of the coccosphere, but this was not observed by Paasche (1968a). A sexual reproductive cycle was observed by Schwarz (1932) in *Orchrosphaera neapolitana*, but has not been noted in recent studies by Parke (1961).

Expulsion of coccoliths from cells and reproduction can not be responsible for the bulk of coccoliths deposited in the sediment. The largest number of coccoliths is probably released through the disintegration of the cell following death. Massive kills of whole populations have been observed in Norwegian fiords during *E. huxleyi* blooms (Birkenes and Braarud, 1952). Such catastrophic events may occur in areas of upwelling or semiclosed basins, but are probably rare in the open ocean. The other way in which cells are killed is by their utilization as food by zooplankton.

Coccoliths released directly into the water will settle slowly to the bottom and be buried. The estimated settling times for coccoliths in oceanic depths range from one to two years according to Smayda (1970) to 10 years (Bramlette, 1961). The rates of sedimentation in the Atlantic Ocean are on the order of 2 cm per 1000 years, or 20 microns per year. A coccolith one micron thick will have a residence time of only 18 days at the sediment water interface before being covered by other coccoliths. A coccosphere twenty microns thick will be covered after one year. The residence time of a coccolith on the sediment surface therefore is comparable to its settling time, during which the coccolith is in contact with the water column. In view of the corrosive nature of deeper oceanic waters to calcium carbonate (Peterson, 1966; Berger, 1967), the preservation of coccoliths at depths of 5000–6000 m requires inhibiting mechanisms.

Factors which may inhibit dissolution of calcareous nannofossils are:

(1) Presence of organic material coating coccoliths which influences the kinetics of their dissolution.

(2) Rapid settling of coccoliths as large aggregate bodies, reducing their exposure to the water column.

(3) Redeposition from shallower sites, both by mass transport (turbidity currents) and by resuspension.

(4) Rapid burial in the sediment.

ORGANIC MEMBRANES AND COATINGS

Wangersky (1969) studied the distribution of suspended carbonate particles at various depths and noticed their stability throughout the water column. He believed that preservation of $CaCO_3$ in the deep sea is largely a result of protective coating by organic material and that the rate of dissolution of the $CaCO_3$ is governed by the rate of destruction of this organic material. Organic inhibition of dissolution was also postulated by Smith *et al.* (1968) to explain discrepancies between expected depth control and the observed $CaCO_3$ content of pelagic sediments. Wangersky and Joensuu (1967) indicated that tests of planktonic foraminifera are resistant to solution so long as their outer organic coating is intact. This was demonstrated by Berger (1967) who found that tests of planktonic foraminifera which had been boiled in H_2O_2 and suspended in the sea on a buoy underwent fragmentation and lost more weight than untreated control tests suspended on the same buoy. McIntyre and McIntyre (1971) were first to suggest that organic membranes might prevent the rapid dissolution of coccoliths.

Organic envelopes, remaining after the dissolution of coccoliths, have been described by Braarud *et al.* (1952) and Honjo (1969). Scales of organic material covering the proximal parts of Recent coccoliths have been observed on suspended material by McIntyre and Bé (1967), Norris (1971) and Black (1968). Studies of coccolith formation have demonstrated a connection between coccolith deposition and presence of organic templates (see Paasche, 1968a and references therein). Manton and Leedale (1969) indicated the presence of organic scales on the proximal shields of *C. pelagicus* and *Cricosphaera* (Syracosphaera) *carterae*, as well as the presence of organic membranes on their distal shields. Decalcified coccoliths of *Hymenomonas* were found to contain proteins and various sacchariids (Isenberg *et al.*, 1966); Green and Jennings (1967) have shown that the scales are composed of carbohydrates. If absorbed organic material inhibits dissolution, then coccoliths with organic envelopes will also be protected.

The proximal surfaces of *C. leptopora* and *C. pelagicus*, observed during this study appeared to be smooth, structureless, and hazy (pl. 3, fig. 6). This is probably an indication of the presence of an organic membrane or scale of the kind described for *C. pelagicus* by Manton and Leedale (1969). The dissolution of the proximal shield seems to be inhibited until the organic membrane has been destroyed.

RAPID SETTLING IN AGGREGATES

Mixed assemblages, containing both well-preserved and corroded specimens were found in several samples. The presence of well-preserved specimens suggests redeposition or rapid transportation inside a protective body. The deposition of coccoliths within faecal pellets would explain such occurrences since these settle rapidly and protect the coccoliths from the corrosive waters (Smayda, 1970). Kullenberg (1953), calculated the amount of fine grained particles suspended in the water column, and concluded that a considerable part of the suspension must be deposited rapidly as aggregates.

Coccolithophores appear to be a major food source for zooplankton.

Lohman (1902) found well-preserved coccolithophores in the guts and faecal pellets of salps, pteropods and appendicularia. He believed that many of the coccoliths in the sea might be transported to the bottom in faecal pellets. Beklemishev (1954) found that in warm seas coccolithophorids are consumed in quantity by copepods. Steeman and Jensen (1957) suggested that in tropical oceans more than 95% of the living algae eaten are consumed in the upper 200 m. The deep living zooplankton and the bottom fauna consume the remainder of the decaying and slowly sinking phytoplankton remains (Sokolova, 1959; Wolff, 1960). Marshall and Orr (1955) have tested *in vitro* utilisation of coccolithophores *Emiliania huxleyi*, *Hymenomonas* (Syracosphaera) *carterae* and *Syracosphaera oblonga* as a food source for the copepod *Calanus*. They indicate a very high utilisation rate, reaching 12 000 cells per copepod per day! During one day, *Calanus* produces 30–40 faecal pellets. Faeces are coated by an organic pellicle, and appear to be firm. The rate of food digestion appears to be high, but many intact cells were found in the pellets. Fretter and Montgomery (1968) have studied the feeding of the pelagic veliger larvae of marine prosobranch gastropods, determining that digestion of *Cricosphaera carterae* took place, but that well-preserved coccospheres and coccoliths with intact organic scales were still found in the faecal pellets. Norris (1971) dissected living salps and found well-preserved coccospheres and coccoliths in their digestive tracts. Some coccoliths even had intact organic scales.

The biological literature indicates that a considerable part of the coccolithophorid population is digested and the coccoliths released to the environment as faecal pellets. Pellets produced by copepods were collected from the water column by Moore (1931). Faecal pellets were found in the deep sea by Moore (1931a, b, 1933, 1939) and Bramlette and Bradley (1946). Griggs *et al.* (1969) studying the sediment-fauna interaction in the Cascadia Abyssal Plain, found pelagic clays with a coarse fraction composed of 95% faecal pellets. Similar pellets also were found by them in clearly redeposited layers.

The faecal pellets deposited in the sediment may be a result of multiple utilisation of the organic material. During descent, coccoliths and faecal pellets may be utilised several times by filter feeding organisms. Material deposited on the bottom is consumed again by various filter and mud feeding organisms. In contrast to faecal pellets of planktonic animals, those of benthic animals are solid and resistant (Moore, 1931b), and can withstand redeposition by turbidity current or bottom currents. The majority of pellets described in the literature were probably excreted by bottom dwelling organisms.

A lack of reports on faecal pellets in modern studies on deep sea sediments probably is due to preparation techniques employed. The common treatment of deep sea samples, particularly those with high organic content, includes both the oxidation of the organic material with hydrogen peroxide and ultrasonic agitation (McIntyre *et al.*, 1967). Moore (1931a) specifically cautioned that treatment with H_2O_2, drying or washing of the material would easily destroy pellets produced by zooplankton. This might be reason for their absence from millipore filtered samples.

During this study, numerous elongate and oval irregular bodies which contain coccoliths were discovered in an untreated sample (V19–303; 12°N; 27°W; 5426 m). These bodies had a hazy appearance in the scanning electron microscope (pl. 1, figs 1–6) which is generally attributed by SEM workers (Hansen *et al.*, 1969) to the presence of organic material. The calcareous nannofossil assemblages extracted from the pellets contained well-preserved coccoliths, including species not otherwise found at such depths. The presence of a partially preserved coccosphere (pl. 1, fig. 4) in one of these pellets may indicate that this is a faecal pellet of some planktonic filter feeding organism—no coccospheres are found in the Recent oceanic sediments. A subsequent search confirmed the presence of such pellets in other samples (pl. 1). It is not clear whether the pellets were deposited from turbidity currents, originating at shallower depths, or settled directly from organisms in overlying waters. Pellets found in several samples (P6303 st 23; P6511 st 20; P6511 st 26; P6511 st 32; P6707B st 16) are probably faeces of bottom dwelling organisms. It is possible, however, that the sediment originally contained more pellets from planktonic animals and that these were destroyed by burrowing organisms, by compaction during coring and sampling, or during sample preparation.

REDEPOSITION

The addition of material from one site to another can also occur through the activity of turbidity currents. In most samples studied, no clear indication of redeposition by turbidity currents occurs. However, in Guiana Basin samples, clearly reworked green clays of the continental slop contain abundant silt, rare diatoms and well-preserved *Gephyrocapsa oceanica* and *Emiliania huxleyi*. In deeper samples, this well-preserved assemblage is intermixed with a poorly-preserved one, composed mainly of *C. leptopora* and *H. kamptneri*. Redeposition from shallow depths was indicated in sample P6511 st 26 and P6511 st 31, where the coccolith assemblage contains *Braarudosphaera bigelowi* (Gran and Braarud) and *Scyphosphaera apsteini*; species not found in any other samples during this study.

In a few samples deposited within the compensation zone, well-preserved assemblages occur intermixed with partly dissolved assemblages of the same species. There is no physical separation between the two kinds of coccolith and it is suggested that the well-preserved assemblage is a result of redeposition from shallower depths. In addition to the downslope introduction of well-preserved coccoliths, clay minerals and small coccoliths seem to be removed from areas of high current velocity, resulting in an increase of the relative proportion of larger coccoliths, as observed in samples CHN75 st 18 (13°′N; 43°′W, 3643 m) and CHN75 st–1 (13°′N; 45°′W, 3148 m) containing 14.2% and 32.8% *Helicopontosphaera kamptneri* respectively. These samples were deposited on the flanks of the Mid Atlantic Ridge, where ripple marks and sediments scouring have been observed (Heezen, 1963).

RAPID BURIAL

Rapid isolation of an assemblage from the ambient waters can occur due to high rates of sedimentation or to burial by a turbidite. The influence of turbidity currents is evident in some Pleistocene samples (core P7008–44 st H) where the coccoliths of the pelagic fraction were more dissolved than coccoliths of the turbidite layer. No additional evidence is available to warrant a discussion of this matter.

The distribution of coccoliths in Atlantic Ocean sediments: selective dissolution and degree of dissolution

McIntyre and McIntyre (1971) have recently reported that coccolith assemblages studied in five closely-spaced cores in the North Atlantic had undergone differential dissolution, resulting in a relative increase in placoliths with depth. Comparing these with other cores McIntyre and McIntyre concluded that their cores, covering the interval of 4524 m to 5466 m represented the maximum depth of preservation, and that coccoliths usually undergo dissolution in shallower depths. They also suggested a relationship between dissolution and skeletal architecture and recorded that placoliths were more resistant to solution than caneoliths. Schneidermann (1971) reported comparable selective dissolution of coccoliths in the Equatorial Atlantic, but recorded coccoliths from depths as great as 6000 m.

The distribution of coccolith species in the sediments approximates the distribution of the parent coccolithophores in the phytoplankton. The greatest diversity is observed in the Equatorial Atlantic sediments. Toward the polar regions the number of species recorded decreases and the assemblage is composed mainly of placoliths. Solution with increasing depth results in an increase in the relative abundance of the more resistant species and a decrease in species diversity.

Table 5 and Fig. 3 record the solution induced depth distribution of Recent coccoliths within the compensation zone of the Equatorial Atlantic. In Fig. 3 *Emiliania huxleyi* and *Gephyrocapsa oceanica* are grouped together because the two species become progressively similar with increased solution and are almost impossible to differentiate with the light microscope in samples from

TABLE 5. Depth distribution of Recent coccoliths in the compensation zone of the Equatorial Atlantic.

Species	Depth (100 m) 37	38	39	40	41	42	43	44	45	46	47	48	49	50	51	52	53	54	55	56	57	58	59	60
Cyclococcolithina leptopora	—	—	—	—	—	—	—	—	—	—	—	—	—	—	—	—	—	—	—	—	—	—	—	—
Emiliania huxleyi	—	—	—	—	—	—	—	—	—	—	—	—	—	—	—	—	—	—	—	—	—	- -	- -	- -
Gephyrocapsa oceanica	—	—	—	—	—	—	—	—	—	—	—	—	—	—	—	—	—	—	—	—	—	- -	- -	- -
Umbilicosphaera mirabilis	—	—	—	—	—	—	—	—	—	—	—	—	—	—	—	—	—	—	—	—	—			
Helicopontosphaera kamptneri	—	—	—	—	—	—	—	—	—	—	—	—	—	—	—	—	—	—	—	—				
Coccolithus pelagicus	—	—	—	—	—	—	—	—	—	—	—	—	—	—	—	—	—	—	—					
Rhabdosphaera claviger	—	—	—	—	—	—	—	—	—	—	—	—	—	—	—	—	—	—						
Syracosphaera pulchra	—	—	—	—	—	—	—	—	—	—	—	—	- -	- -	- -	- -	- -							
Umbilicosphaera tenuis	—	—	—	—	—	—	—	—	—	—	—	- -	- -	- -	- -	- -								
Scapholithus fossilis	—	—	—	—	—	—	—	—	—	—	—	—	—	- -	- -									
Ceratolithus cristatus	—	—	—	—	—	—	—	—	—	—	—	—	—	- -	- -									
Umbilicosphaera irregularis	—	—	—	—	—	—	—	—	—	—	—	- -	- -	- -										
Cyclolithella annula	—	—	—	—	—	—	—	—	—	—														
Thoracosphaera spp.	—	—	—	—	—	—	—	—	—															
Discosphaera tubifera	—	—	—																					
Cyclococcolithus fragilis	—	—	—																					

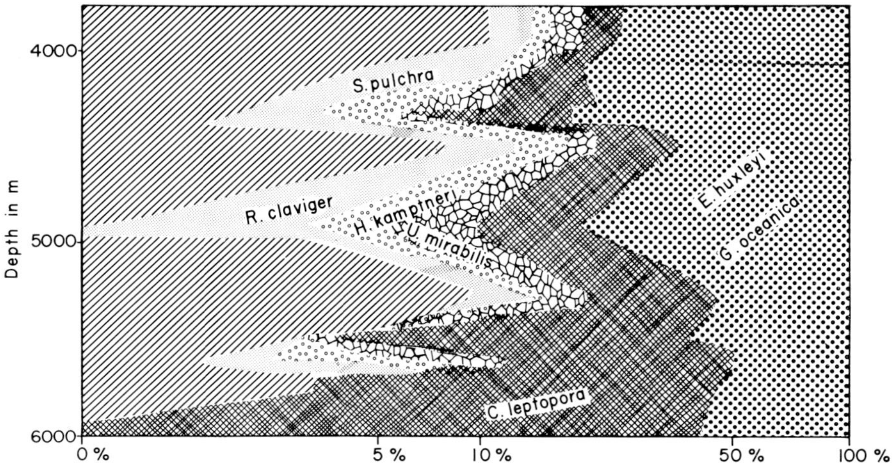

FIG. 3. Depth distribution of coccolith species in bottom sediments of the Equatorial Atlantic (traverse P6303). The unmarked left column indicates the relative proportion of *C. pelagicus*, *U. tenuis* and reworked Tertiary coccoliths. In the upper part, less solution-resistant coccoliths are present.

depths below 5000 m. Changes in the proportion of species due to selective solution are already evident in samples from depths of 3000 m. Holococcoliths (coccoliths constructed of uniform rhombs or prisms) were not observed in the bottom sediments. Caneoliths and scapholiths including *Syracosphaera pulchra* and *Scapholithus fossilis* do not occur below 4900 m, and species of *Thoracosphaera* are not recorded below 4800 m. Most cyrtoliths are absent from sediments below 4000 m but rare specimens of *R. claviger* occur at depths of 5600 m. Below 5000 m the assemblage is mainly composed of a few placoliths species: *Emiliania huxleyi*, *Gephryocapsa oceanica*, and *Cyclococcolithinia leptopora*. *C. leptopora* is the most resistant species (Fig. 4) and rare distal shields are found at depths of 6000 m. A striking result of preferential solution in low latitudes is the gradual introduction of a "cold" aspect to the remaining assemblage. If we exclude rare specimens of *Umbilicosphaera mirabilis* and *H. kamptneri* the remaining assemblage has the characteristics of subarctic or subantarctic assemblage.

In higher latitudes selective solution occurs in shallower depths and reflects active solution by cold bottom waters. The diversity in living assemblages in these regions is small and the sedimentary assemblages are composed mainly of *C. leptopora*, *E. huxleyi* and less commonly *C. pelagicus*. Due to differences in the initial composition of the assemblages, shallower (2000–3000 m depth) samples from the North Atlantic contain virtually the same coccoliths as much deeper (5500–6000 m depth) samples from the Equatorial Atlantic.

The increase in the relative abundance of solution resistant species within a sedimentary assemblage can be used to quantify the degree of dissolution.

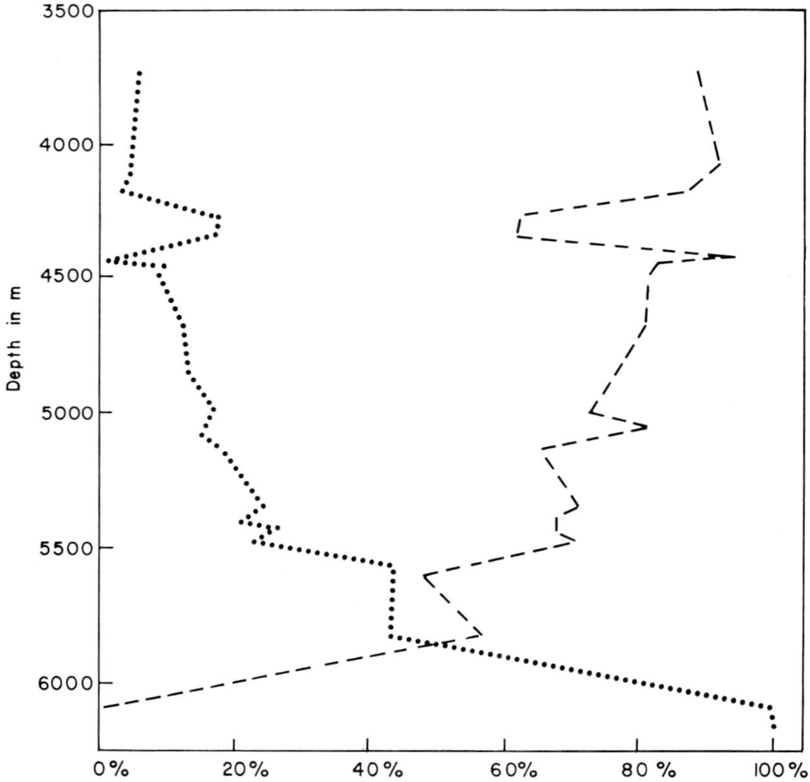

FIG. 4. Changes in the abundance of *Cyclococcolithina leptopora* (· · · · ·) and *Emiliania huxleyi* with *Gephyrocapsa oceanica* (– – – –) with depth; traverse P6511 and lower part of samples from traverse P6303.

In a given living assemblage the initial proportion of resistant species is rather small, and their greater abundance in sedimentary assemblages occurs because non-resistant or less resistant forms are dissolved. Berger (1971) measured the amount of dissolution in foraminiferal assemblages using the formula

$$L = 100\,(1 - Ro/R)$$

where Ro is the initial proportion of resistant species in the living population, R is the proportion of these species in the sediment and L is a measure of the degree of dissolution and is termed the "dissolution index". As is discussed below, this index may be used to characterise and compare relative dissolution among sedimentary assemblages of coccoliths.

C. leptopora, a widely distributed species and apparently the most solution-resistant species, has been chosen to compute dissolution indices. Using the value of $Ro = 1.5\%$ for a living tropical population (McIntyre and Bé, 1967) the following dissolution indices were computed for various assemblages in the central Atlantic:

(*a*) 2626 m (core V 16–191; 27°S; 43°W; McIntyre and Bé 1967). $Ro = 1.5\%$ and $R = 9\%$ indicating 83% dissolution.

(*b*) 3995 m (P 6707 B st. 37; 00°N; 25°W). $Ro = 1.5\%$ and $R = 16\%$ indicating that approximately 90% of the initial assemblage has dissolved.

(*c*) 5426 m (P 6511 st. 21; 29°N; 24°W). $Ro = 1.5$ and $R = 26\%$ indicating 94% dissolution.

(*d*) 5800 m (P 6303 st. 15; 24°N; 35°W). $Ro = 1.5$ and $R = 50\%$ indicating 97% dissolution.

The computation of dissolution indices in the subarctic is difficult owing to a paucity of data concerning the abundance of *C. leptopora* in the living population and hence the value of *Ro*. Assuming that $Ro = 2\%$ (based on the minimum value of this species in bottom sediments) than at 2725 m (V. 27–63; 70°N, 12°W) $Ro = 2\%$ and $R = 20\%$, 90% of the initial population has been dissolved. Similar amounts of dissolution are only recorded from depths of 4000 m and greater in the central and Equatorial Atlantic.

SELECTIVE REMOVAL OF ULTRASTRUCTURAL ELEMENTS

The dissolution of species with depth, discussed in the previous section, does not occur suddenly at a particular level, but through a distinct depth interval. Each coccolith is dissolved in a specific way with certain structural elements removed before others. This selective removal of structural elements provides an additional means of subdividing the compensation zone.

An elaborate terminology has been created by Halldal and Markali (1955); Braarud *et al.* (1955) Hay *et al.* (1966) and Farinacci (1970) to describe the coccolith architecture. The terminology applies mainly to abundant, geographically widespread and solution resistant placoliths. Coccoliths of this group are composed of oval to rounded parallel oriented discs (shields) constructed of radiating elements and connected by a short tube. The external shield is termed "distal," and the usually smaller, inward facing shield is termed "proximal".

Emiliania huxleyi (pl. 2, figs 1–6) is probably the best known Recent placolith. The coccoliths of this species have been described in detail by Black and Barnes

(1961), the coccospheres by Halldal and Markali (1955), and living cells by Watabe and Wilbur (1966), Paasche (1968b) and McIntyre and Bé (1967). This species is highly tolerant to environmental variations, and therefore easily cultured in the laboratory (Isenberg *et al.*, 1965). Phenotypic variations in morphology have been recognised in experiments (Wilbur and Watabe, 1966) and in the field (Lecal, 1966; McIntyre and Bé, 1967; Bartolini, 1970).

The shields of *E. huxleyi* are subequal in size and convex distally. The central column connecting the shields contains a large elliptical central area, covered by a plate or a grill made of interconnected rods (pl. 2, fig. 1). The distal shield is composed of separate elements shaped like the letter "I", with slits between adjacent elements. The proximal shield may be constructed of similar I-shaped elements, or it may be constructed of broader elements which leave no gaps along the sutures.

Well-preserved specimens of *E. huxleyi* were found in the Equatorial Atlantic to depths of 4000 m. Between 4000 m and 5000 m the central plate and most of the I-shaped elements disappear (pl. 2, figs 4–6) though some isolated elements, may be preserved as broken rods, are found down to 5500 m depth. The bulk of the coccoliths of *E. huxleyi*, however, are preserved as tubes or thin rings containing protruberances in the sites originally occupied by I-shaped elements (see also McIntyre and McIntyre, 1971). In the Northern Atlantic, dissolution of the central area and the I-shaped elements occurs in shallower depths, at about 3100 m at 59°N (V23–28; 37°W) latitude and only 2725 m at 70°N (V27–63; 12°W). Samples from the same area (V27–53; 69°N; 02°W) but deposited in deeper waters of 3124 m contain mostly remains of the tubes. A similar situation is observed in the Southern Atlantic Ocean where at 45°S (RC12–290; 20°W; 4276 m) the sediment contains abundant tubes and rings. The asymmetric dissolution rates of $CaCO_3$ about the Mid Atlantic Ridge is evidenced by the dissimilar preservation at comparable depths. In the Western Atlantic (P6303 st 23; 26°N; 51°W; 5224 m) the coccoliths of *E. huxleyi* have lost most of the I-shaped rays and the central grill, whereas east of the Mid Atlantic Ridge (sample P6303 st 11; 21°N; 26°W; 5270 m) at the same depth, many coccoliths of *E. huxleyi* are well preserved (Fig. 6).

Gephyrocapsa oceanica (pl. 3, figs 1–4) and related forms (e.g. *G. caribbeanica*, *G. ericsonii*, see McIntyre and Bé, 1967; Boudreaux and Hay, 1969) are composed of two shields closely connected by a short tube. The shields are composed of pointed elements giving a serrate appearance to the shield margins. A large central area is covered on the proximal side by a very delicate, easily destroyed grill similar to that of *E. huxleyi*. Two bars arising from opposite sides of the central area on the distal side, join together to form a bridge. The overall size, number of elements, size of the central opening and orientation of the bridge with respect to the axes of the shields are used to differentiate the species of *Gephyrocapsa*. Well-preserved specimens are rarely found below 4500 m, although in the Eastern Equatorial Atlantic they occur occasionally at 5000 m. Partial dissolution of the bridge and proximal shield is common in specimens deposited around 5000 m. Below 5000 m it becomes difficult to separate the remains of *G. oceanica* from those of *E. huxleyi*, though the larger size and more imbricate nature of the elements of *G. oceanica* can be used to distinguish between the two species. In the Northern Atlantic, the bridge begins to disappear at 2725 m depth (V27–63; 70°N; 12°W) and is absent at 3124 m (V27–53; 69°N; 02°W). In the Southern Atlantic, the bridge is almost removed at 3037 m (sample V22–106; 46°S; 10°W).

Placoliths of *Cyclococcolithina leptopora* (pl. 3, figs, 5 and 6; pl. 4, figs 1–6) and *Coccolithus pelagicus* (pl. 5, figs 1–4) are built of tightly appressed, imbricate elements with no central pore or only a small central opening. Both are solution resistant, and undergo dissolution in a distinctive manner. Coccoliths of *C. leptopora* have large, distally convex shields, connected by a short central column. The proximal shield has a smaller diameter and often appears featureless in the scanning electron microscope. It is possible that this shield is covered by an organic scale, similar to that of *C. pelagicus*. The small central pore in the proximal side is covered by a thin plate of rhombic crystals. Comprehensive discussions of the structure of *C. leptopora* have been given by Kamptner (1954) and by Gartner (1967).

Well-preserved coccoliths of *C. leptopora* were still found at depths of 5000 m, but, in many assemblages parts of the proximal shield have already dissolved at depths between 4500 and 5000 m. The destruction of the proximal shield is preceded by increasing exposure indicated by sutures between elements and etch marks suggesting removal of the supposed organic scale. (pl. 3, figs 5 and 6). The dissolution seems to proceed more rapidly along the sutures. In some samples, separation of distal and proximal shields occurs due to dissolution or disintegration of the central tube. Coccoliths of *C. leptopora* deposited between 5000 m and 5500 m lose their proximal shields and the central tube (pl. 4, figs 1–6). Distal shields of *C. leptopora* are found to depths of 6000 m. The final destruction occurs through rapid dissolution proceeding along suture lines both from the periphery and the centre (pl. 4, figs 5 and 6).

The proximal shield of *C. pelagicus* is smaller than the distal shield and contains flattened rhomboidal elements covering the central tube. The elongated elements of this shield and the central tube may be covered by an organic membrane, which appears featureless under the scanning electron microscope. Such a membrane was found in *C. pelagicus* by Manton and Leedale (1969). The distal part of the central tube is nearly filled by spirally arranged rhomboidal elements. The placolith is elliptical in comparison to the circular outline of *C. leptopora*.

In the sediments of the Equatorial Atlantic, placoliths of *C. pelagicus* do not constitute an important fraction of the coccolith assemblage. Well-preserved shields are found at 5444 m (P6303 st 28; 26°W; −8°W), although the separation of shields has been observed to begin at 4900 m. The abundance of *C. pelagicus* increases toward the subarctic zone, where a dissolution trend, similar to that described for *C. leptopota*, is observed. Upon destruction of the smooth cover, the proximal shield disintegrates (pl. 5, figs 1 and 2). In sample V23–28 (59°N; 37°W; 3102 m) about 40% of the coccoliths have lost the proximal shield. Further dissolution removes the central tube, so that at 69°N (V27–53; 02°W; −124 m) more than 70% of *C. pelagicus* is represented by distal shields (pl. 5, figs 3 and 4). In the Southern Atlantic coccoliths of *C. pelagicus* are rare, precluding detailed observation of the dissolution sequence. Partially dissolved distal shields are found in an almost pure diatomite (V22–106; 46°S; 10–W; 3037 m).

Coccoliths of *Helicopontosphaera kamptneri* (pl. 5, figs 5 and 6; pl. 6 fig. 1) *Rhabdosphaera claviger* (pl. 6, fig. 2) and *Syracosphaera pulchra* (pl. 6, figs 3 and 4) exhibit an orderly sequence of dissolution. The shallow lopadoliths of *H. kamptneri* are highly resistant, and well-preserved specimens may be found in samples from the vicinity of the CCCD (P6303 st 15; 23°N; 35°W; 5837 m). The initial dissolution removes parts of the broad flange and enlarges the slits of the central area (P6707B st 53; 00°S; 34°W; 4367 m) (pl. 5, figs 5 and 6). Dissolution proceeds to enlarge the slits in the direction of the long axis, so that below 5000 m (P6707B st 20; 00°N; 18°W; 5015 m) coccoliths of this species are commonly separated into two halves (pl. 6, fig. 1).

Rhabdosphaera claviger (pl. 6, figs. 2) is composed of a shield-like base, from which a long, central process (stem, spine, appendix) arises. Coccoliths of this type, termed "cyrtoliths" respond to dissolution by separation of the base from the spine. This occurs in the 4000–4500 m interval although some intact specimens of *R. claviger* are found down to 5000 m. Below this depth, only the spines are found.

Syracosphaera pulchra produces a basket-shaped coccolith, termed a caneolith. The margin of the basket is termed a "wall", and is composed of ring-like distal and proximal shields. The base of the caneolith has a central area covered by a grill of interdigitated rods. The central area shows signs of dissolution even in sediments deposited at 3148 m (CHN75 st 31; 13°N; 45°W), the removal of the grill is complete at about 4500 m (P6707B st 18; 00°S; 18°W; 4380 m). At this level the proximal wall begins to dissolve, and below 4800 m to 5000 m only the ring-like distal part of the wall is present (pl. 6, figs 3 and 4). The differentiation of this wall from ring-like remnants of other coccoliths is difficult. The deepest occurrence found in this study was at 5123 m (P6511 st 39; 20°N; 67°W). In the Northern Atlantic the grill is dissolved at 2765 m depth (V27–78; 75°N; 05°E) whereas in the Southern Atlantic it is absent at a depth of 3393 m (sample RC12–293; 36°S; 13°W).

Surfaces on which members of a species exhibit a similar state of dissolution, e.g. the level at which I-shaped elements are removed in *E. huxleyi* or the surface at which the grill in the central area of *S. pulchra* is removed, rise northward and southward from the Equatorial zone, occupying successively shallower levels. Most dissolution occurs between 4000 m and 5000 m, indicated by the removal of the more readily soluble structural elements of various coccoliths. This depth interval has hydrographic significance, as will be discussed later, corresponding to the mixing zone of Antarctic Bottom Water and North Atlantic Deep Water (Wüst, 1936). The rapid increase in the rate of dissolution of planktonic foraminifera ("lysocline" of Berger, 1968), occurs within the same interval.

Placoliths appear to be the most solution resistant coccoliths. It is not clear why the distal shield is more solution resistant than the proximal. It is noteworthy, however, that the distal shields of *C. leptopora*, and *C. pelagicus* are constructed from calcareous elements whose C-axes parallel the surface, whereas the optical axes of proximal shields and of other coccoliths are inclined to the surface. Since dissolution occurs more readily along the C-axis the parallel orientation of this axis may minimise solution of distal shields. The dissolution sequence of the coccoliths is probably dependant on crystallographic orientation, the degree of imbrication of the elements, protection by organic envelopes, etc. Open structures or central areas are dissolved more readily than walls and shields constructured of imbricate elements. Because the dissolution resistant species *C. leptopora* and *C. pelagicus* have related forms ranging back to the Jurassic, their dissolution sequence may provide a key to determination of proximity to the CCCD in the past.

Subdivision of the compensation zone

Hsu and Anderson (1970) distinguished five dissolution facies in pelagic sediments of the South Atlantic. These facies were based on the percentage of the non-carbonate fraction, which ranged from over 90% to less than 5% and on the percentage and state of preservation of the planktonic foraminifera. As their terminology is equally applicable to the sequence of dissolution and state of preservation of species of calcareous nannoplankton observed by the author (Table 6), Hsu and Anderson's terms are used to subdivide the compensation zone (Fig. 5).

TABLE 6. The bathymetry of dissolution facies in the Equatorial Atlantic based on the preservation of coccolith assemblages.

	Recent calcareous nannoplankton (from this study)	
Facies	Bathymetry related to CCCD	Preservation
Alytic	3500 m or more above the CCCD	Good
Eolytic	3000 m above the CCCD	Initial dissolution of soluble coccoliths
Oligolytic	1500 m above the CCCD	Removal of soluble coccoliths
Mesolytic	500 m above the CCCD	Only solution-resistant coccoliths present
Hololytic	below CCCD	none

MESOLYTIC ZONE (Table 6, Fig. 5)

This is the lowest part of the compensation zone and immediately overlies the CCCD, red clays and diatamaceous or radiolarian ooze of the Hololytic Zone. Sediments deposited in this zone in low latitudes are characterised by: (1) Low $CaCO_3$ content, usually not exceeding 10%, (2) A coccolith assemblage composed mainly of poorly-preserved tubes and shields of *E. huxleyi* and *G. oceanica* and comparatively well-preserved distal shields of *C. leptopora*. Specimens of *H. kamptneri*, *U. mirabilis* and *C. pelagicus* may account for about 10% of the total assemblage. The distal shields of *C. leptopora* are dominant at the base of the zone, (3) The $CaCO_3$ fraction of the sediment is composed almost entirely of coccolith derived carbonate. At higher latitudes, the relative abundance of *C. pelagicus* increases, whereas *H. kamptneri* and *U. mirabilis* may disappear from the assemblage. In the subarctic zone, coccoliths of *G. oceanica* become rare.

The thickness of the mesolytic zone varies: it appears to decrease in the Western and Northern Atlantic, where the increase in solubility may overcome the ultrastructural resistance and where the initial output of $CaCO_3$ is low. In the Southern Atlantic (South of 45°S) the increased thickness of the mesolytic zone to +2000 m, reflects the increasing influence of the Antarctic Bottom Waters and dilution by diatoms.

The sediments of the mesolytic zone rarely contain tests of planktonic foraminifera, although some broken fragments are occasionally found.

OLIGOLYTIC ZONE (Table 6, Fig. 5)

In low latitudes this zone occurs in the depth range 4400–4500 m to 5500 m. At the upper boundary of the zone, which corresponds with Berger's (1968) Lysocline, there is a rapid increase in the rate of dissolution of planktonic foraminifera and calcareous nannoplankton. In the lower part of the zone the following parts of solution resistant species are dissolved: the proximal shield of *C. leptopora* and *C. pelagicus*; the bridge of *G. oceanica*; the central

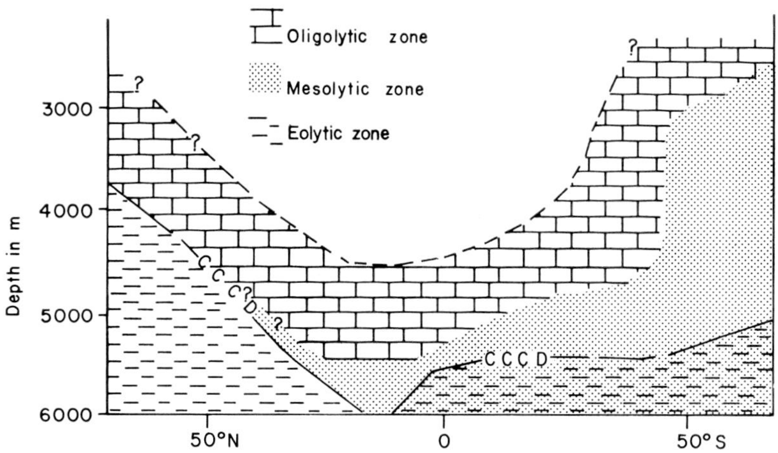

FIG. 5. Dissolution zones of the Atlantic Ocean. Position of CCCD after Lisitzin (1972).

plate and most of the I-shaped rays of *E. huxleyi*. The more soluble coccoliths disappear through the zone, and the sediments contain fragmented foraminifera, well-preserved to partly corroded solution-resistant placcoliths, strongly corroded specimens of *S. pulchra* and spines of *Rhabdosphaera* sp. Planktonic foraminifera disappear at the base of the zone.

In high latitudes increased rates of solution occur at shallower depths. However, since the majority of high latitude samples available for this study were obtained from within the Oligolytic Zone, the upper boundary of the zone can only be approximately defined in these regions (Fig. 5).

EOLYTIC ZONE (Table 6, Fig. 5)

This is the upper part of the compensation zone, i.e. the region of initial dissolution, and is at least 1500 m thick. Pelagic sediments deposited in this zone have a large $CaCO_3$ content and a diversified coccolith assemblage. In the lower part of this zone initial dissolution of *S. pulchra*, *R. stylifera* and *U. tenuis* occurs, whereas, the coccoliths of *U. tubifera* and *U. irregularis* undergo complete disintegration. Solution-resistant coccoliths are well preserved. In comparison with the living population, however, the sediment assemblage is depleted; most caneoliths, lopadoliths, scapholiths and cyrtoliths are gone and placoliths account for more than 60% of the sediment assemblage. This region of initial dissolution is also recognised in the graph of $CaCO_3$ distribution (Turekian, 1964, 1965), where a change in the slope of $CaCO_3$ content is observed at about 3000 m. Much of this carbonate loss may be due to removal of coccoliths, because no appreciable dissolution of planktonic foraminifera was recorded by Ruddiman and Heezen (1967) or Berger (1968) above 4000 m.

The upper boundary of this dissolution zone is not well defined, and may extend to very shallow depths. The sediments deposited in the eolytic zone will contain well-preserved assemblages of planktonic foraminifera and a diversified nannoplankton assemblage depleted in unstable groups and showing signs of initial dissolution. Above the compensation zone well preserved calcareous sediments—chalk oozes and chalks—assigned to the Alytic facies by Hsu and Anderson, (1970) accumulate.

The relationship between the hydrography of the Atlantic and the compensation zone

The bottom water in the North Atlantic around and below a depth of 4000 m is either Antarctic or Arctic in character (Wüst, 1936). In the Southern Atlantic, it is exclusively of Antarctic origin. Cold, polar waters are rich in CO_2 and as such are capable of increased dissolution of carbonate compared with CO_2-depleted warmer waters. It is of interest, therefore, to discuss briefly the distribution of major water masses in the Atlantic.

A mass of cold water is produced in the Weddell Sea. This high density, low temperature water flows as "Antarctic Bottom Water" (ABW) northward. The flow is deflected by the Walvis Ridge, and is excluded from the Angola Basin, but is recognisable in the Western Atlantic as far as 35°N (Wüst, 1936; Wright, 1970). The ABW crosses the Mid Atlantic Ridge through the Romanche Trench into the Cape Verde Basin. The greater part of the Eastern Atlantic is composed of more warm and less dense waters than the Western Atlantic.

In the Northern Atlantic, between Greenland and Iceland, cold water sinks as "North Atlantic Deep Water" (NADW) and flows southward above the ABW (McCoy, 1969; Wright, 1970; Edmond and Anderson, 1971; Reid and Lynn, 1971).

Wright (1970) has detected at the Equator a reduction in rate of northward transport of ABW and an increase in the rate of southward transport of NADW. He recognised a reference level indicated by sharp changes in the shape of temperature-depth and salinity-depth traces, which may be considered the transition between the two water masses. The resulting reference level slopes from about 3400 m depth at 32°S to 4400 m at 16°N. In the Ross Sea, the reference level rises to 500 m depth.

The current regime described above is reflected in the areal distribution of $CaCO_3$ in the bottom sediments illustrated by Correns (1939) and Turekian (1964). Local anomalies, however, do occur in areas of high terrigenous input, or of high surface productivity, e.g. in the Equatorial Atlantic where the CCCD reaches a depth of 600 m.

The distribution of the $CaCO_3$ fraction of bottom sediments is asymmetric, and is generally lower in the Western Atlantic due to the presence of the corrosive Antarctic bottom waters. This asymmetry is also reflected in the dissolution of coccoliths, and in their preservation (Fig. 6). Coccolith assemblages in the Equatorial Eastern Atlantic are better preserved than those from equivalent depths in the Western Atlantic.

The lysocline which defines the level of markedly increased dissolution rates in the planktonic foraminifera (Berger, 1968) and in coccolith assemblages corresponds with the top of the ABW in the South and Central Atlantic. The Mesolytic zone sediments, which are predominantly composed of coccolith derived carbonate, are bathed by the Antarctic Bottom Waters in the Southern and Equatorial Atlantic, and by Arctic Deep Waters (or the lower part of the NADW) in the North Atlantic.

The recognition of compensation zones in ancient calcareous pelagic deposits

Fluctuations in the CCCD in the past (Arrhenius, 1963; Hay, 1970, 1971; Broecker, 1971a; Berger, 1972) preclude the direct application of the Recent dissolution model for absolute palaeobathymetry. Changes in the coccolith populations through geological time will influence the definition of the various dissolution zones; architectural types and their preservation, rather than species composition, must be used. Nevertheless, the long geological range of many Recent placoliths suggests that it should be possible to estimate the depth of deposition relative to the CCCD for most Upper Tertiary, and Pleistocene assemblages. Deposition in the mesolytic zone will be expressed by a coccolith assemblage of low diversity composed mainly of distal shields of placoliths. Relative increases in cyrtoliths and related forms indicates deposition higher above the CCCD. Fossil assemblages composed of abundant cyrtoliths, caneoliths and lopadoliths were probably deposited in the upper part of or above the compensation zone. The relative thickness of each dissolution zone will probably vary with the surface productivity, diversity of the assemblage, chemistry of the bottom waters, position of the saturation and compensation depth and distribution of water masses.

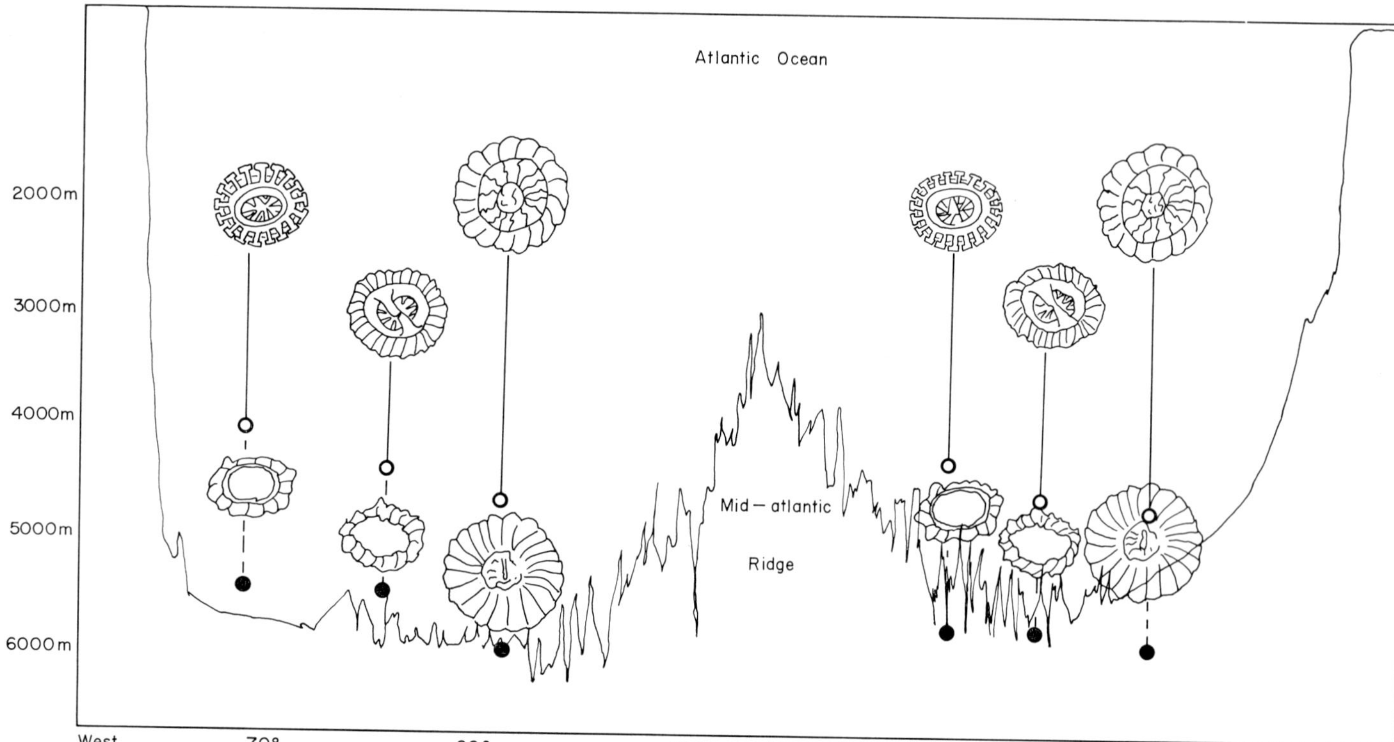

FIG. 6. Assymetric dissolution of *Emiliania huxleyi*, *Gephyrocapsa oceanica* and *Cyclococcolithina leptopora* across the Mid Atlantic Ridge. Initial dissolution indicated by O.

APPENDIX. Core location and core depths of samples used in this study. The origin of the samples is indicated as follows: WHOI, Woods Hole Oceanographic Institution; LDGO, Lamont–Doherty Geological Observatory; RSMAS, Rosenstiel School of Marine and Atmospheric Sciences in Miami.

Core	Latitude	Longitude	Depth (m)	Source
V 23–28	59°48.5′N	37°52′W	3102	LDGO
V 27–45	66°02.9′N	08°35.5′W	1038	LDGO
V 27–53	69°32.7′N	02°49.1′W	3124	LDGO
V 27–63	70°25.4′N	12°04.6′W	2725	LDGO
V 27–71	78°49.9′N	04°37.9′E	2439	LDGO
V 27–76	76°25.4′N	10°48′E	2113	LDGO
V 27–77	74°29.2′N	09°30.6′E	2395	LDGO
V 27–78	75°17.7′N	05°56.5′E	2765	LDGO
V 27–112	56°07.6′N	25°30.9′W	3217	LDGO
P 6511 st 6	35°10.5′N	11°58.1′W	3743	RSMAS
P 6511 st 7	35°30′N	12°55′W	4867	RSMAS
P 6511 st 10	34°50.5′N	13°00′W	3740	RSMAS
P 6511 st 13	34°08.6′N	13°53.2′W	4109	RSMAS
P 6511 st 16	33°36.4′N	14°26.0′W	4173	RSMAS
P 6511 st 17	32°11′N	18°46′W	4418	RSMAS
P 6511 st 18	31°53.2′N	20°01.4′W	4474	RSMAS
P 6511 st 19	30°22′N	22°07′W	5051	RSMAS
P 6511 st 20	30°16.5′N	23°42.1′W	5337	RSMAS
P 6511 st 21	29°52′N	24°56′W	5426	RSMAS
P 6511 st 22	29°53′N	27°51′W	4450	RSMAS
P 6511 st 26	29°42′N	28°43′W	3093	RSMAS
P 6511 st 27	29°24′N	30°20′W	4672	RSMAS
P 6511 st 28	28°33′N	33°44′W	5012	RSMAS
P 6511 st 29	27°42.5′N	37°13′W	4838	RSMAS
P 6511 st 30	27°07.6′N	40°00′W	4340	RSMAS
P 6511 st 31	26°15′N	43°30′W	2826	RSMAS
P 6511 st 32	25°29′N	46°57′W	4266	RSMAS
P 6511 st 34	24°01′N	53°15′W	6115	RSMAS
P 6511 st 35	23°20.6′N	56°07′W	6177	RSMAS
P 6511 st 37	22°06′N	61°08′W	5401	RSMAS
P 6511 st 38	21°23′N	64°04′W	5464	RSMAS
P 6511 st 39	20°36′N	67°08′W	5123	RSMAS
P 6303 st 5	14°55′N	17°55′W	1575	RSMAS
P 6303 st 6	15°00′N	18°00′W	1900	RSMAS
P 6303 st 10	19°54′N	23°02′W	4282	RSMAS
P 6303 st 11	21°57′N	26°32′W	5270	RSMAS
P 6303 st 12	22°42′N	29°49′W	5564	RSMAS
P 6303 st 13	23°01′N	31°20′W	5613	RSMAS

APPENDIX (*contd.*)

Core	Latitude	Longitude	Depth (m)	Source
P 6303 st 15	23°48′N	35°10′W	5837	RSMAS
P 6303 st 16	24°16′N	37°05′W	4905	RSMAS
P 6303 st 17	24°27′N	38°29′W	4451	RSMAS
P 6303 st 19	25°06′N	42°34′W	4169	RSMAS
P 6303 st 20	25°27′N	44°40′W	3950	RSMAS
P 6303 st 21	25°37′N	46°19′W	4389	RSMAS
P 6303 st 22	25°53′N	48°31′W	4333	RSMAS
P 6303 st 23	26°07′N	51°15′W	5224	RSMAS
P 6303 st 28	26°12′N	68°54′W	5444	RSMAS
P 6303 st 30	25°44′N	77°07′W	2777	RSMAS
AII 42 st 13	19°40.0′N	42°44.0′W	3935	WHOI
AII 42 st 14	19°13.0′N	47°27.0′W	4194	WHOI
AII 42 st 15	19°34.0′N	44°57.0′W	3428	WHOI
AII 42 st 16	19°41.5′N	44°33.0′W	3938	WHOI
AII 42 st 22	19°13.0′N	47°27.0′W	4194	WHOI
CHN 75 st 16	13°21.5′N	46°10.5′W	3623	WHOI
CHN 75 st 18	13°22.1′N	43°23.7′W	3643	WHOI
CHN 75 st 19	13°24.0′N	44°38.5′W	3020	WHOI
CHN 75 st 20	13°22.0′N	45°23.7′W	3451	WHOI
CHN 75 st 31	13°00.0′N	45°59.8′W	3148	WHOI
P 6708 st 1	06°35′N	52°00′W	2959	RSMAS
P 6708 st 6	08°45′N	52°57′W	4699	RSMAS
P 6708 st 8	08°37′N	54°25′W	1464	RSMAS
P 6708 st 10	16°35′N	58°04′W	4267	RSMAS
P 6708 st 11	16°41′N	58°03′W	3425	RSMAS
*A 180–69	5°17′N	17°04′W	5100	LDGO
*A 180–78	1°30′S	27°01′W	4261	LDGO
*A 181–6	9°17′N	52°31′W	4862	LDGO
*V 19–298	3°39′N	15°03′W	4792	LDGO
*V 19–303	12°47′N	27°47′W	5426	LDGO
*V 20–237	14°12′N	34°27′W	5978	LDGO
*V 22–194	11°40′N	20°19′W	4903	LDGO
*V 22–195	13°25′N	19°27′W	4210	LDGO
P 6707B st 2	00°59′S	14°00′W	3950	RSMAS
P 6707B st 3	01°20′S	14°47′W	5262	RSMAS
P 6707B st 8	00°24′N	17°05′W	2560	RSMAS

* Indicates samples described by Ruddiman and Heezen (1967) for planktonic foraminifera dissolution.

APPENDIX (*contd.*)

Core	Latitude	Longitude	Depth (m)	Source
P 6707B st 14	01°30′S	18°45′W	4026	RSMAS
P 6707B st 15	01°10′S	18°47′W	4382	RSMAS
P 6707B st 16	00°48′S	18°50′W	4450	RSMAS
P 6707B st 17	00°27′S	18°51′W	3966	RSMAS
P 6707B st 18	00°08′S	18°47′W	4380	RSMAS
P 6707B st 20	00°30′N	18°57′W	5015	RSMAS
P 6707B st 22	01°20′N	18°59′W	4960	RSMAS
P 6707B st 27	00°28′S	21°45′W	4115	RSMAS
P 6707B st 28	00°27′S	22°41′W	4843	RSMAS
P 6707B st 32	00°56.3′S	24°04.7′W	2023	RSMAS
P 6707B st 37	00°52′N	25°05.5′W	3995	RSMAS
P 6707B st 40	01°14′N	26°11′W	3735	RSMAS
P 6707B st 42	00°39′N	26°58′W	2965	RSMAS
P 6707B st 43	00°56′N	29°34′W	1099	RSMAS
P 6707B st 44	00°55′N	29°31′W	2919	RSMAS
P 6707B st 47	01°09′N	31°33′W	3142	RSMAS
P 6707B st 48	00°35′N	32°15.5′W	3526	RSMAS
P 6707B st 53	00°55′S	34°36′W	4362	RSMAS
V 22–103	49°50′S	21°24′W	3871	LDGO
V 22–106	46°08′S	10°54′W	3037	LDGO
RC 11–78	50°52′S	09°52′W	3115	LDGO
RC 12–290	45°38.6′S	20°32.8′W	4276	LDGO
RC 12–293	36°53′S	13°09′W	3393	LDGO
RC 12–300	32°53′S	02°50′E	4652	LDGO

Acknowledgements

Grateful acknowledgement is made to Dr. W. W. Hay, who suggested and supervised this study. Dr. Hay introduced the writer to scanning electron microscopy and aided in securing samples. The writer is indebted to Dr. P. A. Sandberg from the University of Illinois for his continuous help and to Dr. S. Gartner, Jr. from the University of Miami for his review of the manuscript. Dr. A. McIntyre from Lamont–Doherty Geological Observatory kindly provided an unpublished manuscript. This paper was submitted in August 1972.

References

Arrhenius, G. (1963). Pelagic sediments. *In* (Hill, M. N. ed.), *The Sea*, **3,** 655–727. Interscience, New York.

Bartolini, C. (1970). Coccoliths from sediments of the western Mediterranean. *Micropaleont*. **16**(2), 129–154.

Berger, W. H. (1967). Foraminiferal ooze: solution at depths. *Science* **156,** 383–385.

Berger, W. H. (1968). Planktonic Foraminifera: selective solution and paleoclimatic interpretation. *Deep-Sea Res.* **15**(1), 31–43.

Berger, W. H. (1970). Planktonic Foraminifera: selective solution and the lysocline. *Marine Geol.* **8**(2), 111–138.

Berger, W. H. (1971). Sedimentation of planktonic foraminifera. *Marine Geol.* **11,** 325–358.

Berger, W. H. (1972). Deep Sea Carbonates: dissolution facies and age-depth constancy. *Nature* **236**(5347), 392–395.

Berner, R. A. (1965). Activity coefficients of bicarbonate, carbonate and calcium ions in sea water. *Geochim. Cosmochim. Acta* **29,** 947–965.

Birkenes, E. and Braarud, T. (1952). Phytoplankton in the Oslo fjord during a "*Coccolithus huxleyi* summer". *Avh. Norske Vidensk Akad. Oslo, Mat-nat Kl. 1952*, 1–23.

Black, M. (1965). Coccoliths. *Endeavour* **24**, 131–137.

Black, M. (1968). Taxonomic problems in the study of coccoliths. *Paleont.* **11**(5), 793–813.

Black, M. and Barnes, B. (1961). Coccoliths and discoasters from the floor of the South Atlantic Ocean. *J.R. Micr. Soc.* **80,** 137–147.

Boudreaux, J. E. and Hay, W. W. (1969). Calcareous nannoplankton and biostratigraphy of the late Pliocene-Pleistocene-Recent sediments in the Submarex cores. *Rev. Esp. Micropal.* **1**(3), 249–292.

Braarud, T. (1963). Reproduction in the marine coccolithophorid *Coccolithus huxleyi* in culture. *Pubbl. staz. zool. Napoli* **33,** 110–116.

Braarud, T., Deflandre, G., Halldal, P. and Kamptner, E. (1955). Terminology, nomenclature and systematics of the coccolithophoridae. *Micropaleont.* **1,** 157–159.

Bramlette, M. N. (1961). Pelagic sediments. *In* (Sears, M., ed.), *Oceanography* 345–366. Am. Assoc. Adv. Sci. Pub., Washington.

Bramlette, M. N. and Bradley, W. H. (1942). Geology and biology of North Atlantic deep-sea cores between Newfoundland and Ireland. 1. Lithology and geologic interpretations. *U.S. Geol. Surv. Profess. Paper* **196**, 1–54.

Broecker, W. S. (1971a). A kinetic model for the chemical composition of sea water. *Quaternary Research* **1**(2), 188–207.

Bukry, D. (1971). Cenozoic calcareous nannofossils from the Pacific Ocean. *San Diego Soc. Nat. Hist., Trans.* **16**(14), 203–328.

Chen, C. (1964). Pteropod ooze from Bermuda Pedestal. *Science* **144,** 60–62.

Correns, C. W. (1939). Pelagic sediments of the North Atlantic Ocean. *In* (Trask, P. D., ed.), *Recent Marine Sediments*, pp. 373–395. Denver Edition, A.A.P.G.

Edmond, J. M. (1970), What controls the carbonate compensation depth in deep-sea sediments? *Am. Geophys. Union Trans.* **51**(4), 325.

Edmond, J. M. and Gieskes, J. M. T. M. (1970). On the calculation of the degree of saturation of sea water with respect to calcium carbonate uhder *in situ* conditions. *Geochim. Cosmochim Acta.* **34,** 1261–1291.

Farinacci, A. (1971). Round table on calcareous nannoplankton. *In* (Farinacci, A., ed.), *Proc. II Plankt. Conf.* 1343–1360. Tecnoscienza, Roma.

Friedman, G. M. (1965). Occurrence and stability relationships of aragonite, high-magnesium calcite and low magnesium calcite under deep-sea conditions. *Geol. Soc. America Bull.* **76**, 1191–1196.

Gaarder, K. R. (1962). Electron microscope studies on Holococcolithophorids. *Nytt. Mag. Bot.* **10,** 35–51.
Gaarder, K. R. (1970). Three new taxa of Coccolithineae. *Nytt. Mag. Bot.* **17,** 113–126.
Gartner, S., Jr. (1967). Nannofossil species related to *Cyclococcolithus leptoporus* (Murray and Blackman). *Univ. Kansas Paleont. Contr., Paper 28,* 1–4.
Green, J. C. and Jennings, D. H. (1967). A physical and chemical investigation of the scales produced by the Golgi Apparatus within and found on the surface of the cells of *Chrysochromulina chiton* Parke et Manton. *J. Exper. Bot.* **18,** 359–370.
Griggs, G. B., Carey, A. G. and Kulm, L. D. (1969). Deep-sea sedimentation and sediment-fauna interaction in Cascadia Abyssal Plain. *Deep-Sea Res.* **16,** 157–170.
Halldal, P. (1953). Phytoplankton investigations from weather ship M in the Norwegian Sea, 1948–49. Including observations during the Armauer Hansen Cruise, July 1949. *Hvalrad. Skr. norske Vidensk. Akad.* **38,** 5–91.
Hansen, H. J. Reiss, Z. and Schneidermann, N. (1969). Ultramicrostructure of biolamellar walls in forminiferida. *Rev. Espanola Micropaleont.* **1**(3), 293–316.
Hay, W. W. (1970). *In* (Bader, R. G. *et al.*, 1970), *Initial Reports of the Deep Sea Drilling Project*, **4,** 455–502. U.S. Government Printing Office, Washington.
Hay, W. W. (1971). Leg 4 of the Deep Sea Drilling Project. *Science* **172,** 1197–1205.
Hay, W. W., Mohler, H. P., Roth, P. H., Schmidt, R. R. and Boudreuix, J. E. (1967). Calcareous nannoplankton zonation of the Cenozoic of the Gulf Coast and Caribbean-Antillean area, and transoceanic correlation. *Gulf Coast Assoc. Geol. Socs. Trans.* **17,** 428–480.
Heath, G. R. (1969). Carbonate sedimentation in the abyssal equatorial Pacific during the past 50 million years. *Geol. Soc. Am. Bull.* **80,** 689–694.
Heath, G. R. and Culberson, C. (1970). Calcite: degree of saturation, rate of dissolution, and the compensation depth in the deep oceans. *Geol. Soc. Am. Bull.* **81**(10), 3157–3160.
Honjo, S. (1969). Study of fine grained carbonate matrix, sedimentation and diagenesis of micrite. *Paleont. Soc. Japan, spec. paper* **14,** 67–82.
Howley, J. and Pytkowicz, R. M. (1969). Solubility of calcium carbonate in seawater at high pressures and 2°C. *Geochim. Cosmochim. Acta* **33,** 1557–1561.
Hsu, K. J. and Anderson, J. E. (1970). Lithology. *In* (Maxwell A. E. *et al.*, 1970), *Initial Reports of the Deep Sea Drilling Project* **3,** 445–453. U.S. Government Printing Office, Washington,
Hulburt, E. M. (1962). Phytoplankton in the southwestern Sargasso Sea and North Equatorial Current. February 1961. *Limnol. Oceanogr.* **7,** 307–315.
Hulburt, E. M. (1964). Succession and diversity in the phytoplankton flora of the western North Atlantic, *Bull. Mar. Sci. Gulf Caribb.* **14**(1), 33–44.
Hulbert, E. M. (1966). The distribution of phytoplankton and its relationship to hydrography, between southern New England and Venezuela. *J. Mar. Res.* **21**(2), 81–93.
Hulburt, E. M., Ryther, J. H. and Guillard, R. R. L. (1960). The phytoplankton of the Sargasso Sea off Bermuda. *J. Cons. Int. Explor. Mer.* **25,** 115–128.
Isenberg, H. D., Lavine, L. S., Mandell, C. and Weissfellner, H. (1965). Qualitative chemical composition of the calcifying organic matrix obtained from cell-free coccoliths. *Nature* **206,** 1153–1154.
Kamptner, E. (1954). Untersuchungen ueber den Feinbau der Coccolithen. *Archiv. Protistenkunde* **100,** 1–90.

Kuenen, P. H. (1950). *Marine Geology*, 568 pp. Wiley, New York.
Kullenberg, B. (1953). Absolute chronology of deep-sea sediments and deposition of clay on the ocean floor. *Tellus* **5**(3), 302–305.
Lecal-Schlauder, J. (1951). Recherches morphologiques et biologiques sur les Coccolithophorides Nord-Africains. *Ann. Inst. Iceanog.* (*Paris*) **26,** fax. 3, 255–362.
Lecal, J. (1966). Diversification des populations de *Coccolithus huxleyi* (Lohman) Kamptner sous l'influence des facteurs ecologiques. *Protistologica* **11**(4), 57–70.
Li, Y. H., Takahashi, T. and Broecker, W. S. (1969). Degree of saturation of $CaCO_3$ in the oceans. *J. Geophys. Res.* **74**(23), 5507–5525.
Lisitzin, A. P. (1972). Sedimentation in the World Ocean. *Soc. Econ. Paleont. Min., Spec. Publ.* **17,** 218 pp.
Lohman, H. (1902). Die Coccolithophoridae eine Monographie der Coccolithen bildenden Flagellaten, zugleich ein Beitrag zur Kenntnis des Mittelmeerauftriebs. *Arch. Protistenk.* **1,** 89–165.
Manton, J. and Leedale, G. E. (1969). Observations of the microanatomy of *Coccolithus pelagicus* and *Criscosphaera carterae* with special reference to the origin and nature of coccoliths and scales. *J. Marine Biol. Assoc. U.K.* **49,** 1–16.
Markali, J. and Paasche, E. (1955). On two species of *Umbellosphaera*, a new marine coccolithophorid genus. *Nytt. Mag. Bot.* **4,** 95–100.
Marshall, S. M. and Orr, A. P. (1955). On the biology of *Calanus finmarchicus VIII.* Food uptake, assimilation and excretion in adult and Stage V *Calanus. J. Mar. Biol. Ass.* **34,** 495–529.
McCoy, F. W., Jr. (1969). Bottom currents in the western Atlantic Ocean between the Lesser Antilles and the Mid Atlantic Ridge. *Deep-Sea Res.* **16,** 179–184.
McIntyre, A. and Bé, A. W. H. (1967). Modern coccolithophoridae of the Atlantic Ocean. I Placoliths and Cyrtoliths. *Deep-Sea Res.* **14,** 561–597.
McIntyre, A., Bé, A. W. H. and Preikstas, R. (1967). Coccoliths and the Pliocene-Pleistocene boundary. *In* (Sears, M., ed.), *Progress in Oceanography*, **4,** 3–25. Am. Assoc. Advan. Sci. Publ.
McIntyre, A., Bé, A. W. H. and Roche, M. B. (1970). Modern Pacific Coccolithophorida: a paleontological thermometer. *Trans. N.Y. Acad. Sci., ser. 2* **32,** 720–731.
McIntyre, A. and McIntyre, R. (1971). Coccolith concentrations and differential solution in oceanic sediments. *In* (Funnel, B. and Riedel, W. R. eds), *The Micropaleontology of Oceans*, 253–261. Cambridge University Press.
Mjaaland, G. (1956). Some laboratory experiments on the coccolithophorid *Coccolithus huxleyi. Oikos* **7**(2), 251–255.
Moor, H. B. (1931a). The muds of the Clyde area III. Chemical and physical conditions, rate and nature of sedimentaton and fauna. *J. Mar. Biol. Ass.* **17,** 325–358.
Moore, H. B. (1931b). The specific identification of faecal pellets. *J. Mar. Biol. Ass.* **17,** 359–365.
Moore, H. B. (1933). Faecal pellets from marine deposits. *Discovery Repts.* **7,** 17–26.
Moore, H. B. (1939). Faecal pellets in relation to marine deposits. *In* (Trasi, P. D., ed.), *Recent Marine Sediments*, 576–524. Denver Edition, A.A.P.G.
Murray, J. and Renard, A. F. (1891). Deep Sea deposits. *Challenger Expedition Rept.* (*1873–1876*), 525 pp. London.
Murray, J. and Chumley, J. (1924). The deep sea deposits of the Atlantic Ocean. *Royal Soc. Edinburg Trans.* **54**(1), 217 p.

Norris, R. E. (1971). Extant calcareous nannoplankton from the Indian Ocean. *In* (Farinacci, A., ed.), *Proc. II Plankt. Conf.* 899–910. Tecnoscienza, Roma.

Olausson, E. (1965). Evidence of climatic changes in North Atlantic deep-sea cores, with remarks on isotopic paleotemperature analysis. *In* (Sears, M., ed.), *Progress in Oceanography*, **3**, 221–252. Am. Assoc. Advan. Sci. Publ.

Olausson, E. (1971). Quaternary correlations and the geo-chemistry of oozes. *In* (Funnell, B. M. and Riedel, W. R., eds), *The Micropaleontology of Oceans*, 375–398. Cambridge University Press.

Okada, H. and Honjo, S. (1970). Coccolithophoridae distributed in Southwest Pacific. *Pacific Geology* **2**, 11–21.

Osterberg, C., Carey, A. G. and Curl, H. (1963). Acceleration of sinking rates of radionuclides in the ocean. *Nature* **200**, 1276–1277.

Outka, D. E. and Williams, D. C. (1971). Sequential coccolith morphogenesis in *Hymenomonos carterae. Protozool.* **18**(2), 285–297.

Paasche, E. (1967). Marine plankton algae grown with light-dark cycles. I. *Coccolithus huxleyi. Physiol. Plant.* **20,** 946–956.

Paasche, E. (1968a), Biology and physiology of coccolithophorids. *Ann. Rev. Microbiol.* **22,** 71–86.

Paasche, E. (1968b). The effect of temperature, light intensity and photoperiod on coccolith formation. *Limnol. Oceanog.* **13**(1), 178–181.

Parke, M. (1961). Some remarks concerning the class Chrysophyaceae. *Brit. Phyc. Bull* **2**(2), 47–55.

Parke, M. and Adams, I. (1960). The motile (*Crystallolithus hyalinus* Gaarder and Markali) and non-motile phases in the life history of *Coccolithus pelagicus* (Wallich) Schiller. *J. Mar. Biol. Assoc. U.K.* **39,** 263–274.

Peterson, M. N. A. (1966). Calcite rates of dissolution in a vertical profile in the central Pacific. *Science* **154,** 1542–1544.

Pythkowicz, R. M. (1968). The carbon dioxide-carbonate system at high pressures in the oceans. *Oceanogr. Marine Biol. Ann. Rev.* **6,** 83–135.

Pytkowicz, R. M. (1970). On the carbonate compensation depth in the Pacific Ocean. Geochim. *Geochim. Cosmochim. Acta* **34,** 836–839.

Ramsay, A. T. S. (1971). The investigation of Lower Tertiary sediments from the North Atlantic. *In* (Farinacci, A., ed.), *Proc. II Plankt. Conf.* 1039–1055. Tecnoscienza, Roma.

Ramsay, A. T. S. (1974.) Aspects of the distribution of fossil species of calcareous nannoplankton in North Atlantic and Caribbean sediments. *Nature* **236,** 67–70.

Reid, J. L. and Lynn, R. J. (1971). On the influence of the Norwegian-Greenland and Weddell seas upon the bottom waters of the Indian and Pacific Oceans. *Deep-Sea Res.* **18,** 1063–1088.

Revelle, R. and Fairbridge, R. (1957). Carbonates and carbon dioxide, *In* Hedgpeth, I. W. H., ed.), *Treatise on Marine Ecology and Paleoecology*, **1,** 239–295. Geol. Soc. America, Mem. 67.

Ruddiman, W. F. and Heezen, B. C. (1967). Differential solution of planktonic Foraminifera. *Deep-Sea Res.* **14**(6), 801–808.

Schiller, J. (1930). Coccolithineae. *In Rabenorst's Kryptgamenflora von Deutschland, Osterreich und der Schweiz*, **10,** 89–267. Leipzig.

Schneidermann, N. (1971). Selective dissolution of Recent coccoliths in the Atlantic Ocean (abstr.), p. 695. *Geol. Soc. Am. Ann. Meeting*, Washington.

Schott, W. (1935). Die Foraminiferen in dem aequatorialen teil des Atlantischen Ozeans. *Wiss. Ergebn. Deutschen Atlant. Exped. "Meteor", 1925–1927* **3**(1), B.

Schwarz, E. (1932). Beitrage zur Entwicklungsgeshichte der Protophyten IX. Der Formwechsel von *Ochrosphaera neapolitana. Arch. Protistenk.* **77,** 434–462.

Smayda, T. J. (1970). The suspension and sinking of phytoplankton in the sea. *Oceanogr. Mar. Biol. Ann. Rev.* **8,** 353–414.

Smith, S. T., Dygas, J, A. and Chave, K. E. (1968). Distribution of calcium carbonate in pelagic sediments. *Marine Geol.* **6,** 391–400.

Sokolova, M. N. (1959). On the distribution of deep-water bottom animals in relation to their feeding habits and the character of the sediment. *Deep-Sea Res.* **6,** 1–4.

Steemann Nielsen, E. and Aabye Jensen, E. (1957). Primary oceanic production. The autotrophic production of organic matter in the oceans. *Galathea Rept.* **1,** 49–136.

Sverdrup, H. U., Johnson, M. W. and Fleming, R. H. (1942). *The Oceans. Their Physics and General Biology*, 1087 pp. Prentice-Hall, New York.

Turekian, K. K. (1964). The geochemistry of the Atlantic Ocean basin. *Trans. N.Y. Acad. Sci. ser. II* **26,** (3), 312–330.

Turekian, K. K. (1965). Some aspects of the geochemistry of marine sediments. *In* (Riley, J. P. and Skirrow, G. eds), *Chemical Oceanography*, **2,** 81–126. Academic Press. New York.

Wangersky, P. J. (1969). Distribution of suspended carbonate with depth in the ocean. *Limnol. Oceanogr.* **14,** 929–933.

Wangersky, P. J. and Joensuu, O. I. (1967). The fractionation of carbonate deep-sea cores. *J. Geol.* **75,** 148–177.

Watabe, N. and Wilbur, K. M. (1966). Effects of temperature on growth, calcification, and coccolith form in *Coccolithus huxleyi* (Coccolithineae). *Limnol. Oceanogr.* **11**(4), 567–575.

Wilbur, K. M. and Watabe, N. (1963). Experimental studies on calcification in molluscs and the alga *Coccolithus huxleyi. Ann. N.Y. Acad. Sci.* **109**(1), 82–112.

Wolff, T. (1960), The hadal community an introduction. *Deep-Sea Res.* **6,** 95–124.

Worsley, T. R. (1970). The nature of the terminal Cretaceous event as evidenced by calcareous nannoplankton extinctions in Alabama and other areas. Unpublished Ph.D. dissertation, University of Illinois, Urbana.

Wright, W. R. (1970). Northward transport of Antarctic Bottom Water in the Western Atlantic Ocean. *Deep-Sea Res.* **17,** 367-371.

Wüst, G. (1936). Die Stratosphare des Atlantischen Ozean. *Wiss. Ergebn. Deutschen Atlant. Exped. "Meteor", 1925–1927* **6**(1), 105–288.

PLATES

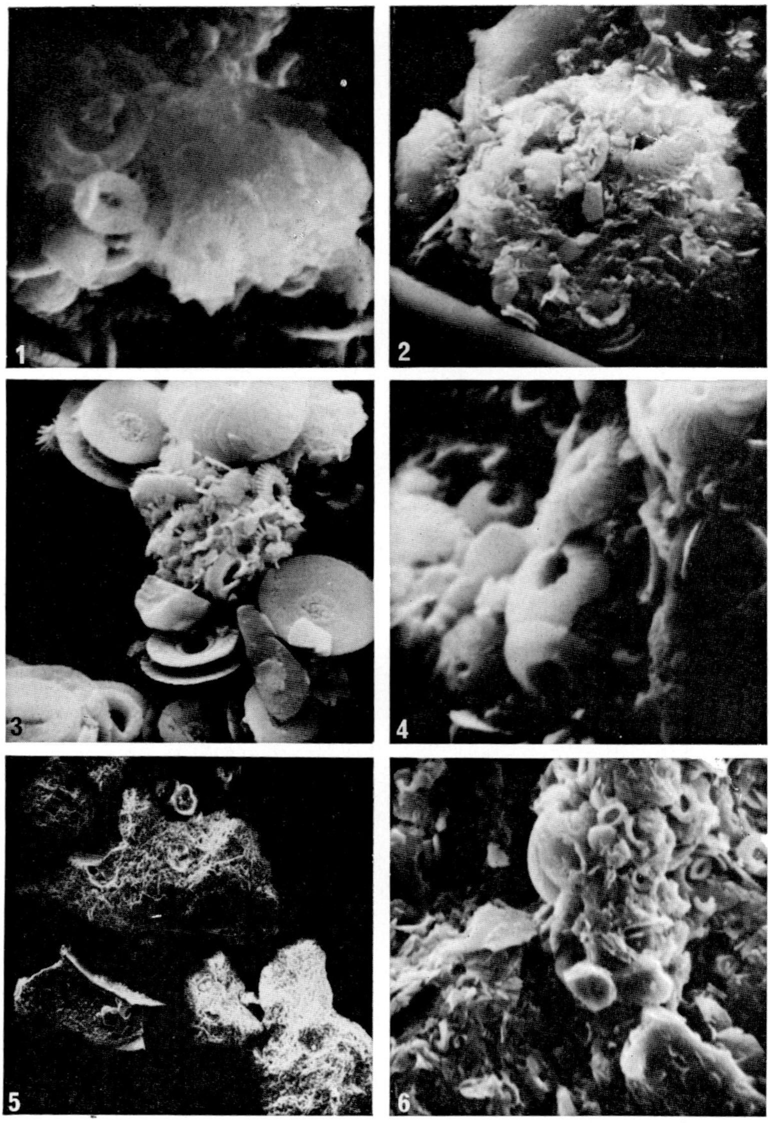

PLATE 1. Pellets.

fig. 1. Sample P6511 st 32, 25°N;46°W 4266 m (×4650). fig. 2. Sample P6303 st 23, 26°N; 51°W 3950 m (×2485). fig. 3. Sample P6511 st 10, 34°N;13°W 3740 m (×4125). fig. 4. Sample V19-303, 12°N;27°W 5426 m, partially preserved coccosphere in a faecel pellet (×3185). fig. 5. Sample V19-303, 12°N;27°W 4526 m (×45); fig. 6. Sample V19-303, 12°N;27°W 5426 m, enlargement of a pellet in fig. 5 (×1817).

PLATE 2. *Emiliania huxleyi* dissolution sequence, distal views.
figs 1–3, Sample P6511 st 10 ,34°N;13°W. 3740 m. fig. 1. (×13 120). fig. 2. (×20 185). fig. 3. (×13 455.). figs 4–6. Sample P. 6511 st 20 30°N;23°W 5337 m. fig. 4. (×13 455). fig. 5. (×13 455). fig. 6. (×7512).

PLATE 3. *Gephyrocapsa oceanica*, dissolution series.
fig. 1. Sample P6511 st 10, 34°N;13°W 3740 m. Distal view (×7512). figs 2, 3. Samples P6511 st 19, 30°N;22°W 5051 m. fig. 2. Proximal and distal views showing effects of initial dissolution (×5400). fig. 3. Proximal view showing effects of initial dissolution (×12 112). fig. 4. Sample P 6511 st 20, 30°N;23°W 5337 m. Distal view showing specimen with a dissolved grill, bridge and partially dissolved distal shield (×7512). *Cyclococcolithina leptopora*, dissolution series. figs 5, 6. Sample P 6511 st 10, 34°N;13°W 3740 m. proximal view of undissolved specimen (×7512). fig. 6. Specimens at various stages of dissolution. The second specimen from the top of photograph, shows the initial stage in the dissolution of the proximal shields (×2855).

PLATE 4. *Cyclococcolithina leptopora*, dissolution sequence.
fig. 1. Sample CHN 78 st 18, 13°N;43°W 3643 m. Proximal view showing a more advanced stage in the dissolution of proximal shield (×7458). fig. 3. Sample V 19–303, 12°N;27°W 5426 m. Proximal view, top left of a specimen with the proximal shield completely dissolved and partially dissolved central tube (×2104). fig. 4. Sample V 10–303, 12°N;27°W 5426 m. Proximal view of specimen showing more advanced dissolution of the central tube (×5250). figs 5 and 6. Sample V 22–106, 46°S;10°W 3037 m. Views of almost completely dissintegrated distal shields.

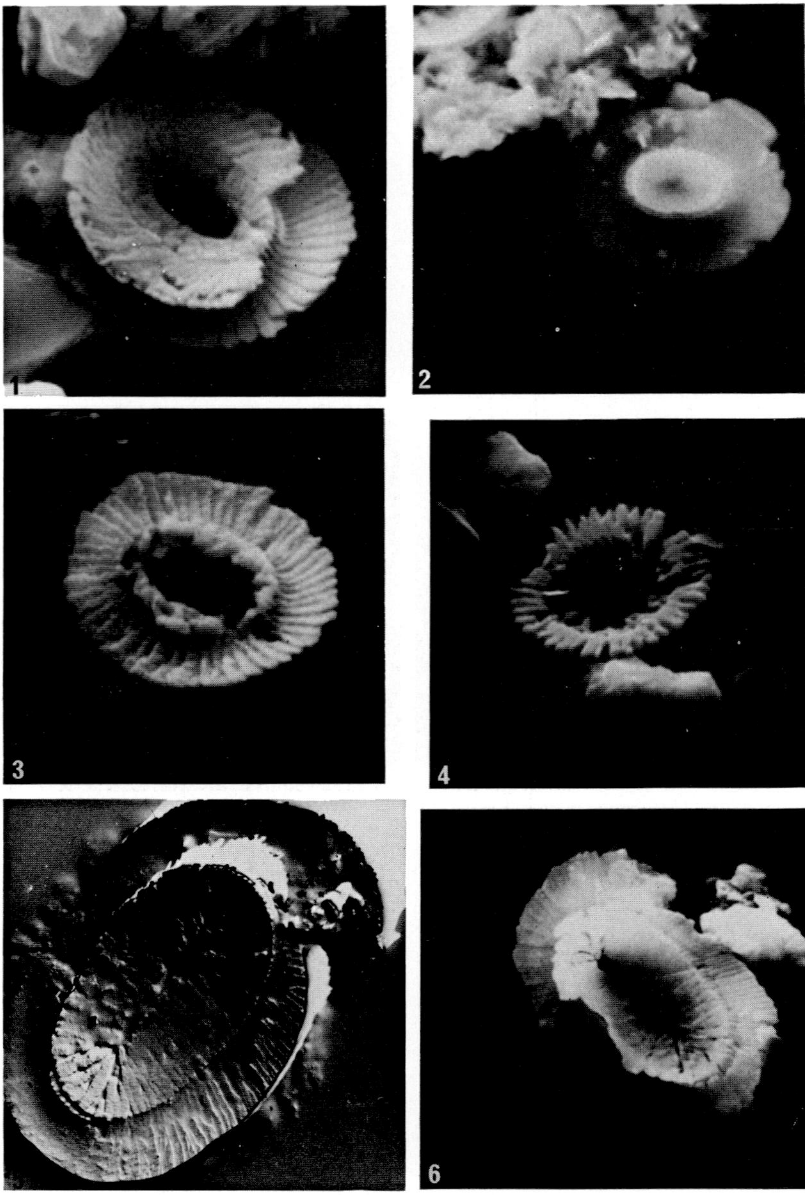

Plate 5. *Coccolithus pelagicus*, dissolution sequence.
figs 1 and 2. Sample V 23–28, 59°N;37°W 3102 m. fig. 1. Proximal view of specimen showing initial dissolution of the proximal shield (×2320). fig. 2. Proximal view of specimen with an almost completely dissolved proximal shield (×4144). figs 3 and 4. Sample P 6603–27, 20°N;68°W 5155 m (*Pseudoemiliania lacunosa* zone). fig. 3. Proximal view of specimen showing an almost completely dissolved central tube. fig. 4. Proximal view of a severely corroded specimen. *Helicopontosphaera kamptneri*, dissolution sequence. fig. 5. Sample P6511 st 26, 29°N;28°W 3093 m. Proximal view of an undissolved specimen (×5517). fig. 6. Sample P 6707B st 14, 01°S;18°W 4026 m. Proximal view showing early stages of dissolution (×3730).

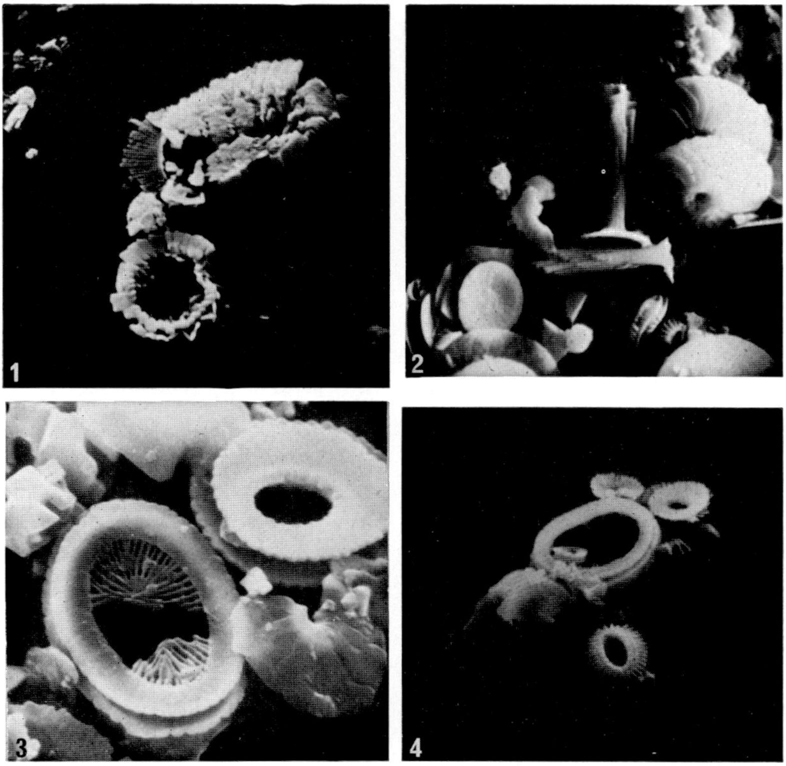

PLATE 6.
Helicopontosphaera kamptneri. fig. 1. Sample V 19–303; 12°N;27°W 5426 m. Intensely corroded specimen (×4200). *Rhabdosphaera claviger*. fig. 2. Sample P 6511 st 10, 34°N; 13°W, 3440 m. Two central specimens, the lower has a partially corroded basal shield (×2754). *Scyphosphaera pulchra*. fig. 3. Sample P 6511 st 10, 34°N;13°W 3740 m. A proximal view of a specimen with a corroded grill (×7344). fig. 4. Sample P 6511 st 26, 29°N; 28°W 3093 m. Strongly corroded specimen (×4125).

13: Calcareous Nannofossils

WILLIAM W. HAY

Division of Marine Geology and Geophysics,
Rosenstiel School of Marine and Atmospheric Sciences,
University of Miami,
Florida 33149, U.S.A.

During the past decade considerable interest in calcareous nannoplankton fossils has developed. Structural analysis has been simplified through use of phase and interference light microscopy in addition to more traditional methods of examination in polarised light, and through development of stereoscopic transmission electron microscopy of replicas, as well as by direct observation of specimens utilising scanning electron microscopy. These techniques permit examination of the same specimen by light and electron microscopy, and promise to resolve many nomenclatural problems.

The taxonomy of calcareous nannoplankton is currently in a state of flux; many new genera and species are still being described, and much work remains to be done before complete lineages can be worked out. The suprageneric classification is gradually emerging as studies proceed. Major orders can be recognised, but it has become evident that in several instances three sets of families are needed, one for Mesozoic nannofossils, another for Tertiary forms, and a third for Quaternary and Recent genera.

The biostratigraphical value of calcareous nannofossils has increased so that they now form the standard against which other groups are compared. Zonations now exist for the Cretaceous—Recent interval.

Outline

Introduction

Although calcareous nannofossils have been known for over a century, it is only during the past decade that they and the algae which produce them have become the objects of widespread interest. An account of the early history of investigation of this group of fossils, and of the attempts to discover their living relatives, has been presented by Gartner (1968).

Detailed scientific investigation and description of calcareous nannoplankton began at the turn of the century, with the discovery of the potential importance of these minute organisms in oceanic plankton surveys. Early studies were dominated by the Viennese, and the descriptions of modern coccolithophores by Lohmann, Schiller, and Kamptner are classic. More recently, Braarud and his colleagues at the Botanical Institute of the University of Oslo have developed a centre for the study of calcareous nannoplankton in culture as well as in the open sea.

Study of the fossil counterparts of the calcareous nannoplankton was sporadic until the stratigraphical usefulness of these minute objects was demonstrated by Bramlette and Riedel (1954). Studies of the stratigraphical distribution of calcareous nannofossils have burgeoned since 1960, along with the introduction of many stratigraphically useful taxa.

The interest in calcareous nannoplankton and calcareous nannofossils has increased as technological advances have made the study of this group easier and more meaningful. Because technology has determined the nature of the information available, major changes in instrumentation have been followed by radical re-evaluations of the criteria used to define species and subsequently have affected interpretation of the ecological and biostratigraphical values of these remarkable fossils. For this reason, technological advances which have affected the study of calcareous nannoplankton will be discussed first, followed by an account of their taxonomy and stratigraphy.

Collection of samples for calcareous nannofossils

Probably the majority of all samples thus far studied for nannofossil content were collected for some other purpose. Unwashed micropalaeontological samples may be good sources of nannofossils. The sediment clinging to the surface of megascopic fossils may also be a useful source of nannofossils from localities which cannot conveniently be visited or which have disappeared. Drill cuttings are liable to be coated with drilling mud, which may contain nannofossils introduced as contaminants during the drilling process. Accordingly, drill cuttings must be processed carefully, and all traces of contaminating mud must be removed from the surface of the pieces to be examined before the sample is prepared for light or electron microscope observation. The drilling mud itself may be interesting to examine as it will acquire assemblages of nannofossils which become mixed together as drilling proceeds. Thus, in the mud from a well which is drilled through a normal sequence of strata, the youngest assemblages, encountered at the top of the well, gradually have older and older assemblages mixed in with them as drilling proceeds. Since the nannofossils are recirculated with the drilling mud, the problem of mixing is more evident than in the case of foraminifera or other larger microfossils.

It is often possible to recover nannofossils from washed micropalaeontological samples by treating them ultrasonically, provided they have not already been ultrasonically cleaned. While this source is generally a last resort, it works well in many instances, and is important since micropalaeontologists in the past frequently washed the entire sample, and did not preserve any unwashed material.

Field collection of samples specifically for calcareous nannofossils is done in a manner different from that employed for larger microfossils. The samples may be very small, a few grams will suffice, and it is possible to sample many thin layers which would be impractical to collect for larger microfossils. The great numbers in which calcareous nannofossils occur in sediments were noted as early as 1873 by Gümbel, who calculated that one cubic metre of the Eocene Stockletten marls contained 8×10^{11} coccoliths. This is an abundance of 800 000 specimens per cubic centimetre, which it is now known is not a particularly rich occurrence. The chief problem to be avoided in collection of field samples is contamination. In nannofossil-rich deposits, an almost imperceptible film of mud on the hands or the collecting instruments may provide a sufficiently high level of contamination to result in errors at the laboratory where the age of the assemblages is to be determined.

Contamination may also occur through natural processes acting on the outcrop. It is easy to imagine how slope wash occurring over the face of an outcrop can transfer nannofossils from higher to lower levels. A more extraordinary case of contamination was noted in the course of examining the lower part of the Schlierenflysch of Switzerland in preparation for the IX European Micropalaeontological Colloquium. A series of samples from the Schlierenflysch below the Gubersandstein was collected. The section lies in a narrow gorge cut by the stream called the Grosse Schliere. The samples were taken as high as practical above water level, so as to avoid contamination by the water of the stream. Sampling was done by cutting a fresh opening into the bank with a hammer or pick, and then carefully extracting several chunks of the fresh marl. In spite of the precautions which had been taken, the assemblages found in these samples were anomalous, and contained species which should be found only near the top of the section far above the Gubersandstein. The assemblages were conspicuous in the relative paucity of calcareous nannofossils, and this aroused suspicion. Very carefully, preparations were made of the material on the surface of the chunks of marl, and of the interior material of the chunks. Calcareous nannofossils were found to occur only in the preparations made from the surface of the fragments, and the assemblages, though very meagre, were typical of the upper part of the Schlierenflysch. During times of high water, the calcareous nannofossils were apparently washed out of the upper part of the section, carried down by the stream, and splashed to considerable heights above the normal level of the stream as it cascades over the Gubersandstein. The water containing the calcareous nannofossils was then drawn into the fine cracks in the marly part of the flysch beds by capillary action, so that the nannofossils penetrated several centimetres into the outcrop. When a highly anomalous distribution of calcareous nannofossils is encountered, the possibility of contamination should be carefully investigated.

There are no infallible rules for determining the lithology which will contain the best preserved calcareous nannofossils. Obviously, the grain size of the sediment must include the size range of calcareous nannofossils (1–25 μm), but beyond this it is difficult to guess in the field which beds will have the richest nannofloras, and which will have the best preserved specimens. Experience has revealed that a few generalities may be made, however:

1. Clay frequently seems to aid in the preservation of the fine structure of coccoliths. The clay particles probably serve to isolate the individual fossils from solutions in the sediment which might cause secondary calcification or etching of the specimen.
2. Chalk, although often composed mostly of coccoliths and thus a very rich source of nannofossils, contains well-preserved specimens in soft layers whilst those in lithified layers are frequently poorly preserved.
3. Well-preserved planktonic foraminifera and well-preserved calcareous nannofossils commonly occur together. An abundance of planktonic foraminifera usually indicates an abundance of calcareous nannofossils as well.
4. Fine grained sediments of the silt and clay range are generally richest in calcareous nannofossils. Samples may be checked in the field with acid to determine whether they contain calcareous material. "Dirty" sandstones are often surprisingly rich in calcareous nannofossils.

Laboratory preparation, washing, cleaning and concentration

For detailed examination in the light microscope and any electron microscope study, fossils should be freed of as much foreign material as possible, concentrated and cleaned but not etched. Samples prepared in this manner are also easier to study and photograph in the light microscope. Light microscopy should be frequently employed as a check during sample preparation to observe the progress of the cleaning and concentration process. Apparently identical samples sometimes behave differently during the cleaning and concentration processes and may need individual adjustment of the processes to obtain the best results.

If the nannofossils are to be preserved from etching during sample preparation, careful attention must be paid to the pH of the solutions used. A saturated solution of $CaCO_3$ in air saturated water has a pH of 8–8.5. The lower limit of these values is probably controlled by the quality of the glassware used, and 8.5 should be taken as a minimum pH value during sample preparation. The best dispersion of clays in a given solution can be expected at a pH of 9.4 (J. Eades and G. McClellan, personal communication). At this value clays are kept in suspension best, and do not flocculate as readily as they do at lower pH values. If the clays are kept dispersed in suspension throughout the sample preparation, they can be almost completely eliminated from the sample. The problems of maintaining a pH level high enough to prevent etching of the nannofossils and to effect maximum dispersion of the clays are solved simultaneously by using a solution made by adding 4 grams of $NaHCO_3$ to twenty litres of distilled water and adjusted to pH 9.4 by the addition of Na_2CO_3 (approx. 3 grams). This solution is stable over a period of several weeks, but after a month or two the pH usually drops, probably due to solution of silica from the surface of the glassware used. Solutions of Na_2CO_3 alone are unstable, the pH dropping rapidly (in a few days) due to uptake of CO_2 from the air. The low concentration of dissolved salts in this solution presents no problems in the preparation of slides for light microscopy or replicas for electron microscopy. Sodium hexametaphosphate [$(Na\ PO_3)_6$; Calgon], widely used for the dispersal of clays in micropalaeontology, should not be used because $Ca(PO_3)_2$ forms as a precipitate which may coat the nannofossils and prevent preparation of a good replica for electron microscopy. While this is not a serious problem for the light microscopist, $Ca(PO_3)_2$ is highly undesirable to the electron microscopist for it is insoluble in HCl and reacts to form insoluble CaF_2 when treated with HF, spoiling the carbon replica. Moreover, a solution of sodium hexametaphosphate has a pH below 8 and is sure to etch the nannofossils if it does not dissolve them completely. Tap water and distilled water are to be avoided, as both not only flocculate clays but often have a pH below 7, due to uptake of CO_2 from the air, resulting in serious etching of ultrastructural detail.

Removal of organic matter, if present, is the first step in sample preparation. This is accomplished through the use of 30% H_2O_2. The hydrogen peroxide solution produces formic acid on contact with organic matter, and unless buffered in advance with $NaHCO_3$ to a pH near 9.4, all the nannofossils may be dissolved. A small amount of the sample is pulverised gently in a mortar with pestle. The sample is placed in a beaker, the buffered 30% H_2O_2 solution is added, and the mixture cautiously heated. After about an hour, a dark organic-rich sample will turn pale grey, indicating that the organic matter has been oxidised. The sample is then centrifuged and given several washings with the sodium carbonate solution. Preparation of samples with a low organic content may be started at this point. Between centrifugings with the sodium carbonate solution, the sample may be treated in an ultrasonic cleaner for periods up to 30 minutes to clean the nannofossils of adhering clay particles. Centrifugation should be stopped while the supernatant contains suspended clays. In this way several cycles of ultrasonic cleaning and centrifuging will serve to remove almost all clay from the sample. The clay may not trouble the light microscopist, as it will commonly be in particles fine enough to be invisible in the light microscope, but clay clinging to the surface of nannofossils obscures detail from the electron microscopist.

The final concentration of the nannofossils is accomplished by gravity settling. Basically the technique is as follows:

1. After straining through 300 mesh screening (40 μm openings), the sample is placed in a beaker and distilled water or sodium carbonate solution added to a depth of 2 cm. The suspension is vigourously agitated and then allowed to settle for $1\frac{1}{2}$ to 2 min.
2. The supernatant is then decanted into a second beaker and allowed to settle for 15 min.
3. The supernatant in the second beaker is decanted into a third and allowed to settle for 1 h before being poured off.

The residue on the floor of the first beaker will contain the largest calcareous nannofossils along with a variety of very small microfossils and silt grains. The residue of the second beaker will contain large and medium-sized calcareous nannofossils and will be suitable for light and electron microscopy. The material remaining on the floor of the third beaker will be useful primarily in electron microscope studies, as most of the nannofossils will be too small to reveal any detail in the light microscope. It may be necessary to repeat the gravity settling several times with each fraction to insure a clean separation, and frequent checks should be made with the light microscope to insure that the settling times are appropriate for the particular sample being processed.

An alternative method, which will result in an even sharper separation of the sample into size fractions, is the use of elutriation. Elutriation apparatus has been described by Joukowski and Charrey (1933), and Noël (1965b). The apparatus essentially consists of a glass tube one or more metres in length. The sample is introduced at the top, and in the course of settling is segregated into size fractions, the coarser material reaching the bottom first. Coccoliths have settling velocities in the order of 1 m/10–20 h, with the smaller forms settling even more slowly. Collection of the different size fractions is accomplished on a series of dishes which can be brought beneath the tube in a preset sequence, or by the less sophisticated method of opening a valve at the bottom of the tube by hand and flushing out the bottom of the tube at selected intervals. This sort of apparatus can offer a very sharp size fractionation if convection currents within the tube can be avoided. Agronomists, who are concerned with precise size analysis of soil materials, sometimes use elutriation tubes surrounded by constant temperature water baths to prevent convection currents. A still sharper fractionation can be accomplished by means of a constant flow elutriator. In this apparatus, the force of gravity is offset by an upward flow of water. The tube has a circular cross-section, but has a conical shape, so that the cross-sectional diameter increases upward with a corresponding decrease in the rate of flow of the water. This apparatus is not so susceptible to the development of convection currents, and the only problem is in maintaining a constant flow of water, accomplished by maintaining a constant head in a reservoir with an overflow always kept running. The sample can be introduced into a constant flow elutriator at any point; commonly it is introduced into the water supply itself. In this case, the particles are carried upward in the tube until the rate of flow drops to the point where their rate of rise is exactly balanced by the force of gravity; i.e. the rate of flow upward of the water = the settling velocity of the coccoliths. The equilibrium rate of flow for coccoliths is in the order of 2 mm/min, and a rate of flow one order of magnitude less should be provided if it is intended to collect the smaller nannofossils. The longer a constant-flow elutriator is allowed to operate on a sample, the more sharply the particles with different hydrodynamic properties will become fractionated. The different fractions will be recognised as cloudy horizontal bands in the elutriator. These are then sampled by means of a pipette for investigation. In some instances it is possible to isolate a single species of nannofossil by this technique if the species is sufficiently different from the others present in the same sample.

The disadvantages of elutriation are:

1. The length of time required for the fractionation of the sample.
2. The fractionation may be too complete to permit preparation of a representative slide for investigation, i.e. the discoasters may be isolated and split into several groups, the coccoliths may be split into a series of fractions each with relatively few species.

Thus elutriation techniques are primarily of value in obtaining a concentration of a few species which might otherwise be so rare as to be nearly impossible to deal with in the electron microscope.

There is yet another technique which may be used to fractionate samples, and this is the quickest of all; a sample may be processed with millepore filters in a matter of seconds. In this case it is advisable to first strain the sample through 400 mesh sieving to remove particles greater than 30 microns.

The strained suspension can then be placed in a syringe with a millepore filter cartridge containing the appropriate filter attached. The filter will clog very quickly, and should not be forced. The coccoliths are easily washed off the filter or the filter can be dissolved, or the coccoliths observed directly on the filter.

Storage

The concentrated and cleaned samples can be stored for long periods of time as a suspension in the sodium carbonate solution in glass or plastic vials. Acid conditions in the vial may arise after a period of time if the organic matter in the sample was not eliminated.

Slides for light microscopy

Two types of slides are commonly in use for light microscopy: permanent mounts and viscous mounts. There are many materials available for permanent mounting, and the user may wish to experiment with media having differing refractive indices. The only important consideration is that the medium itself not be acid, but preferably have a pH about 8–8.5 so that the nannofossils will not dissolve when the slides are stored. There is an advantage to having a mounting medium with a refractive index between the extremes of calcite (1.4864–1.6585) for optimum resolution in the light microscope. The contrast that will be lost by having the indices of refraction close together can be recovered through phase contrast microscopy. Caedax ($n = 1.55$) is an artificial mounting medium which has the advantage of not polymerising when heated, and is always resoluble in xylene. Some calcareous nannofossils, particularly those with a uniform orientation of the c-axis (discoasters, etc.) are most photogenic in media with a very different refractive index. In the case of discoasters, the c-axis of the crystals is perpendicular to the plane of the disc, and the index of refraction is 1.66 in all quadrants. In coccoliths, the c-axes of the crystals commonly have a radial orientation, though the axis may not lie in the plane of the disc. This results in intermediate values of the refractive index, and in a medium such as Caedax, the Becke line may be outside the coccolith in some areas and inside the coccolith outline in others.

Bramlette and Sullivan (1961) have suggested treatment of the samples with hydrofluoric acid to remove clay and alter the calcareous nannofossils to fluorite ($n = 1.43$) so that greater relief is obtained for purposes of photography. Alteration to fluorite is accomplished with little loss in detail, but the fluoritised specimens are isotropic and cannot be studied in polarised light.

Viscous mounts of calcareous nannofossils are particularly useful in light microscope studies, because they permit the specimens to be turned and observed in side view. Silicone oils are commonly used for this purpose, although in principle any non-drying oil with proper pH and refractive index may be used. The cgs unit describing viscosity is the poise, which is 1 dyne cm^2 s. Appropriate viscosities for mounting calcareous nannofossils are in the order of 200 000 poise.

In the preparation of permanent and viscous mounts, a drop of the sample in suspension is spread on a cover slip and dried with warm air, under a heat lamp, or on a warm hot plate. For the permanent mount, the mounting medium may be applied directly to a clean glass slide, which is then inverted and used to pick up the cover slip with the nannofossils on it. In this way there is no chance for the nannofossils to contaminate the reservoir of mounting medium, or the tools used to apply the mounting medium. For viscous mounts, the nannofossils must be loosened with a toothpick, and stirred into the viscous medium, and then the cover slip is joined to a glass slide. The viscous stresses set up may cloud the mounting medium and may take several hours to clear.

Techniques used to study calcareous nannofossils in the light microscope

Ordinary light microscopy has been used in the study of calcareous nannofossils since at least 1854, when Ehrenberg figured a number of fossil examples of this group, assigning them to both organic and inorganic origins. For almost a hundred years simple light microscopy was the sole technique utilised in the investigation of these objects. Ordinary light microscopy is adequate for the study and differentiation of certain kinds of calcareous nannofossil; discoasters are particularly distinctive in simple transmitted light, but many other calcareous nannofossils, such as coccoliths, do not show a sufficient number of distinctive characters to permit consistent differentiation.

Kamptner pioneered the application of new techniques to the study of coccoliths. In 1943, he published the first electron micrograph of a coccolith, a transmission shadowgraph of a specimen of *Emiliania huxleyi*, showing the distinctive shape of the elements constructing the shield, which was not discernible in ordinary light. In 1952, and in a series of subsequent papers, he developed the use of polarised light using crossed nicol prisms. These early descriptions of the appearance of coccoliths between crossed polarisers, and the interpretations of the images, are the most detailed and accurate accounts which have been presented to date.

Phase contrast illumination greatly enhances the images of coccoliths and a number of other kinds of calcareous nannofossil, as well as its widely known use in providing contrast for the study of cytological studies. Deflandre and Fert (1954) were the first to make extensive use of phase contrast illumination in studying coccoliths, and the technique has become standard as a means of illustration since 1964. Of the systems currently available, the condensor after Zernicke seems to produce the best results with calcareous nannofossils.

Several new techniques in microscopy have been applied to the study of calcareous nannofossils with lesser degrees of success. The use of parallel polarisers has been discussed by Hay *et al.* (1967), and pictures made in this way have appeared often, although perhaps inadvertently, in the literature. The effect of plane polarisation is generally to distort the shape of a calcareous nannofossil, and may under certain circumstances produce the optical illusion of perforations at the centre of a radial arrangement of crystallites. Nomarski interference contrast is preferred by some workers to ordinary phase contrast (Gartner, 1967). The effect of Nomarski contrast is to produce an enhancement of relief, apparently increasing the depth of field.

Preparation of replicas for transmission electron microscopy

The development of the carbon replica technique in transmission electron microscopy has been widely used in the study of calcareous nannofossils, especially during the period 1962–1970. Introduced for coccoliths by Deflandre and Durrieu (1957), and subsequently rediscovered by many others, the technique is standard for investigation of electron-dense objects with the transmission electron microscope, and has been described in detail by Kay (1965). For carbon replica studies to be effective and yield good results, a considerable expenditure of time is required. Specimens must be cleaned of debris, concentrated, and dispersed on an appropriate substrate such as mica. Carbon is then evaporated *in vacuo*, and the replica floated off on distilled water. The coccoliths and other foreign materials are removed by treatment with hydrochloric, sulphuric, and hydrofluoric acids, a process which takes at least one day, and may require as long as a week. The results of this technique are excellent, especially when combined with stereomicrography, but the effort and time involved in sample preparation and examination limit its usefulness and introduction of the scanning electron microscope has reduced its popularity. Furthermore, the production of a carbon replica for transmission electron microscopy necessarily entails the dissolution of the specimen, and the replica itself deteriorates during examination by contamination from the electron beam. The following discussion may be useful to those who wish to use this technique with coccoliths and will aid those not familiar with it in evaluating and interpreting the many transmission electron micrographs in the literature.

The first problem in the preparation of replicas is to select a substrate material that will be as structureless as possible. Several types of substrate have been tried. A glass slide is easy to handle, but unless it is very clean, carbon film may not float free of it.

Freshly cleaved sheets of mica are also commonly used, but often the carbon film will not float free of them either. In this case it is generally useful to wipe the surface of the mica with a moist finger before putting the suspension on it. The thin film left on the slide will allow the carbon film to float free very easily. After the dispersion has been placed on a substrate it should be allowed to dry slowly. A heat lamp or a hot air blower can be used to speed the process, but a hot plate is not recommended because convection of the water will cause the sample to dry in clumps and streaks.

The preparation is placed in a vacuum evaporator and a thin film of carbon evaporated onto it *in vacuo*. Carbon has very low self-structure and reproduces details faithfully down to well below 100 Å. It does not granulate under an intense electron beam, and a thin film of carbon is very tough and durable. It is conductive, preventing build up of charge, and minimising distortion. Furthermore, carbon tends to migrate on the surface and will also form a layer in places not in direct "line-of-sight" from the evaporating source. Evaporation of the carbon is commonly done at an angle of 70–80°, and a sturdier replica results if the slide is rotated on a turntable during the evaporation. A drop of glycerin added to the tip of one of the carbon rods may make stripping of the carbon film onto water easier.

Faithful reproduction of detail must also be kept in mind when the replica is shadowed. Shadowing is employed to enhance contrast and thereby increase the amount of detail visible. Gold and palladium are two metals commonly used for shadow casting. Both are very easily evaporated in vacuum, and give a very high contrast without needing to be applied too thickly. Neither of these materials migrates on the surface like carbon, so that their sites of deposition are strictly limited to areas in "line-of-sight" of the evaporation source. Both of these metals also tend to granulate under an intense electron beam, with the result that no detail smaller than the individual grains of the shadowing material can be seen. These grains are responsible for the "sugary" surface often observed in electron micrographs. This problem can be eliminated by simply evaporating carbon at a slight angle to the surface of the substrate, i.e. a self-shadowed replica. The contrast in such a self-shadowed replica will be less than in a metal-shadowed replica and the edges of the shadows will be less sharp but the effect produced is very pleasing.

After the carbon film has been evaporated onto the mica sheet, it is scored into small squares about 2 mm on a side. The mica slide is then slowly submerged in a large Petri dish of distilled water, always keeping the slide inclined about 20° to the surface of the water, with the carbon film on the upper side of the slide. The individual squares of carbon film will float free in the water and all the sediment will remain attached to the film.

Following this stripping process excess water is siphoned out of the Petri dish, and the volume of the remaining water is doubled with 0.2N HCl. After one day all calcareous material has been dissolved, although clays remain and must be dissolved with HF. For this purpose the replicas are transferred into polyethelene containers. This is best accomplished by picking up the replicas on a small piece of 200 mesh wire screen and floating them free again on distilled water in the polyethelene container. The replica should be washed at least once in distilled water so that no free Ca ions are available. If they are present they will react with HF to precipitate fluorite on the replica and may ruin it. When all replicas have been transferred enough HF is added to bring the concentration to 25–40%. The solution of silicates is slow and may take up to a week. A slight amount of H_2SO_4 added with HF will produce a cleaner replica and also seems to speed up solution of the clays.

The clean replicas receive a final washing, and are then picked up on copper grids. If the carbon film is too thin, it should be supported by a collodion or formvar film on the grid. Normal carbon replicas need no supporting film. The replica can be approached by the grid from above or from beneath. With coated grids it is easiest to approach the replica from above, because the coating on the grid, being impervious to the water, makes the replica slip away to one side when approached from beneath. With uncoated grids, the replica will not stick to the grid when approached from above unless the grid has electric charge. The specimen and grid are dried in air on a piece of filter paper.

Observation in the transmission electron microscope

The grid is inserted into the specimen holder of the electron microscope with replica away from the electron gun. This is a standard precaution to avoid distortion of the image by electrons passing close to the copper wires and deviating after having passed through the image-forming carbon film. The image on the screen will be real if the replica was picked up from above, but will be a mirror image if the replica was picked up from below. The mirror image observed when the replica was picked from below presents no problem whatsoever to the recognition of a species, but for purposes of clarity published electron micrographs should be real images.

Stereoscopy of carbon replicas is achieved by tilting the specimen and taking two micrographs. This technique was introduced by Bramlette and Martini (1964), but proper orientation of the pair of photographs is essential in order to achieve correct perception of depth; because a number of stereoscopic micrographs have been published with incorrect orientation, there has been some confusion as to the value of the technique. Depth of field in the transmission electron microscope is so great that interpretation of the micrographs is virtually impossible without a shadowed replica or a stereoscopic pair of micrographs. Many calcareous nannofossils have such great relief that stereoscopic images are uncomfortable to view. A rhabdolith standing on its base looks to the stereo viewer like a nail almost touching his nose. The problem can be overcome by reducing the angle between the stereo views at the time the electron micrographs are taken, but this artificial flattening of the image destroys the value of the stereo pair for obtaining a measure of the true relief.

If the replica has been shadowed at a low angle, then it is easy to calculate the relief of objects by measuring the shadow. It should be borne in mind by those not familiar with shadowed replicas, that the "shadow" cast by an object is an area where no shadowing material is deposited, i.e. a thin area in the replica, and appears lighter than the surrounding area on the screen of the microscope. There are two ways electron micrographs are commonly printed: as positive or negative prints. The positive print is made from a negative. It duplicates the image on the screen of the electron microscope, i.e. shows the objects casting "shadows" which are lighter than the rest of the replica. Many nannaoplankton specialists prefer the positive print, although it may appear grotesque to the micropalaeontologist accustomed to viewing microfossils in reflected light. The positive print shows detail more readily and reproduces better than its counterpart, the negative print.

To produce a negative print, the negative is photographed or copied to make a tone-reversed negative. The reversed negative is printed and the result is a negative print in which the tones are exactly the opposite of what is seen on the screen of the electron microscope. The negative print is easier for the non-electron microscopist to visualise, for the "shadows" are darker than the surroundings as shadows should be. Unfortunately the two kinds of shadow are not really the same. The "shadow" in the electron microscope is an area of non-deposition of carbon or metal particles. The shadow cast by an object being illuminated by light is the area from which the light is blocked by the object. The metal or carbon evaporated to shadow a replica does not behave precisely as light would, however, and bounces off (is reflected from) surfaces that would not reflect light if magnified to the same scale. Thus the interpretation of "shadows" in negative prints is not as easy as it appears, and should be undertaken with caution.

The study of a nannofossil assemblage in the electron microscope is done in a way differing from that used by light microscopists in several important respects. Because the replica reproduces only the surface of a nannofossil, and is usually restricted to the surface facing the carbon source, it is impossible to examine opposite sides of the same specimen and side views are obtained only by chance. The chief problem which has arisen in connection with transmission electron microscopy of carbon replicas is orientation. A mirror image of the coccolith can appear in any one of the following ways: 1. the replica is picked up onto the electron microscope support screen by approaching from above; 2. the replica faces toward the electron gun rather than away from it when inserted in the microscope; 3. a tone-reversed negative is produced by emulsion to emulsion printing of the original negative; or 4. if the negative is printed with the emulsion not facing the emulsion of the paper. Orientation problems may also arise from light microscopes, some of which produce a negative with mirror-image orientation due to an odd number of mirror reflecting surfaces being present in the camera system.

The problems involved in comparing light microscopic views of coccoliths with carbon replicas observed in transmission electron microscopy have plagued researchers in this field. The solution offered by Perch-Nielsen (1967) is a qualified success. In this technique, the coccoliths are concentrated onto a formvar film on a specimen support screen and dispersed as though being prepared for the fabrication of a carbon replica. At this point they are immersed in methylisobutylketone ($n = 1.396$) and examined in the light microscope. Upon removal of the cover slip, the organic embedding medium evaporates, and carbon is then evaporated *in vacuo*. Treatment with acids removes the opaque materials, and the same specimens observed in the light microscope can then be examined in the transmission electron microscope. The disadvantage of this technique is the embedding medium, which has a refractive index very different from balsam or Caedax, the materials usually employed in the light microscopic observation of calcareous nannofossils. The development of ultra-high accelerating voltages offered potential for the study of calcareous nannofossils, because it seemed that they might be sufficiently transparent to the electron beam to obviate the need for replicas. Unfortunately, the results reported by Forchheimer (1968b) indicate that even at 1000 kv the beam is unable to penetrate the thicker parts of coccoliths, and at present ultra-high voltage transmission electron microscopy appears to offer little advantage.

Scanning electron microscopy

Scanning electron microscopy offers a means of directly examining calcareous nannofossils (Black, 1965; Lafitte and Noël, 1967). Hay and Sandberg (1967) have described the scanning electron microscope, compared it to other types of microscopy, and discussed its application by micropalaeontologists. Although the image is not as crisp as that obtained by transmission electron microscopy of a carbon replica, all of the structural details of the nannofossils can be observed if resolution is optimal. Preparation is rapid, requiring only that a conductive coating of a suitable secondary electron emitter (commonly Au–Pd alloy) be applied to the specimens while rotated *in vacuo*. Best results are obtained if a carbon coating is applied first, as this will provide some conductivity over the entire specimen surface. Sample preparation may be greatly simplified because the nannofossils may even be observed directly in the rock if necessary (Noël, 1968, 1969a). Scanning electron microscope investigations have the added advantage that the specimens are not destroyed and do not deteriorate with examination.

Stratigraphical distribution of calcareous nannoplankton

The stratigraphical ranges of the coccolith genera described in the taxonomic section are presented on Tables 1–4. Several genera will be noted to have restricted ranges but in most groups identification must be made to the species level for any detailed stratigraphical analysis. Biostratigraphic zonations based on calcareous nannofossils have been established using the lowest or highest occurrences of species. For the Jurassic, or parts thereof, zonations have been published by Brönnimann (1955), Stradner (1963), and Prins (1969), with the most complete zonation and a discussion of earlier schemes having been presented by Barnard and Hay (1974). For the Early Cretaceous, zonations have been proposed by Brönnimann (1955), Stradner (1963), Worsley (1971) and Thierstein (1971), with a revision of earlier schemes presented by Thierstein (1973). For the Late Cretaceous, a number of zonation schemes have been proposed, most notably by Stradner (1963), Čepek and Hay (1969) and Manivit (1971). That of Manivit is directly related to classic Cretaceous stages. Zonations of the Palaeogene have been developed by Bramlette and Sullivan (1961), Brönnimann and Stradner (1960), Hay (1964), Hay *et al.* (1967), Bramlette and Wilcoxon (1967), Bukry and Bramlette (1969), Roth (1970), Martini (1970, 1971) and Bukry (1973). Zonations of the Neogene, or parts thereof, have been proposed by Hay *et al.* (1967), Bramlette and Wilcoxon (1967), Gartner (1969b), Bukry and Bramlette (1969), Martini and Worsley (1970), Martini (1971) and Bukry (1973). The scheme of zonation presented below and in Tables 5–8 is based chiefly on Barnard and Hay (1974), Thierstein (1973), Manivit (1971) and Martini (1971).

Calcareous nannofossil zones of the Mesozoic and Cainozoic

JURASSIC NANNOFOSSIL ZONES

The ranges of calcareous nannoplankton datum indicators for the Jurassic are shown in Table 5.

Annulithus arkelli Zone

Definition: Interval from the lowest occurrence of *Annulithus arkelli* Rood,

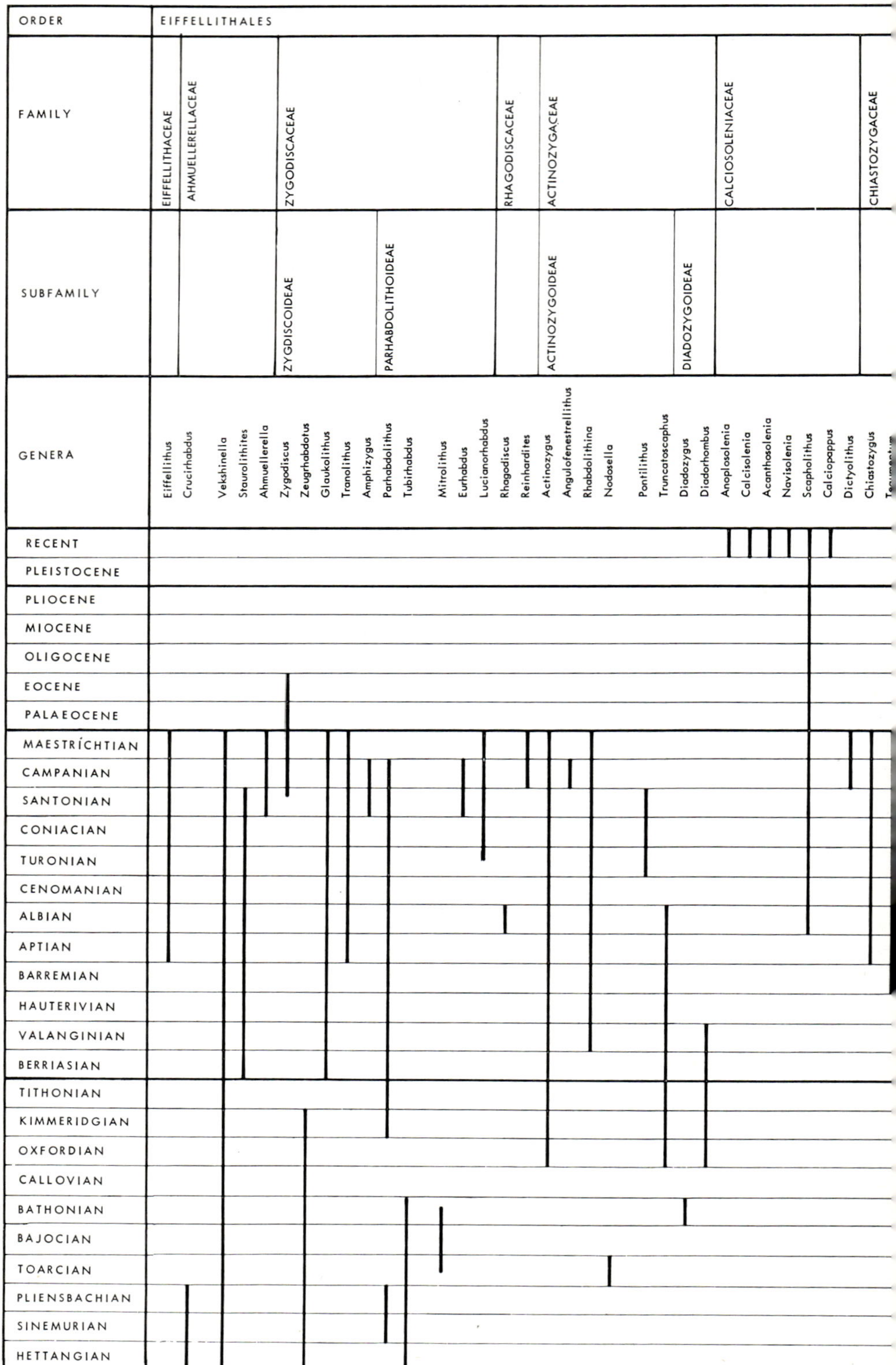

ORDER
EIFFELLITHALES
FAMILY
EIFFELLITHACEAE
AHMUELLERELLACEAE
ZYGODISCACEAE
RHAGODISCACEAE
ACTINOZYGACEAE
CALCIOSOLENIACEAE
CHIASTOZYGACEAE
SUBFAMILY
ZYGDISCOIDEAE
PARHABDOLITHOIDEAE
ACTINOZYGOIDEAE
DIADOZYGOIDEAE
GENERA
Eiffellithus
Crucirhabdus
Vekshinella
Staurolithites
Ahmuellerella
Zygodiscus
Zeugrhabdotus
Glaukolithus
Tranolithus
Amphizygus
Parhabdolithus
Tubirhabdus
Mitrolithus
Eurhabdus
Lucianorhabdus
Rhagodiscus
Reinhardites
Actinozygus
Angulofenestrellithus
Rhabdolithina
Nodosella
Pontilithus
Truncatoscaphus
Diadozygus
Diadorhombus
Anoplosolenia
Calcisolenia
Acanthosolenia
Navisolenia
Scapholithus
Calciopappus
Dictyolithus
Chiastozygus
RECENT
PLEISTOCENE
PLIOCENE
MIOCENE
OLIGOCENE
EOCENE
PALAEOCENE
MAESTRÍCHTIAN
CAMPANIAN
SANTONIAN
CONIACIAN
TURONIAN
CENOMANIAN
ALBIAN
APTIAN
BARREMIAN
HAUTERIVIAN
VALANGINIAN
BERRIASIAN
TITHONIAN
KIMMERIDGIAN
OXFORDIAN
CALLOVIAN
BATHONIAN
BAJOCIAN
TOARCIAN
PLIENSBACHIAN
SINEMURIAN
HETTANGIAN

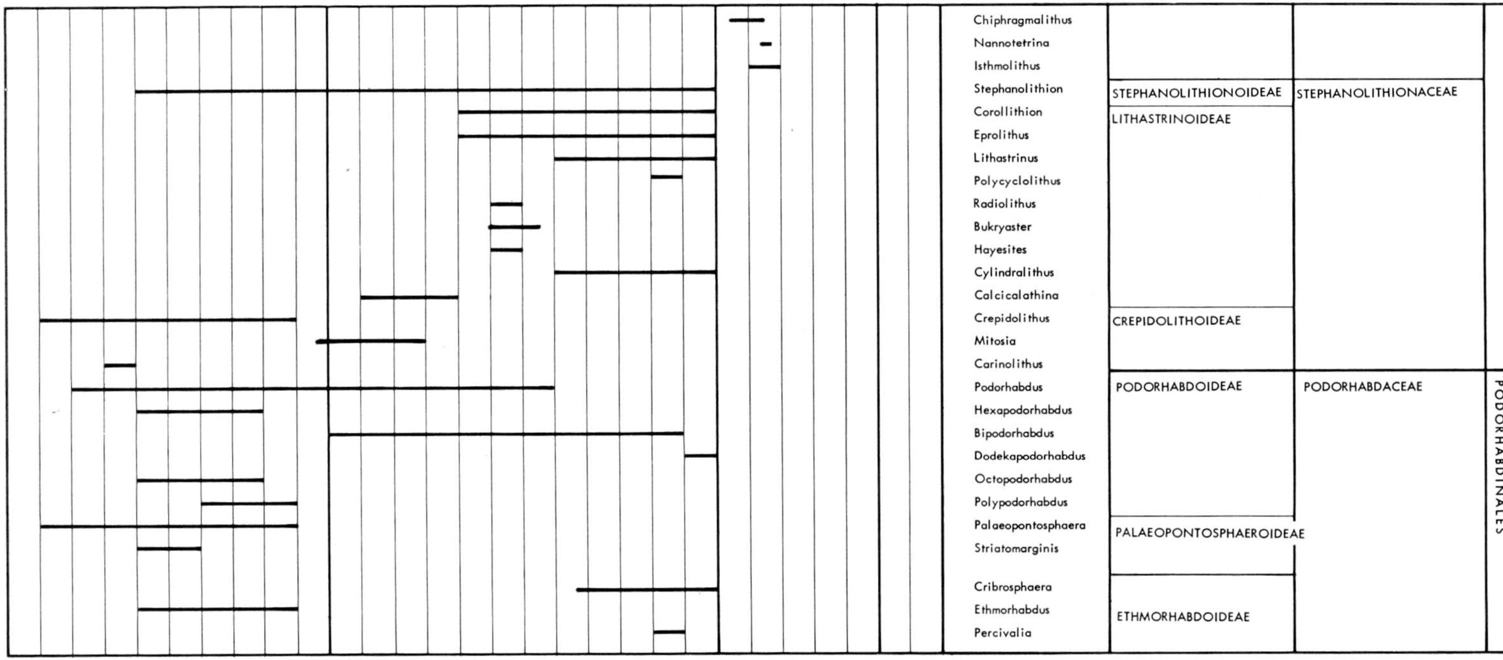

TABLE 1. Known stratigraphical ranges of coccoliths assigned to the orders Eiffelithales and Podorhabdinales.

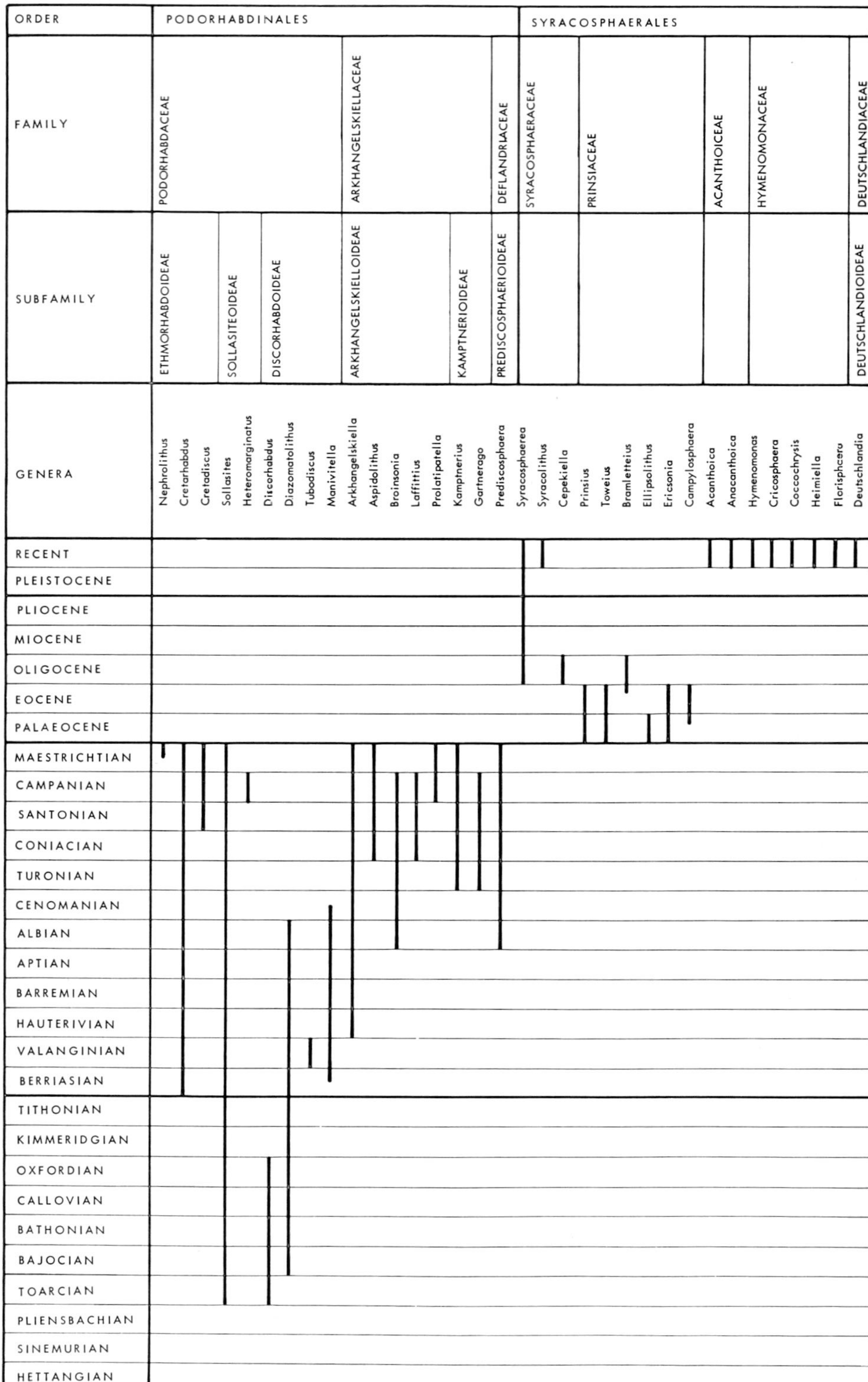
ORDER
PODORHABDINALES
SYRACOSPHAERALES
FAMILY
PODORHABDACEAE
ARKHANGELSKIELLACEAE
DEFLANDRIACEAE
SYRACOSPHAERACEAE
PRINSIACEAE
ACANTHOICEAE
HYMENOMONACEAE
DEUTSCHLANDIACEAE
SUBFAMILY
ETHMORHABDOIDEAE
SOLLASITEOIDEAE
DISCORHABDOIDEAE
ARKHANGELSKIELLOIDEAE
KAMPTNERIOIDEAE
PREDISCOSPHAERIOIDEAE
DEUTSCHLANDIOIDEAE
GENERA
Nephrolithus
Cretarhabdus
Cretadiscus
Sollasites
Heteromarginatus
Discorhabdus
Diazomatolithus
Tubodiscus
Manivitella
Arkhangelskiella
Aspidolithus
Broinsonia
Laffittius
Prolatipatella
Kamptnerius
Gartnerago
Prediscosphaera
Syracosphaerea
Syracolithus
Cepekiella
Prinsius
Toweius
Bramletteius
Ellipsolithus
Ericsonia
Campylosphaera
Acanthoica
Anacanthoica
Hymenomonas
Cricosphaera
Coccochrysis
Heimiella
Florisphcera
Deutschlandia
RECENT
PLEISTOCENE
PLIOCENE
MIOCENE
OLIGOCENE
EOCENE
PALAEOCENE
MAESTRICHTIAN
CAMPANIAN
SANTONIAN
CONIACIAN
TURONIAN
CENOMANIAN
ALBIAN
APTIAN
BARREMIAN
HAUTERIVIAN
VALANGINIAN
BERRIASIAN
TITHONIAN
KIMMERIDGIAN
OXFORDIAN
CALLOVIAN
BATHONIAN
BAJOCIAN
TOARCIAN
PLIENSBACHIAN
SINEMURIAN
HETTANGIAN

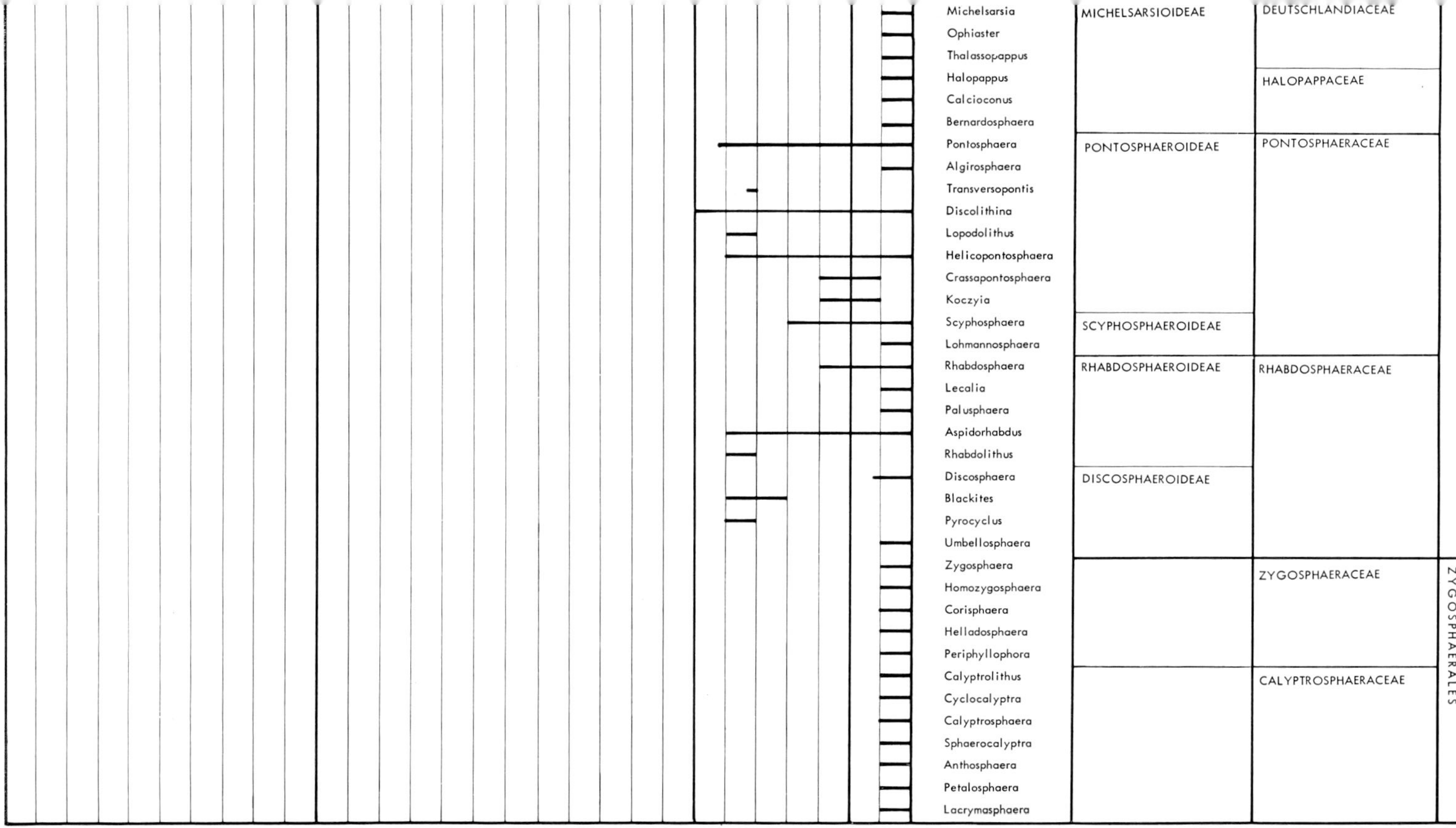

TABLE 2. Known stratigraphical ranges of coccolith genera assigned to the orders Podorhabdinales, Syracosphaerales and Zygosphaerales.

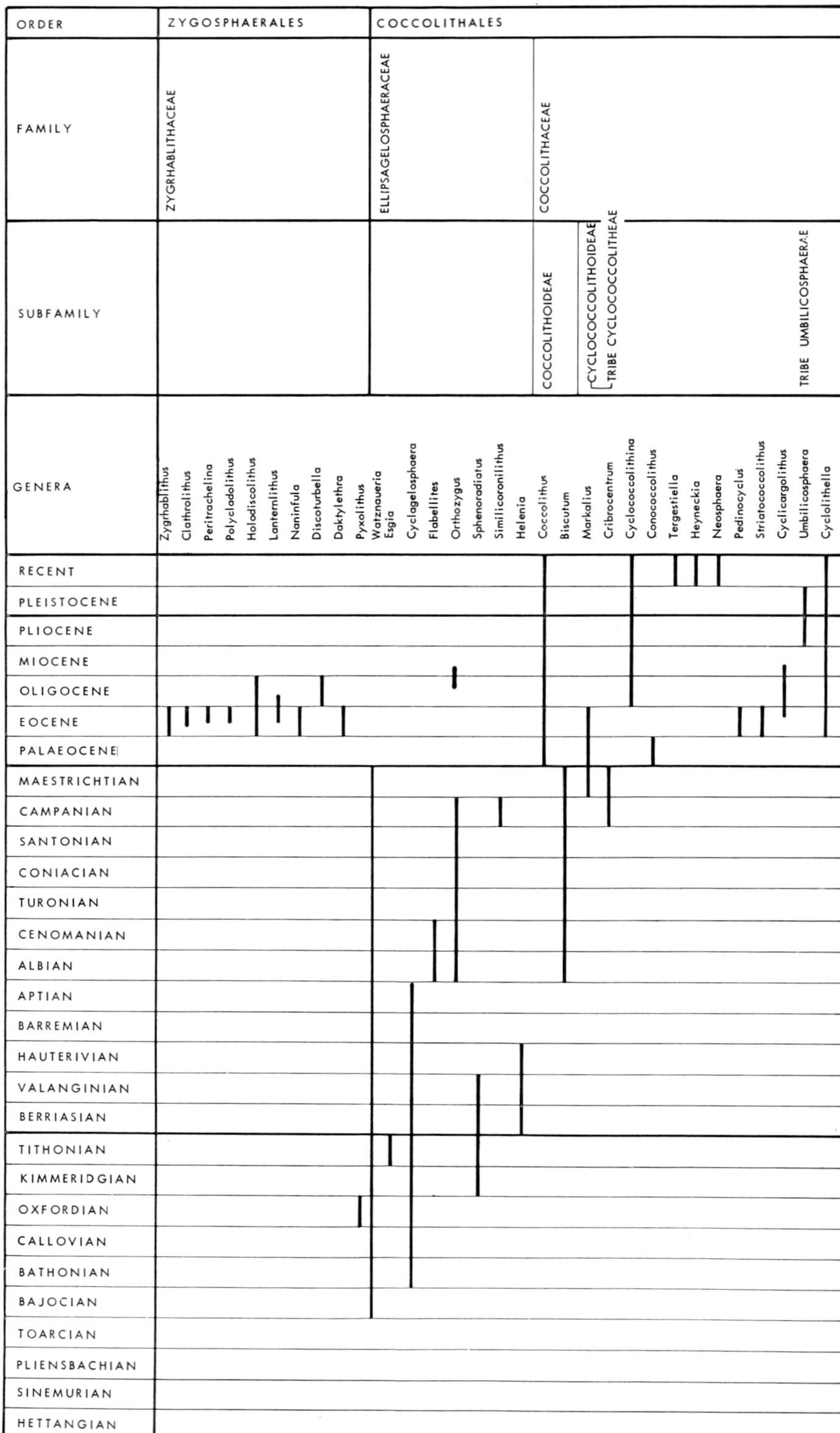
ORDER
ZYGOSPHAERALES
COCCOLITHALES
FAMILY
ZYGRHABLITHACEAE
ELLIPSAGELOSPHAERACEAE
COCCOLITHACEAE
SUBFAMILY
COCCOLITHOIDEAE
CYCLOCOCCOLITHOIDEAE
TRIBE CYCLOCOCCOLITHEAE
TRIBE UMBILICOSPHAERAE
GENERA
Zygrhablithus
Clathrolithus
Peritrachelina
Polycladolithus
Holodiscolithus
Lanternlithus
Naninfula
Discoturbella
Daktylethra
Pyxolithus
Watznaueria
Esgia
Cyclagelosphaera
Flabellites
Orthozygus
Sphenoradiatus
Similicoronilithus
Helenia
Coccolithus
Biscutum
Markalius
Cribrocentrum
Cyclococcolithina
Conococcolithus
Tergestiella
Heyneckia
Neosphaera
Pedinocyclus
Striatococcolithus
Cyclicargolithus
Umbilicosphaera
Cyclolithella
RECENT
PLEISTOCENE
PLIOCENE
MIOCENE
OLIGOCENE
EOCENE
PALAEOCENE
MAESTRICHTIAN
CAMPANIAN
SANTONIAN
CONIACIAN
TURONIAN
CENOMANIAN
ALBIAN
APTIAN
BARREMIAN
HAUTERIVIAN
VALANGINIAN
BERRIASIAN
TITHONIAN
KIMMERIDGIAN
OXFORDIAN
CALLOVIAN
BATHONIAN
BAJOCIAN
TOARCIAN
PLIENSBACHIAN
SINEMURIAN
HETTANGIAN

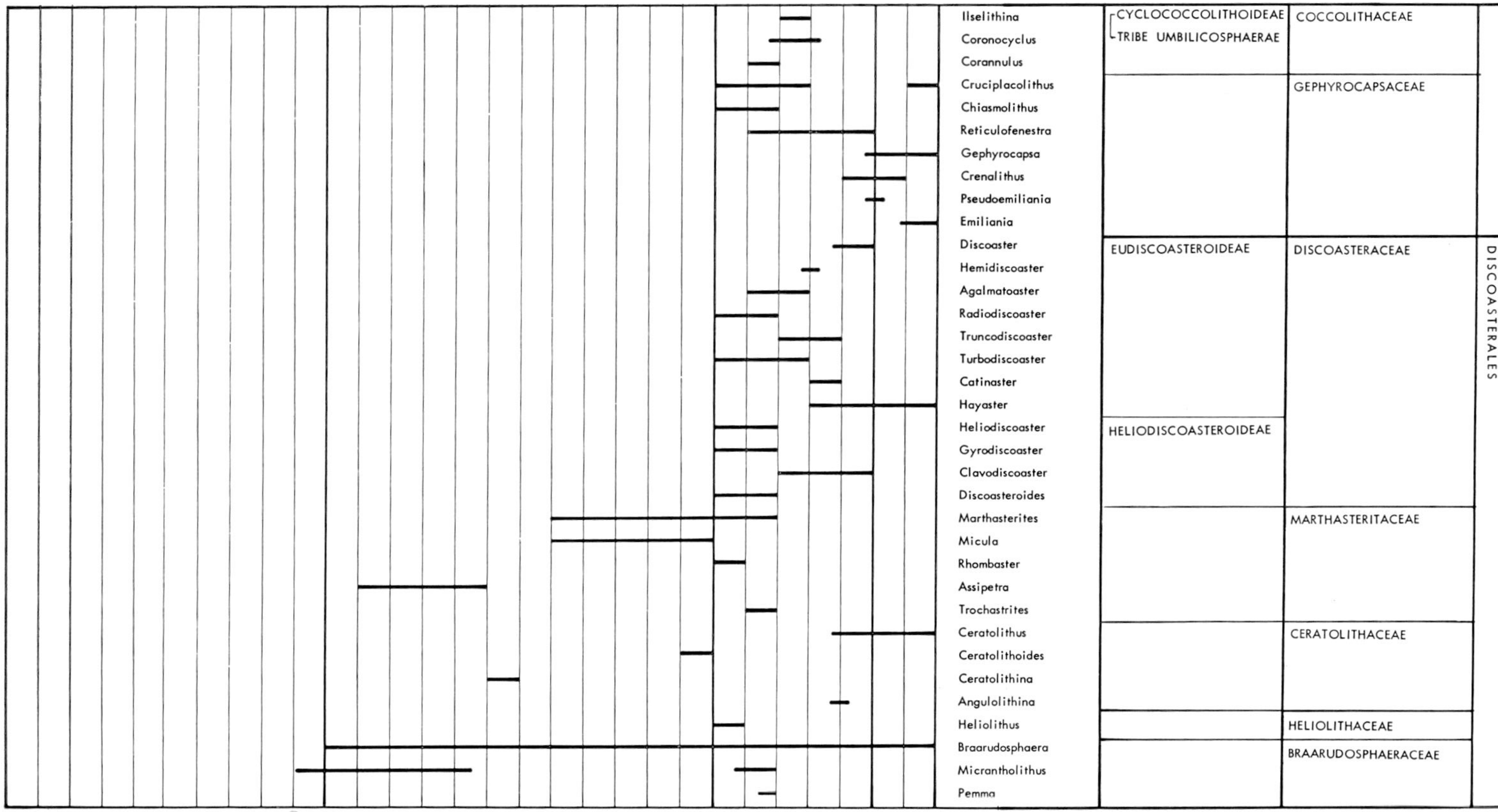

TABLE 3. Known stratigraphical distribution of coccolith genera assigned to the orders Zygosphaerales, Coccolithales and Discoasterales.

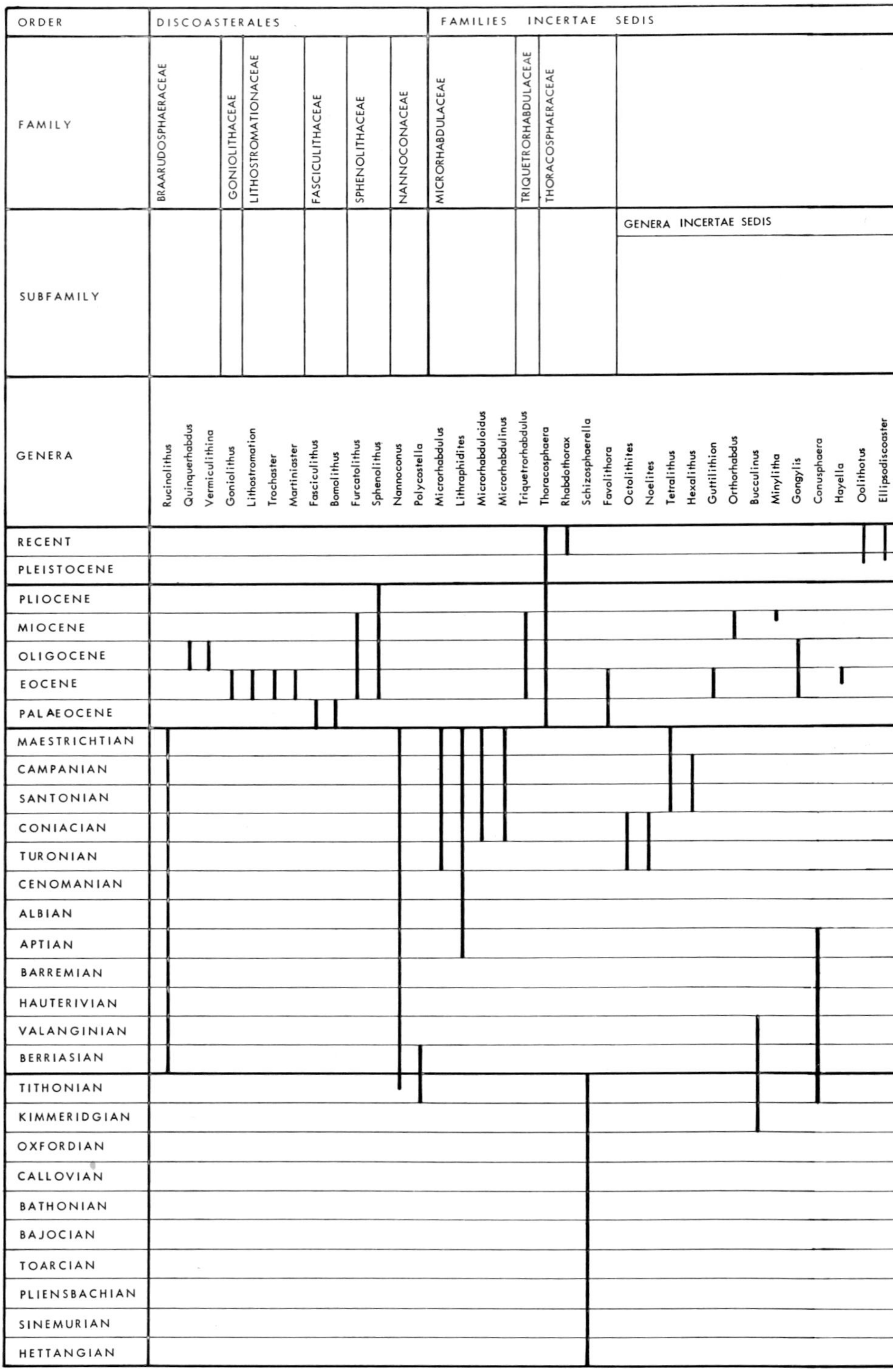

TABLE 4. Known stratigraphical distribution of genera assigned to the orders Discoasterales and Families Incertae Sedis.

Hay and Barnard to the lowest occurrence of *Crucirhabdus primulus* Prins ex Rood, Hay and Barnard.

Important species: *Annulithus arkelli*, *Crepidolithus cavus* Prins ex Rood *et al.*, *Schizosphaerella punctulata* Deflandre.

Authors: Barnard and Hay, 1974.

Age: Hettangian.

Crucirhabdus primulus Zone

Definition: Interval from the lowest occurrence of *Crucirhabdus primulus* to the lowest occurrence of *Parhabdolithus liasicus* Deflandre.

Important species: Those of the *Annulithus arkelli* Zone, plus *Crucirhabdus primulus*, *Vekshinella quadriarculla* (Noël), *Zeugrhabdotus erectus* (Deflandre), *Tubirhabdus patulus* Prins ex Rood *et al.*, *Crepidolithus crucifer* Prins ex Rood *et al.*

Authors: Barnard and Hay, 1974.

Remarks: This corresponds to the upper part of the *Crucirhabdus* Zone of Prins (1969), in which he indicated *Tubirhabdus patulus* and "*Crucirhabdus expansus*" to be present along with an abundance of *Crucirhabdus primulus*.

Age: Latest Hettangian and earliest Sinemurian.

Parhabdolithus marthae Zone

Definition: Interval from the lowest occurrence of *Parhabdolithus liasicus* to the highest occurrence of *Parhabdolithus marthae* Deflandre.

Important species: Those of the *Crucirhabdus primulus* Zone, plus *Parhabdolithus liasicus* and *Parhabdolithus marthae*.

Authors: Barnard and Hay, 1974.

Age: Early Sinemurian.

Parhabdolithus liasicus Zone

Definition: Interval from the highest occurrence of *Parhabdolithus marthae* to the lowest occurrence of *Paleopontosphaera dubia* Noël.

Important species: *Crucirhabdus primulus*, *Tubirhabdus patulus*, *Parhabdolithus liasicus*, *Zeugrhabdotus erectus*, *Vekshinella quadriarculla* and *Crepidolithus cavus*.

Authors: Barnard and Hay, 1974.

Remarks: Prins' (1969) *Parhabdolithus liasicus* Subzone of his *Parhabdolithus* Zone is an approximate equivalent. Stradner (1963) proposed a "nannoplanktonic zone, which might be called '*Liasicus*'-association", characterised by an abundance of *Parhabdolithus liasicus* and *Schizosphaerella punctulata*. It undoubtedly includes the *Parhabdolithus liasicus* Zone and probably portions or all of the overlying and underlying zones.

Age: Middle Sinemurian.

Paleopontosphaera dubia Zone

Definition: Interval from the lowest occurrence of *Paleopontosphaera dubia* to the lowest occurrence of *Crepidolithus crassus* (Deflandre).

Important species: Those of the *Parhabdolithus liasicus* Zone, plus *Paleonpontosphaera dubia.*

Authors: Barnard and Hay, 1974.

Remarks: The *Paleopontosphaera dubia* Zone is an approximate equivalent of the *Crepidolithus cavus* Subzone of Prins (1969).

Age: Late Sinemurian.

Crepidolithus crassus Zone

Definition: Interval from the lowest occurrence of *Crepidolithus crassus* to the lowest occurrence of *Podorhabdus cylindratus* Noël.

Important species: Those of the *Paleopontosphaera dubia* Zone, plus *Crepidolithus crassus.*

Authors: Barnard and Hay, 1974.

Remarks: This interval is probably the closest approximation to the "*Opacus*"-association of Stradner (1963). On his range chart he indicated that *Parhabdolithus liasicus* was a constituent of the assemblage, and this interval occupies the upper part of the range of that species. It is an approximate equivalent of the lower part of the *Crepidolithus crassus* Subzone of Prins' (1969) *Crepidolithus* Zone.

Age: Latest Sinemurian and early Pliensbachian.

Podorhabdus cylindratus Zone

Definition: Interval from the lowest occurrence of *Podorhabdus cylindratus* to the lowest occurrence of *Discorhabdus tubus* Noël.

Important species: Those of the *Crepidolithus crassus* Zone. *Podorhabdus cylindratus* is rare, and must be carefully sought to determine its lowest occurrence.

Authors: Barnard and Hay, 1974.

Remarks: The lower part of the *Podorhabdus cylindratus* Zone might belong to the "*Opacus*"-association of Stradner (1963), because *Parhabdolithus liasicus*, indicated on his range chart to be typical of that unit, has its highest occurrence in this zone. This unit corresponds to part or all of Prins' (1969) *Crepidolithus crassus* Subzone of his *Crepidolithus* Zone and the *Striatococcus opacus* Subzone of his *Striatococcus* Zone.

Age: Early Pliensbachian through Toarcian.

Discorhabdus tubus Zone

Definition: Interval from the lowest occurrence of *Discorhabdus tubus* to the lowest occurrence of *Stephanolithion speciosum* Deflandre.

Important species: Those of the *Podorhabdus cylindratus* Zone, plus *Discorhabdus tubus*, *Podorhabdus macrogranulatus* Prins ex Rood *et al.*, *Striatomarginis primitivus* Rood, Hay and Barnard, *Alvearium dorsetense* Black and *Paleopontosphaera veterna* Prins ex Rood *et al.*, all of which appear at the base of this unit and are common in it so that assemblages of this zone have a strikingly different aspect from the subjacent units.

Authors: Barnard and Hay, 1974.

Age: Latest Toarcian and earliest Bajocian.

Stephanolithion speciosum Zone

Definition: Interval from the lowest occurrence of *Stephanolithion speciosum* s. str. to the lowest occurrence of *Diazomatolithus lehmani* Noël.

Important species: *Zeugrhabdotus erectus*, *Podorhabdus cylindratus*, *Striatomarginis primitivus*, *Paleopontosphaera dubia*, *Tubirhabdus patulus*, *Vekshinella quadriarculla*, *Crepidolithus cavus*, *Paleopontosphaera veterna*, *Discorhabdus tubus*, plus *Stephanolithion speciosum* s. str., *Hexapodorhabdus cuvillieri* Noël, *Octopodorhabdus decussatus* (Manivit), *Watznaueria communis* Reinhardt and *Discorhabdus biperforatus* Rood, Hay and Barnard, all of which appear at the base of this zone.

Authors: Barnard and Hay, 1974.

Age: Early Bajocian to early Bathonian.

Diazomatolithus lehmani Zone

Definition: Interval from the lowest occurrence of *Diazomatolithus lehmani* to the lowest occurrence of *Stephanolithion speciosum* var. *octum* Rood and Barnard.

Important species: Those of the *Stephanolithion speciosum* Zone, plus *Diazomalithus lehmani*. *Striatomarginis primitivus* and *Discorhabdus biperforatus* both have their highest occurrence within this interval; *Diadozygus asymmetricus* Rood, Hay, and Barnard and *Cyclagelosphaera margereli* Noël have their lowest occurrence in this unit.

Authors: Barnard and Hay, 1974.

Age: Early and middle Bathonian.

Stephanolithion speciosum var. *octum* Zone

Definition: Interval from the lowest occurrence of *Stephenolithion speciosum* var. *octum* to the lowest occurrence of *Stephanolithion hexum* Rood and Barnard.

Important species: Those of the *Diazomatolithus lehmani* Zone, plus

Stephanolithion speciosum var. *octum*. *Watznaueria britannica* (Stradner) has its lowest occurrence in this interval.

Authors: Barnard and Hay, 1974.

Age: Late Bathonian.

Stephanolithion hexum Zone

Definition: Interval from the lowest occurrence of *Stephanolithion hexum* to the lowest occurrence of *Stephanolithion bigoti* Deflandre.

Important species: *Watznaueria communis*, *Crepidolithus crassus*, *Stephanolithion hexum*, *Stephanolithion speciosum* var. *octum*, *Zeugrhabdotus erectus*, *Paleopontosphaera dubia*, *Discorhabdus tubus*, *Ethmorhabdus gallicus*, *Diazomatolithus lehmani*, *Vekshinella quadriarculla*, *Hexapodorhabdus cuvillieri*, *Octopodorhabdus decussatus*, *Diadozygus asymmetricus*, *Watznaueria britannica* and *Cyclagelosphaera margereli*. *Stephanolithion speciosum* s. str. has its highest occurrence in the lower part of this unit; *Tubirhabdus patulus* disappears near the top. *Schizosphaerella punctulata* becomes rare and its occurrence sporadic in this interval.

Authors: Barnard and Hay.

Remarks: Stradner (1963) described what he termed "*Decorus*"-associations from Bathonian samples; these probably represent the upper part of the *Stephanolithion hexum* Zone.

Age: Early Callovian.

Stephanolithion bigoti Zone

Definition: Interval from the lowest occurrence of *Stephanolithion bigoti* to the lowest occurrence of *Podorhabdus escaigi* Noël.

Important species: Those of the *Stephanolithion hexum* Zone, plus *Stephanolithion bigoti*. *Discorhabdus patulus* Noël has its lowest occurrence at the base of this unit. *Zeugrhabdotus noeli* Rood, Hay and Barnard has its lowest occurrence near the top of this zone.

Authors: Barnard and Hay, 1974.

Remarks: Stradner (1963) named a unit the "*Bigoti*" Zone ("*Bigoti*-associations" on his chart I) for assemblages "dominated by double rimmed coccoliths" with abundant *Stephanolithion bigoti*. Stradner's unit may correspond to the *Stephanolithion bigoti* Zone or to any of the superjacent units up to the *Vekshinella stradneri* Zone.

Age: Early Callovian.

Podorhabdus escaigi Zone

Definition: Interval from the lowest occurrence of *Podorhabdus escaigi* to the lowest occurrence of *Podorhabdus rahla* Noël.

Important species: *Watznaueria communis, Stephanolithion bigoti, Zeugrhabdotus erectus, Podorhabdus cylindratus, Paleopontosphaera dubia, Ethmorhabdus gallicus, Diadozygus asymmetricus, Cyclagelosphaera margereli, Zeugrhabdotus noeli, Octopodorhabdus decussatus, Watznaueria britannica* and *Vekshinella quadriarculla.*

Authors: Barnard and Hay, 1974.

Age: Middle Callovian.

Podorhabdus rahla Zone

Definition: Interval from the lowest occurrence of *Podorhabdus rahla* to the lowest occurrence of *Discorhabdus jungi* Noël.

Important species: Those of the *Podorhabdus escaigi* Zone, plus *Podorhabdus rahla.*

Authors: Barnard and Hay, 1974.

Age: Middle Callovian.

Discorhabdus jungi Zone

Definition: Interval from the lowest occurrence of *Discorhabdus jungi* to the lowest occurrence of *Diadozygus dorsetense* Rood, Hay and Barnard.

Important species: Those of the *Podorhabdus rahla* Zone, plus *Discorhabdus jungi. Diadozygus minutus* Rood, Hay and Barnard and *Truncatoscaphus delftensis* (Stradner) appear near the top of this unit and are rare.

Authors: Barnard and Hay, 1974.

Age: Late Callovian.

Diadozygus dorsetense Zone

Definition: Interval from the lowest occurrence of *Diadozygus dorsetense* to the lowest occurrence of *Actinozygus geometricus* (Gorka).

Important species: *Watznaueria communis, Zeugrhabdotus erectus, Stephanolithion bigoti, Vekshinella quadriarculla, Podorhabdus cylindratus, Paleontosphaera dubia, Discorhabdus tubus, Ethmorhabdus gallicus, Hexapodorhabdus cuvillieri, Discorhabdus jungi, Zeugrhabdotus noeli, Discorhabdus patulus, Podorhabdus escaigi, Octopodorhabdus decussatus, Watznaueria britannica, Diadozygus asymmetricus, Diadozygus minutus* and *Truncatoscaphus delftensis.*

Authors: Barnard and Hay, 1974.

Age: Early Oxfordian.

Actinozygus geometricus Zone

Definition: Interval from the lowest occurrence of *Actinozygus geometricus* to the lowest occurrence of *Vekshinella stradneri* Rood, Hay and Barnard.

Important species: Those of the *Diadozygus dorsetense* Zone, plus *Actinozygus geometricus.*

Authors: Barnard and Hay, 1974.
Age: Early Oxfordian.

Vekshinella stradneri Zone

Definition: Interval from the lowest occurrence of *Vekshinella stradneri* to the highest occurrence of *Stephanolithion bigoti.*

Important species: Those of the *Actinozygus geometricus* Zone, plus *Vekshinella stradneri. Octopodorhabdus decussatus*, *Podorhabdus rahla* and *Diadozygus minutus* are rare and have their highest occurrence in this unit.

Authors: Barnard and Hay, 1974.

Age: Middle Oxfordian through early Kimmeridgian.

Watznaueria communis Zone

Definition: Interval from the highest occurrence of *Stephanolithion bigoti* to the lowest occurrence of *Parhabdolithus embergeri* (Noël).

Important species: Those of the *Vekshinella stradneri* Zone, less *Stephanolithion bigoti.*

Authors: Barnard and Hay, 1974.

Age: Middle Kimmeridgian.

Parhabdolithus embergeri Zone

Definition: Interval from the lowest occurrence of *Parhabdolithus embergeri* to the lowest occurrence of *Nannoconus colomi* (de Lapparent).

Important species: Those of the *Watznaueria communis* Zone, plus *Parhabdolithus embergeri. Crepidolithus crassus* and *Podorhabdus cylindratus* have their highest occurrence in this interval.

Author: Worsley, 1971.

Age: Late Kimmeridgian.

CRETACEOUS NANNOFOSSIL ZONES

The ranges of calcareous nannoplankton datum indicators for the cretaceous are shown on Table 6.

Nannoconus colomi Zone

Definition: Interval from the lowest occurrence of *Nannoconus colomi* (de Lapparent) to the lowest occurrence of *Cretarhabdus crenulatus* Bramlette and Martini.

Important species: *Polycostella beckmanii* Thierstein, *Cretaturbella rothi* Thierstein, *Parhabdolithus embergeri* (Noël), *Watznaueria barnesae* (Black), *Cyclagelosphaera margereli*, *Diazomatolithus lehmani*, *Watznaueria britannica*, *Biscutum constans* (Gorka), *Nannoconus colomi*, *Nannoconus bronnimanni*

TABLE 5

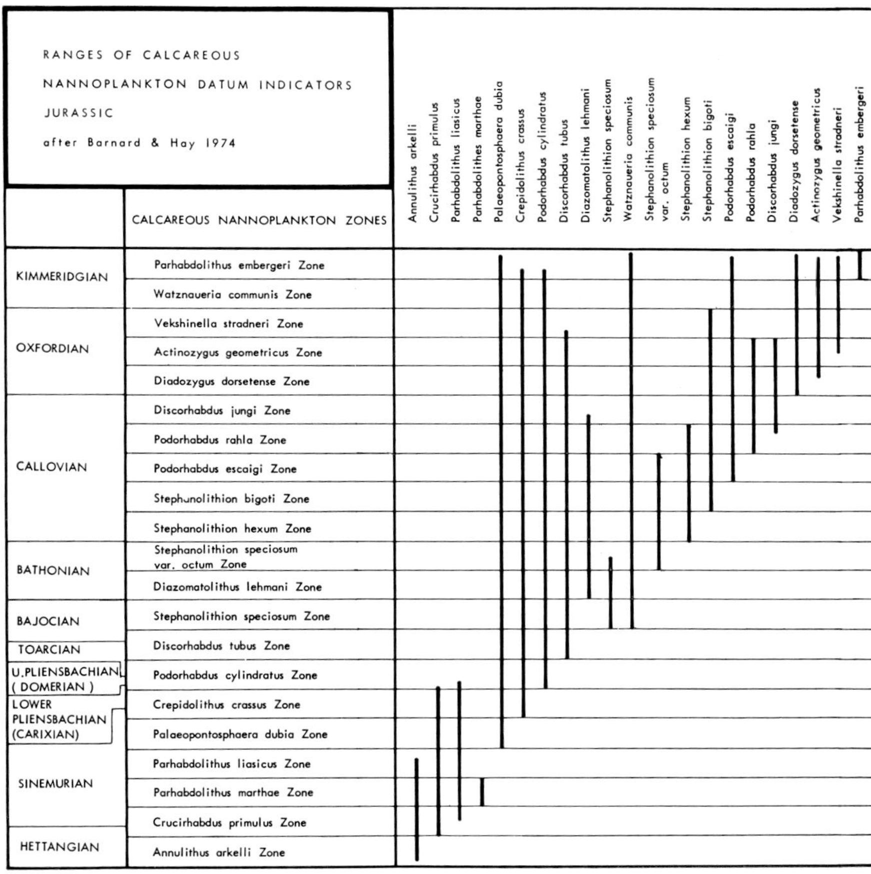
RANGES OF CALCAREOUS
NANNOPLANKTON DATUM INDICATORS
JURASSIC
after Barnard & Hay 1974
CALCAREOUS NANNOPLANKTON ZONES
Annulithus arkelli
Crucirhabdus primulus
Parhabdolithus liasicus
Parhabdolithes marthae
Palaeopontosphaera dubia
Crepidolithus crassus
Podorhabdus cylindratus
Discorhabdus tubus
Diazomatolithus lehmani
Stephanolithion speciosum
Watznaueria communis
Stephanolithion speciosum var. octum
Stephanolithion hexum
Stephanolithion bigoti
Podorhabdus escaigi
Podorhabdus rahla
Discorhabdus jungi
Diadozygus dorsetense
Actinozygus geometricus
Vekshinella stradneri
Parhabdolithus embergeri
KIMMERIDGIAN
Parhabdolithus embergeri Zone
Watznaueria communis Zone
OXFORDIAN
Vekshinella stradneri Zone
Actinozygus geometricus Zone
Diadozygus dorsetense Zone
CALLOVIAN
Discorhabdus jungi Zone
Podorhabdus rahla Zone
Podorhabdus escaigi Zone
Stephanolithion bigoti Zone
Stephanolithion hexum Zone
BATHONIAN
Stephanolithion speciosum var. octum Zone
Diazomatolithus lehmani Zone
BAJOCIAN
Stephanolithion speciosum Zone
TOARCIAN
Discorhabdus tubus Zone
U.PLIENSBACHIAN (DOMERIAN)
Podorhabdus cylindratus Zone
LOWER PLIENSBACHIAN (CARIXIAN)
Crepidolithus crassus Zone
Palaeopontosphaera dubia Zone
SINEMURIAN
Parhabdolithus liasicus Zone
Parhabdolithus marthae Zone
Crucirhabdus primulus Zone
HETTANGIAN
Annulithus arkelli Zone

Trejo, *Lithraphidites carniolensis* Deflandre, *Stephanolithion laffittei* Noël, *Polycostella senaria* Thierstein, *Parhabdolithus asper* (Stradner), *Rucinolithus wisei* Thierstein, *Cruciellipsis chiasta* (Worsley). *Cruciellipsis cuvillieri* (Manivit), and *Micrantholithus hoschulzi* (Reinhardt).

Author: Worsley, 1971, as *Nannoconus steinmanni* Zone; modified by Thierstein, 1971.

Age: Latest Tithonian to late Berriasian.

Cretarhabdus crenulatus Zone

Definition: Interval from the lowest occurrence of *Cretarhabdus crenulatus* to the lowest occurrence of *Calcicalathina oblongata* (Worsley).

Important species: Those of the *Nannoconus colomi* Zone, plus *Cretarhabdus crenulatus*, *Bipodorhabdus roegli* Thierstein, *Reinhardtites fenestratus* (Worsley), and *Tubodiscus verenae* Thierstein. *Cretaturbella rothi* becomes rare near the base of this zone.

Author: Thierstein, 1971.

Age: Late Berriasian to early Valanginian.

Calcicalathina oblongata Zone

Definition: Interval from the lowest occurrence of *Calcicalathina oblongata* to the lowest occurrence of *Lithraphidites bollii* (Thierstein).

Important species: Those of the *Cretarhabdus crenulatus* Zone, plus *Calcicalathina oblongata*, *Glaukolithus diplogrammus* (Deflandre and Fert), *Rhabdolithina splendens* (Deflandre), *Micula infracretacea* Thierstein, *Podorhabdus dietzmanni* (Reinhardt), *Diadorhombus rectus* Worsley, *Nannoconus bucheri* Brönnimann, and *N. wassalli* Brönnimann. *Rucinolithus wisei*, *Diadorhombus rectus* and *Tubodiscus verenae* do not occur above this zone.

Author: Thierstein, 1971.

Age: Early Valanginian–Early Hauterivian.

Lithraphidites bollii Zone

Definition: Interval from the occurrence of *Lithraphidites bollii* to the highest occurrence of *Calcicalathina oblongata.*

Important species: Those of the *Calcicalathina oblongata* Zone, plus *Lithraphidites bollii. Discorhabdus*, *Cruciellipsis cuvillieri*, *Bipordorhabdus roegli*, *Lithraphidites bollii*, and *Calcicalathina oblongata* have their highest occurrence in this unit.

Author: Thierstein, 1971.

Age: Early Hauterivian to early Barremian.

Micrantholithus hoschulzi Zone

Definition: Interval from the highest occurrence of *Calcicalathina oblongata* to the highest occurrence of *Nannoconus colomi* and/or the lowest occurrences of *Chiastozygus litterarius* (Gorka) and/or *Rucinolithus irregularis* Thierstein.

Important species: Those of the *Lithraphidites bollii* Zone, less *Nannoconus colomi*.

Author: Thierstein, 1971.

Age: Early Barremian to Late Barremian.

Chiastozygus litterarius Zone

Definition: Interval from the highest occurrence of *Nannoconus colom* and/or the lowest occurrences of *Chiastozygus litterarius* and/or *Rucinolithus irregularis* to the lowest occurrences of *Parhabdolithus angustus* (Stradner) and/or *Lithastrinus floralis* Stradner.

Important species: Those of the *Micrantholithus hoschulzi* Zone, plus *Chiastozygus litterarius* (Gorka), *Rucinolithus irregularis* Thierstein, *Podorhabdus decorus* (Deflandre) and *Lithastrinus septentrionalis* Stradner. *Nannoconus colomi*, *Micrantholithus hoschulzi*, *Micrantholithus obtusus*, *Cretaturbella rothi*, and *Nannoconus bronnimanni* have their highest occurrences in this Zone.

Author: Thierstein, 1971.

Age: Early Aptian.

Parhabdolithus angustus Zone

Definition: Interval from the lowest occurrences of *Parhabdolithus angustus* and/or *Lithastrinus floralis*, to the lowest occurrence of *Prediscosphaera cretacea* (Arkhangel'ski).

Important species: Those of the *Chiastozygus litterarius* Zone, plus *Parhabdolithus angustus* (Stradner), *Lithastrinus floralis* Stradner, *Corollithion achylosum* (Stover), *Tranolithus gabalus* Stover, and *Cretarhabdus loriei* Gartner. *Nannoconus wassalli*, *Micula infracretacea*, *Cyclagelosphaera margereli*, *Nannoconus bucheri* and *Nannoconus globulus* have their highest occurrences in this Zone.

Author: Manivit, 1971.

Age: Late Aptian–Early Albian.

Prediscosphaera cretacea Zone

Definition: Interval from the lowest occurrence of *Prediscosphaera cretacea* to the lowest occurrence of *Eiffellithus turriseiffeli* (Deflandre and Fert).

Important species: Those of the *Parhabdolithus angustus* Zone, plus *Prediscosphaera cretacea*, *Vagalapilla matalosa* (Stover), *Eiffellithus trabeculatus* (Gorka), *Cretarhabdus coronadventis* Reinhardt, *Broinsonia signata* (Noël), *Tranolithus exiguus* Stover, *Tranolithus orionatus* (Reinhardt),

Broinsonia lata (Noël), *Podorhabdus orbiculofenestrus* (Gartner), and *Prediscosphaera spinosa* (Bramlette and Martini).

Author: Thierstein, 1971.

Age: Early Albian.

Eiffellithus turriseiffeli Zone

Definition: Interval from the lowest occurrence of *Eiffellithus turriseiffeli* to the lowest occurrence of *Hayesites albiensis* Manivit.

Important species: Those of the *Prediscosphaera cretacea* Zone, plus *Eiffellithus turriseiffeli* (Deflandre and Fert), *Cribosphaerella ehrenbergi* (Arkhangel'ski), *Corollithion signum* Stradner, *Broinsonia enormis* (Shumenko), and *Scapholithus fossilis* Deflandre and Fert. *Rucinolithus irregularis*, *Parhabdolithus infinitus*, *Diazomatolithus lehmani* and *Braarudosphaera africana* have their highest occurrences in this unit.

Author: Thierstein, 1971.

Age: Middle Albian.

Hayesites albiensis Zone

Definition: Interval from the lowest occurrence of *Hayesites albiensis* Manivit to the lowest occurrence of *Corollithion rhombicum* (Stradner, Adamiker and Maresch).

Important species: *Zeugrhabdotus erectus* (Deflandre), *Zeugrhabdotus bussoni* (Noël), *Glaukolithus diplogrammus*, *Stuarolithites crux* (Deflandre), *Parhabdolithus embergeri*, *Stephanolithion laffittei*, *Lithastrinus septentrionalis* Stradner, *Lithraphidites carniolensis*, *Nannoconus globulus*, *Nannoconus wassalli*, *Nannoconus minutus* Deflandre, *Predicosphaera columnata* (Stover), *Tetralithus gothicus* Deflandre, *Chiastozygus litteratius*, *Parhabdolithus angustus* (Stradner), *Rhabdolithina splendens*, *Eiffellithus turriseiffeli*, *Lithastrinus floralis* Stradner, *Thoracosphaera deflandrei* Kamptner, *Watznaueria parvidentata* (Deflandre), plus *Hayesites albiensis*.

Author: Manivit, 1971.

Age: Middle Albian.

Corollithion rhombicum Zone

Definition: Interval from the lowest occurrence of *Corollithion rhombicum* to the lowest occurrence of *Staurolithites matalosus* (Stover).

Important species: Those of the *Hayesites albiensis* Zone, plus *Sollasites horticus*, *Cretarhabdus decorus* (Deflandre), *Corollithion signum* Stradner, *Cretarhabdus unicornis* Stover, *Corollithion rhombicum*, *Cretarhabdus helicoides* Stover, *Discorhabdus ignotus* (Gorka). *Nannoconus wassalli*, *Nannoconus globulus*, and *Lithastrinus septentrionalis* have their highest occurrence in this unit.

Author: Manivit, 1971.
Age: Late Albian.

Staurolithites matalosus Zone

Definition: Interval from the lowest occurrence of *Staurolithites matalosus* to the lowest occurrence of *Staurolithites orbiculofenestrus* (Gartner).

Important species: Those of the *Corollithion rhombicum* Zone, plus *Octopodorhabdus decussatus*, *Staurolithites matalosus*, *Tranolithus orionatus*, *Tranolithus exiguus* Stover, *Watznaueria barnesae* (Black).

Author: Manivit, 1971.
Age: Late Albian.

Staurolithites orbiculofenestrus Zone

Definition: Interval from the lowest occurrence of *Staurolithites orbiculofenestrus* to the lowest occurrence of *Gartnerago obliquus* (Stradner).

Important species: Those of the *Staurolithites matalosus* Zone, plus *Staurolithites orbiculofenestrus* (Gartner), *Eiffellithus eximus* (Stover).

Authors: Čepek and Hay, 1969, modified by Manivit, 1971.
Age: Early and Middle Cenomanian.

Gartnerago obliquus Zone

Definition: Interval from the lowest occurrence of *Gartnerago obliquus* to the lowest occurrence of *Corollithion exiguum* Stradner.

Important species: Those of the *Staurolithites orbiculofenestrus* Zone, plus *Gartnerago obliquus*, *Zygodiscus ponticulus* (Deflandre), *Cretarhabdus conicus* Bramlette and Martini.

Author: Manivit, 1971.
Age: Late Cenomian and Early Turonian.

Corollithion exiguum Zone

Definition: Interval from the lowest occurrence of *Corollithion exiguum* to the lowest occurrence of *Micula staurophora* (Gardet).

Important species: Those of the *Gartnerago obliquus* Zone, plus *Lithastrinus grilli* Stradner, *Broinsonia enormis* (Shumenko), *Chiastozygus amphipons* (Bramlette and Sullivan), *Corollithion exiguum* Stradner. *Staurolithites matalosus* and *Watznaueria parvidentata* have their highest occurrence near the base of this zone.

Authors: Čepek and Hay, 1969, modified by Manivit, 1971.
Age: Middle Turonian.

Micula staurophora Zone

Definition: Interval from the lowest occurrence of *Micula staurophora* to the lowest occurrence of *Marthasterites furcatus* Deflandre.

Important species: Those of the *Corollithion exiguum* Zone, plus *Micula staurophora* (Gardet), *Cribrosphaera ehrenbergi* (Arkhangel'ski), *Microrhabdulus decoratus* Deflandre, *Cylindralithus serratus* Bramlette and Martini, *Lucianorhabdus cayeuxi* Deflandre. *Staurolithites orbiculofenestrus*, *Cretarhabdus romani* and *Tetralithus gothicus* have their highest occurrence at the base of this zone.

Author: Manivit, 1971.

Age: Late Turonian and Early Coniacian.

Marthasterites furcatus Zone

Definition: Interval from the lowest occurrence of *Marthasterites furcatus* Deflandre to the lowest occurrence of *Kamptnerius magnificus* Deflandre.

Important species: Those of the *Micula staurophora* Zone, plus *Marthasterites furcatus* Deflandre, *Kamptnerius punctatus* Stradner, *Eiffellithus anceps* (Gorka) and *Arkhangelskiella ethmopora* Bukry. *Markalius circumradiatus* and *Prediscosphaera columnata* both have their highest occurrence near the base of this zone.

Authors: Čepek and Hay, 1969, modified by Manivit, 1971.

Age: Middle and Late Coniacian.

Kamptnerius magnificus Zone

Definition: Interval from the lowest occurrence of *Kamptnerius magnificus* to the lowest occurrence of *Broinsonia parca* (Stradner).

Important species: Those of the *Marthasterites furcatus* Zone, plus *Kamptnerius magnificus* Deflandre, *Tetralithus obscurus* Deflandre, *Microrhabdulinus ambiguus* Deflandre, *Ahmuellerella octoradiata* (Gorka), *Rucinolithus hayi* Stover. *Marthasterites furcatus* Deflandre and the remaining species of *Nannoconus* all have their highest occurrence at the top of this zone.

Authors: Čepek and Hay, 1969, modified by Manivit, 1971.

Age: Santonian.

Arkhangelskiella specillata Zone

Definition: Interval from the lowest occurrence of *Arkhangelskiella specillata* Vekshina to the lowest occurrence of *Tetralithus aculeus* (Stradner).

Important species: Those of the *Kamptnerius magnificus* Zone, plus *Cretarhabdus schizobrachiatus* Gartner, *Arkhangelskiella specillata*, *Staurolithites imbricata* (Gartner), *Broinsonia bevieri* Bukry and *Reinhardites anthophorus* (Deflandre).

Author: Manivit, 1971.

Age: Lower and Middle Campanian.

TABLE 6

RANGES OF CALCAREOUS NANNOPLANKTON DATUM INDICATORS CRETACEOUS

after Thierstein 1973 and Manivit 1971

	CALCAREOUS NANNOPLANKTON ZONES	Nannoconus colomi	Cretarhabdus crenulatus	Calcicalathina oblongata	Lithraphidites bollii	Micrantholithus hoschulzi	Chiastozygus litterarius	Rucinolithus irregularis	Parhabdolithus angustus	Eprolithus floralis	Prediscosphaera cretacea	Eiffellithus turriseiffeli	Staurolithites orbiculofenestrus	Gartnerago obliquus	Corollithion exiguum	Micula staurophora	Marthasterites furcatus	Kamptnerius magnificus	Arkhangelskiella specillata	Tetralithus aculeus	Lithraphidites quadratus	Nephrolithus frequens
MAESTRICHTIAN	Nephrolithus frequens Zone		x				x			x	x	x			x	x		x			x	x
	Lithraphidites quadratus Zone		x				x			x	x	x			x	x		x			x	
CAMPANIAN	Tetralithus aculeus Zone		x				x			x	x	x		x	x	x		x	x	x		
	Arkhangelskiella specillata Zone		x				x			x	x	x		x	x	x	x	x	x			
SANTONIAN	Kamptnerius magnificus Zone		x				x			x	x	x		x	x	x	x	x				
CONIACIAN	Marthasterites furcatus Zone		x				x			x	x	x		x	x	x	x					
TURONIAN	Micula staurophora Zone		x				x		x	x	x	x	x	x	x	x						
	Corollithion exiguum Zone		x				x		x	x	x	x	x	x	x							
CENOMANIAN	Gartnerago obliquus Zone		x				x		x	x	x	x	x	x								
	Staurolithites orbiculofenestrus Zone		x				x		x	x	x	x	x									
ALBIAN	Eiffellithus turriseiffeli Zone		x				x	x	x	x	x	x										
	Prediscosphaera cretacea Zone		x				x	x	x	x	x											
	Parhabdolithus angustus Zone		x				x	x	x	x												
APTIAN	Chiastozygus litterarius Zone		x			x	x	x														
BARREMIAN	Micrantholithus hoschulzi Zone		x			x																
HAUTERIVIAN	Lithraphidites bollii Zone	x	x	x	x	x																
VALANGINIAN	Calcicalathina oblongata Zone	x	x	x		x																
BERRIASIAN	Cretarhabdus crenulatus Zone	x	x			x																
TITHONIAN	Nannoconus colomi Zone	x				x																

Tetralithus aculeus Zone

Definition: Interval from the lowest occurrence of *Tetralithus aculeus* and the lowest occurrence of *Lithraphidites quadratus* Bramlette and Martini.

Important species: Those of the *Arkhangelskiella specillata* Zone, plus *Tetralithus aculeus* (Stradner), *Microrhabdulus stradneri* Bramlette and Martini, *Prediscosphaera spinosa* (Bramlette and Martini), *Marthasterites inconspicuus* Deflandre, *Ceratolithoides kamptneri* Bramlette and Martini, *Staurolithites bochotnicae* (Gorka). *Tetralithus aculeus*, *Broinsonia bevieri*, *Broinsonia enormis*, *Gartnerago obliquus*, *Tetralithus pyramidus*, *Corollithion signum*, *Cretarhabdus schizobrachiatus*, *Microrhabdulinus ambiguus*, *Cylindralithus serratus* and *Tranolithus exiguus* all have their highest occurrence near the top of this zone.

Authors: Čepek and Hay, 1969, modified by Manivit, 1971.

Age: Late Campanian.

Lithraphidites quadratus Zone

Definition: Interval from the lowest occurrence of *Lithraphidites quadratus* to the lowest occurrence of *Nephrolithus frequens* Gorka.

Important species: Those of the *Tetralithus aculeus* Zone, plus *Lithraphidites quadratus* Bramlette and Martini, *Lithraphidites grossopectinatus* Bukry, *Tetralithus nitidus* Martini, *Tetralithus quadratus* Stradner, *Zygodiscus acanthus* (Reinhardt), *Zygodiscus pseudanthophorus* Bramlette and Martini, *Zygodiscus spiralis* Bramlette and Martini, *Rhomboaster cuspis* Bramlette and Martini, *Cylindralithus gallicus* (Stradner).

Authors: Čepek and Hay, 1969.

Age: Early and Middle Maestrichtian.

Nephrolithus frequens Zone

Definition: Interval from the lowest occurrence of *Nephrolithus frequens* Gorka to the level of extinction of most late Cretaceous species.

Important species: Those of the *Lithraphidites quadratus* Zone, plus *Nephrolithus frequens* Gorka, *Tetralithus murus* Martini, and *Markalius astroporus* (Stradner).

Authors: Čepek and Hay, 1969.

Age: Late Maestrichtian.

PALAEOGENE NANNOFOSSIL ZONES

The range of calcareous nannoplankton datum indicators for the Palaeogene are shown in Table 7.

Markalius inversus Zone

Definition: Interval from the highest occurrence of *Arkhangelskiella cymbiformis* Vekshina and other Cretaceous species (Cretaceous extinction plane) to the earliest occurrence of *Cruciplacolithus tenuis* (Stradner).

Important species: *Markalius inversus* (Deflandre), *Zygodiscus sigmoides* Bramlette and Sullivan, *Biantholithus sparsus* Bramlette and Martini, *Thoracosphaera operculata* Bramlette and Martini.

Authors: Mohler and Hay, 1967 emend. Martini, 1971 (syn. *Markalius astroporus* Zone).

Age: Early Danian.

Cruciplacolithus tenüis Zone

Definition: Interval from the lowest occurrence of *Cruciplacolithus tenuis* to the lowest occurrence of *Chiasmolithus danicus* (Brotzen).

Important species: Those of the *Markalius inversus* zone, plus *Cruciplacolithus tenuis*.

Authors: Mohler and Hay, 1967 emend. Martini, 1971.

Age: Early Danian.

Chiasmolithus danicus Zone

Definition: Interval from the lowest occurrence of *Chiasmolithus danicus* to the lowest occurrence of *Ellipsolithus macellus* (Bramlette and Sullivan).

Important species: Those of the *Cruciplacolithus tenuis* Zone, plus *Chiasmolithus danicus*.

Author: Martini, 1971.

Age: Late Danian.

Ellipsolithus macellus Zone

Definition: Interval from the first occurrence of *Ellipsolithus macellus* to the first occurrence of *Fasciculithus tympaniformis* Hay and Mohler.

Important species: Those of the *Chiasmolithus danicus* Zone, plus *Ellipsolithus macellus* and in the upper part *Neococcolithes concinnus* (Martini).

Author: Martini, 1971.

Age: Middle Palaeocene.

Fasciculithus tympaniformis Zone

Definition: Interval from the first occurrence of *Fasciculithus tympaniformis* to the first occurrence of *Heliolithus kleinpelli* Sullivan.

Important species: Those of the *Ellipsolithus macellus* Zone, plus *Fasciculithus tympaniformis* Hay and Mohler, *Ellipsolithus distichus* (Bramlette and Sullivan), *Chiasmolithus bidens* (Bramlette and Sullivan); less *Biantholithus sparsus* Bramlette and Martini. The last occurrence of *Chiasmolithus danicus* (Brotzen) seems to be in the upper part of this zone.

Authors: Mohler and Hay, 1967.

Age. Middle Palaeocene.

Heliolithus kleinpelli Zone

Definition: Interval from the lowest occurrence of *Heliolithus kleinpelli* to the lowest occurrence of *Discoaster gemmeus* Stradner.

Important species: *Heliolithus kleinpelli* Sullivan, *Zygodiscus sigmoides* Bramlette and Sullivan, *Neococcolithes concinnus, Cruciplacolithus tenuis* less *Chiasmolithus danicus.*

Authors: Mohler and Hay, 1967.

Age: Middle Palaeocene.

Discoaster gemmeus Zone

Definition: Interval from the lowest occurrence of *Discoaster gemmeus* to the lowest occurrence of *Heliolithus riedeli* Bramlette and Sullivan.

Important species: *Discoaster gemmeus, Fasciculithus tympaniformis, Zygodiscus sigmoides*, and *Chiasmolithus bidens. Heliolithus kleinpelli* has its highest occurrence at the top of this unit.

Author: Hay, 1964.

Age: Middle Palaeocene.

Heliolithus riedeli Zone

Definition: Interval from the lowest occurrence of *Heliolithus riedeli* to the lowest occurrence of *Discoaster multiradiatus* Bramlette and Riedel.

Important species: *Heliolithus riedeli, Fasciculithus involutus* Bramlette and Sullivan, *Zygodiscus sigmoides, Chiasmolithus bidens, Neococcolithes concinnus, Ellipsolithus distichus.*

Authors: Bramlette and Sullivan, 1961.

Age: Late Palaeocene.

Discoaster multiradiatus Zone

Definition: Interval from the lowest occurrence of *Discoaster multiradiatus* to the lowest occurrence of *Marthasterites bramlettei* Brönnimann and Stradner.

Important species: Those of the *Heliolithus riedeli* Zone, plus *Discoaster multiradiatus* Bramlette and Riedel, *Zygodiscus plectopons* Bramlette and Sullivan, *Discoaster perpolitus* Martini, *Discoaster splendidus* Martini. *Discoaster gemmeus* Stradner probably has its highest occurrence in the upper part of this zone.

Authors: Bramlette and Sullivan, 1961.

Age: Late Palaeocene.

Marthasterites contortus Zone

Definition: Interval from the lowest occurrence of *Marthasterites bramlettei* to the highest occurrence of *Marthasterites contortus* (Stradner).

Important species: *Marthasterites bramlettei*, *Discoaster binodosus* Martini, and except in the lowest part *Marthasterites contortus* and *Discoaster diastypus* Bramlette and Sullivan. *Discoaster multiradiatus* has its highest occurrence above the lowest occurrence of *Marthasterites tribrachiatus* (Bramlette and Riedel) near the top of this zone.

Author: Hay, 1964.

Age: Early Eocene.

Discoaster binodosus Zone

Definition: Interval from the highest occurrence of *Marthasterites contortus* to the lowest occurrence of *Discoaster lodoensis* Bramlette and Riedel.

Important species: *Discoaster binodosus*, *Marthasterites tribrachiatus*, *Discoaster diastypus*, and *Ellipsolithus macellus*. *Chiasmolithus grandis* (Bramlette and Riedel) has its lowest occurrence in middle part of this zone.

Authors: Mohler and Hay, 1967.

Age: Early Eocene.

Marthasterites tribrachiatus Zone

Definition: Interval from the lowest occurrence of *Discoaster lodoensis* to the highest occurrence of *Marthasterites tribrachiatus*.

Important species: Those of the *Discoaster binodosus* Zone, plus *Discoaster lodoensis*, *Discoaster barbadiensis* Tan Sin Hok, *Discoaster distinctus* Martini, *Cyclococcolithina formosa* Kamptner, *Sphenolithus radians* Deflandre, *Discoasteroides kuepperi* (Stradner) and *Neococcolithes dubius* (Deflandre).

Authors: Brönnimann and Stradner, 1960.

Age: Early Eocene.

Discoaster lodoensis Zone

Definition: Interval from the highest occurrence of *Marthasterites tribrachiatus* to the lowest occurrence of *Discoaster sublodoensis* Bramlette and Sullivan.

Important species: Those of the *Marthasterites tribrachiatus* Zone less *Marthasterites tribrachiatus*. *Discoaster diastypus* has its highest occurrence within this zone. *Discoaster nonaradiatus* Klumpp has its lowest occurrence near the top of this zone.

Authors: Brönnimann and Stradner, 1960.

Age: Early Eocene.

Discoaster sublodoensis Zone

Definition: Interval from the lowest occurrence of *Discoaster sublodoensis* to the lowest occurrence of *Chiphragmalithus alatus* (Martini).

Important species: Those of the *Discoaster lodoensis* Zone, plus *Discoaster sublodoensis*, *Discoaster nonaradiatus* and, in the upper part, *Sphenolithus furcatolithoides* Locker.

Author: Hay, 1964.

Age: Middle Eocene.

Chiphragmalithus alatus Zone

Definition: Interval from the lowest occurrence of *Chiphragmalithus alatus* (Martini) to the highest occurrence of *Rhabdosphaera gladius* Locker.

Important species: *Discoaster barbadiensis*, *Discoaster distinctus*, *Neococcolithes dubius*, *Cyclococcolithina formosa*, *Chiphragmalithus alatus*, *Chiphragmalithus cristatus* (Martini), *Chiasmolithus grandis*. *Discoasteroides kuepperi*, *Discoaster sublodoensis* and *Discoaster nonaradiatus* have their highest occurrence in the lower part of this zone. The lowest occurrence of *Reticulofenestra umbilica* (Levin) and *Rhabdosphaera gladius* Locker are within the lower part, and the lowest occurrences of *Discoaster tani nodifer* Bramlette and Riedel, *Discoaster martinii* Stradner and *Lanternithus minutus* Stradner are in the upper part of this zone.

Author: Hay, in Hay *et al.*, 1967 emend. Martini, 1971.

Age: Middle Eocene.

Discoaster tani nodifer Zone

Definition: Interval from the highest occurrence of *Rhabdosphaera gladius* Locker to the highest occurrence of *Chiasmolithus solitus* (Bramlette and Sullivan).

Important species: *Discoaster tani nodifer* Bramlette and Riedel, *Discoaster barbadiensis*, *Cyclococcolithina formosa*, *Chiasmolithus solitus*, *Chiasmolithus grandis*, *Neococcolithes dubius*, *Lanternithus minutus* and *Reticulofenestra umbilica*. *Chiphragmalithus alatus* and *Discoaster martinii* have their highest occurrence in the lower part of this zone. *Discoaster distinctus* has its highest occurrence in the upper part of this zone. *Discoaster saipanensis* Bramlette and Riedel has its lowest occurrence in the upper part of the *Discoaster tani nodifer* zone.

Author: Hay, in Hay *et al.*, 1967, emend. Martini, 1971.

Age: Middle Eocene.

Discoaster saipanensis Zone

Definition: Interval from the highest occurrence of *Chiasmolithus solitus* to the lowest occurrence of *Chiasmolithus oamaruensis* (Deflandre).

Important species: Those listed for the *Discoaster tani nodifer* Zone less *Discoaster distinctus* plus *Discoaster saipanensis* and *Clathrolithus spinosus* Martini. *Helicopontosphaera compacta* (Bramlette and Wilcoxon) has its lowest occurrence in the lower part of this zone.

Author: Martini, 1971.

Age: Middle Eocene.

Chiasmolithus oamaruensis Zone

Definition: Interval from the lowest occurrence of *Chiasmolithus oamaruensis* to the lowest occurrence of *Isthmolithus recurvus* Deflandre.

Important species: *Chiasmolithus oamaruensis* (Deflandre), *Discoaster barbadiensis*, *Chiasmolithus grandis*, *Cyclococcolithina formosa*, *Reticulofenestra umbilica*, *Dictyococcites dictyodus* (Deflandre and Fert).

Author: Martini, 1971.

Age: Late Eocene.

Isthmolithus recurvus Zone

Definition: Interval from the lowest occurrence of *Isthmolithus recurvus* to the lowest occurrence of *Sphenolithus pseudoradians* Bramlette and Wilcoxon.

Important species: Those listed for the *Chiasmolithus oamaruensis* Zone plus *Isthmolithus recurvus*. *Neococcolithes dubius* has its highest occurrence in the lowest part of this zone.

Authors: Hay, Mohler and Wade, 1966, emend. Martini 1971.

Age: Late Eocene.

Sphenolithus pseudoradians Zone

Definition: Interval from the lowest occurrence of *Sphenolithus pseudoradians* to the highest occurrence of *Discoaster saipanensis* Bramlette and Riedel.

Important species: *Isthmolithus recurvus*, *Discoaster saipanensis*, *Dictyococcites dictyodus*, *Reticulofenestra umbilica*, *Cyclococcolithina formosa*, *Lanternithus minutus*. *Corannulus germanicus* Stradner, *Sphenolithus pseudoradians*, *Sphenolithus predistentus* Bramlette and Wilcoxon and *Helicopontosphaera reticulata* (Bramlette and Wilcoxon). The highest occurrence of *Chiasmolithus grandis* is in the lower part of this zone, and that of *Discoaster barbadiensis* below the top of this zone.

Author: Martini, 1971.

Age: Late Eocene.

Ericsonia subdisticha Zone

Definition: Interval from the highest occurrence of *Discoaster saipanensis* to the highest occurrence of *Cyclococcolithina formosa*.

Important species: Those of the *Sphenolithus pseudoradians* Zone, less *Discoaster barbadiensis* and *Discoaster saipanensis*, plus abundant *Ericsonia subdisticha* (Roth and Hay).

Authors: Roth and Hay, 1967, emend. Martini, 1971.

Age: Early Oligocene.

Helicopontosphaera reticulata Zone

Definition: Interval from the highest occurrence of *Cyclococcolithina formosa* to the highest occurrence of *Reticulofenstra umbilica.*

Important species: *Dictyococcites dictyodus*, *Lanternithus minutus*, *Ericsonia subdisticha* and *Helicopontosphaera reticulata. Isthmolithus recurvus* has its highest occurrence near the top of this zone.

Authors: Bramlette and Wilcoxon 1967, emend. Martini, 1971.

Age: Early Oligocene.

Sphenolithus predistentus Zone

Definition: Interval from the highest occurrence of *Reticulofenestra umbilica* to the lowest occurrence of *Sphenolithus ciperoensis* Bramlette and Wilcoxon.

Important species: *Sphenolithus predistentus*, *Dictyococcites dictyodus*, *Cyclicargolithus floridanus* (Roth and Hay), *Reticulofenestra insignita* Roth and Hay, *Helicopontosphaera compacta*, and *Sphenolithus distentus* (Martini), which has its first occurrence near or at the base of this zone.

Authors: Bramlette and Wilcoxon, 1967.

Age: Middle Oligocene.

Sphenolithus distentus Zone

Definition: Interval from the lowest occurrence of *Sphenolithus ciperoensis* to the highest occurrence of *Sphenolithus distentus.*

Important species: Those of the *Sphenolithus predistentus* Zone, plus *Sphenolithus ciperoensis* Bramlette and Wilcoxon, *Discoaster lidzi* Hay, *Coccolithus abisectus* Müller, *Helicopontosphaera recta* (Haq), less *Sphenolithus pseudoradians* and *Reticulofenestra insignita. Sphenolithus predistentus* has its highest occurrence near the top of this zone.

Authors: Bramlette and Wilcoxon, 1967.

Age: Middle Oligocene.

Sphenolithus ciperoensis Zone

Definition: Interval from the highest occurrence of *Sphenolithus distentus* to the highest occurrence of *Helicopontosphaera recta.*

TABLE 7

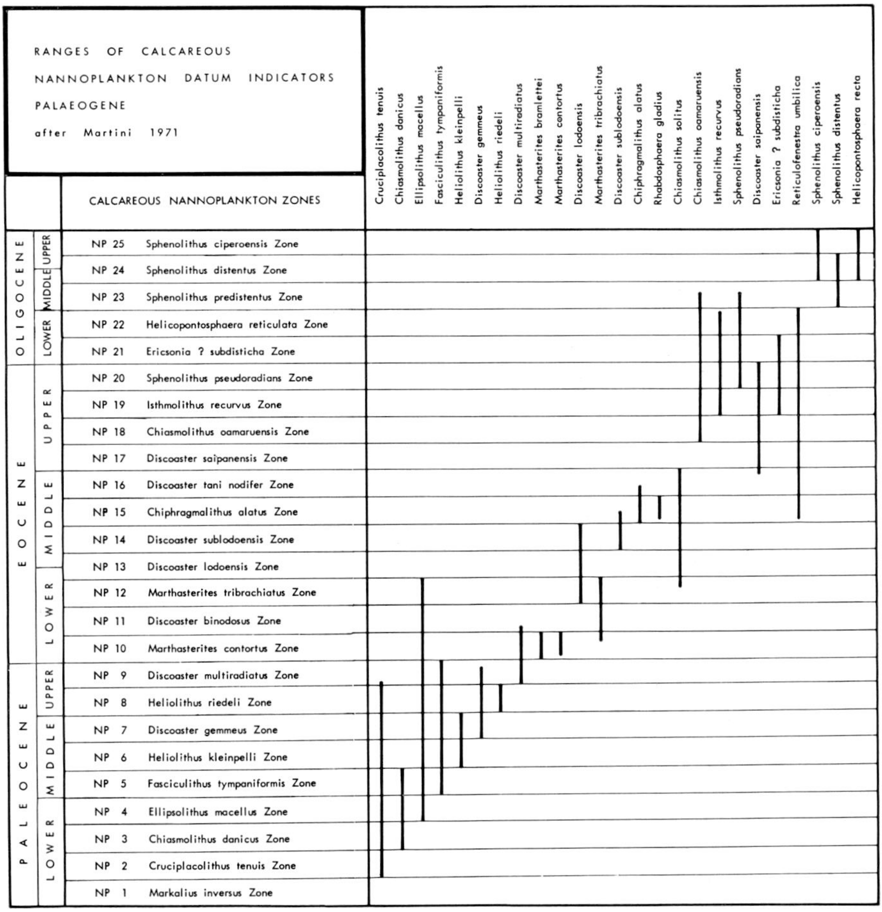
RANGES OF CALCAREOUS NANNOPLANKTON DATUM INDICATORS
PALAEOGENE
after Martini 1971
CALCAREOUS NANNOPLANKTON ZONES
Cruciplacolithus tenuis
Chiasmolithus danicus
Ellipsolithus macellus
Fasciculithus tympaniformis
Heliolithus kleinpelli
Discoaster gemmeus
Heliolithus riedeli
Discoaster multiradiatus
Marthasterites bramlettei
Marthasterites contortus
Discoaster lodoensis
Marthasterites tribrachiatus
Discoaster sublodoensis
Chiphragmalithus alatus
Rhabdosphaera gladius
Chiasmolithus solitus
Chiasmolithus oamaruensis
Isthmolithus recurvus
Sphenolithus pseudoradians
Discoaster saipanensis
Ericsonia ? subdisticha
Reticulofenestra umbilica
Sphenolithus ciperoensis
Sphenolithus distentus
Helicopontosphaera recta
OLIGOCENE
UPPER
MIDDLE
LOWER
EOCENE
UPPER
MIDDLE
LOWER
PALEOCENE
UPPER
MIDDLE
LOWER
NP 25 Sphenolithus ciperoensis Zone
NP 24 Sphenolithus distentus Zone
NP 23 Sphenolithus predistentus Zone
NP 22 Helicopontosphaera reticulata Zone
NP 21 Ericsonia ? subdisticha Zone
NP 20 Sphenolithus pseudoradians Zone
NP 19 Isthmolithus recurvus Zone
NP 18 Chiasmolithus oamaruensis Zone
NP 17 Discoaster saipanensis Zone
NP 16 Discoaster tani nodifer Zone
NP 15 Chiphragmalithus alatus Zone
NP 14 Discoaster sublodoensis Zone
NP 13 Discoaster lodoensis Zone
NP 12 Marthasterites tribrachiatus Zone
NP 11 Discoaster binodosus Zone
NP 10 Marthasterites contortus Zone
NP 9 Discoaster multiradiatus Zone
NP 8 Heliolithus riedeli Zone
NP 7 Discoaster gemmeus Zone
NP 6 Heliolithus kleinpelli Zone
NP 5 Fasciculithus tympaniformis Zone
NP 4 Ellipsolithus macellus Zone
NP 3 Chiasmolithus danicus Zone
NP 2 Cruciplacolithus tenuis Zone
NP 1 Markalius inversus Zone

Important species: Those of the *Sphenolithus distentus* Zone, less *Sphenolithus distentus*. *Zygrhablithus bijugatus* (Deflandre) has its highest occurrence near the top of this zone and *Helicopontosphaera compacta* seems to have its highest occurrence within this zone.

Remarks: Note that *Helicopontosphaera recta* (Haq) is a senior synonym of *Helicopontosphaera truncata* (Bramlette and Wilcoxon).

Authors: Bramlette and Wilcoxon, 1967.

Age: Late Oligocene.

NEOGENE AND QUATERNARY NANNOFOSSIL ZONES

The ranges of selected nanoplankton datum indicators for the Neogene and Quaternary are shown in Table 8.

Triquetrorhabdulus carinatus Zone

Definition: Interval from the highest occurrence of *Helicopontosphaera recta* to the lowest occurrence of *Discoaster druggi* Bramlette and Wilcoxon.

Important species: *Triquetrorhabdulus carinatus* Martini, *Coccolithus abisectus*, *Helicopontosphaera intermedia* (Martini), *Discoaster deflandrei* Bramlette and Riedel, *Sphenolithus pacificus* Martini, *Discoaster nephados* Hay. *Dictyococcites dictyodus* has its highest occurrence in the lowest part of this zone. *Discoaster trinidadensis* Hay has its lowest occurrence in the upper part of this zone.

Authors: Bramlette and Wilcoxon, 1967, emend. Martini and Worsley, 1970.

Age: Late Oligocene–Early Miocene.

Discoaster druggi Zone

Definition: Interval from the lowest occurrence of *Discoaster druggi* to the highest occurrence of *Triquetrorhabdulus carinatus*.

Important species: Those of the *Triquetrorhabdulus carinatus* Zone, plus *Discoaster druggi*, *Orthorhabdus serratus* Bramlette and Wilcoxon, *Sphenolithus belemnos* Bramlette and Wilcoxon, *Helicopontosphaera ampliaperta* (Bramlette and Wilcoxon), *Helicopontispheara kamptneri* Hay and Mohler.

Authors: Martini and Worsley, 1970.

Age: Early Miocene.

Sphenolithus belemnos Zone

Definition: Interval from the highest occurrence of *Triquetrorhabdulus carinatus* to the highest occurrence of *Sphenolithus belemnos*.

Important species: Those of the *Discoaster druggi* Zone, less *Triquetro-*

rhabdulus carinatus. Sphenolithus heteromorphus Deflandre has its lowest occurrence in the upper part of this zone.

Authors: Bramlette and Wilcoxon, 1967.

Age: Early Miocene.

Helicopontosphaera ampliaperta Zone

Definition: Interval from the highest occurrence of *Sphenolithus belemnos* to the highest occurrence of *Helicopontosphaera ampliaperta.*

Important species: Those of the *Sphenolithus belemnos* Zone, less *Sphenolithus belemnos. Discoaster variabilis* Martini and Bramlette has its lowest occurrence in the uppermost part of this zone.

Authors: Bramlette and Wilcoxon, 1967.

Age: Early Miocene.

Sphenolithus heteromorphus Zone

Definition: Interval from the highest occurrence of *Helicopontosphaera ampliaperta* to the highest occurrence of *Sphenolithus heteromorphus.*

Important species: *Sphenolithus heteromorphus, Discoaster variabilis, Discoaster exilis, Discoaster nephados, Discoaster trinidadensis, Helicopontosphaera kamptneri. Discoaster brouweri* Tan Sin Hok has its lowest occurrence in the uppermost part of this zone.

Authors: Bramlette and Wilcoxon, 1967.

Age: Middle Miocene.

Discoaster exilis Zone

Definition: Interval from the highest occurrence of *Sphenolithus heteromorphus* to the lowest occurrence of *Discoaster kugleri* Martini and Bramlette.

Important species: *Discoaster exilis, Discoaster variabilis, Helicopontosphaera kamptneri, Discoaster trinidadensis. Discoaster nephados* has its highest occurrence of this zone. *Triquetrorhabdulus rugosus* Bramlette and Wilcoxon has its lowest occurrence in the lower part of this zone.

Authors: Martini and Worsley, 1970.

Age: Middle Miocene.

Discoaster kugleri Zone

Definition: Interval from the lowest occurrence of *Discoaster kugleri* to the lowest occurrence of *Catinaster coalitus* Martini and Bramlette.

Important species: Those of the *Discoaster exilis* zone plus *Discoaster kugleri. Discoaster trinidadensis* Hay has its highest occurrence in the lower part of this zone. *Discoaster challengeri* Bramlette and Riedel and *Discoaster pseudovariabilis* Martini and Worsley have their lowest occurrences in the upper part of this zone.

Authors: Bramlette and Wilcoxon, 1967.

Age: Middle Miocene.

Catinaster coalitus Zone

Definition: Interval from the lowest occurrence of *Catinaster coalitus* to the lowest occurrence of *Discoaster hamatus* Martini and Bramlette.

Important species: Those of the *Discoaster kugleri* Zone, plus *Catinaster coalitus* Martini and Bramlette. *Discoaster kugleri* has its highest occurrence, in this zone. *Catinaster calyculus* Martini and Bramlette has its lowest occurrence in the lower part and *Discoaster calcaris* Gartner its lowest occurrence in the uppermost part of this zone.

Authors: Bramlette and Wilcoxon, 1967.

Age: Middle Miocene.

Discoaster hamatus Zone

Definition: Interval from the lowest to the highest occurrence of *Discoaster hamatus* Martini and Bramlette.

Important species: *Discoaster hamatus*, *Discoaster variabilis*, *Discoaster challengeri*, *Helicopontosphaera kamptneri*, *Discoaster calcaris*, *Discoaster neohamatus* Bukry and Bramlette and *Discoaster pentaradiatus* Tan Sin Hok have their lowest occurrences in the lower part of this zone. *Discoaster bollii* Martini and Bramlette has its lowest occurrence in the upper part of this zone. *Catinaster coalitus*, *Triquetrorhabdulus rugosus* and *Discoaster pseudovariabilis* have their highest occurrence in the upper part of this zone.

Authors: Bramlette and Wilcoxon, 1967.

Age: Middle–Late Miocene.

Discoaster calcaris Zone

Definition: Interval from the highest occurrence of *Discoaster hamatus* to the lowest occurrence of *Discoaster quinqueramus* Gartner.

Important species: Those of the *Discoaster hamatus* Zone, less *Discoaster hamatus*. *Catinaster calyculus* Martini and Bramlette and *Discoaster bollii* have their highest occurrence and *Sphenolithus abies* Deflandre has its lowest occurrence in the lower part of this zone.

Author: Martini, 1969.

Age: Middle–Late Miocene.

Discoaster quinqueramus Zone

Definition: Interval from the lowest to the highest occurrence of *Discoaster quinqueramus*.

Important species: *Discoaster quinqueramus*, *Discoaster variabilis*, *Discoaster challengeri*, *Discoaster brouweri*, *Discoaster pentaradiatus*, *Sphenolithus abies*, *Helicopontosphaera kamptneri*, *Discoaster calcaris* and *Discoaster neohamatus* have their highest occurrences and *Ceratolithus tricorniculatus* Gartner its lowest occurrence in the lower part of this zone. Rare specimens of *Discoaster surculus* Martini and Bramlette are present and *Reticulofenestra pseudoumbilica* (Gartner) seems to have its lowest occurrence in the upper part of this zone.

Author: Gartner, 1969b.

Age: Late Miocene.

Ceratolithus tricorniculatus Zone

Definition: Interval from the highest occurrence of *Discoaster quinqueramus* to the lowest occurrence of *Ceratolithus rugosus* Bukry and Bramlette.

Important species: Those of the *Discoaster quinqueramus* Zone, less *Discoaster quinqueramus*. *Discoaster surculus* is common.

Author: Gartner, 1969b

Age: Late Miocene–Early Pliocene.

Ceratolithus rugosus Zone

Definition: Interval from the lowest occurrence of *Ceratolithus rugosus* to the lowest occurrence of *Discoaster asymmetricus* Gartner.

Important species: Those of the *Ceratolithus tircorniculatus* Zone plus *Ceratolithus rugosus*.

Author: Gartner, 1969b.

Age: Early Piocene.

Discoaster asymmetricus Zone

Definition: Interval from the lowest occurrence of *Discoaster asymmetricus* to the highest occurrence of *Ceratolithus tricorniculatus*.

Important species: Those of the *Ceratolithus rugosus* Zone, plus *Discoaster asymmetricus* Gartner.

Author: Gartner, 1969b.

Age: Early Pliocene.

Reticulofenestra pseudoumbilica Zone

Definition: Interval from the highest occurrence of *Ceratolithus tricorniculatus* to the highest occurrence of *Reticulofenestra pseudoumbilica*.

Important species: Those of the *Discoaster asymmetricus* zone, less *Ceratolithus tricorniculatus* Gartner, and in the uppermost part less *Sphenolithus abies* Deflandre.

Author: Gartner, 1969b.

Age: Middle Pliocene.

Discoaster surculus Zone

Definition: Interval from the highest occurrence of *Reticulofenestra pseudoumbilica* to the highest occurrence of *Discoaster surculus*.

Important species: *Discoaster surculus*, *Discoaster brouweri*, *Discoaster pentaradiatus*, *Discoaster asymmetricus*, *Ceratolithus rugosus*, *Helicopontosphaera kamptneri*. *Pseudoemiliania lacunosa* (Kamptner) has its lowest occurrence in the lower part of this zone.

Authors: Hay and Schmidt, 1967, emend. Gartner, 1969b.

Age: Middle Pliocene.

Discoaster pentaradiatus Zone

Definition: Interval from the highest occurrence of *Discoaster surculus* to the highest occurrence of *Discoaster pentaradiatus*.

Important species: Those listed for the upper part of the *Discoaster surculus* Zone, less *Discoaster surculus*.

Authors: Martini and Worsley, 1970.

Age: Late Pliocene.

Discoaster brouweri Zone

Definition: Interval from the highest occurrence of *Discoaster pentaradiatus* to the highest occurrence of *Discoaster brouweri*.

Important species: *Discoaster brouweri*, *Helicopontosphaera kamptneri*, *Pseudoemiliania lacunosa*, *Ceratolithus rugosus*. *Ceratolithus cristatus* Kamptner has its lowest occurrence in the upper part of this zone.

Authors: Hay and Schmidt, 1967, emend. Martini and Worsley, 1970.

Age: Late Pliocene.

Pseudoemiliania lacunosa Zone

Definition: Interval from the highest occurrence of *Discoaster brouweri* to the highest occurrence of *Pseudoemiliania lacunosa*.

Important species: *Pseudoemiliania lacunosa*, *Helicopontosphaera kamptneri*, *Cyclococcolithina macintyrei*, *Coccolithus pelagicus*, *Ceratolithus cristatus*. *Ceratolithus rugosus* has its highest occurrence in this zone.

Author: Gartner, 1969b.

Age: Early Pleistocene.

Gephyrocapsa oceanica Zone

Definition: Interval from the highest occurrence of *Pseudoemiliania lacunosa* to the lowest occurrence of *Emiliania huxleyi* (Lohmann).

Important species: *Gephyrocapsa oceanica* Kamptner and related species, *Coccolithus pelagicus*, *Helicopontosphaera kamptner*, *Rhabdosphaera clavigera*

TABLE 8

RANGES OF CALCAREOUS NANNOPLANKTON DATUM INDICATORS NEOGENE / QUATERNARY after Martini 1971

		CALCAREOUS NANNOPLANKTON ZONES
QUATERNARY		NN 21 Emiliania huxleyi Zone
		NN 20 Gephyrocapsa oceanica Zone
		NN 19 Pseudoemiliania lacunosa Zone
PLIOCENE	UPPER	NN 18 Discoaster brouweri Zone
		NN 17 Discoaster pentaradiatus Zone
		NN 16 Discoaster surculus Zone
	LOWER	NN 15 Reticulofenestra pseudoumbilica Zone
		NN 14 Discoaster asymmetricus Zone
		NN 13 Ceratolithus rugosus Zone
MIOCENE	UPPER	NN 12 Ceratolithus tricorniculatus Zone
		NN 11 Discoaster quinqueramus Zone
		NN 10 Discoaster calcaris Zone
	MIDDLE	NN 9 Discoaster hamatus Zone
		NN 8 Catinaster coalitus Zone
		NN 7 Discoaster kugleri Zone
		NN 6 Discoaster exilis Zone
	LOWER	NN 5 Sphenolithus heteromorphus Zone
		NN 4 Helicopontosphaera ampliaperta Zone
		NN 3 Sphenolithus belemnos Zone
		NN 2 Discoaster druggi Zone
		NN 1 Triquetorhabdulus carinatus Zone

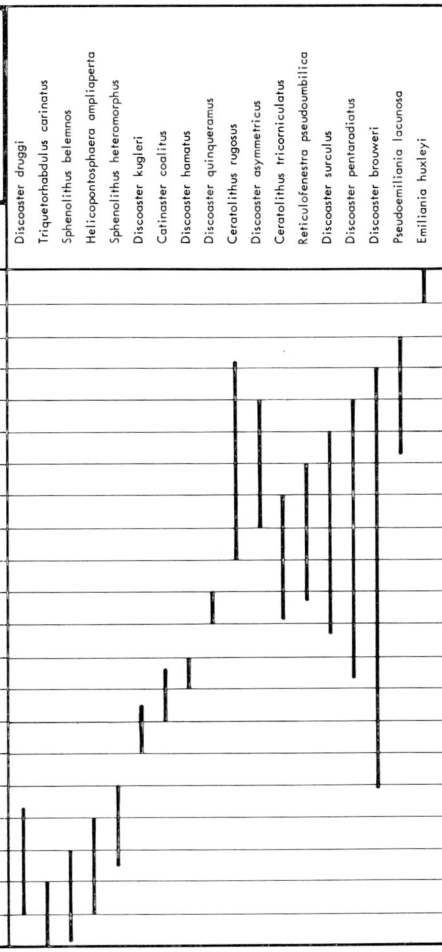

Murray and Blackman, *Cyclococcolithina leptopora* (Murray and Blackman) *Umbilicosphaera mirabilis* (Lohmann), *Ceratolithus cristatus* Kamptner.

Authors: Boudreaux and Hay, 1967, emend. Gartner, 1969b.

Age: Middle Pleistocene.

Emiliania huxleyi Zone

Definition: Interval above the lowest occurrence of *Emiliania huxleyi.*

Important species: Those of the *Gephyrocapsa oceanica* Zone, plus *Emiliania huxleyi* (Lohmann).

Authors: Boudreaux and Hay, 1967.

Age: Late Pleistocene–Holocene.

Coccolith taxonomy

BRIEF HISTORICAL SURVEY OF THE TAXONOMY OF COCCOLITHS

Huxley (1868) distinguished two kinds of coccolith: 1. those consisting of a single shield having a thickened margin, which he termed "discoliths" and placed in the genus *Discolithus*; and 2. those composed of two shields connected by a tube, which he termed "cyatholiths" and for which he erected the genus *Cyatholithus*. Haeckel (1870) made the same differentiation, proposing the term "Monodiske" for discoliths and "Amphidiske" for cyatholiths.

Lohmann (1902) established the first suprageneric classification of the coccolithophores, distinguishing two subfamilies: 1. the Syracosphaerinae, with imperforate coccoliths; and 2. the Coccolithophorinae, with perforate coccoliths. At this stage, all discoliths would belong to the Syracosphaerinae and all cyatholiths, or placoliths as Lohmann termed them, would belong to the Coccolithophorinae, along with the rhabdoliths which had first been described by Schmidt (1870).

During what may be termed the "early oceanographic phase" of investigation of the Coccolithophyceae, between the two World Wars, a number of forms with unique shapes were discovered. Some of the more bizarre forms were placed in new families: Deutschlandiaceae Kamptner, 1927, Halopappaceae Kamptner 1928b, Thoracosphaeraceae Schiller 1930, and the fossil Discoasteridae (=Discoasteraceae) Tan Sin Hok, 1927. The true affinities and phylogenetic position of many of these are still uncertain.

Schiller (1926) observed three different kinds of reproduction in coccolithophorids: 1. production of two offspring of equal size, with loss of flagellae prior to division (observed in *Pontosphaera huxleyi* Lohmann =*Coccolithus huxleyi* (Lohmann) Kamptner =*Emiliania huxleyi* (Lohmann) Hay and Mohler); 2. production of two unequal offspring, with loss of flagellae prior to division, and flagellae produced on the small daughter cell which is then liberated; following this the larger daughter cell expands to fill the original test, and two flagellae are produced (observed in *Calyptrosphaera oblonga* Lohmann); and 3. production of (usually) 16 offspring by multiple fission (gametes?), with loss of flagellae prior to division, flagellae produced on offspring, and the offspring observed in one instance to fuse after leaving the parent body. Schiller also observed that *Rhabdosphaera* and *Discosphaera* never possess flagellae, and he concluded that there are two "Organisation-typen" among the coccolithophores, those which possess flagellae (Flagel-laten–Organisation), and those which are never found with flagellae (Pro-tococcalen–Organisation). Kamptner (1928b) commented on Schiller's observations, suggesting that the loss of flagellae might be of phylogenetic importance. However, the classification which Kamptner proposed at that time was based strictly on that of Lohmann (1902) with the addition of two families (Halopappaceae and Deutschlandiaceae) for highly specialised genera.

Deflandre (1953) noted that polarised light observations of calcareous nannofossils revealed two major groups: one group, Heliolitheae being composed of the fossils which produce an interference cross between crossed polarisers, indicating a "spherolitic" structure; and another group, Ortho-litheae, which appear to be completely extinguished or to show only a few crystal units between crossed polarisers, indicating that the calcite crystallites are oriented with c-axes nearly perpendicular to the plane of disc of the coccolith.

Braarud *et al.* (1955) announced the discovery of a group of coccoliths composed entirely of a single kind of microcrystal, which they termed holo-coccoliths. At present, two kinds of these are known. Almost all holococco-liths are constructed of hexagonal microcrystals, but one genus, *Crystallo-lithus*, utilises rhombic microcrystals. Coccoliths composed of more than one kind of structural element were termed heterococcoliths by Braarud *et al.* (1955). They believed that a complete revision of the classification of the coccolithophores would be needed as more ultrastructural investigations are carried out.

Parke and Adams (1960) discovered two phases in unialgal cultures of "*Coccolithus pelagicus*". One of the phases lacked flagellae and bore typical placoliths; the other possessed flagellae, and produced crystalloliths (disciform holococcoliths composed of rhombic mucrocrystals).

GLOSSARY OF MORPHOLOGICAL TERMS

The terminology of calcareous nannofossils was discussed by a Round Table at the Second Planktonic Conference; many of the terms adopted are illus-trated here.

The following glossary is largely after Hay *et al.*, 1966:

Coccolith (Lohmann, 1902) any chromatophore-bearing protist which at some phase of its life-cycle produces coccoliths.
Coccolith (Huxley, 1868) 1. a general term applied to any calcified skeletal element having heliolithid structure or complexly constructed of minute calcite rhombs; 2. two shields connected by a central tube.
Plate-scales (Parke and Adams, 1960) oval to elliptical organic scales imbedded in the surface layer of the periplast.
Haptonema (Parke *et al.*, 1955; Parke and Adams, 1960) a thread- to club-like organic structure located between the two flagellae; it can be contracted into a spiral or extended and used as an organ of attachment to substrate material; it is more rigid than the flagellae.
Motile phase (Parke and Adams, 1960) the flagellated phase of the life-cycle, producing holococcoliths, caneoliths, cyrtoliths, cricoliths and cribriliths.
Non-motile phase (Parke and Adams, 1960) the non-flagellated phase of the life-cycle, producing scapholiths, rhabdoliths, placoliths, osteoliths and pentaliths.
Coccosphere (Wallich, 1860) the entire spherical test of a coccolithophore, composed of interlocking coccoliths.
Naked flagellar field (Lohmann, 1902) in flagellate coccolithophores lacking a complete cover of coccoliths, the area around the flagellae in which no coccoliths are present.
Covered flagellar field (Lohmann, 1902) in flagellate coccolithophores, coccolith cover complete in flagellar region.
Naked pole (= *pole buccale* Deflandre, 1952a) in non-flagellate coccolithophores, that end which is free of coccoliths.

Terms based on light microscopical studies of coccoliths themselves are generally descriptive, but have little or nothing to do with true coccolith ultrastructure:

Discolith (Huxley, 1868; = Monodiske Haeckel, 1870) a single, apparently imperforate, elliptical or circular shield with thickened margin.
Cyatholith (Huxley, 1868; = Amphidiske Haeckel, 1870) two shields connected by a central tube (never used in the literature).
Placolith (Lohmann, 1902) two shields connected by a central tube, supposedly perforate (frequently used to up 1941, unambiguous, should be used in modern descriptions).
Rhabdolith (Schmidt, 1870) a shield surmounted by a long stem, supposedly perforate.
Tremalith (Lohmann, 1913) 1. centrally perforate coccoliths, i.e. placoliths and rhabdoliths; 2. (redefined Kamptner, 1941) two shields connected by a central tube (= placoliths only).
Lopadolith (Lohmann, 1902) a basket-shaped coccolith opening distally (*Scyphosphaera*).
Calyptrolith (Lohmann, 1902) a basket-shaped coccolith opening proximally (*Calyptrosphaera*).
Zygolith (Kamptner, 1937b) an elliptical ring with a crossbar arching slightly or strongly upward (i.e. distally), and bearing a knob or short spine.

Cricolith (Kamptner, 1958) any elliptical ring coccolith.
Cyclolith (Kamptner, 1948) any elliptical or circular ring coccolith (should be restricted to circular forms).
Sphenolith (Deflandre, 1952a) a nannofossil having a prismatic base, formed by radial elements, surmounted by a cone.
Pentagolith (Farinacci, 1971) a pentalith with heliolithid structure (also called Simple Pentalith).
Pentalith (Gran and Braarud, 1935) coccolith formed of five crystal units.
Porolith (Deflandre, 1952a) a polygonal prism with an axial perforation (a term created for the elements of *Thoracosphaera* which EM studies have shown to be both perforate and imperforate).
Stomatal coccolith (Halldal and Markali, 1955; = Mündungscoccolith (*coccolithi oris*): Kamptner, 1937b) in flagellate coccolithophores exhibiting dimorphism, one of the modified coccoliths surrounding the flagellar field.
Tail coccolith (Halldal and Markali, 1955) in flagellate coccolithophores exhibiting dimorphism, a modified coccolith located at the end opposite the flagellar field (*Calciopappus*).
Pole coccolith (Halldal and Markali, 1955; = Polcoccolith: Kamptner, 1937b) in flagellate coccolithophores exhibiting dimorphism, one of the modified coccoliths found at the flagellar and tail ends (*Acanthoica*).
Polar spine (= *aiguillons polaires*, Deflandre, 1952a) in non-motile fusiform coccolithophores exhibiting dimorphism, modified coccoliths located at the ends (*Calciosolenia*).
Whorl coccolith (Halldal and Markali, 1955) in non-motile coccolithophores exhibiting dimorphism, one of the modified coccoliths forming a whorl about a naked pole (*Ophiaster*).
Float coccolith (Halldal and Markali, 1955) in non-motile coccolithophores exhibiting dimorphism, one of the coccoliths modified to serve as a suspension organ (*Thorosphaera*).
Ordinary coccolith (Halldal and Markali, 1955) in coccolithophores exhibiting dimorphism, one of the unmodified coccoliths.
Micro-coccolith (Markali and Paasche, 1955) in coccolithophores exhibiting dimorphism but with the dimorphic coccoliths irregularly placed; the coccoliths.
Macro-coccolith (Markali and Paasche, 1955) in coccolithophores exhibiting dimorphism but with dimorphic coccoliths irregularly placed; the larger coccoliths.
Asterolith (Deflandre, 1952a, after Sujkowski, 1931) star-shaped or rosette ortholithid elements.

Electron microscopy permits a description of coccoliths based on their ultra-structure; correspondingly, a second group of terms is used for forms observed in the electron microscope:

Holococcolith (Braarud *et al.*, 1955) a coccolith consisting entirely of microcrystals of usual crystallographic shape, identical or not.
Calyptroform holococcolith (Halldal and Markali, 1955) basket-shaped (calotte-shaped) holoccolith. All calyptroliths thus far examined have been found to belong here.
Discoform holoccolith (Halldal and Markali, 1955; = crystallolith Parke and Adams, 1960) discolith-shaped holococcoliths, i.e. with a raised margin two or more cycles of microcrystals high.

Zygoform holoccolith (Halldal and Markali, 1955) zygolith-shaped holococcoliths.

All recent zygoliths thus far examined have been found to belong here.

Heterococcolith (Braarud *et al.*, 1955) a coccolith built of different elements.
Caneolith (Braarud *et al.*, 1955) discoliths: heterococcoliths having a central area with laths, a simple or complex wall, and petaloid upper and lower rims.
Complete caneolith (Halldal and Markali, 1955) a caneolith having upper and lower rim elements and a wall.
Incomplete caneolith (Halldal and Markali, 1955) a caneolith having upper and lower rim elements, but lacking a wall.
Scapholith (Deflandre and Fert, 1954; = rhombolith Halldal, 1954) elongate diamond-shaped heterococcoliths with a central area of parallel laths.
Cyrtolith (Braarud *et al.*, 1955) basket-shaped (calotte-shaped) heterococcoliths with laths and a projecting central structure.
Styliform cyrtolith (Halldal and Markali, 1954) cyrtolith with a long spinose central appendix (pole coccoliths of *Acanthoica*).
Helatoform cyrtolith (Halldal and Markali, 1954) cyrtolith with a nail-shaped central appendix.
Labiatiform cyrtolith (Halldal and Markali, 1955) cyrtolith with a double-lip-shaped central appendix (*Anthosphaera robusta*).
Salpingiform cyrtolith (Braarud *et al.*, 1955) cyrtolith with a trumpet-shaped appendix (*Discosphaera tubifer*).
Placolith s.s. (Braarud *et al.*, 1955) heterococcoliths having two shields connected by a central tube.
Cricolith (Braarud *et al.*, 1955) heterococcoliths having units arranged in a simple ring.
Osteolith (Halldal and Markali, 1955) heterococcoliths having a femur shape, built up of lamellae (*Ophiaster hydroideus*).
Lepidolith (Halldal and Markali, 1955) thin apparently homogenous elliptical plates (for surface coccoliths of *Thorosphaera flagellata*; may represent the surficial organic plate-scales described by Parke and Adams, 1960).
Cribrilith (Halldal and Markali, 1955) discoliths with numerous central perforations and a lamellar rim.
Pentalith (Braarud *et al.*, 1955) a nannofossil with pentagonal outline.
Simple pentalith (Deflandre, 1957) a pentalith with heliolithid structure.
Composite pentalith (Deflandre, 1957) a pentalith composed of five crystals unit.

The terminology of coccolith ultraarchitecture is relatively young. The following terms have been adapted from Halldal and Markali (1954a, 1955), Black and Barnes (1959, 1961) and Hay and Towe (1962):

Holococcoliths

Rhombohedric microcrystals: rhombohedral calcite crystals having an edge length from 430–2500 Å.
Hexagonal microcrystal: found in calyptroform holococcoliths.

Heterococcoliths (general terms)

Segment: any of the parts into which a coccolith naturally separates or is divided.
Rhomb: six-sided, roughly equidimensional.

Tabula: six-sided, two dimensions equal, the third smaller.
Lamina: two dimensions large, the third very small.
Lamella: two dimensions large, the third $\rightarrow 0$.
Lath: one dimension large, one intermediate, one very small.
Rod: one dimension large, two much smaller.
Wedge: five-sided, two dimensions subequal, the third small at one edge, $\rightarrow 0$ at the other.
Central area: the centre of the coccolith.
Margin: the edge of the coccolith.
Cycle: a ring of segments.
Jugum: a transverse structure crossing the centre, connecting one side of the cycle with the other.
Dextral imbrication: each segment overlaps the one to the right when viewed from the centre of the cycle.
Sinistral imbrication: each segment overlaps the one to the left when viewed from the centre of the cycle.
Suture: the boundary between segments.
Clockwise inclination: suture inclined to the right as it proceeds to periphery.
Counterclockwise inclination: suture inclined to the left as it proceeds to the periphery.
Radial: suture corresponds to a radius in a circular form or to a straight line drawn through the nearest focus or the line connecting the foci of an elliptical form.
Flange: a part that spreads out like a rim.

Heterococcoliths (special terms)

(*a*) for caneoliths:
Bottom: (Halldal and Markali, 1954b) the central area.
Wall: (Halldal and Markali, 1954b; = girdle, Halldal and Markali, 1955) the raised margin.
Lower rim: (Halldal and Markali, 1954b) proximal flange peripheral to the wall.
Upper rim: (Halldal and Markali, 1954b) distal flange peripheral to the wall.

(*b*) for placoliths:
Shield: (Black and Barnes, 1959) one of the discs of the placoliths.
Proximal shield: the shield on the concave side of the placolith.
Tube: (Black and Barnes, 1959) the central cylinder connecting the two shields.
Distal shield: the shield on the convex side of the placolith.

Kingdom PLANTAE

DIVISION CHRYSOPHYTA

Class **Coccolithophyceae** *Rothmaler*, 1951

In recent years a number of micropalaeontologists have followed Parke and Dixon (1964) in using the term Class Haptophyceae Christensen, 1962, rather than the older term Class Coccolithophyceae. Classification above the level of family is not subject to the rule of priority, and the selection of terms to be used is left largely to the discretion of the individual taxonomist. The Class Haptophyceae was created to include all unicellular flagellates which possess a haptonema; the Class Coccolithophyceae was created to include all chrysophytes which secrete coccoliths. It is not at all certain that the two terms are synonymous. Only a very few coccolith-bearing flagellates have been found to possess a haptonema (von Stosch, 1955; Rayns, 1962) and there are a number of flagellates which possess a haptonema but never produce coccoliths (Parke, 1971). While recognising the value of the Class Haptophyceae to recent taxonomists attempting to discover relations in this difficult group, the author prefers to retain the older Class Coccolithophyceae because it more objectively includes the criteria used as the basis for classsification.

Order EIFFELLITHALES Rood *et al.*, 1971

Coccoliths having a simple marginal area (eiffellithalid rim) which consists solely of a double cycle of elements; the two cycles are superposed so that they appear to be a single cycle in most proximal and distal views, the suture between them being visible only on the inner surface of the circle, facing the the central area. Central structures are variable, consisting of a single bar, cross, radial bars or more complex feature which may be surmounted by a central spine.

Family EIFFELLITHACEAE Reinhardt, 1965

Coccoliths having an eiffellithalid rim and an asymmetrically arranged x-shaped central structure.

Genus *EIFFELLITHUS* Reinhardt, 1964. Pl. 1, figs 1a–c

Type species: *Zygolithus turriseiffeli* Deflandre, in Deflandre and Fert, 1954.

Synonym: *Clinorhabdus* Stover, 1966; objective junior synonym with the same type species.

Diagnosis: Coccoliths having an expanded elliptical eiffellithalid rim and an inclined stem constructed of numerous small imbricate elements rising from four bars which are asymmetrically arranged with regard to the axes of the elliptical central area.

Range: Aptian–Maestrichtian.

Family AHMUELLERELLACEAE Reinhardt, 1965

Coccoliths with an eiffellithalid rim having a central structure consisting of a cross aligned in the major and minor axes of the ellipse.

Genus *CRUCIRHABDUS* Prins ex Rood *et al.*, 1973, Pl. 1, fig. 2

Type species: *Crucirhabdus primulus* Prins ex Rood *et al.*, 1973.

Diagnosis: Coccoliths with an eiffellithalid rim, in which the central structure is in the form of a symmetrical cross in the major and minor axes of the ellipse supported by one or more diagonal bars in each quandrant.

Range: Hettangian–Pliensbachian.

Genus *VEKSHINELLA* Loeblich and Tappan, 1963, Pl. 1, figs 3a, b

Synonyms: *Ephippium* Vekshina, 1959 (homonym of *Ephippium* Bolten, 1798) objective senior synonym with the same type species; *Valgalapilla* Bukry, 1969, subjective junior synonym (type: *Vekshinella imbricata* Gartner, 1968) established because Vekshina's illustration of the side view of *Ephippium acutiferra* indicated a slight proximal extension of the stem, a feature unknown in any of the species assigned by Gartner to *Vekshinella*, nor reported in any subsequent work. It seems evident that the proximal projection indicated by Vekshina was not an integral part of the coccolith, and replacement of *Vekshinella* by *Vagalapilla* is regarded as unwarranted. *Staurorhabdus*, Noël, 1972, subjective junior synonym with type species *Discolithus quadriarcullus* Noël, 1965, = *Vekshinella*.

Type species: *Ephippium acutiferrus* Vekshina, 1959.

Diagnosis: Coccoliths with an eiffellithalid rim and a central cross aligned in the major and minor axes of the ellipse; the short bar of the cross may be slightly offset at the centre.

Range: Hettangian–Maestrichtian.

Genus *STAUROLITHITES* Caratini, 1960. Pl. 1, fig. 4

Type species: *Staurolithites laffittei* Caratini, 1960.

Synonym: *Zygostephanos* Hoffman, 1970d (type: *Zygolithus crux* Deflandre and Fert, 1954), subjective junior synonym and invalid because the type species of two previously described genera, *Glaukolithus* (*Zygolithus diplogrammus* Deflandre, 1954) and *Zeugrhabdotus* (*Zygolithus erectus* Deflandre, 1954) were included in *Zygostephanos* when originally described.

Diagnosis: An elliptical coccolith with an outer eiffellithid cycle and a secondary inner ring surrounding an open central area spanned by a cross with the axis aligned in the major and minor axes of the ellipse.

Range: Berriasian–Santonian.

Genus *AHMUELLERELLA* Reinhardt, 1964. Pl. 1, fig. 5

Type Species: *Ahmuellerella limbitenuis* Reinhardt, 1964 (=*Discolithus octoradiatus*, Gorka, 1957).

Diagnosis: Elliptical eiffellithid coccoliths with a central area spanned by peripherally bifurcating bars forming a cross in the major and minor axes of the ellipse.

Range: Santonian–Maestrichtian.

Family ZYGODISCACEAE Hay and Mohler, 1967

Coccoliths with an eiffellithalid rim and a central structure having quadripartite construction developed either as a stem with sutures oriented in the equal axes of the ellipse and touching the margin of the ellipse in the minor axis (Subfamily Parhabdolithoideae Gartner, 1968), or as a bar in the minor axis of the ellipse, with or without a stem (Subfamily Zygodiscoideae Bukry, 1969).

Subfamily ZYGODISCOIDEAE Bukry, 1969

Coccoliths with an eiffellithalid rim and a bar oriented in the minor axis of the ellipse. The bar may be complex, and most commonly consists of four optically continuous regions separated along the major and a minor axis of the ellipse. In many species, the bar is surmounted by a short spine having a circular cross section. Coccoliths of this sort superficially resemble the original concept of "zygolith" of Kamptner (1937b, 1941).

Genus *ZYGODISCUS* Bramlette and Sullivan, 1961. Pl. 1, fig. 6; pl. 2, figs 1a, b

Type species: *Zygodiscus adamas* Bramlette and Sullivan, 1961.

Diagnosis: Elliptical coccoliths having a transverse bar displaying some form of quadripartite subdivision of calcite orientation, and commonly some vestige of a structure in the two openings.

Range: Late Cretaceous–Eocene.

Genus *ZEUGRHABDOTUS* Reinhardt, 1965. Pl. 2; figs 2a, b

Type species: *Zygolithus erectus* Deflandre and Fert, 1954.

Synonym: "*Zygolithus*" of many authors. (See Rood *et al.*, 1973 for a discussion of the designation of the type species and confusion over the concept of the genus.)

Diagnosis: Elliptical coccoliths with an eiffellithalid rim, a bar in the minor axis of the ellipse, and a short stem or circular cross section arising from the bar.

Range: Hettangian–Kimmeridgian.

Genus *GLAUKOLITHUS* Reinhardt, 1964. Pl. 2, fig. 3

Type species: *Zygolithus diplogrammus* Deflandre and Fert, 1954.

Diagnosis: Elliptical eiffellithalid coccoliths with a bar oriented along the minor axis of the ellipse. The bar commonly appears to have a longitudinal slit, or may appear as four units which do not meet at the centre.

Range: Berriasian–Maestrichtian.

Genus *TRANOLITHUS* Stover, 1966. Pl. 2, figs 4a, b

Type species: *Tranolithus manifestus* Stover, 1966.

Diagnosis: Elliptical eiffellithalid coccoliths with a bar constructed of four large plates; in extreme cases these may be reduced to only two plates occupying an equal axis of the elliptical central opening.

Range: Aptian–Maestrichtian.

Genus *AMPHIZYGUS* Bukry, 1969. Pl. 2, fig. 5

Type species: *Amphizygus brooksi* Bukry, 1969.

Diagnosis: Coccoliths having a simple rim with a supplementary single cycle visible in proximal view, and a central area having a cross bar in the minor axis of the ellipse; the cross bar and margin of the two perforations are constructed of many small elements.

Range: Santonian–Campanian.

Subfamily PARHABDOLITHOIDEAE Gartner, 1968

Coccoliths with an eiffellithalid rim and a central structure developed as a stem having quadripartite construction with sutures oriented in the equal of the ellipse and touching the margin of the ellipse in the minor axis.

Genus *PARHABDOLITHUS* Deflandre, 1952b. Pl. 2, fig. 6; Pl. 3, figs. 1a-c

Type species: *Parhabdolithus liasicus* Deflandre, 1952.

Diagnosis: Coccoliths having characteristics of the Subfamily and a circular tapering spine.

Range: Sinemurian–basal Toarcian, Kimmeridgian–Maestrichtian.

Genus *TUBIRHABDUS* Prins ex Rood *et al.*, 1973. Pl. 3, fig. 2

Diagnosis: Narrowly elliptical eiffelithalid coccoliths with a broadly oval to circular spinoid central structure in the minor axis of the margin.

Type species: *Tubirhabdus patulus* Prins ex Rood *et al.*, 1973.

Range: Hettangian–Bathonian.

Genus *MITROLITHUS* Deflandre and Fert, 1954. Pl. 3, figs 3a, b

Type species: *Mitrolithus elegans* Deflandre, in Deflandre and Fert, 1954 (by monotypy).

Synonym: *Alvearium* Black, 1967 (Type: *Alvearium dorsetense* Black, 1967, considered by Prins (1969) to be the distal view of *Mitrolithus*).

Diagnosis: Coccoliths with an eiffellithid rim and a central appendage. Coccoliths shaped like a bishop's mitre.

Range: Lias?

Genus *EURHABDUS* Reinhardt, 1965. Pl. 3, fig. 4

Type species: *Eurhabdus luciformis* Reinhardt, 1965.

Diagnosis: Rhabdolith with an elliptical eiffellithalid rim, but with a long cylindrical stem composed of four longitudinal elements.

Range: Santonian–Maestrichtian.

Genus *LUCIANORHABDUS* Deflandre, 1959. Pl. 3, fig. 5

Type Species: *Lucianorhabdus cayeuxi* Deflandre, 1959.

Diagnosis: Straight to curved rod-like elements composed of four longitudinal crystal elements.

Range: Turonian–Maestrichtian.

Family RHAGODISCACEAE, new family

Type genus: *Rhagodiscus* Reinhardt, 1967.

Diagnosis: Elliptical eiffellithalid coccoliths with a rhagodiscid, i.e. granular, centre.

Genus *RHAGODISCUS* Reinhardt, 1967. Pl. 4, figs 1, b

Type species: *Discolithus asper* Stradner.

Diagnosis: Elliptical eiffellithalid coccoliths with a rhagodiscid central area filled by small crystallites.

Range: Albian.

Genus *REINHARDITES* Perch-Nielsen, 1968. Pl. 4, figs 2a, b

Type species: *Rhabdolithus anthophorous* Deflandre, 1959.

Diagnosis: Eiffellithalid coccoliths with many tabular plates in the central area.

Range: Campanian–Maestrichtian.

Family ACTINOZYGACEAE Rood *et al.*, 1971

Elliptical coccoliths with an eiffellithalid rim and a complex central structure consisting of either (1) six or more bars disposed symmetrically with respect to the major and minor axes (Subfamily Actinozygoideae, Rood *et al.*, 1971) or (2) four or more bars disposed asymmetrically in such a way that one end of the coccolith may be duplicated by rotation of the other end through 180 degrees (Subfamily Diadozygoideae) Rood *et al.*, 1971). For a discussion of the evolution of this Family, see Rood and Barnard, 1972.

Subfamily ACTINOZYGOIDEAE Rood *et al.*, 1971

Elliptical coccoliths with an eiffellithalid rim and six or more bars disposed symmetrically with respect to the axes of symmetry of the ellipse. A spine may arise from the point of juncture of the bars in the centre of the coccolith.

Genus *ACTINOZYGUS* Gartner, 1968. Pl. 4, fig. 3

Type: Species *Tremalithus regularis* Gorka, 1957.

Diagnosis: Elliptical eiffellithalid coccoliths with 6–10 symmetrically arranged bars.

Range: Oxfordian–Maestrichtian.

Genus *ANGULOFENESTRELLITHUS* Bukry, 1969. Pl. 4, fig. 4

Type species: *Angulofenestrellithus snyderi* Bukry, 1969.

Diagnosis: Elliptical eiffellithalid coccoliths with a central area having many radial bars and cross struts resulting in a perforate appearance; perforations are surrounded by cycles of elements.

Range: Campanian.

Genus *RHABDOLITHINA* Rheinhardt, 1967. Pl. 4, fig. 5

Type species: *Rhabdolithus splendens* Deflandre.

Diagnosis: Elliptical eiffellithalid coccoliths with many radially disposed bars and a central spine.

Range: Valanginian–Maestrichtian.

Genus *NODOSELLA* Prins ex Rood *et al.*, 1973. Pl. 4, fig. 6

Diagnosis: Elliptical coccoliths with an eiffellithalid rim and a central structure consisting of a longitudinal elongated central island connected to the margin by numerous bars radiating from an axis along the long dimension of the eclipse.

Type species: *Nodosella clatriata* Prins ex Rood *et al.*, 1973

Range: Toarcian?

Genus *PONTILITHUS* Gartner, 1968. Pl. 4, fig. 7

Type species: *Pontilithus obliquicancellatus* Gartner, 1968.

Diagnosis: Elliptical coccoliths with an eiffellithalid rim and a central area divided into four quadrants each having a series of parallel bars lying in the equal axis of the elliptical central area.

Range: Turonian–Santonian.

Genus *TRUNCATOSCAPHUS* Rood *et al.*, 1971. Pl. 5, fig. 1

Type species: *Zygolithus delftensis* Stradner and Adamiker, 1966.

Diagnosis: Elongate coccoliths with an eiffellithalid rim and truncate ends; the longer sides are curved or broadly obtusely angled, the ends are flat so that the overall outline is that of an elongate subhexagonal ring.

Range: Oxfordian–Albian.

Subfamily DIADOZYGOIDEAE Rood *et al.*, 1971

Elliptical or rhomboidal coccoliths with an eiffellithalid rim and a central structure constructed of four or more bars distributed so that rotation of one half of the coccolith by 180 degrees repeats the other half.

Genus *DIADOZYGUS* Rood *et al*., 1971. Pl. 5, fig. 2

Type species: *Diadozygus rotatus* Rood *et al*., 1971

Diagnosis: Elliptical to rhomboidal coccoliths, with six or more bars disposed asymmetrically so that one end of the coccolith may be repeated by diad (180 degree) rotation of the other. A pair of bars extending part of the length of the coccolith but offset at the centre, are diagnostic of most species of the genus.

Range: Bathonian.

Note: Black (1971) proposed a new genus, *Stradnerlithus*, with the type of the species, *Stradnerlithus comptus* Black, based on a fragment of a coccolith. There is some resemblance between Black's type and *Diadozygus dorsetense* Rood, Hay and Barnard, but the specimen designated *S. comptus* does not show the asymmetry characteristic of the latter species; for this reason *Stradnerlithus* and *Diadozygus* are not regarded as synonyms.

Genus *DIADORHOMBUS* Worsley, 1971. Pl. 5, figs 3a, b

Type Species: *Diadorhombus rectus* Worsley, 1971.

Diagnosis: Rhombic coccoliths with an eiffellithalid rim and four or more internal bars supporting a central stem.

Range: Oxfordian–Valanginian.

Family CALCIOSOLENIACEAE Kamptner, 1937b

Fusiform coccolithophores bearing scapholiths, elongate rhomboidal coccoliths having an eiffellithalid rim. Scapholiths are termed scalar if the laths extending from each side meet along the midline of the long axis, or are termed alternate if they are offset along the midline. Long polar spires may arise from scapholiths at one or both ends of the coccolithophore. Most of the generic terms in this family are used for modern coccolithophores and are based on characteristics which require examination of the entire organism.

Genus *ANOPLOSOLENIA* Deflandre, 1952a

Type species: *Cylindrotheca brasiliensis* Lohmann, 1919.

Diagnosis: Spindle-shaped coccolithophores lacking polar spines, with scapholiths of the alternate type.

Range: Recent.

Genus *CALCIOSOLENIA* Gran in Murray and Hjort (1912). Pl. 5, fig. 4

Type species: *Calciosolenia murrayi* Gran, in Murray and Hjort, 1912.

Diagnosis: Spindle-shaped coccolithophores possessing polar spines at one or both ends.

Range: Recent.

Genus *ACANTHOSOLENIA* Bernard, 1939

Type species: *Acanthosolenia mediterranea* Bernard, 1939.

Diagnosis: Spindle-shaped coccolithophores lacking polar spines, with scapholiths of the alternate type.

Range: Recent.

Genus *NAVISOLENIA* Lecal, 1965a

Type species: *Navisolenia aprili* Lecal, 1965a.

Diagnosis: Spindle shaped coccolithophores lacking polar spines, with scapholiths of the scalar type.

Range: Recent.

Genus *SCAPHOLITHUS* Deflandre, in Deflandre and Fert, 1954. Pl. 5, fig. 5

Type species: *Scapholithus fossilis* Deflandre, in Deflandre and Fert, 1954.

Diagnosis: Isolated scapholiths formed by an elongate rhomboidal ring with flat laths extending to the midline in the central area, laths may be either scalar or alternate.

Range: Albian–Recent.

Genus *CALCIOPAPPUS* Gaarder and Ramsfjell, 1954

Type species: *Calciopappus caudatus* Gaarder and Ramsfjell, 1954 (= *Syracosphaera caudata* Braarud, Gaarder and Grøntved, 1953 *nomen nudum*).

Diagnosis: Extremely elongate coccolithophores lacking coccoliths at flagellar pole. Coccoliths, dimorphic, ordinary coccoliths scapholithid, with narrow long elliptical ring and radial laths extending to the mid-length, tail coccoliths typical scapholiths.

Range: Recent.

Genus *DICTYOLITHUS* Gorka, 1957

Type species: *Dictyolithus elegans* Gorka, 1957.

Diagnosis: Isolated rhomboidal plates with a thickened margin and a central area formed of a network of fine laths.

Range: Campanian–Maestrichtian?

Family CHIASTOZYGACEAE, Rood *et al.*, 1973

Coccoliths with an eiffellithalid rim having a central structure consisting of a cross or other structure aligned in the equal axes of the ellipse.

Genus *CHIASTOZYGUS* Gartner, 1968. Pl. 5, figs 6a, b

Type species: *Zygodiscus? amphipons*, Bramlette and Martini, 1964

Diagnosis: Coccolith with an eiffellithalid rim and a central structure consisting of a cross aligned in the equal axes of the ellipse.

Range: Aptian–Maestrichtian.

Genus *TEGUMENTUM* Roth and Thierstein, 1972. Pl. 6, figs 1a, b

Type species: *Tegumentum stradneri* Thierstein, in Roth and Thierstein, 1972.

Diagnosis: Chiastozygid coccoliths with a band of non-imbricate tabular elements covering the inner margin of the rim.

Range: Barremian–Maestrichtian.

Genus *NEOCOCCOLITHES* Sujkowski, 1931. Pl. 6, fig. 2

Type species: *Neococcolithes lososnensis* Sujkowski, 1931.

Synonym: *Zygolithus* Kamptner ex Matthes, 1956, subjective junior synonym because the type species designated by Maslov, 1956, *Zygolithus dubius* Deflandre and Fert, 1954, is considered conspecific with the type species of *Neococcolithes*. *Indumenthalithus* Vekshina, 1959, subjective junior synonym because the type species by original designation, *Coccolithus tetracelluseleganticus* Shamray and Lazareva, 1956, is conspecific with the type species of *Neococcolithes*.

Diagnosis: Isolated coccoliths having an elliptical rim with a peripherally bifurcating transverse bar oriented in the long axis of the ellipse.

Range: ? Maestrichtian–Lower Eocene.

Note: Brönnimann and Stradner, 1960, described a genus, *Heliorthus*, with a type species, *Heliorthus fallax* Brönnimann and Stradner, 1960. This appears to be a monolamellar elliptical rim with an x-shaped central structure in the equal axes of the ellipse. Bramlette and Martini, 1964, have considered this object to be an overgrown specimen belonging to the species *Neococcolithes protenus*.

Genus *CHIPHRAGMALITHUS* Bramlette and Sullivan, 1961. Pl. 6, fig. 6

Type species: *Chiphragmalithus calathus* Bramlette and Sullivan, 1961.

Definition: Basket-like coccoliths with an eiffellithalid rim; the cross bars have been heightened into septa which are higher centrally.

Range: Middle Palaeocene–Middle Eocene.

Genus *NANNOTETRINA* Achuthan and Stradner, 1967. Pl. 6, fig. 3

Type species: *Nannotetraster fulgens* Stradner, in Martini and Stradner, 1960.

Diagnosis: Calcareous nannofossils consisting of four rays, analogous to the central structure of *Chiphragmalithus*.

Range: Eocene.

Genus *ISTHMOLITHUS* Deflandre, in Deflandre and Fert, 1954. P. 6, fig. 4

Type species: *Isthmolithus recurvus* Deflandre, in Deflandre and Fert, 1954 (by original designation).

Diagnosis: Elongate parallelogram with one, two, or three transverse bars.

Range: Eocene.

Family STEPHANOLITHIONACEAE Black, 1968

Elliptical, circular, or polygonal coccoliths with a rim constructed of large, thick, slightly overlapping elements; at least two cycles of elements are present, although one cycle may be very large and the other very small.

Subfamily STEPHANOLITHIONOIDEAE Vekshina, 1959

Circular, hexagonal, or polygonal coccoliths of the Family Stephanolithionaceae with external projections.

Genus *STEPHANOLITHION* Deflandre, 1939. Pl. 6, figs 5a, b

Type species: *Stephanolithion bigoti* Deflandre, 1939.

Diagnosis: Coccoliths with a hexagonal or polygonal rim constructed of one cycle of thin tabular elements and a much thicker cycle of slightly imbricate prismatic elements; lateral projections extend outward from the rim.

Range: Bajocian–Maestrichtian.

Subfamily LITHASTRINOIDEAE Prins, 1971

A tubular or plate-like column generally surrounds a central core which does not protrude above the distal surface of the column; sometimes the central core seems to be lacking; the distal and proximal side of the column often flare so that sometimes the column seems to consist of two separated parts.

Genus *COROLLITHION* Stradner, 1961. Pl. 7, figs 1a–c

Type species: *Corollithion exiguum* Stradner, 1961.

Diagnosis: Broadly round coccoliths with a hexagonal to circular eiffellithalid rim, four to six radial elements extending from the centre to rim; a central knob is present.

Range: Aptian–Maestrichtian.

Genus *EPROLITHUS* Stover, 1966. Pl. 7, figs 2a–e

Type species: *Lithastrinus floralis* Stradner, 1962.

Diagnosis: Short, hollow, thick-walled cylinders with a horizontal partition constructed of 10–12 elements, formed by central extensions of the wall elements.

Range: Aptian–Maestrichtian.

Genus *LITHASTRINUS* Stradner, 1962. Pl. 7, figs 3a–c

Type species: *Lithastrinus grilli* Stradner, 1962.

Diagnosis: Biconcave nannofossils with a deeply incised equator, constructed of 6–9 pairs of asymmetrical rays. One member of each pair lies above and one below the equator; they are connected to each other at a steep angle to the equator.

Range: Turonian–Maestrichtian.

Genus *POLYCYCLOLITHUS* Forchheimer, 1968a

Type species: *Polycyclolithus brotzeni* Forchheimer, 1968.

Diagnosis: Stacks of concentric superposed cycles; at least five such cycles are present.

Range: Campanian.

Genus *RADIOLITHUS* Stover, 1966

Type species: *Radiolithus planus* Stover, 1966.

Diagnosis: Short hollow cylinders with a wall composed of eight or nine segments. At one end is a basal central structure constructed of radiating bars.

Range: Albian.

Genus *BUKRYASTER* Prins, 1971. Pl. 7, fig. 4

Type species: *Discoaster hayi* Bukry, 1969.

Diagnosis: Proximal part of the column with deep incisions between the elements, which curve in a sinistral direction, as is the case in the proximal part of the column in *Lithastrinus*; distal part of the column strongly reduced, probably only consisting of dextrally oriented ridges, which form a continuation of a star-shaped central body.

Range: Cenomanian–Turonian.

Genus *HAYESITES* Manivit, 1971. Pl. 8, fig. 1

Type species: *Hayesites albiensis* Manivit, 1971.

Diagnosis: Starlike coccoliths, consisting of a proximal cycle of 7–8 elongated radial elements meeting centrally, and one or more small proximal cycles of elements.

Range: Albian.

Genus *CYLINDRALITHUS* Bramlette and Martini, 1964. Pl. 8, figs 2a–c

Type species: *Cylindralithus serratus* Bramlette and Martini, 1964.

Diagnosis: Short, slightly tapering cylinders constructed of high wall elements. A thin perforate base with cruciform central structures may be present.

Range: Turonian–Maestrichtian.

Genus *CALCICALATHINA* Thierstein, 1971. Pl. 8, fig. 3; Pl. 31, fig. 7

Type species: *Schizosphaerella oblongata* Worsley, 1971.

Diagnosis: Basket-like elliptical forms with a high wall. Central area filled with calcite crystals of different orientation that rise above the distal margin of the wall. Proximal side flat or slightly concave.

Range: Valanginian–Barremian.

Subfamily CREPIDOLITHOIDEAE Rood *et al.*, 1971

Elliptical coccoliths of the Family Stephanolithionaceae with no external projections.

Genus *CREPIDOLITHUS* Noël, 1965a. Pl. 8, figs 4a–c

Type species: *Discolithus crassus* Deflandre and Fert, 1954.

Diagnosis: Coccoliths with a broad elliptical rim constructed of a proximal cycle of thin tabular elements extending into the centre of the coccolith and a much thicker distal cycle of slightly prismatic elements. As noted by Prins (1969), a vestige of a central cross is usually visible between crossed polarisers.

Range: Sinemurian–Kimmeridgian.

Genus *MITOSIA* Worsley, 1971. Pl. 9, fig. 1

Type species: *Mitosia infinitae* Worsley, 1971.

Diagnosis: Elliptical plates having distinct elevated rims composed of radial elements showing no evident suturing, and a central area composed of randomly to sub-radially oriented granular calcite.

Range: Early Cretaceous.

Genus *CARINOLITHUS* Prins, 1969

Type species: *Rhabdolithus superbus* Deflandre, in Deflandre and Fert, 1954.

Diagnosis: Eiffelithalid coccoliths with a tall spinose structure rising from the inner part of the rim and central area and flaring sharply distally.

Range: Toarcian.

Order PODORHABDINALES Rood *et al.*, 1971

Coccoliths with a marginal area constructed of two or more petaloid cycles of elements which are not at all, or only very slightly, imbricate.

Family PODORHABDACEAE Noël, 1965b

Elliptical coccoliths with a podorhabdid rim, consisting of two cycles of elements.

Subfamily PODORHABDOIDEAE Rood *et al.*, 1971

Elliptical coccoliths with a podorhabdid rim having two cycles of elements and with four or more radial bars supporting a central stem.

Genus *PODORHABDUS* Noël, 1965b. Pl. 9, figs 2a, b

Type species: *Podorhabdus grassei* Noël, 1965b.

Diagnosis: Coccoliths with a podorhabdid rim and a central stem supported by four bars.

Range: Pliensbachian–Cenomanian.

Genus *BIPODORHABDUS* Noël, 1970. Pl. 9, figs 3a, b

Type species: *Bipodorhabdus tesselatus* Noël, 1970.

Diagnosis: Circular coccoliths with a single cross bar in the minor axis of the central area; the cross bar rises centrally to merge with a circular stem.
Range: Berriasian–Campanian.

Genus *HEXAPODORHABDUS* Noël, 1965b. Pl. 9, fig. 4
Type species: *Hexapodorhabdus cuvillieri* Noël, 1965b.
Diagnosis: Elliptical coccoliths with a podorhabdid base having two cycles of elements and a vertical stem composed of numerous small imbricate elements rising from six supports; two of these are in the minor axis and four are arranged symmetrically about the major axis of the ellipse.
Range: Bajocian–Oxfordian.

Genus *OCTOPODORHABDUS* Noël, 1965a. Pl. 9, fig. 5
Type species: *Octopodorhabdus praevisus* Noël, 1965a.
Diagnosis: Elliptical coccoliths with a podorhabdid base having two cycles of elements and a vertical stem arising from eight bars; four of these lie in the major and minor axes of the ellipse and four extend from the margin of the central area to the bars in the major axis.
Range: Bajocian–Oxfordian.

Genus *DODEKAPODORHABDUS* Perch-Nielsen, 1968a. Pl. 9, fig. 6
Type species: *Dodekapordorhabdus noelae* Perch-Nielsen, 1968a.
Diagnosis: Coccoliths with a podorhabdid rim consisting of two cycles of elements and a central area with twelve bars radiating from a central platform.
Range: Maestrichtian.

Genus *POLYPODORHABDUS* Noël, 1965a. Pl. 10, fig. 1
Type species: *Polypodorhabdus escaigi* Noël, 1965a.
Diagnosis: Coccoliths with a podorhabdid rim consisting of two cycles of elements and a vertical stem composed of numerous small imbricate elements rising from the centre of a bar in the major axis of the ellipse; many additional shorter irregularly oriented bars extend from the sides of the central opening, to the longitudinal bar.
Range: Callovian–Kimmeridgian.

Subfamily PALEOPONTOSPHAEROIDEAE Rood *et al.*, 1971
Elliptical coccoliths with a wide podorhabdid rim and a narrow central area lacking prominent bars, but with a central spine commonly present.

Genus *PALEOPONTOSPHAERA* Noël, 1965a. Pl. 10, fig. 2
Type species: *Paleopontosphaera dubia* Noël, 1965a.
Diagnosis: Coccoliths with a podorhabdid margin having only two cycles of elements and a simple elongate central depression with a cross bar in the

short axis of the placolith; the cross bar may be surmounted by a short, stubby spine.

Range: Sinemurian–Kimmeridgian.

Genus *STRIATOMARGINIS* Prins ex Rood *et al.*, 1973. Pl. 10, fig. 3

Type species: *Striatomarginis primitivus* Rood *et al.*, 1973.

Diagnosis: Elliptical coccoliths with a podorhabdid margin having a narrow tube cycle and a central structure consisting of a longitudinal bar from which four groups of diagonally disposed bars extend to the margin; a cross bar in the short axis of the ellipse may be developed.

Range: Toarcian–Bathonian.

Subfamily ETHMORHABDOIDEAE Rood *et al.*, 1971

Coccoliths with a podorhabdid rim and a solid or cribrate centre.

Genus *ETHMORHABDUS* Noël, 1965a. Pl. 10, figs 4a, b

Type species: *Ethmorhabdus gallicus* Noël, 1965a.

Diagnosis: Coccoliths with a podorhabdid rim having two cycles of elements and a cribrate centre.

Range: Bajocian–Kimmeridgian.

Genus *CRIBROSPHAERA* Arkhangel'ski, 1912. Pl. 10, figs 5a, b

Synonym: *Cribrosphaerella* Deflandre, 1952b.

Type species: *Cribrosphaera ehrenbergi* Arkhangel'ski, 1912, objective junior synonym, replacement for animal kingdom, with same type species. *Favocentrum* Black, 1964, subjective junior synonym with type species *Favocentrum laughtoni* Black similar to *Cribrosphaera* spp.

Diagnosis: Elliptical to subcircular coccoliths with a podorhabdid margin having four cycles of petaloid elements and cribrate centre.

Range: Cenomanian–Maestrichtian.

Genus *PERCIVALIA* Bukry, 1969. Pl. 11, figs 1a, b

Type species: *Percivalia porosa* Bukry, 1969.

Diagnosis: Elliptical podorhabdid coccolith having a margin consisting of several cycles of petaloid elements and a cribrate centre.

Range: Campanian.

Genus *NEPHROLITHUS* Gorka, 1957. Pl. 11, fig. 2

Type species: *Nephrolithus frequens* Gorka, 1957.

Diagnosis: Reniform coccoliths having a podorhabdid margin and a cribate centre.

Range: Late Maestrichtian.

Genus *CRETARHABDUS* Bramlette and Martini, 1964. Pl. 11, figs 3a–c

Synonym: *Stradneria* Reinhardt, 1964 (Subjective junior synonym because the type species by original designation, *Stradneria limbicrassa* Reinhardt, 1964, is considered conspecific or very closely related to the type species of *Cretarhabdus*).

Type species: *Cretarhabdus conicus* Bramlette and Martini, 1964.

Diagnosis: Elliptical coccoliths with a podorhabdid margin consisting of two cycles of elements and a cribrate centre with four buttresses aligned in the major and minor axes; a stem arises from the centre.

Range: Berriasian–Maestrichtian.

Genus *CRETADISCUS* Gartner, 1968

Type species: *Cretadiscus polyporus* Gartner, 1968.

Diagnosis: Elliptical to circular coccoliths with a podorhabdid margin having three cycles of petaloid elements and a cribrate centre.

Range: Santonian and Maestrichtian.

Subfamily SOLLASITEOIDEAE Rood *et al.*, 1971

Elliptical coccoliths with a podorhabdid rim, a transverse bar, and three or more longitudinal or sublongitudinal bars. No central stem is present.

Genus *SOLLASITES* Black, 1967. Pl. 11, fig. 4

Type species: *Sollasites barringtonensis* Black, 1967 (=*Coccolithus horticus* Stradner *et al.*, 1966).

Synonyms: *Costacentrum* Bukry, 1969 (Subjective junior synonym with *Coccolithus horticus* Stradner *et al.*, 1966, as type).

Diagnosis: Elliptical coccoliths with a podorhabdid rim, a transverse bar, and three or more longitudinal to sublongitudinal bars.

Range: Toarcian–Maestrichtian.

Genus *HETEROMARGINATUS* Bukry, 1961. Pl. 11, figs 5a, b

Type species: *Heteromarginatus wallacei* Bukry, 1969.

Diagnosis: Elliptical coccoliths with a podorhabdid rim and a central structure consisting of bars aligned in the major and minor axes and four struts connecting the mid-points of the bars.

Range: Campanian.

Subfamily DISCORHABDOIDEAE Noël, 1965a

Circular coccoliths with a circular stem constructed of many short laths.

Genus *DISCORHABDUS* Noël, 1965. Pl. 11, fig. 6

Type species: *Rhabdolithus patulus* Deflandre, in Deflandre and Fert, 1954.

Diagnosis: Coccoliths with a circular podorhabdid base and a circular

stem constructed of numerous small laths. The species of this genus are distinguished by the shape and size of the stem as seen in side view.
Range: Toarcian–Oxfordian.

Genus *DIAZOMATOLITHUS* Noël, 1965. Pl. 12, fig. 1
Type species: *Diazomatolithus lehmani* Noël, 1965.
Diagnosis: Circular coccoliths consisting of a cycle of podorhabdid elements forming a broadly open tube with recessed peripheral groove; the appearance is that of a placolith with wide central opening and a distal shield narrower than the proximal shield.
Range: Bajocian–Albian.

Genus *TUBODISCUS* Thierstein, 1973. Pl. 12, fig. 2
Type species: *Tubodiscus veternae* Thierstein, 1973.
Diagnosis: Two shields are connected by a wide central tube rising distally, producing a distinct collar.
Range: Valanginian.

Genus *MANIVITELLA* Thierstein, 1971. Pl. 12, figs 3a, b
Type species: *Manivitella pemmatoidea* Deflandre ex Manivit, 1965.
Diagnosis: Two closely appressed, narrow shields form an elliptical ring. The distal shield consists of one cycle of non-imbricate elements, the proximal of one or two cycles of imbricated elements.
Range: Berriasian and Cenomanian.

Family ARKHANGELSKIELLACEAE Bukry, 1969
Elliptical coccoliths with a podorhabdid rim consisting of three or four tiers of elements; two or three tiers are visible on the proximal side and two tiers on the distal side.

Genus *ARKHANGELSKIELLA* Vekshina, 1959. Pl. 12, fig. 4
Type species: *Arkhangelskiella cymbiformis* Vekshina, 1959.
Diagnosis: Coccoliths with a margin constructed of three tiers of elements and a central area formed of 10–12 large elements arranged symmetrically about the major and minor axis.
Range: Hauterivian–Maestrichtian.

Genus *ASPIDOLITHUS* Noël, 1969b
Type species: *Aspidolithus angustus* Noël, 1969b.
Diagnosis: Elliptical imperforate to cribrate coccoliths having a margin constructed of three tiers of elements and a central area with a distal layer of irregularly disposed elements and proximal layer of radially arranged elements.
Range: Coniacian–Maestrichtian.

Genus *BROINSONIA* Bukry, 1969. Pl. 12, figs 5a–c

Type species: *Broinsonia dentata* Bukry, 1969.

Diagnosis: Coccoliths with a margin constructed of four tiers of elements and a central area with a solidly constructed cross with arms aligned in the major and minor axes of the ellipse; the four quadrants of the central area have a grille or projecting teeth.

Range: Albian–Campanian.

Genus *LAFFITTIUS* Noël, 1969b. Pl. 13, fig. 1

Type species: *Arkhangelskiella obliqua* Stradner, 1963.

Diagnosis: Coccoliths having a margin constructed of four tiers of elements and an imperforate or cribrate central area with elements arranged about a longitudinal axis in the major axis of the ellipse and a transverse axis slightly oblique to the minor axis of the ellipse.

Range: Coniacian–Campanian.

Genus *PROLATIPATELLA* Gartner, 1968. Pl. 13, figs 2a, b

Type species: *Prolatipatella multicarinata* Gartner, 1968.

Diagnosis: Large coccoliths having a rim constructed of several tiers of elements and an indistinct central area.

Range: Campanian–Maestrichtian.

Subfamily KAMPTNERIOIDEAE Bukry, 1968

Margin of base multilamellar constructed of three, four or five tiers of elements, one of which is very broad.

Genus *KAMPTNERIUS* Deflandre, 1959. Pl. 13, figs 3a, b

Type species: *Kamptnerius magnificus* Deflandre, 1959.

Diagnosis: Asymmetrical coccoliths with a margin of four tiers of elements, one of which forms a distinctive flange, and an imperforate or cribrate central area.

Range: Turonian–Maestrichtian.

Genus *GARTNERAGO* Bukry, 1969. Pl. 13, fig. 4

Type species: *Arkhangelskiella concava* Gartner, 1968.

Diagnosis: Symmetrical coccoliths having a rim of four or five tiers; in proximal view there is a broad distinctive inner tier of narrow elements. The central area is imperforate or cribrate with a subaxial or axial cross bar aligned along the major and minor axes of the ellipse.

Range: Turonian–Campanian.

Family DEFLANDRIACEAE Black, 1967

Circular coccoliths with a podorhabdid rim.

Note: Rood *et al.* (1971) proposed a Family Prediscosphaeraceae intended to be more broadly defined than the Family Deflandriaceae of Black, but the older taxon has priority.

Subfamily PREDISCOSPHAEROIDEAE Gartner, 1968, emended

Circular coccoliths with four buttresses supporting a square stem constructed of a few long laths.

Genus *PREDISCOSPHAERA* Vekshina, 1959. Pl. 13, figs 5a, b; pl. 14, fig. 1

Synonym: *Deflandrius* Bramlette and Martini, 1964, subjective junior synonym because the type species, *Rhabdolithus intercisus* Deflandre, 1954, is considered to be conspecific with the type species of *Prediscosphaera*.

Type species: *Prediscosphaera decorata* Vekshina, 1959 (=*Coccolithophora cretacea* Arkhangel'skii, 1912.

Diagnosis: Circular coccoliths having a podorhabdid base, the proximal cycle being elliptical, the distal cycle circular, each with an x-shaped central structure, that of the distal cycle surmounted by a square appendix flaring distally to form a four-rayed pyramidal tetralith.

Range: Albian–Maestrichtian.

Order SYRACOSPHAERALES, new order

Type family: SYRACOSPHAERACEAE Lemmermann, 1908.

Diagnosis: Coccolithophores, exhibiting a loose cover of delicately constructed, usually perforate coccoliths; coccoliths do not interlock, and may not touch each other, so that areas of coccolith-free cell surface are exposed between crossed polarisers, the coccoliths produce an interference figure. The coccoliths are generally delicate caneoliths, cribriliths, cyrtoliths, and lopodoliths having a marginal area consisting of a cycle of elements forming a tube and proximal and distal rims constructed of delicate petaloid elements. The central area is commonly partially closed by numerous laths.

Family SYRACOSPHAERACEAE Lemmermann, 1908

Coccolithophores bearing caneoliths and cribriliths having the charac-terissics of the order.

Genus *SYRACOSPHAERA* Lohmann, 1902. Pl. 14, figs 2a–f

Type species: *Syracosphaera pulchra* Lohmann, 1902.

Synonym: *Eusyracosphaera* Lohmann, 1902; *Syracorhabdus* Lecal, objective junior synonym with the same type species.

Diagnosis: Coccolithophores with naked flagellar pole, bearing dimorphic caneoliths, those of the circumflagellar cycle bearing a central spine, the others being simple.

Range: Oligocene–Recent.

Genus *SYRACOLITHUS* (Kamptner, 1941) Deflandre, 1952b. Pl. 14, figs 3a, b

Type species: *Syracosphaera dalmatica* Kamptner, 1927.

Diagnosis: Coccolithophores with naked flagellar pole, bearing a single kind of caneolith; caneoliths elliptical, with a central spine surrounded by eight perforations.

Remarks: No electron microscope studies have been undertaken, but the similarity of these supposed caneoliths to the ordinary holococcoliths of *Zygosphaera wettsteini* Kamptner, 1937 suggests that this genus may actualyl bear holococcoliths. (*Syracolithus* was originally proposed as a sub-genus and raised to generic rank by Deflandre.)

Range: Recent.

Genus *CEPEKIELLA* Roth, 1970

Type species: *Cepekiella elongata* Roth, 1970.

Diagnosis: A coccolith consisting of two shields which are connected by short struts. The centre is covered with a cupola, composed of spirally arranged laths which is attached to the distal shield by struts.

Range: Oligocene.

Family PRINSIACEAE Hay and Mohler, 1967

Oval, subcircular or circular placoliths; interference figure between crossed polarisers is dextrogyre in distal view.

Genus *PRINSIUS* Hay and Mohler, 1967. Pl. 14, fig. 4.

Type species: *Coccolithus bisulcus* Stradner, 1963.

Diagnosis: Elliptical placoliths, with simple, solidly constructed distal shield; interference figure between crossed polarisers dextrogyre; proximal shield bright, distal shield somewhat fainter.

Range: Palaeocene–Eocene.

Genus *TOWEIUS* Hay and Mohler, 1967. Pl. 15, figs 5a, b

Type species: *Toweius craticulus* Hay and Mohler, 1967.

Diagnosis: Circular to subcircular placoliths, having shields composed of non-imbricate to slightly imbricate petaloid or trapezoidal segments; central area displaying a reticulate grille; proximal shield bright, distal shield faintly illuminated between crossed polarisers.

Range: Palaeocene–Eocene.

Genus *BRAMLETTEIUS* Gartner, 1969a. Pl. 15, fig. 6

Type species: *Bramletteius serraculoides* Gartner, 1969a.

Diagnosis: Elliptical placoliths having a cruciform central structure with axes in the major and minor axes of the ellipse in top view; in side view the placoliths are seen to bear a wide flaring blade-like spine oriented along the long axis of the ellipse.

Range: Upper Eocene–Oligocene.

Genus *ELLIPSOLITHUS* Sullivan, 1964. Pl. 15, figs 7a, b

Type species: *Coccolithites macellus* Bramlette and Sullivan, 1961.

Diagnosis: Elliptical coccoliths with a margin consisting of two narrow, closely appressed plates surrounding a large central area having elements radiating from two foci and the line joining the foci.

Range: Palaeocene.

Genus *ERICSONIA* Black, 1964. Pl. 15, fig. 1

Type species: *Ericsonia occidentalis* Black, 1964.

Diagnosis: Large elliptical coccoliths having the appearance of a typical coccolithid placolith but with the optical characteristics of the Prinsiaceae.

Range: Eocene–Oligocene.

Genus *CAMPYLOSPHAERA* Kamptner, 1963. Pl. 15, figs 2a–b

Type species: *Campylosphaera bramlettei* Kamptner, 1963.

Diagnosis: Blunt-ended elliptical coccoliths with a central cross aligned in the major and minor axes; in side view the ends are seen to be strongly curved proximally.

Range: Palaeocene–Eocene.

Family ACANTHOICEAE, new family

Type genus: *Acanthoica* Lohmann, 1903.

Diagnosis: Styliform and simple cyrtoliths having the characteristics of the order.

Genus *ACANTHOICA* Lohmann, 1903

Type species: *Acanthoica coronata* Lohmann, 1903.

Diagnosis: Coccolithophores bearing dimorphic cyrtoliths, those of the flagellar and sometimes caudal regions styliform, the others simple.

Range: Recent.

Genus *ANACANTHOICA* Deflandre, 1952

Type species: *Acanthoica acanthos* Schiller, 1925.

Diagnosis: Coccolithophores bearing a single kind of simple cyrtoliths, like shallow calyptroliths.

Range: Recent.

Family HYMENOMONACEAE, new family

Type genus: *Hymenomonas* Stein, 1878.

Diagnosis: Cricolighs having the characteristics of the order.

Genus *HYMENOMONAS* Stein, 1878

Type species: *Hymenomonas roseola* Stein, 1878.

Diagnosis: Spindle to heart-shaped coccolithophores bearing simple cricoliths.

Remarks: Although the type species was described from fresh or slightly brackish water, some marine coccoliths in laboratory culture have been assigned to this genus.

Range: Recent.

Genus *CRICOSPHAERA* Braarud, 1960

Type species: *Syracosphaera carterae* Braarud and Fagerland, 1946.

Diagnosis: Elongate to globular coccolithophores bearing cricoliths composed of anvil-shaped segments.

Remarks: This type species was described from marine water; members of this genus have been cultured in the laboratory but there is virtually no information on their occurrence in nature.

Range: Recent.

Genus *COCCOCHRYSIS* Conrad, 1926

Type species: *Coccochrysis subsala* Conrad, 1926 (by monotypy).

Diagnosis: Coccolithophores bearing a single kind of apparently imperforate caneolith having a central spine.

Remarks: This genus, considered a synonym of *Syracosphaera* by Kamptner (1928a) and Schiller (1930) should be resurrected. The taxon *Syracolithus* Kamptner, 1941, would be a junior synonym if the large perforations of the caneolith found in the type species were not taken into consideration in making a distinction between the two. *C. subsala* is known only from brackish water.

Range: Recent.

Genus *HEIMIELLA* Lohmann, 1913

Type species: *Heimiella excentrica* Lohmann, 1913.

Diagnosis: Bottle-shaped coccolithophores bearing a single kind of shallow caneolith.

Range: Recent.

Family DEUTSCHLANDIACEAE Kamptner, 1928b

Coccolithophores bearing plate-like, tubular or ceratiform coccoliths or osteoliths or jointed spines, constructed of one or more kinds of structural element.

Subfamily DEUTSCHLANDIOIDEAE Kamptner, 1928b

Coccolithophores bearing petaloid, tubular or ceratiform coccoliths.

Genus *FLORISPHAERA* Okada and Honjo, 1973. Pl. 15, fig. 3.

Type species: *Florisphaera profunda* Okada and Honjo, 1973.

Diagnosis: Hemiellipsoidal coccospheres bearing polygonal plate-shaped coccoliths forming a concentric pattern in top view.

Remarks: This genus is very similar to *Deutschlandia* and may be a junior synonym.

Range: Recent.

Genus *DEUTSCHLANDIA* Lohmann, 1912a

Type species: *Deutschlandia anthos* Lohmann, 1912a

Diagnosis: Lenticular coccolithophores bearing dimorphic coccoliths, those of the periphery having a broad petaloid extension.

Range: Recent.

Genus *THOROSPHAERA* Ostenfeld, 1910. Pl. 15, fig. 4

Type species: *Thorosphaera elegans* Ostenfeld, 1910.

Diagnosis: Coccolithophores bearing dimorphic coccoliths; those of the equatorial region are short or long tubular coccoliths which flare distally, the remainder are simple discoliths.

Range: Recent.

Family MICHELSARSIOIDEAE Lecal-Schlauder, 1951

Coccolithophores bearing a whorl of spines of osteolithid coccoliths.

Genus *MICHELSARSIA* Gran, in Murray and Hjort, 1912

Type species: *Michelsarsia elegans* Gran, in Murray and Hjort, 1912.

Diagnosis: Non-flagellate subsphaerical coccolithophores with a flattened apertural face surrounded by a ring of coccoliths supporting long, jointed spines.

Range: Recent.

Genus *OPHIASTER* Gran, in Murray and Hjort, 1912

Type species: *Meringosphaera hydroidea* Lohmann, 1903.

Diagnosis: Spherical coccolithophores possessing a coccosphere of fused coccoliths (discoliths?), the seven about the flagellar pole possessing long sinuous osteoliths.

Range: Recent.

Genus *THALASSOPAPPUS* Kamptner, 1941

Type species: *Thalassopappus pellucidus* Kamptner, 1941.

Diagnosis: Oblong sphaeroidal coccolithophores lacking flagellae, with a circlet of rhabdoliths surrounding the slightly flattened apertural face.

Range: Recent.

Family HALOPAPPACEAE Kamptner, 1928b

Coccolithophores bearing a carapace, with a whorl of coccoliths bearing jointed spines.

Genus *HALOPAPPUS* Lohmann, 1912a

Type species: *Halopappus vahselii* Lohmann, 1912a.

Diagnosis: Flagellate subsphaerical coccolithophores with a flattened apertural face surrounded by a ring of coccoliths supporting long, jointed spines; rest of cell enclosed in a single perforate calcareous shell.

Range: Recent.

Genus *CALCIOCONUS* Schiller, 1925

Type species: *Calcioconus vitreus* Schiller, 1925.

Diagnosis: Conical coccolithophores possessing a flat "flagellar pole" (flagellae not seen) surrounded by long thin adapically directed spines.

Range: Recent.

Genus *BERNARDOSPHAERA* Lecal–Schlauder, 1951

Type species: *Bernardosphaera stellata* Lecal–Schlauder, 1951.

Diagnosis: Subsphaerical coccolithophores with an open apertural area surrounded by 8–10 long jointed sinuous spines; rest of cell covered by discrete simple discoliths.

Range: Recent.

Family PONTOSPHAERACEAE Lemmermann, 1908

Coccolithophores bearing shallow lopodoliths which may or may not possess central spines.

Subfamily PONTOSPHAEROIDEAE Kamptner, 1937

Base of lopodoliths smooth or ornamented, or studded with short spines.

Genus *PONTOSPHAERA* Lohmann, 1902. Pl. 15, figs 5a–b.

Type species: *Pontosphaera syracusana* Lohmann, 1902.

Diagnosis: Coccolithophores bearing simple lopodoliths; depth equal to or a substantial fraction of length; floor solid or perforate.

Range: Palaeocene–Recent.

Genus *ALGIROSPHAERA* Schlauder, 1945

Type species: *Algirosphaera oryza* Schlauder, 1945.

Diagnosis: Coccolithophores bearing lopodoliths about as deep as long; the floor of the lopodoliths is studded with short spines.

Range: Recent.

Genus *TRANSVERSOPONTIS* Hay *et al.*, 1966. Pl. 16, fig. 1

Type species: *Discolithus obliquipons* Deflandre and Fert, 1954.

Diagnosis: Shallow symmetrical elliptical lopodoliths with two large openings in the ends of the central area separated by a bar.

Range: Late Eocene.

Genus *DISCOLITHINA* Loeblich and Tappan, 1963. Pl. 16, figs 2a, b

Type species: *Discolithus vigintiforatus* Kamptner ex Deflandre and Fert, 1954.

Synonym: *Discolithus* Huxley, 1868, non Fortis, 1802.*

Diagnosis: Shallow lopodoliths having various central structures.

Range: Palaeocene–Recent.

Genus *LOPHODOLITHUS* Deflandre, in Deflandre and Fert, 1954. Pl. 16, figs 3a, b

Type species: *Lophodolithus mochlophorus* Deflandre, in Deflandre and Fert, 1954.

Diagnosis: Shallow asymmetrical lopodolithus having a bar spanning the central area in the minor axis of the ellipse.

Range: Eocene.

Genus *HELICOPONTOSPHAERA* Hay and Mohler, 1967. Pl. 16, figs 4a–f;

Type species: *Helicopontosphaera kamptneri* Hay and Mohler, 1967.

Diagnosis: Coccolithophores bearing shallow lopodoliths with a spiral wall.

Remarks: This genus replaced the concept of *Helicosphaera* Kamptner, 1954. The type species of *Helicosphaera*, *Coccolithus carterae* Wallich, 1861, was well illustrated in the original publication, leaving no doubt that it was a species closely related to *Coccolithus pelagicus* Wallich, 1861, and the forms that Kamptner had considered as belonging to *Helicosphaera carteri* (Wallich) were very different from Wallich's original figure. To rectify this confusing situation Hay and Mohler, 1967, proposed the taxa *Helicopontosphaera* and *Helicopontosphaera kamptneri*.

Range: Eocene–Recent.

Genus *CRASSAPONTOSPHAERA* Boudreaux and Hay, 1969

Type species: *Crassapontosphaera jonesi* Boudreaux and Hay, 1969.

Diagnosis: Lopodoliths with an elliptical base, and a low wall formed by inclined laths; distal margin granular.

Range: Pliocene–Pleistocene.

Genus *KOCZYIA* Boudreaux and Hay, 1969

Type species: *Koczyia lepida* Boudreaux and Hay, 1969.

Diagnosis: Lopodoliths having a distal margin which flares peripherally outward.

Range: Pliocene–Pleistocene.

Subfamily SCYPHOSPHAEROIDEAE Boudreaux and Hay, 1969

Coccolithophores bearing large, deep, basket-like lopodoliths.

* See Loeblich and Tappan, 1963.

Genus *SCYPHOSPHAERA* Lohmann, 1902. Pl. 17, figs 1a–c

Type species: *Scyphosphaera apsteini* Lohmann, 1902.

Diagnosis: Coccolithophores bearing dimorphic coccoliths; large, deep, basket-like lopodoliths form an equatorial girdle or circlet; shallow lopodoliths cover most or all of the rest of the cell.

Remarks: In fossil material, only the large, deep, basket-like lopodoliths can be referred to this genus with certainty. The other, shallower lopodoliths found in the recent form would be assigned to genera such as *Pontosphaera*.

Range: Miocene–Recent.

Genus *LOHMANNOSPHAERA* Schiller, 1913

Type species: *Lohmannosphaera adriatica* Schiller, 1913.

Diagnosis: Coccolithophores completely covered by large, deep, basket-like lopodoliths.

Remarks: Some fossil *Scyphosphaera* species might belong to this genus.

Range: Recent.

Family RHABDOSPHAERACEAE Lemmermann, 1908

Coccolithophores bearing rhabdoliths, which may be club, nail, or trumpet-shaped crytoliths.

Subfamily RHABDOSPHAEROIDEAE Kamptner, 1928b, emend. Boudreaux and Hay, 1969

Coccolithophores bearing club or nail-shaped rhabdoliths.

Genus *RHABDOSPHAERA* Haeckel, 1894. Pl. 17, figs 2a–b

Synonyms: *Rhabdolithes* Schmidt, 1870; objective senior synonym with the same type species (designated by Loeblich and Tappan, 1963): petition to suppress this taxon was submitted to the International Commission on Zoological Nomenclature by Braarud *et al.*, 1965. *Clavosphaera* Lecal-Schlauder, 1951. Subjective junior synonym with type species *Clavosphaera furonculata* Lecal-Schlauder, 1951 = *Rhabdosphaera*.

Type species: *Rhabdosphaera clavigera* Murray and Blackman, 1898.

Definition: Rhabdoliths with a circular base.

Range: Pliocene–Recent.

Genus *LECALIA* Loeblich and Tappan, 1963

Type species: *Rhabdocyclus simplex* Lecal, in Bernard and Lecal, 1960.

Synonym: *Rhabdocyclus* Lecal, in Bernard and Lecal, 1960. non Lang and Smith, 1939.*

Diagnosis: Base constructed of a spiral of elements.

Range: Recent.

* See Loeblich and Tappan, 1963.

Genus *PALUSPHAERA* Lecal, 1967
Type species: *Palusphaera vandeli* Lecal, 1967.
Diagnosis: Rhabdoliths with a base consisting of several peripheral cycles of subquadrate plates and a central region with many small irregular elements. The stem is a helix of laths.
Range: Recent.

Genus *ASPIDORHABDUS* Hay and Towe, 1962. Pl. 17, figs 3a, b
Type species: *Aspidorhabdus ovalis* Hay and Towe, 1962.
Diagnosis: Rhabdoliths with an elliptical base constructed of several cycles of elements; stem circular in cross section near the base.
Range: Eocene–Recent.

Genus *RHABDOLITHUS* Kamptner ex Deflandre, 1952b. Pl. 17, figs 4a–d
Type species: *Rhabdolithus perlongus* Deflandre, 1952b.
Diagnosis: Rhabdoliths having a complex base formed of several to many cycles of elements.
Range: Eocene.

Genus *DISCOPHAERA* Haeckel, 1894. Pl. 17, figs 5a, b
Type species: *Discosphaera thomsoni* Ostenfeld, 1899.
Diagnosis: Rhabdoliths developed as recurved salpingiform cyrtoliths having an elliptical base with radial perforations and an entire margin.
Range: Late Pleistocene–Recent.

Subfamily DISCOSPHAEROIDEAE Boudreaux and Hay, 1969
Coccolithophores bearing tube or trumpet-shaped rhabdoliths.

Genus *BLACKITES* Hay and Towe, 1962
Type species: *Discolithus spinosus* Deflandre and Fert, 1954.
Diagnosis: Circular shields constructed of several cycles of elements, and having a circlet of radial perforations. The central cycles are raised to form a boss or may represent the base of a stem.
Range: Eocene–Oligocene.

Genus *PYROCYCLUS* Hay and Towe, 1962
Type species: *Pyrocyclus inversus* Hay and Towe, 1962.
Diagnosis: Subcircular coccoliths with a strongly inclined marginal cycle of elements and a small circular central opening.
Range: Eocene.

Genus *UMBELLOSPHAERA* Paasche, in Markali and Paasche (1955)
Type species: *Coccolithus tenuis* Kamptner, 1937.

Diagnosis: Coccolithophores bearing dimorphic coccoliths; macro-coccoliths: salpingiform cyrtoliths with elliptical base and circular trumpet; micro-coccoliths: simple elliptical caneoliths.
Range: Recent.

Order ZYGOSPHAERALES, new order
Type family: ZYGOSPHAERACEAE, new family.
Diagnosis: Coccolithophores bearing holococcoliths, constructed entirely of minute hexagonal or rhombohedral microcrystals.

Family ZYGOSPHAERACEAE, new family
Type genus: *Zygosphaera* Kamptner, 1937.
Diagnosis: Coccolithophores bearing zygoform holococcoliths constructed of hexagonal microcrystals.

Genus *ZYGOSPHAERA* Kamptner, 1937b. Pl. 18, fig. 1
Type species *Zygosphaera hellenica* Kamptner, 1937.
Diagnosis: Coccolithophores bearing dimorphic holococcoliths, those of circumflagella region being zygoform, the ordinary holococcoliths being disciform, with or without perforations and a central knob.
Range: Recent.

Genus *HOMOZYGOSPHAERA* Deflandre, 1952a. Pl. 18, fig. 2
Type species: *Corisphaera spinosa* Kamptner, 1941.
Diagnosis: Coccolithophores bearing a single kind of zygoform holococcolith, all having a tiny knob on the jugum.
Range: Recent.

Genus *CORISPHAERA* Kamptner, 1937
Type species: *Corisphaera gracilis* Kamptner, 1937.
Diagnosis: Coccolithophores bearing dimorphic zygoform holococcoliths, those of the circumflagellar cycle bearing a knob-like appendix on the jugum.
Range: Recent.

Genus *HELLADOSPHAERA* Kamptner, 1937
Type species: *Syracosphaera cornifera* Schiller, 1913.
Diagnosis: Coccolithophores bearing dimorphic zygoform holococcoliths, those of the circumflagellar region having a petaloid extension arising from the jugum.
Range: Recent.

Genus *PERIPHYLLOPHORA* Kamptner, 1937. Pl. 18, figs 3a–e
Type species: *Calyptrosphaera mirabilis* Schiller, 1925.
Diagnosis: Coccolithophores bearing a single kind of zygoform holococcolith, all with a petaloid extension of the jugum.
Range: Recent.

Family CALYPTROSPHAERACEAE Prins, 1971

Coccolithophores bearing calyptroform holococcoliths, constructed of hexagonal microcrystals.

Genus *CALYPTROLITHUS* Kamptner, 1948

Type species: *Calyptrolithus galerus* Kamptner, 1948.

Diagnosis: Isolated calyptroliths having an elliptical outline.

Remarks: This taxon has yet to be applied to an object other than the type. Kamptner, 1955, designated *Coccolithites detorquatus* Kamptner, 1953 type species of *Calyptrolithus*, but this was an invalid designation because the type species must be selected from those described at the time the genus was erected: *Calyptrolithus galerus* and *Calyptrolithus hemisphaericus*. The latter was designated type species of *Cyclocalyptra* Kamptner by Kamptner, 1958, so that *C. galerus* is the only available form which can be considered as the type species.

Genus *CYCLOCALYPTRA* Kamptner, 1958

Type species: *Calyptrolithus hemisphaericus* Kamptner, 1958.

Diagnosis: Isolated calyptroliths having a circular outline.

Remarks: This taxon has yet to be applied to an object other than the type. It is placed here because all calyptroliths thus far studied in the electron microscope have been found to be holococcoliths.

Range: Recent.

Genus *CALYPTROSPHAERA* Lohmann, 1902

Type species: *Calyptrosphaera globosa* Lohmann, 1902.

Diagnosis: Coccolithophores bearing a single kind of inflated calyptroform holococcolith with a flat distal floor.

Range: Recent.

Genus *SPHAEROCALYPTRA* Deflandre, 1952a

Type species: *Calyptrosphaera quadridentata* Schiller, 1913.

Diagnosis: Coccolithophores bearing dimorphic calyptroform holococcoliths, those of the circumflagellar region being more inflated and having a deep conical distal floor.

Range: Recent.

Genus *ANTHOSPHAERA* Kamptner, 1937b. Pl. 19, fig. 1

Type species: *Anthosphaera fragaria* Kamptner, 1937.

Diagnosis: Coccolithophores bearing dimorphic calyptroform holococcoliths, those of the circumflagellar region being strongly inflated to form a petaloid distal structure.

Range: Recent.

Genus *PETALOSPHAERA* Lohmann, 1913

Type species: *Petalosphaera grani* Lohmann, 1913.

Diagnosis: Coccolithophores with a wide flaring petaloid structure completely surrounding one end of the oblately spheroidal cell.

Remarks: Only one specimen has been observed and that was lost, so that only sketches from Lohmann's notebook were available for the description of this genus.

Range: Recent.

Genus *LACRYMASPHAERA* Lecal, 1965b

Type species: *Lacrymasphaera angelieri* Lecal, 1965b.

Diagnosis: Coccolithophores bearing flattened ceratiform coccoliths having a flat base and pointed tip. Three large perforations are present along the length of the coccolith.

Range: Recent.

Family ZYGRHABLITHACEAE, new family

Type genus: *ZYGRHABLITHUS* Deflandre, 1959

Diagnosis: Holococcoliths constructed of rhomboidal microcrystals.

Genus *ZYGRHABLITHUS* Deflandre, 1959. Pl. 19, fig. 2

Type species: *Zygolithus bijugatus* Deflandre, in Deflandre and Fert, 1954.

Diagnosis: Isolated coccoliths with an elliptical rim and an x-shaped central structure surmounted by a spine.

Range: Eocene.

Genus *CLATHROLITHUS* Deflandre, in Deflandre and Fert, 1954

Type species: *Clathrolithus ellipticus* Deflandre, in Deflandre and Fert, 1954.

Diagnosis: Elliptical reticules.

Range: Eocene.

Genus *PERITRACHELINA* Deflandre, 1952b

Type species: *Peritrachelina ornata* Deflandre, 1952b.

Diagnosis: Crescentic holococcoliths with thickened inner margin and peripherally bifurcating radial ribs.

Range: Eocene.

Genus *POLYCLADOLITHUS* Deflandre, in Deflandre and Fert, 1954

Type species: *Polycladolithus operosus* Deflandre, in Deflandre and Fert, 1954.

Diagnosis: Polyhedral elements of hexagonal outline with ramifying and anastomosing surficial ridges.

Range: Eocene.

Genus *HOLODISCOLITHUS* Roth, 1970. Pl. 19, figs 3a, b

Type species: *Discolithus macroporous* Deflandre, in Deflandre and Fert, 1954.

Diagnosis: Elliptical single plate with perforations composed of small crystal cubes in the manner of holococcoliths; does not exhibit birefringence under crossed nicols.

Range: Eocene–Oligocene.

Genus *LANTERNITHUS* Stradner, 1962. Pl. 19, figs 4a, b

Type species: *Lanternithus minutus* Stradner, 1962.

Diagnosis: Subhexagonal calyptroform holococcoliths having six wedge-shaped segments; in each segment the microcrystals have a slightly different optical orientation.

Range: Middle Eocene–Early Oligocene.

Genus *NANINFULA* Perch-Nielsen, 1968b. Pl. 19, fig. 5

Type species: *Naninfula deflandrei* Perch-Nielsen, 1968b.

Diagnosis: Circular holococcoliths having the form of a pointed hat.

Range: Eocene.

Genus *DISCOTURBELLA* Roth, 1970

Type species: *Discoturbella moori* Roth, 1970.

Diagnosis: A coccolith consisting of one elliptical shield and high central cupola made of a whorl of elongate crystals.

Range: Oligocene.

Genus *DAKTYLETHRA* Gartner, in Gartner and Bukry, 1969

Type species: *Daktylethra punctulata* Gartner, in Gartner and Bukry, 1969.

Diagnosis: Holococcoliths shaped like a helmet and constructed of numerous minute calcite crystallites about 0.1 μm in diameter. The surface bears large shallow pits and pointed conical protuberances.

Range: Eocene.

Genus *PYXOLITHUS* Deflandre, in Deflandre and Fert, 1954

Type species: *Pyxolithus problematicus* Deflandre, in Deflandre and Fert, 1954.

Diagnosis: Minute elongated polyhedral cup with rugose surface.

Range: Oxfordian.

Genus *ORTHOZYGUS* Bramlette and Wilcoxon, 1967. Pl. 19, figs 6a, b

Type species: *Zygolithus aureus* Stradner, 1962.

Diagnosis: Zygoform holococcoliths having a bridge in the minor axis of the ellipse; the bridge is pierced by a number of symmetrically arranged holes.

Range: Late Eocene–Early Oligocene.

Order COCCOLITHALES Rood *et al.*, 1971

Coccoliths with a marginal area constructed of three cycles of elements;

two cycles extend outward peripherally to form proximal and distal shields, connected by an intermediate cycle forming a tube or girdle (coccolithid rim).

Family ELLIPSAGELOSPHAERACEAE Noël, 1965a

Elliptical or circular coccolithids having a coccolithid rim with crystallites oriented so that both shields produce an interference figure between crossed polarisers. Shield elements strongly imbricate.

Genus *WATZNAUERIA* Reinhardt, 1964. Pl. 20, figs 1a, d

Type species: *Watznaueria angustoralis* Reinhardt, 1964.

Synonyms: *Colvillea* Black, 1964 (homonym of *Colvillea* Bojer and Hooker, 1834) subjective senior synonym because the type, *Tremalithus barnesae* Black, 1966 is probably conspecific with *W. angustoralis*; *Maslovella* Tappan and Loeblich, 1966, nomen. subst. pro *Colvillea* Black with the same type, subjective junior synonym; *Ellipsagelosphaera* Noël, 1965a, subjective junior synonym because the type, *Ellipsagelosphaera frequens*, Noël, 1965a, is closely related to *W. angustoralis*; *Actinosphaera* Noël, 1965a, subjective junior synonym because the type, *Actinosphaera deflandrei* Noël, 1965a, is a *Watznaueria* from which the central cycle of elements on the distal shield has been removed; *Calolithus* Noël, 1965a, subjective junior synonym because the type, *Calolithus martelae* Noël, 1965a, is based on a fragment of a species of *Watznaueria*.

Diagnosis: Elliptical coccoliths with a coccolithid rim and one or more supplementary cycles of elements in the central area of the distal shield.

Range: Bajocian–Maestrichtian.

Genus *ESGIA* Worsley, 1971

Type species: *Esgia junior* Worsley, 1971.

Diagnosis: Circular to elliptical placoliths lacking a distinct central area and possessing many long radial to slightly curved elements.

Range: Tithonian.

Genus *CYCLAGELOSPHAERA* Noël, 1965a. Pl. 20, figs 2a, b

Type species: *Cyclagelosphaera margereli* Noël, 1965a.

Diagnosis: Circular coccoliths with a coccolithid rim and one or more supplementary cycles of elements in the central area of the distal shield.

Range: Bathonian–Aptian.

Genus *FLABELLITES* Thierstein, 1973

Type species: *Flabellites biforaminis* Thierstein, 1973.

Diagnosis: Elements of the inner cycle of the distal shield elongate, one or two clusters of these elements often extend beyond the elliptical periphery of the coccolith.

Range: Albian–Cenomanian.

Genus *BIDISCUS* Bukry, 1968. Pl. 22, fig. 3
Type species: *Bidiscus cruciatus* Bukry, 1968.
Diagnosis: Small circular coccoliths constructed of two monocyclic shields composed of radial elements.
Range: Albian–Campanian.

Genus *SPHENORADIATUS* Worsley, 1971
Type species: *Sphenoradiatus pentagonus* Worsley, 1971.
Diagnosis: Circular placoliths possessing two unequal flat shields with radial wedge-shaped elements and a small open central area.
Range: Early Cretaceous.

Genus *SIMILICORONILITHUS* Bukry, 1969. Pl. 21, fig. 1
Type species: *Similicoronilithus primus* Bukry, 1969.
Diagnosis: Circular coccoliths constructed of two or three planar shields, each composed of a single cycle of inclined elements.
Range: Campanian.

Genus *HELENIA* Worsley, 1971. Pl. 21, figs 2a, b
Type species: *Helenia staurolithina* Worsley, 1971.
Synonym: *Cruciellipsis* Thierstein, 1971, subjective junior synonym because the type species *Coccolithus cuvillieri* Manivit, 1965, is similar to *Helenia staurolithina*.
Diagnosis: Elliptical placoliths having solidly constructed shield with radial to slightly inclined intricate elements. The central area has a cross with bars aligned in the major and minor axes of the ellipse.
Range: Berriasian–Hauterivian.

Family COCCOLITHACEAE Kamptner, 1928b, emend. Hay and Mohler, 1967
Coccolithophores bearing placoliths; proximal shield dark or faintly illuminated in two quadrants between crossed polarisers. Shield elements of placothis strongly imbricate.

Subfamily COCCOLITHOIDEAE Kamptner, 1928b, emend. Hay and Mohler, 1967
Coccolithiphores bearing elliptical placoliths; interference figure of proximal shield laevogyre in distal view.

Genus *COCCOLITHUS* Schwarz, 1894. Pl. 21, figs 3a, b

Synonyms: *Cyatholithus* Huxley, 1868 (objective senior synonym with the same type species (designated by Loeblich and Tappan, 1963): petition to suppress this taxon was submitted to the International Commission on Zoological Nomenclature by Braarud *et al*., 1964, and a petition to suppress this taxon in the Kingdom Plantae also is in preparation); *Coccosphaera* Wallich, 1877 (not *Coccosphaera* Perty, 1862)*; *Cyathosphaera* Haeckel, 1894 (objective junior synonym with the same type species (designated by Hay and Towe, 1962); *Coccolithophora* Lohmann, 1902 (objective junior synonym with the same type species (designated by Loeblich and Tappan, 1963)).

Type species: *Coccosphaera pelagica* Wallich, 1877.

Definitions: Elliptical placoliths, proximal shield smaller than the distal shield; distal shield with calcite elements having c-axes arranged perpendicular to the plane of the shield; interference figure observed in polarised light produced by proximal shield only.

Range: Palaeocene–Recent.

Genus *BISCUTUM* Black, in Black and Barnes, 1959. Pl. 21, figs 4a–c

Type species: *Biscutum testudinarium* Black, in Black and Barnes, 1959 (by original designation).

Diagnosis: Isolated elliptical placoliths with the two shields closely appressed, tube apparently lacking.

Range: Albian–Maestrichtian.

Subfamily CYCLOCOCCOLITHOIDEAE Hay and Mohler ex Boudreaux and Hay, 1969

Placoliths circular, interference figure of proximal shield radial to subradial.

Tribe CYCLOCOCCOLITHEAE Boudreaux and Hay, 1969

Placoliths imperforate.

Genus *MARKALIUS* Bramlette and Martini, 1964. Pl. 22, figs 1a, b

Type species: *Cyclococcolithus leptoporus inversus* Deflandre, in Deflandre and Fert, 1954.

Diagnosis: Placoliths with both proximal and distal shields extinguished between crossed polarisers; the interference figure is produced only by the tube cycle.

Range: Maestrichtian–Eocene.

Genus *CRIBROCENTRUM* Perch-Nielsen, 1972. Pl. 22, fig. 2

Type species: *Cribrocentrum reticulatum* Perch-Nielsen, 1972.

* See Loeblich and Toppan, 1963.

Diagnosis: Circular placoliths with a cribrate or reticulate centre.
Range: Campanian–Maestrichtian.

Genus *CYCLOCOCCOLITHINA* Wilcoxon, 1970. Pl. 22, figs 3a–c

Type species: *Coccosphaera leptopora* Murray and Blackman, 1898.

Synonyms: *Cyclococcolithus* Kamptner, 1954: objective senior synonym with same type species but illegitimate because the type species of *Umbilicosphaera*, Lohmann, 1902, *U. mirabilis*, Lohmann, 1902, was included among the species indicated to belong to the genus; *Calcidicus* Kamptner, 1952: subjective senior synonym because the type species *Calcidicus quadriforatus* Kamptner = fragment of *Cyclococcolithina*; *Tiarolithus* Kamptner, 1958: subjective senior synonym because the type species *Calcidicus medusoides* Kamptner = fragment of *Cyclococcolithina.*

Diagnosis: Coccolithophores bearing circular placoliths having a distal shield larger than the proximal shield, imperforate, but with a central depression on the distal shield, producing the illusion of a perforation in the light microscope.

Range: Oligocene–Recent.

Genus *CONOCOCCOLITHUS* Hay and Mohler, 1967. Pl. 22, fig. 4

Type species: *Conococcolithus minutus* Hay and Mohler, 1967.

Diagnosis: Circular placoliths having a conical outline in side view.

Range: Palaeocene.

Genus *TERGESTIELLA* Kamptner, 1941

Type species: *Tergestiella adriatica* Kamptner, 1941.

Diagnosis: Coccolithophores bearing circular placoliths having a central boss rather than an apparent pore when seen in the light microscope; placoliths probably complex, constructed of at least four cycles of elements.

Range: Recent.

Genus *HEYNECKIA* Lohmann, 1913

Type species: *Heyneckia barkowi* Lohmann, 1913.

Diagnosis: Coccolithophores bearing simple circular placoliths having a broad distal shield and a much smaller proximal shield; the distal shield is concave distally.

Range: Recent.

Genus *NEOSPHAERA* Lecal-Schlauder, 1950

Type species: *Neosphaera coccolithomorpha* Lecal-Schlauder, 1950.

Diagnosis: Coccolithophores bearing circular placoliths having a large proximal shield, a short tube closed proximally, and a very narrow distal shield.

Range: Recent.

Genus *STRIATOCOCCOLITHUS* Bukry, 1971

Type species: *Striatococcolithus pacificanus* Bukry, 1971.

Diagnosis: Circular to subcircular placoliths composed of two simple shields connected at the centre by a small tube; each shield is composed of a single cycle of narrow, essentially radial elements. In cross-polarised light either both the shields and the small central area are dark to faintly visible, or the shields are dark but a tiny central area is bright. The distal shield is distinctly smaller than the proximal shield.

Range: Eocene.

Genus *PEDINOCYCLUS* Bukry and Bramlette, 1971

Type species: *Leptodiscus larvalis* Bukry and Bramlette, 1969.

Diagnosis: Circular coccoliths constructed as a single thin, nearly flat shield, composed of slightly inclined to radial elements that are not clearly imbricate. Central opening relatively small.

Range: Eocene.

Genus *CYCLICARGOLITHUS* Bukry, 1971. Pl. 23, figs 1a, b

Type species: *Coccolithus floridanus* Roth and Hay, in Hay *et al.*, 1967.

Diagnosis: Circular to subcircular placoliths constructed of two shields connected by a central tube that may be closed or open. In plan view, the distal shield is bright between crossed polarisers.

Range: Late Eocene–Early Miocene.

Tribe UMBILICOSPHAEREAE Boudreaux and Hay, 1969

Placoliths broadly perforate.

Genus *UMBILICOSPHAERA* Lohmann, 1902. Pl. 23, fig. 2

Type species: *Umbilicosphaera mirabilis* Lohmann, 1902.

Synonyms: *Craspedolithus* Kamptner, 1963: subjective junior synonym with type species *Craspedolithus declivus* Kamptner = fragment of *Umbilicosphaera*.

Diagnosis: Circular placoliths having a very short tube connecting the larger proximal shield with the smaller distal shield; central perforation large.

Range: Pliocene–Pleistocene.

Genus *CYCLOLITHELLA* Loeblich and Tappan, 1963. Pl. 23, fig. 3

Type species: *Cyclolithus inflexus* Kamptner ex Deflandre, 1952b.

Synonym: *Cyclolithus* Kamptner ex Deflandre (1952b) (not *Cyclolithus* Koenig, 1825).

Diagnosis: Elliptical to subcircular placoliths with a large central opening.

Range: Eocene–Recent.

Genus *ILSELITHINA* Stradner, in Stradner and Adamiker, 1966. Pl. 23, fig. 4

Type species: *Ilselithina iris* Stradner, in Stradner and Adamiker, 1966.

Synonym: *Hayella* Roth, 1969: Subjective junior synonym with type species: *Hayella elegans* Roth = *Ilselithina riis*.

Diagnosis: Circular coccoliths with I-shaped shield elements.

Range: Oligocene.

Genus *CORONOCYCLUS* Hay *et al.*, 1966. Pl. 23, fig. 5

Type species: *Coronocyclus serratus* Hay, Mohler and Wade, 1966.

Diagnosis: Circular coccoliths of serrate outline having two narrow shields and a large open central area.

Range: Upper Eocene–Lower Miocene.

Genus *CORANNULUS* Stradner, 1962. Pl. 23, fig. 6; pl. 24, fig. 1

Type species: *Corannulus germanicus* Stradner, 1962.

Synonym: *Diademopetra* Hay *et al.*, 1966: Subjective junior synonym with type species *Diademopetra luma* Hay *et al.*, 1966 = *Corannulus germanicus*.

Diagnosis: Coccoliths consisting of a single shield forming an open circlet with spines or two circlets having open spaces between the inner and outer cycle.

Range: Eocene.

Family GEPHYROCAPSACEAE, new family

Type genus: *Gephyrocapsa* Kamptner, 1943.

Diagnosis: Elliptical placoliths constructed of elements which are not or only slightly imbricate.

Genus *CRUCIPLACOLITHUS* Hay and Mohler, 1967. Pl. 24, figs 2a, b

Type species: *Heliorthus tenuis* Stradner, 1961.

Diagnosis: Elliptical placoliths having a cruciform central structure aligned in the major and minor axes of the ellipse.

Range: Palaeocene–Oligocene (Recent?).

Genus *CHIASMOLITHUS* Hay *et al.*, 1966. Pl. 24, figs 3a, b

Type species: *Tremalithus oamaruensis* Deflandre, in Deflandre and Fert, 1954.

Diagnosis: Elliptical placoliths having a central cross in the form of a Greek letter χ.

Range: Palaeocene–Eocene.

Genus *RETICULOFENESTRA* Hay *et al.*, 1966. Pl. 24, figs 4a, b

Type species: *Reticulofenestra caucasica* Hay *et al.*, 1966.

Synonym: *Apertapetra* Hay *et al.*, 1966: Subjective synonym because the type species, *Apertapetra samodurovi* Hay *et al.*, 1966 = *Reticulofenestra*; *Dictyococcites* Black, 1967: Subjective junior synonym because the type species *Dictyococcites danicus* Black, 1967 = *Reticulofenestra*.

Diagnosis: Elliptical placoliths having a central area which is either open or spanned by many small laths which may form a reticulum.

Range: Eocene–Pliocene.

Genus *DICTYOCOCCITES* Black, 1967. Pl. 25, fig. 4; pl. 26, fig. 1

Type species: *Dictyococcites danicus* Black, 1967.

Diagnosis: Coccoliths having elliptical shields constructed of many non-imbricate elements. The shields are joined together at the margin of the central area. The centre is spanned by a grid of bars which meet along the major axis of the ellipse. Each bar is continuous with one or more of the elements of the proximal shield.

Range: Late Eocene–Early Miocene.

Genus *GEPHYROCAPSA* Kamptner, 1943. Pl. 25, fig. 1

Type species: *Gephyrocapsa oceanica* Kamptner, 1943.

Diagnosis: Elliptical placoliths with non-imbricate elements, having the central opening spanned by a bridge not aligned with either the major or minor axis of the ellipse.

Range: Late Pliocene–Recent.

Genus *CRENALITHUS* Roth, 1973

Type species: *Coccolithus doronicoides* Black and Barnes, 1961.

Diagnosis: Small elliptical placoliths with non-imbricate or only slightly overlapping elements. Margins of shields serrate. Central area open.

Range: Pliocene–Pleistocene.

Genus *PSEUDOEMILIANIA* Gartner, 1969. Pl. 25, fig. 2

Type species: *Pseudoemiliania lacunosa* Kamptner ex Gartner, 1969.

Diagnosis: Elliptical placoliths with gaps between some of the elements of the distal shield; the elements are joined centrally and may or may not touch peripherally; the gaps are irregularly distributed.

Remarks: Kamptner (1963, Plate 9, Figure 50) described and figured the species *E. lacunosus*, type of the provisionally proposed *Ellipsoplacolithus*. Because provisional names are not valid under the International Code of Botanical Nomenclature (Article 34), the genus and species proposed under it were invalid. Gartner (1969), apparently unaware that the species had never been validated or that *E. lacunosus* had been intended as the type species of *Ellipsoplacolithus*, proposed the new genus *Pseudoemiliania*, designating *E. lacunosus* as its type species. From Gartner's discussion, it is evident that he did not consider the new genus *Pseudoemiliania* to be in any way provisional, and his citation and discussion of the species *P. lacunosus* attributed to Kamptner may be construed to have validated the species. The genus *Ellipsoplacolithus* was never validated, so that Gartner's validation of the species and its assignment to *Pseudoemiliania* are correct.

Range: Late Pliocene–Early Pleistocene.

Genus *EMILIANIA* Hay and Mohler, in Hay *et al.*, 1967. Pl. 25, figs 3a–c

Type species: *Pontosphaera huxleyi* Lohmann, 1902.

Diagnosis: Elliptical placoliths, with the distal shield slightly larger than the proximal shield; elements of the distal and/or proximal shield 'I' shaped.

Range: Late Pleistocene–Recent.

Order DISCOASTERALES, new order

Type family: DISCOASTERACEAE Vekshina, 1959

Diagnosis: All tubular, and disc-, dome-, star-, tower-, or umbrella-shaped nannofossils.

Remarks: Prins (1971) considered these to be descendents of the Early Cretaceous representatives of the genus *Eprolithus* Stover, 1966.

Family DISCOASTERACEAE Tan Sin Hok, 1927

Outline circular or star-shaped; in cross section plate- or umbrella-shaped, composed of radially arranged large crystal units, sometimes with a central stem. In plan view not birefringent except in a few species, where the stem shows a strong birefringence.

Subfamily EUDISCOASTEROIDEAE Prins, 1971

Specimens in side view plate-like to conical, lacking a stem; ray tips are often ornamented.

Genus *DISCOASTER* Tan Sin Hok, 1927. Pl. 26, figs 2a–b

Type species: *Discoaster pentaradiatus* Tan Sin Hok, 1927.

Synonym: *Eudiscoaster* Tan Sin Hok, 1927 objective synonym with same type species.

Diagnosis: *Discoaster sensu strictu*: rayed asteroliths constructed of five segments each having a slightly different optical orientation. *Discoaster sensu lato*: asteroliths.

Range: *S. s.* Late Miocene–Pliocene, *s. l.* Tertiary.

Genus *HEMIDISCOASTER* Tan Sin Hok, 1927. Pl. 26, fig. 3

Type species: *Discoaster molengraaffi* Tan Sin Hok, 1927.

Diagnosis: Six-rayed asteroliths having rays arranged as two sets of three.

Range: Upper Oligocene–Lower Miocene.

Genus *AGALMATOASTER* Klumpp, 1953. Pl. 26, fig. 4

Type species: *Agalmatoaster septemradiatus* Klumpp, 1953.

Diagnosis: Ray tips ornamented; distal sutures straight, proximal sutures sinistrally curved.

Range: Eocene–Oligocene.

Genus *RADIODISCOASTER* Prins, 1971. Pl. 26, figs 5a, b

Type species: *Discoaster helianthus* Bramlette and Sullivan, 1961.

Diagnosis: Rays without ornamentation; distal sutures straight, often oriented in a dextral direction; proximal sutures curved in a sinistral direction.

Range: Palaeocene–Eocene.

Genus *TRUNCODISCOASTER* Prins, 1971. Pl. 26, fig. 6

Type species: *Discoaster adamanteus* Bramlette and Wilcoxon, 1967.

Diagnosis: Rather shallow incisions between the rays; no ornamentation; ray tips blunt; sutures on both sides straight and running from the interray area to the centre of the disc.

Range: Oligocene–Miocene.

Genus *TURBODISCOASTER* Prins, 1971. Pl. 26, fig. 7

Type species: *Discoaster tani* Bramlette and Riedel, 1954.

Diagnosis: Deep incisions between the rays: sutures on the distal side of the disc straight, on the proximal side curved sinistrally, with straight sutures on both sides of the disc; rays a little offset.

Range: Palaeocene–Oligocene.

Genus *CATINASTER* Martini and Bramlette, 1963. Pl. 28, fig. 8

Type species: *Catinaster coalitus* Martini and Bramlette, 1963.

Diagnosis: Asteroliths having rays curved out of the plane of the disc to form a basket-like structure.

Range: Miocene.

Genus *HAYASTER* Bukry, 1973. Pl. 27, fig. 1

Type species: *Discoaster perplexus* Bramlette and Riedel, 1954.

Diagnosis: Flat discoidal fossils of approximately circular outline modified by straight peripheral edges of several triangular-outlined crystallites composing the disc. Crystallites radiate from a simple centre point and are essentially equant. Intercrystallite sutures are simple and straight, although they may be curved on one side of the disc. The disc is non-birefringent and is dark in cross-polarised light in normal plan view. A small, thin, accessory, central disc on well-preserved specimens is birefringent.

Range: Miocene–Recent.

Subfamily HELIODISCOASTEROIDEAE Prins, 1971

Type genus *HELIODISCOASTER* Tan Sin Hok, 1927

Specimens in side view umbrella-shaped; stem generally present; ray tips never ornamented.

Genus *HELIODISCOASTER* Tan Sin Hok, 1927. Pl. 27, fig. 2

Type species: *Discoaster barbadiensis* Tan Sin Hok, 1927.

Diagnosis: Asteroliths having a circular or subcircular outline.
Range: Palaeocene–Eocene.

Genus *GYRODISCOASTER* Stradner, 1961. Pl. 27, figs 3a, b
Type species: *Discoaster saipanensis* Bramlette and Riedel, 1954.
Diagnosis: Asterolith having rays curved in the plane of the disc.
Range: Palaeocene–Eocene.

Genus *CLAVODISCOASTER* Prins, 1971. Pl. 27, fig. 4
Type species: *Discoaster aulakos* Gartner, 1967.
Diagnosis: Ray ornamentation confined to the ray tips; sutures—if visible—straight on both sides of the disc; one star-shaped central knob in the centre of each side of the disc; points of the distal central knob are directed to the interray areas, points of the proximal central knob in line with the rays; sides of the rays more or less parallel; ornamentation at the end of the rays rather heavy.
Range: Oligocene–Pliocene.

Genus *DISCOASTEROIDES* Bramlette and Sullivan, 1961. Pl. 27, figs 5a, b
Type species: *Discoaster kuepperi* Stradner, 1959.
Diagnosis: Asteroliths having a subcircular outline and a stem which produces an interference figure between crossed polarisers.
Range: Palaeocene–Eocene.

Family MARTHASTERITACEAE, new family

Type genus: *MARTHASTERITES* Deflandre, 1959
Rhomboidal to triradiate ortholithid elements.

Genus *MARTHASTERITES* Deflandre, 1959. Pl. 27, figs 6a–c
Type species: *Discoaster* (?) *furcatus* Deflandre, in Deflandre and Fert, 1954.
Diagnosis: Triradiate ortholithid elements lacking structures.
Range: Turonian–Eocene.

Genus *MICULA* Vekshina, 1959. Pl. 27, fig. 8
Type species: *Micula decussata* Vekshina, 1959.
Synonym: *Nannotetraster* Martini and Stradner, 1960, objective junior synonym with type species *Discoaster staurophorus* Gardet, synonymous with the type species of *Micula*.
Diagnosis: Isolated ortholithid nannofossils having an equidimensional rhomboidal outline and displaying a cross with arms ending at the corners of the rhomb when viewed in transmitted light.
Range: Cenomanian–Maestrichtian.

Genus *RHOMBASTER* Bramlette and Sullivan, 1961
Type species: *Rhombaster cuspis* Bramlette and Sullivan, 1961.
Diagnosis: Forms with the basic symmetry of a rhombohedron but with the faces strongly depressed, resulting in extended spine-like corners.
Range: Palaeocene.

Genus *ASSIPETRA* Roth, 1973
Type species: *Micula infracretacea* Thierstein, 1973.
Diagnosis: Subrectangular to suboval calcareous nannofossils characterised by having two sets of flat crystal plates, one set piercing the other at an obtuse angle.
Range: Valanginian–Aptian.

Genus *TROCHASTRITES* Stradner, 1961
Type species: *Discoaster bramletti* Martini, 1958.
Diagnosis: Triradiate ortholithid elements lacking sutures but with peripherally bifurcating ribs.
Range: Eocene.

Family CERATOLITHACEAE Norris, 1965
Coccolithophores bearing ceratoliths.

Genus *CERATOLITHUS* Kamptner, 1950. Pl. 27, figs 9a–b
Type species: *Ceratolithus cristatus* Kamptner, 1950.
Diagnosis: Horseshoe-shaped objects with or without a spine and which may have a blade-like structure on one of the arms; both arms have the same optical orientation.
Range: Late Miocene–Recent.

Genus *CERATOLITHINA* Martini, 1967
Type species: *Ceratolithina hamata* Martini, 1967.
Diagnosis: Ortholithid ceratoliths with one strongly recurved arm.
Range: Albian.

Genus *CERATOLITHOIDES* Bramlette and Martini, 1964
Type species: *Ceratolithoides kamptneri* Bramlette and Martini, 1964.
Diagnosis: Crescentic nannofossils with two curved arms having different optical orientation of the calcite; and with a small knob at their juncture.
Range: Late Cretaceous.

Genus *ANGULOLITHINA* Bukry, 1973
Type species: *Angulolithina arca* Bukry, 1973.
Diagnosis: Two thick calcite limbs joined in the form of a V and making

an angle of less than 90 degrees. In plan view, shows birefringence in cross-polarised light, acting as a single optical unit.
Range: Late Miocene–Early Pliocene.

Family HELIOLITHACEAE Hay and Mohler, 1967
Coccoliths constructed of two abutting shields, lacking a connecting tube; one shield has subradial sutures, the other steeply inclined sutures.

Genus *HELIOLITHUS* Bramlette and Sullivan, 1961. Pl. 27, figs 9a–b
Type species: *Heliolithus riedeli* Bramlette and Sullivan, 1961.
Diagnosis: Coccoliths having two closely appressed conical shields.
Range: Palaeocene.

Family BRAARUDOSPHAERACEAE Deflandre, 1947
Coccolithophores bearing pentaliths.

Genus *BRAARUDOSPHAERA* Deflandre, 1947. Pl. 27, figs 10a–c; pl. 31, fig. 1
Type species: *Pontosphaera bigelowi* Gran and Braarud, 1935.
Synonym: *Eodiscoaster* Martini, 1961: Subjective junior synonym with type species *Eodiscoaster danicus* Martini = *Braarudosphaera*.
Diagnosis: Coccolithophores with pentagonal to subpentagonal pentaliths, each pentalith constructed of five crystal units; pentaliths smooth, united to form a dodecahedral coccosphere.
Range: Cretaceous–Recent.

Genus *MICRANTHOLITHUS* Deflandre 1950. Pl. 28 figs. 2a–e
Type species: *Micrantholithus flos* Deflandre, 1950.
Diagnosis: Pentaliths constructed of five or more elements; sutures between elements extend to corners or tips of the pentalith.
Range: Early Cretaceous, Palaeocene–Eocene.

Genus *PEMMA* Klumpp, 1953. Pl. 28, figs 3a–c
Type species: *Pemma rotundum* Klumpp, 1953.
Diagnosis: Subcircular pentaliths having a pit or depression near the centre of each segment.
Range: Middle–Late Eocene.

Genus *RUCINOLITHUS* Stover, 1966. Pl. 28, fig. 4
Type species: *Rucinolithus hayi* Stover, 1966.
Diagnosis: Coccoliths constructed of five or six overlapping segments.
Range: Cretaceous.

Genus: *QUINQUERHABDUS* Bukry and Bramlette, 1971
Type species *Quinquerhabdus colossicus* Bukry and Bramlette, 1969.
Diagnosis: Pentaliths with five long, blade-like elements tapering abruptly

upward from the heavy and centrally depressed base. The characteristic optical orientation of the five segments in basal view shows the typical orientation of braarudosphaerids, although not clearly because of the high upward taper.

Range: Oligocene.

Genus *VERMICULITHINA* Bukry and Percival, 1971

Type species: *Vermiculithina arca* Bukry and Percival, 1971.

Diagnosis: Curved cylindrical calcareous nannofossils which taper slightly toward the ends. The curving cylindrical wall is constructed of many crystallites.

Range: Oligocene.

Family GONIOLITHACEAE Deflandre, 1957

Pentoliths having a margin of radial elements and a centre of small rhomboidal laths.

Genus *GONIOLITHUS* Deflandre, 1957. Pl. 28, fig. 5

Type species: *Goniolithus fluckigeri* Deflandre, 1957.

Diagnosis: Pentaliths having a double cycle of radial marginal elements and a central area of rhomboidal laths arranged around hexagonal perforations.

Range: Eocene.

Family LITHOSTROMATIONACEAE Deflandre, 1959

Thick triangular nannofossils ornamented by pits.

Genus *LITHOSTROMATION* Deflandre, 1942. Pl. 28, fig. 6

Type species: *Lithostromation perdurum* Deflandre, 1942.

Diagnosis: Thick triangular nannofossils ornamented by pits of varying size.

Range: Eocene.

Genus *TROCHASTER* Klumpp, 1953. Pl. 29, figs 1a, b

Type species: *Trochaster simplex* Klumpp, 1953.

Diagnosis: Hexagonal to six-pointed lenticular elements with surficial ridges and depressions starting from a central triradiate pattern.

Range: Eocene.

Genus *MARTINIASTER* Loeblich and Tappan, 1963

Synonym: *Coronaster* Martini, 1961 (non *Coronaster* Perrier, 1884).

Type species: *Coronaster fragilis* Martini, 1961.

Diagnosis: Multirayed lenticular elements with radial surface sculpture.

Range: Eocene.

Family FASCICULITHACEAE Hay and Mohler, 1967
Short solid cylinders constructed of triangular segments.

Genus *FASCICULITHUS* Bramlette and Sullivan, 1961. Pl. 29, fig. 2
Type species: *Fasciculithus involutus* Bramlette and Sullivan, 1961.
Diagnosis: Short solid cylinders constructed of triangular segments; one end is concave, the other convex; sides are convex and may be ornamented by pits and ridges.
Range: Palaeocene.

Genus *BOMOLITHUS* Roth, 1973
Type species: *Bomolithus elegans* Roth, 1973.
Diagnosis: Fasciculiths consisting of three cycles of elements. The upper cycle is taller than the others and bears a central depression.
Range: Palaeocene.

Family SPHENOLITHACEAE Deflandre, 1952b
Beehive- to spine-shaped sphenoliths having a base constructed of radial elements, and which may have a spine.

Genus *FURCATOLITHUS* Martini, 1965. Pl. 29, fig. 3
Type species: *Furcatolithus distentus* Martini, 1965.
Diagnosis: Sphenoliths with a bifurcating spine.
Range: Eocene–Miocene.

Genus *SPHENOLITHUS* Deflandre, 1952b. Pl. 29, figs 4a–c
Type species: *Sphenolithus radians* Deflandre, 1952b.
Synonym: *Nannoturbella* Brönnimann and Stradner, 1960: Subjective junior synonym with type species *N. moriformis* Brönnimann and Stradner = *Sphenolithus*.
Diagnosis: Beehive-shaped or simple spinose sphenoliths.
Range: Eocene–Pliocene.

Family NANNOCONACEAE
Conical to spheroidal calcareous nannofossils constructed of elements radiating from the axis.

Genus *NANNOCONUS* Kamptner, 1931. Pl. 30, fig. 1
Type species: *Nannoconus steinmanni* Kamptner, 1931.
Diagnosis: Conical, prolate or spheroidal calcareous nannofossils constructed of elements radiating from the axis of symmetry. The axis may be a canal.
Range: Portlandian–Maestrichtian.

Genus *POLYCOSTELLA* Thierstein, 1971

Type species: *Polycostella senaria* Thierstein, 1971.

Diagnosis: Flat cone of radially arranged elements sloping slightly towards the proximal niche and forming six to eight radial ridges on the distal side. Proximal side flat with a niche in the centre.

Range: Tithonian–Berriasian.

FAMILIES *Incertae Sedis*

Family MICRORHABDULACEAE Deflandre, 1963

Cylindrical to spindle-shaped calcareous nannofossils.

Genus *MICRORHABDULUS* Deflandre, 1959. Pl. 29, figs 4a, b

Type species: *Microrhabdulus decoratus* Deflandre, 1959.

Diagnosis: Canaliculate rods constructed of longitudinal circlets of aligned elements.

Range: Turonian–Maestrichtian.

Genus *LITHRAPHIDITES* Deflandre, 1963. Pl. 30, figs 2a–c

Type species: *Lithraphidites carniolensis* Deflandre, 1963.

Diagnosis: Cruciform rods constructed of four blades.

Range: Aptian–Maestrichtian.

Genus *MICRORHABDULOIDES* Deflandre, 1963

Type species: *Microrhabdulus rugosus* Bouche, 1962.

Diagnosis: Canaliculate or non-canaliculate rods having a rounded or angular cross section, constructed of homogenous or heterogenous irregular elements.

Range: Coniacian–Maestrichtian.

Genus *MICRORHABDULINUS* Deflandre, 1963

Type species: *Microrhabdulinus ambiguus* Deflandre, 1963.

Diagnosis: Canaliculate rods having rounded or polygonal cross section and constructed of elements arranged in spirals.

Range: Coniacian–Maestrichtian.

Family TRIQUETRORHABDACEAE Lipps, 1969

Trigonal rods, pointed at both ends.

Genus *TRIQUETRORHABDULUS* Martini, 1965. Pl. 30, figs 3a, b

Type species: *Triquetrorhabdulus carinatus* Martini, 1965.

Diagnosis: Spindle-shaped rods constructed of three longitudinal blades so that the cross section resembles a triangle with concave sides.

Range: Eocene–Miocene.

Family THORACOSPHAERACEAE Deflandre, 1952a

Coccolithophores producing a spherical or subspherical test of fused coccoliths.

Genus *THORACOSPHAERA* Kamptner, 1927. Pl. 30, figs 4a, b

Type species: *Thoracosphaera pelagica* Kamptner, 1927. (=*Scyphosphaera heimi* Lohmann, 1919).

Diagnosis: Coccolithophores with a spherical test composed of interlocking polygonal elements derived from coccoliths.

Range: Maestrichtian–Recent.

Genus *RHABDOTHORAX* Kamptner, 1958. Pl. 30, figs 5a, b

Type species: *Rhabdothorax erinaceus* Kamptner, 1958.

Diagnosis: Coccolithophores bearing interlocking asteroliths, each asterolith surmounted by a spine.

Range: Recent.

Genus *SCHIZOSPHAERELLA* Deflandre and Dangeard, 1938. Pl. 30, fig. 6; pl. 31, fig. 1

Type species: *Schizosphaerella punctulata* Deflandre and Dangeard, 1938.

Synonym: *Nannopatina* Stradner, 1961: Subjective junior synonym with type species *N. grandaeva* Stradner, 1961 = *Schizosphaerella*.

Diagnosis: Pitted subsphaerical calcareous nannofossils consisting of two unequal parts joined along a girdle.

Range: Jurassic.

Genus *FAVOLITHORA* Stradner, 1961

Type species: *Favolithora cyclopia* Stradner, 1961.

Diagnosis: Complexly interlocking calcareous nannofossil fragments.

Remarks: Entire specimens of this genus have not been found or, if they are known, are referred to *Thoracosphaera* Kamptner.

Range: Palaeocene–Eocene.

GENERA *Incertae Sedis*

Genus *OCTOLITHITES* Caratini, 1960. Pl. 31, fig. 2

Type species: *Octolithites lecali* Caratini, 1960.

Diagnosis: Monolamellar disc with lobate outline composed of eight segments with radial sutures.

Range: Late Cretaceous.

Genus *NOELITES* Caratini, 1960

Type species: *Noelites sectus* Caratini, 1960.

Diagnosis: Pentagonal rings composed of numerous segments separated by radial sutures.

Range: Late Cretaeceous.

Genus *TETRALITHUS* Gardet, 1955. Pl. 31, figs 3a, b

Type species: *Tetralithus pyramidus* Gardet, 1955.

Diagnosis: Skeletal elements made of four crystal units.

Range: Upper Cretaceous.

Genus *HEXALITHUS* Gardet, 1955

Type species: *Hexalithus lecali* Gardet, 1955.

Diagnosis: Skeletal elements made of six crystal units.

Range: Santonian–Campanian.

Genus *GUTTILITHION* Stradner, 1962. Pl. 31, fig. 4

Type species: *Guttilithion cassum* Stradner, 1962.

Diagnosis: Irregularly elliptical calcareous nannofossils with a spinose margin and an elongate central opening.

Range: Eocene.

Genus *ORTHORHABDUS* Bramlette and Wilcoxon, 1967. Pl. 31, fig. 5

Type species: *Orthorhabdus serratus* Bramlette and Wilcoxon, 1967.

Diagnosis: Elongate rhabdolith-like objects, tapering to a point. The entire nannofossil is ortholithid, appearing like a single crystal in polarised light.

Range: Miocene.

Genus *BUCCULINUS* Noël, 1957

Type species: *Bucculinus algeriensis* Noël, 1957.

Diagnosis: Subrectangular calcareous plates pierced by one or two elongate to crescentic openings.

Range: Kimmeridgian–Valanginian.

Genus *MINYLITHA* Bukry, 1973

Type species: *Minylitha convallis* Bukry, 1973.

Diagnosis: Thick plate-like individual calcite elements with polygonal outline, depressed central area, and raised rim. Only moderately bright in cross-polarised light, with slight birefringence when rotated with respect to polarisation directions.

Range: Late Miocene.

Genus *GONGYLIS* Hoffmann, 1970a

Type species: *Gongylis salzwedelensis* Hoffmann, 1970a.

Diagnosis: Conical nannofossils with several ribs which flare outward at the base of the cone.

Range: Eocene–Oligocene.

Genus *CONUSPHAERA* Trejo, 1969. Pl. 31, fig. 6

Type species: *Conusphaera mexicana* Trejo, 1969.

Synonym: *Cretaturbella* Roth, 1971, subjective junior synonym because the type species, *Cretaturbella rothi* Thierstein, 1971 = *Conusphaera mexicana* Trejo, 1969.

Diagnosis: Elongate truncate cone consisting of dextrally inclined plates arranged into a dextrally twisted spindle when viewed from the smaller end. A central canal is present. The cone is covered by long, narrow, non-imbricated plates which extend from one end of the cone to the other, i.e. separated by sutures parallel to the median axis.

Range: Tithonian–Aptian.

Genus *HAYELLA* Gartner, 1969. Pl. 31, fig. 7

Type species: *Hayella situliformis* Gartner, 1969.

Diagnosis: A heliolithid calcareous object in the shape of a truncated cone with the smaller end closed. Peripheral rimlike flanges are present at both ends, modified by a constricted lip at the wider, open end.

Range: Late Eocene.

Genus *OOLITHOTUS* Reinhardt, 1968. Pl. 31, fig. 8

Type species: *Oolithotus antillarum* Cohen, 1964.

Diagnosis: Subcircular asymmetrical cocoliths having one large shield and one small shield connected at an excentric point. Optic axes of both shields are approximately at right angles to the plane of the shields.

Range: Late Pleistocene–Holocene.

Genus *ANNULITHUS* Rood *et al.*, in Barnard and Hay, 1974

Type species: *Annulithus arkelli* Rood *et al.*, in Barnard and Hay, 1974.

Diagnosis: A coccolith-like object consisting of a simple ring of a few large, coarse, irregular elements.

Range: Hettangian–Sinemurian.

Genus *ELLIPSODISCOASTER* Boudreaux and Hay, 1969. Pl. 31, fig. 9

Type species: *Ellipsodiscoaster lidzi* Boudreaux and Hay, 1969.

Diagnosis: Elliptical discoaster-like object constructed of numerous petaloid rays radiating from a central area. Reinforcing flanges extend along the rays. The central area is an elevated ellipse composed of elements having heliolithid optical orientation.

Range: Late Pleistocene–Holocene.

UNRECOGNISABLE GENERA

FUSELLINUS Noël, 1957

PSEUDORHABDOSPHAERA, Lecal and Bernheim 1960

AETHIOPITES Caratini, 1963

FIBRILLINUS Noël, 1957
MENNERIUS Lyul'eva, 1967
BRACIOLITHUS Noël, 1957

NAMES APPLIED TO ISOLATED RINGS

CRICOLITHUS Kamptner, 1958
Type species: *Cricolithus multiradiatus* Kamptner, 1958.
CALCITREMA Kamptner, 1963
Type species: *Calcitrema praeceps* Kamptner, 1963.
NOELITES Caratini, 1963
Type species: *Noelites sectus* Caratini, 1963.
LOXOLITHUS Noël, 1965
Type species: *Cyclolithus armilla* Black, in Black and Barnes, 1959
CYCLODISCOLITHUS Kamptner, 1958
Type species: *Cyclodiscolithus leptos* Kamptner, 1958
ELLIPTICOLITHITES Caratini, 1963
Type species: *Cyclolithus ellipticus* Kamptner, 1948

References

Achuthan, M. V. and Stradner, H. (1967). Calcareous nannoplankton from the Wemmelian Stratotype. *Abstracts Papers, Plankt. Conf.*, Geneva, 1.

Arkhangel'ski, A. D. (1912). Verkhnemelovyya otolozheniya vostoka evropeiskoi Rossii (Upper Cretaceous deposits of east European Russia) *Mater. Geol. Ross.* **25,** n. 4, 1–631, pls. 10.

Barnard, T. and Hay, W. W. (1974). On Jurassic coccoliths: a tentative zonation of the Jurassic of southern England and north France. *Eclog. geol. Helvet.* **67,** 563–585, 6 pls., 3 tables, 2 text-figs.

Bernard, F. (1939). Coccolithophorides nouveaux ou peu connus observés à Monaco en 1938. *Arch. Zool. Expér. et Gén.* **81** (Notes et Revue), 33–44, 2 figs.

Black, M. (1964). Cretaceous and Tertiary coccoliths from Atlantic seamounts: *Paleontology* **7,** n. 2, 306–316, pls. 50–53.

Black, M. (1965). Coccoliths. *Endeavour* **24,** n. 93, 131–137, 25 figs.

Black, M. (1967). New names for some coccolith taxa. *Geol. Soc. London, Proc.*, n. 1640, 139–145, 4 text-figs.

Black, M. (1968). Taxonomic problems in the study of coccoliths. *Palaeontology* **11,** 793–813, pls. 143–154.

Black, M. (1971). Coccoliths of the Speeton clay and Sutterby marl. *Proc. York. geol. Soc.* **38,** 381–424, pls. 30–34.

Black, M. and Barnes, B. (1959). The structure of coccoliths from the English Chalk. *Geol. Mag.* **96,** 321–328, pls. 8–12.

Black, M. and Barnes, B. (1961). Coccoliths and discoasters from the floor of the South Atlantic Ocean. *Roy. Micr. Soc.* (*London*), *Jour.*, *ser.* 3, **80,** 137–147, pls. 19–26.

Bouche, P. M. (1962). Nannofossiles calcaires du Lutetien du bassin de Paris. *Rev. Micropaleont.* **5,** 75–103, 4 pls., 32 text-figs., figs A–C.

Boudreaux, J. E. and Hay, W. W. (1969). Calcareous nannoplankton and biostratigraphy of the Late Pliocene–Pleistocene–Recent sediments in the Submarex cores. *Revista Esp. Micropaleont.* **1,** 249–292, pls. 1–10, 1 fig.

Braarud, T. (1960). On the coccolithophorid genus *Cricosphaera* n. gen. *Nytt Mag. Bot.* **8,** 211–212.

Braarud, T., Bramlette, M. N., Deflandre, G., Kamptner, E., Loeblich, Jr., A. R., Martini, E. and Tappan, H. (1964). *Rhabdosphaera* Haeckel, 1894 (Coccolithophorida): proposed validation under the plenary powers and designation of a lectotype for *Coccolithus oceanicus* Schwarz, 1894. Z. N. (S.) 1658. *Bull. Zool. Nomencl.* **21,** 397–400.

Braarud, T., Deflandre, G., Halldal, P. and Kamptner, E. (1955). Terminology, nomenclature, and systematics of the Coccolithophoridae. *Micropaleont.* **1,** 157–159.

Braarud, T. and Fagerland, E. (1946). A coccolithophoride in laboratory culture, *Syracosphaera carterae* n. sp. *Avh. Norske Vid.-Akad. Oslo, Mat.-Naturv. Kl.* **1946**(2), 10, pl. 1.

Braarud, T., Gaarder, K. R. and Grøntved, J. (1953). The phytoplankton of the North Sea and adjacent waters in May 1948. *Rapp. et Proc.-Verb. Cons. Int. Explor. Mer*, **133,** 1–87, pls. A & B, 18 text-figs.

Bramlette, M. N. and Martini, E. (1964). The great change in calcareous nannoplankton fossils between the Maestrichtian and Danian. *Micropaleont.* **10,** 291–322, incl. 7 pls., 1 text-fig.

Bramlette, M. N. and Riedel, W. R. (1954). Stratigraphic value of discoasters and some other microfossils related to Recent Coccolithophores. *J. Paleont.* **28,** 385–403, pls. 38–39, 3 text-figs.

Bramlette, M. N. and Sullivan, F. R. (1961). Coccolithophorids and related nannoplankton of the early Tertiary in California. *Micropaleont.* **7,** 129–188, incl. 14 pls., 1 text-fig.

Bramlette, N. N. and Wilcoxon, J. A. (1967). Middle Tertiary calcareous nannoplankton of the Cipero section, Trinidad, W.I. *Tulane Studies Geol. Paleont.* **5,** 93–131, pls. 1–10.

Brönnimann, P. (1955). Microfossils *incertae sedis* from the Upper Jurassic and Lower Cretaceous of Cuba. *Micropaleont.* **1,** 28–51, pls. 1–2, 10 text-figs.

Brönnimann, P. and Stradner, H. (1960). Die Foraminiferen- und Discoasteridenzonen von Kuba und ihre interkontinentale Korrelation. *Erdoel-Zeitschr.* **76,** 364–369, 44 figs, 2 schemata.

Bukry, D. (1969). Upper Cretaceous coccoliths from Texas and Europe: *Univ. Kansas Paleont. Contr.* **Art. 51** (Protista 2), 79 pp., 40 pls. 1 text-fig., 2 tables.

Bukry, D. (1971). Cenozoic calcareous nannofossils from the Pacific Ocean. *San Diego Soc. Nat. Hist., Trans.* **16,** 303–327, pls. 1–7.

Bukry, D. (1973). Low latitude coccolith biostratigraphic zonation. *In* (Edgar, N. T., *et al.*), *Initial Rep. Deep Sea Drilling Project* **15,** 685–703, 1 table, 2 text-figs. U.S. Government Printing Office, Washington.

Bukry, D. and Bramlette, M. N. (1969). Some new and stratigraphically useful calcareous nannofossils of the Cenozoic. *Tulane Studies Geol. Paleont.* **7,** 131–142, pls. 1–3.

Bukry, D. and Bramlette, M. N. (1971). Validation of *Pedinocyclus* and *Quinquerhabdus*, new calcareous nannoplankton genera. *Tulane Studies Geol. Paleont.* **8,** 122.

Bukry, D. and Percival, Jr., S. F. (1971). New Tertiary calcareous nannofossils. *Tulane Studies Geol. Paleont.* **8,** 123–146, pls. 1–7.

Caratini, C. (1960). Sur la Decouverte de Nannoconides dans le Cenomanien et le Turonian du bassin de Paris. *Soc. Geol. France, C.R.*, 106–7, 6 text-figs.

Caratini, C. (1963). Contribution a l'étude des coccolithes du Cénomanien supérieur et du Turonien de la région de Rouen. *Thèse, Faculté des Sciences, Université d'Alger*. 61 pp., 5 pls., 2 + 7 text-figs.

Čepek, P. and Hay W. W. (1969). Calcareous nannoplankton and biostratigraphic subdivision of the Upper Cretaceous. *Trans. Gulf Coast Assoc. Geol. Socs.* **19,** 323–336, 4 figs.

Christensen, T. (1962). *Botanik, Bind II: Systematisk Botanik, Nr.* 2, Alger. Munksgaard, Copenhagen, 1–178.

Cohen, C. L. D. (1964). Coccolithophorids from two Caribbean deep-sea cores. *Micropaleont.* **10,** 231–250, 6 pls.

Conrad, W. (1926). Recherches sur les flagellates de nos eaux saumatres. 2e partie. *Chrysomonadines. Arch. Protistenk*, **56,** 167–231, pls. 7–9, 28 text-figs.

Deflandre, G. (1939). Les stéphanolithes, représentants d'un type nouveau de coccolithes du Jurassique supérieur. *C. R. Acad. Sci.* (*Paris*) **208,** 1331–1333, 14 figs.

Deflandre, G. (1942). Possibilités morphogenetiques comparées du calcaire et de la silice à propos d'un nouveau type de microfossile calcaire de structure complexe, *Lithostromation perdurum* n.g. n. sp. *C. R. Acad. Sci.* (*Paris*) **214,** 917–919, 9 figs.

Deflandre, G. (1947). *Braarudosphaera* nov. gen., type d'une famille nouvelle de Coccolithophoridés actuels à éléments composites. *Ibid.* **225,** 439–441, 5 figs.

Deflandre, G. (1950). Observations sur les Coccolithophoridés, à propos d'un nouveau type de Braarudosphaeridé, *Micrantholithus*, à éléments clastiques. *C. R. Acad. Sci.* **231,** 1156–1158, 11 figs.

Deflandre, G. (1952a). Classe des Coccolithophoridés (Coccolithophoridae Lohmann, 1902). *In* (Grasse, P. P.), *Traité de zoologie*, **1,** 439–470, figs 339–364 bis. Masson, Paris.

Deflandre, G. (1952b). Classe des Coccolithophoridés: *In* (Piveteau, J.) *Traité de paléontologie* **1,** 107–115, figs 31–79. Masson, Paris.

Deflandre, G. (1953). Hétérogénéité intrinsèque et pluralité des éléments dans les coccolithes actuels et fossiles. *Ibid.* **237,** 1785–1787, 7 figs.

Deflandre, G. (1957). *Goniolithus* nov. gen., type d'une famille nouvelle de Coccolithophoridés fossiles, à éléments pentagonaux non composites. *C. R. Acad. Sci.* (*Paris*) **244,** 2539–2541, 4 figs.

Deflandre, G. (1959). Sur les nannofossiles calcaires et leur systematique. *Rev. Micropaleont.* **2,** 127–152, 4 pls.

Deflandre, G. (1963). Sur les Microhabdulides, famille nouvelle de nannofossiles calcaires. *C. R. Acad. Sci.* (*Paris*) **256,** 3484–3486, 25 text-figs.

Deflandre, G. and Dangeard, L. (1938). *Schizosphaerella*, un nouveau microfossile méconnu du Jurassique moyen et supérieur. *C. R. Acad. Sci.* (*Paris*) **207,** 1115–1117, 6 figs.

Deflandre, G. and Durrieu, L. (1957). Application de la technique d'empreintes de carbone à la systématique des Coccolithophoridés fossiles. *C. R. Acad. Sci.* (*Paris*) **244,** 2948–2951, 2 figs.

Deflandre, G. and Fert, C. (1953a). Application du microscope électronique à l'étude des Coccolithophoridés. Technique et résultats préliminaires. *Bull. Soc. Hist. Nat. Toulouse* **88,** 301–313, 4 pls.

Deflandre, G. and Fert, C. (1953b). Étude des Coccolithophoridés des vases actuelles au microscope électronique. Orientation des particules élémentaires de calcaire en rapport avec les notions d'Heliolithae et d'Ortholithae. *Ibid.* **326,** 328–330, 9 figs.

Deflandre, G. and Fert, C. (1954). Observations sur les Coccolithophoridés actuels et fossiles en microscopie ordinaire et électronique. *Ann. Paléont.* **40,** 115–176, 15 pls., 127 text-figs.

Ellis, C. H., Lohmann, W. H. and Wray, J. L. (1972). Upper Cenozoic calcareous nannofossils from the Gulf of Mexico (Deep Sea Drilling Project, Leg. 1, Site 3) *Quart. J. Colorado Schl. Mines* **67,** n. 3, xii + 103, 18 pls., 1 table, 5 text-figs.

Ehrenberg, C. G. (1854). *Mikrogeologie. Das Erden und Felsen schaffende Wirken des unsichtbar kleinen selbständigen Lebens auf der Erde.* **374,** 31 (index to names); 88 (index to atlas), 40 pls. Voss, Leipzig.

Farinacci, A. (1971). Round table on calcareous nannoplankton. *In* (Farinacci, A., ed.), *Proc. II Plankt. Conf.* **2,** 1343–1369. Tecnoscienza, Roma.

Forchheimer, S. (1968a). Die Coccolithen des Gault-Cenoman, Cenoman und Turon in der Bohrung Höllviken I, Südwest-Schweden. *Sver. Geol. Undersök. Ser. C., Nr.* 635, *Årsbok* **62,** n. 6, pp. 1–84, pls. 1–9, 23 figs.

Forchheimer, S. (1968b). JEM-1000, a new tool in geological science. *JEOL News* **6M,** p. 18.

Gaarder, K. R. and Heimdal, B. R. (1973). Light and scanning electron microscope observations on *Rhabdothorax regale* (Gaarder) Gaarder nov. comb. *Norwegian J. Bot.* **20,** 89–97, figs 1–19.

Gaarder, K. R. and Markali, J. (1956). On the coccolithophorid *Crystallolithus hyalinus* n. gen., n. sp. *Nytt Mag. Bot.* **5,** 1–5, 1 pl., 1 text-fig. ("1957", separate 1956).

Gaarder, K. R. and Ramsfjell, E. (1954). A new coccolithophorid from northern waters. *Calciopappus caudatus* n. gen. n. sp. *Nytt Mag. Bot.* **2,** 155–156, 1 fig.

Gardet, M. (1955). Contribution à l'étude des coccolithes des terrains Néogènes de l'Algérie. *Publ. Serv. Carte Géol. Algérie, ser.* 2, *Bull.* **5,** 477–550, incl. 11 pls., 1 text-fig.

Gartner, S., Jr. (1967). Calcareous nannofossils from Neogene of Trinidad, Jamaica, and Gulf of Mexico. *Univ. Kansas Paleont. Contr., Art.* **48** (Protista 1), 1–56, pls. 1–10.

Gartner, S., Jr. (1968). Coccoliths and related calcareous nannofossils from Upper Cretaceous deposits of Texas and Arkansas. *Univ. Kansas Paleont. Contr. Protista,* **Art 1,** 56 pp., 28 pls., 5 text-figs.

Gartner, S., Jr. (1969a). Two new calcareous nannofossils from the Gulf Coast Eocene. *Micropaleont.* **15,** 31–34, pl. 1.

Gartner, S., Jr. (1969b). Correlation of Neogene planktonic foraminifera and calcareous nannofossil zones. *Trans. Gulf Coast Assoc. Geol. Soc.* **19,** 585–599, pl. 1, 2; 7 figs.

Gartner, S., Jr., and Bukry, D. (1969). Tertiary holoccoliths. *J. Paleont.* **43,** 1213–1221, pl. 139–142.

Gohrbandt, K. (1963). Zur Gliederung des Paläogen im Helvetikum nördlich Salzburg nach planktonischen Foraminiferen. *Mitt. Geol. Ges. Wien* **56,** 1–116, 11 pls.

Gorka, H. (1957). Coccolithophoridae z gornego mastrychtu Polski srodkowej (Les Coccolithophoridés du Maestrichtien supérieur de Pologne) *Acta Paleont. Polon.* **2,** 235–284, 5 pls.

Gran, H. H. and Braarud, T. (1935). A quantitative study of the phytoplankton in the Bay of Fundy and the Gulf of Maine (including observations on hydrography, chemistry and turbidity). *J. Biol. Board Canada* **1,** 279–467, 69 figs.

Gümbel, C. W. (1873). Coccolithen im Eocänmergel. *Neues Jahrb. Min. Geol. u. Paleont.* **1873,** 299–304.

Haeckel, E. (1870). Beiträge zur Plastidentheorie. *Jena Zeitschr. Med. u. Naturw.* **5,** 482–550, pls. 17–18.

Haeckel, E. (1894). *Systematische Phylogenie der Protisten und Pflanzen.* xv, 400. Reimer, Berlin.

Halldal, P. (1954). Comparative observations on coccolithophorids in light and electron microscopes and their taxonomical significance. *Rapp. et Commun. VIII Congr. Int. Bot.* (*Paris*) **17,** 122–124.

Halldal, P. and Markali, J. (1954a). Morphology and microstructure of coccoliths studied in the electron microscope. Observations on *Anthosphaera robusta* and *Calyptrosphaera papillifera. Nytt Mag. Bot.* **2,** 117–119, 2 pls.

Halldal, P. and Markali, J. (1954b). Observations on coccoliths of *Syracosphaera mediterranea* Lohm., *S. pulchra* Lohm., and *S. molischi* Schill. in the electron microscope. *J. Cons. Internat. Expl. Mer.* **19,** 329–336, pls. 1–6.

Halldal, P. and Markali, J. (1955). Electron microscope studies on coccolithophorids from the Norwegian Sea, the Gulf Stream and the Mediterranean. *Avh. Norske Vid.-Akad. Oslo, Mat.-Naturv. Kl.* n, 1, 30, 27 pls.

Hay, W. W. (1964). Utilisation stratigraphique des Discoasterides pour la zonation du Paleocène et de l'Eocéne inferieur: *Mem. Bur. Rech. Geol. et Min.* **28,** 885–889.

Hay, W. W. and Towe, K. M. (1962). Electron microscopic examination of some coccoliths from Donzacq (France). *Eclog. geol. Helv.* **55,** 497–517, 10 pls., 2 text-figs.

Hay, W. W., Mohler, H. P. and Wade, M. E. (1966). Calcareous nannofossils from Nal'hik (Northwest Caucasus). *Eclog. geol. Helvet.* **59,** 379–400, 13 pls.

Hay, W. W. and Mohler, H. P. (1967). Calcareous nannoplankton from Pont Labau, France, and zonation of the Paleocene and lower Eocene. *J. Paelont.* **41** n. 6, 1505–1541, pls. 196–206, 5 text-figs.

Hay, W. W., Mohler, H. P., Roth, P. H., Schmidt, R. R. and Boudreaux, J. E. (1967). Calcareous nannoplankton zonation of the Cenozoic of the Gulf Coast and Caribbean–Antillean area, and trans-oceanic correlation: *Trans. Gulf Coast Assoc. Geol. Soc.* **17,** 428–480, pls. 1–13.

Hay, W. and Sandberg, P. A. (1967). The scanning electron microscope, a major breakthrough for micropaleontology. *Micropaleont.* **13,** 407–418, pls. 1–2.

Hoffmann, N. (1970a). Elektronenmikroskopische Untersuchungen an "stabförmigen" Nannofossilien aus der Kreide und dem Paläogen Norddeutschlands. *Hercynia* **7,** 131–162.

Hoffmann, N. (1970b). *Placozygus*, n. gen. (Coccolithineen) aus der Oberkreide des nördlichen Mitteleuropas. *Geologie* **19,** 1004–1009.

Hoffmann, N. (1970c). Taxonomische Untersuchungen an coccolithineen aus der Kreide Norddeutschlands anhand elektronenmikroskopischer Aufnahmen. *Hercynia, N.F.* **7,** 163–198, pls. 1–10.

Hoffmann, N. (1970d). Coccolithineen aus der weissen Schreibkreide (Unter-Maastricht) von Jasmund auf Rügen. *Geologie* **19,** 846–879, pls. 1–7, 4 figs.

Huxley, T. H. (1868). On some organisms living at great depths in the North Atlantic Ocean. *Quart. J. Micr. Sc.*, Ser. 2, **8,** 203–212, pl. 4.

Joukowski, E. and Charrey, L. (1933). Un hydroclasseur à milieu liquide immobile. *C. R. Soc. Phys. Hist. Nat. Genève* **30,** 209–211.

Kamptner, E. (1927). Beitrag zur Kenntnis adriatischer Coccolithophoriden. *Arch. Protistenk.* **58,** 173–184, 6 figs.

Kamptner, E. (1928a). Über eine Coccolithophoride aus der "Alten Donau" bei Wien, nebst einigen systematischen Bemerkungen. *Arch. Protistenk.* **61,** 38–44, 2 figs.

Kamptner, E. (1928b). Über das System und die Phylogenie der Kalkflagellaten: *Ibid.* **64,** 19–43.

Kamptner, E. (1931). *Nannoconus steinmanni* nov. gen., nov. spec. ein merkwürdiges gesteinsbildendes Mikrofossil aus dem jüngeren Mesozoikum der Alpen. *Paläont. Z.* **13,** 288–298, 3 figs.

Kamptner, E. (1937). Neue und bemerkenswerte coccolithineen aus dem Mittelmeer. *Arch. Protistenk.* **89,** 279–316, pls. 14–17.

Kamptner, E. (1941). Die Coccolithineen der Südwestküste von Istrien. *Ann. Naturh. Mus. Wien.* **51,** 54–149, 15 pls.

Kamptner, E. (1943). Zur Revision der Coccolithineen-Spezies *Pontosphaera huxleyi* Lohm. *Akad. Wiss. Wien, Math.-Naturw. Kl., Anz.* **80,** 43–49.

Kamptner, E. (1948). Coccolithen aus dem Torton des Inneralpinen Wiener Beckens. *Sitzber. Österr. Akad. Wiss., Math.-Naturw. Kl., Abt. I.* **157,** 1–16, 2 pls.

Kamptner, E. (1950). Über den submikroskopischen Aufbau der Coccolithen. *Anz. Osterr. Akad. Wiss. Math.-Naturw. Kl.* **87,** 152–158.

Kamptner, E. (1952). Das mikroskopische Studium des Skelettes der Coccolithineen (Kalkflagellaten). Überischt der Methoden und Ergebnisse. I. Die Gestalt des Gehäuses und seiner Bauelemente. *Mikroskopie* **7,** 232–244, figs. 1–16, II. Der Feinbau der Coccolithen. *Ibid.* **7,** 375–386, figs 17–27.

Kamptner, E. (1953). Zur Frage des geologischen Alters der Coccolithineen und ihrer Eignung für fossile Erhaltung. *Anz. Österr. Akad. Wiss., Math.-Naturw. Kl.* **90,** 184–188.

Kamptner, E. (1954). Untersuchungen über den Feinbau der Coccolithen. *Arch. Protistenk.* **100,** 1–90, 50 figs.

Kamptner, E. (1955). Fossile Coccolithineen-Skelettreste aus Insulinde. Eine mikropaläontologische Untersuchung. *K. Neder. Akad. Wet., Afd. Natuurk. ser.* **2, 50,** n. 2, 1–105, 9 pls.

Kamptner, E. (1958). Betrachtungen zur Systematik der Kalkflagellaten, nebst Versuch einer neuen Gruppierung der Chrysomonadales. *Arch. Protistenk.* **103,** 54–116.

Kamptner, E. (1963). Coccolithineen-Skelettreste aus Tiefseeablagerungen des Pazifischen Ozeans. *Naturhist. Mus. Wien, Ann.* **66,** 139–204, 9 pls., 39 text-figs.
Kay, D. (1965). *Techniques for Electron Microscopy*, xiv, 560 pp. F. A. Davis, Co., Philadelphia.
Klumpp, B. (1953). Beitrag zur Kenntnis der Mikrofossilien des Mittleren und Oberen Eozän. *Paleontolographica* **130A,** 377–406, pls. 16–20, 5 text-figs.
Lafitte, R. and Noël, D. (1967). Sur la formation des calcaires lithographiques. *C. R. Acad. Sci. (Paris)* **264,** 1379–1382, pls. 1, 2.
Lecal, J. (1965a). *Navisolenia.* n.g. *aprili* n.s.p. nouveau genre de calciosolenides. *Bull. Soc. Hist. Nat. Toulouse* **100,** 422–426, 3 pls.
Lecal, J. (1965b). A propos des modalités d'élaboration des formations épineuses des coccolithophoridés. *Protistologica* **1,** 63–70, 2 pls., schema A-L.
Lecal, J. (1967). Le nannoplancton des côtes d'Israël. *Hydrobiologia* **28,** 305–387.
Lecal, J. and Bernheim, A. (1960). Microstructure du squelette de quelques Coccolithophorides. *Bull. Soc. Hist. Nat. Afr. Nord.* **51,** 273–297, 22 pls.
Lecal-Schlauder, J. (1950). Notes préliminaires sur les Coccolithophoridés d'Afrique du Nord. *Bull. Soc. Hist. Nat. Afr. Nord.* **39,** 15–21.
Lecal-Schlauder, J. (1951). Recherches morphologiques et biologiques sur les Coccolithophoridés nord-africains: *Ann. Inst. Océanogr. (Monaco)*, ser. 2, **26,** 255–362, pls. 9–13, 47 text-figs.
Lemmermann, E. (1908). XXI: Flagellatae, Chlorophyceae, Coccosphaerales und Silicoflagellatae. *In* (Brandt, K. and Apstein, C., eds.), *Nordisches Plankton. Botanischer Teil*, 40 pp., 35 figs. Lipsius and Tischer, Kiel and Leipzig.
Lipps, J. H. (1969). *Triquetorhabdulus* and similar calcareous nannoplankton. *J. Paleont.* **43,** 1029–1032, pl. 126.
Loeblich, A. R., Jr. and Tappan, H. (1963). Type fixation and validation of certain calcareous nannoplankton genera. *Proc. Biol. Soc. Wash.*, **76,** 191–196.
Lohmann, H. (1902). Die Coccolithophoridae, eine Monographie der Coccolithen bildenden Flagellaten, zugleich ein Beitrag zur Kenntnis des Mittelmeerauftriebs. *Arch. Protistenk.* **1,** 89–165, pls. 4–6.
Lohmann, H. (1903). Neue Untersuchungen über den Reichthum des Meeres an Plankton und über die Brauchbarkeit der verschiedenen Fangmethoden. Zugleich auch ein Beitrag zur Kenntniss des Mittelmeerauftriebs. *Wiss. Meeresuntersuch. Kiel* **7,** 1–87, pls. 1–4.
Lohmann, H. (1912). Untersuchungen über das Pflanzen und Tierleben der Hochsee. Zugleich ein Bericht über die biologischen Arbeiten auf der Fahrt der "Deutschland" von Bremerhaven nach Buenos Aires in der Zeit von 7. Mai bis 7. September 1911. *Veröff. Inst. Meeresk. Univ. Berlin, N.F., Geogr. Naturw. Reihe* 1. viii + 92 pp., 2 pls., 14 text-figs.
Lohmann, H. (1913). Über Coccolithophoriden. *Verh. Deutsch. Zool. Ges.* **23,** 143–164, 19 figs.
Lohmann, H. (1919). Die Bevölkerung des Ozeans mit Plankton nach den Ergebnissen der Zentrifugenfänge während der Ausreise der "Deutschland" 1911. *Arch. Biont.* **4,** n. 3, 1–617, pp., 16 pls., 113 text-figs.
Lyul'eva, S. A. (1967). Kokolitoforidi turons'kikh vidkladiv Dniprovs'ko-Donets'-koy zapadini. *Geol. Zhurnal* **27,** 91–98, pls. 1–4.
Manivit, H. (1965). Nannofossiles calcaires de l'Albo-Aptien. *Rev. Micropaléont.* **8,** 189–201, 2 pls.

Manivit, H. (1971). *Nannofossiles calcaires du Cretace français (Aptien-Maestrichtien). Essai de biozonation appuyée sur les stratotypes*, 187 pp., 32 pls. Ph.D. Thesis, University Paris, Orsay.
Markali, J. and Paasche, E. (1955). On two species of *Umbellosphaera*, a new marine coccolithophorid genus. *Nytt Mag. Bot.* **4,** 95–100, 6 pls.
Martini, E. (1958). Discoasteriden und verwandte Formen im NW-deutschen Eozän (Coccolithophorida). 1. Taxionomische Untersuchungen. *Senckenb. leth.* **38,** 353–388, incl. 6 pls.
Martini, E. (1961). Nannoplankton aus dem Tertiaer und der obersten Kreide von SW-Frnakreich. *Senck. Leth.* **42,** 1–32, 5 pls., 3 text-figs.
Martini, E. (1965). Mid-Tertiary calcareous nannoplankton from Pacific deep sea cores. *In* (Whittard, W. F. and Bradshaw, R. B., eds.), *Submarine geology and geophysics*, 393–411, pls. 33–37, text-fig. 153. Butterworths, London.
Martini, E. (1967). *Ceratolithina hamata* n. g., n. sp., aus dem Alb von N-Deutschland (Nannoplankton, *incertae sedis*). *Neues Jahrb. Geol. Palaeont., Abh.* **128,** 284–298, pl. 30.
Martini, E. (1969). Nannoplankton aus dem Miozan von Gabon (Westafrika). *N. Jb. Geol. Paleont. Abh.* **132,** 285–300, 26–29, 1 fig.
Martini, E. (1970). Standard Paleogene calcareous nannoplankton zonation. *Nature* **226,** 560–561, 1 fig.
Martini, E. (1971). Standard Tertiary and Quaternary calcareous nannoplankton zonation. *In* (Farinacci, A., ed.), *Proc. II Plankt. Conf.* **2,** 739–785, 4 pls., 6 tables. Tecnoscienza, Roma.
Martini, E. and Bramlette, M. N. (1963). Calcareous nannoplankton from the experimental Mohole drilling. *J. Paleont.* **37,** 845–856, pls. 102–105, 2 text-figs.
Martini, E. and Stradner, H. (1960). *Nannotetraster*, eine stratigraphisch bedeutsame neue Discoasteridengattung. *Erdoel-Zeitschr.* **76,** 266–270, 19 figs.
Martini, E. and Worsley, T. R. (1970). Standard Neogene calcareous nannoplankton zonation. *Nature* **225,** 289–290.
Maslov, V. P. (1956). Iskopaemye izvestkovye vodorosli SSSR (Fossil calcareous algae of the USSR). *Trudy Inst. Geol. Akad. Nauk SSSR* **160,** 1–301, 86 pls., 136 text-figs.
Matthes, H. W. (1956). *Einführung in die Mikropaläontologie*, 348 pp., 1050 figs Hirzel, Leipzig.
Murray, G. and Blackman, V. H. (1898). On the nature of coccospheres and rhabdospheres. *Phil. Trans. Roy. Soc. London* **190B,** 427–441, pls. 15–16.
Murray, J. and Hjort, J. (1912). *The Depths of the Ocean. A general account of the modern science of oceanography based largely on the scientific researches of the Norwegian steamer* Michael Sars *in the North Atlantic*, xx, 821 pp., 9 pls. Macmillan, London.
Murray, J. and Renard, A. F. (1891). Report on deep-sea deposits based on specimens collected during the voyage of H.M.S. *Challenger* in the years 1872 to 1876, pt. **3.** Deep-sea deposits, vii–xxix + 1–525, 29 pls., 35 text-figs.
Noël, D. (1957). Coccolithes des terrains jurassiques de l'Algérie. *Publ. Serv. Carte Géol. Algérie, ser.* 2, *Bull.* **8,** 303–345, incl. 8 pls. (1A, 1B, 2–7). ("1956").
Noël, D. (1965a). Note préliminaire sur les Coccolithes jurassiques. *Cahiers Micropaléont. sér.* 1, no. **1** (Arch. Orig. Centre Document., C.N.R.S. 408), 1–12, figs. 1–60.
Noël, D. (1965b). Sur les coccolithes du Jurassique Européen et d'Afrique du Nord.

Essai de classification des coccolithes fossiles. *Paris, Centre Nat. Rech. Sci.* 209 pp., 29 pls., 74 figs.

Noël, D. (1968). Nature et genèse des alternances de marnes et de calcaires du Barrémien supérieur d'Angles (Fosse Vocontienne, Basses Alpes). *C. R. Acad. Sci. (Paris)* **266,** 1223–1225, pls. 1–2.

Noël, D. (1969a). Etude des coccolithes in situ dans les roches: la notion de nannofacies. *In* (Brönnimann, P. and Renz, H. H., eds.), *Proc. I. Plankt. Conf.* **2,** 486–491, pls. 1–4. E. J. Brill, Leiden.

Noël, D. (1969b). *Arkhangelskiella* (Coccolithes Crétacés) et formes affinés du Bassin de Paris. *Rev. Micropaleont.* **11,** 181–204, pls. 1–3, 3 figs.

Noël, D. (1970). Coccolithes crétacés de la craie campanienne du Bassin de Paris. *Paris, Centre Nat. Rech. Sci.* 128 pp., 48 pls., 23 figs.

Noël, D. (1971). Quelgues coccolithes du Crétace superieur du Bassin de Paris. *In* (Farinacci, A., ed.), *Proc. II Plankt. Conf.* **2,** 475–485, 3 pls. Tecnoscienza, Roma.

Noël, D. (1972). Nannofossiles calcaires de sediments jurassiques finement laminés. *Bull. Mus. Nat. Hist. Nat.*, ser. 3. **75,** 95–155, 15 pls., 9 figs.

Okada, H. and Honjo, S. (1973). The distribution of oceanic coccolithophorids in the Pacific. *Deep-Sea Res.* **20,** 355–374.

Ostenfeld, C. H. (1899). Über Coccosphaera und einige neue Tintinniden im Plankton des nördlichen Atlantischen Oceans. *Zool. Anz.* **22,** 433–439, 2 figs.

Ostenfeld, C. H. (1910). *Thorosphaera*, eine neue Gattung der Coccolithophoriden. *Ber. Deutsch. Bot. Ges.* **28,** 397–400, 5 figs.

Parke, M. (1971). The production of calcareous elements by benthic algae belonging to the Class Haptophyceae (Chrysophyta). *In* (Farinacci, A., ed.), *Proc. II Plankt. Conf.* **2,** 929–937. Tecnoscienza, Roma.

Parke, M. and Adams, I. (1960). The motile (*Crystallolithus hyalinus* Gaarder and Markali) and non-motile phases in the life history of *Coccolithus pelagicus* (Wallich) Schiller. *J. Mar. Biol. Assoc. U.K.* **39,** 263–274, 4 pls.

Parke, M., Manton, I. and Clarke, B. (1955). Studies on marine flagellates II. Three new species of *Chrysochromulina*. *J. Mar. Biol. Assoc. U.K.* **34,** 579–609, 9 pls.

Parke, M. and Dixon, P. S. (1964). A revised check-list of British marine algae. *J. Mar. Biol. Assoc. U.K.* **44,** 499–542, 1 fig.

Parke, M., Manton, J. and Clarke, B. (1955). Studies on marine flagellates. II. Three new species of *Chrysochromulina*. *J. Mar. Biol. Ass. U.K.* **34,** 579–609.

Perch-Nielsen, K. (1967). Eine Präparationstechnik zur Untersuchung von Nannoplankton im Lichtmikroskop und im Elektronenmikroskop. *Medd. Dansk Geol. Foren., Kobenhavn* **17,** 129–130, 1 pl.

Perch-Nielsen, K. (1968a). Der Feinbau und die Klassifikation der Coccolithen aus dem Maastrichtien von Daenemark. *Kong. Danske Videnskab. Selskab., Biol. Skr.* **16,** 1–96, 32 pls., 44 text-figs.

Perch-Nielsen, K. (1968b). *Naninfula*, genre nouveau de Nannofossiles calcaires du Tertiare danois. *C. R. Acad. Sci. (Paris)* **267,** 2298–2300, pls. 1, 2.

Perch-Nielsen, K. (1971). Elektronenmikroskopische Untersuchungen an Coccolithen und Verwandten Formen aus dem Eozän von Dänemark. *Kongl. Danske Videnskab. Sedskab. Biol. Skr.* **18,** n. 3, pp. 1–76, pls. 1–61.

Perch-Nielsen, K. (1972). Remarks on Late Cretaceous to Pleistocene coccoliths from the North Atlantic. *In* (Laughton, A. S. and Berggren, W. A. *et al.*), *Initial*

Rep. of the Deep Sea Drilling Project **14,** 1003–1069, pls. 1–22, 7 tables, 1 text-fig. U.S. Gov. Printing Office, Washington.

Prins, B. (1969). Evolution and stratigraphy of coccolithinids (*sic*) from the Lower and Middle Lias. *In* (Brönnimann, P. and Renz, H. H., eds.), *Proc. I. Plankt. Conf.* **2,** 547–558, pls. 1–3, 1 table. E. J. Brill, Leiden.

Prins, B. (1971). Speculations on relations, evolution, and stratigraphic distribution of discoasters. *In* (Farinacci, A., ed.). *Proc. II Plankt. Conf.* 1017–1037, pls. 1–7. Tecnoscienza, Roma.

Rayns, D. G. (1962). Alternation of generations in a coccolithophorid, *Cricosphaera carterae* (Braarud and Fagerl) Braarud. *J. Mar. Biol. Assoc. U.K.* **42,** 481–484, 2 pls.

Reinhardt, P. (1964). Einige Kalkflagellaten-Gattungen (Coccolithophoriden, Coccolithineen) aus dem Mesozoikum Deutschlands: *Deutsch. Akad. Wiss. Berlin, Monatsber.* **6,** 749–759, 2 pls., 8 text-figs.

Reinhardt, P. (1965). Neue Familien für fossile Kalkflagellaten (Coccolithophoriden, Coccolithineen). *Deutsch. Akad. Wiss. Berlin, Monatsber.* **7,** 30–40, 3 pls., 6 text-figs.

Reinhardt, P. (1967). Fossile Coccolithen mit rhagoidem Zentralfeld (Fam. Ahmuellerellaceae, Subord. Coccolithineae). *Neues Jb. Geol. Paleont., Monatshefte* **163–178,** 12 text-figs.

Rood, A. P. and Barnard, T. (1972). On Jurassic coccoliths: *Stephanolithion, Diadozygus* abd related genera. *Eclog. geol. Helv.* **65,** 327–342, 2 pls., 2 figs.

Rood, A. P., Hay, W. W. and Barnard, T. (1971). Electron microscope studies of Oxford Clay coccoliths. *Eclog. geol. Helv.* **64,** 245–272, pls. 1–5.

Rood, A. P., Hay, W. W. and Barnard, T. (1973). Electron microscope studies of Lower and Middle Jurassic coccoliths. *Eclog. geol. Helv.* **66,** 365–382, pls. 1–3.

Roth, P. H. (1969). Calcareous nannoplankton zonation of Oligocene sections in Alabama (U.S.A.), on the islands of Trinidad and Barbados (W.I.), and the Blake Plateau (E. coast of Florida, U.S.A.). *Eclog. geol. Helv.* **61,** 459–465, 1 pl., 3 figs.

Roth, P. H. (1970). Oligocene calcareous nannoplankton biostratigraphy. *Eclog. geol. Helv.* **63,** 799–881, pl. 1–14, 17 figs.

Roth, P. H. (1973). Calcareous nannofossils—Leg 17, Deep Sea Drilling Project. *In* (Winterer, E. L., Ewing, J. I. *et al.*), *Init. Repts. Deep Sea Drilling Proj.* **17,** 695–795, pls. 1–27.

Roth, P. H. and Thierstein, H. (1972). Calcareous nannoplankton: Leg 14 of the Deep Sea Drilling Project. *In* (Hayes, D. E., Pimm, A. C. *et al.*), *Init. Repts. Deep Sea Drilling Proj.* **14,** 421–485, pls. 1–16.

Roth, P. H., Franz, H. E. and Wise, S. W., Jr. (1971). Morphological study of selected members of the genus *Sphenolithus* Deflandre (*Incertae Sedis*, Tertiary). *Proc. II Plankt. Conf.* 1099–1119, 6 pls., 5 text-figs. Tecnoscienza, Roma.

Schiller, J. (1913). Vorläufige Ergebnisse der Phytoplankton-Untersuchungen auf den Fahrten S.M.S. "Najade" in der Adria 1911/12. I. Die Coccolithophoriden. *SitzBer. K. Akad. Wiss. (Wien), Math.-Naturw. Kl.*, Abt. I, **122,** 597–617, 3 pls.

Schiller, J. (1925). Die planktonisch Vegetationen des adriatischen Meeres. A. Die Coccolithophoriden-Vegetation in den Jahren 1911–14. *Arch. Protistenk.* **51,** 1–130, pls. 1–9, text-figs. A–Y.

Schiller, J. (1926). Über Fortpflanzung, geissellose Gattungen und die Nomenklatur der Coccolithophoraceen nebst Mitteilung über Copulation bei *Dinobryon. Arch. Protistenk.* **53,** 326–342, figs. A–H.

Schiller, J. (1930). Coccolithineae. *In Dr. L. Rabenhorst's Kryptogamen-Flora von Deutschland, Österreich und der Schweiz.* 10. Band, **2,** Abt. 89–267, 137 figs, figs A–F. Akademische Verlagsgesellschaft, Leipzig.

Schlauder, J. (1945). Recherches sur les Flagellés calcaires de la baie d'Alger. *Diplôme Faculté des Sciences, Université d'Alger*, 51 pp., 9 pls.

Schmidt, O. (1870). Über Coccolithen und Rhabdolithen. *SitzBer. K. Akad. Wiss. (Wien), Math.-Naturw. K.* **62** (Abt. I), 669–682, 2 pls.

Schwarz, E. H. L. (1894). Coccoliths. *Ann. Mag. Nat. Hist., ser.* 6 **14,** 341–346, 27 figs.

Shamray, I. A. and Lazareva, E. P. (1956). Paleogenovye Coccolithophoridae i ikh stratigraficheskoe znachenie (Paleogene Coccolithophoridae and their stratigraphic significance). *Dokl. Akad. Nauk SSSR*, **108,** 711–714. 1 pl.

Stein, F. Ritter von (1878). *Der Organismus der Infusionsthiere nach eigenen Forschungen in systematischer Reihenfolge bearbeitet.* 3 Abt. 1 Hälfte, x + 154, 24 pls. Engelmann, Leipzig.

Stosch, H. A. von (1955). Ein morphologischer Phasenwechsel bei einer Coccolithophoridae. *Naturwissenschaften* **42,** 423.

Stover, L. (1966). Cretaceous coccoliths and associated nannofossils from France and the Netherlands. *Micropaleont.* **12,** 133–167, 9 pls., 3 text-figs.

Stradner, H. (1959). Die fossilen Discoasteriden Osterreichs. II Teil. *Erdoel Erdgas Zeitschr.* **75,** 472–488, 77 figs.

Stradner, H. (1961). Vorkommen von Nannofossilien im Mesozoikum und Alttertiaer. *Erdoel Erdgas Zeitschr.* **77,** 77–88, 99 text-figs.

Stradner, H. (1962). Über neue und wenig bekannte Nannofossilien aus Kreide und Alttertiär. *Geol. Bundesanst. Wien., Verh.* **1962,** 363–377, 3 pls.

Stradner, H. (1963). New contributions to Mesozoic stratigraphy by means of nannofossils. *Proc. 6th World Petrol. Congr., Sect.* 1, *Paper* 4, (preprint), 16 pp., 6 pls.

Stradner, H. and Adamiker, D. (1966). Nannofossilien aus Bohrkernen und ihre elektrononmikroskopische Bearbeitung. *Erdoel Erdgas Zeitschr.* **82,** 330–341, pls. 1–3, 16 text-figs.

Stradner, H., Adamiker, D. and Maresch, O. (1968). Electron microscope studies on Albian calcareous nannoplankton from Delft 2 and Leidschenden 1 deepwells, Holland. *Verh. K. Nederl. Akad. Wetensch., Afd. Natuurk., Eerst Reeks,* **24,** n. 4, 1–107, 48 pls. 11 text-figs., 1 table.

Stradner, H. and Edwards, A. R. (1968). Electron microscopic studies on Upper Eocene coccoliths from the Oamaru Diatomite, New Zealand. *Jahrb. Geol. Bundesanstalt, Sonderband* **13,** 1–66, 48 pls., 10 figs.

Stradner, H. and Papp, A. (1961). Tertiäre Discoasteriden aus Osterreich und deren stratigraphischen Bedeutung mit Hinweisen auf Mexico, Rumänien und Italien. *Jahrb. Geol. Bundesanstalt, Sonderband* **7,** 1–159, 42 pls.

Sujkowski, Z. (1931). Petrografja kredy Polski. Kreda z glebokiego wiercenia w Lublinie w porownaniu z kreda niektorych innych obszarow Polski: *Spraw. Polsk. Inst. Geol.* **6,** 485–628, pls. 6–13, 4 text-figs.

Sullivan, F. R. (1964). Lower Tertiary nannoplankton from the California Coast Ranges. I. Paleocene. *Univ. California Publ. Geol. Sci.* **44,** 163–227, 12 pls., 2 figs.

Tan Sin Hok (1927). Discoasteridae *incertae sedis*: *Sect. Sci. K. Akad. Wet. Amsterdam, Proc.* **30,** 411–419, 14 figs.

Tappan, H. and Loeblich, A. R., Jr. (1966). *Maslovella* nom. nov. *Taxon* **15,** 43.

Thierstein, H. (1971). Tentative Lower Cretaceous calcareous nannoplankton zonation. *Eclog. geol. Helv.* **64,** pp. 459–488, 8 pls.
Thierstein, H. R. (1973). Lower Cretaceous calcareous nannoplankton biostratigraphy. *Abh. Geol. Bundesanst.* **28,** 1–52, 6 pls.
Thierstein, H. R., Franz, H. E. and Roth, P. H. (1971). Scanning electron and light microscopy of the same small object. *Micropaleont.* **17,** 501–502.
Trejo, M. (1969). *Conusphaera mexicana*, un nuevo coccolitoforido del Jurasico Superior de Mexico. *Revista Inst. Mexicano Petrol.* **1,** n. 4, 5–15, 5 pls.
Vekshina, V. N. (1959). Kokkolitoforidy maastrikhtskikh otlozhenii Zapado-Sibirskoi nizmennosti (Coccolithophoridae of the Maastrichtian deposits of the west Siberian lowland). *Trudy Sibir. Nauch-Issled. Inst. Geol. Geofiz. i Min. Syr'ya* (*SNIGGIMS*) **2,** 56–81, 2 pls, 1 text-fig.
Wallich, G. C. (1860). Results of soundings in the North Atlantic. *Ann. Mag. Nat. Hist., ser.* 3 **6,** 457–458.
Wallich, G. C. (1861). Remarks on some novel phases of organic life, and on the boring powers of minute annelids, at great depths in the sea. *Ann. Mag. Nat. Hist. ser.* 3 **8,** 52–58, 4, 3 figs.
Wallich, G. C. (1877). Observations on the coccosphere. *Ann. & Mag. Nat. Hist.*, ser. 4, **18,** 342–350, pl. 17.
Wilcoxon, J. A. (1970). *Cyclococcolithina* Wilcoxon nom. nov. (nom. subst. pro *Cyclococcolithus* Kamptner, 1954). *Tulane Studies Geol. Paleontol.* **8,** 82–83.
Worsley, T. R. (1971). Calcareous nannofossil zonation of Upper Jurassic and Lower Cretaceous sediments from the western Atlantic. *In* (Farinacci, A., ed.), *Proc. II Plankt. Conf.* 1301–1321. Tecnoscienza, Roma.

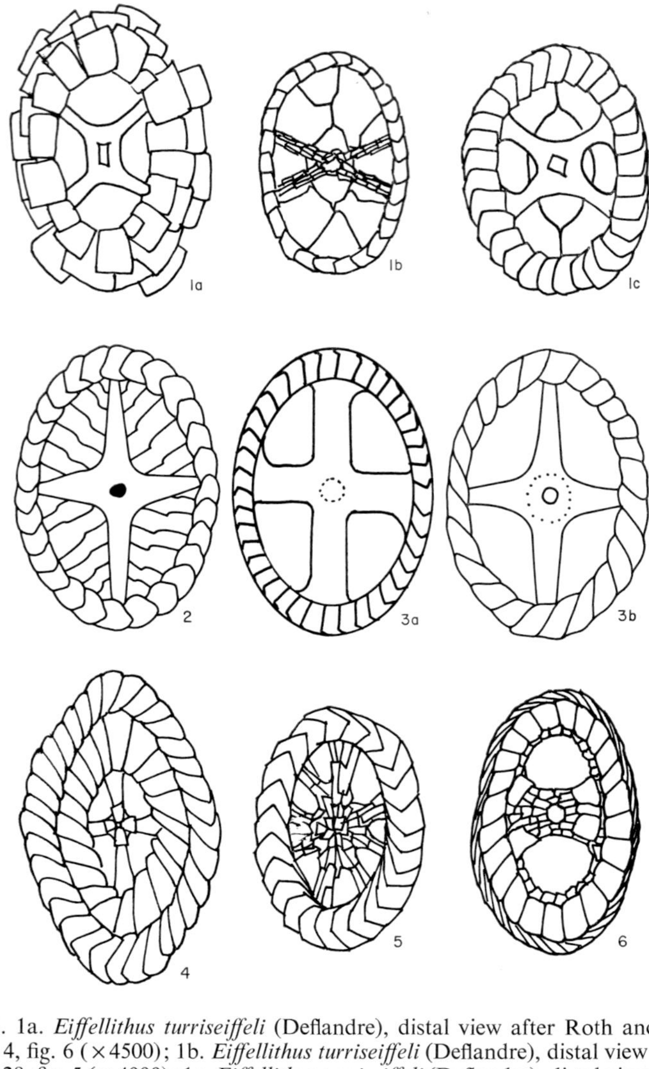

PLATE 1. 1a. *Eiffellithus turriseiffeli* (Deflandre), distal view after Roth and Thierstein (1972), pl. 4, fig. 6 (× 4500); 1b. *Eiffellithus turriseiffeli* (Deflandre), distal view after Bukry (1969), pl. 28, fig. 5 (× 4000); 1c. *Eiffellithus turriseiffeli* (Deflandre), distal view after Thierstein (1971), pl. 7, fig. 8 (× 5000); 2. *Crucirhabdus primulus* Prins ex Rood *et al.* distal view after Rood *et al.* (1973), pl. 1, fig. 1 (× 14 000); 3a. *Vekshinella stradneri* Rood *et al.*, distal view after Rood *et al.* (1971), pl. 1, fig. 2 (× 12 000); 3b. *Vekshinella quadriarculla* Noël, distal view after Rood *et al.* (1971), pl. 1, fig. 1 (× 12 000); 4. *Staurolithites matalosus* (Stover), distal view after Roth and Thierstein (1971), pl. 13, fig. 7 (× 35 000); 5. *Ahmuellerella octoradiata* (Gorka), distal view after Bukry (1969), pl. 33, fig. 5 (× 4500); 6. *Zygodiscus deflandrei* Bukry, distal view, after Bukry (1969), pl. 34, fig. 4 (× 5000).

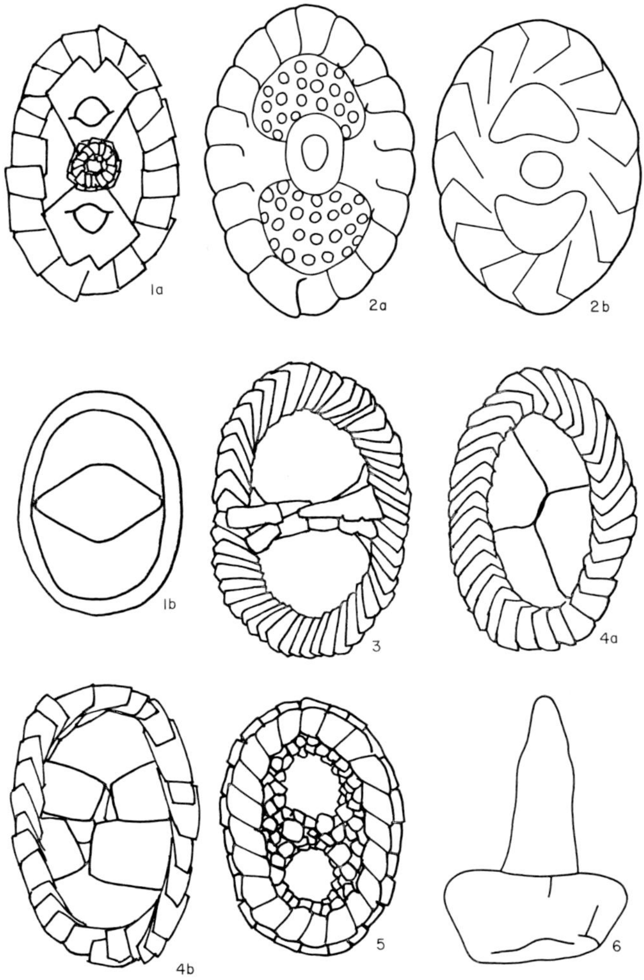

PLATE 2. 1a. *Zygodiscus macleodae* Bukry, distal view after Bukry (1969), pl. 35, fig. 1 (×10 000); 1b. *Zygodiscus adamas* Bramlette and Sullivan, distal view, after Bramlette and Sullivan (1961), pl. 4, fig. 10a (×2500); 2a. *Zeugrhabdotus erectus* (Deflandre), distal view after Rood *et al.* (1971), pl. 1, fig. 3 (×9000); 2b. *Zeugrhabdotus noeli* Rood *et al.*, distal view after Rood *et al.* (1971), pl. 1, fig. 4 (×13 500); 3. *Glaukolithus diplogrammus* (Deflandre), distal view after Roth and Thierstein (1972), pl. 11, fig. 2 (×5000); 4a. *Tranolithus gabalus* Stover, distal view after Roth and Thierstein (1972), pl. 10, fig. 3 (×10 000); 4b. *Tranolithus orionatus* (Reinhardt), distal view after Roth and Thierstein (1972), pl. 10, fig. 12 (×5500); 5. *Amphizygus brooksi brooksi* Bukry, proximal view after Bukry (1969), pl. 25, fig. 3 (×4500); 6. *Parhabdolithus liasicus* Deflandre, side view after Rood *et al.* (1937), pl. 2, fig. 1 (×5000).

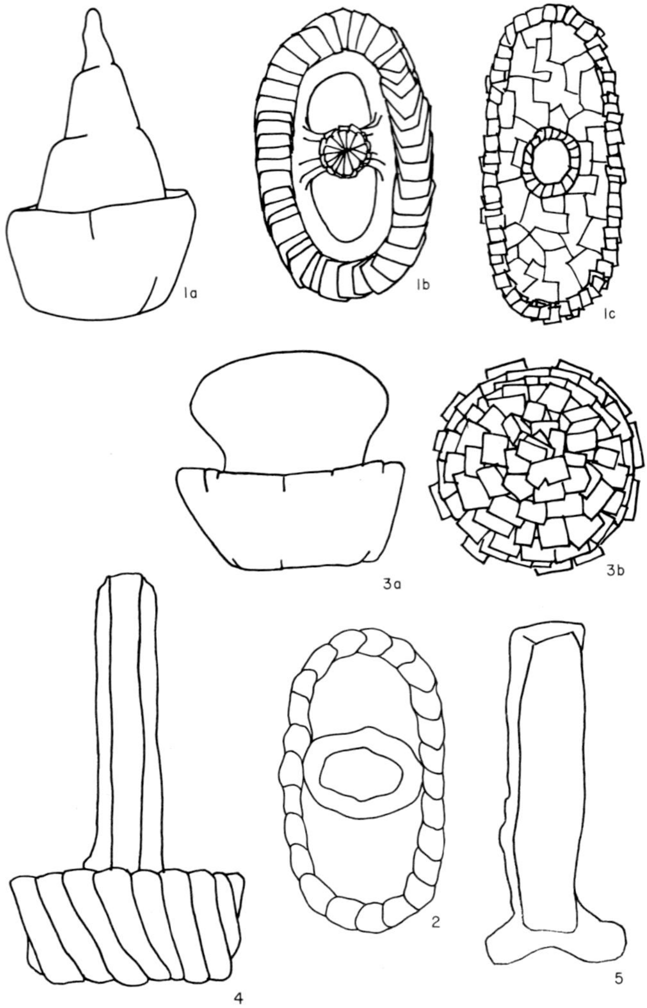

PLATE 3. 1a. *Parhabdolithus marthae* Deflandre and Fert, side view after Rood *et al.* (1973), pl. 2, fig. 2 (×5000); 1b. *Parhabdolithus embergeri* (Noël), distal view after Roth and Thierstein (1972), pl. 9, fig. 6 (×2500); 1c. *Parhabdolithus angustus* (Stradner), distal view after Bukry (1969), pl. 28, fig. 11 (×6500); 2. *Tubirhabdus patulus* Prins ex Rood *et al.*, distal view after Rood *et al.* (1973), pl. 2, fig. 3 (×10 000); 3a. *Mitrolithus elegans* Deflandre, side view after Deflandre and Fert (1954), fig. 67 (×5000); 3b. "*Alvearium dorsetense*" Black, distal view (?) after Black (1965), fig. 5 (×7500); 4. *Eurhabdus luciformis* Reinhardt, side view after Reinhardt (1965), pl. 1, fig. 5 (×8000); 5. *Lucianorhabdus cayeuxi* Dflandre, side view after Gartner (1968), pl. 16, fig. 3 (×2600).

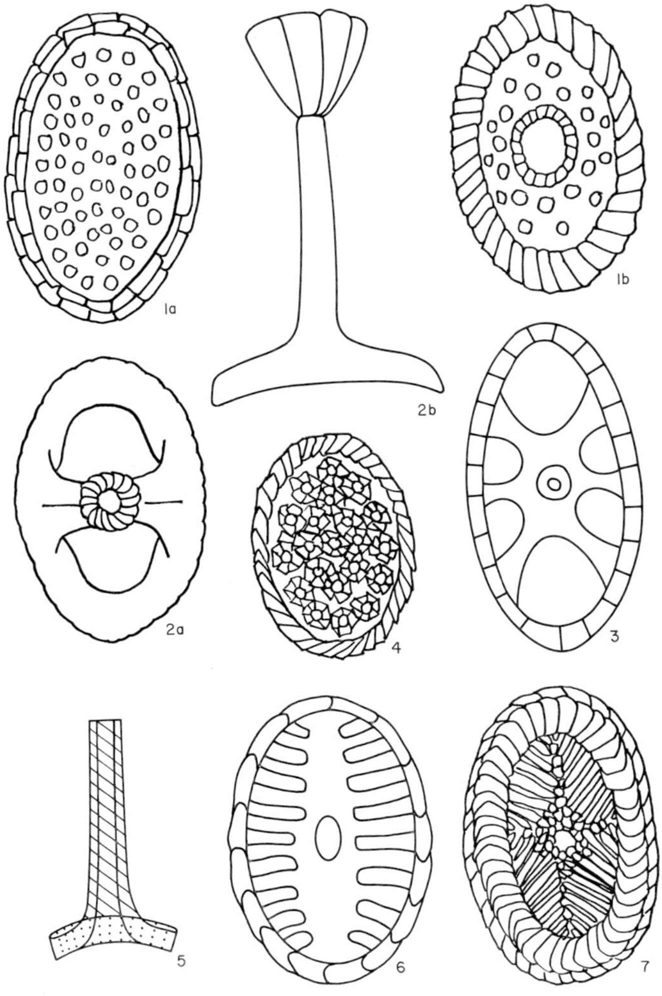

PLATE 4. 1a. *Rhagodiscus asper* (Stradner), distal view after Stradner *et al.* (1968), pl. 24, fig. 1 (×10 000); 1b. *Rhagodiscus asper* (Stradner), distal view after Roth and Thierstein (1972), pl. 7, fig. 12 (×5000); 2a. *Reinhardites anthophorous* (Deflandre), distal view after Stradner (1963) pl. 5, fig. 48 (×8000); 2b. *Reinhardites anthophorus* (Deflandre), side view after Stradner (1963), pl. 5, fig. 46 (×8000); 3. *Actinozygus geometricus* (Gorka) distal view after Rood *et al.* (1971), pl. 1, fig. 6 (×12 500); 4. *Angulofenestrellithus snyderi* Bukey, distal view after Bukry (1969), pl. 26, fig. 1 (×8000); 5. *Rhabdolithina splendens* (Deflandre), side view after Deflandre and Fert (1954), fig. 88 (×4000); 6. *Nobosella clatriata* Prins ex Rood *et al.*, distal view after Rood *et al.* (1973), pl. 1, fig. 9 (×15 000); 7. *Pontilithus obliquicancellatus* Gartner, distal view after Gartner (1968), p. 23, fig. 25 (×5500).

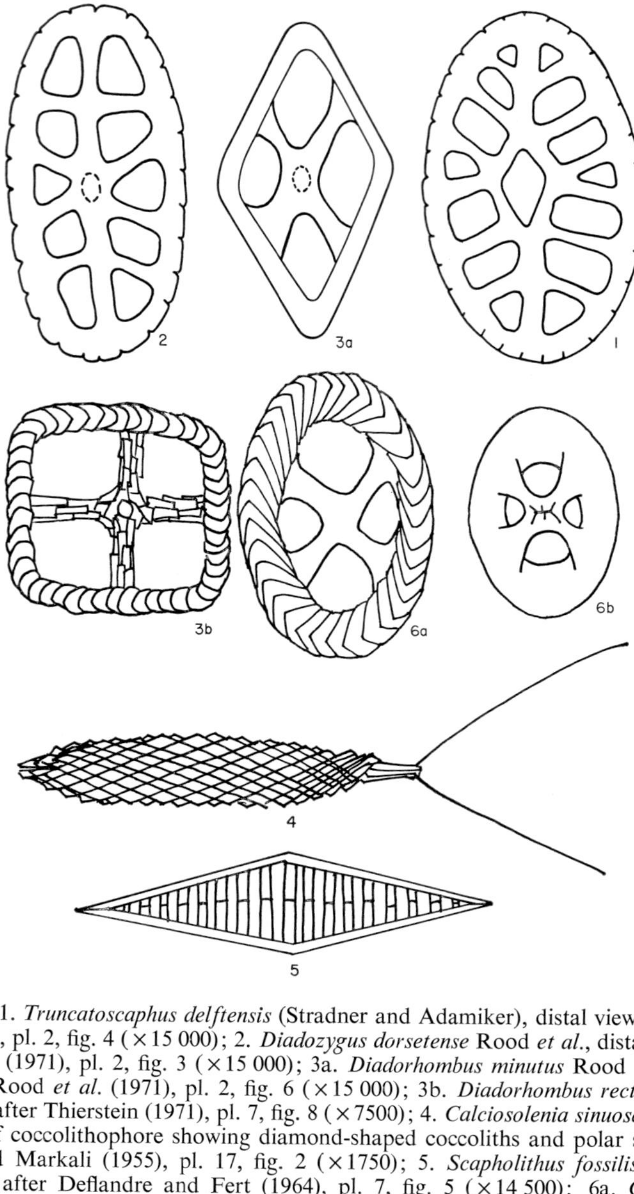

PLATE 5. 1. *Truncatoscaphus delftensis* (Stradner and Adamiker), distal view after Rood *et al.* (1971), pl. 2, fig. 4 (×15 000); 2. *Diadozygus dorsetense* Rood *et al.*, distal view after Rood *et al.* (1971), pl. 2, fig. 3 (×15 000); 3a. *Diadorhombus minutus* Rood *et al.*, distal view after Rood *et al.* (1971), pl. 2, fig. 6 (×15 000); 3b. *Diadorhombus rectus* Worsley, distal view after Thierstein (1971), pl. 7, fig. 8 (×7500); 4. *Calciosolenia sinuosa* Schlauder, side view of coccolithophore showing diamond-shaped coccoliths and polar spines, after Halldal and Markali (1955), pl. 17, fig. 2 (×1750); 5. *Scapholithus fossilis* Deflandre, distal view after Deflandre and Fert (1964), pl. 7, fig. 5 (×14 500); 6a. *Chiastozygus cuneatus* Lyul'eva, proximal view after Roth and Thierstein (1972), pl. 12, fig. 1 (×7500); 6b. *Chiastozygus litterarius* (Gorka), distal view after Thierstein (1971), pl. 2, fig. 21 (×4000).

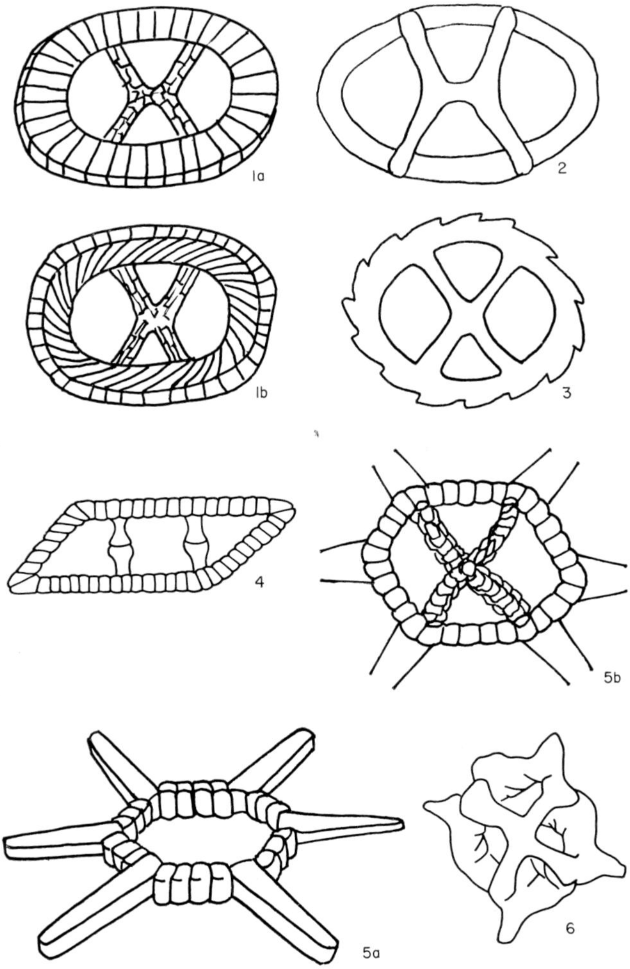

PLATE 6. 1a. *Tegumentum stradneri* Thierstein, proximal view after Roth and Thierstein (1972), pl. 1, fig. 12 (×5000); 1b. *Tegumentum stradneri* Thierstein, distal view after Roth and Thierstein (1972), pl. 1, fig. 11 (×5000); 2. *Neococcolithes concinnus* (Martini), distal view after Perch-Nielsen (1968a), pl. 1, fig. 3 (×1250); 3. *Nannotetrina pappi* Stradner, distal view after Perch-Nielsen (1972), pl. 15, fig. 2 (×3000); 4. *Isthmolithus recurvus* Deflandre, plan view after Hay *et al.* (1966), pl. 12, fig. 1 (×10 500); 5a. *Stephanolithion bigoti* Deflandre, oblique view after Noël (1965), fig. 12 (×2500); 5b. *Stephanolithion bigoti* Deflandre, distal view after Noël (1965), fig. 13 (×2500); 6. *Chiphragmalithus acanthodes* Bramlette and Sullivan, distal view after Bramlette and Sullivan (1961), pl. 10, fig. 1 (×2500).

PLATE 7. 1a. *Corollithion achylosum* (Stover), distal view after Thierstein (1971), pl. 7, fig. 13 (×5000); 1b. *Corollithion signum* Stradner, proximal view after Thierstein (1971), pl. 8, fig. 19 (8000); 1c. *Corollithion acutum* Thierstein, proximal view after Roth and Thierstein (1972), pl. 2, fig. 1 (×5000); 2a. *Eprolithus floralis* (Stradner), plan view after Stradner (1963), pl. 2, fig. 8a (×6000); 2b. *Eprolithus floralis* (Stradner), side view after Stradner (1963), pl. 2, fig. 8a (×6000); 2c. *Eprolithus floralis* (Stradner), plan view after Thierstein (1971), pl. 7, fig. 1 (×6000); 2d. *Eprolithus floralis* (Stradner), plan view after Thierstein (1971), pl. 7, fig. 5 (×6000); 2e. *Eprolithus floralis* (Stradner), oblique view after Thierstein (1971), pl. 7, fig. 5 (×6000); 3a. *Lithastrinus grilli* Stradner, plan view after Stradner (1963), pl. 2, fig. 9 (×6000); 3b. *Lithastrinus grilli* Stradner, side view after Stradner (1963), pl. 2, fig. 9a (×6000); 3c. *Lithastrinus grilli* Stradner, plan view after Bukry (1969), pl. 21, fig. 4 (×7500); 4. *Bukryaster hayi* (Bukry), distal view after Bukry (1969), pl. 38, fig. 12 (×4500).

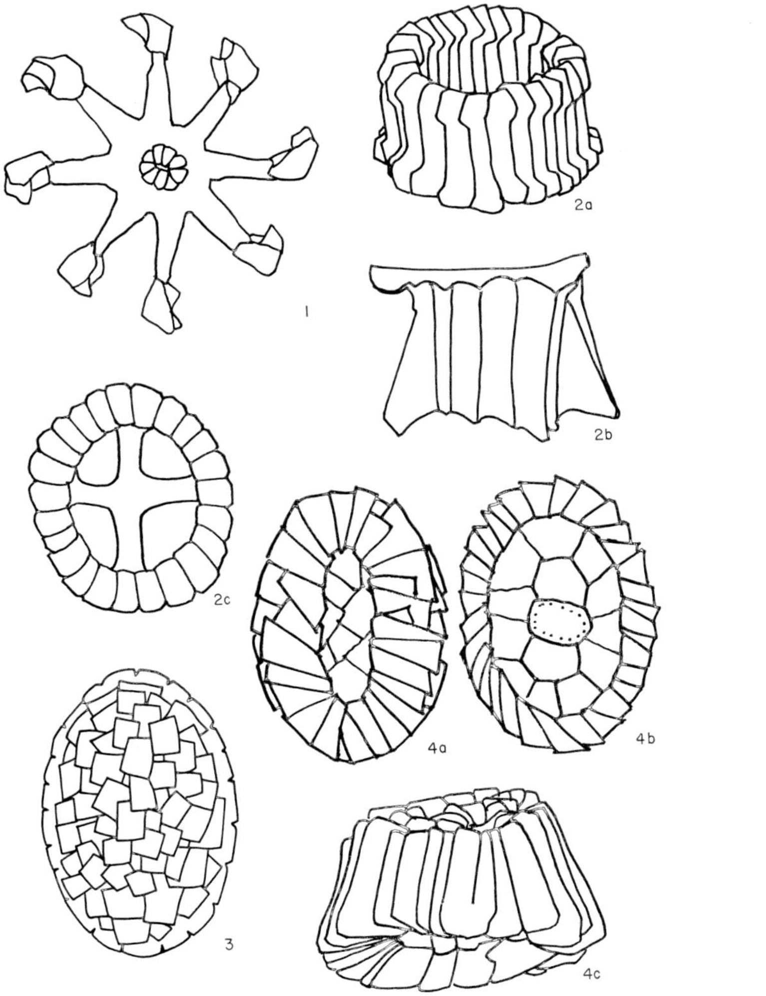

PLATE 8. 1. *Hayesites bulbus* Thierstein, distal view after Thierstein and Roth (1972), pl. 2, fig. 20 (×5000); 2a. *Cylindralithus gallicus* (Stradner), oblique view after Manivit (1971), pl. 10, fig. 13 (×3500); 2b. *Cylindralithus serratus* Bramlette and Martini, side view after Bukry (1969), pl. 20, fig. 12 (×5000); 2c. *Cylindralithus coronatus* Bukry, distal view after Roth and Thierstein (1972), pl. 13, fig. 2 (×5000); 3. *Calcicalathina oblongata* (Worsley), distal view after Thierstein (1971), pl. 4, fig. 10 (×3500); 4a. *Crepidolithus crucifer* Prins, distal view after Rood *et al.* (1973), pl. 2, fig. 4 (×6500); 4b. *Crepidolithus cavus* Prins, distal view after Rood *et al.* (1973), pl. 2, fig. 5 (×8000); 4c. *Crepidolithus crassus* (Deflandre), oblique view after Noël (1965b), fig. 20 (×7500).

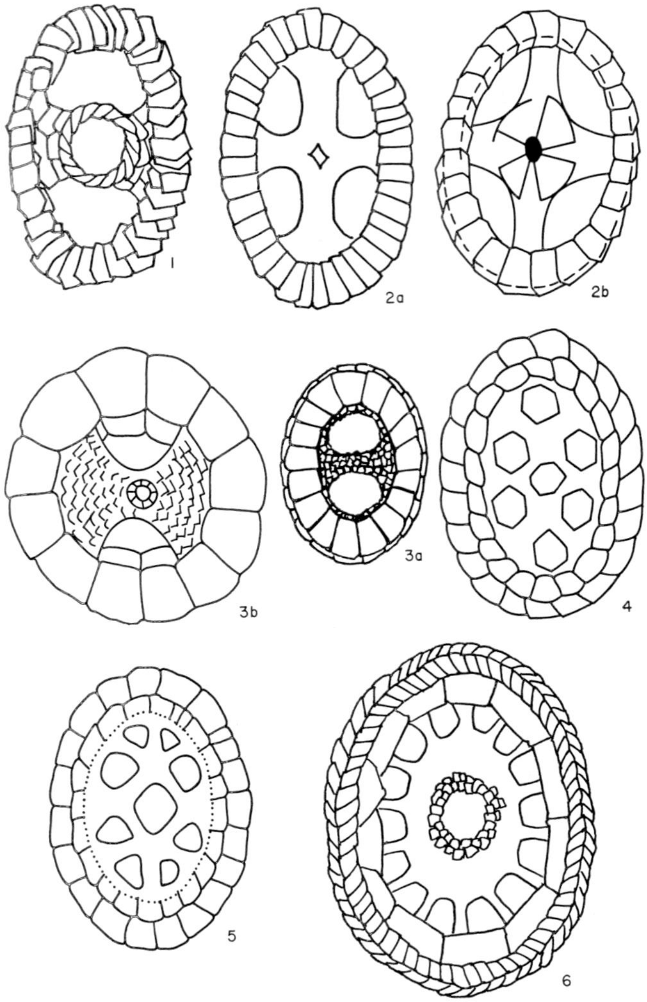

PLATE 9. 1. *Mitosia infinitae* Worsley, distal view after Roth and Thierstein (1972), pl. 8, fig. 7 (×3500); 2a. *Podorhabdus orbiculofenestrus* (Gartner), distal view after Roth and Thierstein (1972), pl. 6, fig. 1 (×5500); 2b. *Podorhabdus cylindratus* Noël, distal view after Rood *et al.* (1973), pl. 3, fig. 1 (×6500); 3a. *Bipodorhabdus roegli* Thierstein, proximal view after Thierstein (1971), pl. 1, fig. 8 (×3500); 3b. "*Discorhabdus*" *biperforatus* Rood *et al.*, distal view after Rood *et al.* (1973), pl. 3, fig. 7 (×5000). The circular base of this species distinguished it from the type of *Bipodorhabdus*, but it may more properly belong to that genus rather than *Discorhabdus*; 4. *Hexapodorhabdus cuvillieri* Noël, distal view after Rood *et al.* (1971), pl. 3, fig. 3 (×6000); 5. *Octopodorhabdus decussatus* (Manivit) proximal view after Rood *et al.* (1971), pl. 3, fig. 4 (×35 000); 6. *Dodekapodorhabdus noelae* Perch-Nielsen, distal views after Perch-Nielsen (1968a), pl. 9, fig. 5 (×3000).

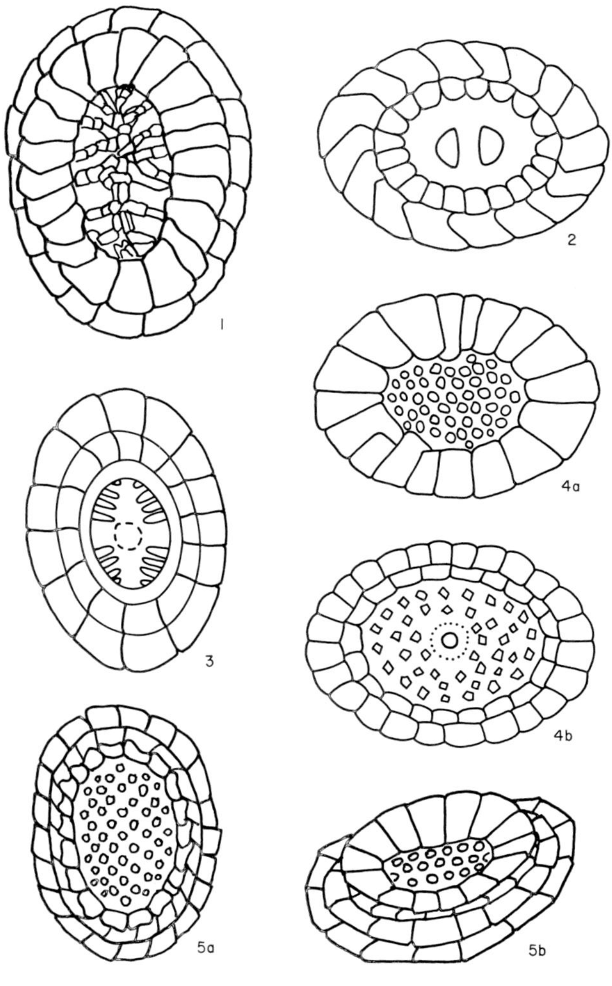

PLATE 10. 1. *Polypodorhabdus escaigi* Noël, proximal view after Noël (1965b), fig. 32; 2. *Paleopontosphaera dubia* Noël, distal view after Rood *et al.* (1971), pl. 4, fig. 9 (×10 500); 3. *Striatomarginis primitivus* Prins ex Rood *et al.*, distal view after Rood *et al.* (1973), pl. 3, fig. 4 (×12 500); 4a. *Ethmorhabdus anglicus* Rood *et al.*, proximal view after Rood *et al.* (1971), pl. 3, fig. 8 (×8000); 4b. *Ethmorhabdus gallicus* Noël, distal view after Rood *et al.* (1971), pl. 3, fig 7 (×6000); 5a. *Cribosphaera ehrenbergi* (Arkhangel'ski), proximal view after Bukry (1969), pl. 22, fig. 11 (×5000); 5b. *Cribosphaera ehrenbergi* (Arkhangel'ski), oblique proximal view after Manivit (1971), pl. 7, fig. 6 (×3500).

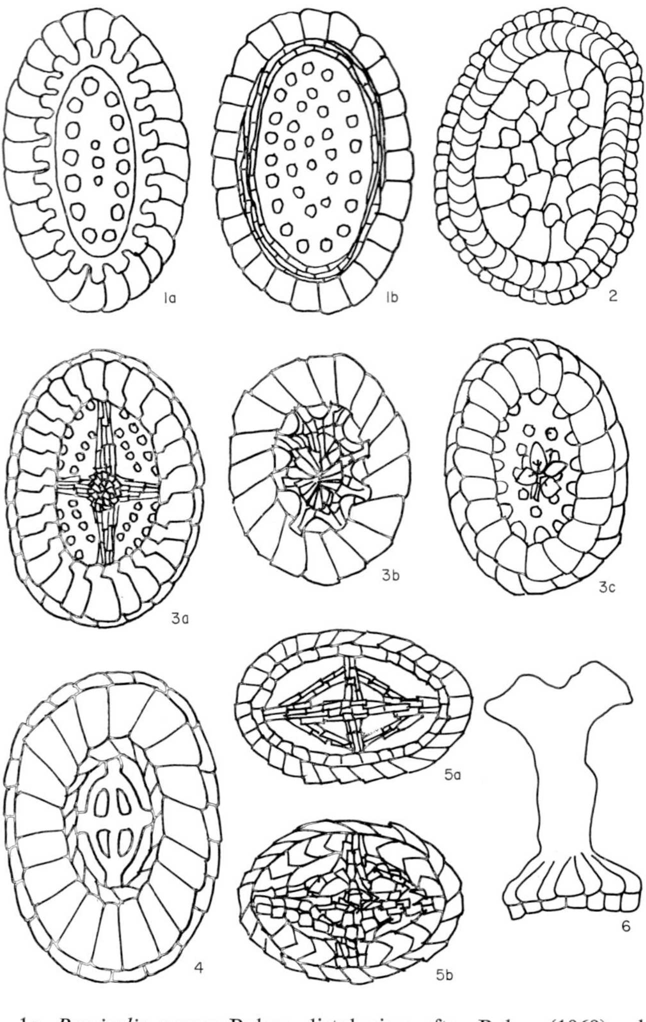

PLATE 11. 1a. *Percivalia porosa* Bukry, distal view after Bukry (1969), pl. 31, fig. 2 (×5000); 1b. *Percivalia porosa* Bukry, proximal view after Bukry (1969), pl. 31, fig. 3 (×5000); 2. *Nephrolithus frequens* Gorka, distal view after Perch-Nielsen (1968a), pl. 18 fig. 6 (×7500); 3a. *Cretarhabdus loriei* Gartner, distal view after Bukry (1969), pl. 15, fig. 3 (×5000); 3b. *Cretarhabdus crenulatus* Bramlette and Martini, distal view after Thierstein (1971), pl. 5, fig. 13 (×4000); 3c. *Cretarhabdus conicus* Bramlette and Martini, proximal view after Thierstein (1971), pl. 6, fig. 7 (×5000); 4. *Sollasites lowei* (Bukry), distal view after Bukry (1969), pl. 22, fig. 6 (×7500); 5a. *Heteromarginatus wallacei* Bukry, proximal view after Bukry (1969), pl. 29, fig. 7 (×4500); 5b. *Heteromarginatus wallacei* Bukry, distal view after Bukry (1969), pl. 29, fig. 6 (×7000); 6. *Discorhabdus patulus* Noël, side view after Noël (1965b), fig. 57 (×12 000).

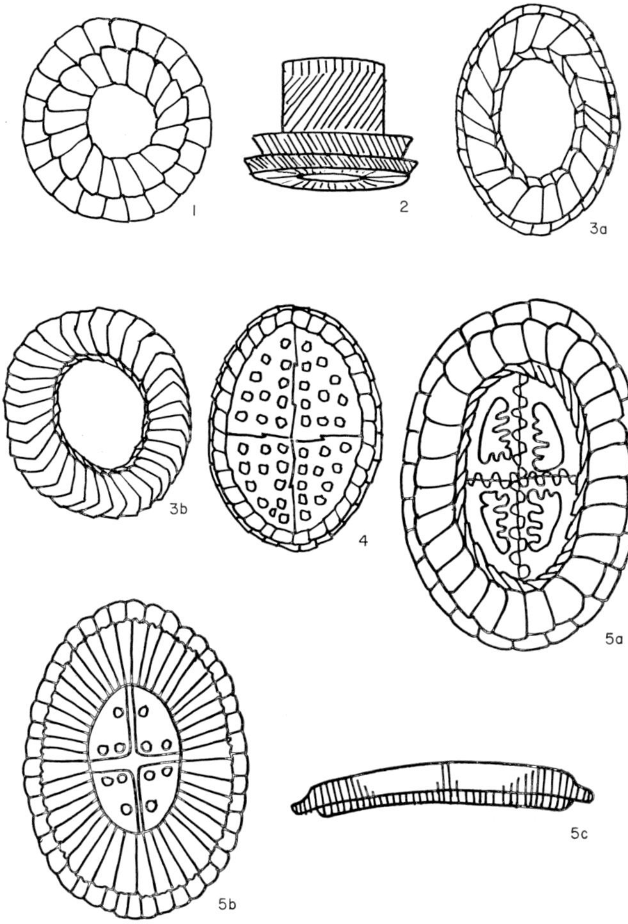

PLATE 12. 1. *Diazomatolithus lehmani* Noël, plan view after Thierstein (1971), pl. 3, fig. 11 (×7000); 2. *Tubodiscus veternae* Thierstein, side view after Thierstein (1973), pl. 2, fig. 3 (×2500); 3a. *Manivitella pemmatoidea* (Deflandre ex Manivit), proximal view after Thierstein (1971), pl. 5, fig. 1 (×2500); 3b. *Manivitella gronosa* (Stover) n. comb (=*Cyclolithus gronosus* Stover, 1966), distal view after Bukry (1959), pl, 6, fig. 6 (×2000); 4. *Arkhangelskiella specillata ethmopora* Bukry, distal view after Bukry (1969), pl. 1, fig. 6 (×3500); 5a. *Broinsonia dentata dentata* (Bukry), distal view after Bukry (1969), pl. 32, fig. 3 (×10 000); 5b. *Broinsonia parca* (Stradner), distal view after Stradner (1963), pl. 1, fig. 3 (×8000); 5c. *Broinsonia parca* (Stradner), side view after Stradner (1963), pl. 1, fig. 3a (×8000).

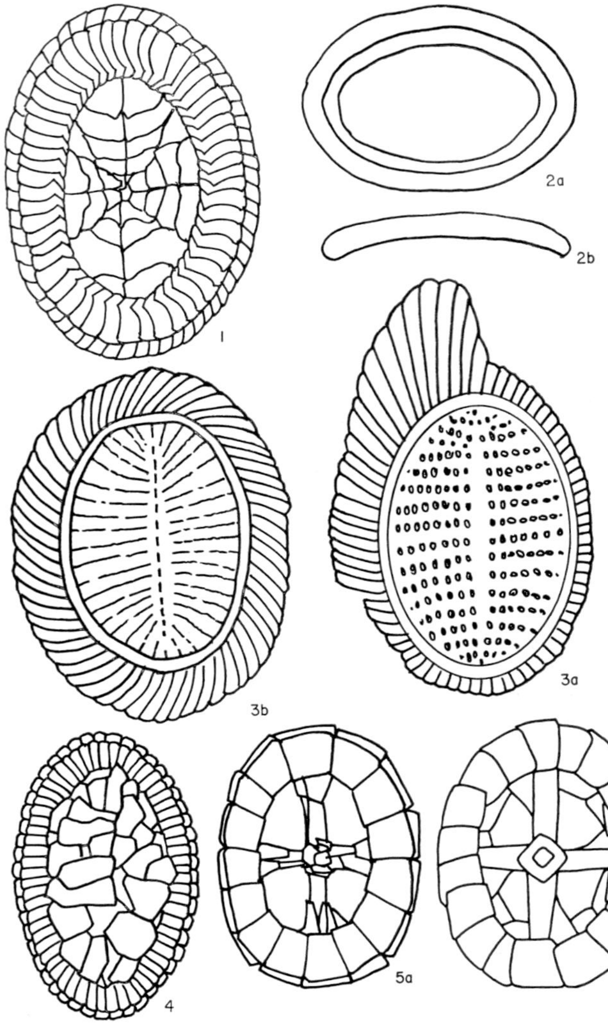

PLATE 13. 1. *Laffittius obliquus* (Stradner), distal view after Perch-Nielsen (1968a), pl. 19, fig. 3 (×4000); 2a. *Prolatipatella multicarinata* Gartner, distal view after Gartner (1968), pl. 7, fig. 11 (×3750); 2b. *Prolatipatella multicarinata* Gartner, side view after Gartner (1968), pl. 7, fig. 10d (×3750); 3a. *Kamptnerius punctatus* Stradner, distal view after Stradner (1963), pl. 2, fig. 3 (×10 000); 3b. *Kamptnerius magnificus* Deflandre, distal view after Stradner (1963), pl. 2, fig. 2 (× 10 000); 4. *Gartnerago concavum* (Gartner), distal view after Bukry (1969), pl. 4, fig. 2 (×4500); 5a. *Prediscosphaera spinosa* (Bramlette and Martini), distal view after Bukry (1969), pl. 18, fig. 7 (×6500); 5b. *Prediscosphaera cretacea* (Arkhangel'ski), distal view after Perch-Nielsen (1968a), pl. 13, fig. 1 (×3000).

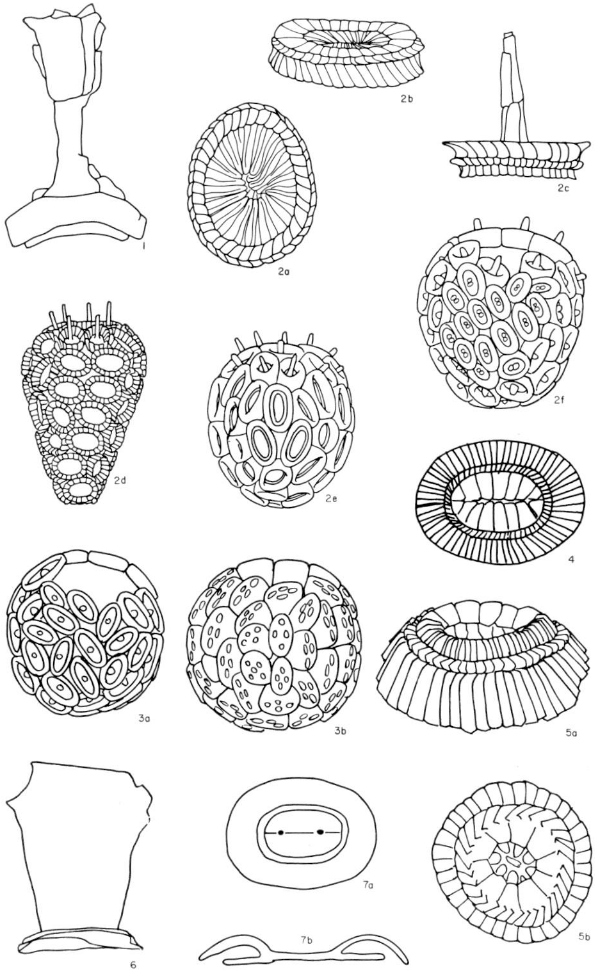

PLATE 14. 1. *Prediscosphaera cretacea* (Arkhangel'ski), side view after Perch-Nielsen (1968a), pl. 14, fig. 1 (×2620); 2a. *Syracosphaera pulchra* Lohmann, proximal view after Ellis *et al.* (1972), pl. 8, fig. 4 (×3500); 2b. *Syracosphaera pulchra* Lohmann, oblique proximal view after Ellis *et al.* (1972), pl. 8, fig. 5 (×4200); 2c. *Syracosphaera pulchra* Lohmann, side view after Ellis *et al.* (1972), pl. 8, fig. 6 (×4900); 2d. *Syracosphaera pulchra* Lohmann, entire coccosphere, after Kamptner (1937), pl. 15, fig. 12 (×1400); 2e. *Syracosphaera tuberculata* Kamptner, entire coccosphere, after Kamptner (1937), pl. 15, fig. 17 (×1750); 2f. *Syracosphaera binodata* Kamptner, entire coccosphere, after Kamptner (1937), pl. 14, fig. 7 (×1750); 3a. *Syracolithus catilliferus* (Kamptner), entire coccosphere, after Kamptner (1937), pl. 14, fig. 10 (×1750); 3b. *Syracolithus quadriperforatus* (Kamptner), entire coccosphere, after Kamptner (1937), pl. 15, fig. 15 (×1750); 4. *Prinsius bisulcus* (Stradner), distal view after Hay and Mohler (1967), pl. 197, fig. 6 (×5500); 5a. *Toweius callosus* Perch-Nielsen, oblique view after Perch-Nielsen (1972), p. 7, fig. 6 (×3500); 5b. *Toweius craticulus* Hay and Mohler, distal view after Hay and Mohler (1967), pl. 197, fig. 2 (×4250); 6. *Bramletteius serraculoides* Gartner, side view after Gartner (1969), pl. 1, fig. 1 (×3150); 7a. *Ellipsolithus macellus* (Bramlette and Sullivan), distal view after Bramlette and Sullivan (1961), pl. 7, fig. 13d (×1750); 7b. *Ellipsolithus macellus* (Bramlette and Sullivan), side view after Stradner in Gohrbandt (1963), text fig. 3/36 (×2800).

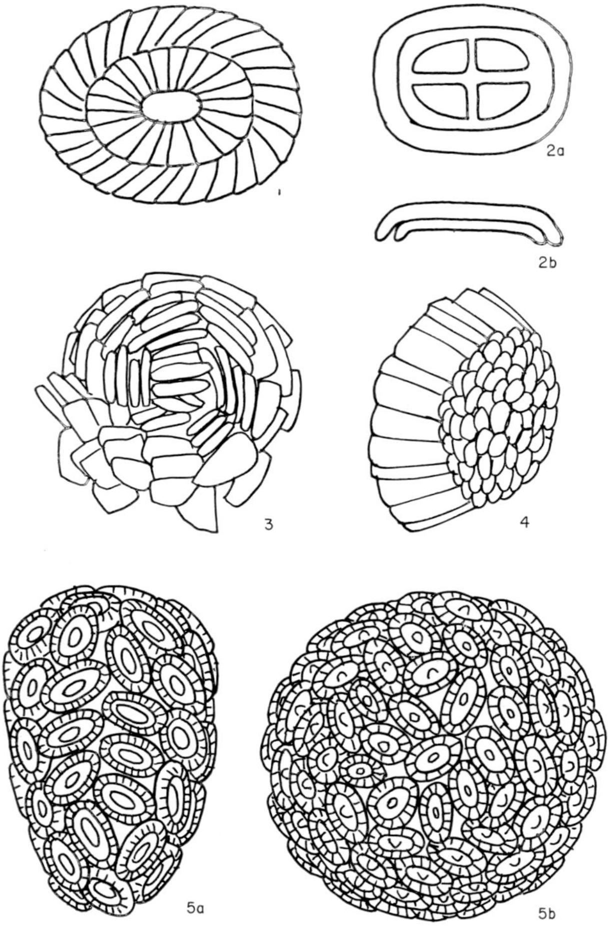

PLATE 15. 1. *Ericsonia ovalis* Black (1964), distal view after Perch-Nielsen (1971), fig. 5 (×5000); 2a. *Campylosphaera dela* (Bramlette and Sullivan), distal view after Bramlette and Sullivan (1961), p. 7, fig. 1a (×2500); 2b. *Campylosphaera dela* (Bramlette and Sullivan), side view after Bramlette and Sullivan (1961), pl. 7, fig. 1b (×2500); 3. *Florisphaera profunda* Okada and Honjo, distal view after Okada and Honjo (1973), pl. 2, no. 4 (×2000); 4. *Thorosphaera flabellata* Okada and Honjo, oblique view after Okada and Honjo (1973), pl. 2, no. 3 (×2000); 5a. *Pontosphaera pietschmanni* Kamptner, entire coccosphere after Kamptner (1937), pl. 14, fig. 5 (×4000); 5b. *Pontosphaera achillea* Kamptner, entire coccosphere after Kamptner (1937), pl. 14, fig. 1 (×3500).

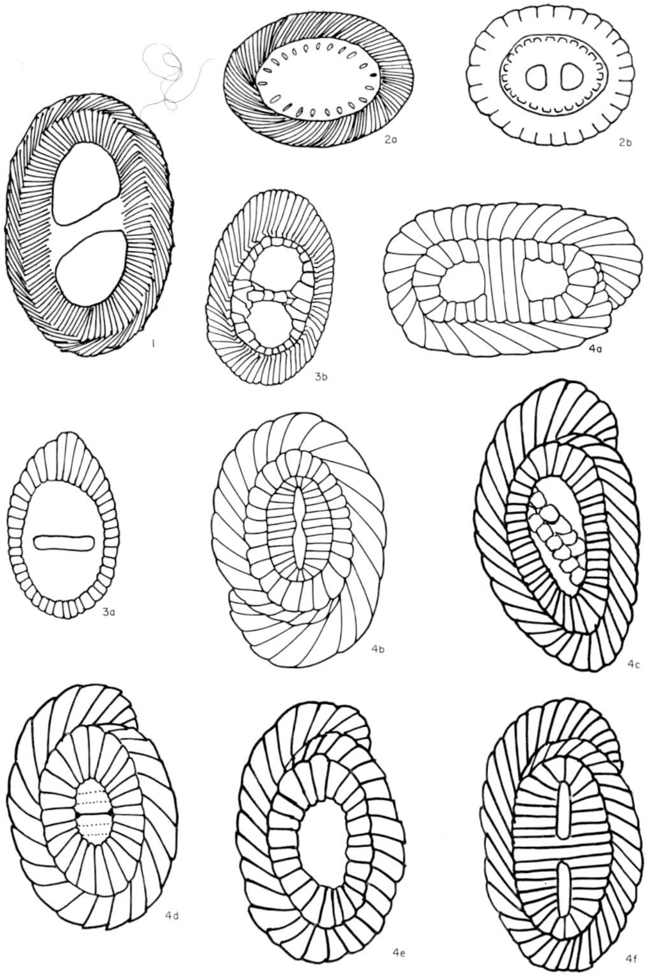

PLATE 16. 1. *Transversopontis obliquipons* (Deflandre), distal view after Hay *et al.* (1966), pl. 8, fig. 5 (×3500); 2a. *Discolithina anisotrema* (Kamptner), distal view after unpublished electron micrograph (×1750); 2b. *Discolithina fimbriata* (Bramlette and Sullivan), distal view after Bramlette and Sullivan (1961), pl. 3, fig. 1d (×1400); 3a. *Lophodolithus mochlophorus* Bramlette and Sullivan, distal view after Bramlette and Sullivan, (1961), pl. 4, fig. 62 (×1750); 3b. *Lophodolithus nascens* Bramlette and Sullivan, distal view, after Perch-Nielsen (1972), pl. 19, fig. 2 (×2800); 4a. *Helicopontosphaera recta* (Haq), proximal view after Perch-Nielsen (1971), fig. 14 (×4200); 4b. *Helicopontosphaera compacta* (Bramlette and Wilcoxon) proximal view after Perch-Nielsen (1971), fig. 14 (×4200); 4c. *Helicopontosphaera reticulata* (Bramlette and Wilcoxon), proximal view after Perch-Nielsen (1971), fig. 14 (×4800); 4d. *Helicopontosphaera ampliaperta* (Bramlette and Wilcoxon), proximal view after Perch-Nielsen (1971), fig. 14 (×4800); 4e. *Helicopontosphaera euphratis* (Haq), proximal view after Perch-Nielsen (1971), fig. 14 (×4800); 4f. *Helicopontosphaera kamptner*. Hay and Mohler in Hay *et al.*, proximal view after Perch-Nielsen (1971), fig. 14 (×4800).

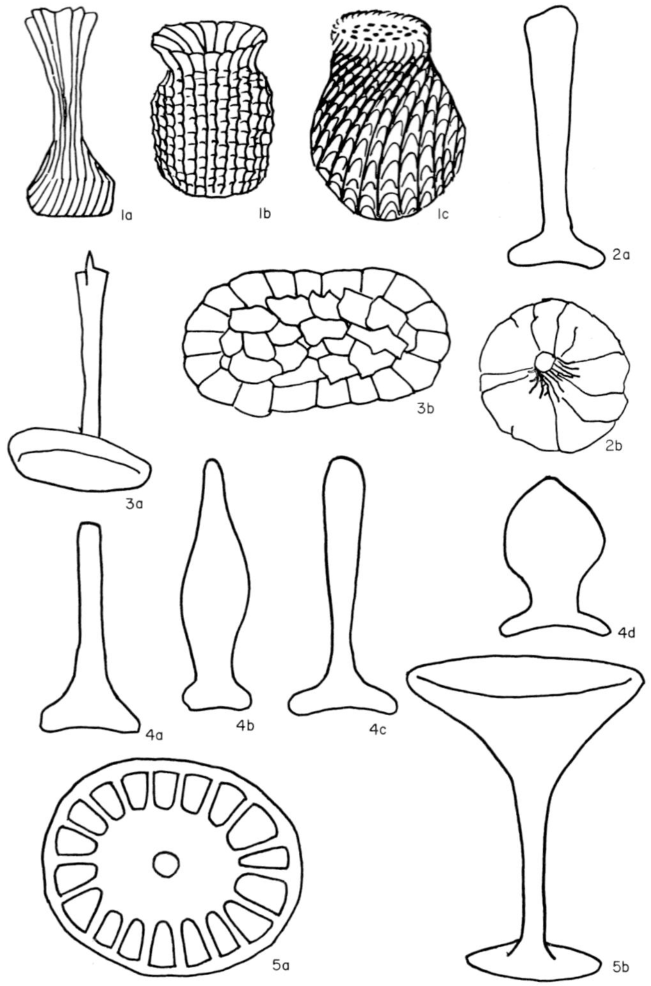

PLATE 17. 1a. *Scyphosphaera intermedia* Deflandre, side view after Ellis *et al.* (1972), pl. 7, fig. 5 (×1500); 1b. *Scyphosphaera pulcherrima* Deflandre, oblique view after Ellis *et al.* (1972), pl. 7, fig. 6 (×2000); 1c. *Scyphosphaera recurvata* Deflandre, oblique view after Ellis *et al.* (1972), pl. 8, fig. 1 (×2000); 2a. *Rhabdosphaera clavigera* Murray and Blackman, side view after Boudreaux and Hay (1969), pl. 4, fig. 6 (×4000); 2b. *Rhabodosphaera clavigera* Murray and Blackman, distal view after Boudreaux and Hay (1969), pl. 4, fig. 7 (×5000); 3a. *Aspidorhabdus stylifer* (Lohmann), side view after Boudreaux and Hay (1969), pl. 4, fig. 11 (×5500); 3b. *Aspidorhabdus stylifer* (Lohmann), proximal view after Boudreaux and Hay (1969), pl. 4, fig. 13 (×12 500); 4a. *Rhabdolithus creber* Deflandre, side view after Bramlette and Sullivan (1961), pl. 5, fig. 1 (×2500); 4b. *Rhabdolithus inflatus* (Bramlette and Sullivan), side view after Bramlette and Sullivan (1961), pl. 5, fig. 4 (×2500); 4c. *Rhabdolithus perlongus* Deflandre, side view after Bramlette and Sullivan (1961), pl. 5, fig. 7 (×2500); 4d. *Rhabdolithus morionum* Deflandre, side view after Bramlette and Sullivan (1961), pl. 5, fig. 12 (×2500); 5a. *Discosphaera tubifera* Murray and Blackman, proximal view after Halldal and Markali (1955), pl. 22, fig. 3 (×1500); 5b. *Discosphaera tubifera* Murray and Blackman, side view after Halldal and Markali (1955), pl. 22, fig. 3 (×7500).

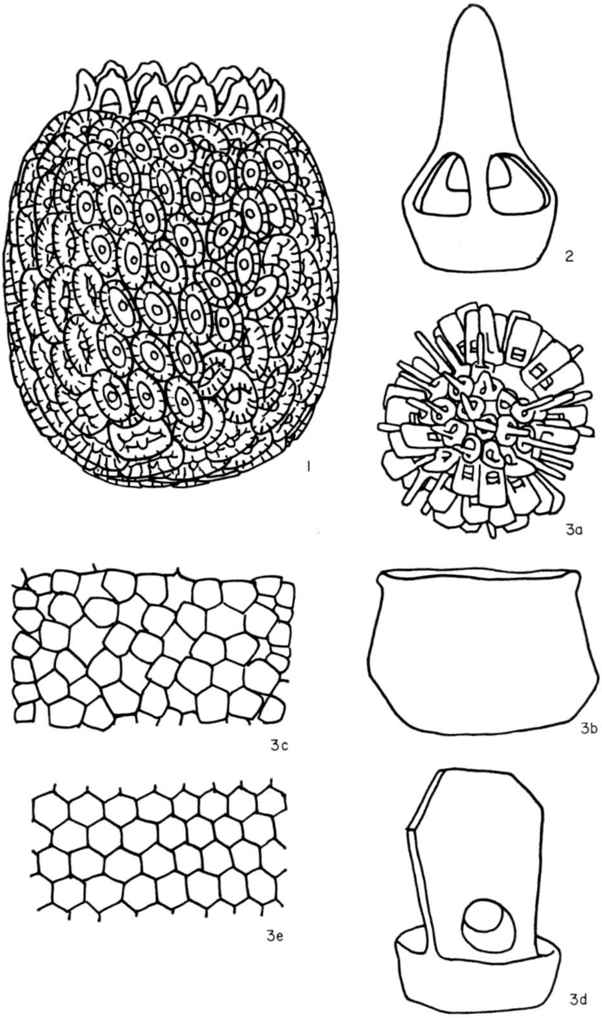

PLATE 18. 1. *Zygosphaera quadriperforata* (Kamptner), entire coccosphere after Kamptner (1937), pl. 16, fig. 27 (×3500); 2. *Homozygosphaera triarcha* Halldal and Markali, side view after Halldal and Markali (1955), pl. 4, fig. 4 (×12 500); 3a. *Periphyllophora mirabilis* (Schiller), entire coccosphere after Kamptner (1937), pl. 17, fig. 39 (×2000); 3b. *Calyptrosphaera oblonga* Lohmann, side view after Halldal and Markali (1955), pl. 1, fig. 1 (×18 000) 3c. *Calyptrosphaera oblonga* Lohmann, part of calyptrolith, showing microcrystals after Halldal and Markali (1955), pl. 1, fig. 3 (×3500); 3d. *Periphyllophora mirabilis* (Schiller), side view after Halldal and Markali (1955), pl. 3, fig. 1 (×12 500); 3e. *Periphyllophora mirabilis* (Schiller), part of calyptrolith, showing microcrystals, after Halldal and Markali (1955), pl. 3, fig. 4 (×37 500).

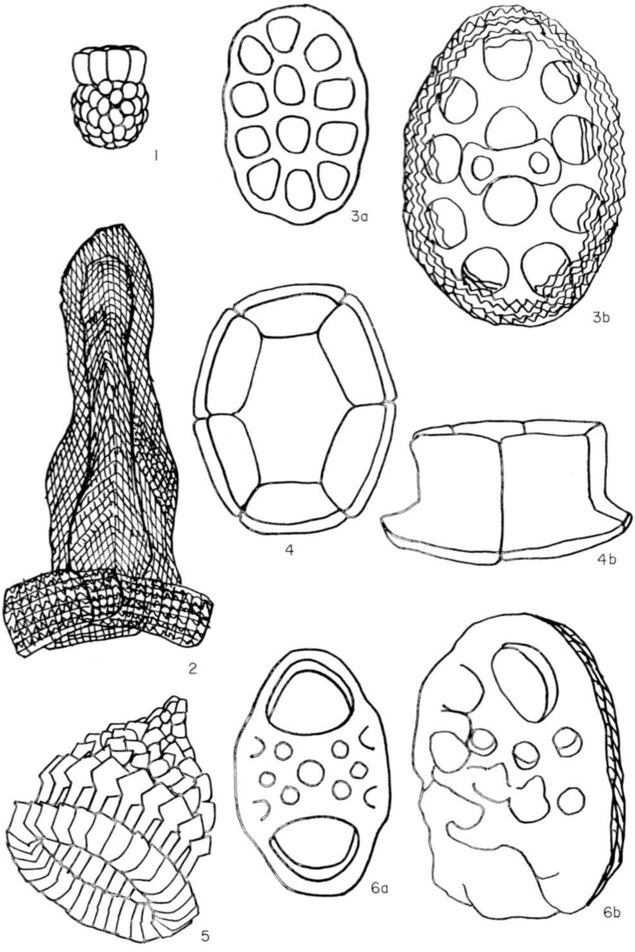

PLATE 19. 1. *Anthosphaera fragaria* Kamptner, entire coccosphere after Kamptner (1937), pl. 15, fig. 20 (×1875); 2. *Zygrhablithus bijugatus* (Deflandre), side view after Gartner and Bukry (1969), pl. 140, fig. 5 (×6000); 3a. *Holodiscolithus macroporus* (Deflandre), distal view after Stradner (1962), pl. 1, fig. 12 (×6000); 3b. *Holodiscolithus macroporus* (Deflandre), distal view after Gartner and Bukry (1969), pl. 140, fig. 2 (×12 000); 4a. *Lanternithus minutus* Stradner, distal view after Gartner and Bukry (1969), pl. 139, fig. 4 (×10 000); 4b. *Lanternithus minutus* Stradner, side view after Gartner and Bukry (1969), pl. 139, fig. 5 (×10 000); 5. *Naninfula deflandrei* Perch-Nielsen, oblique view after Perch-Nielsen (1968), pl. 1, fig. 6 (×6000); 6a. *Orthozygus aureus* Stradner, distal view after Stradner (1962), pl. 1, fig. 33 (×4500); 6b. *Orthozygus aureus* Stradner, oblique distal view after Gartner and Bukry (1969), pl. 139, fig. 1 (×6000).

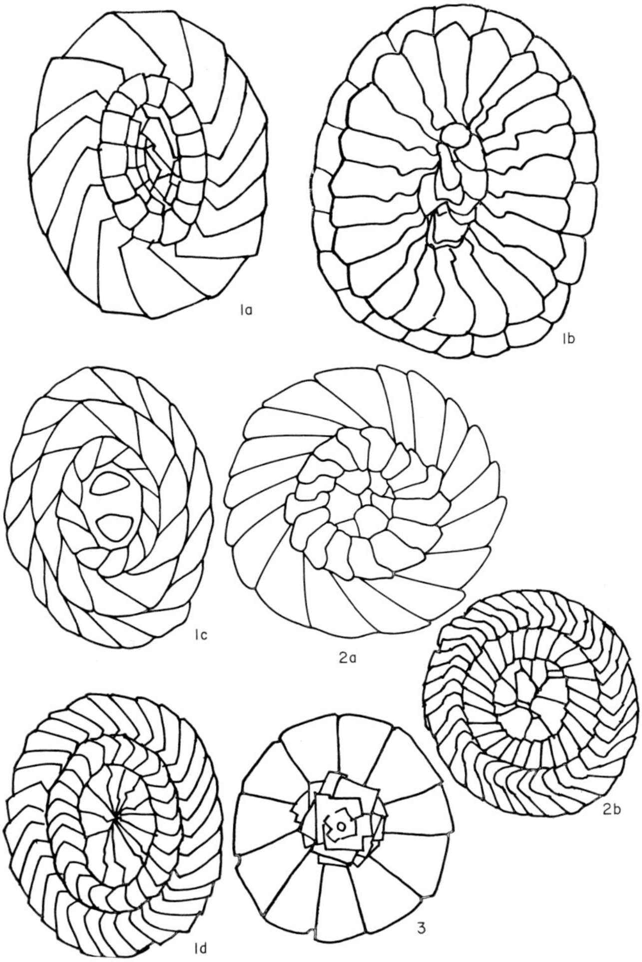

PLATE 20. 1. *Watznaueria frequens* (Noël), distal view after Noël (1965b), fig. 36A (× 18 000); 1b. *Watznaueria frequens* (Noël), proximal view after Noël (1965b) fig. 37 (× 18 000); 1c. *Watznaueria biporta* Bukry, distal view after Bukry (1969), pl. 10, fig. 9 (× 3500); 1d. *Watznaueria coronata* (Gartner), distal view after Bukry (1969), pl. 11, fig. 1 (× 6000); 2a. *Cyclagelosphaera margereli* Noël, distal view after Rood *et al.* (1971), pl. 5, fig. 9 (× 8000); 2b. *Cyclagelosphaera specioclypeata* Bukry, disal view after Bukry (1969), pl. 9, fig. 9 (× 4000); 3. *Bidicus cruciatus multicruciatus* Bukry, distal view after Bukry (1969), pl. 6, fig. 12 (× 10 000).

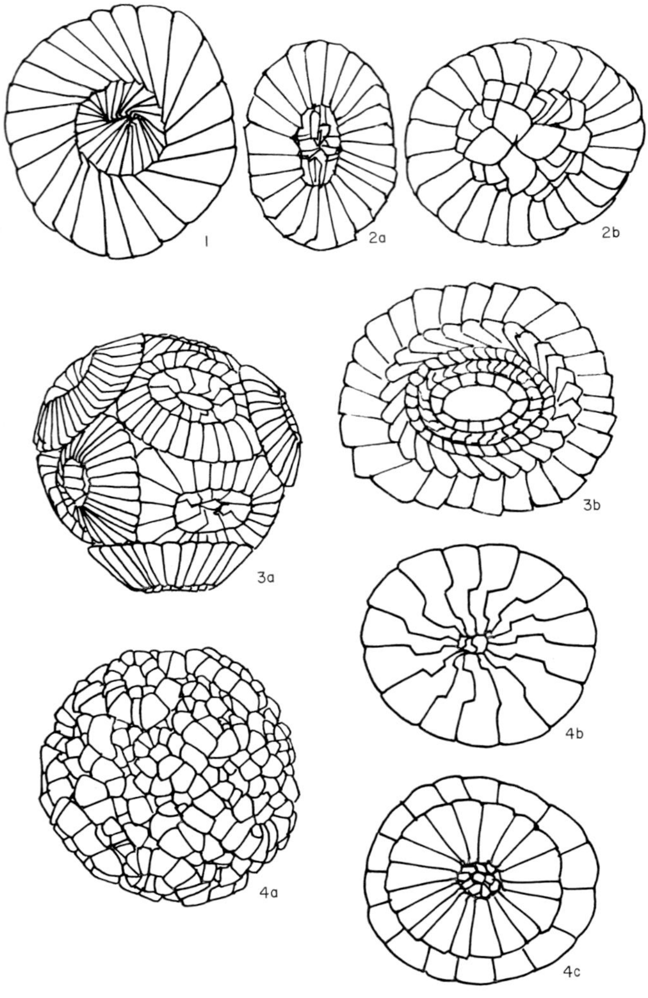

PLATE 21. 1. *Similicoronolithus primus* Bukry, distal view after Bukry (1969), pl. 9, fig. 11 (×3500); 2a. *Helenia cuvillieri* (Manivit), distal view after Thierstein (1971), pl. 5, fig. 8 (×3500); 2b. *Helenia chiasta* Worsley, distal view after Roth and Thierstein (1972), pl. 6, fig. 9 (×5000); 3a. *Coccolithus pelagicus* (Schwarz), coccosphere after Perch-Nielsen (1972), pl. 1, fig 6 (×4000); 3b. *Coccolithus pelagicus* (Schwarz), distal view, after unpublished electron micrograph (×6000); 4a. *Biscutum dimorphosus* Perch-Nielsen, coccosphere after Perch-Nielsen (1972), pl. 8, fig. 2 (×6000); 4b. *Biscutum testudinarium* Black, distal view after Bukry (1969), pl. 9, fig. 10 (×4500); 4c. *Biscutum testudinarium* Black, proximal view after Bukry (1969), pl. 8, fig. 11 (×4500).

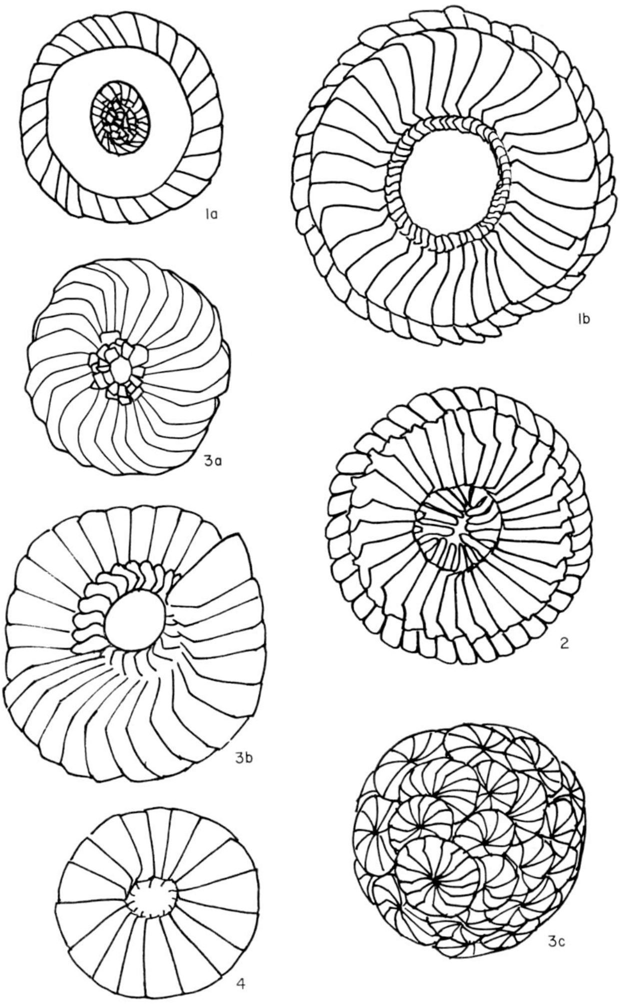

PLATE 22. 1a. *Markalius circumradiatus* (Stover), proximal view after Thierstein (1971), pl. 4, fig. 5 (×5000); 1b. *Markalius inversus* (Deflandre), proximal view after Perch-Nielsen (1972), pl. 5, fig. 6 (×5000); 2. *Cribrocentrum reticulatum* Perch-Nielsen, proximal view after Perch-Nielsen (1972), pl. 7, fig. 5 (×6000); 3a. *Cyclococcolithina macintyrei* Bramlette and Bukry, distal view after Perch-Nielsen (1972), pl. 5, fig. 1 (×3750); 3b. *Cyclococcolithina macintyrei* Bramlette and Bukry, distal view of coccolith with part of distal shield removed, after Stradner (1972), pl. 10, fig. 3 (×7500); 3c. *Cyclococcolithina fragilis* (Lohmann), entire coccosphere after Okada and Honjo (1973), pl. 2, no. 2 (×4000); 4. *Conococcolithus minutus* Hay and Mohler, distal view after Hay and Mohler (1967), pl. 197, fig. 1 (×4500).

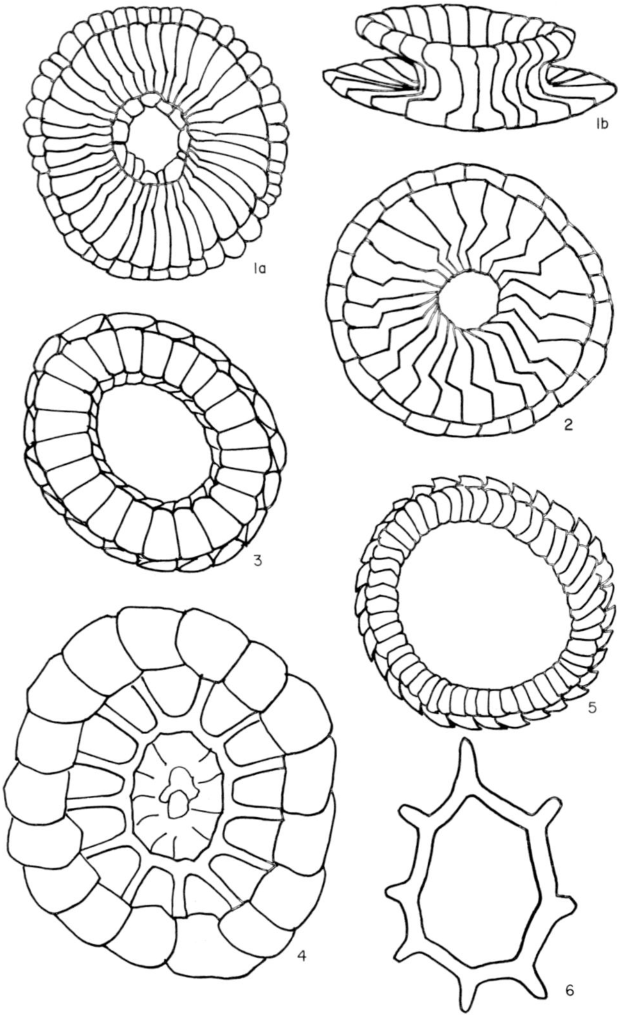

PLATE 23. 1a. *Cyclicargolithus floridanus* (Roth and Hay) proximal view after Roth (1973), pl. 11, fig. 1 (×5000); 1b. *Cyclicargolithus floridanus* (Roth and Hay), oblique side view after Roth (1973), pl. 11, fig. 2 (×5000); 2. *Umbilicosphaera mirabilis* Lohmann, distal view after Boudreaux and Hay (1969), pl. 3, fig. 8 (×4000); 3. *Cyclolithella annula* (Cohen), distal view after Boudreaux and Hay (1969), pl. 1, fig. 10 (×7000); 4. *Ilselithina iris* Stradner, after Stradner and Edwards (1968), pl. 28, fig. 5 (×15 000); 5. *Coronocyclus nitescens* (Kamptner), distal view after Perch-Nielsen (1972), pl. 21, fig. 3 (×6000); 6. *Corannulus germanicus* Stradner, plan view after Stradner (1962), pl. 1, fig. 21 (×6000).

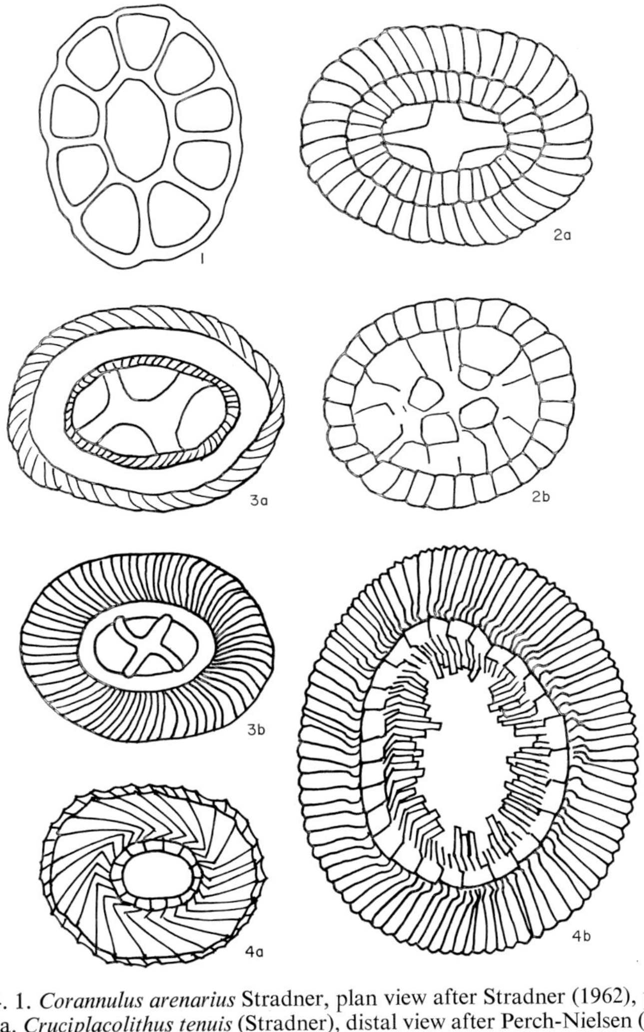

PLATE 24. 1. *Corannulus arenarius* Stradner, plan view after Stradner (1962), pl. 1, fig. 14 (×6000); 2a. *Cruciplacolithus tenuis* (Stradner), distal view after Perch-Nielsen (1972), pl. 4, fig. 4 (×5000); 2b. *Cruciplacolithus tenuis* (Stradner), proximal view after Hay and Mohler (1967), pl. 198, fig. 1 (×5500); 3a. *Chiasmolithus grandis* (Bramlette and Riedel), distal view after Perch-Nielsen (1972), pl. 3, fig. 4 (×2500); 3b. *Chiasmolithus gigas* (Bramlette and Sullivan), distal view after Bramlette and Sullivan (1961), pl. 1, fig. 6 (×1250); 4a. *Reticulofenestra samodurovi* (Hay *et al*), distal view after Hay *et al.* (1966), pl. 6, fig. 6 (×5000); 4b. *Reticulofenestra caucasica* Hay *et al.*, distal view after Hay *et al.* (1966), pl. 4, fig. 2 (×5000).

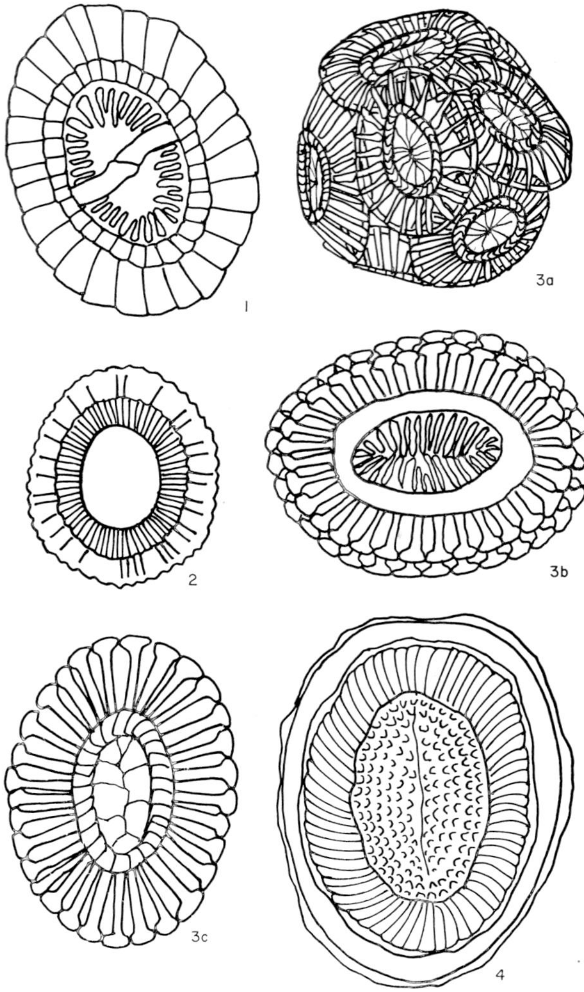

Plate 25. 1. *Gephyrocapsa oceanica* Kamptner, distal view after Boudreaux and Hay (1969), pl. 1, fig. 20 (×9000); 2. *Pseudoemiliania lacunosa* Kamptner, distal view after Ellis, Lohmann and Wray (1972), pl. 2, fig. 6 (×5000); 3a. *Emiliania huxleyi* (Lohmann) entire coccosphere after Okada and Honjo (1973), pl. 1, no. 3 (×4000); 3b. *Emiliania huxleyi* (Lohmann), proximal view after Boudreaux and Hay (1969), pl. 2, fig. 11 (×12 000); 3c. *Emiliania huxleyi* (Lohmann), distal view after an unpublished electron micrograph (×12 000); 4. *Dictyococcites oamaruensis* (Deflandre), proximal view after Stradner and Edwards (1968), pl. 16 (×4500).

PLATE 26. 1. *Dictyococcites placomorpha* (Kamptner), distal view after Stradner and Edwards (1968) pl. 21, fig. 1 (×10 000); 2a. *Discoaster pentaradiatus* Tan Sin Hok, plan view after Stradner and Papp (1961), pl. 21, fig. 3 (×2000); 2b. *Discoaster pentaradiatus* Tan Sin Hok, side view after Stradner and Papp (1961), pl. 21, fig. 6 (×2000); 3. *Hemi-discoaster molengraaffi* (Tan Sin Hok), plan view after Murray and Renard (1891), pl. 9, fig. 4 (×2000); 4. *Agalmatoaster gemmifer* (Stradner), plan view after Stradner and Papp (1961), pl. 8, fig. 52 (×2500); 5a. *Radiodiscoaster helianthus* (Bramlette and Sullivan), plan view after Bramlette and Sullivan (1961), pl. 11, fig. 18a (×2000); 5b. *Radiodiscoaster helianthus* (Bramlette and Sullivan), side view after Bramlette and Sullivan (1961), pl. 11, fig. 18b (×2000); 6. *Truncodiscoaster adamanteus* (Bramlette and Wilcoxon), plan view after Gartner (1967), pl. 3, fig. 1 (×3000); 7. *Turbodiscoaster tani* (Bramlette and Riedel), plan view after Prins (1971), pl. 4, fig. 21 (×2000); 8. *Catinaster coalitus* Martini and Bramlette, plan view after Bramlette and Wilcoxon (1967), pl. 8, fig. 10 (×2500).

PLATE 27. 1. *Hayaster perplexus* (Bramlette and Riedel), plan view after Boudreaux and Hay (1969), pl. 9, fig. 15 (×2880); 2. *Heliodiscoaster multiradiatus* (Bramlette and Riedel), plan view after Bramlette and Riedel (1954), pl. 38, fig. 10 (×1150); 3a. *Gyrodiscoaster lodoensis* (Bramlette and Riedel), plan view after Stradner and Papp (1961), pl. 26, fig. 1a (×2000); 3b. *Gyrodiscoaster saipanensis* (Bramlette and Riedel), plan view after Bramlette and Riedel (1954), pl. 39, fig. 4 (×1400); 4. *Clavodiscoaster aulakos* (Gartner), plan view after Gartner (1967), pl. 29, fig. 4 (×1400); 5a. *Discoasteroides kuepperi* Stradner, plan view after Stradner and Papp (1961), pl. 27, fig. 5 (×2800); 5b. *Discoasteroides kuepperi* Stradner, side view after Stradner and Papp (1961), pl. 27, fig. 41 (×2100); 6a. *Marthasterites furcatus* Deflandre, plan view after Bukry (1969), pl. 39, fig. 4 (×2100); 6b. *Marthasterites contortus* Stradner, plan view after Stradner and Papp (1961), pl. 36, fig. 6 (×1400); 6c. *Marthasterites tribrachiatus* (Bramlette and Riedel), plan view after Stradner and Papp (1961), pl. 35, fig. 4 (×1400); 7. *Micula staurophora* (Gardet), oblique view after Stradner (1963), pl. 4, fig. 12c (×2800); 8a. *Ceratolithus eristatus* Kamptner, plan view after Boudreaux and Hay (1969), pl. 10, fig. 17 (×2880); 8b. *Ceratolithus tricorniculatus* Gartner, plan view after Gartner (1967), pl. 10, fig. 5b (×1750); 9a. *Heliolithus kleinpelli* Sullivan, plan view after Hay and Mohler (1967), pl. 200, fig. 1 (×1220); 9b. *Heliolithus riedeli* Bramlette and Sullivan, oblique view after Bramlette and Sullivan (1961), pl. 14, fig. 9b (×2100); 10a. *Braarudosphaera bigelowi* (Gran and Braarud), distal view after Stradner and Papp (1961), pl. 37, fig. 1 (×1400); 10b. *Braarudosphaera bigelowi* (Gran and Braarud), entire coccosphere after Stradner and Papp (1961), pl. 37, fig. 2 (×1400); 10c. *Braarudosphaera undata* Stradner, plan view after Stradner and Papp (1961), pl. 37, fig. 6 (×1400).

PLATE 28. 1. *Braarudosphaera africana* (Stradner), distal view after Stradner and Papp (1961), pl. 37, fig. 4 (×2000); 2a. *Micrantholithus flos* Deflandre, plan view after Stradner and Papp (1961), p. 39, fig. 3b (×2000); 2b. *Micrantholithus vesper* Deflandre, plan view after Stradner and Papp (1961), pl. 39, fig. 5a (×2000); 2c. *Micrantholithus procerus* Bukry and Bramlette, side view after Bukry and Bramlette (1969), pl. 2, fig. 14 (×2500); 2d. *Micrantholithus procerus* Bukry and Bramlette, plan view after Bukry and Bramlette (1969), pl. 2, fig. 15 (×2500); 2e. *Micrantholithus obtusus* Stradner, plan view after Thierstein (1971), pl. 5, fig. 9 (×4000); 3a. *Pemma rotundum* Klumpp. plan view after Stradner and Papp (1961), pl. 38, fig. 1 (×2000); 3b. *Pemma papillatum* Martini, plan view after Stradner and Papp (1961), pl. 38, fig. 4 (×2000); 3c. *Pemma papillatum* Martini, side view after Stradner and Papp (1961), pl. 38, fig. 5 (×2000); 4. *Rucinolithus wisei* Thierstein, plan view after Thierstein (1971), pl. 4, fig. 11 (×6000); 5. *Goniolithus fluckigeri* Deflandre, plan view after Stradner and Edwards (1968), pl. 41, fig. 1 (×6000); 6. *Lithostromation perdurum* Deflandre, plan view after Stradner and Papp (1961), pl. 41, fig. 4 (×1000).

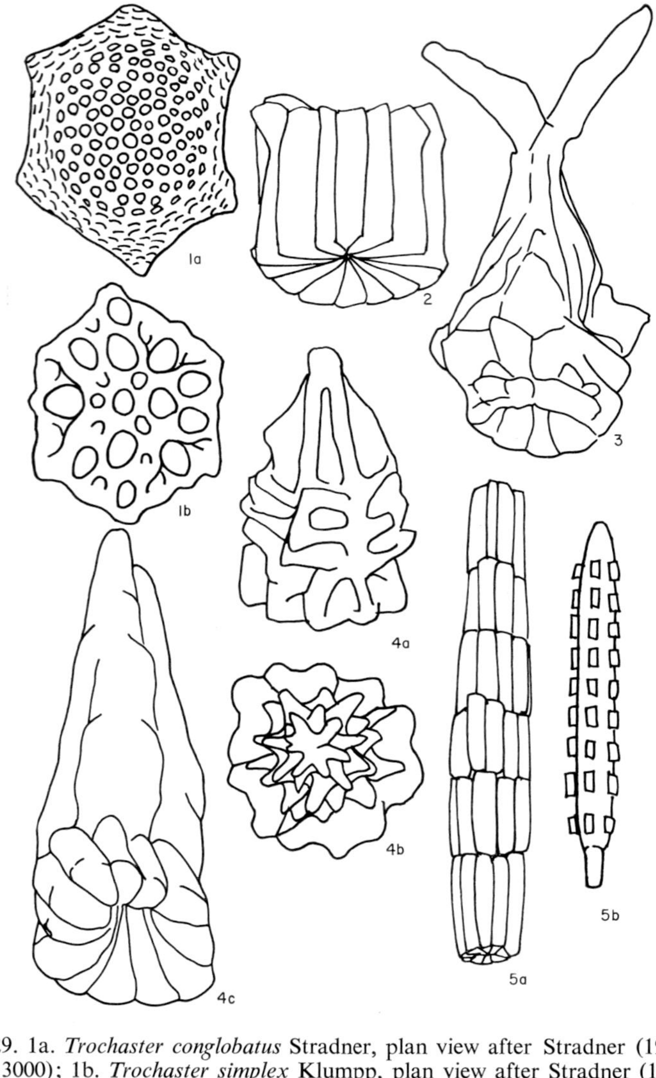

PLATE 29. 1a. *Trochaster conglobatus* Stradner, plan view after Stradner (1962), pl. 2, fig. 18 (×3000); 1b. *Trochaster simplex* Klumpp, plan view after Stradner (1962), pl. 2, fig. 17 (×3000); 2. *Fasciculithus tympaniformis* Hay and Mohler, oblique view after Hay and Mohler (1967), pl. 205, fig. 7 (×3500); 3. *Furcatolithus tribulosus* (Roth), oblique view after Roth *et al.* (1971), pl. 1, fig. 5 (×8000); 4a. *Sphenolithus radiaus* Deflandre, side view after Perch-Nielsen (1972), pl. 17, fig. 4 (×6000); 4b. *Sphenolithus abies* Deflandre, top view after unpublished electron micrograph; 4c. *Sphenolithus predistentus* Bramlette and Wilcoxon, oblique view after Roth *et al.* (1971), pl. 1, fig. 9 (×8000); 5a. *Microrhabdulus decoratus* Deflandre, side view after Perch-Nielsen (1968a), pl. 30, fig. 11 (×6500); 5b. *Microrhabdulus belgicus* Hay and Towe, side view after Bukry (1969), pl. 39, fig. 10 (×3500).

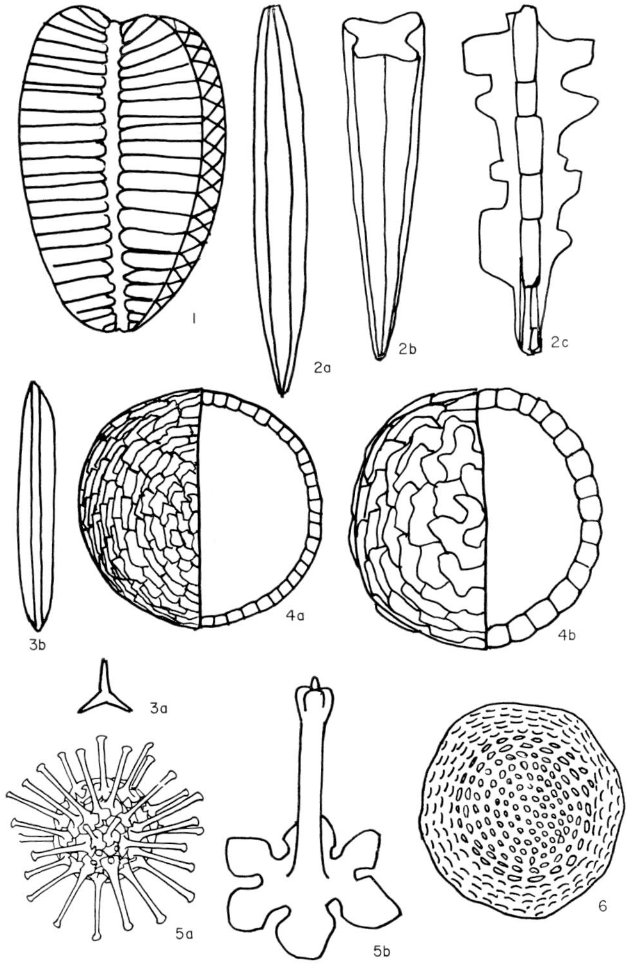

PLATE 30. 1. *Nannoconus steinmanni* Kamptner, cut-away drawing showing arrangement of crystallites, after Brönnimann (1955), text fig. 1 (×2500); 2a. *Lithraphidites carniolensis* Deflandre, side view after Deflandre (1963), fig. 1 (×1750); 2b. *Lithraphidites carniolensis* Deflandre, oblique view showing cross section, after Deflandre (1963), fig. 5 (×3000); 2c. *Lithraphidites grossopectinatus* Bukry, side view after Bukry (1969), pl. 40, fig. 3 (×6000); 3a. *Triquetrorhabdulus carinatus* Martini, cross section (×3500); 3b. *Triquetrorhabdulus carinatus* Martini, side view after Perch-Nielsen (1972), pl. 15, fig. 5 (×3500); 4a. *Thoracosphaera deflandrei* Kamptner, cut-away view after Stradner (1963, pl. 3, fig. 2 (×4000); 4b. *Thoracosphaera saxea* Stradner, cut-away view, after Stradner (1963), pl. 3, fig. 3 (×4000); 5a. *Rhabdothorax regale* (Gaarder), entire coccosphere, after Gaarder and Heimdal (1973), fig. 9 (×1000); 5b. *Rhabdothorax regale* (Gaarder), distal view of coccolith after Gaarder and Heimdal, fig. 12 (×6000); 6. *Schizosphaerella punctulata* Deflandre, end view after Stradner (1963), pl. 3, fig. 1 (×4000).

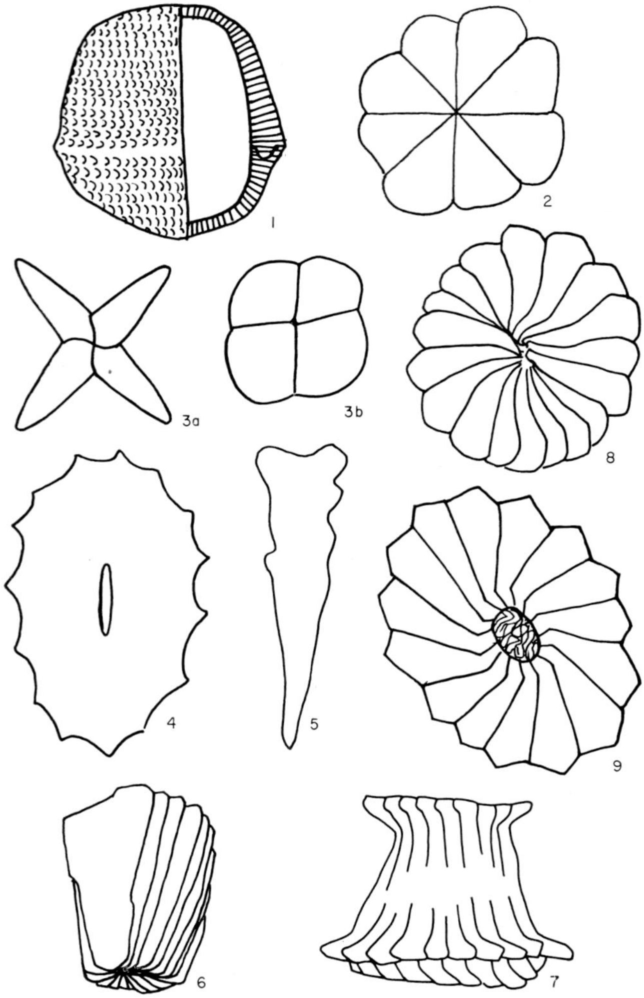

PLATE 31. 1. *Schizosphaerella punctulata* Deflandre, cut away side view after Stradner (1963), pl. 3, fig. 12 (×4000); 2. *Octolithites lecali* Caratini, after Caratini (1960), pl. 5, fig. 90 (×2000); 3a. *Tetralithus gothicus* Deflandre, plan view after Stradner and Papp (1961), pl. 40, fig. 14a (×2000); 3b. *Tetralithus copulatus* Deflandre, plan view after Stradner and Papp (1961), pl. 40, fig. 14a (×2000); 4. *Guttilithion cassum* Stradner, plan view after Stradner (1962), pl. 3, figs. 1–4 (×5000); 5. *Orthorhabdus serratus* Bramlette and Wilcoxon, side view after Bramlette and Wilcoxon (1967), pl. 9, fig. 7 (×3000); 6. *Conusphaera mexicana* Trejo, oblique view after Thierstein (1971), pl. 3, fig. 1 (×4500); 7. *Hayella situliformis* Gartner, side view after Gartner (1969), pl. 1, fig. 6 (×6000); 8. *Oolithotus antillarum* (Cohen), plan view after Boudreaux and Hay (1969), pl. 9, fig. 20 (×4000); 9. *Ellipsodiscoaster lidzi* Boudreaux and Hay (1969), plan view after Boudreaux and Hay, pl. 10, fig. 4 (×6000).

14. Automated Biostratigraphy

THOMAS R. WORSLEY and MAURICE L. JORGENS

Department of Oceanography,
University of Washington, Seattle,
Washington 98195, U.S.A.

In recent years, biostratigraphical data for many fossil groups has been accumulating at an ever accelerating rate. The volume now available is so large that it has become expedient to computer store, reduce and analyse it. Toward this end we have developed probability-based computer programs which greatly simplify collection, manipulation, and reduction of biostratigraphical data.

Input requires only occurrence-abundance lists of species within samples from sections plus locality information of the sections. Because keypunching has proven to be the ultimate bottleneck in most computer-based systems, we have adapted our system so that it does not require keypunched cards, but instead can use direct input from mark-sense sheets which can be directly fed into a computer via an optical scanning system.

Output includes range charts, automatic taxonomic reassignment to preferred names, biostratigraphical subdivision at any given level of confidence, and distribution maps of taxa or combinations of taxa.

It is envisioned that this system or similar ones will ultimately lift palaeontology from the realm of "art" into that of "science".

Outline

Introduction

With the advent of deep-sea drilling, micropalaeontology has captured the limelight of geological oceanography. It has also begun to leave the realm of "art" and is fast becoming a "science". As ever more oceanic cores are recovered, the problem of storing, processing and recalling the results of micropalaeontological studies becomes increasingly difficult. Ultimately, the volume of data will grow so large that only computers will be able to efficiently store and process it. The purpose of this paper is to show several ways in which computer based systems can greatly increase the capabilities of biostratigraphers while at the same time eliminating much of the tedium inherent in data collection.

The major difficulty one encounters in attempting to make biostratigraphical data amenable to machine processing lies in the realm of accounting for the palaeontological uncertainty in any given situation in terms that are programmable. The frustration becomes extreme when one considers, for example, the relative ease with which competent biostratigraphers recognise most "Oppel" or "assemblage" zones despite the high degree of ambiguity in the definitions of these zones. Of course, the ambiguity and subjectivity of many of these units, coupled with the knowledge of experienced palaeontologists, is exactly what has made them useful for international correlation. Unfortunately, this same ambiguity has made communication among ever larger numbers of investigators very difficult and has apparently set a limit to biostratigraphical resolution by these methods which is now being approached. In addition it is hindering successful adaptation of biostratigraphy to computer based subdivision systems so that for the most part, the rapid developments being made in machine processing of data have completely by-passed the field of biostratigraphy.

Hay (1972) in a paper furthering the ideas of Shaw (1964) has sought to rectify this situation by suggesting methods for using probabilistic techniques to quantify uncertainty, thereby avoiding the ambiguity inherent in current biostratigraphical practice. Worsley *et al.* (1973) and Worsley and Jorgens (1974) have developed and used machine processing techniques based in large part on the ideas of Shaw (1964) and Hay (1972) for synthesising raw biostratigraphical data (presence and abundance of taxa in samples) into probabilistically defined biostratigraphical units. This system was used to describe the distribution of calcareous nannofossils in strata of Oligocene age. However, in these initial studies, situations were encountered for which adequate description was not possible using common biostratigraphical terminology. Therefore, we must outline a simple system for describing most situations likely to be encountered in the stratigraphic column and then show how automated data processing techniques may be used in the framework of this system to eliminate much of the tedium involved in biostratigraphical data collection while retaining the information useful to synthesis.

As we have already done considerable work on Oligocene nannofossils, we will use these for illustrative purposes throughout this paper. It should be apparent that those techniques would be equally useful for any other fossil group or time period. In fact, they are already being applied to the study of planktonic foraminifera and radiolaria in studies now in progress.

Fossils in space and time

The distribution of fossil groups in the geological record can be represented in two main ways; within an area–stratigraphical volume or an area–time volume. Area–stratigraphical space can be represented by block or fence diagrams in which all three of the axes are defined in terms of physically measurable lengths. Usually the two horizontal axes are geographical and the vertical is stratigraphical. Area–time space on the other hand, is defined with time, rather than stratigraphical thickness as its vertical axis. The way stratigraphical (including biostratigraphical) information can be represented in each of these frameworks can look very different, as demonstrated by Krumbein and Sloss (1963). Basically, this is because unconformities are represented by surfaces in Area–stratigraphical space whereas they are represented by lacunas in area–time space (Fig. 1). While it is possible to

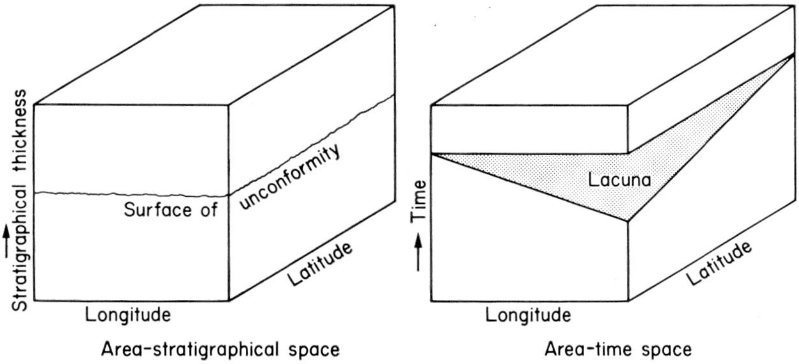

FIG. 1. Area–stratigraphical space and area–time space.

construct a set of parallel terminologies to describe each of these systems, it is far simpler to use the same unmodified geometrical terms for either as long as no difference in interpretation would result. Appropriate modifiers or context can be used to clarify which system is being used in cases where the difference does matter. In the following, the basic definitions necessary to successful machine processing of biostratigraphical information are presented.

Basic definitions

A *datum* is the lowest or highest occurrence of a fossil group in an area–time or area–stratigraphical section. As such it represents a point in area–time or area–stratigraphical space. Its plural is defined as *datums* because of the spurious connotative meaning attached to "data". Hay (1972) used the term event to express the concept here defined as a datum. However, we feel it wiser to retain the term datum for this purpose because of its already widespread use in a very similar sense. Admittedly, little confusion would result from the use of both.

A *surface* is the locus in area–stratigraphical (or area–time) space of a specified datum for all sections in which that datum occurs. Therefore, surfaces may be defined in either area–stratigraphical or area–time space and may be planar or irregular depending on the distribution of the points upon which they are defined. Two or more surfaces may commonly coincide locally in area–stratigraphical space but rarely do so in area–time space (see Fig. 1). Many types of surfaces are available to the biostratigrapher and they need only be defined by an adequate descriptive term preceding the word surface. The two specific types of most concern to the biostratigrapher are the *highest occurrence surface* (HOS) and the *lowest occurrence surface* (LOS) as used by Hay (1971). These represent highest and lowest occurrences of fossil groups. The vertical separation between any two surfaces is an *increment*. It can either be a stratigraphical or a time increment, and its length is a function of both evolutionary processes and the locality at which it is measured. It shall be noted that the length of stratigraphical increments can be influenced by sedimentation rates whereas time increments cannot. The term "interval" would seem appropriate for this concept, but its previous use as an approximate synonym for assemblage zone renders it inappropriate. The simplest way to describe any given increment would be to name the two surfaces from oldest to youngest in the stratigraphical order in which they occur. This avoids the use of any formal zonal terminology (and accompanying emotion) while at the same time retaining the time or stratigraphical value of that pair of surfaces. In addition, it defines the area for which the sequence is valid. For example; the increment represented by the two surfaces in Fig. 2 most

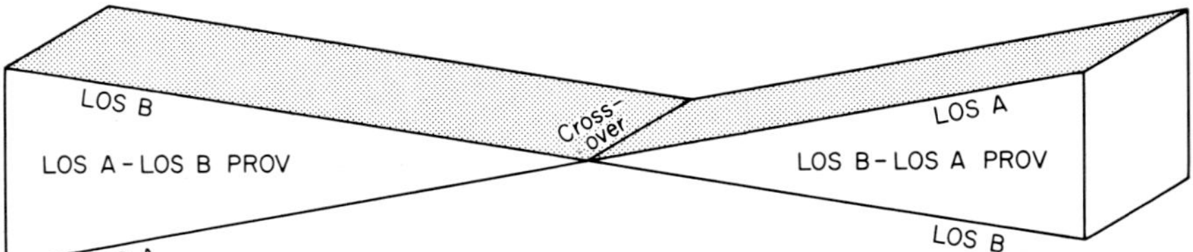

FIG. 2. Areal relationship between crossing datum surfaces.

certainly depends upon the location of where it is measured. As such, the interrelationship among increments can easily be used to define a type of biostratigraphical province. This type of province defines a volume that has no formal name, is not necessarily occupied by any particular fossil group, and need not be bounded above or below by synchronous surfaces. It is simplest to refer to such a province by the stratigraphical or temporal sequence which defines it. In Fig. 2 the two provinces shown would be the LOS A-LOS B province on the left and the LOS B-LOS A province on the right. A convenient name for the line of intersection between the two provinces (the place where the sequence reverses) is a *crossover*.

In terms of the above, we can now define continuous and discontinuous sections. A *continuous section* is one that can be sampled in such a way that no two datums occur in the same sample (Hay, 1972). Conversely, a *discontinuous section* is one that cannot. A *discontinuity* is therefore a position in a section where two or more datums not at crossover cannot be separated. The term discontinuity conveys only the idea that a section is not biostratigraphically continuous and offers no hint of the reason. In many cases discontinuities represent unconformities, but they may also represent vertical facies changes, covered intervals, "barren" intervals, uncored intervals, or other situations for which it is impossible to separate all datums. Of course, other definitions can be added to augment those presented above, but as most would be corollaries or logical outgrowths of those above, they are better left to definition as needed.

The construction of range charts

Having these basic definitions at hand, how does one put them to use for unambiguously collecting and processing biostratigraphical information? The answer to this question lies in a system which is capable of unambiguously comparing ranges of species. In biostratigraphy, with or without a computer, the basic comparative unit is the range chart, on which the presence and abundance of fossil taxa and therefore their stratigraphical ranges are recorded for a section. In the past, range charts were looked upon as the end result of research endeavours. Synthesis of the biostratigraphical information they contained (i.e. zonations or province maps) was done by visual comparison of the charts. However, in machine processing, range charts themselves can be regarded as the basic framework in which to compile raw data. Their synthesis into probabilistically defined stratigraphical or geographical subdivisions are the results. Therefore, the first requirement of any machine processing system is a means of recording, storing and recalling range charts.

It should be realised that although the basic building blocks of biostratigraphy are range charts, the elements of range charts are taxa.

As yet computers can add little to the art of describing and identifying taxa defined in Linnaean terms so that palaeontologists can feel assured of their role in biostratigraphy for as long as Linnaean taxonomy persists. Perhaps the remarks of Worsley *et al.* (1973, p. 77) best summarise our feelings on this matter.

With regard to taxonomy, it should be apparent that the value of a fossil taxon to a biostratigrapher is essentially a function of its utility in stratigraphic or environmental differentiation. This requires taxa which are common, easy to identify, and which have discrete vertical and lateral distributions. It is suggested here that most such fossils have already been discovered and described. Unfortunately, many of these have been independently named and described by several authors so that each has several available names. Naturally, the rules of taxonomy dictate that each taxonomic entity may have only one valid name so that much research endeavor is spent trying to ascertain who really was first to validly describe a taxon. The result of this "noble quest" has been a proliferation of taxonomic literature, which, in many cases has obscured rather than clarified taxonomic classifications at the expense of time which might have been better devoted to delineating the precise ranges of the taxa in question regardless of what their valid names are.

Ultimately, this situation will most likely not be resolved by exclusive appeal to a set of complex rules, but rather by the dictates of common usage among-biostratigraphers, as is the case for most flexible terminologies or for that matter any type of living language.

For our own taxonomic purposes, we have compiled a master list of species names which is periodically revised. This list contains the names of all the species known to occur in the part of the column under investigation, in this case the Oligocene. In addition it contains our opinions on the synonomy of species names in such a way that no matter under what name a species is entered into the machinery memory, it is stored and recalled under its preferred name. Of course, preferred names can be updated as taxonomic fashion changes. The preferred names can be printed out in alphabetical order by species name with synonyms of each printed to the right of the preferred names. Using this system any number of taxonomic recombinations are possible and groups of species may be combined into larger taxonomic units so that genera or even larger entities may be used as single taxonomic units. Table 1 contains the list of species whose datums we have chosen to evaluate for this paper.

The program also makes selective use of species possible. If, for instance, out of the list on Table 1 we wish to use only 10 species, the others are arbitrarily made junior synonyms to a "dead letter" species which is never used. In that way, any number of species may be selectively eliminated from as long a list as desired.

When taxonomic standardisation has been completed, the next step is to compile the range chart. Initially we compiled range charts by keypunching all pertinent information about each sample in a section. We soon found, however, that keypunching is very inefficient and have subsequently adopted optical scanning procedures. Optical scanning is the term used for the technique based on machine recognition of marks or characters. It has received

TABLE 1. Alphabetical listing of species used for this paper and the number assigned the to HOS and LOS of each.

OLIGOCENE 50 SECTIONS -- 51 FOSSILS

NUMBER OF CORES 50 NUMBER OF FOSSILS 51 NUMBER OF <DATUMS> 102

EV.	EXT.	
1	2	COCCOLITHUS ABISECTUS
3	4	HELICOPONTOSPHAERA AMPLIAPERTA
5	6	DISCOASTER AULAKOS
7	8	DISCOASTER BARBADIENSIS
9	10	SPHENOLITHUS BELEMNOS
11	12	ZYGRHABLITHUS BIJUGATUS
13	14	DISCOASTER BINODOSUS
15	16	RETICULOFENESTRA BISECTA
17	18	TRIQUETRORHABDULUS CARINATUS
19	20	SPHENOLITHUS CIPEROENSIS
21	22	SYRACOSPHAERA CLATHRATA
23	24	HELICOPONTOSPHAERA COMPACTA
25	26	BLACKITES CREBER
27	28	SPHENOLITHUS DISTENTUS
29	30	DISCOASTER DRUGGI
31	32	HELICOPONTOSPHAERA EUPHRATIS
33	34	ERICSONIA FENESTRATA
35	36	CYCLOCOCCOLITHINA FLORIDANA
37	38	ERICSONIA FORMOSA
39	40	RETICULOFENESTRA FOVEOLATA
41	42	CORANNULUS GERMANICUS
43	44	CHIASMOLITHUS GRANDIS
45	46	SPHENOLITHUS HETEROMORPHUS
47	48	RETICULOFENESTRA HILLAE
49	50	HELICOPONTOSPHAERA INTERMEDIA
51	52	CYCLOCOCCOLITHUS INVERSUS
53	54	HELICOPONTOSPHAERA KAMPTNERI
55	56	CYCLOCOCCOLITHUS KINGI
57	58	RETICULOFENESTRA LAEVIS
59	60	CYCLOCOCCOLITHUS LEPTOPORUS
61	62	DISCOASTER LIDZI
63	64	CYCLOCOCCOLITHUS MARGARITAE
65	66	LANTHERNITHUS MINUTUS
67	68	PONTOSPHAERA MULTIPORA
69	70	RETICULOFENESTRA OAMARUENSIS
71	72	CHIASMOLITHUS OAMARUENSIS
73	74	TRANSVERSOPONTIS OBLIQUIPONS
75	76	HELICOPONTOSPHAERA PARALLELA
77	78	DISCOASTER PERPLEXUS
79	80	SPHENOLITHUS PREDISTENTUS
81	82	SPHENOLITHUS PSEUDORADIANS
83	84	HELICOPONTOSPHAERA RECTA
85	86	ISTHMOLITHUS RECURVUS
87	88	DISCOASTER SAIPANENSIS
89	90	DISCOASTER SAUNDERSI
91	92	COCCOLITHUS SCISSURUS
93	94	BRAMLETTEIUS SERRACULOIDES
95	96	ORTHORHABDUS SERRATUS
97	98	ERICSONIA SUBDISTICHA
99	100	DISCOASTER TANI
101	102	RETICULOFENESTRA UMBILICA

wide use in the banking field for cheque identification and for examination purposes in college courses so that most large American universities now use optical scanning for machine grading of multiple-choice tests. Keypunching is still preferable to optical scanning for small projects when it is not worthwhile developing optical scanning techniques. Figures 3a and b show the

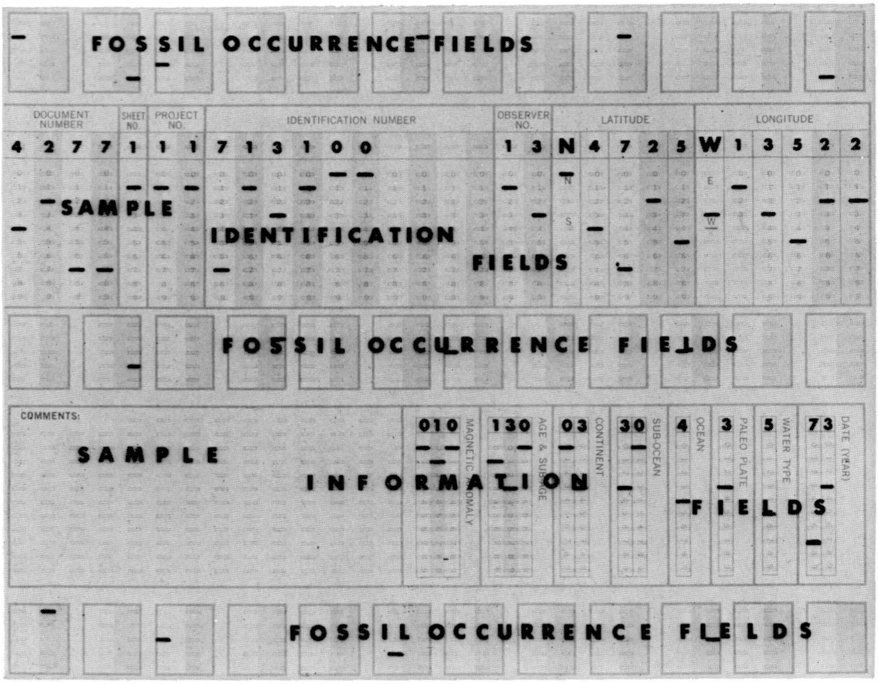

FIG. 3a. Mark sense sheet without overlain template.

main features of the optical scanning system we have adopted. Basically a template having pictures of fossil groups is placed over a mark-sense sheet on which abundance information can be recorded for machine processing. We have found that 72 is the maximum number of taxonomic units that can be accommodated on a template with pictures without overcrowding. It is possible to include as many as 200 if pictures are not needed. Information about abundance and diversity of groups in a sample is recorded simply by marking the sheets in the appropriate fields. For our purposes the abundance scale is logarithmic to the base 10 and runs from abundant (>10 specimens/field of view) in the left column to very rare (<1 specimen/1000 fields of view), on

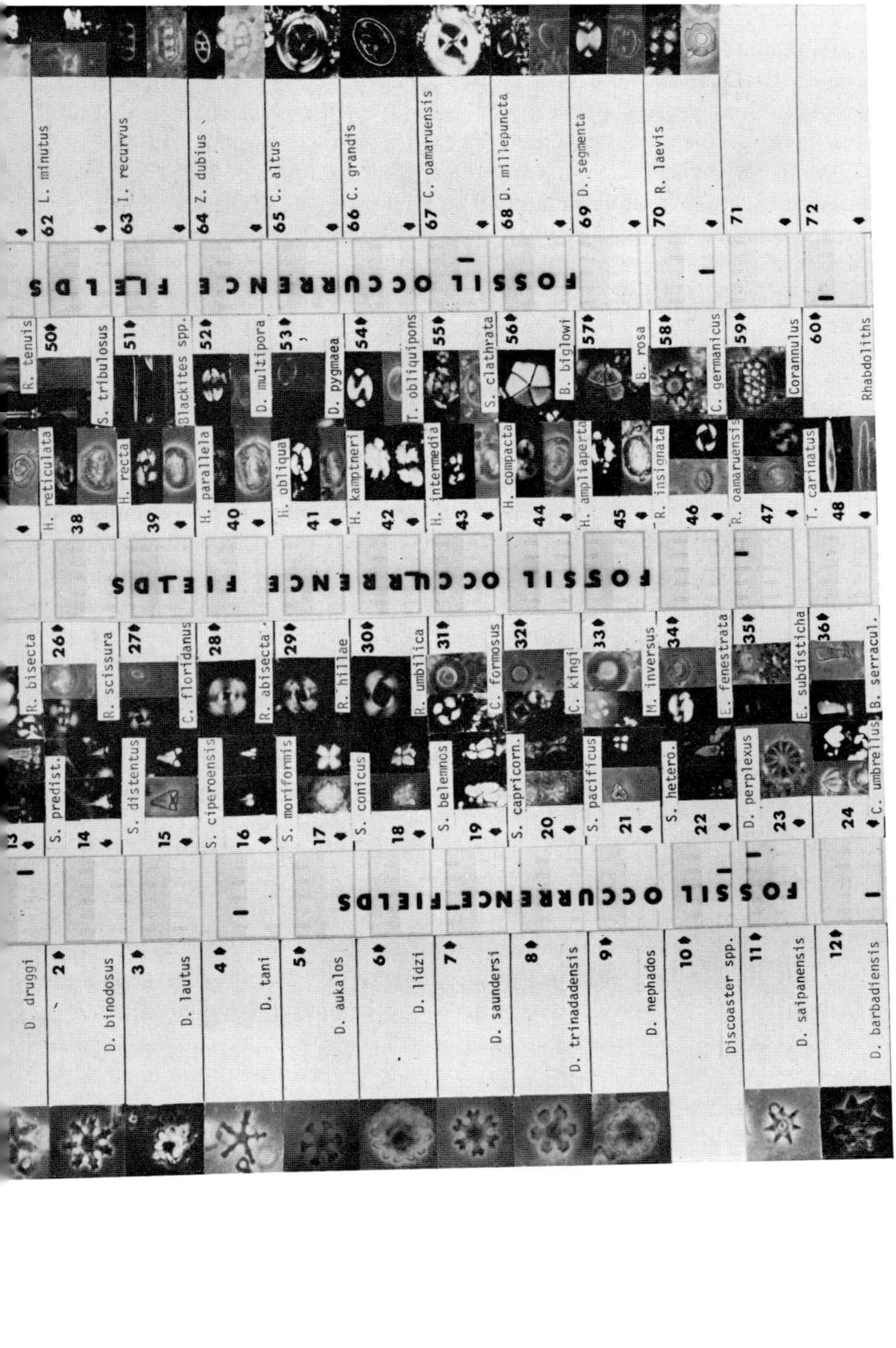

FIG. 3b. Mark-sense sheet with overlain template.

the right-hand column. Specimen counts could just as well be used if they are better suited to a particular purpose or fossil group. The difference in purpose and results between visual estimation and counting have been discussed in detail by Dennison and Hay (1967) and Worsley (1974). The basic determinant for choosing between the two is whether one wants an accurate estimate of the relative abundance of the more common taxa without a reasonable chance of consistently detecting rare ones, or a good chance of detecting rare taxa while having only a crude estimate of the abundance of the common ones. Biostratigraphy for the most part, is more concerned with the latter, as demonstrated by Shaw (1964).

In order to construct a range chart, the abundance and diversity of a flora or fauna in each sample from a section is recorded on a mark-sense sheet (i.e. the one in Figs 3a and b). When information on all the samples studied for a section has been recorded in the above manner, the mark-sense sheets may then be sequentially read by means of an optical reader, and the stratigraphical information they contain transcribed onto a magnetic tape. This tape may then be computer interpreted to produce a range chart (Table 2) which also contains the number of datums occurring in each sample level. This allows for preliminary recognition of discontinuities and will give an idea of whether additional sampling is necessary. Although we present this format, many others are possible. The above procedure for producing a range chart can be repeated for as many sections as are available.

It should be noted that the recording of stratigraphical data from published range charts is easily accomplished using this system, and has been found to be much faster, cheaper, and more error free (i.e. more efficient) than key-punching for this purpose. If punched cards are desired, they can be obtained by simply running the interpreted tape through the computer system and having it punch them.

When all range charts from both primary inspection of samples and published information have been recorded, checked for errors, and corrected if necessary, several additional preliminary operations can be performed on them prior to using them for synthesis purposes.

First, the number of samples from a section in which any given taxon occurs can be tallied to give a crude idea of its distribution throughout a section. Next, the total number of sections in which a taxon occurs can also be determined. This allows for deletion of forms too rare to be profitably analysed statistically. Generally we have found that it is not fruitful to consider taxa that occur in fewer than four or five sections; for conclusions about their distribution have little statistical significance.

When taxonomic selection and exclusion is complete, the remaining taxa are alphabetised by species name as in Table 1. It was found easier to alphabetise

TABLE 2. Range chart for a section compiled from mark-sense sheets.

```
                                                                                ^
                                                  1111111111111111122222222222222
                    CORE  #         112223456667890112234455667889901234455667777
                                    2AAAAAAAAAAAAAAAAAAAAAAAAAAAAAAAAAAAAAAAAAAAA
                    SECTION #       3313C33113C313336363363C3633C3C33333C3C3C369C

                    CM FROM TOP       1    11 1 1          2    2 2     3 3 3
                    OF SECTION        2    45 6 7          4    6 7     0 1 2
                                      2    98 8 7          7    9 5     5 1 0

N06.20 W140.22 JOIDES HOLE  70
HAQ B.. AND J. LIPPS  INITIAL REPORTS  LEG 8

   DISCOASTER BARBADIENSIS                                                     11
   SPHENOLITHUS BELEMNOS            11111
   RETICULOFENESTRA BISECTA            111111111111111111111111111111111111111111
   TRIQUETRORHABDULUS CARINATUS     1111XXXXXX11
   SPHENOLITHUS CIPEROENSIS            1111111
   HELICOPONTOSPHAERA COMPACTA             1            1111111111111111111111111
   SPHENOLITHUS DISTENTUS                   111111111111111111111
   CYCLOCOCCOLITHUS FORMOSUS                                                   11
   CHIASMOLITHUS OAMARUENSIS                                                  111
   SPHENOLITHUS PREDISTENTUS                1111111111111111111111111111111111111
   SPHENOLITHUS PSEUDORADIANS                            11111111
   DISCOASTER TANI                                  1X111111111111111111111111111
   RETICULOFENESTRA UMBILICA                                       1XXXXX11111111
```

by species rather than by generic names because the frequent taxonomic reassignment common to calcareous nannofossils generally has less effect upon the alphabetical position of species names than generic ones (e.g. *Ellipsolithus subdistichus* (ROTH and HAY) is now called *Ericsonia subdisticha* MARTINI). Upon alphabetisation of a species list, code numbers are assigned sequentially to the alphabetical order with each taxon receiving an odd number for its LOS and the next highest even number for its HOS. This is shown in Table 1, which is a print-out of the list we have analysed for this chapter.

Probabilitistic stratigraphy

When the master list of species to be analysed and all range charts to be included in a study have been reduced to these species, the data is ready for synthesis into a *most probable sequence* (Hay, 1972) of datum surfaces for that collection of sections. This is the step in which computers are truly superior for decision making. The collection of sections may be from a single continent or ocean, or part thereof, or may be worldwide. In determining the most probable sequence the matrix programme first scans the range chart for each section and determines the highest or lowest occurrence of each taxon in that section. If two or more highest or two or more lowest occurrences are found in a sample, a discontinuity is noted. However, if a highest occurrence of one taxon and a lowest occurrence of another are contained in a single sample, they are by definition separate and the former is recorded as being above the latter for that section. This should need no further explanation, because the fact that the two fossils occur in the same sample is indicative that their ranges overlap. When sequential information has been compiled for all sections, the next step is to compare sequences among the sections. This is by far the most subtle and difficult problem in biostratigraphy because of the tremendous number of comparisons that must be made simultaneously in order to synthesise a sequence. In the computer this is accomplished by comparing the position of each datum in a section with all the others in that section, repeating the process for each section, then taking a "weighted average" of the results. This is most easily accomplished by constructing a matrix and using a probabilistic framework to solve it as outlined by Hay (1972). The procedure is as follows. Initially, the input data is broken into occurrences within samples from levels within each section. An occurrence matrix is generated and filled one section at a time. An entry is made in the matrix each time one datum point (highest or lowest occurence of a species) occurs at a distinctly separate level from another in a section.

After the raw data to be used is sorted, a matrix array is generated in which any pair of matrix elements X_{IJ} and X_{JI} denotes the number of occurrences of J which precede I and the number of occurrences of I which precede J, respectively. The sum of the pair is the total number of separate occurrences of the two datums.

The final order of species for the sections under consideration is derived by ordering the matrix so that the larger of the pair of numbers is above the main diagonal (the one with X's) in all cases. The order is simply the I indices read off along one side of the matrix. For example, the matrix (X-array) of Fig. 4, when ordered, becomes that of Fig. 5, so that the final order is CADB.

	A	B	C	D
A	X	2	3	2
B	1	X	2	1
C	4	4	X	3
D	0	7	1	X

Fig. 4.
Simple unordered
4 × 4 matrix.

	C	A	D	B
C	X	4	3	4
A	3	X	2	2
D	1	0	X	7
B	2	1	1	X

FIG. 5.
Ordered matrix of
Fig. 4.

This order can be verified for the pair (CA. $X_{CA} = 4$ $X_{AC} = 3$, indicating that C must come before A in the order, because it occurs first 4 out of 7 times. This means X_{CA} must be above the main diagonal, which is true. Note that there are some instances which cannot be ordered in this manner. If the initial matrix had been the one in Fig. 6 the order would be nondeterministic, for the matrix indicates that C must come before A, A before D, D before B, and B before C. This is clearly circular. To eliminate this inconsistency in the data (caused by improper identification, contamination, or non-mutual occurrence among pairs in different sections), a sequential probability is assigned to each pair in the following way. Assuming the two occurrences are random events, the degree of correlation between them can be determined by a modification of the binomial expansion. The probability that a pair occurs in a preferred sequence is given by the formula:

$$P_{IJ} = 1 - \left[\left(\tfrac{1}{2}\right)^{N-1} \sum_{r=n}^{N} \frac{N!}{r!(N - r!)} \right]$$

where $N = X_{ij} + X_{JI}$, $n =$ the larger of H_{IJ} or X_{JI}, and r is a variable running from n to N, a detailed explanation of the use of this formula may be found in Hay (1972). A probability of 1 denotes absolute certainty that the pairing is in that order. A zero indicates no significant separation for that pair.

If P_{IJ} is close or equal to P_{JI}, pairing can be shown to have a probability close or equal to zero. This can happen in two ways; Pair CD ($N = 7$, $n = 4$) in Fig. 6 demonstrates the first (a coin tossed 7 times will most likely come up heads either 3 or 4 times). The second is to never have the pair occur in sequence in any section. This is shown by the pair CA. Such pairs can be eliminated by zeroing X_{CA} and X_{AC}. No information is lost if the probability of pairing in that order is zero. In this manner the array shown in Fig. 6 then becomes the one in Fig. 7.

	A	B	C	D
A	X	2	3	2
B	1	X	5	1
C	4	2	X	3
D	0	7	4	X

FIG. 6.
Indeterminate 4 × 4 matrix, unordered.

	A	D	B	C
A	X	2	2	0
D	0	X	7	4
B	1	1	X	5
C	0	3	2	X

FIG. 7.
Ordered matrix of Fig. 6 after zeroing pairs with very low sequential probabilities.

Our research has shown that pairs having lower probability values are of little use in determining a most probable sequence. Therefore, any pair with probability less than a specified value (predetermined by the investigator) can be zeroed in this manner, so that it can be ignored in the ordering process. Table 3 (after Hay, 1972) shows the values of various combinations of n and N

TABLE 3. Probability values of combinations of n and N.

	>0.5	>0.75	>0.9	>0.95	>0.99
No. preferred seq./total no. in seq.	2/2	3/3	5/5	6/6	8/8
	4/5	5/6	7/8	7/7	9/9
	5/7	7/9	9/11	8/9	10/10
	6/8	8/11	10/13	9/10	10/11
	7/10	9/12	11/14	10/12	11/11
	8/12	9/13	12/15	11/13	11/12

calculated by the above-mentioned formula. Any pair with a probability less than a specified sum (a probability so low as not to be stratigraphically useful) will be placed at the bottom of the order, and a note stating the number still in the order will be printed. In the past, we have zeroed pairs having probabilities of 0.5 and lower. The programme will print any part of the original occurrence array or the probability array desired; both arrays can be printed either before or after ordering, so that all recorded occurrences can be studied. Furthermore, all occurrences of any fossil can be deleted from the matrix before ordering.

In its present state, our program can analyse 170 species (340 datum surfaces) simultaneously. Insofar as the number of datum surfaces compared is approximately twice the number of species analysed to the $2\frac{1}{2}$ power, cost increase exponentially as additional species are added. For example, if 170 species were actually used, computer storage requirements would exceed 2 million bytes.

Figure 8, the Oligocene matrix compiled by Worsley and Jorgens (1974), is an example of an unordered matrix. It compares the 42 datums of 21 of the more significant species in Table 1 for the 51 sections studied. The 102 datum matrix is available but is too large to display conveniently. As explained above, such a matrix contains all the information necessary for obtaining the most probable sequence. However, such large matrices are exceedingly difficult to solve by hand. Furthermore, as indicated above the computer space necessary, and therefore the cost, increases in proportion to twice the number of taxa to the $2\frac{1}{2}$ power.

After the occurrence matrix has been ordered, the most probable sequence is printed as shown in Table 4. The position number is sequential from the top down in the left half of the column. The bottom number will equal twice the number of taxa analysed. The datum numbers shown correspond to the initial list (Table 1). The adjacent link expectation column on the right, shows the probability that any datum is in the proper sequence with the one just below it, based on the information given. It should be realised that only a portion of the results shown in the most probable sequence are valid. In this case, only the Oligocene was being analysed so that raw occurrence data would primarily be collected only from Oligocene sections or the Oligocene parts of longer sections. Inasmuch as the highest occurrence of *Sphenolithus ciperoensis* (position 7) approximates the Oligocene–Miocene boundary (Martini, 1971) so that data collection above this level was artificially truncated, results are incomplete above this datum. *Discoaster saipanensis* (position 54) approximates the Eocene–Oligocene boundary so that data collected below this position is likewise not complete. Therefore, in this example, valid positions would include 7 through 54, suggesting that 47 of the 102 datum surfaces of the 51 taxa analysed occur in Oligocene strata. This does not necessarily mean that 47 subdivisions are available for the Oligocene based

I/J	7	8	9	10	11	12	15	16	17	18	19	20	23	24	27	28	37	38	51	52	65	66	67	68	71	72	73	74	79	80	81	82	85	86	87	88	97	98	99	100	83	84	101	102
7		0	1	1	1	1	0	0	0	0	0	0	1	0	0	0	0	0	0	0	0	0	0	0	0	1	1	0	2	0	1	0	0	1	0	1	1	0	0	0	0	0	0	0
8	21		1	1	0	1	18	0	0	0	0	0	10	0	2	0	19	0	0	0	7	0	0	0	5	0	0	0	11	0	0	0	7	1	15	0	6	0	17	1	0	0	17	0
9	4	4		0	0	0	7	0	6	1	5	0	5	0	7	0	6	0	0	0	0	0	0	0	0	0	0	0	8	0	4	4	0	0	0	0	0	0	7	0	0	0	6	0
10	4	4	11		0	0	8	0	9	0	6	0	5	4	7	6	6	5	0	0	0	0	0	0	0	0	0	0	8	6	4	4	0	0	0	0	0	0	7	5	0	0	6	0
11	4	0	0	0		0	7	0	1	1	1	0	0	0	1	0	4	0	0	0	0	0	0	0	0	0	5	1	0	0	1	0	0	0	0	0	0	0	2	0	0	0	0	0
12	8	7	0	0	22		18	3	4	0	5	0	11	0	0	2	12	8	0	0	11	8	11	0	5	4	12	5	16	0	7	5	13	8	7	6	13	0	14	0	0	0	17	9
15	0	1	1	1	1	0		0	1	1	1	0	0	0	0	0	0	0	0	0	0	0	2	0	0	0	1	0	1	0	0	0	0	0	0	1	0	0	1	0	0	0	0	0
16	19	15	0	0	18	8	36		11	2	13	3	20	13	20	9	24	17	0	0	9	7	13	0	6	5	12	7	30	14	15	11	12	9	15	14	13	9	23	15	8	5	27	18
17	10	10	1	1	4	1	13	5		0	7	2	10	8	15	0	11	10	0	0	0	0	0	0	4	4	0	0	15	11	9	9	4	4	7	7	4	4	13	13	0	0	11	11
18	10	10	9	0	4	0	15	11	23		15	7	10	9	15	14	11	11	0	0	0	0	4	0	4	4	0	0	15	14	9	9	4	4	7	7	4	4	13	13	4	4	11	11
19	9	9	1	0	5	1	13	3	3	0		0	9	8	11	2	12	12	0	0	4	4	0	0	0	0	0	0	14	0	10	10	4	4	6	6	4	4	13	9	0	1	12	12
20	9	9	0	0	6	0	16	7	13	1	19		10	9	14	9	12	12	0	0	4	4	4	0	0	0	0	0	15	13	10	10	4	4	6	6	4	4	13	13	4	0	12	12
23	6	2	1	1	0	0	9	0	1	1	1	0		0	1	0	11	5	0	0	5	2	0	0	0	0	0	1	0	1	3	2	4	1	6	3	0	2	8	1	0	0	10	4
24	12	11	0	1	11	0	20	2	3	2	2	1	21		11	2	19	14	0	0	9	7	8	0	4	0	8	6	21	4	11	0	8	4	10	9	12	5	17	0	4	0	17	9
27	12	11	2	2	9	0	17	1	1	1	1	0	9	4		0	16	14	0	0	6	6	7	1	4	4	7	0	18	3	7	5	7	7	10	9	8	0	15	4	0	0	17	16
28	13	13	0	2	11	6	21	3	0	2	12	2	15	9	25		16	15	0	0	6	6	9	0	4	4	9	8	25	12	14	11	7	7	10	10	9	6	16	13	6	0	17	17
37	0	0	0	0	0	0	0	0	0	0	0	0	0	0	0	0		0	0	0	0	0	0	0	0	0	0	0	0	0	0	0	0	0	0	0	0	0	0	0	0	0	0	0
38	19	11	0	1	12	2	24	1	1	0	0	0	14	3	2	1	28		0	0	11	0	5	1	5	0	9	0	21	0	0	1	10	1	14	11	14	0	23	5	0	0	26	1
51	0	0	0	0	0	0	0	0	0	0	0	0	0	0	0	0	0	0		0	0	0	0	0	0	0	0	0	0	0	0	0	0	0	0	0	0	0	0	0	0	0	0	0
52	0	0	0	0	4	0	0	0	0	0	0	0	0	0	0	0	0	0	4		0	0	0	0	0	0	0	0	0	0	0	0	4	0	0	0	0	0	0	0	0	0	4	0
65	0	0	0	0	0	0	0	0	0	0	0	0	1	0	0	0	0	0	0	0		0	0	0	0	0	0	0	1	0	1	0	0	0	0	0	1	0	1	0	0	0	0	0
66	7	6	0	0	11	0	9	0	0	0	0	0	7	1	0	0	11	0	0	0	11		5	1	4	0	8	0	10	0	7	0	10	2	6	5	11	1	11	0	0	0	10	0
67	5	5	0	0	0	1	6	0	0	0	0	0	0	0	1	0	5	2	0	0	4	1		0	0	0	4	3	0	1	0	0	4	2	5	5	4	0	0	1	0	0	5	2
68	5	5	0	0	12	0	13	0	0	0	4	0	8	0	8	0	7	4	0	0	6	4	13		0	0	11	6	12	5	4	0	6	5	5	5	9	6	8	5	0	0	10	5
71	0	0	0	0	0	1	0	0	0	0	0	0	0	0	0	0	0	0	0	0	0	0	0	0		0	0	0	0	0	0	0	0	0	0	0	0	0	0	0	0	0	0	0
72	4	0	0	0	6	1	6	0	0	0	0	0	0	0	0	0	5	0	0	0	4	0	0	0	7		0	0	6	0	4	0	6	0	4	0	4	0	6	1	0	0	7	1
73	4	0	0	0	2	0	5	0	0	0	0	0	0	0	1	0	5	1	0	0	0	0	1	0	0	0		0	0	0	0	0	0	0	4	0	0	2	0	0	0	0	4	1
74	6	5	0	0	11	1	12	1	0	0	0	0	7	1	0	0	10	0	0	0	8	0	8	0	0	0	14		12	0	0	0	6	0	5	5	11	0	9	0	0	0	11	2
79	13	7	1	1	0	2	15	0	1	1	1	0	0	0	1	0	14	3	0	0	6	0	0	1	0	0	0	1		0	3	1	7	1	12	0	7	1	9	1	0	0	15	3
80	18	16	0	2	18	0	30	3	5	2	0	1	20	10	22	2	24	20	0	0	10	8	12	2	6	4	13	8	35		16	12	11	8	15	14	15	9	24	13	0	0	26	19
81	10	0	1	1	6	1	13	0	1	1	1	1	10	2	3	0	12	0	0	0	5	0	0	0	4	1	0	0	11	1		0	4	0	9	0	5	0	12	1	0	0	11	0
82	13	11	1	1	8	2	15	1	1	1	1	1	11	0	9	0	15	11	0	0	7	0	0	0	5	0	0	0	16	1	17		5	0	10	9	7	1	17	3	0	0	14	0
83	0	1	0	0	0	0	0	0	0	0	0	0	1	0	0	0	0	1	0	0	0	0	1	1	0	0	0	1	1	0	1	1		0	0	1	1	0	1	0	0	0	0	0
84	7	5	0	0	13	1	12	0	0	0	0	0	7	1	0	0	11	7	4	0	10	6	5	1	6	0	7	0	10	0	6	0	14		7	6	11	0	12	0	0	0	14	1
85	0	0	0	0	0	1	0	0	0	0	0	0	0	0	0	0	0	0	0	0	0	0	0	0	0	0	0	0	1	0	0	0	0	0		0	0	0	0	0	0	0	0	0
86	14	0	0	0	0	1	14	0	0	0	0	0	7	0	1	0	14	0	0	0	6	0	0	0	0	0	0	0	0	0	0	0	6	0	15		0	0	11	0	0	0	12	0
87	6	2	0	0	0	0	8	0	0	0	0	0	0	0	0	0	9	0	0	0	6	0	1	0	0	0	0	0	3	0	0	0	5	0	6	0		0	0	0	0	0	8	1
88	8	7	0	0	13	0	13	0	0	0	0	0	10	1	0	0	14	0	0	0	11	7	0	2	4	0	9	0	14	2	7	5	11	0	7	6	16		12	6	0	0	14	0
97	0	2	1	1	6	1	7	0	1	1	1	1	2	0	1	0	6	0	0	0	4	0	0	0	0	0	0	0	4	0	0	0	4	0	0	2	0	0		0	1	1	6	0
98	19	13	0	2	15	0	23	1	1	1	5	1	16	0	14	0	23	12	4	0	11	0	7	1	6	4	9	0	23	4	16	7	12	0	13	11	12	2	30		0	0	23	12
99	0	0	0	0	4	0	6	0	0	1	0	1	0	0	5	1	0	0	0	0	0	0	0	0	0	0	4	4	5	0	0	0	0	0	0	0	4	0	5	0		0	0	0
100	0	0	0	0	6	0	8	1	0	1	4	0	4	0	7	0	0	0	0	0	0	0	0	0	0	0	4	4	8	0	0	0	0	0	0	0	4	0	5	0	10		0	0
101	0	1	0	0	0	0	0	0	0	0	0	0	1	0	0	0	0	0	0	0	0	0	1	0	0	0	1	0	2	0	0	0	0	0	0	2	0	0	1	1	0	0		0
102	18	12	0	0	17	4	27	0	0	0	0	0	13	5	1	0	26	12	4	4	10	7	8	2	7	5	10	6	23	1	0	0	14	8	14	12	13	0	24	6	0	0	32	

TABLE 4. The most probable sequence for the raw data of Fig. 8.

SITION JMBER	DATUM NUMBER		FOSSIL NAME	LINK: EXPECTATION	
1	2	EXT.	COCCOLITHUS ABISECTUS	EXPECTATION ↓	0.00000
2	10	EXT.	SPHENOLITHUS BELEMNOS	EXPECTATION ↓	0.00000
3	18	EXT.	TRIQUETRORHABDULUS CARINATUS	EXPECTATION ↓	0.00000
4	36	EXT.	CYCLOCOCCOLITHINA FLORIDANA	EXPECTATION ↓	.87300
5	9	EV.	SPHENOLITHUS BELEMNOS	EXPECTATION ↓	0.00000
6	16	EXT.	RETICULOFENESTRA BISECTA	EXPECTATION ↓	0.00000
7	20	EXT.	SPHENOLITHUS CIPEROENSIS	EXPECTATION ↓	0.00000
8	90	EXT.	DISCOASTER SAUNDERSI	EXPECTATION ↓	.93750
9	28	EXT.	SPHENOLITHUS DISTENTUS	EXPECTATION ↓	0.00000
10	50	EXT.	HELICOPONTOSPHAERA INTERMEDIA	EXPECTATION ↓	.54688
11	12	EXT.	ZYGRHABLITHUS BIJUGATUS	EXPECTATION ↓	0.00000
12	76	EXT.	HELICOPONTOSPHAERA PARALLELA	EXPECTATION ↓	.87300
13	17	EV.	TRIQUETRORHABDULUS CARINATUS	EXPECTATION ↓	0.00000
14	68	EXT.	PONTOSPHAERA MULTIPORA	EXPECTATION ↓	0.00000
15	84	EXT.	HELICOPONTOSPHAERA RECTA	EXPECTATION ↓	.54688
16	19	EV.	SPHENOLITHUS CIPEROENSIS	EXPECTATION ↓	0.00000
17	58	EXT.	RETICULOFENESTRA LAEVIS	EXPECTATION ↓	0.00000
18	62	EXT.	DISCOASTER LIDZI	EXPECTATION ↓	.62500
19	80	EXT.	SPHENOLITHUS PREDISTENTUS	EXPECTATION ↓	.97754
20	24	EXT.	HELICOPONTOSPHAERA COMPACTA	EXPECTATION ↓	0.00000
21	26	EXT.	BLACKITES CREBER	EXPECTATION ↓	0.00000
22	32	EXT.	HELICOPONTOSPHAERA EUPHRATIS	EXPECTATION ↓	0.00000
23	64	EXT.	CYCLOCOCCOLITHUS MARGARITAE	EXPECTATION ↓	0.00000
24	61	EV.	DISCOASTER LIDZI	EXPECTATION ↓	0.00000
25	89	EV.	DISCOASTER SAUNDERSI	EXPECTATION ↓	0.00000
26	98	EXT.	ERICSONIA SUBDISTICHA	EXPECTATION ↓	0.00000
27	34	EXT.	ERICSONIA FENESTRATA	EXPECTATION ↓	.87500
28	100	EXT.	DISCOASTER TANI	EXPECTATION ↓	.65625
29	82	EXT.	SPHENOLITHUS PSEUDORADIANS	EXPECTATION ↓	0.00000
30	83	EV.	HELICOPONTOSPHAERA RECTA	EXPECTATION ↓	.99219
31	27	EV.	SPHENOLITHUS DISTENTUS	EXPECTATION ↓	.99973
32	102	EXT.	RETICULOFENESTRA UMBILICA	EXPECTATION ↓	0.00000
33	57	EV.	RETICULOFENESTRA LAEVIS	EXPECTATION ↓	.93750
34	86	EXT.	ISTHMOLITHUS RECURVUS	EXPECTATION ↓	.71094
35	38	EXT.	ERICSONIA FORMOSA	EXPECTATION ↓	0.00000
36	40	EXT.	RETICULOFENESTRA FOVEOLATA	EXPECTATION ↓	0.00000
37	56	EXT.	CYCLOCOCCOLITHUS KINGI	EXPECTATION ↓	0.00000
38	66	EXT.	LANTHERNITHUS MINUTUS	EXPECTATION ↓	0.00000
39	74	EXT.	TRANSVERSOPONTIS OBLIQUIPONS	EXPECTATION ↓	.62500
40	63	EV.	CYCLOCOCCOLITHUS MARGARITAE	EXPECTATION ↓	.87300
41	33	EV.	ERICSONIA FENESTRATA	EXPECTATION ↓	.54688
42	25	EV.	BLACKITES CREBER	EXPECTATION ↓	0.00000
43	67	EV.	PONTOSPHAERA MULTIPORA	EXPECTATION ↓	.93750
44	8	EXT.	DISCOASTER BARBADIENSIS	EXPECTATION ↓	0.00000
45	94	EXT.	BRAMLETTEIUS SERRACULOIDES	EXPECTATION ↓	.93750
46	1	EV.	COCCOLITHUS ABISECTUS	EXPECTATION ↓	0.00000
47	31	EV.	HELICOPONTOSPHAERA EUPHRATIS	EXPECTATION ↓	0.00000
48	42	EXT.	CORANNULUS GERMANICUS	EXPECTATION ↓	0.00000
49	52	EXT.	CYCLOCOCCOLITHUS INVERSUS	EXPECTATION ↓	0.00000
50	72	EXT.	CHIASMOLITHUS OAMARUENSIS	EXPECTATION ↓	.62500
51	81	EV.	SPHENOLITHUS PSEUDORADIANS	EXPECTATION ↓	.87300
52	49	EV.	HELICOPONTOSPHAERA INTERMEDIA	EXPECTATION ↓	0.00000
53	73	EV.	TRANSVERSOPONTIS OBLIQUIPONS	EXPECTATION ↓	.62500
54	88	EXT.	DISCOASTER SAIPANENSIS	EXPECTATION ↓	0.00000

on these species, because many adjacent link expectations in Table 4 are zero. This means that the positions of those adjacent taxa are interchangeable, i.e. their preferred order is not known for the sections analysed. Also note that on the print-out of the most probable sequence (Table 4), it is not possible to compare the sequential probabilities of non-adjacent links. For example, Table 4 does not show the link probability of the LOS (datum 19) and HOS (datum 20) of *S. ciperoensis* even though *S. ciperoensis* obviously evolved before it became extinct. However, such probabilities can be found by knowing the preferred sequence and consulting the unordered *X*-array (Fig. 8). For example, the table shows that datum 19 occurs below 20 in 19 sections without reversal, corresponding to a probability of 1.00. However, the process of calculating sequential probabilities of non-adjacent links can become very time consuming, considering that for just this sample table 1722 ($=42^2-42$) pairs are available. Obviously, it is inefficient to hand calculate so many relationships. It should be mentioned that non-adjacent link probabilities can also be found by consulting the ordered *X*-array. However, we have found little practical use for the ordered *X*-array because the alphabetical ordering and numbering system of fossils it contains has been destroyed, thereby making it difficult to locate taxa.

For the above reasons, it was found expedient to machine construct most probable sequences for which the probability of the weakest acceptable link could be specified in advance, thereby eliminating the need to sort through links of little or no significance. For instance, one might desire a most probable sequence at the 0.99 level from the 47 available datums shown in Table 4. In this case, the computer would start with position 7 (the top of the Oligocene) and check downward link probabilities with positions 8, 9, 10, etc. until it found a link at or above 0.99; in this case, position 13. From position 13, it would sequentially check lower positions until it found another link above 0.99; namely, position 28. From position 28, links exist with positions 42, 44, 51, 53, and 54, but none of these positions links with one another. As position 54 is by definition the bottom of the Oligocene, the search is therefore complete starting from position 7 and four links have been found at the 0.99 level. These links are stored and a search would then initiate downward from position 8, recording all links above 0.99 and storing these. The process would continue until searches had been initiated from all the positions through 54, the bottom of the Oligocene. With searching complete, the longest chains available at the 0.99 level are printed. We have found it convenient when synthesising a subdivision to have chains generated at the 0.99, 0.95, 0.9, 0.8, 0.7, and 0.5 levels. These have been condensed and are shown in Figs 9, 10, 11, 12 and 13. The 0.5 sequence has been deleted because of its complexity and inherent weakness. It should be mentioned here that if a worker wishes

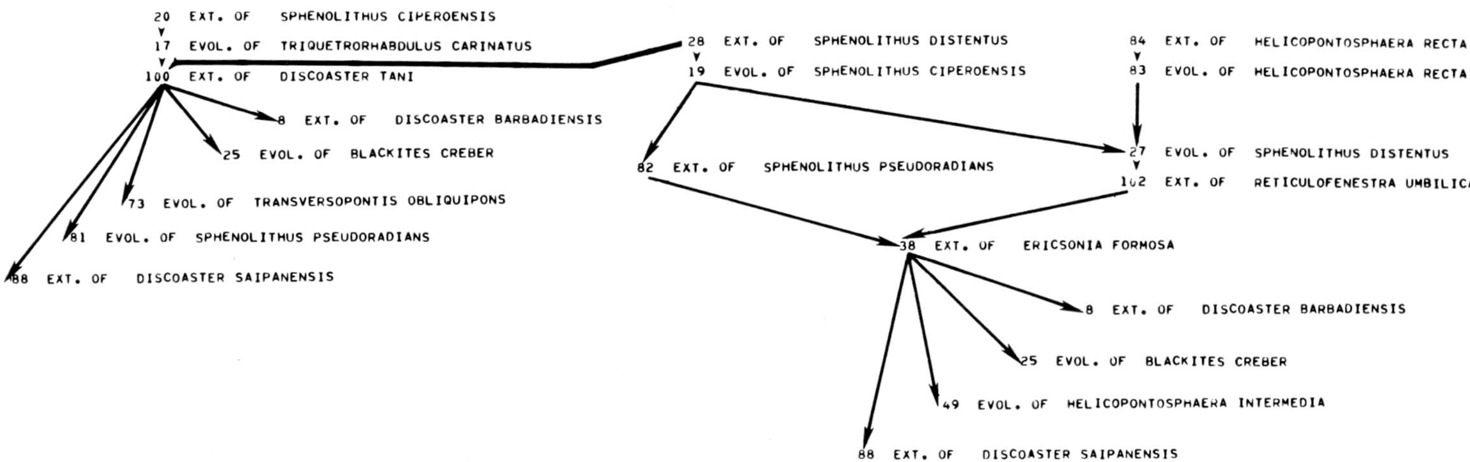

FIG. 9. The most probable sequence of the datums in Table 4 at minimum link value 0.99.

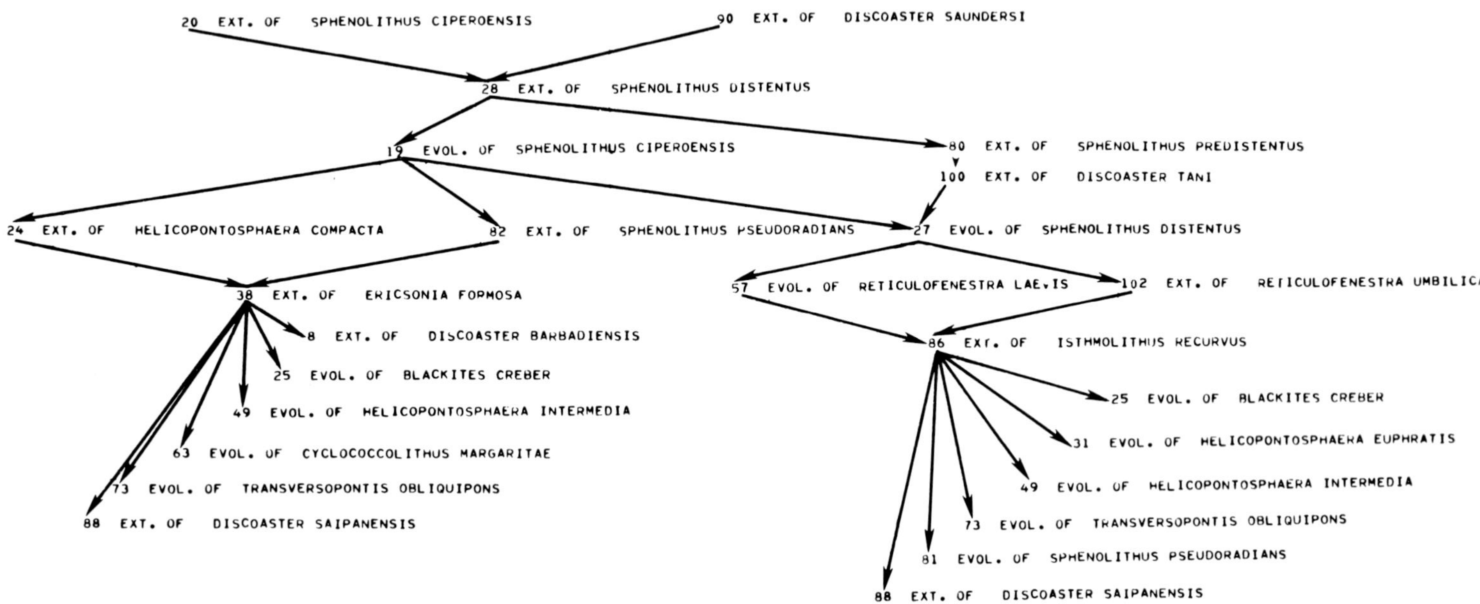

FIG. 10. The most probable sequence of the datums in Table 4 at minimum link value 0.95.

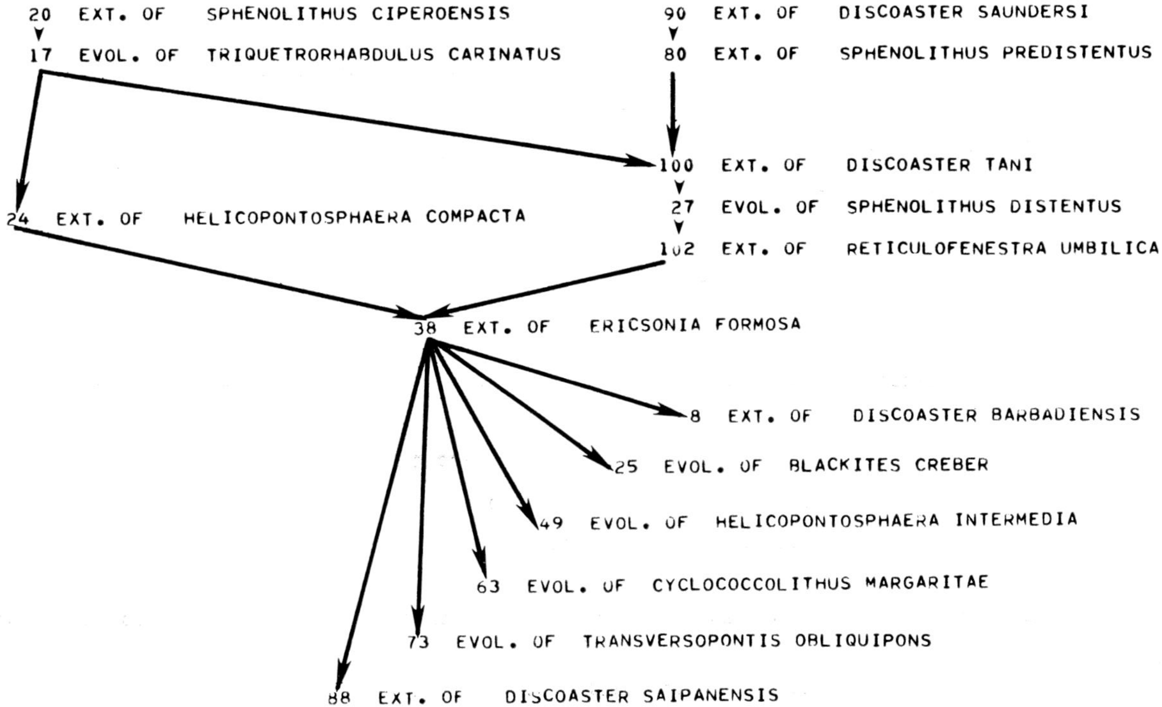

FIG. 11. The most probable sequence of the datums in Table 4 at minimum link value 0.9.

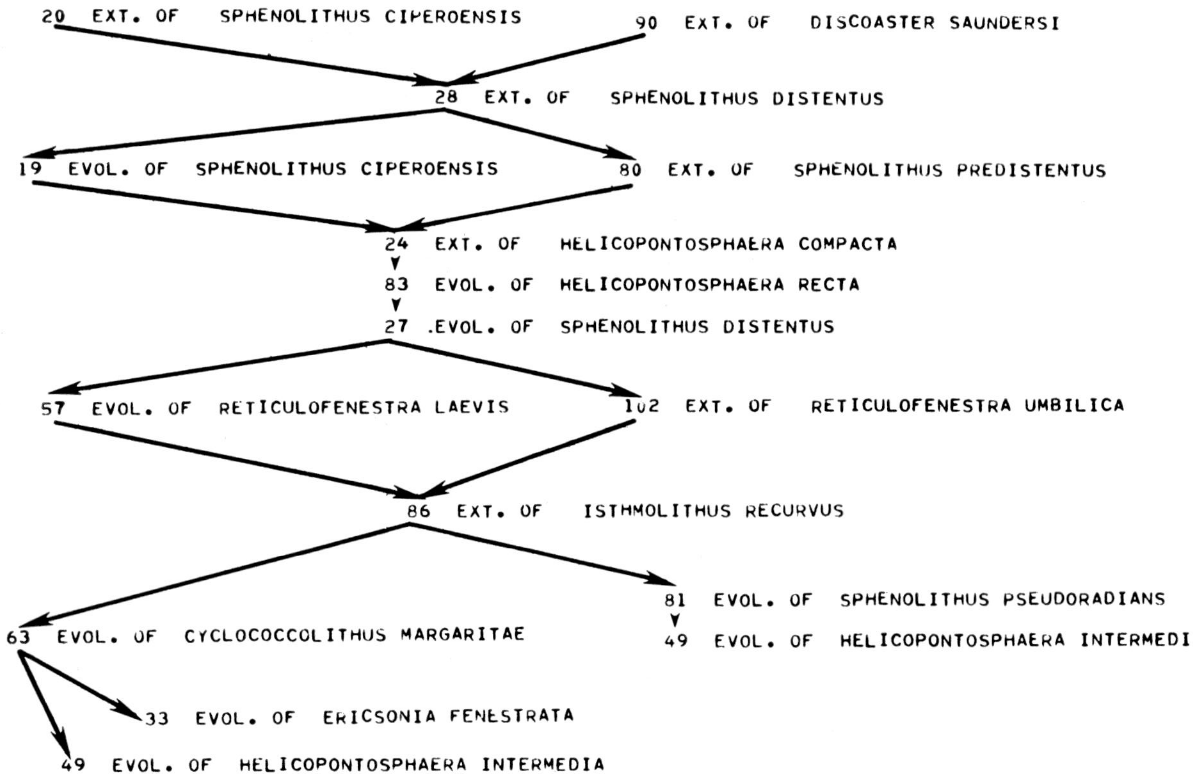

FIG. 12. The most probable sequence of the datums in Table 4 at minimum link value 0.8.

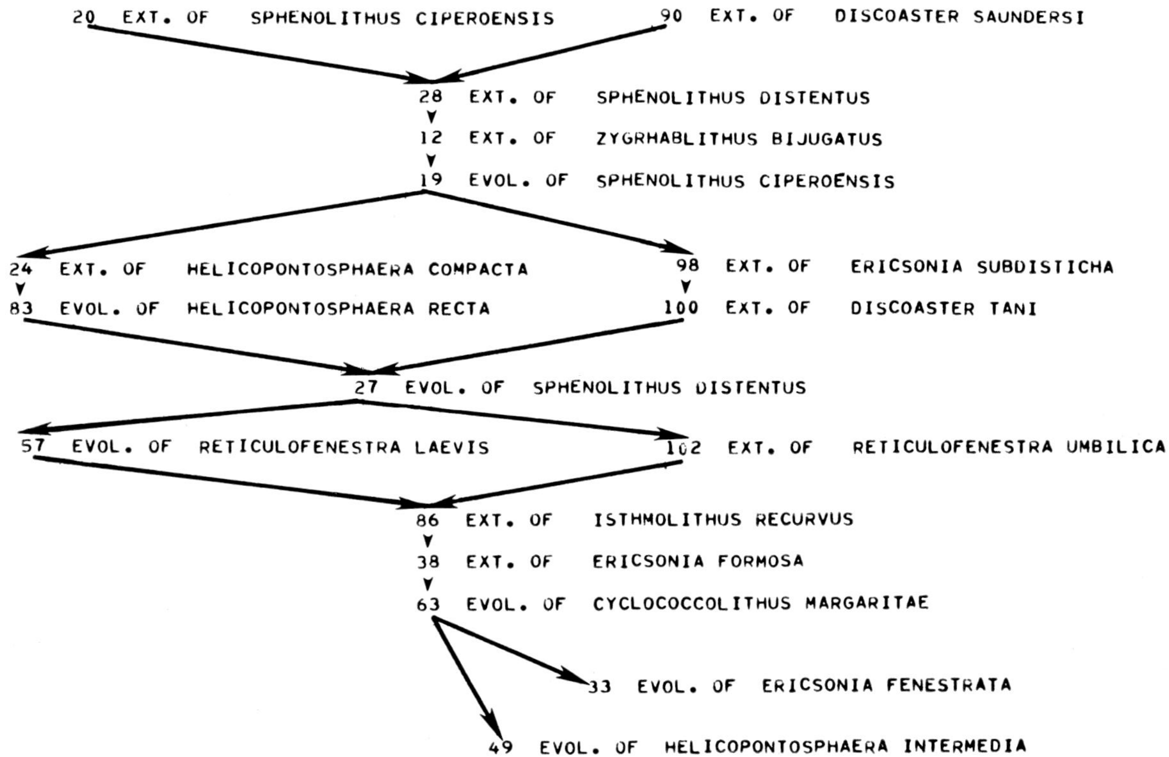

FIG. 13. The most probable sequence of the datums in Table 4 at minimum link value 0.7.

to subdivide only a part of the column, he simply selects the desired interval by specifying the top position and the bottom position.

Suppose, however, that instead of a most probable sequence for all sections, one wishes to construct one only for a subgroup of sections. This can easily be done by hand sorting the sections, or if the parameters for sorting are known, machine selecting them and then finding a most probable sequence for that set. For instance, using our programme, it is possible to sort sections by latitude and/or longitude, water depth, author, tectonic plate, ocean or continent name. For example, one can request a sequence at the 0.90 level for the entire Pacific Ocean or another for the North Pacific Ocean at the 0.80 level, or perhaps a 0.95 subdivision of all the neritic sections studied by a specific investigator. Many other combinations are possible, but it must always be remembered that the number of links obtained at a particular level of probability is generally directly proportional to the number of sections available and inversely proportional to the minimum acceptable probability of the links. We have found that a minimum of about 10 sections is required before probabilistic solutions are superior to the intuitions of experienced biostratigraphers; a prerequisite which it is not always possible to meet.

After subdivisions have been established for different geographical, sedimentary, or climatic regimes, they can be compared. This problem is not as yet readily programmable, and so, comparisons are still in part done by hand. The fence diagram in Fig. 14 represents one possible way of illustrating the relationships among sequences for different regimes. In this case, the length–width relationships on the diagram do not represent geographical coordinates but, rather are end members of subgroups of climatic and water–depth regimes. Time represents the vertical axis. Much information can be condensed into such a diagram and we expect that similar diagrams will rapidly become widely used in biostratigraphy for presentation of complex results. The reader should be aware that the numbers of datums on Fig. 14 do not correspond to those of the rest of this paper.

Careful examination of Fig. 14 should reveal that many of the relationships among the surfaces should be amenable to mapping on true geographical coordinates. Of special interest are those which seem to cross each other in a manner similar to that shown in Fig. 2; i.e. those which seem to reverse position between "provinces". Fortunately, it is possible to machine-pick possible candidates and make "calcomp maps" of them without resort to fence diagrams. The basis for choosing lies in the fact that links having a high probability of being in sequence occur many times and have few reversals. Conversely, links with low probabilities of being in sequence either result because they occur in few or no sections or they occur in many sections but have frequent reversals.

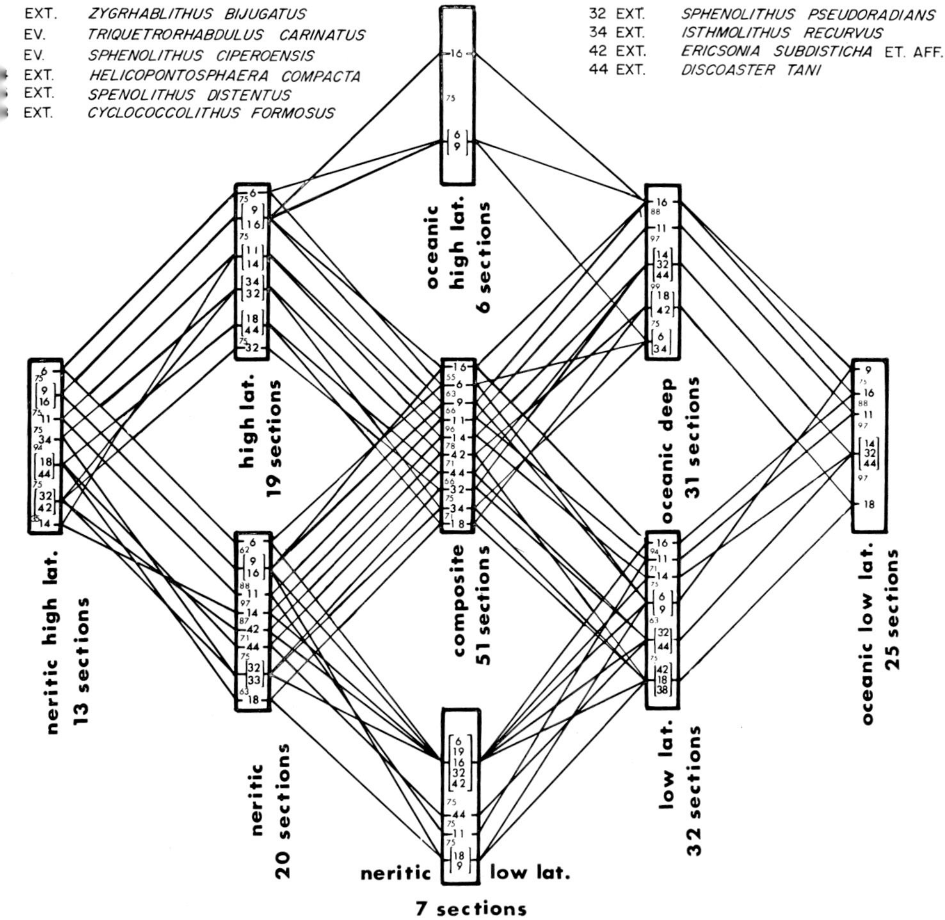

FIG. 14. Fence diagram showing the relationship of some Oligocene datum surfaces (after Worsley *et al.*, 1973).

The first case is of little interest, but the second is important because in sorting for such pairs, one has the potential of machine recognition of the type of fossil provinces discussed earlier in this chapter. We have elected to sort for pairs occurring in sequence at least eight times but having sequential probabilities below 0.8. Consultation of Table 3 shows that this would require reversals in at least 25% of those sections. For example, the HOS of *S. pseudoradians* and the HOS of *R. umbilica* occur in sequence 12 times but the probability of a preferred sequence for the two is only 0.61. However, the

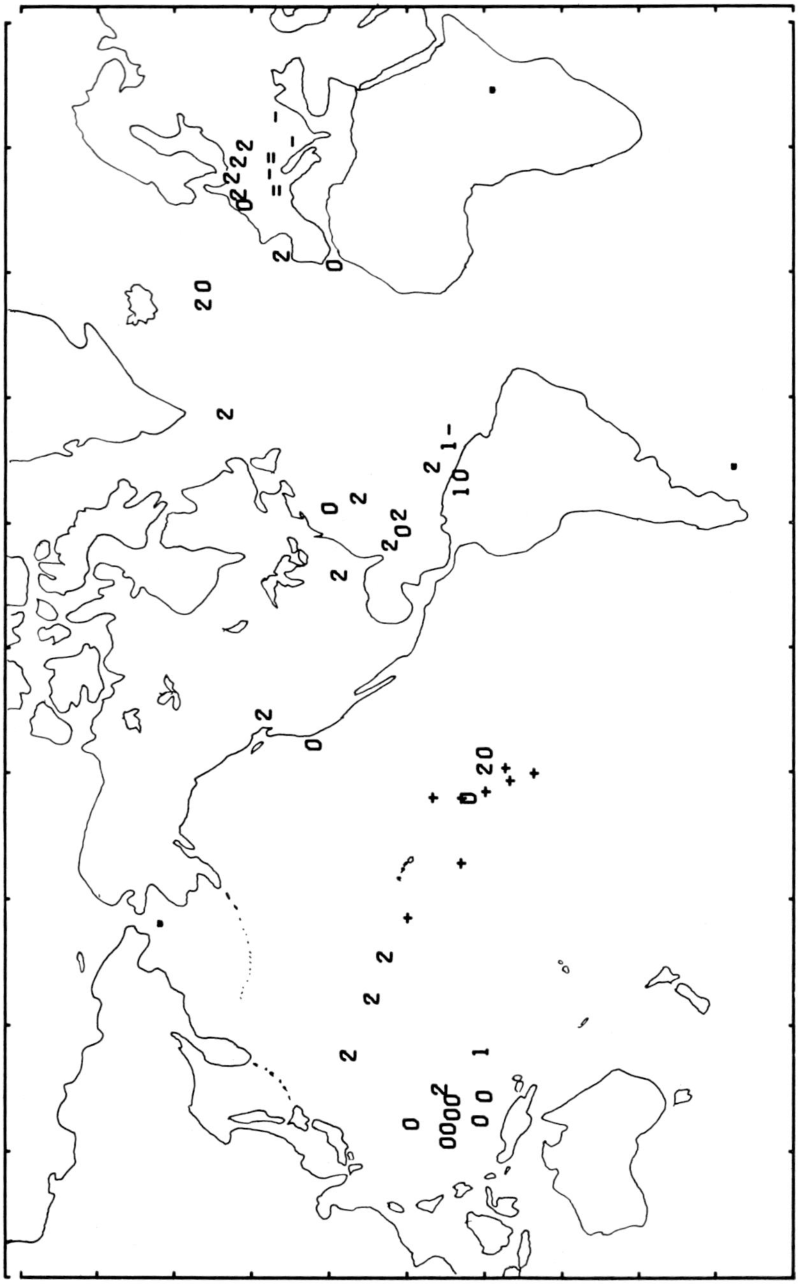

FIG. 15. Stratigraphic relationship between the HOS of *Sphenolithus pseudoradians* and that of *Reticulofenestra umbilica*.

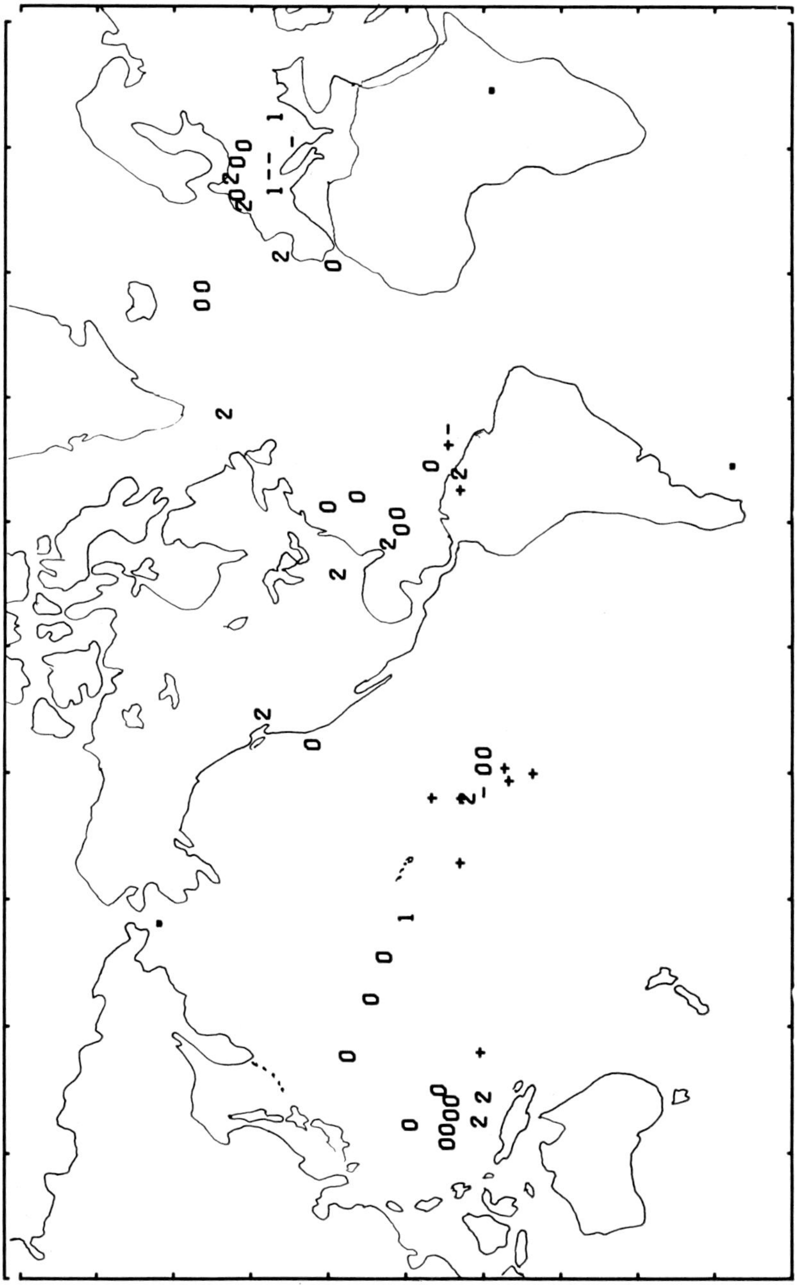

FIG. 16. Stratigraphic relationship between the HOS of *Sphenolithus pseudoradians* and the LOS of *Sphenolithus distentus*.

HOS of *S. pseudoradians* occurs below that of *R. umbilica* at 0.99 in the Pacific Ocean whereas the opposite sequence exists in Europe and Central America at 0.88. Figure 15 is a map of this situation, and shows that the sequence of these two surfaces is reversed between Europe and the Pacific Ocean. Figure 14 shows a similar relationship for these two surfaces between high latitude neritic and low latitude oceanic section also but the map gives a far more accurate idea of the areal configuration of this reversal pattern. Of course, it is plausible to have a pair occur in sequence many times, have a low probability of preferential sequence, and still not have a "province" defined. Figure 16 illustrates a case for which the reversals appear random. This can sometimes be caused because the two surfaces are so nearly congruent and so very closely appressed so that the sequence one finds for any given section is strictly a matter of randomly finding one above the other. In other cases, gradual evolution may be the culprit so that it becomes very difficult to define the exact point at which one species evolves into another, or in other words, the endpoints of their stratigraphic ranges. In any case, such closely appressed surfaces can be used interchangeably in most practical biostratigraphical applications. A well-known example of such a pair would be the HOS of both *D. saipanensis* and *D. barbadiensis*, which, while not identical, approximate each other closely enough in many regimes for most practical problems.

Conclusion

If a conclusion is to be drawn from this chapter it is that studies of this kind have left the experimental stage and have moved into the developmental stage. The approach they use may soon dominate palaeontological thinking.

At the time of this writing, further studies are progressing on calcareous nannofossils from the entire Cainozoic and Cretaceous together with similar studies for planktonic foraminifera and radiolaria. Ultimately it will be possible to combine information obtained for all of these microfossil groups into a single scheme offering an almost limitless flexibility for both biostratigraphical subdivision and provincial differentiation among the taxa within these groups for their entire geological record.

Acknowledgments

This chapter is the outgrowth of several years' research effort among numerous persons at the University of Washington. We are especially indebted to Bill Snow and Craig Suchland who devised the computer programs used in this study and to Gretchen Blechschmidt for stimulating scientific discussion and technical assistance in researching this topic.

This research was supported by National Science Foundation grants: GU-2655, GA-36628X, and by the University of Washington Department of Oceanography. This is contribution number 934 of the Department of Oceanography, University of Washington.

References

Dennison, J. M. and Hay, W. W. (1967). Estimating the needed sampling area for subaquatic ecologic studies. *J. Paleont.* **41,** 706–708.

Dennison, J. M. and Hay, W. W. (1972). Probabilistic Stratigraphy. *Eclog. Geol. Helv.* **65,** 255–266.

Krumbein, W. C. and Sloss, L. L. (1963). *Stratigraphy and Sedimentation*, 660 pp. Freeman and Co., San Francisco.

Martini, E. (1971). Standard Tertiary and Quaternary calcareous nannoplankton zonation. *In* (Farinacci, A., ed.) *Proc. II Plankt. Conf.* **2,** 739–786. Tecnoscienza, Roma.

Shaw, A. B. (1964). *Time in Stratigraphy.* McGraw Hill, New York.

Worsley, T. R. (1974). The Cretaceous–Tertiary boundary event in the ocean. *In* (Hay, W. W., ed.) *Studies in Palaeo-oceanography. Soc. Econ. Paleont. Mineral Spec. Publ.* **20,** 94–125.

Worsley, T. R. and Jorgens, M. (1974). Oligocene calcareous nannoplankton provinces. *In* (Ross, C. A., ed.) *Palaeogeographic Provinces and Provinciality. Soc. Econ. Paleont. Mineral. Spec. Publ.* **21,** 85–108.

Worsley, T. R., Jorgens, M. , Blechcchmidt, G., Ralston, S. and Snow, B. (1973). Probability-based analysis of the area–time distribution of Oligocene calcareous nannofossils. *In* (Smith, L. and Hardenbol, J., eds.) G.C.A.G.S.–S.E.P.M. Symposium on calcareous nannofossils.

15: Dinocysts

Their classification, biostratigraphy and palaeoecology

GRAHAM L. WILLIAMS

Atlantic Geoscience Center,
Geological Survey of Canada,
Bedford Institute of Oceanography,
Dartmouth,
Nova Scotia, Canada

The comprehensive studies of fossil dinoflagellates have confirmed their significance as biostratigraphical markers. Thirty Concurrent Range Zones and Twelve Subzones, named for index species, delineate the Late Triassic, Jurassic, Cretaceous and Tertiary. A distinctive suite of species characterises each zone or subzone, which is usually cosmopolitan in the Jurassic, Cretaceous, Palaeocene and Eocene, and regional in the Oligocene, Miocene and Pliocene. Most fossil dinoflagellates fall into one of seven lineages, defined according to their morphology. Evolutionary trends may be defined in several morphogenetic series of dinocyst taxa. Fossil dinoflagellates appear to be viable palaeoecological indicators, being sensitive to such parameters as temperature, water depth, salinity, ocean current patterns and shoreline proximity.

Outline

Introduction

Dinoflagellates (from the Greek *dinos*, a whirling rotating eddy; *flagellates*, to scourge or whip) are unicellular protists, which are enclosed in a cellular covering, the amphiesma. They have a size range of 5–2000 microns. Dinoflagellates are placed by botanists in the Class Dinophyceae within the Division Pyrrhophyta of the Algae. They possess two flagella, both of which arise from pores on the ventral surface. The transverse flagellum encircles the body in an equatorial position and lies in the transverse furrow. The longitudinal flagellum extends posteriorly (Fig. 1). The two flagella provide both the motive force and the directional control of the motile cells. Dinoflagellates have a varied habitat. The vast majority form an important constituent of the marine phytoplankton, being a vital link in the food chain and second in importance to the diatoms as producers of organic matter. Others are marine psammophytes, brackish or fresh water dwellers, parasites or symbionts.

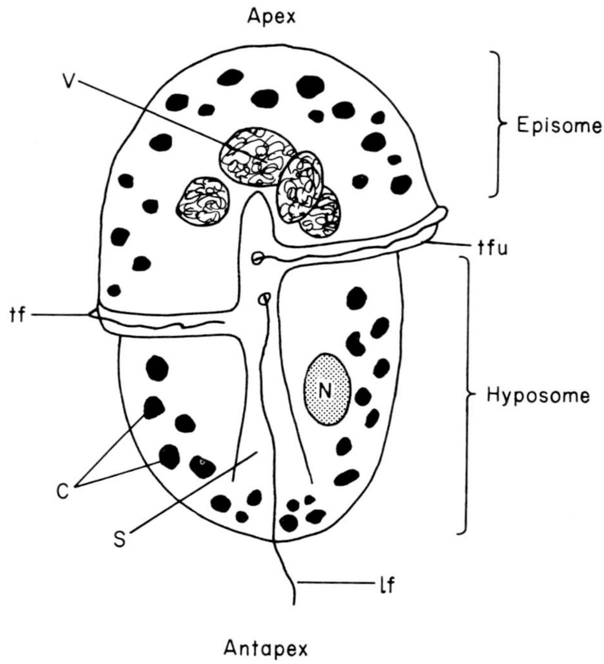

FIG. 1. Vegetative Stage of a Gymnodiniale. This is colloquially known as a "naked" dinoflagellate or gymnodinioid stage. C, chromatophore; lf, longitudinal flagellum; N, nucleus; S, sulcus; tf, transverse flagellum; tfu, transverse furrow; V, vacuole.

The majority of dinoflagellates are autotrophic, synthesising organic compounds from inorganic substances and containing chlorophyll like the green plants; others are heterotrophic, that is they are devoid of chromatophores and feed like animals by ingesting their food since they do not have the ability to synthesise organic compounds from inorganic compounds. Many dinoflagellates possess a complex life cycle including a vegetative stage popularly known as the motile stage, and an encysted stage (Fig. 2). In the encysted stage the organic wall of the dinocyst is extremely tough being readily preserved in the fossil record. The vegetative stage in contrast is readily broken down into discrete components and is rarely, if ever, preserved for posterity Unquestionable fossil dinocysts have been recorded from the Silurian. They are common in marine sediments of Permian to Recent age and are being recorded increasingly from fresh water sediments.

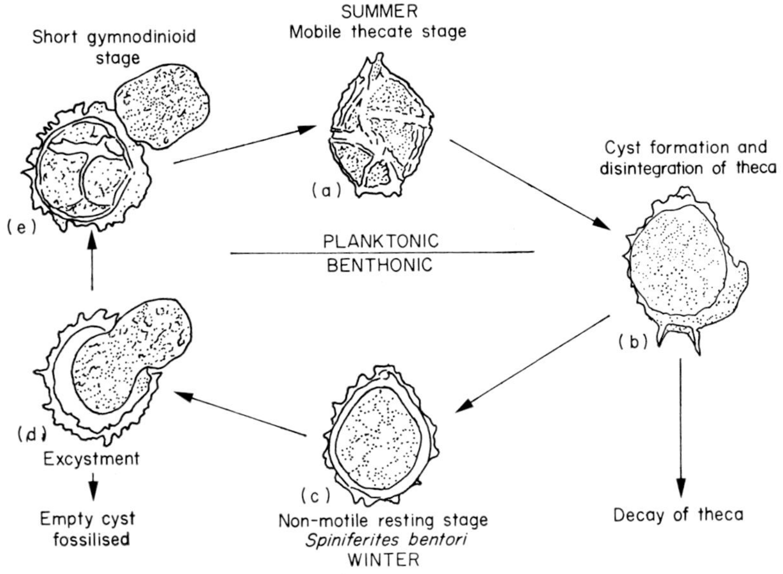

FIG. 2. Encystment-excystment cycle of *Gonyaulax digitalis* (Pouchet) Kofoid. (a), motile thecate stage; (b), cyst formation and ultimate decay of the theca; (c), non-motile resting stage (*Spiniferites bentori*); (d) excystment stage; (e) gymnodinioid stage. (From Wall and Dale, 1968.)

The encysted stage of the dinoflagellate, the dinocyst, may be similar in appearance to the vegetative stage as presumably in the fossil species formerly included in the extant genus *Gonyaulax* Diesing (Fig. 3) but now attributed to *Gonyaulacysta* Deflandre (Fig. 4), or dissimilar, when processes of varying length are commonly developed (Fig. 5). This latter type of cyst was formerly termed a hystrichosphere. Many hystrichospheres are now known to be dinocysts. others whose natural affinities are questionable have been placed in the Group Acritarcha. The acritarchs are organic walled microfossils allegedly of unknown origin. Current research indicates that many acritarchs are dinocysts of naked (unarmoured) and thecate (armoured) dinoflagellates. If this is correct the known geological range of the dinocysts would extend back into the Cambrian.

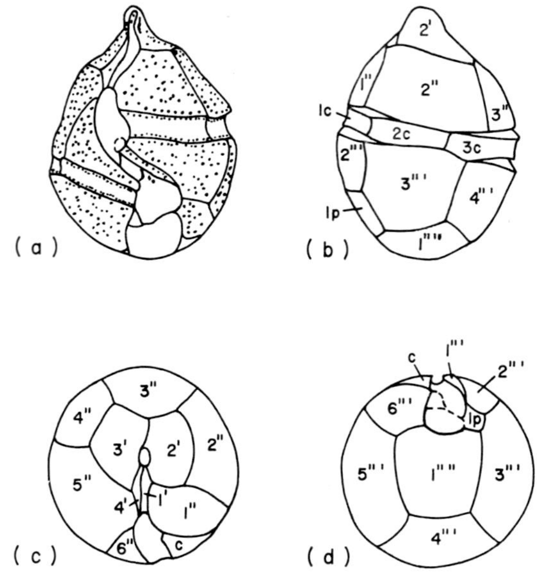

FIG. 3. Theca of *Gonyaulax spinifera* (Claparede and Lachmann). (a), ventral view; (b), lateral view; (c), apical view; (d), antapical view. (From Wall and Dale, 1970.) Kofoid's alphanumeric system has been adopted for denoting the tabulation.

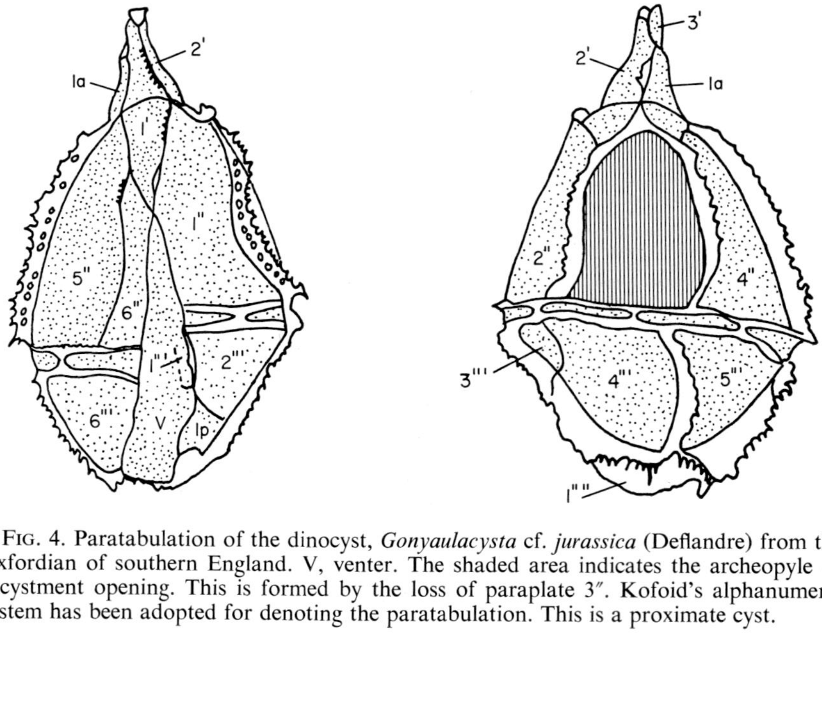

FIG. 4. Paratabulation of the dinocyst, *Gonyaulacysta* cf. *jurassica* (Deflandre) from the Oxfordian of southern England. V, venter. The shaded area indicates the archeopyle or excystment opening. This is formed by the loss of paraplate 3″. Kofoid's alphanumeric system has been adopted for denoting the paratabulation. This is a proximate cyst.

History of the study of fossil dinoflagellates

Fossil dinoflagellates were first observed by Ehrenberg (1836) but his findings did not appear in print until 1838. Ehrenberg, during a microscopical examination of flint flakes from the Late Cretaceous, observed two distinct types of microfossils. The first had an ovoidal to polygonal test with two furrows; one almost completely encircling the test, and obviously equating with the transverse furrow in the vegetative stage, and the other bisecting the first at right angles and being confined to one surface. The specimens of this group were morphologically similar to Recent species of dinoflagellates with which Ehrenberg was familiar. Consequently he placed them in the extant dinoflagellate *Peridinium* Ehrenberg, and presumed that they represented the vegetative stage. The second type had a spherical main body, apparently lacking "furrows" and bearing numerous, seemingly haphazardly orientated, spines. Ehrenberg believed these were the siliceous zygospores of the freshwater desmid genus *Xanthidium* Ehrenberg.

Mantell (1845) proved conclusively that the "xanthidia" were of organic composition and formally erected the genus *Spiniferites* in a later publication (1850) to include these organisms. His work, largely overlooked until the last decade, has been fully vindicated in the universal recognition of the validity of the generic name *Spiniferites*. Subsequent studies of fossil dinocysts were largely superficial until O. Wetzel (1933a, 1933b) published a monograph of Late Cretaceous microplankton. He erected the genus *Hystrichosphaera* to include spiny spheres formerly attributable to the xanthidia and proposed the family Hystrichosphaeridae to accommodate them. In addition, he described several fossil dinocysts which he placed in the extant genera *Ceratium* Schrank, and *Peridinium*. Later workers, notably Deflandre (1935, 1937), and Lejeune Carpentier (1938), delineated new genera within the Order Hystrichosphaeridea. Members of this order came to be known as "hystrichospheres" or, even "hystrix".

The affinites of the hystrichospheres and the relationship of the dinocyst to the life cycle of the dinoflagellate remained enigmas until the studies of Evitt (1961b, 1963), Evitt and Davidson (1964) and Wall and Dale (1968a, 1970). Evitt (1961b) called attention to the presence of regular openings in the epicyst, that part of the dinocyst equivalent to the epitheca of the vegetative stage, of fossil dinocysts. Such an opening which was formed by the loss of a single paraplate or group of paraplates he termed an archeopyle. This, he postulated correctly, was an excystment aperture for release of the protoplasm. The recognition of archeopyles and paratabulation (reflected tabulation) in many hystrichospheres has resulted in their transfer to the dinoflagellates and the erection of the Group Acritarcha for species formerly included within the hystrichospheres but whose affinities are not demonstrably dinophycean.

Today most Mesozoic "hystrichospheres" are classed as dinoflagellates, while the majority of Palaeozoic "hystrichospheres" are placed in the Acritarcha.

Most "typical" fossil dinocysts which were formerly regarded as theca, are now known to possess an archeopyle, whose position and paraplate composition are constant for a species. The vegetative stage undergoes degradation immediately after death of the organism and is not preserved in sediments. Fossil dinoflagellates are believed to represent the encysted stage which is a more resistant structure than the vegetative stage. Recently more attention to paratabulation, archeopyle type and ornamentation has led to an increase of fossil dinocyst genera to over 250.

Morphology of the cyst

The dinocyst may closely resemble the corresponding vegetative stage or be totally dissimilar, and possess processes as in *Gonyaulax digitalis* (Pouchet).

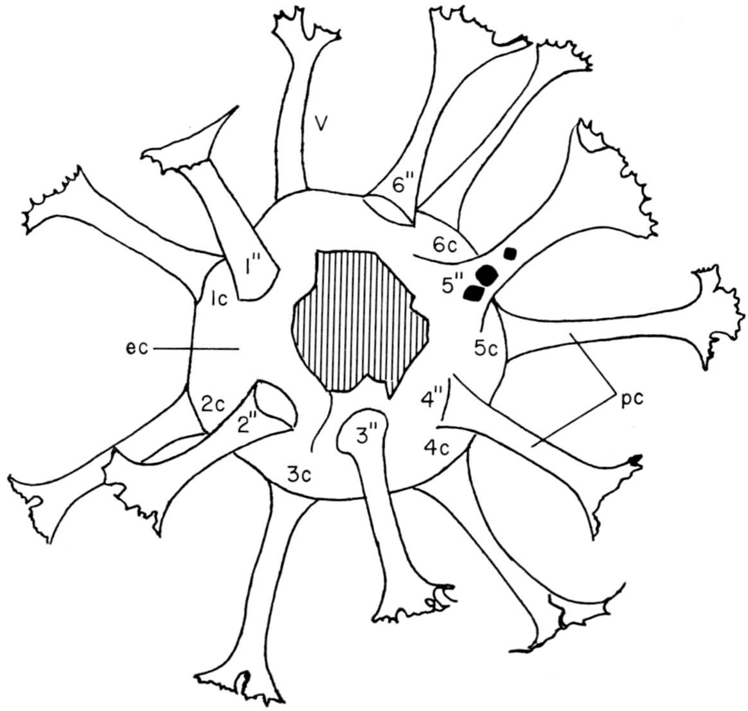

FIG. 5. Paratabulation of the epicyst of *Hystrichosphaeridium tubiferum* (Ehrenberg), as inferred from the number and position of the intratabular processes. Visible are the apical tetratabular archeopyle (shaded), precingular processes (1″, etc.), cingular processes (1c, etc.), and a ventral process (V). ec, endocoel; pc, pericoel. This is a chorate cyst. (×900)

The presence or absence of processes is reflected in the spatial relationship of the cyst and theca. This permits the recognition of *proximate* and *chorate* cysts (Downie and Sarjeant, 1966). In proximate cysts, or those devoid of processes the cyst closely approaches the theca in shape and morphology (Fig. 4). It presumably is formed in close contact with, or slightly separated from, the theca. Proximate cysts were long mistaken for thecae. Chorate cysts possess a main body which has processes, as in *Hystrichosphaeridium* Deflandre (Fig. 5), or other prominently developed ornamentation. Downie and Sarjeant believed that in chorate cysts the tips of the processes are in contact with the inner wall layer of the theca. The overall size of the cyst and theca should therefore be approximately equal. This assumption has been questioned by Wall (1971) who believes that cyst growth could occur after sloughing off of the theca.

The third major cyst type recognised by Downie and Sarjeant (1966) is the *cavate* cyst (Fig. 6). In this the two walls of the cyst do not come into contact. This group includes many cysts attributable to the Peridiniaceae such as *Deflandrea* Eisenack. The reader is referred to Sarjeant (1969) for a detailed discussion of the subdivisions of these three major cyst types. While these groups are useful for morphological separation they should not be taken to

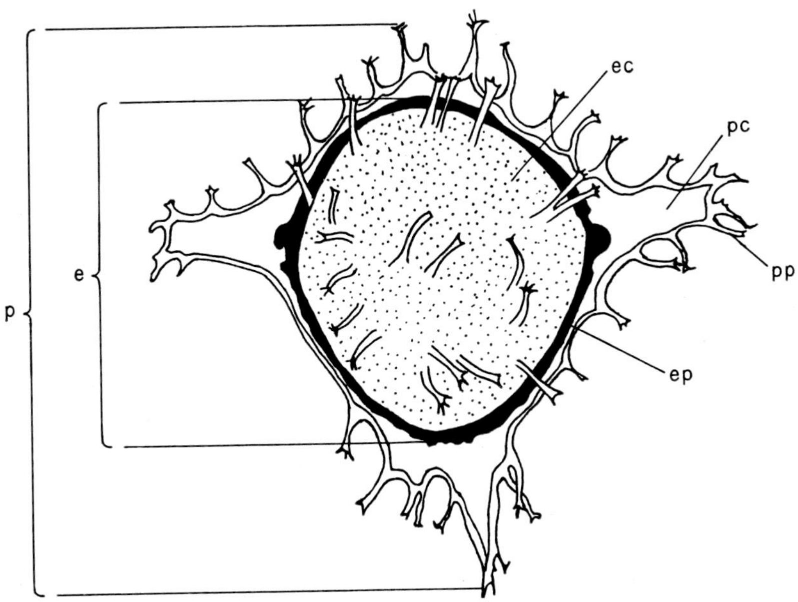

FIG. 6. *Wetzeliella lunaris* Gocht, a cavate cyst from the Early Eocene, England. e, endocyst; ec, endocoel; ep, endophragm; p, pericyst; pc, pericoel; pp, periphragm. (×550)

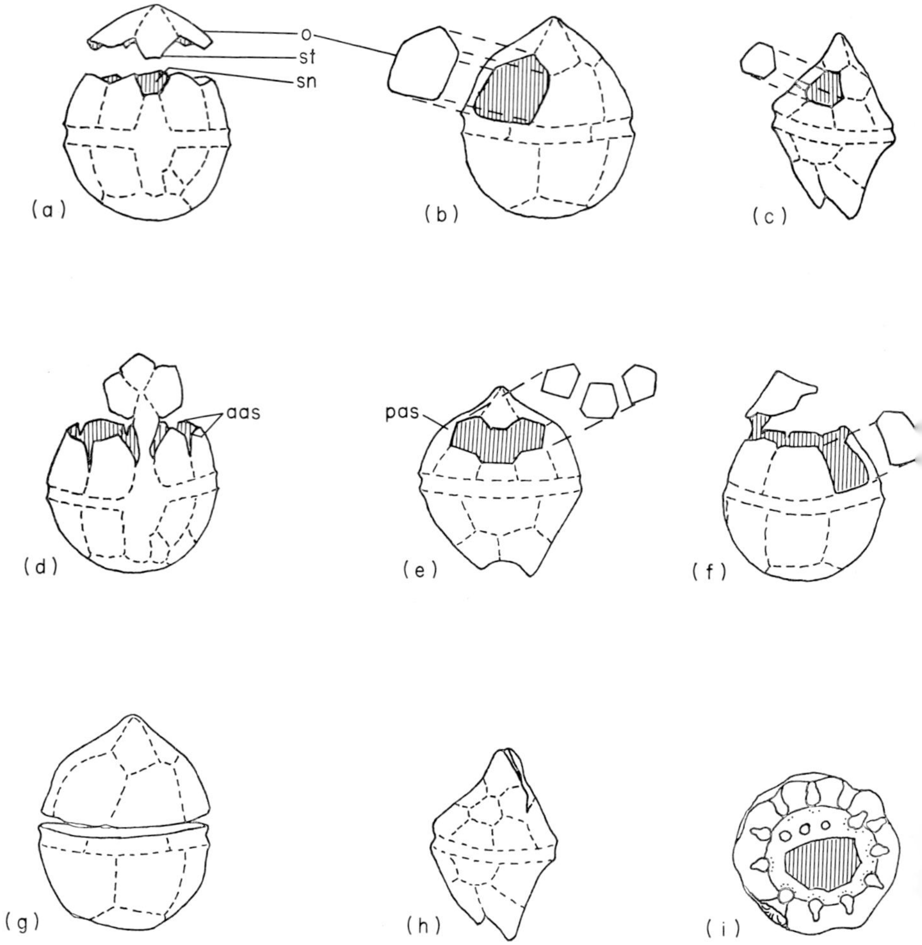

FIG. 7. Archeopyle types: (a), apical archeopyle with simple free operculum which corresponds to all the apical paraplates. Ventral view; (b), precingular archeopyle with simple free operculum which corresponds to the third precingular paraplate (3″). Dorsolateral view; (c) intercalary archeopyle with simple free operculum which corresponds to the second anterior intercalary paraplate (2a). Dorsolateral view; (d), apical archeopyle with simple attached operculum which corresponds to all four apical paraplates (1′, 2′, 3′ and 4′). Ventral view; (e), intercalary archeopyle with compound, free operculum which corresponds to all three intercalary paraplates (1a, 2a, and 3a). Dorsal view; (f), combination archeopyle with compound operculum, the apical corresponding to all four apical paraplates (1′, 2′, 3′ and 4′) is attached; the precingular piece, corresponding to the third precingular paraplate (3″) is free. Dorsolateral view; (g), combination archeopyle with simple free operculum which corresponds to the epicyst; this type of archeopyle in which the archeopyle margin lies immediately anterior to the paracingulum is also known as an epicystal archeopyle. Dorsolateral view; (h), Archeopyle which results from schism along

indicate genetic affinities among members. It is a common fallacy that the theca exerts a controlling influence on the dinocyst. Both are genetically controlled and therefore can assume a similar, or dissimilar form.

Much of the terminology for dinocyst morphology was originally applied to modern forms and has been adopted without change in meaning. Recently, however, the realisation that fossil dinoflagellates are dinocysts has led to a revision of the terminology for the fossils. The reader is referred to Williams *et al.* (1973) for a glossary of dinoflagellate terminology.

The terminology for archeopyles was originally based on fossil dinocysts. The archeopyle has been shown to be an excystment aperture in the wall of a dinocyst. It almost always results from the complete or partial removal of one or more than one paraplate (*the operculum*). There are five major categories of archeopyles, apical intercalary, precingular, combination and miscellaneous (Fig. 7). The category names indicate which series of paraplates are lost in archeopyle formation. Evitt (1967) has devised an alphanumeric system for denoting the archeopyle. By use of this system, it is possible to delineate the series of paraplates involved in archeopyle formation in the different cyst wall layers, the number of opercular pieces, whether the operculum or opercular pieces are free or attached and are composed of more than one paraplate, and the number of wall layers involved in archeopyle formation. The system is finding widespread acceptance.

Classification of dinocysts

Wall (1971) has observed an alternation of a planktonic, vegetative (motile) stage with an encysted benthic, nonmotile stage in the life cycles of over 40 extant species of dinoflagellates. Freshwater and neritic naked and thecate species are known to be capable of producing resting cysts. In the past, neontologists have concentrated their studies on the vegetative stage, while the palynologists of necessity, have exclusively studied the dinocyst. A few neontologists have briefly mentioned fossil dinoflagellates; some palynologists have attempted to assign extinct dinoflagellates to modern taxa. Generally,

---

FIG. 7 (*continued*)

the margins of adjacent lateral paraplates on the epicyst. This schism, the transapical suture, crosses the apex and extends in a right lateral and left lateral direction almost to the paracingulum. The archeopyle is in effect a slit-like opening formed by the transapical suture only, and therefore lacks an operculum. Dorsolateral view; (i), hypocystal archeopyle with compound, free operculum which corresponds to three posterior antapical paraplates. This is characteristic of *Pyrophacus vancampoae*. According to Evitt's (1967) classification this would be a miscellaneous archeopyle. Antapical view.

aas, accessory archeopyle parasuture; o, operculum; pas, principal accessory parasuture; sn, parasulcal notch; st, parasulcal tongue. (A-H, from Evitt, 1967; I, from Wall and Dale, 1971.)

however, two distinct and independent classification schemes have evolved with the consequence that the vegetative stage and encysted stage of a dinoflagellate have been given different specific and even generic names.

EXISTING CLASSIFICATIONS

Palynologists have adopted two approaches to dinocyst classification. Sarjeant and Downie (1966) recognised cyst taxa up to the familial level even where the equivalent vegetative stage was known. Other palynologists, including Vozzhennikova (1967), have assigned some fossi ldinocyst species to extant genera and families, while placing others in genera and families known only from the fossil record. Wall and Dale (1968a) found serious drawbacks to the proposed classification of Sarjeant and Downie, which they said was nonbiological and nonevolutionary. They believed that every effort should be made to combine the classification of modern dinoflagellates with that of fossil dinoflagellates. The "semi" natural classification of Vozzhennikova was criticised because of the exclusion of hystrichospheres and the failure to recognise that most fossil dinoflagellates are cysts and not thecae. This led to a disregard for paratabulation and the excystment aperture.

Schemes of classification based only on the vegetative stage of living dinoflagellates fail to take into account the cyst stage of encysting species. Thus they overlook the polyphyletic nature of some of the genera including *Gonyaulax*. A unified system of classification for dinoflagellates based on, if possible, a knowledge of all phases of the life cycle of the organism must be developed. When the vegetative stage of a species can be equated with a fossil dinocyst previously named, Wall and Dale have argued that nomenclatural priority should preferably be given to the motile stage. Extinct taxa should be defined with special reference to those morphologic features which are of most taxonomic importance in living dinocysts. As with foraminifera too much emphasis may have been placed on gross morphological features, which are often adaptations to a planktonic mode of life. This would have been equally as critical to the dinocyst as to the motile stage if excystment was to take place. Certainly more attention should be given to the micro-structure of the phragma (wall), which is known to be of importance in the classification of certain taxa, as in the genus *Cordosphaeridium* Eisenack (Pl. 6, fig. 2).

Morphological features utilised for generic classification include *paratabulation*, archeopyle type and ornamentation, including horns and processes.

ORDER PERIDINIALES

The majority of fossil dinoflagellates appear to be attributable to the Order Peridiniales, the thecate dinoflagellates. In these forms the theca is composed of a series of plates numbered alphanumerically according to the system of

TABLE 1. Alphanumeric system of plate designation in dinoflagellates (after Kofoid, 1907, 1909).

Vegetative stage, plate series	Encysted stage, paraplate series	Location	Symbol	Individual plate or paraplate: Position	Individual plate or paraplate: Symbol
Apical	Apical	Touching the apex but not the cingulum or paracingulum	′	Second Apical	2′
Anterior Intercalary	Anterior Intercalary	Between the precingulars and apicals but not touching the apex or cingulum (paracingulum); can be ventrally or dorsally located	a	Third Anterior Intercalary	3a
Precingular	Precingular	Immediately anterior to the cingulum or paracingulum and not touching the apex	″	Fifth Precingular	5″
Cingular	Paracingular or girdle	Cingulum or Paracingulum	c or pc	Fourth Cingular or Paracingular	4c 4pc
Postcingular	Postcingular	Immediately posterior to the cingulum or paracingulum and not touching the antapex	″ ′	Sixth Postcingular	6″ ′
Posterior Intercalary	Posterior Intercalary	Between the postcingulars and antapicals(s) and not touching the cingulum (paracingulum) or antapex. Usually adjacent to the sulcus or parasulcus	p	First Posterior Intercalary	1p
Antapical	Antapical	Touching the antapex but not the cingulum or paracingulum	″ ″	First Antapical	1″ ″
Sulcal	Parasulcal or Ventral	Occupying a position on the sulcus (parasulcus) or venter	s v	Anterior Sulcal (Parasulcal) or Ventral	1s 1v

nomenclature devised by Kofoid (1907, 1909) (Table 1 and Fig. 8). The fossil Peridiniales showed maximum development in periods when epicontinental seas were most widespread as in the Callovian–Oxfordian, the Cretaceous and the Paleocene–Eocene. Over 200 fossil genera are included in this order. The

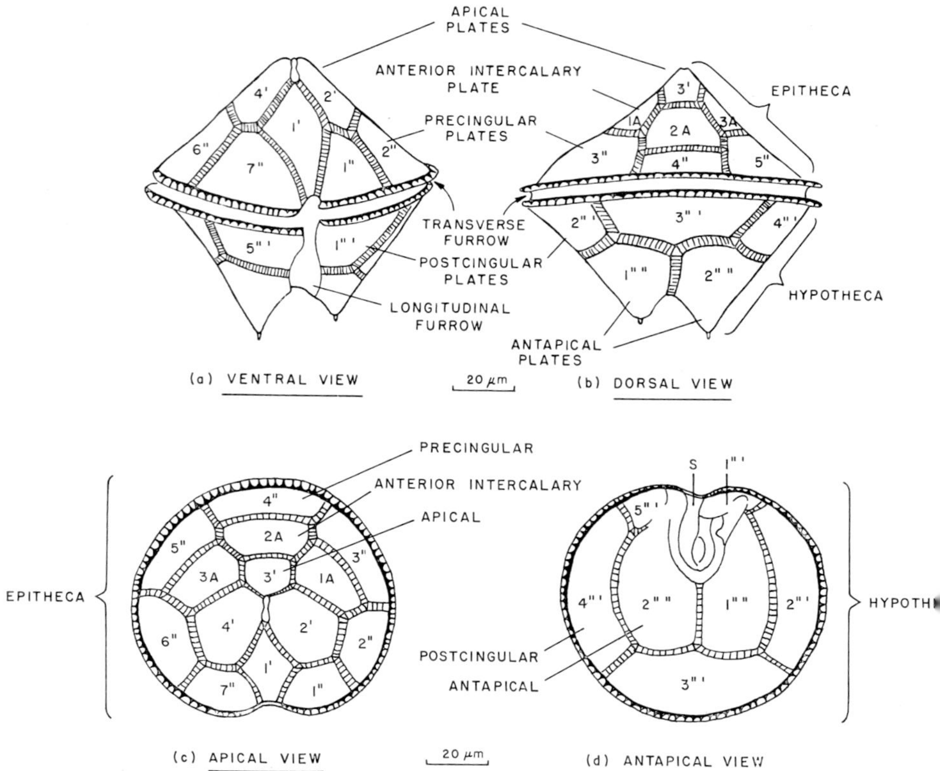

FIG. 8. Theca of *Peridinium leonis* Pavillard, showing tabulation and the alphanumeric symbols for denoting the individual plates. The tabulation or plate formula for this specimen may be expressed thus, 4′, 3A, 7″, 5″, ′, 2″ ″. This signifies that there are four apical plates, three anterior intercalary plates, seven precingular plates, five postcingular plates, and two antapical plates. Note that the ventral, dorsal, apical and antapical views are always drawn as though they are the upper surface. This gives an inverted aspect to the plate numbers. S = sulcus. (After Lebour, 1925.)

majority of these genera exhibit a shape and paratabulation that appears to relate them to one or another of the extant genera, *Peridinium*, *Gonyaulax* or *Ceratium*.

Included in *Peridinium* (Fig. 8) are over 600 species, most of which possess the plate formula 4′, 7″, 5′, 2″. Variations in plate number may occur in any of the prime series, with the exception of the postcingular series. Species differentiation is commonly based on variations in the number and arrangement of apical cingular, sulcal or anterior intercalary plates. Particularly important is the shape of the second anterior intercalary plate, 2a. This can be "quadra", "penta" (two types), or "hexa" (Fig. 9). This is also true of fossil genera, *Deflandrea* Eisenack and *Wetzeliella* Eisenack (Pl. 6, Fig. 4) respectively having a quadra and a hexa anterior intercalary. Species of the genus *Peridinium* have a typically peridinioid outline (with one apical and two antapical horns), but can also be spherical or ellipsoidal. Fossil genera with a peridinioid paratabulation include *Australiella* Vozzhennikova, *Deflandrea*, *Hexagonifera* Cookson and Eisenack, *Pareodinia* Deflandre, *Phthanoperidinium* Drugg and Loeblich, *Svalbardella* Manum and *Wetzeliella* (Fig. 6). Archeopyle formation commonly results from the loss of one or more than one anterior intercalary in the dinocyst.

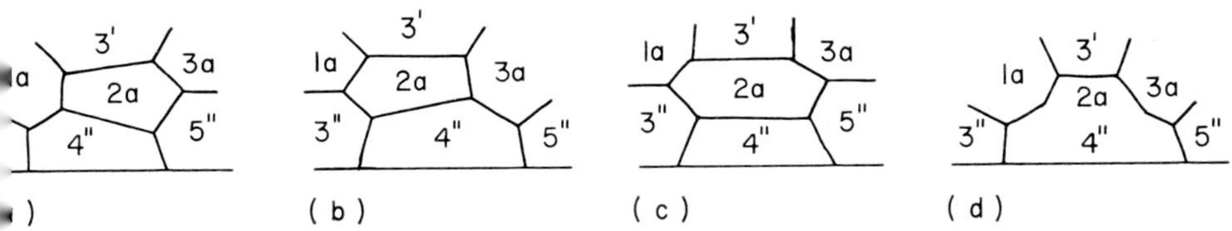

Fig. 9. Variation in the second anterior intercalary plate (2a) and its relationship to bordering plates in *Peridinium*: (a) and (b), penta; (c), hexa; (d), quadra. (From Graham, 1942.)

Gonyaulax (Fig. 3) has the basic tabulation 3–4′, 0–3a, 6″, 6c, 5–10s, 5–6″′, 1p, 1″″. All species possess six precingulars and one antapical. The number and disposition of the anterior intercalaries largely controls the number and size of the apical plates. When six postcingulars are present 1″′ is usually very reduced in size. Species of *Gonyaulax* are generally spherical to ellipsoidal and possess, at the most, one horn, which is apical in position. Fossil representatives include *Ctenidodinium* Deflandre emend. Sarjeant, *Gonyaulacysta*, *Meiourogonyaulax* Sarjeant and many typical hystrichospheres e.g. *Oligosphaeridium* Davey and Williams and *Spiniferites* (Pl. 5, fig. 6). Archeopyle formation commonly results from the loss of apical or precingular paraplates and never from the loss of intercalary paraplates.

The genus *Ceratium* (Fig. 10) is one of the most widely distributed of modern dinoflagellates, with species abundant in both freshwater and marine plankton. All the species are holophytic. The basic tabulation is 4′, 6″, 5–6c, 6″′, 1p, 1″″. Jörgensen (1911) stated that all species of *Ceratium* have five precingular, four to five cingular, five postcingular and two antapical plates.

Wall and Evitt (1975) believe that the genus has six precingular, six cingular, six postcingular, one posterior intercalary and one antapical plates. Individuals are asymmetric, compressed dorsoventrally and produced into one apical, one antapical and one or two long, hollow postcingular horns. According to Wall the postcingular horns characterise this genus. The cingular is almost circular. The sulcus lies to the left of the midventral line. Fossil genera,

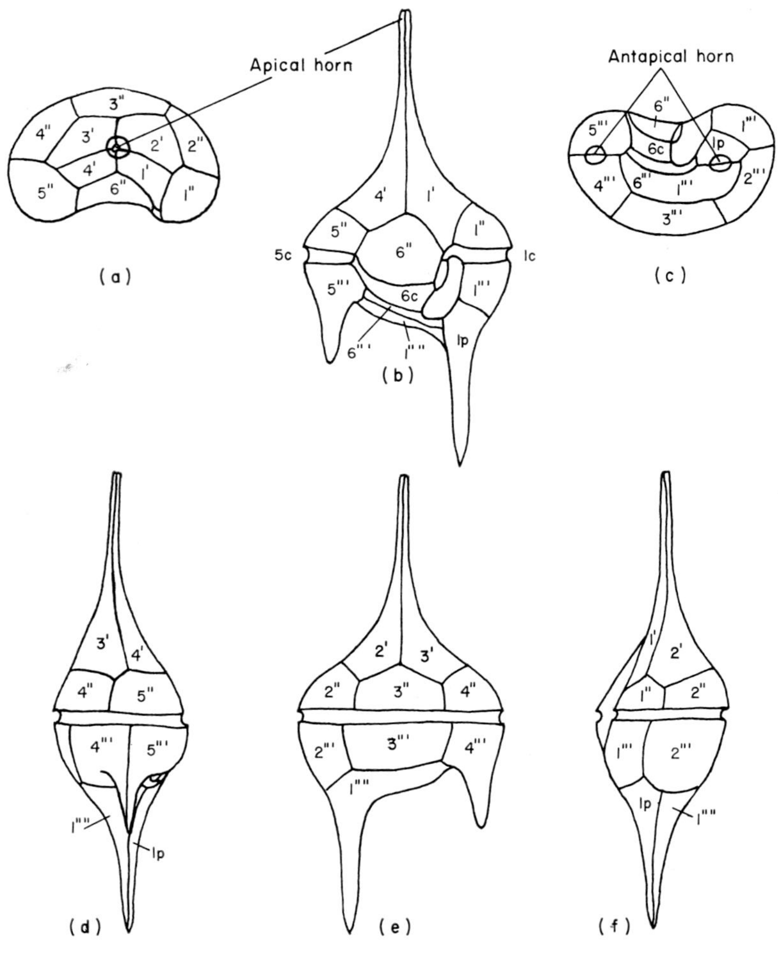

FIG. 10. *Ceratium* sp., showing tabulation of the theca or vegetative stage. The tabulation may be expressed thus; 4′, 6″, 6c, 6″ ′, 2″ ″. (a), apical view. The circle denotes the apical horn; (b), ventral view. Note the offset sulcus; (c), antapical view. The circles denote the antapical horns; (d), right lateral view; (e), dorsal view; (f), left lateral view. (From Evitt's Teaching Manual.)

which appear to be related to *Ceratium* include *Muderongia* Cookson and Eisenack, *Odontochitina* Deflandre, *Phoberocysta* Millioud and *Pseudoceratium* Gocht.

PROPOSED LINEAGES

In this chapter the fossil dinocysts have been placed into seven lineages (Fig. 11), based on paratabulation and archeopyle formation. These are the ceratioid, *Cyclonephelium* Deflandre and Cookson, *Dinogymnium* Evitt, Clarke and Verdier, *Gonyaulacysta, Nannoceratopsis* Deflandre, peridinioid, and *Tuberculodinium* Wall lineages. The peridinioid and *Gonyaulacysta* lineages are further subdivided, the former into the proto-peridinioid group with more than three intercalary paraplates all of which are lost in archeopyle formation, the *Deflandrea* group with three intercalary paraplates and intercalary archeopyles of the I or 3I type, the *Pareodinia* Deflandre group with four intercalary paraplates and an archeopyle of the 2I type; the *Palaeoperidinium* Deflandre ex Sarjeant group wih combination archeopyle involving the loss of occasionally one apical, one to three anterior intercalaries and up to three precingulars (this group includes *Peridinium limbatum* (Stokes) which opens by schism); and the *Ovoidinium* Davey group characterised by combination archeopyles of the AI type.

The *Gonyaulacysta* lineage is subdivided into the *Gonyaulacysta* group, proximate cysts with simple or compound precingular archeopyle; the *Meiourogonyaulax* group, proximate cysts with apical (A) or combination (AP) archeopyle; the *Valensiella* Eisenack, group with reticulate or areolate periphragm, elongate main body and apical archeopyle; the *Ctenidodinium* group, proximate cysts with epicystal archeopyle; the *Spiniferites* group with precingular archeopyle and bearing *gonal* and *intergonal* possesses; the *Hystrichosphaeridium* Deflandre group with apical archeopyle and intratabular processes not exceeding one per paraplate in the apical precingular, postcingular and antapical series; the *Systematophora* Klement, group with annulate complexes on the precingular and postcingular paraplates and apical archeopyle; the *Polysphaeridium* Davey and Williams, group with apical archeopyle and nontabular or intratabular processes numbering more than 50 per individual; the *Cordosphaeridium* group with precingular archeopyle with simple or compound operculum and intratabular processes not exceeding one per paraplate in the precingular series; the *Operculodinium* Wall, group with intratabular or nontabular processes numbering more than 50 per individual and precingular archeopyle with simple or compound operculum; the *Hemicystodinium* Wall, group with epicystal archeopyle and compound operculum; and the *Callaiosphaeridium* Davey and Williams, group which possess an epicystal archeopyle and both gonal and intratabular processes. In the *Cyclonephelium* lineage three branches are recognised, the *Cyclonephelium* group characterised by processes of length less than 10 μm and which only rarely delineate paratabulation; the *Areoligera* Lejeune–Carpentier group with processes greater than 10 μm and which usually delineate paratabulation; and the *Canningia* Cookson and Eisenack group, which are devoid of processes.

Stratigraphical record

Population sizes of organisms do not fluctuate haphazardly. There is a recognisable population production cycle in algal cultures. An initial *lag phase*, in which the cells may undergo a size increase but do not reproduce, gives way to the *exponential growth* phase, a period of maximum cell division and hence increase of population. The third phase is the *balanced growth* phase in which the population remains stationary presumably when it has reached maximum concentration. The cycle terminates in the *death phase*, when there is a rapid population decline (Fig. 12). Zeuner (1946) believed that all organisms exhibit a similar cycle in their evolution. A lag phase with relatively insignificant diversification is followed successively by an explosive phase (in which the influx of new species exceeds extinction), a climax or stationary phase, and a

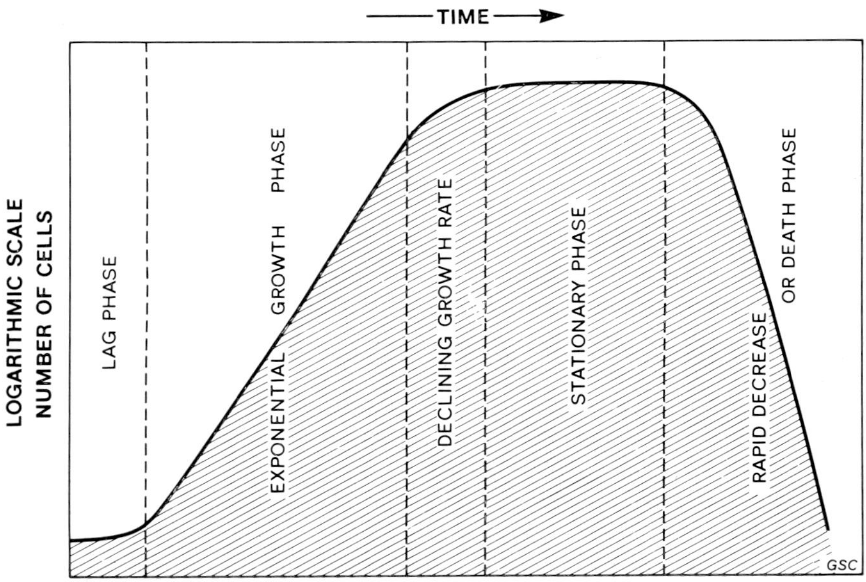

FIG. 12. Growth phases of algal protists in a limited culture, from inoculation to death. (From Tappan and Loeblich, 1970.)

decline phase (in which the influx of new species is exceeded by extinction). This cycle appears to have been time controlled, lasting for approximately fifty million years regardless of the organisms involved.

Tappan and Loeblich (1970) have applied Zeuner's cycle to the phytoplankton evolution in the geologic column (Fig. 13). They postulate that the cycle is particularly applicable to organisms which reproduce asexually. Also they believe the explosive phase is characterised by small dinocysts, the decline phase by large dinocysts. The dinocyst lag phase is recognised in the Silurian to Early Jurassic. This is followed by an explosive phase in the Toarcian to

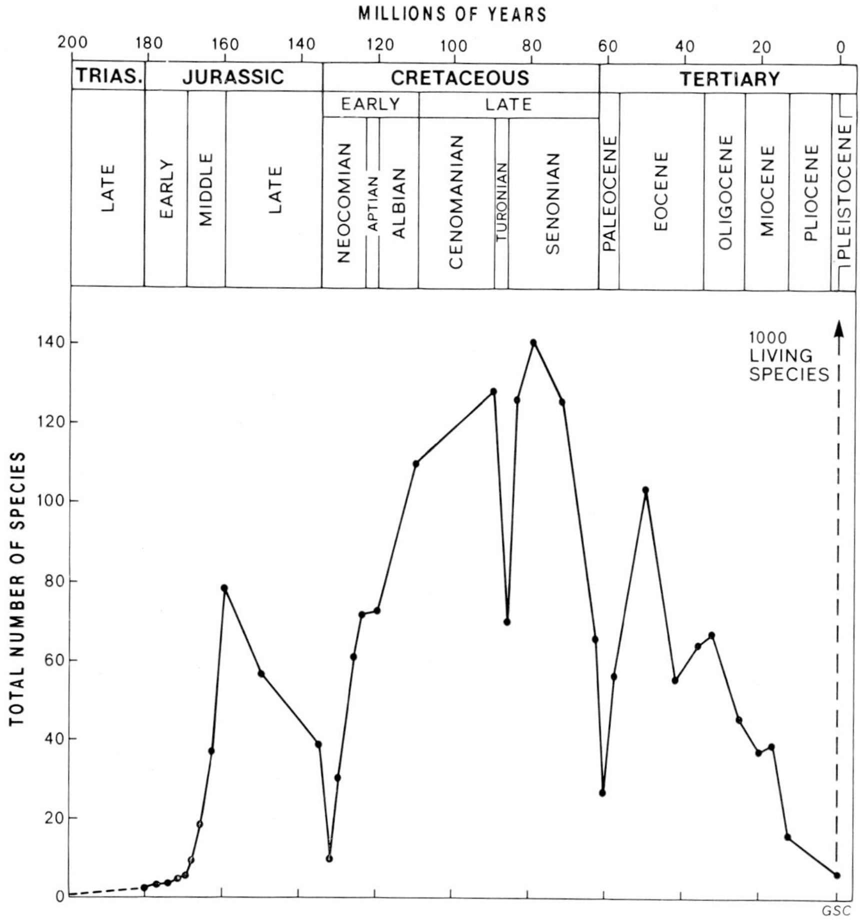

FIG. 13. Variations in number of dinocyst species in the Mesozoic-Cainozoic. Points denote end of period for which species were totaled. (From Tappan and Loeblich, 1970.)

Oxfordian (24 million years) terminating in a climax and succeeded by a Late Jurassic to Lower Cretaceous (Berriasian) decline. A second cycle containing a separate evolutionary group commenced its explosive phase in the Early Cretaceous and continued for 25 million years when it reached its climax. The Late Cretaceous marked a decline phase which was particularly accentuated in the Maestrichtian and Danian. A third explosive phase commencing in the Late Palaeocene and extending into the Early Eocene, was followed by a decline phase which extends to the Recent. If we assume a cycle occupies 50 million years, it is possible that Recent dinoflagellates are either experiencing or approaching the evolutionary phase of a fourth cycle.

The death phase of one cycle tends to be succeeded by the explosive phase of another cycle. These explosive cycles often seem to be related to maximum development of epicontinental seas. Other major environmental changes may also act as triggering mechanisms.

By compiling the data generated by Sarjeant (1967), Tappan and Loeblich have estimated that a new fossil dinocyst genus appeared every 1.4 million years and the average duration of this genus was 42.4 million years. Inclusion of the modern genera increases the post Palaeozoic evolution rate to one genus every 460 000 years. Such a rapid turnover suggests that dinocysts are ideally suited for biostratigraphical zonation.

Tappan and Loeblich (1972) pointed out that there are approximately 265 extant genera of dinoflagellates. Since many of these are aberrant parasitic or symbiotic forms and others are oceanic taxa which do not encyst, the greater proportion of these living dinoflagellates could not be preserved as fossils. In actuality only forty-odd species living today are known to encyst. This compares favourably with approximately twelve in the Pleistocene but contrasts with the numbers recorded from the Jurassic–Late Cretaceous and Early Tertiary. The fossil record indicates that the rate of generic extinctions was 34% in the Jurassic, dropping to 23% in the Early Cretaceous and increasing to 61% in the Late Cretaceous. Both extinction and evolutionary rates dropped in the Palaeocene, the few taxa being cosmopolitan. The percentage of generic extinctions increased throughout the Cainozoic, from 31% in the Palaeocene, to 43% in the Eocene, 57% in the Oligocene and 65% in the Miocene, culminating in the increasingly restricted diversity of the later Neogene.

The abundance in modern seas of dinocysts with a gonyaulacacean or peridinioid paratabulation is reflected in the fossil record. A few fossil genera including *Muderongia*, *Odontochitina*, *Phoberocysta* and *Pseudoceratium* appear to belong to the *Ceratium* group. Other fossil genera that are possibly related to *Ceratium*, notably *Areoligera*, *Chiropteridium* Gocht, *Cyclonephelium* and *Tenua* Eisenack (Pl. 4, fig. 11) are placed by Wall and Dale

(1968a) in the gonyaulacacean lineage. The majority of fossil dinocyst genera, however, are readily assignable to either the gonyaulacacean or the peridinioid lineages.

Mesozoic and Cainozoic biostratigraphy

It is proposed in this chapter to palynologically subdivide the Mesozoic–Cainozoic into a series of concurrent range zones, each zone bearing the name of a dinocyst (Fig. 14). These dinocysts zones will perforce constitute biostratigraphical units. The definitions of a concurrent range zone and biostratigraphical unit as formulated by the American Commission on Stratigraphic Nomenclature are the ones adhered to in this chapter. Each concurrent range zone is thus defined by the overlapping ranges of specified dinocysts from one of which it takes its name.

TRIASSIC BIOZONES

Known occurrences of dinocysts are few and are restricted to the Carnian and Rhaetian Stages of the Late Triassic. The general paucity of dinocysts is probably attributable to the unfavourable environments and lithologies present in the areas studied palynologically, such as Germany and England. Knowledge should increase with study of Triassic sediments from the Arctic and South America.

The Triassic is dominated by members of the proto-peridinioid group which are characterised by numerous paraplates and an archeopyle which results from the loss of more than three intercalaries. The assemblages described permit delineation of the *Shublikodinium arcticum*, Zone of Carnian age and the *Rhaetogonyaulax rhaetica*, Zone of Rhaetian age (Fig. 14). The ranges of selected species for the Triassic is shown on Table 2.

Shublikodinium arcticum Zone (=Carnian)

Wiggins (1973) described a Late Carnian dinocyst assemblage from the Arctic North Slope of Alaska. All of the ten species are assigned to the genus *Shublikodinium* Wiggins. This has a peridinioid paratabulation having seven precingulars and allegedly five anterior intercalaries and an A + 5I archeopyle. From the accompanying illustrations, the archeopyle appears to be of type 5I, with the apex sometimes being accidentally broken off. *Shublikodinium* belongs to the proto-peridinioid group. The zone has as yet not been recognised elsewhere.

Rhaetogonyaulax rhaetica Zone (=Rheatian)

The Rhaetic of Gloucestershire England contains two species of dinocysts of very similar morphology to *Shublikodinium* and which Sarjeant (1963) placed in a new genus *Rhaetogonyaulax*. *R. rhaetica* is a paratabulate species. *R. chaloneri* lacks paratabulation apart from the presence of a paracingulum and parasulcus. The archeopyle appears identical to that in *Shublikodinium* which may be a junior synonym. *Rhaetogonyaulax* falls within the proto-peridinioid group. The genus has also been recorded from the Rhaetic by Fisher (1972), and Morbey and Neves (1974).

JURASSIC BIOZONES

Throughout the Jurassic, the dinocysts become increasingly important for biostratigraphical zonation until the Tithonian, when there is a marked decline in number of species and their abundances. Evolutionary trends become stabilised with individuals tending to have fewer paraplates and the simple precingular archeopyle becoming the dominant type of excystment opening. During the Jurassic, process bearing dinocysts evolve although they do not attain such high concentrations as in the Cretaceous and Cainozoic. The Early Jurassic is dominated by the *Nannoceratopsis* lineage and proto-peridinioid group. In the Middle Jurassic, the *Ctenidodinium* and *Pareodinia* groups appear and the *Gonyaulacysta* group shows a marked expansion. The Late Jurassic marks the influx of the important Cretaceous ceratioid lineage and *Oligosphaeridium*. The extensive literature coverage of Jurassic dinocyst assemblages permits the recognition of nine zones and five subzones (Fig. 14). The ranges of selected Jurassic dinocyst species are shown in Table 2.

Dapcodinium priscum Zone (=Hettangian–Sinemurian)

Dapcodinium priscum was described from the Lias Alpha (Hettangian) of Denmark by Evitt (1961a). Morbey and Neves (1974) have also recorded this species from the Rhaetic of Austria. *D. priscum* has a peridinioid paratabulation with four anterior intercalaries and seven precingulars. The archeopyle appears to be 4I resulting from the loss of all the anterior intercalary paraplates, the apical paraplates remaining attached in the ventral area. *Dapcodinium* Evitt represents a continuation of the proto-peridinioid group in the Lower Jurassic.

Nannoceratopsis gracilis Zone (=Pliensbachian–Toarcian)

Dinocyst assemblages which are assigned to the *Nannoceratopsis gracilis* Zone can be recognised in Denmark (Evitt, 1961a), Germany (Morgenroth, 1970), England and the Grand Banks (Williams, this Volume) and in Arctic Canada (Johnson and Hills, 1973).

In the late Pliensbachian-Toarcian of the Grand Banks the author has recorded *Mancodinium semitabulatum* and *Nannoceratopsis gracilis*. *N. gracilis*, has also been observed in the early Toarcian of England.

Beju (1971) recognised a *Nannoceratopsis gracilis* Zone to which he assigned a Bajocian–Bathonian age. This age is questionable in view of the dinocyst assemblage which contains *Nannoceratopsis gracilis*, *N. pellucida*, *Fromea elongata* Beju and *Canningia* cf. *minor* Cookson and Hughes. Also since Beju's samples were obtained from wells the possibility of caving cannot be excluded.

The late Pliensbachian marks the onset of the first exponential growth phase of the dinoflagellates and the distribution of dinocysts on a regional scale. Two species have well developed processes, *Valvaeodinium armatum* with intratabular processes, and *Luehndea spinosa* with gonal processes. This evolutionary trend which may represent an adaptation to flotation becomes widespread in the dinocysts of the Mesozoic–Cainozoic.

Pareodinia ceratophora Zone (=Bajocian)

The Bajocian dinoflagellates are poorly studied mainly because it was a period of widespread carbonate deposition. Since *Pareodinia ceratophora* (Pl. 2, fig. 12) has not as yet been recorded from pre-Bajocian sediments it has been designated as the zonal index species. Dinocyst assemblages which are assigned to this zone are recorded from the Scotian Shelf and Grand Banks (Williams, this volume) and from Western Canada (Pocock, 1972).

Personal observations on the dinoflagellates from the Grand Banks and Scotian Shelf have revealed the presence of numerous specimens of species of *Canningia* together with forms which show characteristics of the *Polysphaeridium* and *Meiourogonyaulax* groups *Ctenidodinium pachydermum*, *Dictyopyxidia reticulata*, *Pareodinia ceratophora*, *Tenua rioulti* and *Valensiella ovulum* are also recorded.

The ranges of selected species for this zone are shown in Table 2.

Ctenidodinium ornatum Zone (=Bathonian)

Assemblages which can be assigned to the *C. ornatum* (Pl. 1, fig. 10) Zone are recorded from Germany (Gocht, 1970), France (Valensi, 1953), Southern England, the Scotian Shelf and Grand Banks (Williams, this paper), Johnson and Hills (1973) recognised a *Pareodinia tripartita* Subzone of Middle to Late Bathonian age in the Jurassic sediments of the Sverdrup Basin, Arctic Canada. This subzone may be of limited geographical extent. The Bathonian of southern England is distinguished by a high relative percentage of *Pareodinia*. Gocht (1970) has included within *P. ceratophora* forms with a paracingulum, forms with a kalyptra and forms with an antapical horn. In the British material nine morphotypes of *P. ceratophora* are recognised (Fig. 15). Forma 1 lacks a paracingulum and has a hollow apical horn.

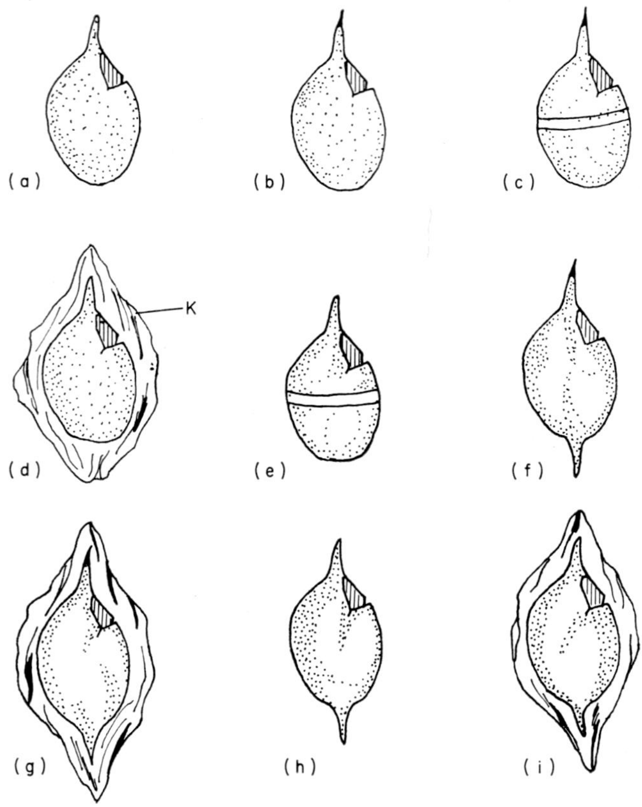

FIG. 15. The nine morphotypes of *Pareodinia ceratophora* Deflandre recognised in sediments from the ammonite *Morrisiceras morrisi* Zone of the Bathonian of southern England. (a), Forma 1, this lacks a paracingulum and has a hollow apical horn; (b), Forma 2, this lacks a paracingulum and possesses a solid extension (spine) of the apical horn; (c), Forma 3, this has a clearly delineated paracingulum and possesses a solid extension to the apical horn; (d), Forma 4, this lacks a paracingulum, has a hollow apical horn, and a surrounding outer membrane, the kalyptra (K); (e), Forma 5, this has a hollow apical horn and a well-developed paracingulum; (f), Forma 6, this possesses an apical, and an antapical horn, the former being produced into a solid spine; (g) Forma 7, this possesses an apical horn produced into a solid extension, an antapical horn, and a paracingulum; (h), Forma 8, this has a hollow apical and a hollow antapical horn; (i), Forma 9, this has a hollow apical horn, a hollow antapical horn and a kalyptra. The shaded area indicates the archeopyle which is intercalary.

Forma 2 possesses a solid extension of the apical horn; Forma 3 possesses a solid extension to the apical horn and a clearly delineated paracingulum. Forma 4 lacks a paracingulum but possesses a kalyptra. Forma 5, the most abundant has a hollow apical horn and a well-developed paracingulum.

Forma 6 has an apical and an antapical horn, the former being produced into a solid spine. Forma 7 differs from Forma 6 in having a kalyptra. Forma 8 has a hollow apical and a hollow antapical horn. Forma 9 differs from 8 in possessing a kalyptra. Forma 4 is identical to *Kalyptea monoceras* Cookson and Eisenack; Forma 7 and 9 would be *Kalyptea diceras* Cookson and Eisenack. In specimens with an antapical horn, a paracingulum is never developed. The relative abundances vary considerably. Formae 1–4 are confined to the *Morrisiceras morrisi* Zone. Forma 5 is the dominant form and extends into the *Prohecticoceras retrocostatum* Zone where it usually develops tubercles. Formae 6, 7, 8 and 9 are not found in the *Morrisi* Zone. Other species present in the Bathonian of Southern England are a species of *Cleistosphaeridium* Davey *et al.* possessing some trifurcate processes (*Polysphaeridium group*), *Ctenidodinium continuum*, *C. ornatum*, *C. pachydermum*, *Endoscrinium eisenacki*, *Gonyaulacysta aldorfensis*, *Leptodinium subtile* subsp. *pectinigerum*, *Lithodinia jurassica*, *Nannoceratopsis pellucida*, *Valensiella ampulla* and *V. vermiculata*.

In the Scotian Shelf–Grand Banks region *Ctenidodinium ornatum*, *C. Pachydermum*, *Endoscrinium eisenacki*, *Gonyaulacysta filapicata*, *leptodinium regali* Gocht, *L. subtile* subsp. *pectinigerum* Gocht, *Lithodinia jurassica*, *Valensiella ovulum* and *V. vermiculata* (Pl. 5, fig. 3) are recorded. The chorate cyst *Hystrichosphaeridium polonicum* first appears in the Bathonian of the Grand Banks.

Adnatosphaeridium caulleryi Zone (=Early Callovian) *and Adnatosphaeridium aemulum* Zone (=Late Callovian)

The Callovian has been the subject of detailed studies by Sarjeant (1965, 1968). It is a period of expansion in the number of dinocyst species and Callovian assemblages generally show a closer affinity with Oxfordian dinocysts rather than with Bathonian species.

The *Adnatosphaeridium caulleryi* (Pl. 1, fig. 7) Zone is known from Southern England where *A. caulleryi*, *A. filamentosum*, *Gonyaulacysta ambigua*, *G. cladophora*, *G. jurassica*, *Meiourogonyaulax cristulata* (Sarjeant), *M. rioulti* and *Polystephanephorus paracalathus* (Sarjeant) are recorded. In the Scotian Shelf–Grand Banks region the zone is distinguished by the first occurrence of *Adnatosphaeridium caulleryi*, *Cleistosphaeridium tribuliferum*, *Gonyaulacysta jurassica* and *Scriniodinium crystallinum*.

The *Adnatosphaeridium aemulum* Zone is known from Poland, where Gorka (1970) described an assemblage of forms which is indicative of a boreal climate, France and Southern England.

In the late Callovian of England *A. aemulum*, *Cleistosphaeridium lumectum* (Sarjeant), *Ctenidodinium ornatum*, *Ellipsoidictyum cinctum*, Klement, *Fromea*

Cookson and Eisenack, *Gonyaulacysta dangeardi* Sarjeant, *Meiourogonyaulax deflandrei* Sarjeant, *Spiniferites* sp., *Stephanelytron caytonense*, *S. scarburghense*, *Surculosphaeridium vestitum* (Deflandre), *Systematophora* cf. *areolata* Klement, *Tenua pilosa*, *T. rioulti*, *T. verrucosa*, Sarjeant, *T. villersense* Sarjeant, *Wanaea digitata* and *W. fimbriata* are recorded.

Species first appearing in this zone in the Scotian Shelf–Grand Banks sediments include *Adnatosphaeridium aemulum*, *Hexagonifera jurassica*, *Stephanelytron scarburghense*, *Ctenidodinium pachydermum*, *Gonyaulacysta aldorfensis* and *Leptodinium regale: Lithodinia jurassica*, *Valensiella ampulla* and *V. ovulum* do not range into the Oxfordian.

Johnson and Hills (1973) recognised two dinocyst zones in the Callovian of the Canadian Arctic, the *Paragonyaulacysta calloviense* Johnson and Hills subzone of Early Callovian age and the *Acanthaulax* spp. Zone which extends from the Middle Callovian into the Early Oxfordian. Pocock (1972) recorded *Chytroeisphaeridia scabrata* Pocock, *Cyclodictyon minor* Pocock, *Evansia granulata* Pocock, *Opaeopsomus wapellensis* Pocock, *Scriniodinium klementii* Pocock and *S. reticulatum* Pocock both of which appear very similar to *S. dictyotum* Cookson and Eisenack and *Tenua evittii* from the Callovian of western Canada. The difference between the Canadian and English material presumably indicates provincialism.

Gonyaulacysta jurassica Zone (=Oxfordian)

The *G. jurassica* (Pl. 2, fig. 7) Zone is recognised throughout western Europe, on the Scotian Shelf and Grand Banks and in Arctic Canada. This zone is provisionally assigned an Oxfordian age but probably extends into the early Kimmeridgean. Studies of Oxfordian dinocysts undertaken in England (Sarjeant, 1961a, 1962, 1968) France, (Deflandre, 1938, 1947) Roumania (Beju, 1971) and Canada (Pocock, 1972; Johnson and Hills, 1973) confirm the cosmopolitan nature of *Gonyaulacysta jurassica*.

In Arctic Canada, Johnson and Hills (1973) defined a *Gonyaulacysta jurassica* Range Zone to which they assigned an age of ?Middle Jurassic to ?Tithonian. Within this zone they recognised a *Gonyaulacysta jurassica* var. *longicornis* Deflandre Subzone of Oxfordian age.

In Southern England the *G. jurassica* Zone can be subdivided into three subzones:

1. The *Polystephanephorus calathus–Hystrichosphaeridium polonicum* Subzone (=Early Oxfordian).

Within this subzone species chorate cysts constitute a low relative percentage of the dinocyst count, and the following taxa are recorded: *Acanthaulax paliuros*, *Apteodinium* sp., *Cleistosphaeridium tribuliferum*, *Eisenackia*

(Deflandre and Cookson), *Gonyaulacysta jurassica*, *G. nuciformis*, *Hystrichogonyaulax nealei*, *Hystrichosphaeridium polonicum* (Pl. 3, fig. 4), *Pareodinia ceratophora*, Forma 1, *Polystephanephorus calathus*, *Prolixosphaeridium granulosum* (Deflandre), *Scriniodinium dictyotum* subsp. *pyrum* Gitmez, *Senoniasphaera* sp., *Stephanelytron redcliffense* Sarjeant and *Tenua pilosa*.

2. *Adnatosphaeridium filamentosum* Subzone (=Middle Oxfordian)

In the Middle Oxfordian *Adnatosphaeridium aemulum* remains common. There is a marked influx of *A. filamentosum* and related species with solid processes, and species of *Ctenidodinium* Deflandre which lack gonal and border processes. *Pluriarvalium osmingtonense* Sarjeant, *Systematophora fasciculigera* and *S. orbifera* appear for the first time, and *Ctenidodinium ornatum* does not range into younger sediments. *Endoscrinium eisenacki* and *Gonyaulacysta cladophora* were also observed.

3. *Systematophora orbifera* Subzone (=Late Oxfordian)

The *S. orbifera* (Pl. 4, fig. 5) Subzone embraces the late Oxfordian (Upper Corallian beds) of Southern England, These contain a prolific dinocyst assemblage in which species of Gonyaulacysta dominate both in their abundance and variability. *Gonyaulacysta jurassica*, *Pareodinia ceratophora* Forma 1 and *P. ceratophora* Forma 4 are common. Also present are *Chytroeisphaeridia chytroeides* (Sarjeant), *Cleistosphaeridium tribuliferum*, *Gonyaulacysta cladophora*, *Lanterna bulgarica*, *Occisucysta* Gitmez, *Palaeostomocystis* Deflandre (of unknown affinities), *Prolixosphaeridium deirense* Davey et al., *Scriniodinium crystallinum*, *S. dictyotum*, *Systematophora orbifera*, *Tenua rioulti* and *Verrucosphaera lukoviensis*. The zonal species *Systematophora orbifera* is common in the Late Oxfordian of the Scotian Shelf and southern England.

Gonyaulacysta cladophora Zone (=Kimmeridgian)

The Kimmeridgian marks the onset of the decline of the Jurassic dinocyst taxa, with a few Cretaceous morphotypes appearing towards the top. The clay, shaley lithologies dominating this stage in several areas have yielded excellent assemblages which have been described by Klement (1960), Gitmez (1970), Beju (1971), Gitmez and Sarjeant (1972).

Beju (1971) erected a *Gonyaulacysta cladophora* Zone for subsurface Kimmeridgean rocks of Rumania. Assemblages which can be assigned to this zone are recognised from Southern England and the Scotian Shelf–Grand Bank region.

In southern England the *G. cladophora* (Pl. 3, fig. 1) Zone can be subdivided into two subzones.

1. *Endoscrinium luridum–Stephanelytron scarburghense* Subzone (=Early Kimmeridgian)

The dinocysts from the basal Kimmeridge (*Baylei* zone) of England, Scotland and France have recently been figured and described by Gitmez (1970). Species believed to be restricted to the Basal Kimmeridge are *Cryptarchaeodinium calcaratum* Deflandre, *Gonyaulacysta granuligera* and *Leptodinium amabilis*. Other species present include *Belodinium dysculum*, *Chytroeisphaeridia chytroeides*, *Endoscrinium eisenacki*, *Leptodinium arcuatum* Klement, *Nannoceratopsis pellucida*, *Scriniodinium playfordi* Cookson and Eisenack, *Systematophora orbifera*, *Taeniophora iunctispina*, *Tenua pilosa* and alledgedly *Oligosphaeridium pulcherrimum*. The specimen identified as *O. pulcherrimum* does not appear to belong to this species. Gitmez observed that the Australian specimens are considerably larger than the European specimens of the same species. She postulated that the size of the dinocysts was palaeoecologically controlled. It could also reflect climatic control or provincialism, and delineate an Australian and a European biogeographical province respectively. In the Dorset section *Endoscrinium luridum* (Pl. 2, fig. 6) and *Scriniodinium crystallinum* do not range above the Basal Kimmeridgian.

The Early Kimmeridgian of Dorset England is characterised by the presence of *Acanthaulax venusta* (Klement) Sarjeant, *Stephanelytron scarburghense* (Pl. 4, fig. 3) and *Tenua villersense*. Klement (1960) has recorded *Endoscrinium galeritum*, *E. luridum*, *Histiophora ornata* and *Systematophora areolata* from the Early Kimmeridgian.

2. *Gonyaulacysta perforans* Subzone (=Middle to Late Kimmeridgian)

Several of the ancestral Cretaceous dinocysts first appear in the Middle and Late Kimmeridgian. Particularly significant is the chorate cyst *Oligosphaeridium pulcherrimum* (Middle Kimmeridgian–Coniacian), presumably the ancestral form of the *Oligosphaeridium* species which often dominate Cretaceous assemblages. The ceratioid lineage, another dominant Cretaceous morphotype, first appears in the Late Kimmeridgian in the form of *Muderongia simplex* Alberti. Other new species are *Lanterna sportula* and *Sirmiodinium grossi*. *Hexagonifera jurassica* is restricted to the Middle and Late Kimmeridgian. Diagnostic species of the Late Kimmeridgian are *Chlamydophorella wallala* Cookson and Eisenack, *Gonyaulacysta perforans*, *Lanterna bulgarica*, *Systematophora areolata* and *Wanaea fimbriata* Sarjeant.

Lanterna sportula–Belodinium dysculum Zone (=Portlandian)

Assemblages of dinocysts which can be assigned to this zone have been described from Papua (Cookson and Eisenack, 1968), Dorset, England (Norris, 1965), Moscow region, U.S.S.R. (Vozzhennikova, 1967), California

(Warren, 1967), Bulgaria (Dodekova, 1969). On the Scotian Shelf the *Ctenidodinium panneum* Zone provisionally dated Portlandian, is considered as the coeval equivalent in part of the *L. sportula* (Pl. 3, fig. 8)–*B. dysculum* (Pl. 2, fig. 11) Zone.

On the Scotian Shelf *Ctenididinium culmulum*, *C. panneum* and *Lanterna sportula* are described from Portlandian sediments.

CRETACEOUS BIOZONES

The Cretaceous Period is characterised by the rise to prominence of the ceratioid lineage and the *Canningia*, *Cyclonephelium*, *Deflandrea*, *Hystrichosphaeridium*, *Operculodinium*, *Ovoidinium* and *Palaeoperidinium* groups. In the Early Cretaceous, the *Gonyaulacysta* group is characterised by the appearance of thicker-walled species which represent adaptation to a neritic environment. The *Dinogymnium* lineage is restricted to the Late Cretaceous. This may be a reflection of the widespread marine transgression and the broad climatic belts which are characteristic of this epoch. The presumably cosmopolitan neritic environments also seem to have stimulated the evolution of the *Areoligera* and *Deflandrea* groups. Throughout the Cretaceous paraplate number stabilised so that most dinocysts can be readily assigned a gonyaulacacean or peridinioid paratabulation. This is not always accompanied by archeopyle standardisation. The *Gonyaulacysta* lineage includes predominantly apical or simple precingular archeopyles and the ceratioid lineage is almost exclusively apical. The peridinioid lineage, however, shows a diversification, particularly evident in the Albian and Late Cretaceous that should permit detailed biostratigraphic control when fully documented.

Zonation of the Cretaceous utilising dinocysts permits recognition of all the stages (Fig. 14); Nine dinocyst zones and five subzones are recognised. The ranges of selected Cretaceous dinocyst species are shown on Table 2.

Phoberocysta neocomica Zone (=Neocomian)

The dinocysts of the Neocomian (taken to include the Barremian) stratotypes have been studied and described by Millioud (1969). His findings are discussed under the subzone headings. Alberti (1961) plotted the ranges of 70 species in the Cretaceous. Of six species of *Broomea* Cookson and Eisenack (all six of which belong to the ceratioid lineage) five are found only in the Neocomian. *Operculodinium* species lacking the reticulate wall appear in this epoch. Subdivision of the Neocomian into Berriasian–Valanginian, Hauterivian and Barremian has been achieved using dinocysts.

1. *Biorbifera johnewingi–Cribroperidinium orthoceras* Subzone (Berriasian–Valanginian)

Dinocyst assemblages which can be assigned to this subzone are recorded from Russia (Vozzhennikova, 1967), California (Warren, 1967), France (Millioud, 1969), in Mesozoic sediments cored during Leg 11 of JOIDES (Habib, 1972) and in the Scotian–Shelf–Grand Banks region where the *Phoberocysta neocomica* Zone is its coeval equivalent.

Millioud (1969) noted that *Cribroperidinium orthoceras*, *Cyclonephelium distinctum*, *Muderongia tomaszowensis* and *Phoberocysta neocomica* (Pl. 3, fig. 11) appear in the Berriasian. Only two species *Cordosphaeridium eoinodes* and *Spiniferites ramosus* (Ehrenberg) appear in the Valanginian.

The Speeton Clay of Yorkshire yields beautifully preserved dinocyst assemblages. At the base in the Ryazanian *Deflandrea pirnaensis* Alberti, *Dingodinium albertii*, *Eopseudoceratium gochti*, *Gonyaulacysta* cf *serrata* (Cookson and Eisenack), *Heliodinium voigti* Alberti and *Imbatodinium Kondratjeri* occur. The Valanginian marks the influx of *Apteodinium ciliatum* Gocht, *?Broomea longicornuta*, *Chlamydophorella nyei*, *Cordosphaeridium eoinodes*, *?C. fasciatum*, *Cribroperidinium orthoceras* (Pl. 2, fig. 3), *C. sepimentum*, *Endoscrinium campanula*, *Hystrichodinium ramoides* Alberti, *Meiourogonyaulax stoveri*, *Muderongia staurota* Sarjeant, *?Oligosphaeridium asterigerum*, *?Spiniferites dentatus*, *S. ramosus* and *Systematophora complicata* Neale and Sarjeant.

Habib (1972) recorded several dinocysts from Mesozoic sections cored during leg 11 of the JOIDES Project. The Early Cretaceous (presumably pre-Valanginian) is characterised by the incoming of *Microdinium deflandrei* Millioud, *Scriniodinium dictyotum* and *Wallodinium krutzschi* (Alberti). In the Valanginian *Cyclonephelium distinctum*, *Deflandrea pirnaensis* and *Endoscrinium campanula* appear for the first time while *Biorbifera johnewingi* (Pl. 2, fig. 1), *Diacanthum hollisteri* Habib and *Scriniodinium dictyotum* have their tops.

Warren (1967) described a species of *Hystrichosphaeridium* from the Berriasian. This the oldest known occurrence of this genus. He also noted the incoming of *?Oligosphaeridium asterigerum* in the Late Valanginian.

In the Scotian Shelf–Grand Banks region, the *Phoberocysta neocomica* Zone is dated Berriasian–Valanginian. Other species present in this zone but not ranging into the Hauterivian include *Achomosphaera neptuni* and *Biorbifera johnewingi*.

Evans (1966a, b) proposed a dinocyst zonation for the Cretaceous of Australia, with each zone taking the name of the dominant species. The Early Neocomian is the *Cannosphaeropsis mirabilis* Cookson and Eisenack, *Scriniodinium attadalense* (Cookson and Eisenack) Eisenack Zone.

2. *Ctenidodinium elegantulum* Subzone (=Hauterivian)

The Hauterivian is a time of influx of Cretaceous species. In the stratotype, according to Millioud (1969), *Coronifera oceanica, Ctenidodinium elegantulum, Deflandrea pirnaensis, Meiourogonyaulax stoveri, Muderongia staurota, Odontochitina operculata* and *Oligosphaeridium complex* (White) appear.

Assemblages of dinocysts which can be assigned to this subzone are described from Germany (Alberti, 1961), Switzerland (Millioud, 1969), Northern England and the Scotian Shelf–Grand Banks region (Williams, this chapter).

In the Speeton Clay, the Hauterivian marks the incoming of *Achomosphaera neptuni, Broomea jaegeri* Alberti, *Callaiosphaeridium asymmetricum, Coronifera oceanica, Ctenidodinium elegantulum, Cyclonephelium distinctum* Deflandre and Cookson, *Endoscrinium eisenacki* (Deflandre), *Gardodinium eisenacki* (Alberti), *Gonyaulacysta serrata, Muderongia mcwhaei* (Cookson and Eisenack), *Oligosphaeridium complex* and *Sirmiodinium grossi* Alberti.

The *Ctenidodinium elegantulum* (Pl. 1, fig. 11) Zone is recognised on the Scotian Shelf and Grand Banks. Within this Zone *Pseudoceratium pelliferum* attains its maximum abundance. Other species include *Broomea jaegeri, Callaiosphaeridium asymmetricum, Gonyaulacysta serrata* (Cookson and Eisenack), *Oligosphaeridium complex, Systematophora complicata* and *Wanaea fimbriata* Sarjeant.

3. *Tenua anaphrissa* Subzone (=Barremian)

Dinocyst assemblages described from Germany (Alberti, 1961), France (Millioud, 1969), northern England (Davey, 1974) and the Scotian Shelf and Grand Banks are assigned to this subzone.

The type Barremian of southern France is characterised by *?Cordosphaeridium fasciatum*. Species not extending into the Aptian are *Ctenidodinium elegantulum, Muderongia tomaszowensis* and *Pareodinia ceratophora*.

In the *Tenua anaphrissa* (Pl. 4, fig. 11) Zone on the Scotian Shelf–Grand Banks, *T. anaphrissa* attains a relative abundance of up to 15% of the palynomorph assemblage. Several species, notably *?Cordosphaeridium fasciatum, Meiourogonyaulax stoveri, Muderongia simplex* Alberti, *M. tomaszowensis, Oligosphaeridium asterigerum, Polystephanephorus sarjeanti* Gitmez and *Pseudoceratium pelliferum* do not range into the Aptian.

The presence of the *Tenua anaphrissa* Subzone in eastern North America suggests that this area lay within the European biogeographical province during the Barremian.

Systematophora schindewolfi–Deflandrea perlucida Zone (=Aptian)

Dinocyst assemblages which are assigned to this zone are recorded from southern France, northern England, the Scotian Shelf and Grand Banks (Williams, this chapter).

The dinocysts from the Aptian stratotypes of southeast France include

Cribroperidinium sepimentum which is restricted to the lower Aptian, *Astrocysta cretacea*, *Callaiosphaeridium asymmetricum Canningia colliveri*, *Cleistosphaeridium*, *polypes ?Cyclonephelium attadalicum*, *C. distinctum*, *Deflandrea perlucida* (Pl. 2, fig. 9), *Dingodinium alberti*, *Gardodinium trabeculosum*, *Hystrichodinium pulchrum* Deflandre, *Hystrichosphaeridium tubiferum* Davey and Williams, *Odontochitina operculata*, *Oligosphaeridium complex*, *Protoellipsodinium spinosum* and *Systematophora schindewolfi* (Pl. 5, fig. 4).

The Aptian of the Speeton Clay is characterised by the presence of *Achomosphaera neptuni*, *Broomea jaegeri*, *B. longicornuta*, *Callaiosphaeridium asymmetricum*, *Cauca parva*, *Cleistosphaeridium polypes*, *Cordosphaeridium eoinodes*, *Dingodinium albertii*, *Endoscrinium campanula*, *E. eisenacki*, *Eopseudoceratium gochti*, *Gardodinium eisenacki*, *Heliodinium voigti* Alberti, *Hystrichodinium ramoides* Alberti, *Hystrichosphaeridium recurvatum*, *H. readei* Davey, *Muderongia staurota*, *Oligosphaeridium anthophorum* (Gocht) *O. perforatum* (Gocht), *Pareodinia ceratophora* Forma 3, *Polysphaeridium laminaspinosum*, *Prolixosphaeridium deirense* Davey *et al.*, *Pseudoceratium pelliferum*, *Sirmiodinium grossi*, *Spiniferites perforatus* (Davey and Williams) Sarjeant, *Systematophora schindewolfi*, *Tanyosphaeridium* sp. and *Wallodinium glaessneri* (Cookson and Eisenack).

The *Deflandrea perlucida* (Pl. 2, fig. 9)–*Systematophora schindewolfi* (Pl. 5, fig. 4) Zone delineates the Aptian of the Scotian Shelf and Grand Banks. Neither species extends into the Albian. Other species not found in younger sediments include *Aptea polymorpha* Eisenack, *Canningia colliveri* and *Dingodinium cerviculum*. In part equivalent to the *Deflandrea perlucida* Zone is the *?Cyclonephelium attadalicum* (Pl. 2, fig. 2) Zone, which appears to be confined to the Early Aptian in some of the Scotian Shelf wells. The species *Cordosphaeridium eoinodes*, *Cribroperidinium sepimentum*, *?Cyclonephelium attadalicum*, *Pareodinia ceratophora* and *Tenua anaphrissa* are not found in post-Aptian and possibly post-Early Aptian sediments.

Spinidinium cf *vestitum* Zone (=Albian)

Dinocyst assemblages which are assigned to this zone have been described from the Paris Basin, France (Davey and Verdier, 1971, 1973), southern England (Cookson and Hughes, 1964), and the Scotian Shelf–Grand Banks region (Williams, this chapter).

The Albian represents a significant phase in the evolution of the archeopyle types in the peridinioid lineage, and particularly the *Deflandrea*, *Ovoidinium* and *Palaeoperidinium* groups. First recorded from this stage are the combination I3P (*Spinidinium* cf. *vestitum*, Pl. 3, fig. 6) and combination AI (*Ovoidinium*) 3I3P (as yet unnamed), A3I3P (*Scriniodinium eurypylum* Manum and Cookson and *Astrocysta cretacea*) archeopyle types. These diagnostic

Subzone	Zone
	Shublikodinium arcticum
	? ? ? ? *Rhaetogonyaulax rhaetica*
	Dapcodinium priscum
	Nannoceratopsis gracilis
	Pareodinia ceratophora
	Ctenidodinium ornatum
	Adnatosphaeridium caulleryi
	Adnatosphaeridium aemulum
Polystephanephorus calathus *Hystrichosphaeridium polonicum*	*Gonyaulacysta jurassica*
Adnatosphaeridium filamentosum	
Systematophora orbifera	
Endoscrinium luridum *Stephanelytron scarburghense*	*Gonyaulacysta cladophora*
Gonyaulacysta perforans	
	Lanterna sportula *Belodinium dysculum*
Biorbifera johnewingi *Cribroperidinium orthoceras*	*Phoberocysta neocomica*
Ctenidodinium elegantulum	
Tenua anaphrissa	
	Systematophora schindewolfi *Deflandrea perlucida*
Protoellipsodinium spinocristatum	*Spinidinium* cf. *vestitum*
Xenascus ceratioides *Carpodinium obliquicostatum*	
	Cleistosphaeridium polypes
	Surculosphaeridium longifurcatum
	Callaiosphaeridium asymmetricum *Oligosphaeridium pulcherrimum*
	Hystrichosphaeridium truncigerum
	Odontochitina operculata
	Dinogymnium euclaensis
	Deflandrea diebeli *Palaeoperidinium pyrophorum*
	Deflandrea speciosa *Wetzeliella parva*
	Homotryblium tenuispinosum *Spiniferites septatus*
	Membranilarnacia ursulae *Tanyosphaeridium ellipticum*
	Areosphaeridium diktyoplokus *Pentadinium laticinctum*
Deflandrea heterophlycta	*Chiropteridium dispersum* *Samlandia chlamydophora*
Cordosphaeridium funiculatum *Thalassiphora pelagica*	
	Cordosphaeridium cantharellum
	Lejeunia fallax *Xenicodinium hispidum*
	Hystrichosphaeropsis obscurum *Polysphaeridium pastielsi*
	Achomosphaera ramulifera
SUBZONE	ZONE

Tanyosphaeridium ellipticum (Cookson)
Cyclopsiella vieta Drugg and Loeblich
Pentadinium laticinctum Gerlach
Cordosphaeridium cantharellum (Brosius)
Eisenackia ornata Cookson and Eisenack
Hemiplacophora semilunifera Cookson and Eisenack
Deflandrea heterophlycta Deflandre and Cookson
Leptodinium incompositum (Drugg)
Lingulodinium machaerophorum Deflandre and Cookson
Chiropteridium dispersum Gocht
Chiropteridium lobospinosum (Gocht)
Deflandrea spinulosa Alberti
Hystrichogonyaulax coreoides Benedek
Phthanoperidinium amoenum Drugg and Loeblich
Tanyosphaeridium paradoxum (Brosius)
Lejeunia fallax Morgenroth
Lejeunia paratenella Benedek
Tuberculodinium rossignolae Drugg
Xenicodinium hispidum Drugg
Hystrichosphaeropsis obscura Habib
Leptodinium aculeatum Wall
Leptodinium patulum Wall
Leptodinium strialatum Wall
Operculodinium giganteum Wall
Tectatodinium pellitum Wall
Leptodinium churchillii Harland

morphotypes plus the several detailed studies of Albian assemblages usually permit ready recognition of this stage.

1. *Protoellipsodinium spinocristatum* Subzone (=Early Albian)

Species believed by Davey and Verdier (1971) to first occur in the early Albian are *Achomosphaera ramulifera*, *Cassiculosphaeridia reticulata* Davey, *Hystrichosphaeridium radiculatum* Davey and Williams, *H. stellatum*, *H*, *tubiferum*, *Ovoidinium scabrosum* and *Protoellipsodinium spinocristatum* (Pl. 4. fig. 2). This subzone is currently recognised only in the Paris Basin.

2. *Xenascus ceratioides–Carpodinium obliquicostatum* Subzone (=Middle to Late Albian)

Albian dinocyst assemblages which are assigned to the *X. ceratioides* (Pl. 5, fig. 10; Pl. 6, fig. 5)–*C. obliquicostatum* Subzone are described from the Middle and Late Albian of the Paris Basin (Davey and Verdier, 1971), the highest Albian of France and Late Albian (Vraconian) of Switzerland (Davey and Verdier, 1973), the Upper Gault (Albian) of England (Cookson and Hughes, 1964), and Alberta, Canada (Singh, 1971; Brideaux, 1971b). Species first appearing in this Zone include *Carpodinium obliquicostatum*, *Dinopterygium cladoides*, *Litosphaeridium siphoniphorum*, *Palaeohystrichophora infusorioides*, *Psaligonyaulax deflandrei*, *Xenascus ceratioides* and *Xiphophoridium alatum*. *Gonyaulacysta helicoidea*, *Hystrichosphaeridium arundum* Eisenack and Cookson, *Meiourogonyaulax stoveri* and *Protoellipsodinium spinocristatum* do not range into the Cenomanian. Detailed discussions of the species recorded from these areas and their ranges are contained in the works of the authors mentioned above.

Middle to Late Albian sediments in Alberta commonly contain the species *Scriniodinium eurypylum*. This species first described from the Cretaceous of Arctic Canada has a very distinctive AIP archeopyle. *S. eurypylum* has not as yet been recorded from Europe or the Scotian Shelf and was probably restricted to the Canadian Arctic biogeographical province during the Albian. This presumably included part of Alberta. The Late Albian of the Paris Basin, Alberta and central U.S.A. contains the species *Ovoidinium verrucosum*. This species may be diagnostic of the Late Albian–Early Cenomanian on a worldwide scale.

Cleistosphaeridium polypes Zone (=Cenomanian)

Assemblages which are assigned to this zone have been described from the Isle of Wight, England (Clarke and Verdier, 1967), southern England (Davey, 1969, 1970) and the Scotian Shelf and Grand Banks where the zone can be recognised in eight out of nine wells.

In their detailed study of the chalk (Cenomanian–Senonian) from the Isle of Wight, Clarke and Verdier (1967) included the Cenomanian in a single zone, the *Litosphaeridium siphoniphorum* Zone, which they subdivided into three subzones. These are from oldest to youngest: the *Pterodinium perforatum* Subzone (described as the *Dinopterygium perforatum* Subzone); the *Epelidosphaeridia spinosa* Subzone (described as the *Palaeoperidinium spinosum* Subzone) and the *Cleistosphaeridium huguonioti* Subzone.

Davey (1969, 1970) stated that the Cenomanian could be adequately defined by the first occurrence of *Palaeohystrichosphora infusorioides* at the base and the last appearance of *Cleistosphaeridium huguonioti* at the top. The two subzones characterised by *E. spinosa* and *C. huguonioti* could be recognised in England and France.

On the Scotian Shelf and Grand Banks species which do not range upwards into the Turonian include *Ascodinium pontis-mariae*, *Cleistosphaeridium polypes* (Plate 1, fig. 8), *Cribroperidinium intricatum*, *C. orthoceras*, *Epelidosphaeridia spinosa*, *Hystrichosphaeridium cooksonii* and *Oligosphaeridium totum*. *Cyclonephelium vannophorum* is abundant throughout this stage.

Surculosphaeridium longifurcatum Zone (=Turonian)

The Turonian, where studied, has yielded relatively poor assemblages. On the Grand Banks–Scotian Shelf it is delineated by the *Surculosphaeridium longifurcatum* Zone. *S. longifurcatum* (Pl. 4, fig. 10) ranges up into the Santonian but never exceeds two or three specimens per sample in that stage. It is abundant in the Turonian where it is found in association with *Aiora fenestrata* (Deflandre and Cookson) *Callaiosphaeridium asymmetricum*, *Coronifera oceanica*, *Cyclonephelium vannophorum* and *Litosphaeridium siphoniphorum* (Pl. 4, fig. 7). *Deflandrea pirnaensis* is sometimes common.

Clarke and Verdier (1967) subdivided the Turonian into the *Endoscrinium campanula* and *Cyclonephelium membraniphorum* Cookson and Eisenack (*Cyclonephelium distinctum*) Zones, the latter extending into the basal Coniacian. The *E. campanula* Zone is characterised by the presence of the zonal species plus *Cyclonephelium hughesii* Clarke and Verdier and *Xiphophoridium alatum* (as *Pyramidium alatum*). Unfortunately both zones are based on poor sample coverage.

The Turonian dinocysts of Kazakhstan (Vozzhennikova, 1967) include *Albertia recticornis* Vozzhennikova, *Cooksoniella manumi* Vozzhennikova and *C. vnigri* Vozzhennikova, *Diconodinium rhombiformis* Vozzhennikova, *Dinogymnium digitus*, *D. digitus* subsp. *crassus* (Vozzhennikova), *D. mitratum* (Vozzhennikova) (as *Amphidinium*), *D. vozzhennikovae* Lentin and Williams (as *Gymnodinium albertii*) and *Microdinium ornatum* Cookson and Eisenack. The presence of *Cooksoniella* Vozzhennikova and *Dinogymnium digitus* cast doubt on the Turonian age of these sediments. They would appear to be Campanian or Maestrichtian.

The *Surculosphaeridium longifurcatum* Zone is recognised in seven wells on the Scotian Shelf and Grand Banks. Elsewhere knowledge of Turonian assemblages is insufficient for delineation of zonal species.

Callaiosphaeridium asymmetricum–Oligosphaeridium pulcherrimum Zone (= ? Coniacian–Early Santonian)

The *C. asymmetricum* (Pl. 1, fig. 9)–*O. pulcherrimum* (Pl. 3, fig. 9) Zone is recognised on the Grand Banks–Scotian Shelf and in the Isle of Wight. On the Grand Banks, the Coniacian is marked by the incoming of *Deflandrea acuminata* Cookson and Eisenack, *Hystrichosphaeridium readei* (as *H. stellatum* var. A. Williams and Brideaux, 1975) and *Leptodinium striatum* (Clarke and Verdier). On the Scotian Shelf, the Coniacian is characterised by the last occurrence of *Chlamydophorella nyei*, *Cyclonephelium vannophorum* and *Oligosphaeridium pulcherrimum*. Clarke and Verdier (1967) note that *Callaiosphaeridium asymmetricum* (as *Hexasphaera asymmetrica*) extends into the Early Santonian. In the *C. asymmetricum* Zone, *Ellipsodinium rugulosum* becomes extinct. It is possible that in the Scotian Shelf sediments the top of this zone falls also within the Early Santonian.

In Australia the *Deflandrea cretacea* Zone of Evans (1966a, b) has been assigned a Late Coniacian–Early Santonian age.

Hystrichosphaeridium truncigerum Zone (=Santonian)

Assemblages of dinocysts which are assigned to this zone are described from the Santonian of Khatanga (Vozzhennikova, 1967), the Senonian of the Isle of Wight, England (Clarke and Verdier, 1967), the Scotian Shelf (Williams, this chapter) and the Grand Banks (Williams and Brideaux, 1975).

The Senonian of the Isle of Wight (Clarke and Verdier, 1967) is divided into the *Deflandrea echinoidea* Zone (Santonian) and the *Xenascus ceratioides* (as *Pseudoceratium ceratioides*) Zone Santonian–Campanian). The *X. ceratioides* Zone is subdivided into a lower *Callaiosphaeridium asymmetricum* (as *Hexasphaera asymmetrica*) Subzone (Santonian) and an upper *Pterospermopsis spinosa* Clarke and Verdier Subzone (Santonian–Campanian).

The *Hystrichosphaeridium truncigerum* (Pl. 3, fig. 5) Zone demarcates the Santonian of the Scotian Shelf–Grand Banks and the Isle of Wight, England. In the Grand Banks the base of the Santonian is delinated by the appearance of *Australiella victoriensis*, *Dinogymnium acuminatum*, and *Hystrichosphaeridium truncigerum*. A much more significant change palynologically occurs at the top of the Early Santonian. Species having their tops are *Callaiosphaeridium asymmetricum*, *Hystrichosphaeridium readei* (as *H. stellatum* var.

A Williams and Brideaux, 1975), *Oligosphaeridium anthophorum*, *O. complex*, *O. pulcherrimum*, *Senoniasphaera rotundata* and *Tenua hystrix* Eisenack. Species coming in include *Hexagonifera chlamydata* and *Senoniasphaera protrusa*. Within the Santonian *Canningia reticulata*, *Dinopterygium cladoides*, *Hystrichokolpoma ferox* (Deflandre), *Hytrichosphaeridium truncigerum*, *Palaeohystrichophora infusorioides* and *Senoniasphaera protrusa* become extinct.

On the Scotian Shelf the *Hystrichosphaeridium truncigerum* Zone has been recognised in five of the wells examined for palynomorphs. Species first appearing in this zone include *Australiella victoriensis*, *Deflandrea sverdrupiana* Manum, *Dinogymnium acuminatum*, *D. undulosum* Cookson and Eisenack and *Hystrichosphaeropsis ovum* Deflandre. Species not extending upwards into the Campanian are *Dinopterygium cladoides*, *Hystrichokolpoma ferox*, *Hystrichosphaeridium readei*, *Senoniasphaera protrusa* and *Stephodinium coronatum*. *Australiella victoriensis* attains its maximum abundance in this zone whilst *Palaeohystrichophora infusorioides* is not present in any abundance in younger sediments.

The *Nelsoniella aceras* Cookson and Eisenack (includes *Hexagonifera vermiculata* Cookson and Eisenack) zone of Evans (1966a, b) has been assigned a Santonian age.

The Santonian of Khatanga (Vozzhennikova, 1967) is distinguished by the presence of *Australiella granulifera* (Manum) Vozzhennikova, *A. micracantha* (Cookson and Eisenack) Lentin and Williams, *A. tripartita* (Cookson and Eisenack) Vozzhennikova, *Chatangiella niiga* Vozzhennikova, *Cooksoniella manumi*, *C. vnigri*, *Hexagonifera laticaudata* Vozzhennikova and *H. suspecta*, all of the *Deflandrea* group, and *Dinogymnium muticum* (Vozzhennikova) Lentin and Williams (as *Gymnodinium muticum*).

Odontochitina operculata Zone (=Campanian–Earliest Maestrichtian)

Dinocyst assemblages from the Isle of Wight, England (Clarke and Verdier, 1967), Belgium and Holland (Wilson, 1971), Zone A of the Navarro Formation Texas (Zaitzeff and Cross, 1970), Alberta (Harland, 1973), the Scotian Shelf (Williams, this chapter) and the Grand Banks (Williams and Brideaux, 1975; Williams, this chapter) can all be assigned to this zone.

On the Isle of Wight, the Campanian dinocyst assemblages constitute the upper part of the *Xenascus ceratioides* Zone and *Pterospermopsis spinosa* Subzone.

According to Wilson (1971) the Campanian of northern Europe is characterised by *Odontochitina operculata*, *Odontochitina* sp., *O.* aff. *costata* and *X. ceratioides*. On the Scotian Shelf the *O. operculata* (Pl. 1, fig. 2) Zone is characterised by the last occurrence of *Australiella victoriensis*, *Canningia reticulata, Gardodinium deflandrei, Hystrichosphaeridium recurvatum, Odontochitina costata, O. operculata, Oligosphaeridium anthophorum, Senoniasphaera rotundata* and *Xenascus ceratioides. Cannosphaeropsis utinensis* O. Wetzel, *Horologinella apiculata* and *Microdinium veligerum* are not found in pre-Campanian sediments. *Deflandrea echinoidea, Palaeohystrichophora infusorioides* and *Trichodinium castanea* do not appear to range up into the Late Campanian. *Deflandrea belfastensis* Cookson and Eisenack reaches its maximum abundance within this zone.

On the Grand Banks the *Odontochitina operculata* Zone is characterised by the dominance of species of *Deflandrea* and *Dinogymnium*, and the presence of *Deflandrea magnifica, D. sverdrupiana, Dinogymnium digitus, D. digitus* var. A Williams and Brideaux (1975), *Palaeotetradinium silicorum* and *?Svalbardella* sp. Wilson, and the last appearance of *Australiella victoriensis, Odontochitina costata* and *Xenascus ceratioides.*

Zaitzeff and Cross (1970), Williams and Brideaux (1975), Williams (1975) and Wilson (1971) all have a palynological top where *Odontochitina costata, O. operculata* and *Xenascus ceratioides* become extinct. Williams and Brideaux, Wilson, and Williams equate this with the Campanian–Maestrichtian boundary. Only Zaitzeff and Cross place it within the Early Maestrichtian. This may be provincialism, indicating closer proximity of the Grand Banks to Europe than to Texas during the Campanian–Maestrichtian.

The Campanian of Kazakhstan contains only one species of *Dinogymnium* (*D. kasachstanicum*, Vozzhennikova) Lentin and Williams. Other species include *Albertia curvicornis* Vozzhennikova, *Australiella cooksoniae* (Alberti) Vozzhennikova, *A. granulifera* and *Cooksoniella manumi* Vozzhennikova, all belonging to the *Deflandrea* group. The absence of other species of *Dinogymnium* and the abundance of species of *Australiella* and *Cooksoniella* in the Campanian may indicate a shallow water nearshore environment or alternatively mean that this area represents a distinct biogeographical province.

The Campanian of Australia (Evans, 1966a, b) is delineated by the *Xenikoon australis* Cookson and Eisenack (of unknown affinities) Zone.

The *Odontochitina operculata* Zone is recognisable in Alberta, Texas, the Scotian Shelf, Grand Banks, southern England, Belgium and Holland. It is a very distinctive a zone which is as cosmopolitan as the equivalent planktonic foraminiferal zone.

Dinogymnium euclaensis Zone (=Maestrichtian)

Dinocyst assemblages which can be assigned to this zone are described from the Maestrichtian of Belgium, Holland and Denmark (Wilson, 1971), Zones Band C of the Navarro Formation of Texas (Zaitzeff and Cross, 1970),

California Drugg (1967), the Scotian Shelf–Grand Banks (Williams, this chapter) and Grand Banks (Williams and Brideaux, 1975).

No species of *Dinogymnium* are known from post-Maestrichtian sediments, this genus thus permitting separation of the Mesozoic and Cainozoic Eras. In the Scotian Shelf sediments, species of *Dinogymnium* becoming extinct in the *D. euclaensis* Zone include *D. acuminatum*, *D. digitus*, *D. euclaensis* (Pl. 1, fig. 6), *D. microgranulosum* Clarke and Verdier and *D. undulosum*. Other species which do not range into the Cainozoic are *Australiella granulifera*, *Cyclonephelium distinctum*, *Deflandrea acuminata*, and *D. sverdrupiana*, *Deflandrea diebeli* (as *Ceratiopsis*), *Deflandrea magnifica* and *Spongodinium delitiense* do not extend downwards into pre-Maestrichtian sediments. *Areoligera senonensis* Lejeune–Carpentier is often abundant.

In the Maestrichtian of the Grand Banks (Williams and Brideaux, 1975) species of *Deflandrea* and *Dinogymnium* remain dominant elements of most dinocyst assemblages. The stage is distinguished by the presence of *Deflandrea acuminata*, *D. asymmetrica* Wilson, *D. pannucea* Stanley, *Dinogymnium digitus*, *Hystrichodinium pulchrum* Deflandre, *Spiniferites cingulatus* subsp. *cingulatus*, *Svalbardella* sp. Wilson, and *Trigonopyxidia ginella*.

According to Wilson (1971) who has studied the type Maestrichtian dinocyst assemblages from Belgium and coeval assemblages from Holland and Denmark, this stage has a large dinocyst population including *Areoligera senonensis*, *Cannosphaeropsis utinensis*, *Deflandrea balmei* Cookson and Eisenack, *D. diebeli*, *Gonyaulacysta wetzeli* (Lejeune Carpentier) Sarjeant, *Hexagonifera chlamydata*, *Hystrichokolpoma sequanaportus*, *Hystrichosphaeropsis ovum*, *Membranilarnacia* aff. *tenella* Morgenroth (*Hystrichosphaeridium* group), *Microdinium veligerum*, *Palaeoperidinium pyrophorum*, *Palynodinium grallator* Gocht, *Renidinium membraniferum* Morgenroth (*Areoligera* group), *Triblastula nuda*, *T. utinensis* and *Trigonopyxidia ginella*. In the Early Maestrichtian species of *Deflandrea* and *Dinogymnium microgranulosum* are common. *Thalassiphora pelagica* (Eisenack) Eisenack and Gocht comes in in the Late Maestrichtian. Species of *Areoligera* and *Cyclonephelium* are common throughout.

The youngest Cretaceous dinocyst zone recognised by Evans (1966a, b) in his study of the Australian Cretaceous is the *Deflandrea pellucida* (Deflandre and Cookson) Cookson and Eisenack Zone. This presumably is the coeval equivalent of the *Dinogymnium euclaensis* Zone.

TERTIARY BIOZONES

Within the Tertiary Period members of the *Gonyaulacysta* lineage generally dominated assemblages. In the Palaeogene the *Deflandrea* group of the peridinioid lineage is cosmopolitan in distribution and locally attains prominence particularly in inner neritic or brackish water environments. All the species of *Deflandrea* have a simple 2A archeopyle and two antapical horns of more or less equal length. Other members of the *Deflandrea* group include *Lejeunia*, *Palaeocystodinium* Alberti, *Phthanoperidinium*, *Svalbardella* and *Wetzeliella*. The *Gonyaulacysta* lineage in the Palaeogene is represented by species of the *Cordosphaeridium*, *Gonyaulacysta*, *Operculodinium* and *Spiniferites* groups, all with precingular archeopyles; species of the *Hystrichosphaeridium*, *Polysphareidium* and *Systematophora* groups all with apical archeopyles; and species of the *Hemicystodinium* group with epicystal archeopyles. The *Tuberculodinium* lineage appears. The *Cyclonephelium* lineage is represented by the *Areoligera* group, while the *Cyclonephelium* group does not extend into the Palaeocene. The absence of the ceratioid lineage in the Palaeogene and Neogene fossil record suggests that species assignable to this lineage did not encyst during this period.

In the Oligocene, the dinocyst species diversity showed a marked decline so that in the Neogene, of the marine peridinioid dinocysts, only *Palaeocystodinium golzowense* Alberti and a species attributed to *Xenicodinium* Klement survived. The *Gonyaulacysta* lineage dominates Neogene assemblages with species of *Leptodinium*, *Lingulodinium* Wall, *Operculodinium* and *Spiniferites* fluctuating in relative abundances. These fluctuations appear to be primarily controlled by climatic cooling or warming, so that within the Plio-Pleistocene it is possible to recognise glacial and interglacial stages. Throughout the Tertiary these stable long-ranging dinocyst morphotypes remain in the ascendency.

The detailed studies of Tertiary dinocyst assemblages permit the recognition of ten zones and two subzones (Fig. 14); the stratigraphical distribution of selected species is shown on Table 2.

Deflandrea diebeli—Palaeoperidinium pyrophorum Zone (=Early Palaeocene)

Early Palaeocene (Danian) assemblages of dinocysts which have been described from Siberia (Vozzhennikova, 1967), Denmark and North Germany (Morgenroth, 1968), Wilson, 1971, California (Drugg, 1967), Alabama (Drugg, 1970b), the Scotian Shelf and Grand Banks (Williams, this chapter) and Australia (Cookson and Eisenack, 1965c) are assigned to this zone.

The Lower Danian type section (Wilson, 1971) contains *Cordosphaeridium* spp., *Danea mutabilis* (*Gonyaulacysta* group) and *Spongodinium delitiense*. The basal Danian at Svens Klint section in Denmark contains *Danea mutabilis*, *Deflandrea cretacea*, *D. denticulata* Alberti, *D. diebeli* and a species of *Svalbardella* with spinules covering the distal parts of both horns. Absent from the Danian but present in the youngest Maestrichtian are *Hystrichosphaeropsis ovum*, *Palynodinium grallator* and *Triblastula utinensis*.

The *Deflandrea diebeli* (Pl. 2, fig. 4)–*Palaeoperidinium pyrophorum* (Pl. 4, fig. 1) Zone delineates the Early Palaeocene of the Scotian Shelf. Neither of these species is recorded from younger sediments. Other species which do not range above this zone are *Deflandrea delineata*, *Eisenackia circumtabulata* Drugg, *Hystrichosphaeropsis ovum*, *Palaeocystodinium benjaminii* Drugg, *Subtilidinium minutum* Morgenroth, and *Svalbardella* sp. Wilson (1971). Species which occur for the first time are *Cordosphaeridium inodes*, *Cyclonephelium ordinatum* and *Spiniferites septatus*.

The occurrence of *Cordosphaeridium inodes* and *Deflandrea delineata* in the Upper Moreno Formation, California Drugg (1967) are indicative of an early Palaeocene age.

In the Danian assemblage of Alabama described by Drugg (1970b) the form assigned to *Palaeoperidinium basilium* (Drugg) appears to be identical with the zonal index species *Palaeoperidinium pyrophorum*.

The presence of *Deflandrea delineata* and *Eisenackia crassitabulata* indicate age for the Pebble Point.

The Palaeocene of Siberia (Vozzhennikova, 1967) is characterised by *Deflandrea diebeli* (as *Ceratiopsis*), *Eisenackia crassitabulata*, *Lejeunia hyalina* Gerlach (*Deflandrea* group), *Palaeocystodinium hyperxantha* (Vozzhennikova) Vozzhennikova (probably *P. golzowense*), *Palaeoperidinium deflandrei* subsp. *larjakiensis* (as *Peridinium conicum* var. *larjakienis*), *Pentagonum marginatum* Vozzhennikova, *P. sibiricum* Vozzhennikova (both probably synonymous with *Palaeoperidinium pyrophorum*), *Trigonopyxidia ginella* (of unknown affinities) and *Uvatodinium marginatum* Vozzhennikova (*Deflandrea* group). The presence of *Deflandrea diebeli* and *Palaeoperidinium pyrophorum*, together with *Palaeocystodinium hyperxantha* and *Eisenackia crassitabulata* indicate an Early Palaeocene age.

The *Deflandrea diebeli–Palaeoperidinium pyrophorum* Zone delineates the Early Palaeocene of Siberia, Denmark, Germany, the Scotian Shelf, Albama California and Australia. It is the most cosmopolitan of dinocyst zones. In the Early Palaeocene the remaining Cretaceous morphotypes die out to be replaced in the Late Palaeocene by several of the diagnostic Palaeocene species.

In south-east Australia the presence of *Deflandrea delineata* and *Eisenackia crassitabulata* in an assemblage described from the Pebble Point formation by Cookson and Eisenack (1965c) indicate an early Palaeocene age.

During the early Palaeocene the remaining Cretaceous morphotypes die out and are replaced in the late Palaeocene by more typical Tertiary forms.

Deflandrea speciosa–Wetzeliella parva Zone (=Late Palaeocene)

This zone is diagnostic of the late Palaeocene and assemblages have been described from southeast England (Hussain, 1967), northern Spain (Caro,

1973), Australia (Cookson and Eisenack, 1965b, 1967b), the JOIDES coreholes, offshore Florida and the Scotian Shelf–Grand Banks (Williams, this chapter).

On the Scotian Shelf *Deflandrea speciosa* (Pl. 2, fig. 8), *Cyclonephelium pastielsi*, *Polysphaeridium pastielsi*, *Spiniferites pseudofurcatus* (Klumpp) and *Wetzeliella homomorpha* Deflandre and Cookson are recorded for the first time.

In the Late Palaeocene of the JOIDES Coreholes, offshore Florida, species present include: *Adnatosphaeridium reticulense*, *Areoligera senonensis*, *A. senonensis* sensu Gocht, 1969, *Batiacasphaera compta* Drugg, *Comasphaeridium* cf. *cometes* (Valensi) (acritarch), *Cordosphaeridium cracenospinosum* Davey and Williams, *C. gracilis*, *Cyclonephelium divaricatum* (*Areoligera* group), *Deflandrea denticulata* Alberti, *D. wetzeli* Morgenroth, *Diphyes colligerum*, *Eisenackia scrobiculata* Morgenroth, *Eocladopyxis peniculatum*, *Hystrichokolpoma unispinum* Williams and Downie, *Hystrichosphaeridium pseudorecurvatum* Morgenroth, *Lanternosphaeridium lanosum*, *L.* sp. 2 Gocht, 1969, *Operculodinium centrocarpum*, *O.* cf. *hirsutum* Gocht, *Spiniferites cornutus* (Gerlach), *S. pseudofurcatus*, *S. septatus*, *Wetzeliella homomorpha* and *W. parva*.

Caro (1973) recognised two zones for the Palaeocene of northern Spain. The *Deflandrea speciosa* Zone assigned an ?early to Middle Palaeocene age, and the younger *Wetzeliella hyperacantha* Cookson and Eisenack Zone which is believed to be Middle to Late Palaeocene (Late Landinian or Early Sparnacian). Both these zones appear to equate with the *Deflandrea speciosa–Wetzeliella parva* Zone erected in this chapter.

Hussain (1967) completed a study of the Late Palaeocene–Early Eocene of southeast England. Only three species *Areoligera medusettiformis* (O. Wetzel) Lejeune Carpentier, *Tanyosphaeridium variecalamum* Davey and Williams and *Wetzeliella parva* were restricted to the Late Palaeocene (Thanetian).

The presence of *Lanternosphaeridium bipolare* (Cookson and Eisenack) and *Wetzeliella hyperacantha* in the Dartmoor Formation of southeast Australia (Cookson and Eisenack, 1965b) indicates a Late Palaeocene age for this formation. These workers also record these species from the Palaeocene Rivernook Bed of Victoria, together with *Muratodinium fimbriatum* (as *Kenleyia fimbriata*) which is also diagnostic for the Late Palaeocene.

The Late Palaeocene assemblages are distinctly Tertiary and are dominated by species of the *Gonyaulacysta* lineage; species of *Wetzeliella* and *Deflandrea* are locally common.

Homotryblium tenuispinosum–Spiniferites septatus Zone (=Early Eocene)

Early Eocene dinocyst assemblages which are assigned to the *H. tenuispinosum*

(Pl. 3, fig. 7)–*S. septatus* Zone are described from southeastern England (Williams and Downie, 1966a, b, c), Belgium (De Coninck, 1969), Germany (Morgenroth, 1966a; Gocht, 1969), Northern Spain (Caro, 1973), the Scotian Shelf and Grand Banks (Williams and Brideaux, 1975; Williams, this paper), and the JOIDES coreholes offshore Florida (Williams, this paper).

In the Illerdian stratotype section of northern Spain Caro defined three zones on the basis of species of *Wetzeliella*. These are: the *W. articulata* Eisenack Zone (earliest Ypresian), the *W. unicaudalis* Caro Zone (Ypresian) and the *W. coleothrypta* Zone (Latest Ypresian).

Williams and Brideaux (1975) recognise early Eocene sediments on the Grand Banks. These conclusions are based on the presence of *Achomosphaera alcicornu* (Pl. 6, fig. 3), *Areosphaeridium multicornutum* Eaton (*Areoligera* group), *Comasphaeridium cometes* (as *C.* cf. *cometes*), *Cordosphaeridium gracilis* (Pl. 6, fig. 2), *Deflandrea wardenensis*, *Hemicystodinium zoharyi* (Rossignol), *Hystrichokolpoma unispinum* Williams and Downie, *Hystrichosphaeridium salpingophorum* Deflandre and *Spiniferites septatus*.

On the Scotian Shelf, Lower Eocene sediments fall within the *Areoligera senonensis* (Pl. 6, fig. 1) sensu Gocht (1969) Zone. This taxon whose morphology has been described in considerable detail by Gocht, attains dominance in the Early Eocene, although sporadic specimens have been found as high as the Late Eocene. Species not extending into the Middle Eocene include *Comasphaeridium cometes*, *Deflandrea hialina* (Baltes) Lentin and Williams, *Duosphaeridium nudum* (Cookson) Loeblich and Loeblich, *Hystrichosphaeridium salpingophorum* and *Spiniferites septatus*. Species appearing for the first time include *Homotryblium pallidum* and *Wetzeliella coleothrypta* (Pl. 6, fig. 4). *Wetzeliella homomorpha* frequently attains dominance in Early Eocene assemblages.

Stover (1973) has undertaken a detailed study of the species of *Deflandrea* in the Palaeocene, Eocene of southeastern Australia. Species restricted to the Early Eocene are *Deflandrea dartmooria* Cookson and Eisenack, *D. flounderensis* Stover, *D. obliquipes* Deflandre and Cookson, *D. robusta* Deflandre and Cookson and *D. truncata* Stover. *Deflandrea medcalfii* Stover and *D. pachyceros* Deflandre and Cookson, also present in the Palaeocene, do not range into the Middle Eocene. *Deflandrea heterophlycta* Deflandre and Cookson and *D. phosphoritica* are absent from Early Eocene sediments, first appearing in the Middle Eocene. Thus recognition of the Early Eocene is feasible using only species of *Deflandrea*.

The evolutionary expansion of the dinocysts in the Early Eocene is marked by the appearance of several new species. These include *Dinopterygium cladoides* sensu Morgenroth (1966a), *Hemicystodinium zoharyi*, *Homotryblium pallidum*, *H. tenuispinosum*, *Perisseiasphaeridium pannosum* Davey and Williams, *Samlandia chlamydophora* (according to Morgenroth), *Wetzeliella clathrata* Eisenack, *W. coleothrypta*, *W. lunaris* Gocht, *W. meckelfeldensis* Gocht, *W. reticulata* and *W. tenuivirgula*, *Membranilarnacia ursulae* first appears within the Ypresian, here taken as the base of the Late Ypresian. This has been confirmed in four independent studies. Species becoming extinct in the Early Eocene include *Comasphaeridium cometes* (an acritarch), *Hystrichokolpoma unispinum*, *Hystrichosphaeridium salpingophorum*, *H. tubiferum*, *Lanternosphaeridium lanosum*, *Spiniferites septatus* and *Wetzeliella* (*Wetzeliella*) *parva*.

Membranilarnacia ursulae—Tanyosphaeridium ellipticum Zone (=Middle Eocene)

Dinocyst assemblages which may be assigned to this Zone are recorded from southern England (Eaton, 1969, 1971), the Grand Banks (Williams and Brideaux, 1975, and Williams, this chapter) and the JOIDES coreholes, offshore Florida (Williams, this chapter).

Eaton (1971) recognised five zones in the Bracklesham Beds (Lutetian–Auversian) of the Isle of Wight, England. His study showed that there is an overlap in the range of two stratigraphically important species, *Membranilarnacia ursulae* (Pl. 3, fig. 2) which extends to the top of his Zone 3, and *Areosphaeridium diktyoplokus* which first appears towards the top of Zone 2 or within Zone 3. Other stratigraphically important species *Areosphaeridium arcuatum* var. A Eaton, *A. arcuatum* var. B Eaton, and *A. multicornutum* came in either in Zone 4 or 5.

Areosphaeridium diktyoplokus is first recorded from the Middle Eocene of the Isle of Wight (Eaton, 1971) and Grand Banks (Williams and Brideaux, 1975). The only unequivocal record of this species in the early Eocene is by Morgenroth (1966a). The extension of the range of this species into the Oligocene is based on the occurrence of eight specimens in three samples (Maier, 1959). According to Eaton *A. diktyoplokus* may well have attained "world-wide" distribution during the Late Eocene. Evitt (1972) has noted provincial variation in process numbers in this species; European and Grand Banks forms are devoid of the paracingular processes which are present in South American specimens.

The Middle Eocene of the Grand Banks marks the incoming of *Areosphaeridium diktyoplokus*, *Rhombodinium intermedium*, *Wetzeliella tenuivirgula*. *Chiropteridium aspinatum* and *Pentadinium laticinctum* subsp. *granulatum* first appear within the Middle Eocene. *Adnatosphaeridium reticulense*, *Membranilarnacia ursulae*, *Muratodinium fimbriatum* and *Wetzeliella homomorpha* subsp. *quinquelata* (Williams and Downie) have their tops at the top of the Middle Eocene. *Areosphaeridium arcuatum* var. B is restricted to this horizon. *Hemicystodinium zoharyi* and *Thalassiphora pelagica* show a marked increase in relative percentage of the dinocyst population.

The Middle Eocene of JOIDES core holes 1 and 2, offshore Florida is characterised by the first appearance of *Areosphaeridium arcuatum*, *A. multicornutum*, *Cordosphaeridium cantharellum*, *Cyclopsiella vieta*, *Homotryblium plectilum*, *Pentadinium laticinctum* subsp. *granulatum* and *Tanyosphaeridium ellipticum*. *Tanyosphaeridium ellipticum* (Pl. 5, fig. 1) first appears in the Middle Eocene and attains some prominence in the Late Eocene and Oligocene before dying out in the Miocene. *Cordosphaeridium cantharellum* appears towards the top of the Middle Eocene.

Areosphaeridium diktyoplokus–Pentadinium laticinctum Zone (=Late Eocene)

This Zone is cosmopolitan and is recognised in Late Eocene dinocyst assemblages recorded from Southern England (Eaton, 1971), the JOIDES coreholes, offshore Florida, the Scotian Shelf–Grand Banks (Williams and Brideaux, 1975; Williams, this chapter), Argentina (Arkhangelsky, 1969) and southeast Australia (Cookson and Eisenack, 1965a).

The *Areosphaeridium diktyoplokus–Pentadinium laticinctum* Zone is recognised in Upper Eocene sediments of the JOIDES core holes offshore Florida. In samples from these coreholes the Late Eocene contains *Areoligera senonensis* (Pl. 6, fig. 1) sensu Gocht (1969), *Areosphaeridium multicornutum* (Pl. 5, fig. 2), *Cordosphaeridium funiculatum*, *C. gracilis*, *Cyclonephelium intricatum* (Eaton), *C. pastielsi*, *Cyclopsiella vieta* Drugg and Loeblich, *Dinopterygium cladoides* sensu Morgenroth (1966), *Diphyes colligerum*, *Homotryblium plectilum*, *Hystrichokolpoma* cf. *rigaudae* (Pl. 6, fig. 6), *Lingulodinium machaerophorum* (Deflandre and Cookson) Wall, *Operculodinium centrocarpum* (Deflandre and Cookson,) *Pentadinium laticinctum*, *P. taeniagerum* Gerlach, *Polysphaeridium pastielsi*, *Rhombodinium intermedium*, *Samlandia chlamydophora*, *Tanyosphaeridium ellipticum*, *Wetzeliella coleothrypta* and *W. tenuivirgula* subsp. *crassoramosa* Williams and Downie.

The Late Eocene, possibly in part Early Oligocene, dinocyst Zone in the Scotian Shelf sediments is the *Diphyes colligerum* Zone. Species present in this zone and not extending into the Early Oligocene *Deflandrea heterophlycta* Zone include *Areosphaeridium diktyoplokus*, *Cyclonephelium intricatum* and *Rhombodinium draco*, *Pentadinium laticinctum* subsp. *granulatum* is not present in pre-Upper Eocene sediments. *Leptodinium incompositum* (Drugg) Lentin and Williams is restricted to this horizon.

In the Late Eocene of the Grand Banks *Cordosphaeridium cantharellum* and *Pentadinium laticinctum* (Pl. 3, fig. 10) first appear, while *Areosphaeridium diktyoplokus* (Pl. 1, fig. 4), *A. multicornutum*, *Cordosphaeridium biarmatum* Morgenroth, *C. gracilis*, *Cyclonephelium ordinatum*, *Deflandrea wardenensis*,

Diphyes colligerum, *Homotryblium tenuispinosum*, *Rhombodinium draco*, *R. intermedium*, *Wetzeliella homomorpha*, *W. reticulata* and *W. tenuivirgula* become extinct. Species restricted to this horizon include *Eisenackia ornata* and *Leptodinium incompositum*.

According to Benedek (1972) *Cordosphaeridium gracilis* and *Cyclonephelium ordinatum* range upwards into the Middle Oligocene and Late Oligocene respectively. The basis for the age determinations are not given by Benedek. Also the illustration of *Cyclonephelium ordinatum* appears to be of another species. Further evidence is needed before extending the ranges of these species.

The dinocysts of the Browns Creek Clays of Late Eocene age from Australia have been studied by Cookson (1965) and Cookson and Eisenack (1965a). Species present include *Cordosphaeridium capricornum* Cookson and Eisenack (figs 1 and 2, Pl. 15) appear to be specimens of *Areosphaeridium arcuatum* and fig. 4, Pl. 15 a specimen of *A. multicornutum*. Also recorded are specimens of *Eisenackia ornata*, *Schematophora speciosa* Deflandre and Cookson and *Tanyosphaeridium ellipticum* which typify the Late Eocene.

Archangelsky (1969) in a study of Middle and Late Eocene dinocysts of Argentina recorded *Areosphaeridium diktyoplokus* which indicates a Middle to Late Eocene Age.

Chiropteridium dispersum–Samlandia chlamydophora Zone (=Oligocene)

Assemblages of dinocysts which can be assigned to this Zone are recorded from Germany (Gerlach, 1961, Gocht, 1969; Benedek, 1972), Denmark and Germany (Morgenroth, 1966b), Scotian Shelf and Grand Banks (Williams and Brideaux, 1975; Williams, this chapter) and the JOIDES coreholes, offshore Florida (Williams, this paper). In addition Oligocene dinocysts are also described in studies by Maier (1959), Vozzhennikova (1967) and Drugg and Loeblich (1967).

The Oligocene is characterised by species of *Chiropteridium* Gocht (*C. aspinatum*, *C. dispersum*, *C. lobospinosum*), *Cordosphaeridium* Eisenack (*C. cantharellum*, *C. funiculatum*), *Cyclonephelium* Deflandre and Cookson (*C. pastielsi*, *C. reticulosum* Gerlach, *C. semicirculatum* Morgenroth), *Lejeunia* Gerlach (*L. aechmophora* Benedek, *L. hyalina* Gerlach, *L. paratenella*, *L. spatiosa* Morgenroth, *L. tenella* Morgenroth), *Phthanoperidinium* Drugg and Loeblich (*P. amoenum*) and *Tanyosphaeridium* (*T. ellipticum*, *T. paradoxum*, *T. tenerum* Benedek), together with *Thalassiphora pelagica*. Remnants of the Eocene dinocysts represented by species of *Deflandrea* and *Wetzeliella* are also present, especially in the Early Oligocene.

In the Oligocene provincial variations are noticeable in the dinocyst assemblages. Samples from the Oligocene of Germany and the Grand Banks usually contain species of *Chiropteridium*, *Lejeunia*, *Pentadinium* Gocht, *Systematophora* Klement, *Tanyosphaeridium* and *Thalassiphora* Eisenack and Gocht. The Oligocene of the Scotian Shelf differs in the absence of species of *Lejeunia*. The JOIDES samples off Florida contain high relative percentages of *Apteodinium conjunctum* Eisenack and Cookson sensu Benedek, 1972), *Cyclopsiella* Drugg and Loeblich, *Homotryblium plectilum* Drugg and Loeblich, *Pentadinium laticinctum* subsp. *granulatum* Gocht, *Systematophora ancyrea* Cookson and Eisenack, *Thalassiphora pelagica* (Eisenack) and, in the Early Oligocene, *Chiropteridium aspinatum*.

The abundance of *Homotryblium plectilum* in the Oligocene of the JOIDES core holes and Mississippi and its paucity in the Grand Banks and apparent absence in Germany may be due to climatic control if *Homotryblium* is a tropical form related to *Hemicystodinium zoharyi*.

In several areas the *Chiropteridium dispersum* (Pl. 1, fig. 5)–*Samlandia chlamydophora* (Pl. 5, fig. 7) Zone can be subdivided into the *Deflandrea heterophlycta* Subzone (Early Oligocene) and the *Cordosphaeridium funiculatum–Thalassiphora pelagica* Subzone (Middle to Late Oligocene).

Deflandrea heterophlycta Subzone (=Early Oligocene)

On the Scotian Shelf, the *Deflandrea heterophlycta* (Pl.2, fig. 5) Subzone marks the last occurrence of several species of *Wetzeliella*. Species not ranging up into the Middle Oligocene include *Areosphaeridium arcuatum*, *Cordosphaeridium funiculatum*, *Deflandrea heterophlycta*, *D. wardenensis*, *Eocladopyxis peniculatum*, *Thalassiphora pelagica*, *Wetzeliella articulata* Eisenack, *W. coleothrypta*, *W. lunaris* and *W. ovalis* Eisenack. This is a period of decline in species diversity with no corresponding influx of new species.

On the Grand Banks the Early Oligocene is characterised by *Apteodinium* sp. Gocht. *Chiropteridium dispersum*, *C. lobospinosum*, *Diphyes colligerum*, *Tuberculodinium rossignolae*, *Wetzeliella coleothrypta* and *W. meckelfeldensis* Gocht, *Diphyes colligerum*, *Wetzeliella coleothrypta*, and *W. meckelfeldensis* do not range up into the Middle–Late Oligocene. Species restricted to this horizon include *Deflandrea spinulosa* Alberti, and *Homotryblium plectilum*. In the Early Oligocene the genera *Deflandrea* and *Wetzeliella* become extinct, with the exception of three specimens of *D. phosphoritica* subsp. *phosphoritica* Eisenack in one core hole. Species appearing for the first time include *Tuberculodinium rossignolae*, which was described by Drugg (1970a) from the Miocene of Sumatra. The occurrence in the Late Eocene–Early Oligocene extends the known range of the genus *Tuberculodinium* Wall originally described from the Pleistocene.

The *D. heterophlycta* Subzone of the Scotian Shelf is in part the coeval equivalent of the *Diphyes colligerum* Zone of the JOIDES corehole.

In the JOIDES core holes, offshore Florida *Areosphaeridium arcuatum*, *Deflandrea wardenensis*, *Diphyes colligerum*, *Wetzeliella coleothrypta* and *W. symmetrica* Weiler become extinct during the Early Oligocene.

1. *Cordosphaeridium funiculatum–Thalassiphora pelagica* Subzone (=Middle to Late Oligocene)

Provincialism in dinocyst assemblages increases throughout the Oligocene while the cosmopolitan nature of the zones decreases. Thus the *Cordosphaeridium funiculatum–Thalassiphora pelagica* (Pl. 4, fig. 4) Zone is recognised in Germany (Benedek, 1972) the JOIDES coreholes, offshore Florida (Williams, this chapter) may be present in part on the Grand Banks (Williams and Brideaux, 1975; Williams, this chapter) but is as yet unknown on the Scotian Shelf.

Benedek (1972) studied the dinocysts from 116 core samples of a corehole in the Lower Rhine region. These were assigned a Middle and Late Oligocene age by Kempf (1968). The assemblages described by Benedek contain a high percentage of Eocene and Early Oligocene species. Either reworking is a problem in this core or the age determinations are incorrect. Caution should therefore be exercised in comparing information from other areas with Benedek's data.

The Middle–Late Oligocene of the Grand Banks is characterised by a marked reduction in species diversity with the number of species dying out exceeding those appearing for the first time. This marks the peak of the decline phase according to Tappan and Loeblich (1970). Species becoming extinct include *Areosphaeridium multicornutum*, *Chiropteridium aspinatum*, *Cyclonephelium exuberans* Deflandre and Cookson, *Hemicystodinium zoharyi*, *Pentadinium laticinctum* subsp. *granulatum*, *Samlandia chlamydophora* and *Thalassiphora pelagica*. Species restricted to the Oligocene are *Chiropteridium dispersum* and *C. lobospinosum*. *Hemicystodinium zoharyi* according to Wall and Dale (1969) ranges in age from Early Eocene to Recent and is restricted to tropical waters. It is abundant in the Miocene and Pliocene of the JOIDES samples from offshore Florida (Drugg and Williams, in preparation) and occurs in the Early and Middle Miocene of the Scotian Shelf. Its absence from post-Oligocene sediments on the Grand Banks and post-Middle Miocene on the Scotian Shelf is indicative of a cooling trend initiated in the Late Oligocene and Early Miocene in the Grand Banks spreading southwest to reach the Scotian Shelf by Late Miocene time.

The *Chiropteridium dispersum* Zone spans Middle and Late Oligocene time on the Scotian Shelf. The dinocysts continue to show the decrease in species diversity first noted in the Early Oligocene. *Chiropteridium aspinatum*, *C. dispersum*, *Deflandrea phosphoritica*, *D. spinulosa*, *Pentadinium laticinctum* subsp. *granulatum* and *Rhombodinium intermedium* are recorded for the last time. *Homotryblium plectilum* which on the Grand Banks is restricted to the Early Oligocene ranges throughout the Oligocene and into the Middle Miocene. This species is abundant in the Miocene of the JOIDES core holes offshore Florida. It would appear therefore that *H. plectilum* is a subtropical to tropical species. The presence of this species in sediments as young as Middle Miocene on the Scotian Shelf and its disappearance in the Early Oligocene on the Grand Banks indicates climatic control and is the first indication of climatic differences between the two regions in the Cainozoic.

In the JOIDES coreholes, offshore Florida the Middle Oligocene can be separated from the Late Oligocene. *Tuberculodinium rossignolae* appears at the base of the Middle Oligocene whose top is marked by the extinction of *Leptodinium incompositum* and *Hystrichogonyaulax coreoides*. *Deflandrea phosphoritica* is last recorded from the Middle Oligocene.

In the Late Oligocene *Cordosphaeridium fibrospinosum*, *C. funiculatum*, *Cyclonephelium reticulosum*, *Heteraulacacysta fehmarnensis* Lentin and Williams, *Phthanoperidinium amoenum*, *Samlandia chlamydophora* and *Thalassiphora pelagica* become extinct.

Cordosphaeridium cantharellum Zone (=Early Miocene)

The dinocysts of the Miocene show some conformity with Recent assemblages. There is a marked reduction in species diversity and provincialism becomes more marked than in the Oligocene so that the zones delineated are generally regional. The *Cordosphaeridium cantharellum* (Pl. 1, fig. 3) Zone is recognised in the JOIDES coreholes, offshore Florida (Williams, this paper), the Scotian Shelf and the Grand Banks (Williams and Brideaux, 1975; Williams, this paper).

In the JOIDES coreholes the Miocene is delineated by the presence of *Hystrichokolpoma rigaudae* Deflandre and Cookson, *Hystrichosphaeridium pseudorecurvatum* Morgenroth, *Lingulodinium machaerophorum*, *Palaeocystodinium golzowense* and *Xenicodinium hispidum*. Species not ranging above the Early Miocene include *Apteodinium conjunctum* sensu Benedek, 1972, *Chiropteridium dispersum*, *Cordosphaeridium cantharellum*, *C. fibrospinosum*, *Homotryblium plectilum*, *Pentadinium laticinctum* subsp. *granulatum* and *Systematophora ancyrea* (Cookson and Eisenack).

Within the Early Miocene of the Scotian Shelf *Apteodinium* sp. B. Williams and Brideaux, 1975, *Cannosphaeropsis* sp. A. Williams and Brideaux, 1975, *Cordosphaeridium cantharellum* and *Tanyosphaeridium paradoxum* become extinct. *Hemicystodinium zoharyi* is occasionally present.

On the Grand Banks, the Miocene is subdivided into Early, Middle and Late Miocene, mainly on extinctions of species. In the Early Miocene the species *Cordosphaeridium cantharellum* and *Dinopterygium cladoides* sensu Morgenroth (1966a) die out. *Cannosphaeropsis* sp. A and *Xenicodinium hispidum*, the latter described by Drugg (1970a) from the Middle Miocene and Pliocene of Sumatra, appear for the first time.

The *Cordosphaeridium cantharellum* Zone has not, as yet, been delineated in Europe.

Lejeunia fallax–Xenicodinium hispidum Zone (=Middle Miocene)

This Zone is so far only recognised on the Grand Banks and eastern Scotian Shelf (Williams and Brideaux, 1975; Williams, this chapter). The dinocyst assemblages of the Middle Miocene of other areas have not been studied in sufficient detail to determine their zonal index species.

In the Middle Miocene of the Grand Banks (Williams and Brideaux, 1975) *Impletosphaeridium transfodum* Morgenroth, *Tanyosphaeridium* sp. A, *Thalassiphora delicata* and *Tuberculodinium rossignolae* (Pl. 5, fig. 9) appear for the last time. *Hystrichosphaeropsis obscurum*, previously known also from the Middle Miocene of Germany (Gerlach, 1961) and the Late Miocene of Italy (Habib, 1971) is first recorded from this horizon.

The *Pentadinium laticinctum* Zone of the Scotian Shelf is believed to be of Middle Miocene age although *P. laticinctum* has a known range of Late Eocene–Late Miocene on the Grand Banks. Species not ranging up into the Late Miocene include *Lejeunia fallax*, *L. paratenella*, *Spiniferites crassipellis* Deflandre and Cookson, *Systematophora ancyrea* and *Tanyosphaeridium* sp. A Williams and Brideaux, in press.

Hystrichosphaeropsis obscurum–Polysphaeridium pastielsi Zone (=Late Miocene)

This Zone is apparently widespread in occurrence. Habib (1971) has recognised it in the Tabiano Section, Italy, where outcrop sediments of late Miocene and Pliocene age. Although most of the species range throughout the section, a few have a very restricted range. *Chiropteridium partispinatum* (Gerlach), *Cordosphaeridium inodes* (a questionable identification) and *Hystrichosphaeropsis obscurum* (as *H.* sp. cf., *H. ovum*) do not range into the Pliocene.

Within the Late Miocene of the Grand Banks (Williams and Brideaux, 1975) two zones are recognised. The older, Late Miocene A is characterised by the last appearance of *Hystrichokolpoma rigaudae* Deflandre and Cookson, *Hystrichosphaeridium pseudorecurvatum*, *Hystrichosphaeropsis obscurum* (Pl. 3, fig. 3), *Palaeocystodinium golzowense* (Pl. 5, fig. 8), *Pentadinium laticinctum* and *Systematophora ancyrea*. This extends the range of *P. golzowense* which previously had not been reported in sediments younger than Oligocene.

In the Late Miocene B *Apteodinium conjunctum* sensu Benedek, 1972, and *Polysphaeridium pastielsi* (Pl. 4, fig. 9) appear for the last time. A distinct form, *Nematosphaeropsis* sp. A (Williams and Brideaux, 1975), is restricted to this zone. Many of the dinocysts which are known to extend into the Recent in other regions have their last appearances in the Late Miocene in the Grand Banks. This may be a result of climatic control, a general cooling of the area causing species to migrate to warmer waters.

The Late Miocene of the eastern Scotian Shelf falls within the *Cannosphaeropsis* sp. A Zone. The zonal index species together with *Lingulodinium machaerophorum*, *Nematosphaeropsis balcombiana* (Deflandre and Cookson) and *Tuberculodinium vancompoae* (Rossignol) Wall do not extend into the Plio-Pleistocene.

Achomosphaera ramulifera Zone (Habib, 1971) (=Pliocene)

This Zone is presently unique to the Tabiano Section, Italy. In his study of the Tabiano Section Habib noted that the following species range throughout the section from Late Miocene to Pleistocene, *Hystrichokolpoma rigaudae*, *Leptodinium aculeatum*, *L. paradoxum* Wall, *L. patulum*, *Lingulodinium machaerophorum* Deflandre and Cookson, *Nematosphaeropsis balcombiana* Deflandre and Cookson, *Operculodinium centrocarpum* Deflandre and Cookson, *Spiniferites bentori* (Rossignol), *S. ramosus*, *Tectatodinium pellitum* and *Tuberculodinium vanccmpoae* (Rossignol). Habib believed that *Achomosphaera ramulifera* (Pl. 1, fig. 1) is a good marker for the top of the Pliocene since it has not been recorded from Pleistocene sediments.

On the Grand Banks an assemblage of dinocysts containing the species *Hystrichosphaeridium pseudorecurvatum*, *Thalassiphora delicata* and *Xenicodinium hispidum* are recorded.

Baltes (1971) has described a very distinctive dinocyst assemblage from a brackish water early Pliocene (Pontian) deposit in Romania.

QUATERNARY DINOFLAGELLATES

Since dinoflagellate biozones are climatically controlled within the Quaternary and can rarely be correlated from one region to another, no formal zonation is proposed for this period. Several major studies indicate however that the glacial, interglacial stages can be differentiated even in areas in front of the maximum ice advance. Invariably a zone is based on quantitative rather than qualitative data so that care must be exercised in projecting data, even over relatively short distances.

Assemblages of marine dinocysts have been described from the Calabrian, Sicilian Tyrrhenian and Holocene of Israel (Fig. 16) by Rossignol (1961, 1962,

1964) the Pleistocene, Holocene of the Red Sea (Wall, 1967), the Early Pleistocene of England (Wall and Dale, 1968b), the Holsteinian of England (Harland and Downie, 1969), the Holocene of England and Ireland (Harland, 1968; Downie and Singh, 1969) the Late Glacial and Holocene of the Red Sea (Wall and Warren, 1969) and of Australia (Harland and Sarjeant, 1970).

Freshwater dinocysts have been described from the Holocene of Australia (Churchill and Sarjeant, 1963) Minnesota (Eisenack and Fries, 1965, Evitt and Wall, 1968) and Massachusetts (Evitt and Wall, 1968). Genera indigenous to the marine Quaternary are *Hemicystodinium* (1 species), *Hystrichokolpoma* Klumpp (1 species), *Leptodinium* (1 species), *Nematosphaeropsis* Deflandre and Cookson (1 species), *Operculodinium* (4 species), *Spiniferites* (15 species), *Tectatodinium* Wall (1 species) and *Tuberculodinium* (1 species), plus some forms which can be assigned to the extant genus *Peridinium*.

The species diversity in the Quaternary is low with 10–15 being an average. Biostratigraphical zonation within the Quaternary is based primarily on dinocyst associations determined largely from relative and absolute percentages of the dinocysts. Associations which are temperature controlled permit recognition of glacial and interglacial associations. Rossignol (1962) recognised seven zones in the Ashdod Borehole, Israel (Fig. 16). Wall and Dale (1968b) delineated five species associations which correspond with local pollen stages in the Ludham Borehole (Fig. 20). Wall and Warren (1969) recorded repetitive associations dominated by one of the following species: *Hemicystodinium zoharyi*, *Leptodinium aculeatum* and *Nematosphaeropsis balcombiana*. *Hemicystodinium zoharyi* is the dominant species of a recurrent association which occurs throughout the Pleistocene of the Gulf of Mexico. In some samples 99 out of every 100 dinocysts are assignable to this species.

Lateral variation in Quaternary marine dinocyst associations presumably reflects climatic control. Species of *Leptodinium* which are abundant in the Caribbean are rare or absent in the Gulf of Mexico. In Israel Rossignol (1962) found an alternation of assemblages dominated by either *Hemicystodinium* or *Spiniferites*, *Operculodinium* and *Lingulodinium*; whereas in England according to Wall and Dale (1968b) the assemblages are dominated by *Leptodinium multiplexum* Wall and Dale, *Operculodinium israelianum* (Rossignol) Wall and *Tectatodinium pellitum* respectively.

Harland (1971) has described a Holocene (Flandrian) assemblage from Lake Gnotuk, Victoria, Australia. In the species of *Peridinium* archeopyle formation was apparently variable. One specimen appeared to lose paraplate 2′, another appeared to lose paraplate 1c plus some sulcal paraplates. Variability in archeopyle formation seems to be a characteristic of *Peridinium*. This has been discussed by Evitt and Wall (1968), Norris and McAndrews (1970) and Norris and Hedlund (1972).

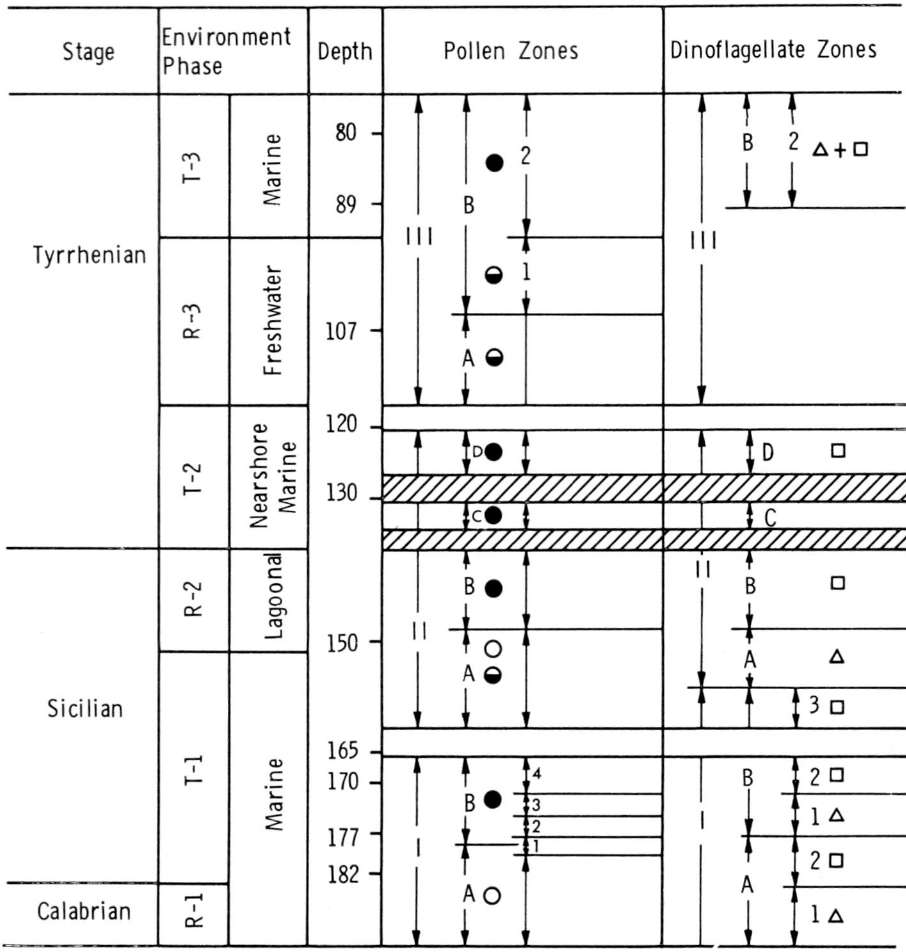

FIG. 16. Palynological zonation of the Plio–Pleistocene of the Ashdod 15/0 Borehole, Israel. R–1, R–2 and R–3 are regressive phases; T–1, T–2 and T–3 equate with marine transgressions. The symbol legend is; O, Chenopodiaceae; ◒, Compositae; ●, Gramineae. These are all pollen. △, *Spiniferites bentori*, *S. ramosus* and *Lingulodinium machaerophorum*. □, *Hemicystodinium zoharyi*. These are dinocysts. (Adapted from Rossignol, 1962.)

RECENT DINOFLAGELLATES

Modern dinocysts have been described by Evitt and Davidson (1964), Wall (1971), Wall and Dale (1966, 1968a, 1969, 1970) and Evitt and Wall (1968). Wall (1971) defined four groups which have living species that produce resting cysts. These are the *Gonyaulax* group, the *Peridinium* group, the *Ceratium*

SAMPLE DEPTH (Ft.)	LOCAL STAGES		CLIMATE	VEGETATION	COMMON DINOFLAGELLATES	NATIONAL STAGES	
	POLLEN	FORAMINIFERA				ENGLAND	NETHERLANDS
44.7	L_5		Temperate	Forest	Tectatodinium pellitum Operculodinium centrocarpum		
64	L_4	LVII LVI	Glacial Cold	Oceanic Heath	Tectatodinium pellitum Operculodinium israelianum	Baventian	Menapian
76	L_3	LV	Temperate	Forest (with Tsuga, Pterocarya)	Spiniferites bulloideus S. mirabilis S. ramosus Operculodinium israelianum	Antian	Waalian
91 99.5 99.7	L_2	LIV	Glacial	Oceanic Heath	Operculodinium israelianum	Thurnian	Eburonian
126.8 146.5	L_{1b} L_{1a}	LIII LII LI	Temperate	Forest (with Tsuga, Pterocarya)	Leptodinium multiplexum Operculodinium centrocarpum Leptodinium multiplexum Tectatodinium pellitum Spiniferites bulloideus	Ludhamian	Tiglian

GSC

FIG. 17. The stratigraphy and palaeontology of the Early Pleistocene at Ludham, Norfolk, England. Within each pollen and Foraminifera zone are listed the predominant dinocysts. This permits recognition of five dinocyst species-associations (facies-associations) whose vertical distribution parallels that of the pollen and foraminifers, and is environmentally controlled, probably climatically. (From Wall and Dale, 1968.)

group and the *Pyrophacus* Stein group (Table 3). The theca of the *Gonyaulax* group is characterised by the possession of six precingular plates and one antapical plate. It includes as well as *Gonyaulax* the genera *Protoceratium* Bergh and *Pyrodinium* Plate. Members of this group produce hystrichosphaerid cysts varying from paratabulate with gonal processes and a simple precingular archeopyle resulting from the loss of paraplate 3″ (*Nematosphaeropsis*, *Spiniferites*); to cysts with numerous intratabular? processes and an archeopyle resulting from the loss of paraplate 3″ (*Operculodinium*); to cysts with numerous intratabular processes and an archeopyle resulting from the loss of more than one precingular paraplate (*Gonyaulax polyhedra* Stein); to cysts with numerous intratabular processes and an archeopyle resulting from the loss of the epicyst (*Pyrodinium bahamense* Plate); to cysts with thick phragma and devoid of processes and simple precingular (3″) or rarely more complex precingular (type 2P) archeopyle, (*Tectatodinium*).

The theca of the *Peridinium* group is characterised by the possession of 4 apicals, 7 precingulars, 5 postcingulars and 2 antapicals with possible variations in all the major series except the postcingular. Wall recognised twelve different types of cyst in this group but listed only four: 1, acapsulate cysts with a dorsal intercalary archeopyle resulting from the loss of paraplate 1a

TABLE 3. Modern cyst-producing dinoflagellates. (From Wall, 1971.)

Group	Cyst-producing taxon	Archeopyle	Fossil equivalent
Gonyaulax	*Gonyaulax digitalis* *G. scrippsae* *G. spinifera*	Precingular (P)	*Nematosphaeropsis* *Spiniferites*
	Protoceratium reticulatum	Precingular (P or 3P)	*Operculodinium*
	Gonyaulax polyedra	Precingular (P to 5P)	*Lingulodinium*
	Pyrodinium bahamense	Epicystal (2A + 6P)	*Hemicystodinium*
	Gonyaulax sp.	Precingular (P or 2P)	*Tectatodinium*
Peridinium	*Diplopsalis lenticula* *Peridinium claudicans* *P. latissimum* *P. leonis* *P. stellatum*	Intercalary (I)	*Lejeunia* *Selenopemphix*
	Peridinium limbatum	Transapical Parasuture	*Palaeoperidinium*
	P. wisconsinense	Apical ($\overline{\text{A}}$a)	
	Diplopeltopsis minor	Epicystal ($\overline{\text{AP}}$a)	
	Diplopsalopsis orbicularis	Combination ($\overline{\text{AI}}$a)	
	Peridinium trochoideum *Scrippsiella sweenyae*	Apical ($\overline{\text{A}}$) or or Combination ($\overline{\text{AI}}$)	*Ovoidinium*
	Ensiculifera sp.	Combination ($\overline{\text{AP}}$)	
Ceratium	*Ceratium hirundinella*	Lateral parasuture?	
Pyrophacus	*Pyrophacus horologium*	Hypocystal	*Tuberculodinium*

or 2a (many marine species of *Peridinium*); 2, cysts of freshwater species possessing an endocyst and pericyst, included here are cysts with two different types of excystment aperture, a transapical parasuture as in *Peridinium limbatum* and an apical archeopyle with attached operculum as in *P. wisconsinense* Eddy; 3, oblate peridinioid cysts with combination archeopyles as in *Diplopsalopsis* Meunier and *Diplopeltopsis* Pavillard; and 4, a number of small modern marine dinoflagellates which produce calcareous cysts.

In the *Ceratium* group only freshwater species appear to produce cysts. The resulting cyst has a ceratioid outline with a lateral parasuture as excystment aperture. In the *Pyrophacus* group the cysts possess a hypocystal archeopyle resulting from the loss of several antapical paraplates.

Evolution of dinoflagellates

At best the fossil record provides an incomplete evolutionary history of the dinoflagellates and contains many problems.

Modern gymnodinioid resting cysts are rapidly destroyed when subjected to mild oxidation and it is possible that they would be incapable of fossilisation. Resting cysts described for the modern Gymnodiniales appear to lack clearly defined archeopyles or paratabulation. Some, however, do bear close resemblance to acritarchs of the acanthomorphitic and herkomorphitic types, and it is possible that the acritarchs which dominate the Palaeozoic microplankton are the cysts of naked dinoflagellates.

The Cretaceous genus *Dinogymnium* appears to be referable to the Gymnodiniales on morphological grounds. The fact that it does possess a resistant phragma and a small apical archeopyle, however, is taken to indicate that it is a cyst. The similarity of this form to *Gymnodinium* Stein may indicate a close relationship or may be an example of homeomorphy.

The absence of gymnodinioid cysts in the Cainozoic may be due to a possible extinction of cyst producing species at the close of the Maestrichtian or in the early Danian, a change in habitat of species of *Gymnodinium* from the neritic to the oceanic during the Palaeogene and consequently a cessation in the production of cysts; or the production of "acanthomorphic" cysts by many Cainoic species which have been overlooked by palynologists.

Other problems in the fossil record are: the sudden disappearance of *Nannoceratopsis* in the Late Jurassic, the demise of the ceratioid lineage in the Cainozic, the disappearance, in the Miocene, of the apical archeopyle in the *Gonyaulax* lineage, the absence of the *Hystrichosphaeridium* group in post Miocene sediments, the relationship of the different kinds of intercalary archeopyles, the function of the endocyst, and the development of the transapical parasuture as an excystment aperture.

PRIMITIVE FEATURES

Wall (1970) introduced the concept of "primitive" features in dinocysts. He is of the opinion that the endocyst, found only in freshwater species of *Peridinium*, is such a feature. Primitive features in such relatively simple unicellular organisms are difficult to determine.

Postulation on the evolutionary trend in paratabulation must in the light of present knowledge be somewhat conjectural. Lister (1970) has argued that there are four broad categories of peridinialean cysts; 1, proximate cysts; 2, proximochorate cysts; 3, chorate cysts and; 4, nontabulate cysts recognised solely on the excystment aperture. Lister stated that these represent a morphological series (a cyst modification series) representing various degrees of differentiation from the theca. Whether, however, evolution goes from 1 to 4, 4 to 1, or independently in each category, cannot be ascertained. Thus we cannot say if *Cleistosphaeridium* Davey *et al.* a nontabulate cyst with an apical archeopyle, is a primitive, or advanced form. Certainly, too much significance should not be attached to proximate, chorate or cavate which merely represent an artificial grouping of convenience.

Primitive features of paratabulation appear to be observable in some lineages. In the peridinioid lineage genera with four anterior intercalaries (e.g. *Pareodinia*) are present in the Lower Jurassic. The presence of more than three intercalaries may therefore be remarkably constant. The tendency for 1″′ to be reduced and incorporated into the parasulcal region is found in the *Gonyaulax* group throughout its known stratigraphic range. The secondary bilateral symmetry in the *Cyclonephelium* lineage is not found in modern cysts. Modern cysts do not possess intratabular processes as in the *Hystrichosphaeridium* group, which became extinct in the Miocene.

ARCHEOPYLE EVOLUTION

The recognition of evolutionary trends in archeopyle types must also be largely conjectural although Lister (1970) has recognised a sequential development. In the most primitive stage, the excystment aperture is defined by a parasuture which has no surface manifestation before dehiscence (*Palaeohystrichophora infusorioides*). Thus the transapical parasuture present in *Peridinium limbatum* is a primitive feature. A more advanced stage is present in species with an archeopyle margin which is not polygonal. Species with a polygonal archeopyle margin represent the most advanced stage in development of the archeopyle. This last category would presumably include apical archeopyles. Unfortunately such sweeping generalities concerning evolution in dinoflagellates give a very distorted viewpoint. This is apparent from the absence of the apical archeopyle in present day *Gonyaulax* species indicating that in this group it may be the obverse of an advanced feature.

LINEAGES

A fascinating development in dinocyst studies in recent years has been the recognition of evolutionary lineages. Eaton (1971) has recognised a morphogenetic series of dinocysts in the Early and Middle Eocene of southern England. The distinctive species *Membranilarnacia ursulae* is first recorded in the Early Eocene (Fig. 18). This species is characterised by a reticulate membrane which units the processes distally and completely encircles the

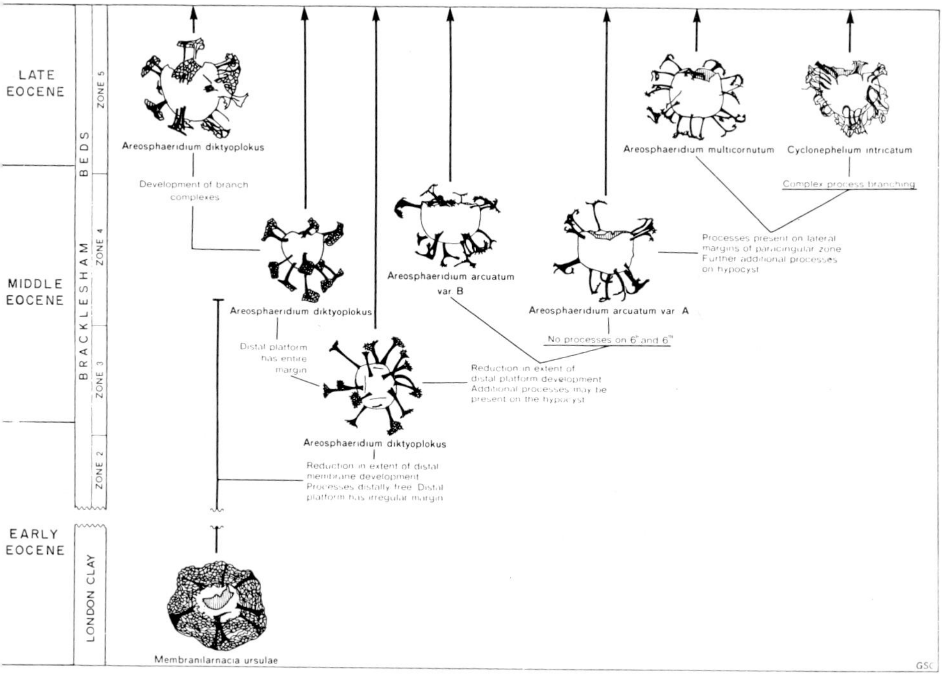

FIG. 18. The postulated evolution of the *Membranilarnacia ursulae* lineage in the Early and Middle Eocene of southeast England. Species include *Membranilarnacia ursulae*, *Areosphaeridium diktyoplokus*, *A. arcuatum*, *A. multicornutum* and *Cyclonephelium intricatum*. (From Eaton, 1971.)

dinocyst. In Zone 3 of the overlying Bracklesham Beds an apparent reduction in the extent of the distal membrane gives rise to *Areosphaeridium diktyoplokus* which has free processes with an irregular distal margin. This root stock gives rise to two further lineages. In the first, specimens of *A. diktyoplokus* with entire distal margins evolve. At the top of the Bracklesham Beds these forms have evolved into specimens of *A. diktyoplokus* possessing process complexes. In the second lineage a further reduction in the extent of the distal platform gives rise to *A. arcuatum*. This may be accompanied by an increase in the number of processes on the hypocyst. Two varieties of *A. arcuatum* are recognised. *A. arcuatum* var. B has processes on all six precingular and all six postcingular paraplates and occasionally on the paracingulum. In *A. arcuatum* var. A, paraplates 6″ and 6‴ and the paracingulum are devoid of processes. *A. arcuatum* var. A evolves into *A. multicornutum* by the development of paracingular processes and an increase in the number of hypocystal processes, and into *Cyclonephelium intricatum* by the development of complex process branching. All three species lack processes on paraplates 6″ and 6‴.

Stover (1973) has determined trends in the variability and relationships of various morphological features in species of *Deflandrea* from the Palaeocene–Eocene of southeastern Australia (Fig. 19). These include the shape of the

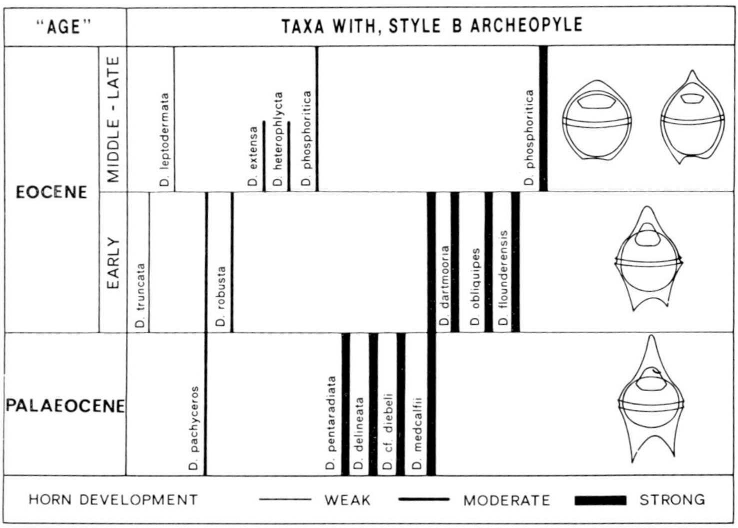

FIG. 19. Stratigraphic distribution of species of *Deflandrea* with the Style B archeopyle and arranged according to the prominence of the apical and antapical horns, in the Palaeocene–Eocene of southeast Australia. The Style B archeopyle is an intercalary archeopyle resulting from the loss of paraplate 2a. The periarcheopyle is roundly triangular or irregularly hexagonal with the width being greater than the height. The perioperculum is completely detached. There is always an endoarcheopyle, the endoperculum remaining attached to the perioperculum. (From Stover, 1973.)

pericyst, apical and antapical horn development, degree and clarity of paratabulation, archeopyle, ornamentation and the position and size of the endocyst relative to the pericyst. In taxa with free operculae the following changes occur with decreasing geological age: the length to width ratio of the archeopyle decreases, the clarity of the paratabulation and the prominence of the paracingulum and parasulcus decrease, apical and antapical horn development tends to become subdued and concomitantly the outline tends to become more rotund with the endocyst occupying a greater proportion of the pericoel, and the younger taxa are more coarsely sculptured. Study of these trends should provide greater preciseness in determining the relative ages of *Deflandrea*-bearing assemblages.

Malloy (1972) has described an Upper Cretaceous dinocyst lineage in sidewall cores from an offshore well, Gabon, West Africa. The lineage (Fig. 20) is believed to be a morphologically transitional evolutionary series. The ancestral form *Deflandrea laevigata* (Malloy) has a subcircular to subpentagonal pericyst with one short apical and two reduced antapical horns. The endophragm and periphragm are closely appressed apart from in the vicinity of the horns. This root stock gives rise to three distinct morphotypes. *Svalbardella polymorpha* (Malloy) is evolved from *D. laevigata* by an elongation of the pericyst and an increase in length of the apical and one of the

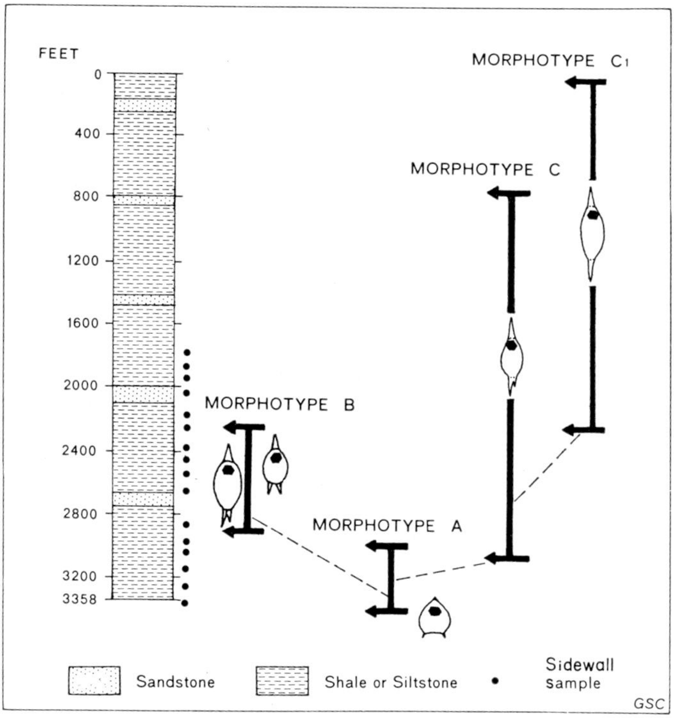

FIG. 20. The postulated evolution of the *Deflandrea laevigata* lineage in the Senonian–Palaeocene of Gabon. Morphotype A, *Deflandrea laevigata* Malloy; Morphotype B, *Svalbardella polymorpha* Malloy; Morphotype C, *Svalbardella australina* Cookson; Morphotype C_1, *Palaeocystodinium golzowense* Alberti, (from Malloy, 1972).

antapical horns. Independently *Svalbardella australina* (Cookson) Malloy evolves from *D. laevigata* by an elongation of the pericyst and an increase in length of the apical and one of the antapical horns. The second antapical horn is reduced to a minute secondary spur. The remaining antapical and the apical horn are approximately equal in length. *Palaeocystodinium golzowense* is derived from *S. australina* by the disappearance of the secondary antapical spur. Unfortunately biostratigraphic control does not permit the precise delineation of the lineage in geological time.

The recognition of lineages in dinocyst studies is obviously of considerable importance. It will permit more precise speciation and more importantly be an invaluable aid to refinement of biostratigraphic zonation and correlation. Care must however be exercised in not stretching morphogenetic series beyond the bounds of scientific probability.

Ecology

VEGETATIVE STAGE

The ecology of living dinoflagellates is imperfectly known and permits only generalisations concerning distribution patterns. The data are based almost exclusively on the motile stage of oceanic forms. Dinoflagellates are found in all environments from freshwater to open oceanic. The motile stage of autotrophic forms must stay above the compensation depth of the photic zone to survive. This depth varies with the species. In marine species the greatest species diversity is found in tropical areas, in colder waters absolute numbers can be high but there is a reduction in species number.

Distribution

Temperature is of primary importance in dinoflagellate distribution. Species may be grouped according to their response to temperature, into temperature tolerant species which are cosmopolitan in distribution and temperature intolerant species which have a somewhat restricted distribution, usually to warmer waters.

The distribution patterns of only one genus, *Ceratium*, have been studied in any detail. Peters (1932) using the species, subspecies and varieties of this genus, recognised warm, temperate and cold water forms. Of fifty-five species, thirty-three species were restricted to warm water, twenty-one in warm to cool waters and only one to cold water. Peters also noted that tropical species of *Ceratium* had thinner amphiesmae and longer horns than cold water species; this according to D. B. Williams (1971a) is an adaptation to reduced viscosities and densities. Peters could not distinguish any variations which were directly relatable to salinity. He did observe that where there is a high species diversity, absolute numbers are small, where there are few species, absolute numbers are generally high. Graham and Bronikowsky (1944) confirmed Peters findings and noted that variation in phosphate content affected distribution patterns of *Ceratium*. The dominant factors in *Ceratium* distribution appear to be the current systems and their associated water masses.

Productivity

The productivity of phytoplankton is controlled by factors often of little importance in their distribution. D. B. Williams (1971a) categorises the agencies affecting phytoplankton production into incident radiation, effective radiation, temperature, salinity, major nutrients, hydrodynamics and other physical controls and minor nutrients. Incident radiation is a function of both the latitude and time of year. The relationship of this function to photosynthesis is a factor in the seasonal fluctations of productivity. The depth of effective radiation limits productivity to the photic zone and varies according to productivity. Where productivity is highest the compensation depth is at its shallowest. The effect of temperature and productivity is variable since the areas of highest productivity are in the higher latitudes. The evidence for the effects of salinity is questionable. Of the major nutrients, oxygen, carbon dioxide, carbon, nitrogen, oxygen, and silicon, only silicon is of no importance to dinoflagellates. Oxygen and carbon dioxide are always present in sufficient amounts. The limiting factor on productivity appears to be nitrogen concentration. Since nitrogen is removed from the photic zone when phytoplankton sink in deeper water, agencies recirculating this lost nitrogen will constitute the dominant influence on productivity. Thus the effect of hydrodynamics and other physical controls is paramount. In regions of divergence of surface waters, upwelling of deep water replenishes the nitrogen content of the photic zone. This results in areas of increased productivity. The zone between two parallel opposed surface currents is often one of increased productivity, as are the eastern margins of major oceanic basins where the combination of ocean current and wind causes an offshore transport of water which is replaced by upwelling of deep water. The seasonally related high productivity in the subpolar regions is due to replenishment of the nutrients in the photic zone. In the sub-Arctic, cooling of the surface waters in winter causes them to sink thus setting up convection currents which bring the deep water to the surface. In spring the warming of this surface water causes stratification and retains the dinoflagellates above the compensation depth. This produces the spring bloom. In the sub-Antarctic the sinking of cold water brings deeper water enriched in nutrients to the surface. The importance of minor nutrients has only recently been realised. Vitamins are of some importance not only in total productivity but also in species distribution.

Seasonal fluctuations control the abundance of dinoflagellates. Commonly they bloom in late spring or summer, following the maximum diatom bloom. At this period the concentration of dinoflagellates which can be as high as 100 million per litre, often imparts a red or yellow colouration to the water. Some dinoflagellates secrete a compound toxic to fish or shellfish and cause mass mortalities amongst these animals. Other dinoflagellates are bioluminescent.

Salinity tolerance

As observed by D. B. Williams (1971a) the effects of salinity fluctuations are imperfectly known. Individual species appear to be salinity sensitive although genera such as *Gymnodinium* and *Peridinium* can be found in freshwater and open oceanic environments. Individual species or variants tend to be restricted to oceanic, neritic, brackish or freshwater environments. Hulbert (1963) in a comparative study of phytoplankton assemblages in open oceanic, continental shelf and estuarine environments in the western North Atlantic Ocean, observed definite trends in species diversity and dominance. Hulbert found that in the open ocean, population sizes are usually small, on the continental shelf they increase in size and attain their maximum in estuarine areas. In the latter environment there is a reduction in species diversity with sometimes one species dominating an assemblage. Superimposed on this empirical picture are seasonal variations. This makes it extremely difficult to apply Hulbert's parameters to fossil dinocysts.

CYST STAGE

Distribution patterns of the living vegetative stage are concerned with the biocoenose (an assemblage of organisms that live together as an interrelated community). Most studies of cyst assemblages are dealing with a thanatocoenose (a group of organisms brought together after death). Further difficulty results from the fact that we are dealing with separate stages of the life cycle and that not all motile stages produce cysts. We are also dealing with a benthonic nonmotile stage as opposed to a planktonic motile stage. The dinoflagellate genus whose distribution pattern is most understood is *Ceratium*. Unfortunately marine species of this genus do not appear to produce cysts.

To make palaeoecological interpretations based on dinocyst distribution it is necessary to know which species encyst and the conditions necessary for encystment. The majority of the more than 40 species possessing an encysted stage are temperate or eurythermal neritic marine, naked and thecate species. A few are freshwater forms. All fall within one of the four groups defined by Wall (1971). The dominant two groups are the *Gonyaulax* group which produce hystrichosphaerid cysts and the *Peridinium* group which produce capsulate, acapsulate and calcareous cysts. *Gonyaulax* and *Peridinium* are heterogeneous genera since cysts of such dissimilar form are produced by individual species.

The causes of encystment remain uncertain. Temperature is an important external factor influencing both encystment and excystment. Other external controlling factors are largely conjectural. Internal factors play some part. In several species there appears to be a relationship between the period of exponential growth and encystment. In this case temperature does not directly control encystment.

Studies of the distribution patterns of modern cysts are few. Dinocysts do not appear to be restricted to any one sedimentary facies or depositional environment. This is to be expected since transportation after death must play an important part in distribution. Whether the cysts are autochthonous or allochthonous must first be determined before conclusions concerning

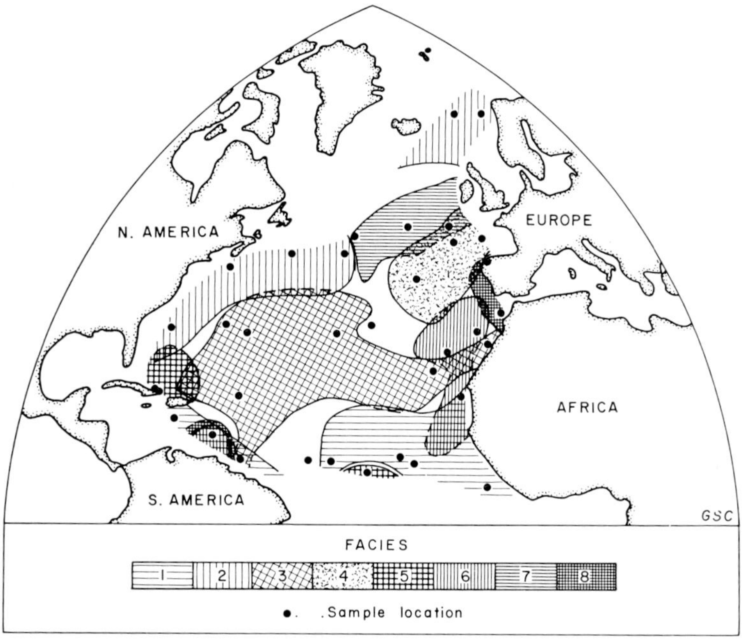

FIG. 21. Biofacies map of dinocyst assemblages in the North Atlantic. This has been obtained from superimposing plots of similarity between real assemblages and ideal assemblages 1 to 8, resulting from Principal Components Analysis of assemblage data. Facies 1, high proportion of Organism A Williams, 1971; Facies 2, high proportion of *Operculodinium centrocarpum*; Facies 3, high proportion of Organism B Williams, 1971; Facies 4, high proportion of *Spiniferites mirabilis*; Facies 5, undifferentiated *Spiniferites* species; Facies 6, high proportion of *Leptodinium aculeatum*; Facies 7, high proportion of *Nematosphaeropsis* sp. C, Williams, 1971. (From D. B. Williams, 1971b.)

palaeoecology, palaeoclimatology, palaeogeography or palaeobathymetry are drawn.

There is an immediate need for detailed studies of the distribution of cysts in modern sediments to see how they relate to the planktonic community, and to determine if they show any climatic zonation or permit the recognition of varying environments such as freshwater, estuarine, neritic or open marine.

A pioneer study has recently been completed by D. B. Williams (1971b) who plotted the distribution of cysts in Recent North Atlantic sediments (Fig. 21). He recognised 44 taxa but only used species comprising more than 25 per cent of at least one sample in his frequency calculations. On the basis of the dinocyst content Williams recognised eight facies, generally relatable to ocean currents. Facies I appears to be an area of low salinity corresponding in part with the Equatorial Counter Current. Facies II is an assemblage of cold-water dinoflagellates and especially *Operculodinium centrocarpum*. It is concentrated along the line of contact of the Gulf Stream and North Atlantic Drift. Facies III is in an area of high salinity (the Sargasso Sea). Facies VIII is off the coast of Gibralter and is dominated by *L. machaerophorum*. This may represent an assemblage transported from the Mediterranean. It would appear that the distribution of marine dinocysts is related to the conditions of the surface environment. Absolute frequency is a factor of productivity. Relative frequencies are hydrodynamically controlled. The weakness of this study lies in the low number (35) of samples analysed and the fact that some were from abyssal and bathyal depths. Whether the cysts are autochthonous or allochthonous was not investigated.

Palaeoecology

QUATERNARY

Recent studies of fossil communities appear to question the traditional interpretation that dinocysts indicate a marine environment. Evitt and Wall (1968) have demonstrated that cysts can be produced by freshwater species. They described the encystment of the freshwater species *Peridinium limbatum* which produces a cyst having a pericyst and endocyst and thus of a similar morpholopy to that of *Deflandrea*. The excystment aperture however results from separation along a transapical parasuture (Fig. 7h). Another freshwater species *P. wisconsinense* has an attached apical archeopyle. Churchill and Sarjeant (1963) have recorded the presence of freshwater cysts in Quaternary sediments.

Studies of marine Quaternary sediments indicate that climatic variations are reflected in the dinocyst communities. Rossignol (1961, 1962) studied dinocysts, pollen and spore assemblages from marine Pleistocene sediments penetrated in the Ashdod 15/0 borehole in southern Israel. She observed three principal palynological zones, I, II and III in the Plio-Pleistocene (Calabrian–Tyrrhenian) (Fig. 16). The cyclic occurrence of dinocyst species permitted each principal zone to be subdivided. Throughout the marine sediments a dinocyst assemblage dominated by *Hemicystodinium zoharyi* (as *Hystrichosphaeridium zoharyi*) alternated with an assemblage dominated by *Lingulodinium machaerophorum*, *Spiniferites bentori* and *S. ramosus* (as *Hystrichosphaera*). Rossignol concluded that no simple parallelism exists between the respective dominances of the two groups of microfossils, dinocysts and pollen. She suggested that subnormal salinities caused by increased freshwater run off favoured the development of a *Hemicystodinium zoharyi* association and that the alternating *Lingulodinium machaerophorum*, *Spiniferites bentori*, *S. ramosus* association accumulated under conditions of normal salinity for the area and time.

Wall (1967) extended Rossignol's studies in an examination of three piston cores from the Caribbean, two from the abyssal plain of the Yucatan Basin in the West Caribbean and the third from the Cariaco Trench. The age of the sediments encountered in the cores range from Sangamon to Post Glacial. According to Wall neritic and epicontinental assemblages in the Quaternary appear liable to considerable and rapid fluctuations in composition involving a small number of common species, especially where there are nonmarine intercalations. Individual assemblages are often poor in species. Wall does not recognise temperature fluctuations in the assemblages encountered. He suggests the maximum absolute abundances of microplankton expressed in terms of specimens per gram of lutite may indicate the warmer climatic intervals and secondly an increase in the abundance of *Leptodinium aculeatum* may be indicative of climatic change.

Further extensive research on Pleistocene assemblages of dinoflagellates has been carried out by Wall and Dale (1968b) who studied samples from the Royal Society borehole at Ludham, Norfolk (Fig. 17). West (1961) had originally conducted a palynological investigation of this borehole while Funnell (1961) studied the distribution of foraminifera in a parallel pilot borehole lying a few feet to the West. West recognised five stages (pollen zones), based on the pollen spectra, representing alternations of temperate mixed coniferous/deciduous forest and glacial oceanic heath vegetation. He also noted the presence of several dinoflagellate species and observed that they exhibited numerical fluctations going down the hole. However, West did not formally name any of these species. Wall and Dale selected untreated samples from each of West's stages L1–5 and ran analysis of the dinoflagellates only. They recognised five dinocyst species associations (facies associations) which tied in with West's and Funnell's works. The time relationship of the Ludham borehole to the Ashdod 15/0 (see previous discussion of

Rossignol's work) is imperfectly known. Wall and Dale believe that the Ludham sequence is probably equivalent to the Calabrian (ml), Sicilian (mla) and perhaps the lowest sections of the Ante Tyrrhenian (m2) at Ashdod. The Ludham assemblages exhibit marked differences from the Ashdod assemblages despite some similarity in age. The two alternating associations of dinoflagellates noted in the Ashdod borehole are not present in the Ludham section. Conversely, none of the Ludham associations are found in Israel. However, a *Lingulodinium machaerophorum–Spiniferites bentori* association has been found in the Holocene sediments in the Vilaine Valley by Morzadec-Kerfourn (1966).

From the equation with the foraminiferal and pollen data Wall and Dale (1968b) concluded that the dinoflagellates found in the Ludhamian (L1), Antian (L3) and stage L5 are temperate stages (interglacial associations). These associations are the *Leptodinium multiplexum* dominated assemblage of the Ludhamian, the association of *Spiniferites* species (*S. bulloideus, S. furcatus, S. mirabilis*) in the Antian, and a *Tectatodinium pellitum* dominated association in stage L5. In contrast, *Operculodinium israelianum* appears to prefer colder water conditions. Wall (personal communication) has subsequently plotted the present day occurrence of *O. israelianum* and discovered that it is a tropical species. It therefore appears that the pollen data, used as the basis for climatic determinations in the Ludham sequence, have been misinterpreted. As is the case today, in the Ludham borehole, the number of species present in the temperate stages is greater than that found in the glacial or cooler stages. The alternating dinocyst associations are believed to represent stadials and interstadials.

Wall and Dale (1969) interpreted Quaternary sequences in which sediments containing *Pyrodinium bahamense* alternate with sediments containing *Gonyaulax polyhedra* resting spores and cysts of the *Spiniferites* type, as indicative of oscillations of sea level. Sediments of the littoral facies are dominated by cysts of *P. bahamense* (a form which today occupies tropical and subtropical bays with bordering mangrove swamps), while cysts of *Gonyaulax* dominate sediments of the neritic facies.

Davey (1971) studied the distribution of palynomorphs in Quaternary continental shelf sediments off the coast of southwest South Africa. In this region two oceanic currents, the cold northward flowing Benguela current and the warm westward flowing Mozambique–Algulhas current converge. The former is nutrient rich, the latter deficient in nutrients. The absolute numbers of palynomorphs is related to the lithology of the sample and distance from shore. The dinocysts are dominated by two species, *Operculodinium centrocarpum* and *Spiniferites ramosus* subsp. *ramosus* (Ehrenberg), whose distribution patterns are related to the two ocean currents. *S. ramosus* subsp. *ramosus* is dominant in sediments beneath the Benguela current while *O. centrocarpum* is dominant in sediments beneath the Agulhas current. Other species present include *Spiniferites membranaceus* (Rossignol) Sarjeant, *S. nodosus* (Wall) Sarjeant and species of *Leptodinium* and *Peridinium*. Absolute numbers seem to be a reflection on lithology type, whereas the dominance of various taxa is a factor of the hydrodynamics and the proximity of land. The minimum values for abundances tend to be along the edge of the continental shelf increasing on the upper part of the slope and in the neritic zone. This is apparently controlled by deposition, the areas where clay and coarse silt are being deposited denoting the highest concentration of palynomorphs.

TERTIARY

Scull *et al.* (1966), in a detailed palaeoenvironmental study of the subsurface Oligocene Vicksburg, Texas, illustrated the usefulness of dinoflagellate cysts in the delineation of marine zones. They categorised the hystrichosphaerids into two artificial groups, those with barbed processes (probably *Cleistosphaeridium* sp.) and those with forked processes (possibly *Spiniferites pseudofurcatus*) and found that there was a consistent relationship between water depth and salinity of the habitat and the process type. The barbed forms appeared to be restricted to brackish water, whereas the forked variety was indicative of a shallow marine environment. They concluded that the hystrichosphaerids (dinocysts) were particularly good facies indicators and were useful in recognising transgressive and regressive phases.

Dinocyst associations have been recognised in the Early Eocene of southeast England by Downie *et al.* (1971). In addition to those dinocyst species which were used to erect a palynologic zonation, there are several taxa which ranged throughout the time interval covered. Some of these long-ranging species display marked variations in abundances (based on counts of 400 specimens per sample), thus permitting the recognition of distinctive assemblages termed associations (Fig. 22). These associations are named for the genus to which the dominant species belong. Four associations are recognised. These are:

1. The *Spiniferites* (as *Hystrichosphaera*) association-dominated by species belonging to the genera *Achomosphaera* Evitt, *Cordosphaeridium*, *Hystrichosphaeridium* and *Spiniferites.*
2. The *Micrhystridium*, Deflandre emend. Lister, association-dominated by the arcritarch genera *Comasphaeridium* Staplin *et al.* and *Micrhystridium.*
3. The *Areoligera* association-dominated by the genera *Areoligera* and *Cyclonephelium.*
4. The *Wetzeliella* association-dominated by species of *Wetzeliella* and (perhaps) *Deflandrea.*

In the majority of samples analysed, one or the other of the above associations is dominant. In a few examples, however, there are transitional associations. Less frequently, different individual species attain dominance within an assemblage. The *Areoligera* association of Whitecliff Bay, Isle of Wight, is dominated by *Cyclonephelium pastielsi* while species of *Areoligera* predominate in the Thanet Sands and *C. pastielsi* is relatively less abundant. There does not appear to be any correlation between the dominant association and pollen percentage in a particular sample.

The presence of an association commonly relates to a particular lithology so that:

1. The Thanet Sands have the *Areoligera* association.
2. The Woolwich Beds have the *Areoligera* and *Wetzeliella* associations.
3. The Oldhaven Beds have the *Areoligera* association.
4. The London Clay Basement Beds are barren of dinoflagellates but contain pollen and spores.
5. The Lower London Clay has the *Micrhystridium* association.
6. The remainder of the London Clay has the *Spiniferities* association.

The presence of an association in a particular lithological unit has more than local significance. The *Areoligera* association has been found to characterise the Thanet Sands of Ipswich, East London, Herne Bay (Kent) and Pegwell Bay (Kent). *Micrhystridium* associations occur in the Lower London Clay at Ipswich, London, Herne Bay (Kent) and the Isle of Wight. *Spiniferites* associations generally predominate in the Middle and Upper London Clay (Fig. 22).

Evolutionary change does not appear to be the main factor controlling the assemblage composition of the samples analysed since the assemblages reflect fluctuations in the numbers of common species that range throughout the sequence under review.

The associations are interpreted as indicating open marine (*Areoligera*, *Spiniferites*), inner neritic (*Micrhystridium*) and lagoonal, estuarine or brackish water environments (*Wetzeliella*). The *Micrhystridium* association presumably marks the initial and closing stages of a marine transgression. This makes it possible to recognise the transgressive, regressive phases of the cycles of sedimentation in the Early Eocene.

Gruas-Cavagnetto (1967, 1968) presented the results of a palynological analysis of Sparnacian sediments from the Paris Basin. The Sparnacian of the Paris Basin is represented by sediments of lagoonal, lacustrine and fluviatile origin. Five dinoflagellate/acritarch associations, which were given a numerical connotation, were recognised in the lagoonal sediments. Associations 1 through 4 were dominated by species of *Wetzeliella* (*W. homomorpha* or *W.*

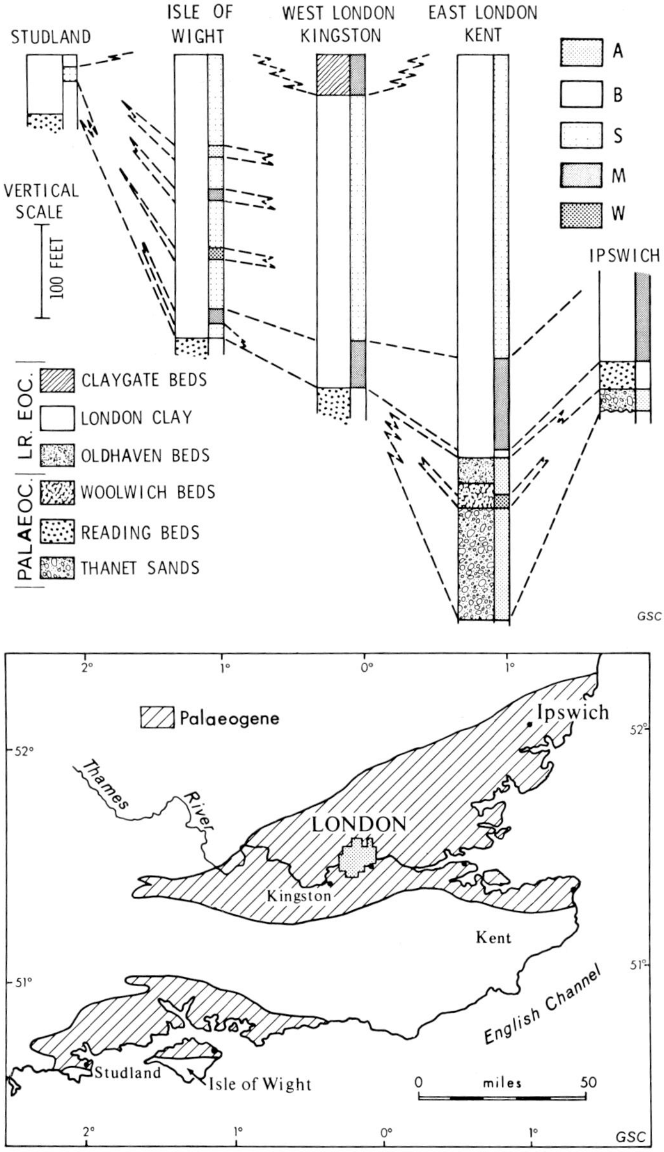

FIG. 22. Correlation of the dinocyst associations recognised in the Palaeocene-Lower Eocene of southeast England. In each section the wide column (on the left) represents lithology and the narrow column (on the right) indicates the association. A, *Areoligera* association; B, barren; S, *Spiniferites* association; M, *Micrhystridium* association; W, *Wetzeliella* association. (From Downie *et al.*, (1971.)

parva) or an unnamed species of *Palaeocystodinium*. In association 5 *Lingulodinium machaerophorum* (as *Baltisphaeridium machaerophorum*) constituted 25–65% of the total dinocyst assemblage. The first three associations were recovered from a lagoonal facies, with no connection with the sea, the molluscs present being lagoonal–continental forms. The No. 5 association (*Lingulodinium machaerophorum*) predominated in a lagoonal–marine deposit.

Gruas-Cavagnetto's data shows close agreement with the findings of Downie *et al.* (1971). Gruas-Cavagnetto found a typical *Wetzeliella* association in most of the Sparnacian samples examined, which in the marine Lower Ypresian gave way to what appears to be either a *Spiniferites* association or a *Micrhystridium–Spiniferites* association. In the early stages of a transgression, the assemblages tended to be uniform in composition and dominated by acanthomorphids. Establishment of open-sea conditions resulted in a greater variety of species, and the marine regression was marked by a reduction in the diversity and abundance of microplankton, with acanthomorphids again becoming dominant.

Other workers in the Tertiary have rarely presented their data quantitatively. Morgenroth (1966a) described assemblages from Eocene clays of North Germany which are clearly *Micrhystridium* associations which appear to be correlatable with the London Clay. An assemblage which he described from the Ypres Clay of Belgium was dominated by *Cyclonephelium* but does not appear to be closely similar to the *Areoligera* association found in the Thanet Sand. This may represent an association not yet described from England.

Gocht (1969) ran counts on 32 samples from 2 boreholes located near Hamburg, Germany, which penetrated Palaeocene, Lower Eocene, questionable Upper Eocene and Oligocene (Fig. 23a and 23b). Special emphasis was placed on the Lower Eocene. He noted definite trends in the composition of the assemblages. The dominant form in the Upper Palaeocene was *Areoligera senonensis*. In the Lower Eocene *Cordosphaeridium* predominated, forming more than 50% of the total assemblages in some samples. Other important elements in the Lower Eocene were species of *Achomosphaera*, *Cyclonephelium*, *Spiniferites*, and *Wetzeliella*. Environmental influences on the local, often monotypical, assemblages were not recognised. However, tabulation of Gocht's data gives a typical *Areoligera* association for the Palaeocene and the *Spiniferites* association for the Lower Eocene. The absence of the *Micrhystridium* association may be related to distance from shoreline.

The recognition of associations in the fossil dinoflagellate cysts and acritarchs has not detracted from their usefulness as zonal indices. Rather, the association may be the key to correlation in the Palaeogene of the Anglo–Paris–Belgian Basin. Confirmation of this is provided by an analysis of the

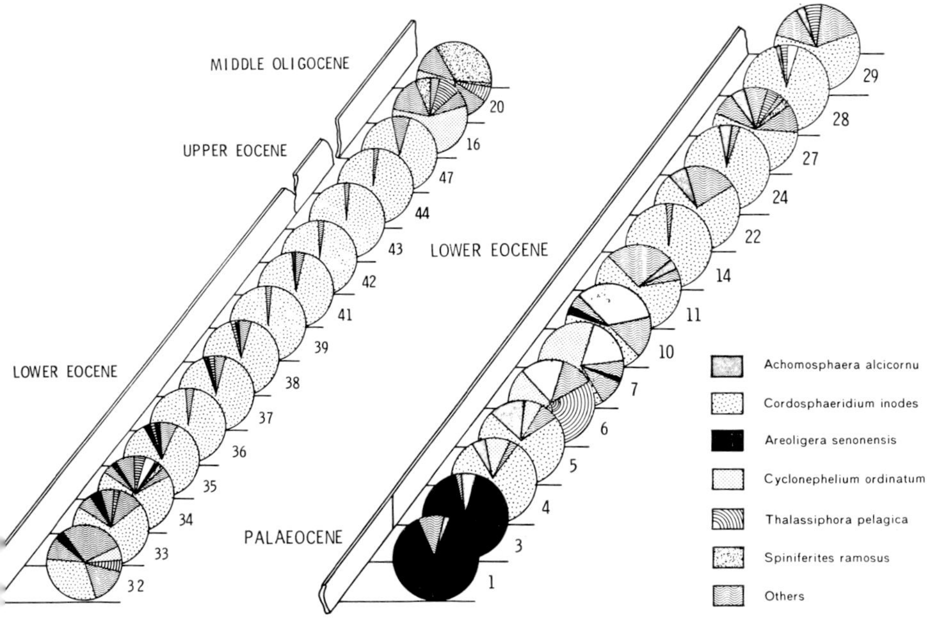

FIG. 23. Relative abundances of fossil dinocysts in the Palaeogene of the Meckelfeld Bore holes 86 and 87 located near Hamburg, northwest Germany. In the Palaeocene *Areoligera senonensis* Lejeune-Carpentier is the dominant species. Throughout the Eocene *Cordosphaeridium gracilis* (Eisenack) Davey and Williams is the most abundant species, except for in samples 6 and 7. *Cordosphaeridium cantharellum* (Brosius) Gocht dominates the two Oligocene samples. (From Gocht, 1969.)

Ypresian assemblages described by De Coninck (1969) from a borehole near Antwerp, Belgium. These assemblages can be classified into one or the other of the four associations present in southeast England (Fig. 24).

CRETACEOUS

Davey (1970) noted that the Albian–Cenomanian of Saskatchewan was characterised by a high relative percentage of *Deflandrea* which was absent from sediments of the same age from England, Texas and France. *Deflandrea* is also found in sediments of similar age in northern Germany, Arctic Canada and eastern Australia. Davey postulated that *Deflandrea* had a preference for cooler water, being a boreal species, whose distribution was thus climatically controlled in the Albian–Cenomanian. Davey also suggested that the decrease in density as water increases in temperature is reflected in the dinocyst assemblage, chorate cysts being predominant in warmer waters. This would explain why the Cenomanian assemblages from England and France are dominated

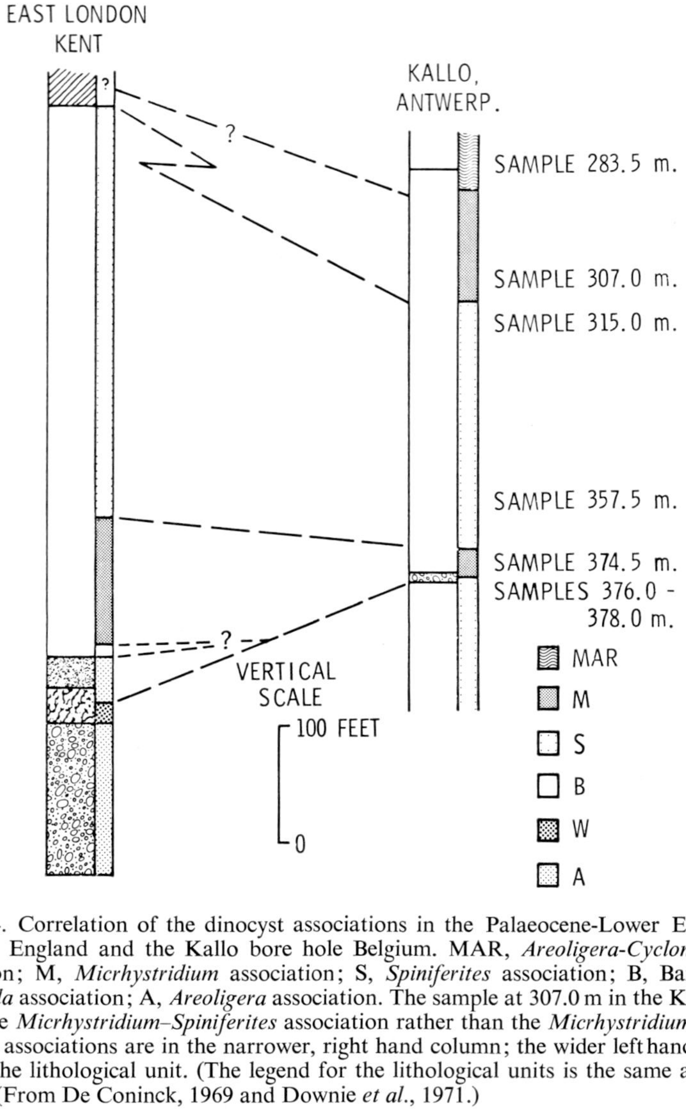

FIG. 24. Correlation of the dinocyst associations in the Palaeocene-Lower Eocene of southeast England and the Kallo bore hole Belgium. MAR, *Areoligera-Cyclonephelium* association; M, *Micrhystridium* association; S, *Spiniferites* association; B, Barren; W, *Wetzeliella* association; A, *Areoligera* association. The sample at 307.0 m in the Kallo bore hole is the *Micrhystridium–Spiniferites* association rather than the *Micrhystridium* association. The associations are in the narrower, right hand column; the wider left hand column signifies the lithological unit. (The legend for the lithological units is the same as that in Fig. 22.) (From De Coninck, 1969 and Downie *et al.*, 1971.)

by chorate cysts and contain very few proximate and cavate cysts, while the Saskatchewan assemblages are composed predominantly of the cavate genus, *Deflandrea*, with a few proximate cysts. As a corollary to this reasoning Vozzhennikova (1965) suggested that morphological differences in dinocysts could be utilised for predicting depositional environments. Thick-walled (not defined) dinocysts were concentrated in the neritic zone, thin-walled forms which developed flotation devices (e.g. processes) were largely confined to open marine environments.

Harland (1973) in a study of the Bearpaw Formation of southern Alberta recognised a parameter, which he termed the gonyaulacacean ratio. This is the number of gonyaulacacean cysts divided by the number of peridiniacean cysts. This was taken as a guide to the salinity, the higher the ratio, the more open marine the environment. Harland noted that the gonyaulacacean ratio for dinocysts in deep sea cores in the Caribbean (Wall, 1967) is 18.0 whereas for a near shore cyst population at Woods Hole, Mass. (Wall and Dale, 1968a) the gonyaulacacean ratio is 0.44. Highs in the gonyaulacacean ratio are therefore taken to denote an open marine environment. This confirms the findings of Downie *et al.* (1971) since their open marine *Spiniferites* association is dominated by gonyaulacacean dinocysts while the brackish water *Wetzeliella* association is predominantly composed of peridiniacean dinocysts.

In the Lethbridge area shallow subsaline near shore environments appear to be indicated by species of *Deflandrea*.

Brideaux (1971), by using the binomial test demonstrated the presence of recurrent species groups of dinocysts and acritarchs in assemblages from the Albian section encountered in two wells in Alberta. The recurrent species groups contain three to five species which can be further grouped into one of two sections. Section I is dominated by dinocysts and in particular species of *Gonyaulacysta* and *Oligosphaeridium*. Section II is dominated by acritarchs. The recurrent species groups in one section are not associated with the recurrent species groups in the other. Brideaux believed that the recurrent groups have palaeoecological significance and reflect response to changes in the sedimentary or environmental regime. The sequential appearance and development of the groups appears to be related to changes in lithology or proximity to shoreline. The groups of Section I appear to denote a transgressive phase; the groups of Section II represent a stable phase.

Reworking of dinocysts and proximity to shoreline

Reworked dinocysts can be used to indicate proximity to shoreline. G. L. Williams (1964) observed the presence of derived dinocysts (Pl. 3, fig. 12; Pl. 4, fig. 8) and acritarchs, primarily from the Middle and Upper Jurassic, in Lower Eocene (London Clay) sediments of southeast England (Fig. 25). Their state of preservation was indistinguishable from that of the indigenous species. However, whereas the indigenous specimens contained pyrite inclusions, the reworked forms did not. Either pyrite was not formed in the reworked specimens during diagenesis of the Jurassic sediments, or if formed it had

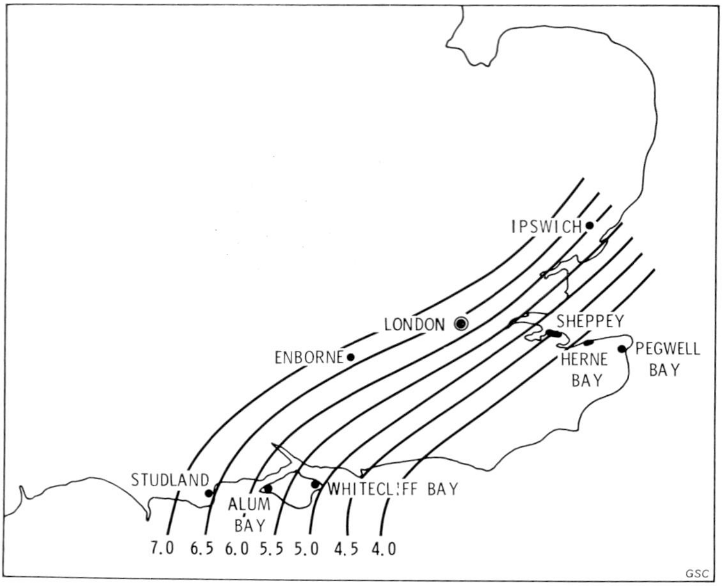

FIG. 25. Relative percentage of derived dinocysts in the London Clay (Early Eocene) of southern England. One hundred per cent represents the total dinocyst count per sample. The majority of derived species are from the Middle and Upper Jurassic.

been removed by subsequent oxidation during reworking. The percentage of reworked dinocysts and acritarchs attained a peak of 17.1% in the vicinity of the presumed shoreline and declined to 4% at the locality farthest from the shoreline. This method may have application in studies where derived material can be identified.

McLean (1968) recorded the Late Cretaceous dinocyst *Dinogymnium acuminatum* and Late Cretaceous species of *Deflandrea*, plus *Gonyaulacysta jurassica* from the Late Jurassic, in the Palaeocene Naheole Formation of southwest Alabama. These were concluded to be reworked from older sediments.

Dinocysts should permit recognition of cycles of sedimentation and transgressive, regressive phases but not in the classical empirical method whereby presence denotes a marine environment. The cyst types and their relative abundances are equally as important. Only interpretations based on a study of all these factors can hope to be reliable.

Conclusions

Since the recognition of the significance of the archeopyle, there has been a proliferation of new taxa particularly at the generic level. In the next few years this should hopefully level off and more thought given to the relative importance of this morphological feature in classification. How much emphasis should be placed on paratabulation, archeopyle type, nature of the operculum and shape at the generic level? Should archeopyle type alone be diagnostic for generic differentiation? Immediate needs are for a unified classification system applicable to cysts and the vegetative stage. This will require a compromise on the part of palynologist and planktonologist. A suprageneric classification taking into account evolutionary trends is urgently needed. The thecate stages of the heterogenous genera *Gonyaulax* and *Peridinium* need to be studied, the genera redefined and if necessary sub-divided. Detailed studies of the ecology of freshwater, brackish water, neritic and oceanic dinoflagellates must be undertaken and the areas of cyst formation determined. The factors causing cyst formation need to be more fully understood. Evolutionary trends need to be determined. Life cycles must be studied in much more detail with the hope that ultimately the neontologist and palynologist will become competent in each other's fields.

Acknowledgements

This chapter has been written in its entirety at the Bedford Institute of Oceanography, Dartmouth, Nova Scotia. I am grateful to the Geological Survey of Canada for all the assistance I have received throughout this venture. I am extremely grateful to Maisie Trapnell and Florrie MacAusland for typing the several versions of the manuscript and to Gary Cook and Gary Grant for drafting some of the figures.

For permission to reproduce Figs 2 and 3 I am indebted to Dr. David Wall, and the Editor of *Micropaleontology*; Figs 7 and 10 are reproduced by kind permission of Professor W. R. Evitt; Fig. 8 is reproduced by courtesy of Dr. E. J. Denton, Director, the Marine Biological Association of the United Kingdom; Fig. 9 is reproduced with the permission of the Carnegie Institute of Washington; Figs 12 and 13 are reproduced with the kind permission of Dr. A. R. Loeblich, Jr.; Fig. 16 is a redrawn version of a figure by Dr. M. Rossignol produced with her permission. Figure 17 is reproduced by courtesy of Dr. David Wall; Figure 18 is reproduced from the original by G. L. Eaton with his permission and that of Edizioni Tecnoscienza, Rome; Fig. 19 is a redrawn version of a figure by Dr. L. E. Stover, with the kind consent of the author and the Geological Society of Australia; Fig. 20 is redrawn from an original by Dr. R. E. Malloy, with his permission and that of the Editor of *Geoscience and Man*; Fig. 21 is reproduced by courtesy of Dr. D. B. Williams and Cambridge University Press; Figs 22 and 24 are by permission of the Editor of *Geoscience and Man* and Fig. 23 is reproduced with the permission of Dr. Hans Gocht.

References

Alberti, G. (1959). Kenntnis der Gattung *Deflandrea* Eisenack (Dinoflag.) in der Kreide und im Alttertiär Nord- und Mitteldeutschlands. *Mitt. geol. StInst. Hamb.* **28,** 93–105. Pls. 8–9.

Alberti, G. (1961). Zur Kenntnis Mesozoischer und Alttertiärer Dinoflagellaten und Hystrichosphaerideen von Nord- und Mitteldeutschland sowie einigen anderen Europäischen Gebieten. *Palaeontographica* Abt. **A, 116,** 1–58. Pls. 1–12.

American Commission on Stratigraphic Nomenclature. (1961). Code of stratigraphic nomenclature. *Bull. Am. Ass. Petrol. Geol.* **45,** 645–660.

Archangelsky, S. (1969). Estudio del paleomicroplancton de la Formacion Rio Turbio (Eoceno), Provincia de Santa Cruz. *Ameghiniana* **6,** 181–218, Pls. 1–5.

Baltes, N. (1971). Pliocene Dinoflagellata and Acritarcha in Romania. *In* (Farinacci, A., ed.) *Proc. II Plankt. Conf.*, **1,** 1–19. Pls. 1–5. Tecnoscienza, Roma.

Beju, D. (1971). Jurassic microplankton from the Carpathian foreland of Romania. *Annls. Inst. geol. publ. hung.* **54,** 275–302, Pls. 1–8.

Benedek, P. N. (1972). Phytoplanktonten aus dem Mittel- und Oberoligozän von Tonisberg (Niederrheingebiet). *Palaeontographica* Abt. **B, 137,** 1–71. Pls. 1–16.

Brideaux, W. W. (1971a). Recurrent species groupings in fossil microplankton assemblages. *Palaeogeogr. Palaeoclimatol. Palaeoecol.* **9,** 101–122. Pls. 1–4.

Brideaux, W. W. (1971b). Palynology of the Lower Colorado Group, central Alberta Canada. I. Introductory remarks. Geology and microplankton studies. *Palaeontographica* Abt. **B, 135,** 53–114. Pls. 21–30.

Burgess, J. D. (1971). Palynological interpretation of Frontier environments in central Wyoming. *Geoscience and Man* **3,** 69–82. Pl. 1.

Caro, Y. (1973). Contribution à la connaissance des dinoflagellés du Paléocène–Eocène inferieur des Pyrénées espagnoles. *Rev. Españ. Micropaleontol.* **5,** 329–372. Pls. 1–5.

Churchill, D. M. and Sarjeant, W. A. S. (1963). Freshwater microplankton from Flandrian (Holocene) peats of southwestern Australia. *Grana Palynologica* **3,** 29–53. Pls. 1–2.

Clarke, R. F. A. and Verdier, J. P. (1967). An investigation of microplankton assemblages from the Chalk of the Isle of Wight, England. *Verh. K. ned. Akad. Wet., Afd. natuurk., Eerste reeks* **24,** 1–96. Pls. 1–17.

Cookson, I. C. (1965). Cretaceous and Tertiary microplankton from south-eastern Australia. *Proc. R. Soc. Vict.* **78,** 85–93. Pls. 9–11.

Cookson, I. C. and Eisenack, A. (1958). Microplankton from Australian and New Guinea Upper Mesozoic sediments. *Proc. R. Soc. Vict.* **70,** 19–79. Pls. 1–12.

Cookson, I. C. and Eisenack, A. (1960a). Microplankton from Australian Cretaceous sediments. *Micropaleont.* **6,** 1–18. Pls. 1–3.

Cookson, I. C. and Eisenack, A. (1960b). Upper Mesozoic microplankton from Australia and New Guinea. *Palaeontology* **2,** 243–261. Pls. 37–39.

Cookson, I. C. and Eisenack, A. (1962a). Some Cretaceous and Tertiary microfossils from Western Australia. *Proc. R. Soc. Vict.* **75,** 269–273. Pls. 37.

Cookson, I. C. and Eisenack, A. (1962b). Additional microplankton from Australian Cretaceous sediments. *Micropaleont.* **8,** 485–507. Pls. 1–7.

Cookson, I. C. and Eisenack, A. (1965a). Microplankton from the Browns Creek Clays, S.W Victoria. *Proc. R. Soc. Vict.* **79,** 119–131. Pls. 11–15.

Cookson, I. C. and Eisenack, A. (1965b). Microplankton from the Dartmoor Formation, SW. Victoria. *Proc. R. Soc. Vict.* **79,** 133–137. Pls. 16–17.

Cookson, I. C. and Eisenack, A. (1965c). Microplankton from the Paleocene Pebble Point Formation, south-western Victoria. *Proc. R. Soc. Vict.* **79,** 139–146. Pls. 18–19.

Cookson, I. C. and Eisenack, A. (1967a). Some early Tertiary microplankton and pollen grains from a deposit near Strahan, western Tasmania. *Proc. R. Soc. Vict.* **80,** 131–140. Pls. 17–21.

Cookson, I. C. and Eisenack, A. (1967b). Some microplankton from the Paleocene Rivernook Bed, Victoria. *Proc. R. Soc. Vict.* **80,** 247–257. Pls. 39–42.

Cookson, I. C. and Hughes, N. F. (1964). Microplankton from the Cambridge Greensand (mid-Cretaceous). *Palaeontology* **7,** 37–59. Pls. 5–11.

Davey, R. J. (1969). Non-calcareous microplankton from the Cenomanian of England, northern France and North America. Part I. *Bull. Br. Mus. nat. Hist.* (*Geol.*) **17,** 103–180. Pls. 1–11.

Davey, R. J. (1970). Non-calcareous microplankton from the Cenomanian of England, northern France and North America, Part II. *Bull. Br. Mus. nat. Hist.* (*Geol.*) **18,** 333–397. Pls. 1–10.

Davey, R. J. (1971). Palynology and palaeo-environmental studies with special reference to the continental shelf sediments of South Africa. *In* (Farinacci, A., ed.) *Proc. II Plankt. Conf.* **1,** 331–347. Tecnoscienza, Roma.

Davey, R. J. (1974). Dinoflagellate cysts from the Barremian of the Speeton Clay, England. *In* (Sah, S. C. D., ed.) Symposium on Stratigraphical Palynology. *Birbal Sahni Institute of Palaeobotany, Special Publication,* No. **3,** 41–75. Pls. 1–9.

Davey, R. J. and Verdier, J. P. (1973). An investigation of microplankton assemblages from the Albian of the Paris Basin. *Verh. K. ned. Akad. Wet., Afd. natuurk., Eerste reeks* **26,** 1–58. Pls. 1–7.

Davey, R. J. and Verdier, J. P. (1973). An investigation of microplankton assemblages from latest Albian (Vraconian) sediments. *Revi. Españ. Micropaleont.* **5,** 173–212. Pls. 1–5.

Davey, R. J. and Williams, G. L. (1966). The genus *Hystrichosphaeridium* and its allies. *In* (Davey, R. J., Downie, C., Sarjeant, W. A. S. and Williams, G. L. *Studies on Mesozoic and Cainozoic dinoflagellate cysts. Bull. Br. Mus. nat. Hist.* (*Geol.*) Suppl. **3,** 53–106.

De Coninck, J. (1969). Dinophyceae et Acritarcha de l'Yprésien du Sondage de Kallo. *Mém. Inst. r. Sci. nat. Belg.* **161,** 1–67. Pls. 1–17.

Deflandre, G. (1935). Considérations biologique sur les microorganisms d'origine planctonique conservés dans les silex de la craie. *Bull. biol. Fr. Belg.* **69,** 213–244. Pls. 5–9.

Deflandre, G. (1937). Microfossiles des silex crétacés. Deuxième partie. Flagellés *incertae sedis* Hystrichosphaeridés. Sarcodinés organisms divers. *Annls. Paléont.* **26,** 51–103. Pls. 8–18.

Deflandre, G. (1938). Microplancton des mers Jurassiques conservé dans les marnes de Villers-sur-Mer (Calvados). Étude luminaire et considérations générales. *Strat. Zool. Wimereux Trav.* **13,** 147–200. Pls. 5–11.

Deflandre, G. (1947). Sur quelques microorganismes planctoniques des silex Jurassiques. *Bull. Inst. océanogr. Monaco* **921,** 1–10.

Deflandre, G. and Cookson, I. C. (1955). Fossil microplankton from Australian

Late Mesozoic and Tertiary sediments. *Aust. J. mar. Freshwat. Res.* **6,** 242–313. Pls. 1–9.

Dodekova, L. (1969). Dinoflagellés et acritarches du Tithonique aux environs de Pleven, Bulgarie Central du Nord. *Bulgarian Acad. Sci., Committee Geol., Bull. Inst., Ser. Paleontol.* **18,** 13–24. Pls. 1–5.

Dodekova, L. (1971). Dinoflagelati i acritarchi ot Tithona v tsentralni severna Bulgariya. *Bulgarian Acad. Sci., Committee Geol., Bull. Geol. Inst., Ser. Paleontol.* **20,** 5–22. Pls. 1–5. (Dinoflagellates and acritarchs of the Tithonian of North–Central Bulgaria.)

Downie, C. and Sarjeant, W. A. S. (1966). The morphology, terminology and classification of fossil dinoflagellate cysts. *In* (Davey, R. J., Downie, C., Sarjeant, W. A. S. and Williams, G. L. *Studies on Mesozoic and Cainozoic dinoflagellate cysts. Bull. Br. Mus. nat. Hist.* (*Geol.*), *Suppl.* **3,** 10–17.

Downie, C. and Singh, G. (1969). Dinoflagellate cysts from estuarine and raised beach deposits at Woodgrange, Co. Down, N. Ireland. *Grana Palynologica* **9,** 124–132.

Downie, C., Hussain, M. A. and Williams, G. L. (1971). Dinoflagellate cyst and acritarch associations in the Paleogene of southeast England. *Geoscience and Man* **3,** 29–35. Pls. 1–2.

Drugg, W. S. (1967). Palynology of the Upper Moreno Formation (Late Cretaceous–Paleocene) Escarpado Canyon, California. *Palaeontographica* Abt. **B, 120,** 1–71. Pls. 1–9.

Drugg, W. S. (1970a). Two new Neogene species of *Tuberculodinium* and one of *Xenicodinium* (Pyrrhophyta). *Proc. biol. Soc. Wash.* **83,** 115–122.

Drugg, W. S. (1970b). Some new genera, species, and combinations of phytoplankton from the Lower Tertiary of the Gulf Coast, U.S.A. *North Amer. Paleontol. Convention, Chicago*, 1969 Proc. **G,** 809–843.

Drugg, W. S. and Loeblich, A. R., Jr. (1967). Some Eocene and Oligocene phytoplankton from the Gulf Coast, U.S.A. *Tulane Stud. Geol.* **5,** 181–194. Pls. 1–3.

Eaton, G. L. (1969). Dinoflagellate cysts and acritarchs from the Bracklesham Beds of the Isle of Wight. Ph.D. Thesis, Univ. of Sheffield, unpublished.

Eaton, G. L. (1971). A morphogenetic series of dinoflagellate cysts from the Bracklesham Beds of the Isle of Wight, Hampshire, England. *In* (Farinacci, A., ed.) *Proc. II Plankt. Conf.* **1,** 355–379. Pls. 1–4. Tecnoscienza, Roma.

Ehrenberg, C. G. (1838). Über das Massenverhältniss der jetzt lebenden Kiesel-Infusorien und über ein neues Infusorien-Conglomerat als Polirschiefet von Jastraba in Ungarn. *Abh. preuss. Akad. Wiss.* **1836,** 109–135. Pls. 1.

Eisenack, A. and Fries, M. (1965). *Peridinium limbatum* (Stokes) verglichen mit der tertiären *Deflandrea phosphoritica* Eisenack. *Geol. För. Stockh. Förh.* **87,** 239–248.

Evans, P. R. (1966a). Mesozoic stratigraphic palynology in Australia. *Australas. Oil Gas J.* **12,** 58–63.

Evans, P. R. (1966b). Mesozoic stratigraphic palynology of the Otway Basin. *Rec. Bur. Miner. Resour. Geol. Geophys. Aust.* 1966/69. Unpublished.

Evitt, W. R. (1961a). *Dapcodinium priscum* n. gen., n. sp., a dinoflagellate from the Lower Lias of Denmark. *J. Paleont.* **35,** 996–1002. Pl. 119.

Evitt, W. R. (1961b). Observations on the morphology of fossil dinoflagellates. *Micropaleont.* **7,** 385–420. Pls. 1–9.

Evitt, W. R. (1963). A discussion and proposals concerning fossil dinoflagellates, hystrichospheres, and acritarchs. *Proc. natn. Acad. Sci. U.S.A.* **49,** 158–164, 298–302.

Evitt, W. R. (1967). Dinoflagellate studies. II. The archeopyle. *Stanford Univ. Publ., Geol. Sci.* **10,** 1–83. Pls. 1–11.

Evitt, W. R. (1972). Significance of process number and arrangement in some Lower Tertiary hystrichospheres. Abstr. *Geoscience and Man* **4,** 130.

Evitt, W. R. (1975). *Manual for a Teaching Conference on fossil dinoflagellates.* Private publication, Stanford University.

Evitt, W. R. and Davidson, S. E. (1964). Dinoflagellate studies. I. Dinoflagellate cysts and thecae. *Stanford Univ. Publ., Geol. Sci.* **10,** 1–12. Pl. 1.

Evitt, W. R. and Wall, D. (1968). Dinoflagellate studies. IV. Theca and cyst of Recent freshwater *Peridinium limbatum* (Stokes) Lemmermann. *Stanford Univ. Publ., Geol. Sci.* **12,** 1–15. Pls. 1–4.

Fisher, M. J. (1972). The Triassic palynofloral succession in England. *Geoscience and Man* **4,** 101–109. Pls. 1–2.

Funnell, B. M. (1961). The Palaeogene and Early Pleistocene of Norfolk. *Trans. Norfolk Norwich Nat. Soc.* **19,** 340.

Gerlach, E. (1961). Mikrofossilien aus dem Oligozän und Miozän Nordwestdeutschlands, unter besonderer Berücksichtigung der Hystrichospharen und Dinoflagellaten. *Neues Jb. Geol. Paläont., Abh.* **112,** 143–228. Pls. 25–29.

Gitmez, G. U. (1970). Dinoflagellate cysts and acritarchs from the basal Kimmeridgian (Upper Jurassic) of England, Scotland and France. *Bull. Br. Mus. nat. Hist. (Geol.)* **21,** 171–257. Pls. 1–14.

Gitmez, G. U. and Sarjeant, W. A. S. (1972). Dinoflagellate cysts and acritarchs from the Kimmeridgian (Upper Jurassic) of England, Scotland and France. *Bull. Br. Mus. nat. Hist. (Geol.)* **21,** 171–257. Pls. 1–17.

Gocht, H. (1969). Formengemeinschaften Alttertiären Mikroplanktons aus Bohrproben des Erdölfeldes Meckelfeld bei Hamburg. *Palaeontographica* Abt. **B, 126,** 1–100. Pls. 1–11.

Gocht, H. (1970). Dinoflagellaten-Zysten aus dem Bathonium des Erdölfeldes Aldorf (NW-Deutschland). *Palaeontographica* Abt. **B, 129,** 125–165. Pls. 26–35.

Gorka, H. (1965). Les microfossiles du Jurassique supérieur de Magnuszew (Pologne). *Acta palaeont. pol.* **10,** 291–327. Pls. 1–5.

Gorka, H. (1970). Dinoflagellate cysts from Callovian of Kukow (Poland). *Acta palaeont. pol.* **15,** 479–498. Pls. 1–6.

Graham, H. W. and Bronikowsky, N. (1944). The genus *Ceratium* in the Pacific and North Atlantic Oceans. *Publs. Carnegie Instn.* **542,** 1–209.

Gruas-Cavagnetto, C. (1967). Complexes sporo-polliniques du Sparnacien du bassin de Paris. *Review Palaeobotan. Palynol.* **5,** 243–261.

Gruas-Cavagnetto, C. (1968). Étude palynologique des divers gisements du Sparnacien du bassin de Paris. *Mém. Soc. géol. Fr.,* (N.S.) **47,** 1–44. Pls. 1–18.

Gruas-Cavagnetto, C. (1970). Microflore et microplancton des Woolwich Beds (Swanscombe, Kent). *Pollen Spores* **12,** 71–82. Pls. 1–2.

Habib, D. (1971). Dinoflagellate stratigraphy across the Miocene–Pliocene boundary, Tabiano stratotype Section. *In* (Farinacci, A., ed.) *Proc. II Plankt. Conf.* **1,** 591–598, Pls. 1–4. Tecnoscienza, Roma.

Habib, D. (1972). Dinoflagellate stratigraphy Leg II, Deep Sea Drilling Project. *In* (Hollister, C. D., Ewing, J. I., *et al.*) *Initial Rep. Deep Sea Drilling Project* **9,** 367–425. Pls. 1–22. U.S. Govt. Printing Office, Washington.

Harland, R. (1968). A microplankton assemblage from the post-Pleistocene of Wales. *Grana Palynologica* **8,** 536–554.

Harland, R. (1971). Fossil dinoflagellate cysts from Lake Gnotuk, Victoria, Australia. *Proc. R. Soc. Vict.* **84,** 245–254. Pls. 15–16.

Harland, R. (1973). Dinoflagellate cysts and acritarchs from the Bearpaw Formation (Upper Campanian) of southern Alberta, Canada. *Palaeontology* **16,** 665–706, Pls. 84–88.

Harland, R. and Downie, C. (1969). The dinoflagellates of the interglacial deposits at Kirmington, Lincolnshire. *Proc. Yorks. geol. Soc.* **37,** 231–237. Pls. 7.

Harland, R. and Sarjeant, W. A. S. (1970). Fossil freshwater microplankton (dinoflagellates and acritarchs) from Flandrian (Holocene) sediments of Victoria and Western Australia. *Proc. R. Soc. Vict.* **83,** 211–234. Pls. 21–22.

Haskell, T. R. (1969). Dinoflagellate species *Dingodinium cerviculum*, *Odontochitina operculata*, and *Muderongia tetracantha* in Lower Cretaceous strata of the Great Artesian Basin, Australia. *Proc. R. Soc. Qd.* **81,** 57–68. Pl. 1.

Haskell, T. R. (1974). Foraminiferal affinities exhibited by the dinoflagellate *Nannoceratopsis* Deflandre. *Geology* **2,** 579–583.

Hulbert, E. M. (1963). The diversity of phytoplanktonic populations in oceanic, coastal, and estuarine regions. *J. mar. Res.* **21,** 81–93.

Hussain, M. A. (1967). Dinoflagellates and acritarchs from the Eocene and Palaeocene of southeast England. Ph.D. Thesis, University of Sheffield, England, unpublished.

Jansonius, J. (1962). Palynology of Permian and Triassic sediments, Peace River area, western Canada. *Palaeontographica* Abt. **B, 110,** 35–98. Pls. 11–16.

Johnson, C. D. and Hills, L. V. (1973). Microplankton zones of the Savik Formation (Jurassic), Axel Heiberg and Ellesmere Islands, District of Franklin. *Bull. Can. Petrol. Geol.* **21,** 178–218. Pls. 1–3.

Jörgensen, E. (1911). Die Ceratien. Eine kurze Monographie der Gattung *Ceratium* Schrank. *Int. Revue ges. Hydrobiol. Hydrogr.* **4,** *Biol. Suppl.* **1,** 1–124. Pls. 1–10.

Kempf, E. K. (1968). Geologische und paläontologische Ergebnisse der Abteufung von Schacht Tönisberg. *Der Niederrhein, Krefeld* **35,** 12–17.

Klement, K. W. (1960). Dinoflagellaten und Hystrichosphaerideen aus dem unteren und mittleren Malm Südwestdeutschlands. *Palaeontographica* Abt. **A, 114,** 1–104. Pls. 1–10.

Kofoid, C. A. (1907). The plates of *Ceratium* with a note on the unity of the genus. *Zool. Anz.* **32,** 177–183.

Kofoid, C. A. (1909). On *Peridinium steini* Jörgensen, with a note on the nomenclature of the skeleton of the Peridinidae. *Arch. Protistenk.* **16,** 25–47. Pl. 2.

Lejeune-Carpentier, M. (1938). L'étude microscopique des *Areoligera*: nouveau genre d'Hystrichosphaeridée (Sixième note). *Annls. Soc. géol. Belg.* **62,** B163–B174.

Lister, T. R. (1970). A monograph of the acritarchs and chitinozoa from the Wenlock and Ludlow Series of the Ludlow and Millichope areas, Shropshire. *Palaeontogr. Soc. (Monogr.)* **124,** 1–100. Pls. 1–13.

Maier, D. (1959). Planktonuntersuchungen in tertiären und quartären marinen Sedimenten. Ein Beitrag zur Systematik, Stratigraphie and Ökologie der Coccolithophorideen, Dinoflagellaten und Hystrichosphaerideen vom Oligozän bis zum Pleistozän. *Neues Jb.Geol. Paläont. Abh.* **107,** 278–340. Pls. 27–33.

Malloy, R. E. (1972). An Upper Cretaceous dinoflagellate cyst lineage from Gabon, West Africa. *Geoscience and Man* **4,** 57–65. Pl. 1.

Mantell, G. A. (1845). Notes of a microscopical examination of the Chalk and Flint of southeast England, with remarks on the Animalculites of certain Tertiary and modern deposits. *Ann. Mag. nat. Hist.* **16,** 73–88.
Mantell, G. A. (1850). *A pictorial atlas of fossil remains consisting of coloured illustrations selected from Parkinson's "Organic remains of a former world", and Artis's "Antediluvian phytology"*, xii, 207 pp. Pls. 1–74. Henry G. Bohn, London.
McLean, D. M. (1968). Reworked palynomorphs in the Paleocene Naheola Formation of southwest Alabama. *J. Paleont.* **42,** 1478–1485, Pls. 187–188.
McLean, D. M. (1971). Organic-called phytoplankton from the lower Tertiary lamunky group of Virginia, Maryland. Ph.D. Thesis, University of Stanford, California, U.S.A., unpublished.
Millioud, M. E. (1969). Dinoflagellates and acritarchs from some western European Lower Cretaceous type localities. *In* (Brönnimann, P. and Renz, H. H., eds), *Proc. I. Plankt. Conf.*, **2,** 420–434, Pls. 1–3. E. J. Brill, Leiden.
Morbey, S. J. and Neves, R. (1974). A scheme of palynologically defined concurrent-range zones and subzones for the Triassic Rhaetian Stage (sensu lato). *Review Palaeobot. Palynol.*, **17,** 161–173, Pls. 1–3.
Morgenroth, P. (1966a). Mikrofossilien und Konkretionen des nordwesteuropäischen Untereozäns. *Paleontographica* Abt. **B, 119,** 1–53, Pls. 1–11.
Morgenroth, P. (1966b). Neue in organischer Substanz erhaltene Mikrofossilian des Oligozäns. *Neues Jb. Geol. Palaont. Abh.* **127,** 1–12, Pls. 1–2.
Morgenroth, P. (1968). Zur Kenntnis der Dinoflagellaten und Hystrichosphaeridien des Danien. *Geol. Jb.* **86,** 533–578, Pls. 41–48.
Morgenroth, P. (1970). Dinoflagellate cysts from the Lias Delta of Luhnde Germany. *Neues Jb. Geol. Palaont. Abh.* **136,** 345–359, Pls. 9–13.
Morzadec-Kerfourn, M. T. (1966), Étude des Acritarches et Dinoflagellés des sédiments vaseux de la Vallée de la Vilaine. *Bull. Soc. géol. minér. Bretagne*, N. S. 137–146, Pls. 1–4.
Norris, G. (1965). Archeopyle structures in Upper Jurassic dinoflagellates from southern England. *N. Z. J. Geol. Geophys.* **8,** 792–806.
Norris, G. and Hedlund, R. W. (1972). Transapical sutures in dinoflagellate cysts. *Geoscience and Man* **4,** 49–56, Pls. 1–3.
Norris, G. and McAndrews, J. H. (1970). Dinoflagellate cysts from post-Glacial muds, Minnesota (U.S.A.). *Review Palaeobot. Palynol.* **10,** 131–156, Pls. 1–3.
Peters, N. (1932). Die Bovölkerung des Südatlantischen Ozeans mit Ceratien. *Wiss. Ergebn. dt. atlant. Exped. "Meteor"* **11,** 1–69.
Pocock, S. A. J. (1972). Palynology of the Jurassic sediments of western Canada. Part 2. Marine Species. *Palaeontographica* Abt. **B, 137,** 85–153. Pls. 22–29.
Rossignol, M. (1961). Analyse pollinique de sédiments marins Quaternaires en Israël. I: Sédiments Récents. *Pollen Spores* **3,** 303–324. Pls. 1–2.
Rossignol, M. (1962). Analyse pollinique de sédiments marins Quaternaires en Israël. II: Sédiments Pleistocenes. *Pollen Spores* **4,** 121–148. Pls. 1–2.
Rossignol, M. (1964). Hystrichosphères du Quaternaire en Méditerranée orientale, dans les sédiments Pléistocenes et les boues marines actuelles. *Revue Micropaléont.* **7,** 83–99. Pls. 1–3.
Sarjeant, W. A. S. (1960a). New hystrichospheres from the Upper Jurassic of Dorset. *Geol. Mag.* **97,** 137–144. Pl. 6.
Sarjeant, W. A. S. (1960b). Microplankton from the Corallian rocks of Yorkshire. *Proc. Yorks. geol. Soc.* **32,** 389–408. Pls. 12–14.

Sarjeant, W. A. S. (1961). Microplankton from the Kellaways Rock and Oxford Clay of Yorkshire. *Palaeontology* **4,** 90–118. Pls. 13–15.

Sarjeant, W. A. S. (1962). Upper Jurassic microplankton from Dorset, England. *Micropaleont.* **8,** 255–268. Pls. 1–2.

Sarjeant, W. A. S. (1963). Fossil dinoflagellates from Upper Triassic sediments. *Nature* **199,** 353–354.

Sarjeant, W. A. S. (1965). Microplankton from the Callovian (*S. calloviense Zone*) of Normandy. *Revue Micropaleont.* **8,** 175–184, Pl. 1.

Sarjeant, W. A. S. (1967). The stratigraphical distribution of fossil dinoflagellates. *Review Palaeobot. Palynol.* **1,** 323–343.

Sarjeant, W. A. S. (1968). Microplankton from the Upper Callovian and Lower Oxfordian of Normandy. *Revue Micropaléont.* **10,** 221–242. Pls. 1–3.

Sarjeant, W. A. S. (1969). Microfossils other than pollen and spores in palynological preparations. *In* (Erdtman, G.) *Handbook of palynology*, 165–208. Hafner Publishing Co., New York.

Sarjeant, W. A. S. and Downie, C. (1966). The classification of dinoflagellate cysts above generic level. *Grana Palynologica* **6,** 503–527.

Scull, B. J., Felix, C. J., McCaleb, S. B. and Shaw, W. G. (1966). The inter-discipline approach to paleoenvironmental interpretations. *Trans. Gulf-Cst Ass. geol. Socs.* **16,** 81–117. Pls. 1–9.

Singh, C. (1971). Lower Cretaceous microfloras of the Peace River area, northwestern Alberta. *Bull. Res. Coun. Alberta* **28,** 1–542. Pls. 1–80.

Stover, L. E. (1973). Palaeocene and Eocene species of *Deflandrea* (Dinophyceae) in Victorian coastal and offshore basins, Australia. *Spec. Pubs geol. Soc. Aust.* **4,** 167–188. Pls. 1–5.

Tappan, H. and Loeblich, A. R., Jr. (1970). Geobiologic implications of fossil phytoplankton evolution and time-space distribution. *In* (Kosanke, R. M. and Cross, A. T., eds) *Symposium on palynology of the Late Cretaceous and Early Tertiary. Spec. Pap. geol. Soc. Am.* **127,** 247–340. Pl. 1.

Tappan, M. and Loeblich, A. R. (1972). Fluctuating rates of protistan evolution, diversification and extinction. *In* (Mamet, B. L. and Westermann, G. E. G., eds) Section 7 *Paleontology*, *24th Int. Geol. Congress*, *Montreal*, *Canada* 205–213.

Valensi, L. (1953). Microfossiles des silex du Jurassique moyen. Remarques pétrographiques. *Mém. Soc. géol. Fr.* **68,** 1–100. Pls. 1–16.

Vozzhennikova, T. F. (1965). Vvedenie v izuchenie iskopaemykh vodorosley. *Akad. Nauk SSSR, Sib. Otd., Inst. Geol. Geofiz., Tr.* 156 pp. (Introduction to the study of fossil peridinian algae).

Vozzhennikova, T. F. (1967). Iskopaemye peridinei yurskikh, melovykh i paleogenovykh otlozheniy SSSR. *Akad. Nauk SSSR, Sib. Ord., Inst. Geol. Geofiz., Tr.*, 347 pp. Pls. 1–121. (Fossil peridinians of the Jurassic, Cretaceous and Paleogene deposits of the USSR).

Wall, D. (1967). Fossil microplankton in deep-sea cores from the Caribbean Sea. *Palaeontology* **10,** 95–123. Pls. 14–16.

Wall, D. (1970). Quaternary dinoflagellate micropaleontology: 1959 to 1969. *North Amer. Paleontol. Convention, Chicago*, 1969, Proc. **G,** 844–866.

Wall, D. (1971). Biological problems concerning fossilizable dinoflagellates. *Geoscience and Man* **3,** 1–15. Pls. 1–2.

Wall, D. and Dale, B. (1966). "Living fossils" in western Atlantic plankton. *Nature* **211,** 1025–1026.

Wall, D. and Dale, B. (1968a). Modern dinoflagellate cysts and evolution of the Peridiniales. *Micropaleont.* **14,** 265–304. Pls. 1–4.

Wall, D. and Dale, B. (1968b). Early Pleistocene dinoflagellates from the Royal Society borehole at Ludham, Norfolk. *New Phenol.* **67,** 315–326. Pl. 1.

Wall, D. and Dale, B. (1969). The "hystrichosphaerid" resting spore of the dinoflagellate *Pyrodinium bahamense*, Plate, 1906. *J. Phycology* **5,** 140–149.

Wall, D. and Dale, B. (1970). Living hystrichosphaerid dinoflagellate spores from Bermuda and Puerto Rico. *Micropaleont.* **16,** 47–58. Pl. 1.

Wall, D. and Evitt, W. R. (1975). A comparison of the modern genus *Ceratium* Schrank 1873 with certain marine dinoflagellates. *Micropaleont.* **21,** 14–44.

Wall, D. and Warren, J. S. (1969). Dinoflagellates in Red Sea piston cores. *In* (Degens, E. T. and Ross, D. A., eds) *Hot brines and Recent heavy metal deposits in the Red Sea*, 317–328. Springer-Verlag, New York.

Warren, J. S. (1967). Dinoflagellates and acritarchs from the Upper Jurassic and Lower Cretaceous rocks on the west side of the Sacramento Valley, California. Ph.D. thesis, University of Stanford, California, U.S.A., unpublished.

Warren, J. S. (1973). Form and variation of the dinoflagellate *Sirmiodinium grossi* Alberti, from the Upper Jurassic and Lower Cretaceous of California. *J. Paleont.* **47,** 101–114. Pls. 1–3.

West, R. G. (1961). Vegetational history of the Early Pleistocene of the Royal Society borehole at Ludham, Norfolk. *Proc. R. Soc., Ser. B,* **155,** 437–453.

Wetzel, O. (1933a). Die in organischer Substanz erhaltenen Mikrofossilien des baltischen Kreide-Feuersteins mit einem sedimentpetrographischen und stratigraphischen Anhang. *Palaeontographica* Abt. **A, 77,** 141–188.

Wetzel, O. (1933b). Die in organischer Substanz erhaltenen Mikrofossilien des baltischen Kreide-Feuerstein mit einem sedimentpetrographischen und stratigraphischen Anhang. *Palaeontographica* Abt. **A, 78,** 1–110. Pls. 1–7.

Wiggins, V. D. (1973). Upper Triassic dinoflagellates from arctic Alaska. *Micropaleont.* **19,** 1–17. Pls. 1–5.

Williams, D. B. (1971a). The distribution of marine dinoflagellates in relation to physical and chemical conditions. *In* (Funnell, B. M. and Riedel, W. R., eds) *Micropalaeontology of the Oceans*, 91–95. Cambridge University Press.

Williams, D. B. (1971b). The occurrence of dinoflagellates in marine sediments. *In* (Funnell, B. M. and Riedel, W. R., eds) *Micropaleontology of the Oceans*, 231–243. Cambridge University Press.

Williams, D. B. and Sarjeant, W. A. S. (1966). Organic-walled microfossils as depth and shoreline indicators. *Marine Geol.* **5,** 389–412.

Williams, G. L. (1964). Organic-walled microplankton of the London Clay. Ph.D. thesis, University of Sheffield, England, unpublished.

Williams, G. L. (1975). Dinoflagellate and spore stratigraphy of the Mesozoic–Cenozoic offshore Eastern Canada. *Geol. Surv. Pap. Can.* **2,** 107–161, Pls. 1–8.

Williams, G. L. and Brideaux, W. W. (1975). Palynological analyses of Late Mesozoic–Cenozoic rocks of the Grand Banks of Newfoundland. *Bull. geol. Surv. Can.* **236,** 1–162, Pls. 1–47.

Williams, G. L. and Downie, C. (1966a). The genus *Hystrichokolpoma*. *In* (Davey, R. J., Downie, C., Sarjeant, W. A. S. and Williams, G. L.) *Studies on Mesozoic and Cainozoic dinoflagellate cysts. Bull. Br. Mus. nat. Hist.* (*Geol.*) *Suppl.* **3** 176–181.

Williams, G. L. and Downie, C. (1966b). *Wetzeliella* from the London Clay.

In (Davey, R. J., Downie, C., Sarjeant, W. A. S. and Williams, G. L.) *Studies on Mesozoic and Cainozoic dinoflagellate cysts. Bull. Br. Mus. nat. Hist.* (*Geol.*) *Suppl.* **3,** 182–198.

Williams, G. L. and Downie, C. (1966c). Further dinoflagellate cysts from the London Clay. *In* (Davey, R. J., Downie, C., Sarjeant, W. A. S. and Williams, G. L.) *Studies on Mesozoic and Cainozoic dinoflagellate cysts. Bull. Br. Mus. nat. Hist.* (*Geol.*) *Suppl.* **3,** 215–235.

Williams, G. L., Sarjeant, W. A. S. and Kidson, E. J. (1973). A glossary of the terminology applied to dinoflagellate amphiesmae and cysts and acritarchs. *American Association of Stratigraphic Palynologists, Contr. Ser.* no. **2,** 1–222. Pls. 1–14.

Wilson, G. J. (1971). Observations on European Late Cretaceous dinoflagellate cysts. *In* (Farinacci, A., ed.) *Proc. II Plankt. Conf.* **2,** 1259–1275. Pls. 1–4. Tecnoscienza, Roma.

Zaitzeff, J. B. and Cross, A. T. (1970). The use of dinoflagellates and acritarchs for zonation and correlation of the Navarro Group (Maestrichtian) of Texas. *Spec. Pap. geol. Soc. Am.* **127,** 341–377. Pls. 1–6.

Zeuner, F. E. (1946). *Dating the past, an introduction to geochronology,* 444 pp. Methuen and Co., Ltd., London.

PLATES

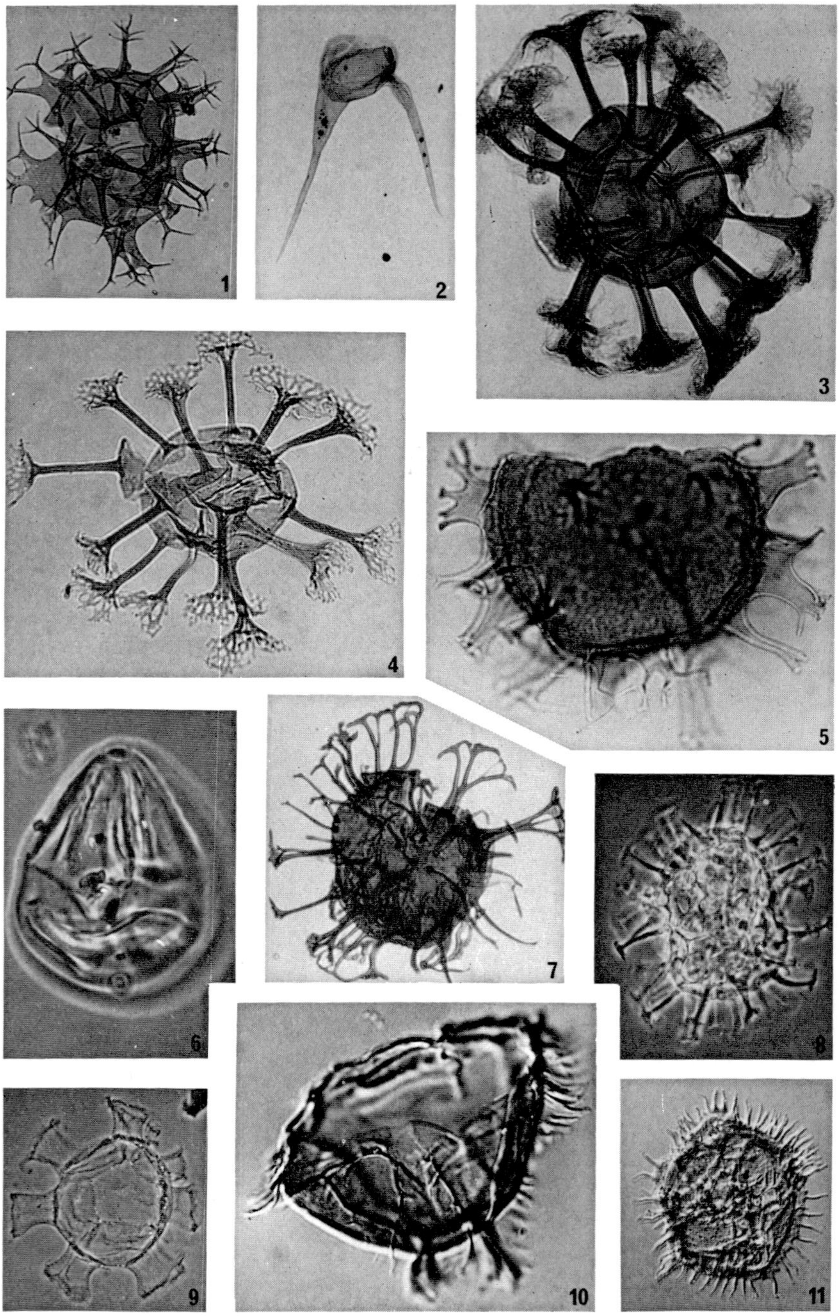
1
2
3
4
5
6
7
8
9
10
11

Plate 1

fig. 1. *Achomosphaera ramulifera* (Deflandre) Evitt. Early Eocene, southern England. This is the species after which the Pliocene dinocyst zone is named (×260).

fig. 2. *Odontochitina operculata* (O. Wetzel) Deflandre and Cookson. Santonian, Grand Banks of Newfoundland, Canada. The archeopyle is apical. This is the species after which the Campanian dinocyst zone is named (×130).

fig. 3. *Cordosphaeridium cantharellum* (Brosius) Gocht. Early Miocene, JOIDES corehole, offshore Florida, U.S.A. This is the species after which the Early Miocene dinocyst zone is named (×325).

fig. 4. *Areosphaeridium diktyoplokus* (Klumpp) Eaton. Late Eocene, Grand Banks of Newfoundland, Canada. Archeopyle apical. The apical archeopyle is formed by the loss of all four apical paraplates. This is one of the species after which the Late Eocene dinocyst zone is named (×260).

fig. 5. *Chiropteridium dispersum* Gocht. Oligocene, Grand Banks of Newfoundland, Canada. Archeopyle apical. This is one of the species after which the Oligocene dinocyst zone is named (×360).

fig. 6. *Dinogymnium euclaensis* Cookson and Eisenack. Maestrichtian, Grand Banks of Newfoundland, Canada. This is the species after which the Maestrichtian dinocyst zone is named (×550).

fig. 7. *Adnatosphaeridium caulleryi* (Deflandre) Williams and Downie. Callovian, southern England. This is the species after which the Early Callovian dinocyst zone is named (×290).

fig. 8. *Cleistosphaeridium polypes* (Cookson and Eisenack) Davey. Albian, Nova Scotian Shelf, Canada. This is the species after which the Cenomanian dinocyst zone is named (×260).

fig. 9. *Callaiosphaeridium asymmetricum* (Deflandre and Courteville) Davey and Williams. Coniacian, Grand Banks of Newfoundland, Canada. Archeopyle epicystal. This is one of the species after which the Coniacian dinocyst zone is named (×195).

fig. 10. *Ctenidodinium ornatum* (Eisenack) Deflandre. Oxfordian, southern England. Archeopyle epicystal. This is the species after which the Bathonian dinocyst zone is named (×425).

fig. 11. *Ctenidodinium elegantulum* Millioud. Hauterivian, Nova Scotian Shelf, Canada. This is the species after which the Hauterivian dinocyst subzone is named (×230).

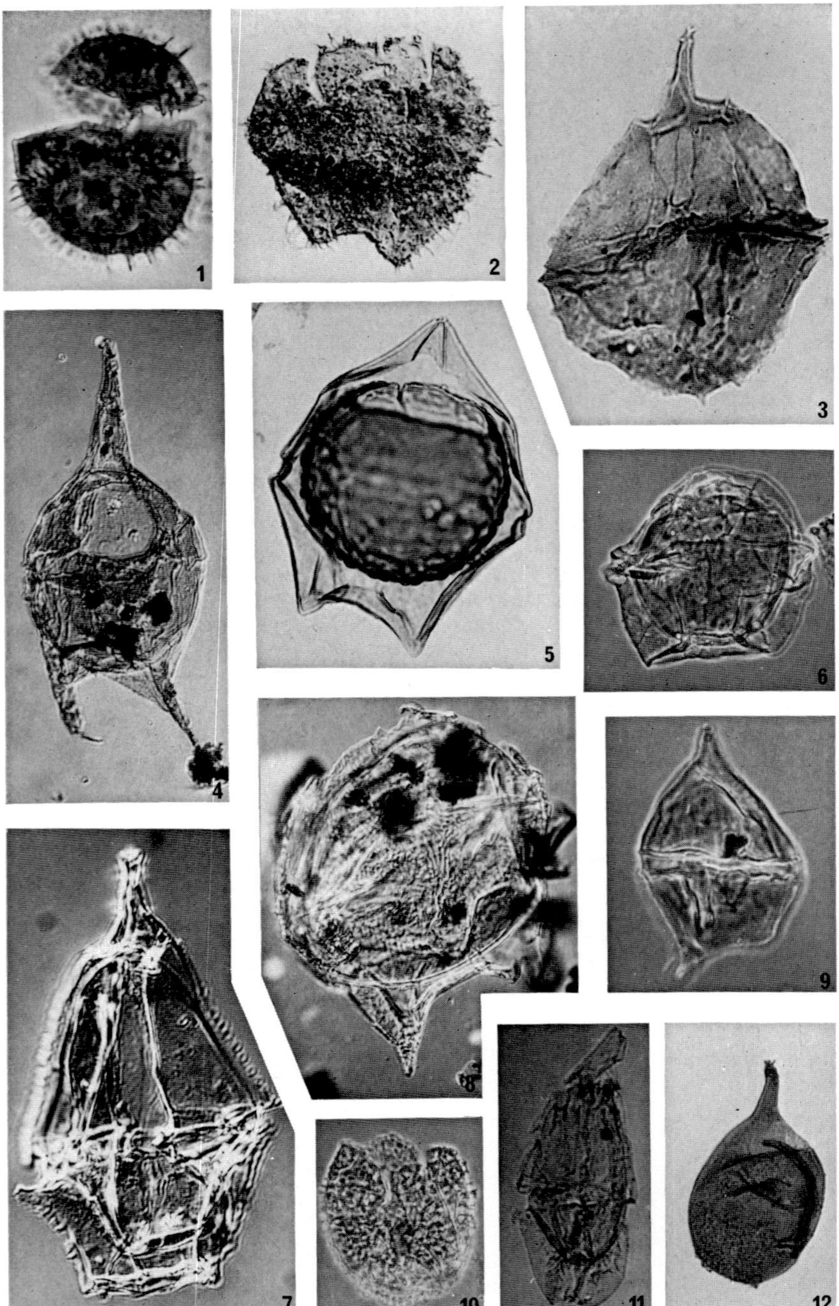
1
2
3
4
5
6
7
8
9
10
11
12

Plate 2

fig. 1. *Biorbifera johnewingi* Habib. Berriasian-Early Valanginian, Grand Banks of Newfoundland, Canada. This is one of the species after which the Berriasian-Early Valanginian dinocyst subzone is named (×550).

fig. 2. *Cyclonephelium attadalicum* Cookson and Eisenack. Aptian, Nova Scotian Shelf, Canada. Archeopyle apical (×360).

fig. 3. *Cribroperidinium orthoceras* (Eisenack) Davey. Hauterivian, Nova Scotian Shelf, Canada. This is one of the species after which the Berriasian-Early Valanginian dinocyst subzone is named (×360).

fig. 4. *Deflandrea diebeli* Alberti. Early Palaeocene, Nova Scotian Shelf, Canada. Archeopyle intercalary, resulting from the loss of paraplate 2a. This is one of the species after which the Early Palaeocene dinocyst zone is named (×230).

fig. 5. *Deflandrea heterophlycta* Deflandre and Cookson. Early Oligocene, Grand Banks of Newfoundland, Canada. Archeopyle intercalary, resulting from the loss of paraplate 2a. This is the species after which the Early Oligocene subzone is named (×260).

fig. 6. *Endoscrinium luridum* (Deflandre) Gocht. Oxfordian, Nova Scotian Shelf, Canada. Archeopyle precingular, resulting from the loss of paraplate 3″. This is one of the species after which the Early Kimmeridgian dinocyst subzone is named (×230).

fig. 7. *Gonyaulacysta jurassica* (Deflandre) Norris and Sarjeant. Oxfordian, southern England. This is the species after which the Oxfordian dinocyst zone is named (×390).

fig. 8. *Deflandrea speciosa* Alberti. Palaeocene, Nova Scotian Shelf, Canada. This is one of the species after which the Late Palaeocene dinocyst zone is named (×230).

fig. 9. *Deflandrea perlucida* Alberti. Barremian, Grand Banks of Newfoundland, Canada. This is one of the species after which the Aptian dinocyst zone is named (×260).

fig. 10. *Lanterna* sp. Kimmeridgian, Nova Scotian Shelf, Canada (×230).

fig. 11. *Belodinium dysculum* Cookson and Eisenack. Portlandian, Western Australia. This is one of the species after which the Portlandian dinocyst zone is named (×230).

fig. 12. *Pareodinia ceratophora* Deflandre. Bathonian, Grand Banks of Newfoundland, Canada. Archeopyle intercalary, resulting from the loss of two anterior intercalary paraplates. This is the species after which the Bajocian dinocyst zone is named (×260).

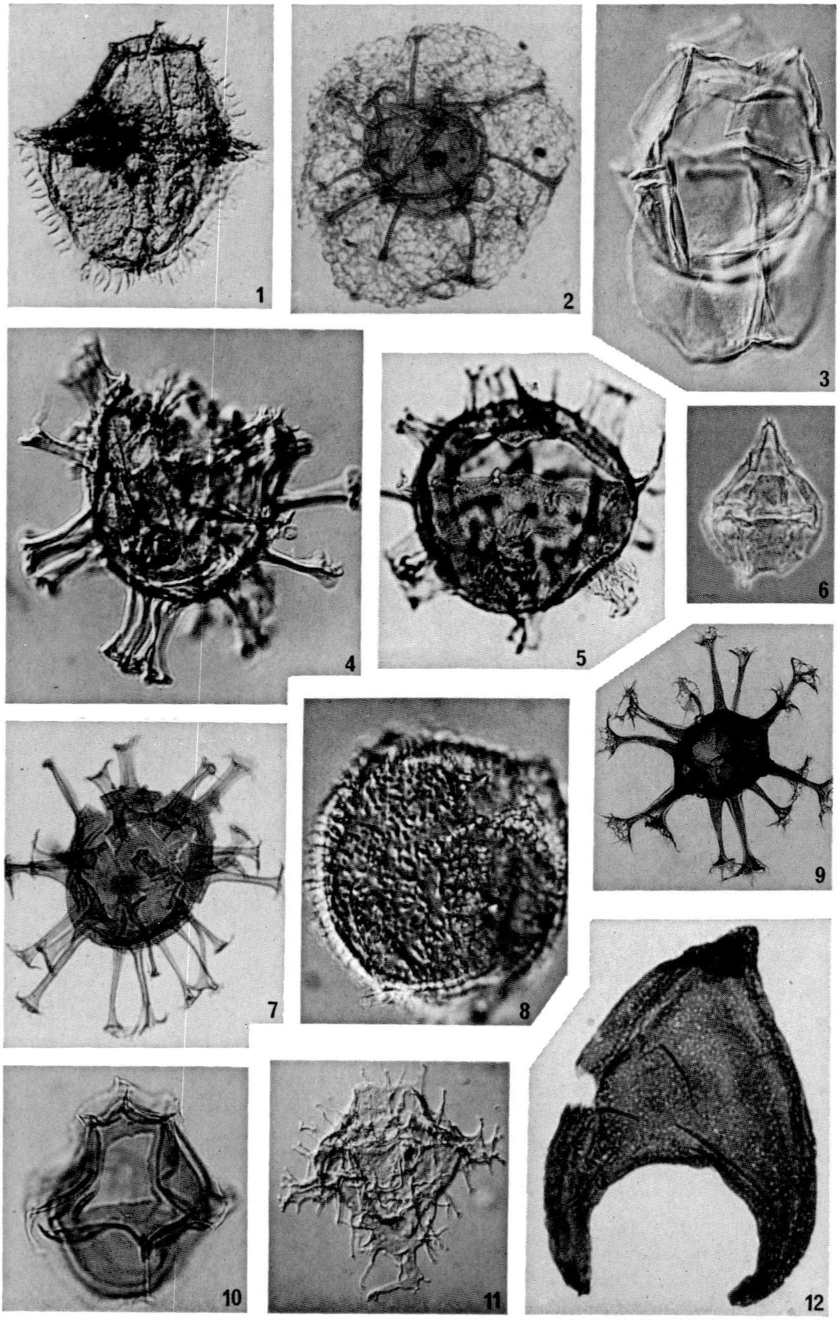
1
2
3
4
5
6
7
8
9
10
11
12

Plate 3

fig. 1. *Gonyaulacysta cladophora* (Deflandre) Dodekova. Kimmeridgian. Archeopyle precingular, resulting from the loss of paraplate 3″. This is the species after which the Kimmeridgian dinocyst zone is named (×230).

fig. 2. *Membranilarnacia ursulae* (Morgenroth) De Coninck. Early Eocene, southern England. This is one of the species after which the Middle Eocene dinocyst zone is named (×260).

fig. 3. *Hystrichosphaeropsis obscurum* Habib. Late Miocene, Grand Banks of Newfoundland, Canada. Archeopyle precingular, resulting from the loss of paraplate 3″. This is one of the species after which the Late Miocene dinocyst zone is named (×325).

fig. 4. *Hystrichosphaeridium polonicum* Gorka. Oxfordian, southern England. This is one of the species after which the Early Oxfordian dinocyst subzone is named (×260).

fig. 5. *Hystrichosphaeridium truncigerum* Deflandre. Santonian, Grand Banks of Newfoundland, Canada. Archeopyle precingular, resulting from the loss of two paraplates. This is the species after which the Santonian dinocyst zone is named (×260).

fig. 6. *Spinidinium* cf. *vestitum* Brideaux. Albian, Nova Scotian Shelf, Canada. This is the taxon after which the Albian dinocyst zone is named (×260).

fig. 7. *Homotryblium tenuispinosum* Davey and Williams. Early Eocene, southern England. Archeopyle epicystal. This is one of the species after which the Early Eocene dinocyst zone is named (×260).

fig. 8. *Lanterna sportula* Dodekova. Kimmeridgian, southern England. Archeopyle apical. This is one of the species after which the Portlandian dinocyst zone is named (×390).

fig. 9. *Oligosphaeridium pulcherrimum* (Deflandre and Cookson) Davey and Williams. Coniacian, Grand Banks of Newfoundland, Canada. This is one of the species after which the Coniacian dinocyst zone is named (×180).

fig. 10. *Pentadinium laticinctum* Gerlach. Late Miocene, Grand Banks of Newfoundland, Canada. Archeopyle precingular, resulting from the loss of paraplate 3″. This is one of the species after which the Late Eocene dinocyst zone is named (×260).

fig. 11. *Phoberocysta neocomica* (Gocht) Millioud. Berriasian-Valanginian, Grand Banks of Newfoundland, Canada. Archeopyle apical. This is the species after which the Neocomian dinocyst zone is named (×230).

fig. 12. *Nannoceratopsis pellucida* Deflandre. Reworked specimen Early Eocene, southern England. The closely related *Nannoceratopsis gracilis* Alberti is the species after which the Pliensbachian–Toarcian dinocyst zone is named (×360).

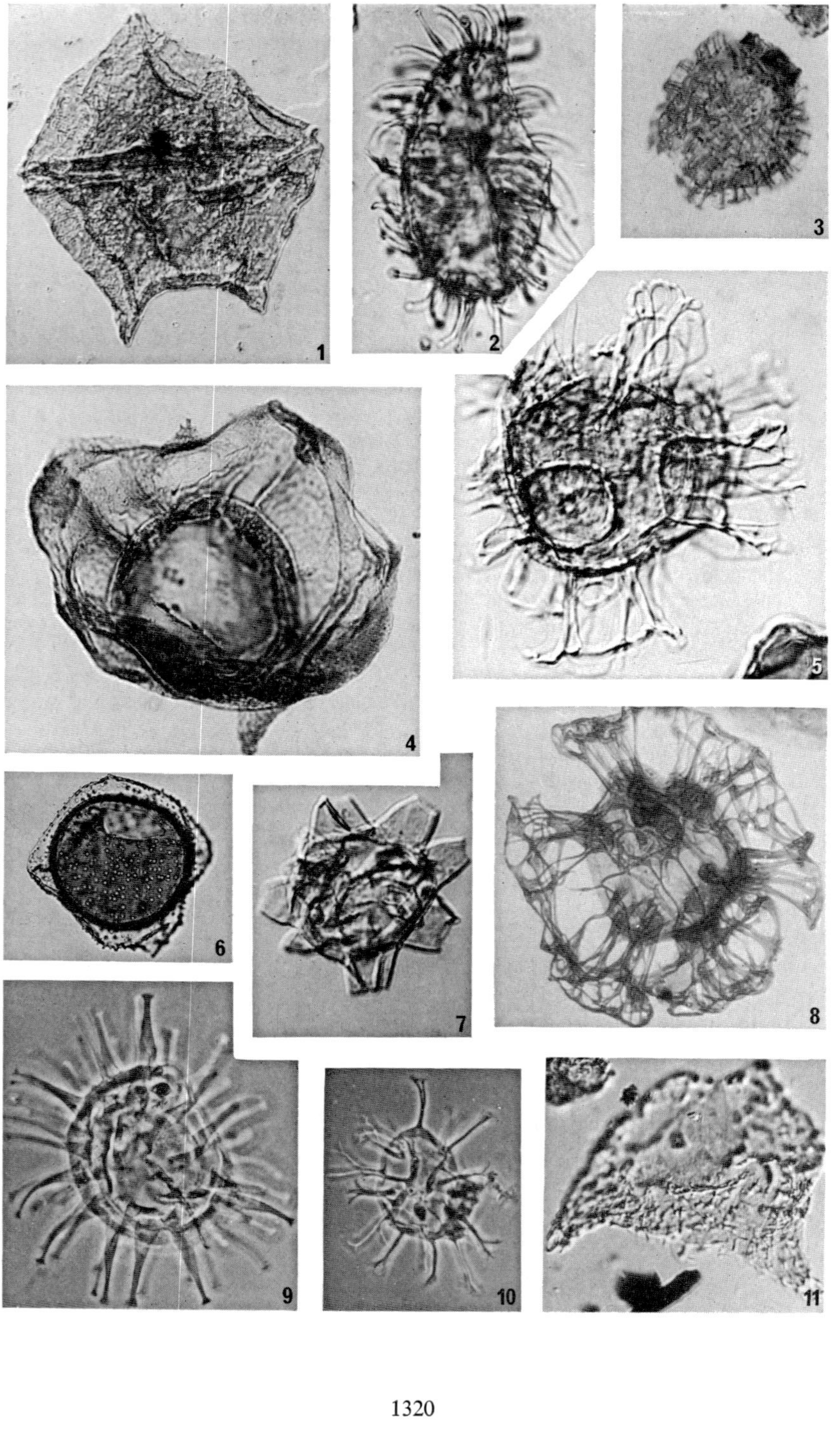

PLATE 4

fig. 1. *Palaeoperidinium pyrophorum* (Ehrenberg) Sarjeant. Early Palaeocene, Nova Scotian Shelf, Canada. This is one of the species after which the Early Palaeocene dinocyst zone is named (×230).

fig. 2. *Protoellipsodinium spinocristatum* Davey and Verdier. Albian, Nova Scotian Shelf, Canada. Archeopyle precingular. This is the species after which the Early Albian dinocyst subzone is named (×360).

fig. 3. *Stephanelytron scarburghense* Sarjeant. Reworked specimen Early Eocene, southern England. This is one of the species after which the Early Kimmeridgian dinocyst subzone is named (×260).

fig. 4. *Thalassiphora pelagica* (Eisenack), Eisenack and Gocht. Oligocene, JOIDES corehole, offshore Florida, U.S.A. Archeopyle precingular, resulting from the loss of one paraplate. This is one of the species after which the Middle-Late Oligocene dinocyst subzone is named (×195).

fig. 5. *Systematophora orbifera* Klement. Kimmeridgian, southern England. Archeopyle apical. This is one of the species after which the Late Oxfordian dinocyst subzone is named (×290).

fig. 6. *Wetzeliella condylos* Williams and Downie. Early Eocene, southern England. Archeopyle intercalary, resulting from the loss of anterior intercalary paraplate 2a (×180).

fig. 7. *Litosphaeridium siphoniphorum* (Cookson and Eisenack) Davey and Williams. Turonian, Grand Banks of Newfoundland, Canada (×325).

fig. 8. *Polystephanephorus paracalathus* (Sarjeant) Downie and Sarjeant. Reworked specimen Early Eocene, southern England. The closely related *Polystephanephorus calathus* (Sarjeant) Downie and Sarjeant, which lacks the inter-connecting trabeculae between process complexes, is one of the species after which the Early Oxfordian dinocyst subzone is named (×260).

fig. 9. *Polysphaeridium pastielsi* Davey and Williams. Middle Miocene, Grand Banks of Newfoundland, Canada. This is one of the species after which the Late Miocene dinocyst zone is named (×325).

fig. 10. *Surculosphaeridium longifurcatum* (Firtion) Davey *et al.* Turonian, Nova Scotian Shelf, Canada. This is the species after which the Turonian dinocyst zone is named (×260).

fig. 11. *Tenua anaphrissa* (Sarjeant) Benedek. Barremian, Nova Scotian Shelf, Canada. Archeopyle apical. This is the species after which the Barremian dinocyst subzone is named (×230).

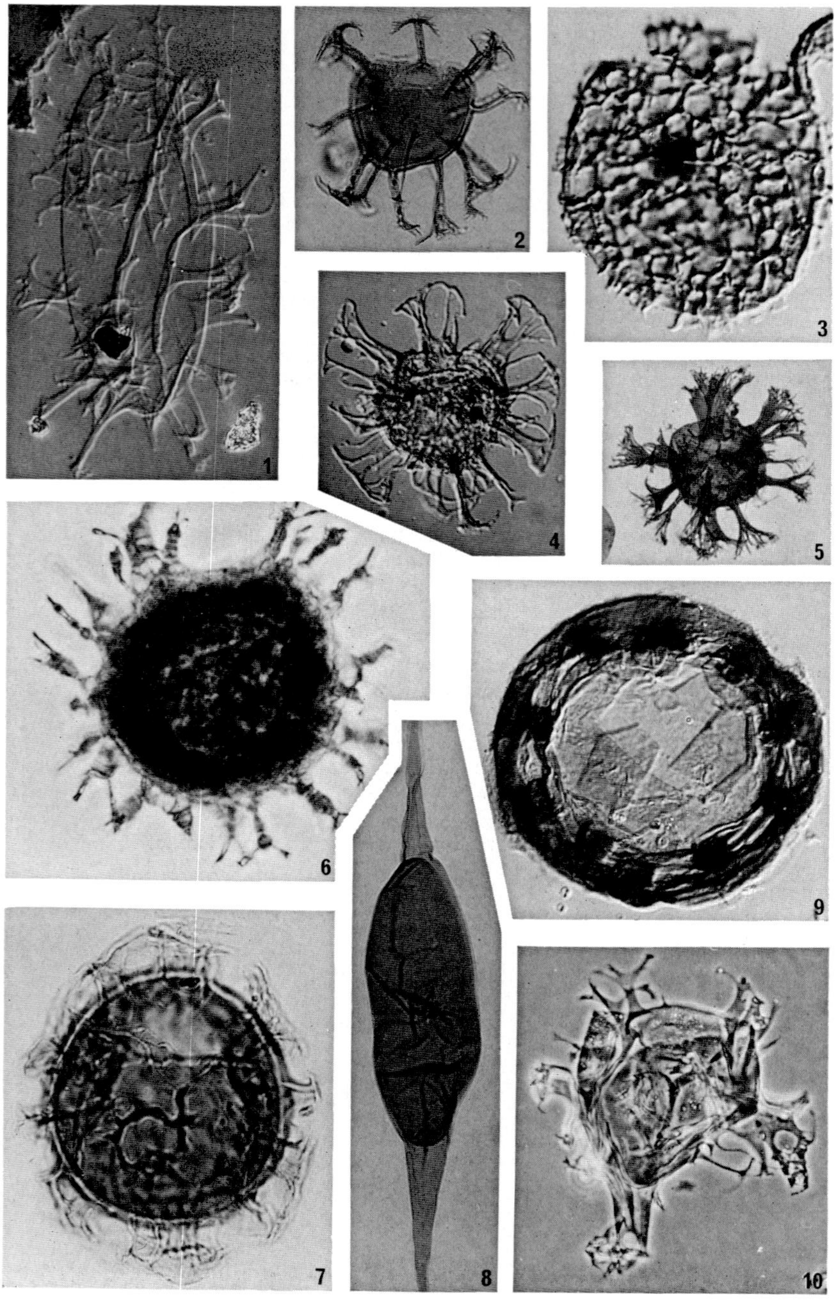
1
2
3
4
5
6
7
8
9
10

PLATE 5

fig. 1. *Tanyosphaeridium ellipticum* (Cookson) Davey and Williams. Late Eocene, JOIDES corehole, offshore Florida, U.S.A. This is one of the species after which the Middle Eocene dinocyst zone is named (×425).

fig. 2. *Areosphaeridium multicornutum* Eaton. Late Eocene, Grand Banks of Newfoundland, Canada (×230).

fig. 3. *Valensiella ovulum* (Deflandre) Eisenack. Bathonian, Grand Banks of Newfoundland, Canada (×585).

fig. 4. *Systematophora schindewolfi* (Alberti) Downie and Sarjeant. Hauterivian, Nova Scotian Shelf, Canada. This is one of the species after which the Aptian dinocyst zone is named (×260).

fig. 5. *Perisseiasphaeridium pannosum* Davey and Williams. Early Eocene, southern England (×195).

fig. 6. *Spiniferites septatus* (Cookson and Eisenack) McLean. Early Eocene, Grand Banks of Newfoundland, Canada. This is one of the species after which the Early Eocene dinocyst zone is named (×460)

fig. 7. *Samlandia chlamydophora* Eisenack. Late Eocene, Grand Banks of Newfoundland, Canada. Archeopyle precingular, resulting from the loss of paraplate 3″. This is one of the species after which the Oligocene dinocyst zone is named (×325).

fig. 8. *Palaeocystodinium golzowense* Alberti. Early Miocene, Grand Banks of Newfoundland, Canada. Archeopyle intercalary, resulting from detachment of anterior intercalary paraplate 2a, which is still present (×260).

fig. 9. *Tuberculodinium rossignolae* Drugg. Late Miocene, Grand Banks of Newfoundland, Canada. Archeopyle resulting from the loss of three antapical paraplates (×245).

fig. 10. *Xenascus ceratioides* (Deflandre) Lentin and Williams. Santonian, Grand Banks of Newfoundland, Canada. This is one of the species after which the Middle-Late Albian dinocyst subzone is named (×290).

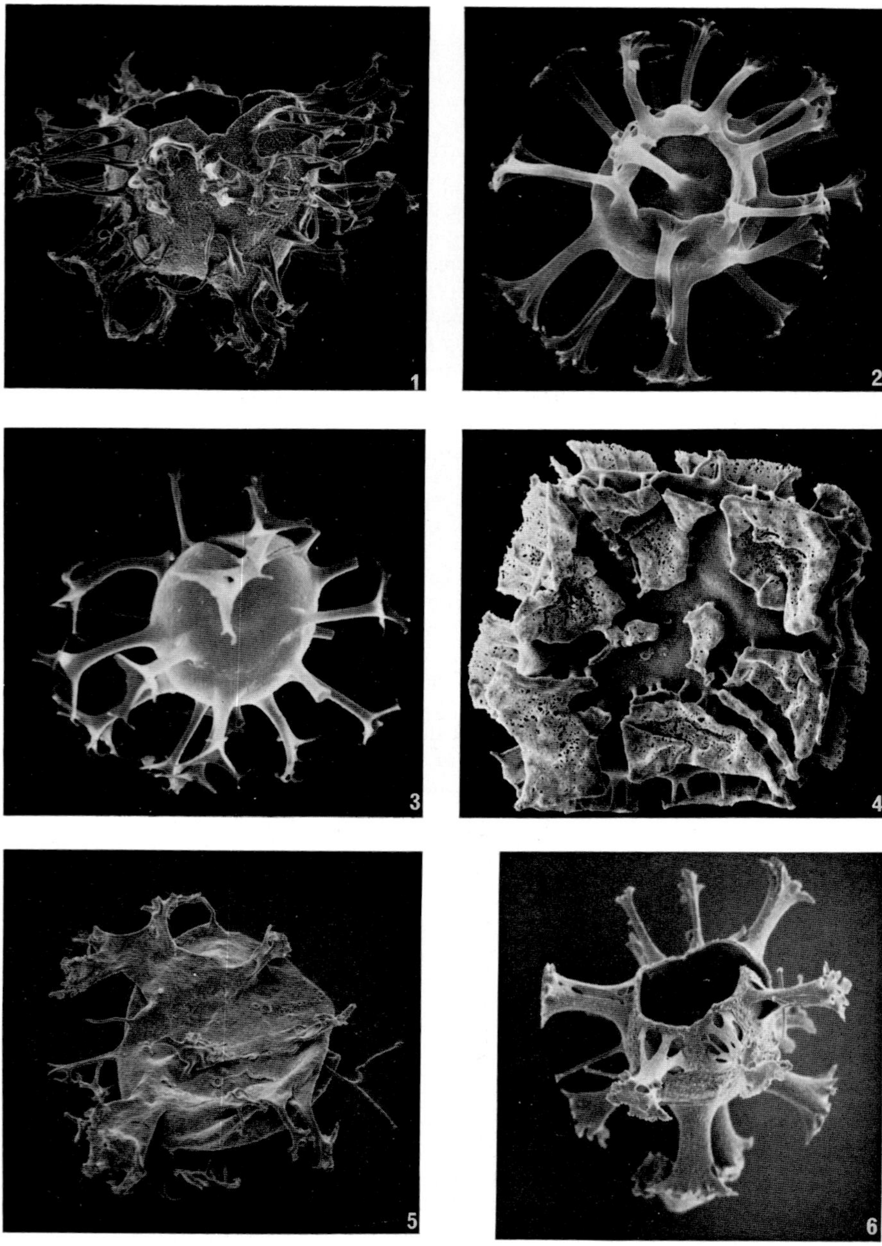
1
2
3
4
5
6

Plate 6. Scanning electron micrographs

fig. 1. *Areoligera senonensis* Lejeune-Carpentier. Early Eocene, Grand Banks of Newfoundland, Canada. Archeopyle apical (×390).

fig. 2. *Cordosphaeridium gracilis* (Eisenack) Davey and Williams. Early Eocene, Grand Banks of Newfoundland, Canada. Archeopyle precingular (×260).

fig. 3. *Achomosphaera alcicornu* (Eisenack) Davey and Williams. Early Eocene, Grand Banks of Newfoundland, Canada (×275).

fig. 4. *Wetzeliella coleothrypta* Williams and Downie. Early Eocene, Grand Banks of Newfoundland, Canada. Ventral view showing parasulcus (×885).

fig. 5. *Xenascus ceratioides* (Deflandre) Lentin and Williams. Santonian, Grand Banks of Newfoundland, Canada. Archeopyle apical (×390).

fig. 6. *Hystrichokolpoma* cf. *rigaudae* Deflandre and Cookson. Late Eocene, JOIDES corehole, offshore Florida, U.S.A. Archeopyle apical (×390).

16: Systematics, Distribution and Stratigraphical Application of Silicoflagellates

E. MARTINI

Geologisch-Paläontologisches Institut der Johann Wolfgang Goethe Universität, Frankfurt-am-Main, Germany

The biology, systematics, phylogeny, ecology and biostratigraphy of the silicoflagellates are considered.

The silicoflagellates are subdivided into three families which contain nine genera. Three genera are classed as *incertae sedis*. Our understanding of the phylogeny of silicoflagellates is incomplete and further studies are necessary. Optimal conditions for the distribution of these organisms occur in fertile regions of the oceans at salinities ranging from 30–40%. High percentages of aberrant forms occur in nearshore assemblages.

Fifteen zones are based on silicoflagellate events are recognised for the Tertiary and Quaternary.

Outline

Introduction

The first silicoflagellates were described by Ehrenberg (1837) who assigned two fossil species to a new genus *Dictyocha*. Both species, however, were nomina nuda. In the following years Ehrenberg described many new taxa from Recent as well as from fossil material sent to him from Russia, California, Maryland, Greece, Haiti and other countries. He discussed the distribution of modern silicoflagellates and listed geological and ecological data. Unfortunately the geological age of many samples he received was not very well known and some locality-names were rather broadly defined. Most of the Ehrenberg collection has survived and is currently being restudied in Berlin.

The oceanographic expeditions of various countries added much to the knowledge of living silicoflagellates found in plankton tows and bottom samples. Lemmermann in 1901 and Gemeinhardt in 1930 summarised the knowledge of silicoflagellate distribution and systematics of those days. With the work of Schulz (1928) interest in fossil silicoflagellates increased, and fossil material was described in numerous articles, especially from France, Germany, Russia, Austria, Japan, and Yugoslavia. Most of the authors of these articles, however, made no attempt to apply silicoflagellates to stratigraphy. In the last few years the stratigraphical value of silicoflagellates has been demonstrated by Stradner (1961), Glezer (1966), Bachmann (1970), and Martini (1971). An index of fossil and Recent silicoflagellates with descriptions and illustrations of validly proposed taxa, and a bibliography of all published works up to 1967, was compiled by Loeblich *et al.* (1968). Most present-day investigations are concerned with establishing a biozonation based on silicoflagellates, with the evolution of the group and with the palaeogeographic distribution of certain species.

Morphology and reproduction

Silicoflagellates are unicellular flagellates possessing a siliceous skeleton (Fig. 1). Their size commonly ranges from 20 to 100 μm, but may reach 150 μm in some species. The protoplasm is generally clear and hyaline, and occupies the space enclosed by the basal ring and, if present, by the apical apparatus and supporting spines. It normally extends in a thin film beyond the skeleton. The many chromatophores are yellowish to greenish brown, and a colourless rounded to oval nucleus is present. Pseudopodia which extend from the spines of the skeleton have been described in the living species, their presence suggests that they may possibly also rely on captured food for nutrition. A single flagellum extends from one corner, and with respect to the direction of movement is anterior in position.

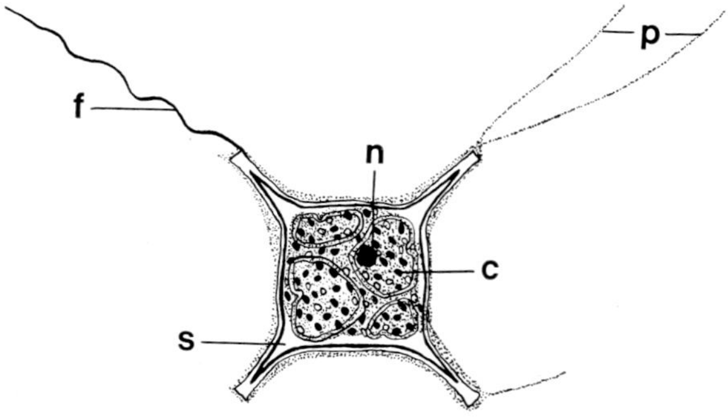

FIG. 1. Organisation of living specimen of *Dictyocha fibula* Ehrenburg, showing the siliceous skeleton, s; flagellum, f; nucleus, n; chromatophores, c and pseudopodia, p. (After Marshall, 1934.)

The siliceous skeletons of the various species range from simple triradiate spicules to complex dome-like structures (Fig. 2), and are typically tubular. Most genera show a basal ring or polygon, from which an apical structure arise on one side. The apical structure may consist of a few bars, an apical ring supported by bars or may be built of a meshwork of rods forming a hemispherical to spherical structure. The basal ring commonly has spines, and additional supporting spines are present in many species. The siliceous skeleton shows a remarkable degree of variation within large populations, and aberrant forms which occur due to unfavourable conditions were often regarded as genuine species. Studies of the skeleton with the electron microscope show that it can be smooth, rough or reticulate. The reticulations are formed by a fine network of ridges and small knobs (Pl. 1, fig. 1), and their function is unknown.

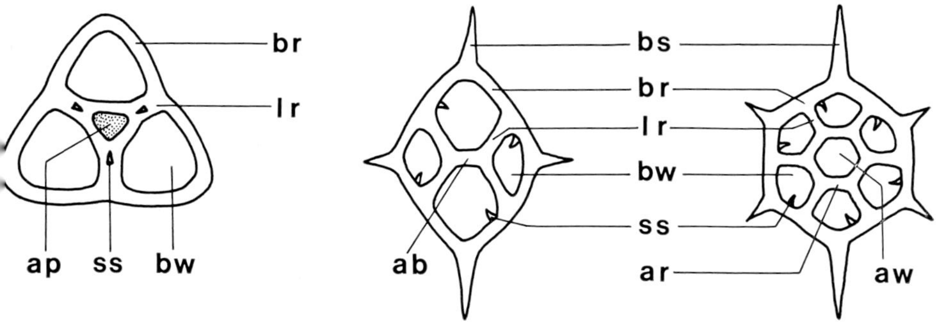

FIG. 2. Main elements in silicoflagellate skeleton morphology: ab, apical bar; ap, apical plate; ar, apical ring; aw, apical window; br, basal ring; bs, basal spine; bw, basal window; lr, lateral rod; ss, supporting spine.

The reproduction of silicoflagellates has been described as a simple division (Gemeinhardt, 1930). First a daughter skeleton is developed, which is a mirror image of its parent. Then the nucleus divides, and the daughter cell separates from the maternal cell. Firmly bound double and triple skeletons have been observed in many species, and these were sometimes considered as evidence to support the view that the silicoflagellate skeleton like the diatoms consists of two halves (Gemeinhardt, 1930) or they were regarded as a copulation stage (Schulz, 1928). Further studies in cultures are needed to improve the limited knowledge of silicoflagellate reproduction.

Systematics and phylogeny

SYSTEMATICS

The systematics of silicoflagellates is in a rather confused state particularly at the species and subspecies level. This confusion has arisen because of the narrow limits of morphological variation which are used by many authors to define taxa, and because the significance of aberrant forms has been overemphasised. There are also some doubts as to the present usage of names of some silicoflagellates which were described by early investigators, and which were illustrated by schematic drawings only. The present restudy of the Ehrenberg collection may help to improve the situation in the near future.

Other forms like *Clathrium* Frenguelli, *Lutetianella* Filipescu, *Nothyocha* Deflandre, *Pseudomesocena* Hovasse and *Rocella* Hanna, which were formerly included with the silicoflagellates can be excluded, as they were proved to be parts of radiolaria or diatoms (Lipps, 1970).

Class **CHRYSOPHYCEAE**

Subclass **Silicoflagellatophycidae**

Order CHRYSOMONADALES

Three families, to which nine genera can be assigned, may be recognised together with forms which are better classed as *incertae sedis*.

Family VALLACERTACEAE Deflandre, 1950

Genus *VALLACERTA* Hanna, 1928

Polygonal ring with spines projecting from the corners, and a convex sculptured apical plate (Santonian to Maestrichtian; early Palaeocene?).

Family CORNUACEAE Gemeinhardt, 1930

Genus *CORNUA* Schulz, 1928

Three to four lateral rods with terminal horns, indicating a rudimentary basal ring. Terminal horns commonly with supporting spines at their bases (Santonian to Maestrichtian).

Genus *LYRAMULA* Hanna, 1928

"U" or "Y"-shaped tubular rods, probably representing a primitive apical apparatus (Santonian to Maestrichtian; early Palaeocene?).

Family DICTYOCHACEAE Lemmermann, 1901

Genus *CORBISEMA* Hanna, 1928

Basal ring triangular, connected by three bars to a central hyaline, or in some cases perforated structure. Rounded corners or with short processes. Supporting spines may be present on lateral rods (Albian to early Miocene).

Genus *NAVICULOPSIS* Frenguelli, 1940

Basal ring boat-shaped, usually with a spine at each end, and without supporting spines. The apical part consists of an arched bar connecting the two

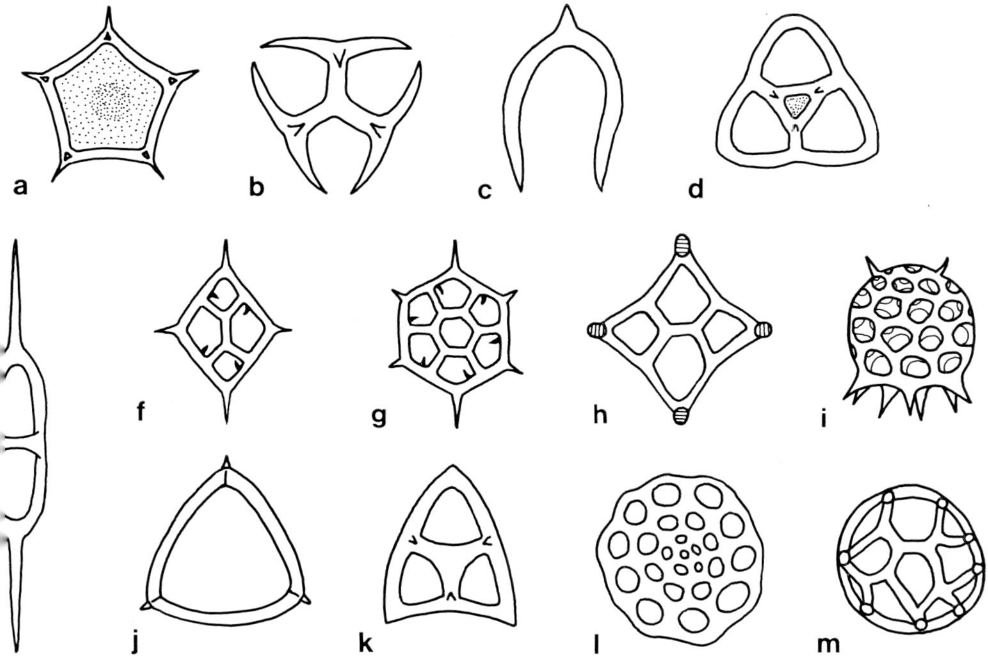

FIG. 3. Schematic diagrams of representatives of the genera *Vallacerta* (a), *Cornua* (b), *Lyramula* (c), *Corbisema* (d), *Naviculopsis* (e), *Dictyocha* (f), and (g), *Hannaites* (h), *Cannopilus* (i), *Mesocena* (j), *Phyllodictyocha* (k), *Pseudorocella* (l), *Clathropyxidella* (m).

sides of the basal "ring" at their central parts (Palaeocene to early Miocene). *Note:* aberrant naviculid forms occur in some species of the genus *Dictyocha.*

Genus *DICTYOCHA* Ehrenberg, 1837

Triangular to polygonal basal ring with or without radial spines. Lateral rods fused at the centre or support an apical bar or ring. Instead of an apical ring a meshwork of bars occur in some species. Supporting spines commonly are present on the basal ring below the lateral rods or to their sides (Albian to Recent). *Note: Distephanus* Stöhr, 1880 (Fig. 3g), *Deflandryocha* Jerković, 1963, and part of the genus *Paradictyocha* Frenguelli, 1940 are included here.

Genus *HANNAITES* Mandra, 1969

Quadrate basal ring with spherical protuberances at each corner. Apical bar connected by rods to the basal ring. Supporting spines missing or rudimentary (Eocene). According to recent studies this may be a diseased *Dictyocha.*

Genus *CANNOPILUS* Haeckel, 1887

Basal ring polygonal or circular with spherical apical structure, which may have spines on the apical rods. Spines of the basal ring bent downwards forming an angle with the plane of the basal ring (Oligocene to Miocene). *Note: Cannopilus* Haeckel, 1887 is regarded a junior synonym of *Halicalyptra* Ehrenberg, 1847 by some authors (Lipps, 1970), but as most species of the latter are radiolaria, a reinvestigation of the type-species is necessary.

Genus *MESOCENA* Ehrenberg, 1843

Elliptical, triangular to polygonal basal ring, with or without radial spines. Apical structure and supporting spines are absent (Early Palaeocene? to Pleistocene). *Note:* Aberrant forms and broken pieces of the genus *Dictyocha* may be mistaken for representatives of *Mesocena*. Part of the genus *Paradictyocha* Frenguelli, 1940 may be placed with *Mesocena*.

INCERTAE SEDIS

Genus *PHYLLODICTYOCHA* Deflandre, 1947

Triangular nontubular basal ring with three convex central bars joined at the centre. Small supporting spines occur on the bars near the basal ring (Eocene).

Genus *PSEUDOROCELLA* Deflandre, 1938

Rounded plates with numerous circular perforations and an arched central area, with much variation as to number of perforations and outline of plate (Eocene to Miocene).

Genus *CLATHROPYXIDELLA* Deflandre, 1938

More or less circular basal ring without spines and irregular semispherical apical structure (Eocene).

PHYLOGENY

The phylogenetic relationships of the silicoflagellates have been studied be various authors, among them Deflandre (1950) and Glezer (1966). Although the generic status of Lower Cretaceous silicoflagellates must be treated with reservation, the genus *Dictyocha* appears to be the main stock from which the genera *Mesocena*, *Naviculopsis*, *Hannaites*, *Cannopilus* and probably *Corbisema* have developed. Glezer places the genus *Lyramula* at the base of a phylogenetic scheme for the silicoflagellates, followed by the genera *Cornua* and *Dictyocha*, according to their morphological similarities and development. As the Cretaceous occurrences of silicoflagellates are rather sparse and direct development from one to the other form has not yet observed this part of the phylogenetic tree is somewhat obscured and needs further study. Also there are difficulties in relating the genus *Vallacerta* to the rest of the silicoflagellates as indicated by Lipps (1970).

Ecology

Silicoflagellates are typical inhabitants of the marine plankton. The adaptability to salinity and temperature is relatively great. The optimal salinity for these planktonic organisms is 30–40‰, but they may tolerate a salinity lower than 10‰ as in the eastern part of the Baltic Sea, where small numbers of *Dictyocha speculum* still occur (Schulz, 1928).

High percentages of aberrant specimens in a silicoflagellate assemblage seem to occur only in assemblages living under abnormal conditions in nearshore waters or basins. In open ocean deposits aberrant silicoflagellates generally account for less than 3% of the total number of specimens. Isolated stocks may develop into local "new genera" or "species" as indicated by members of the genus *Deflandryocha* (Fig. 4) in the Sarmatian of Yugoslavia (Jerković, 1969), where a development from a *Dictyocha crux* stock is most likely. It is interesting to note that the calcareous nannoplankton in the same samples show an assemblage dominated by species of *Reticulofenestra*, which are very resistant to abnormal conditions, and nearshore species like *Braarudosphaera bigelowi* and *Holodiscolithus macroporus*.

Reported water temperatures for living silicoflagellates show a wide range from approximately 0°C to 30°C, but certain species show preference for cool waters or warmer waters. Ehrenberg (1875) was the first worker to discuss the distribution of silicoflagellates in modern seas. Gemeinhardt (1934) working on samples obtained during the "Meteor"-Expedition in the Atlantic

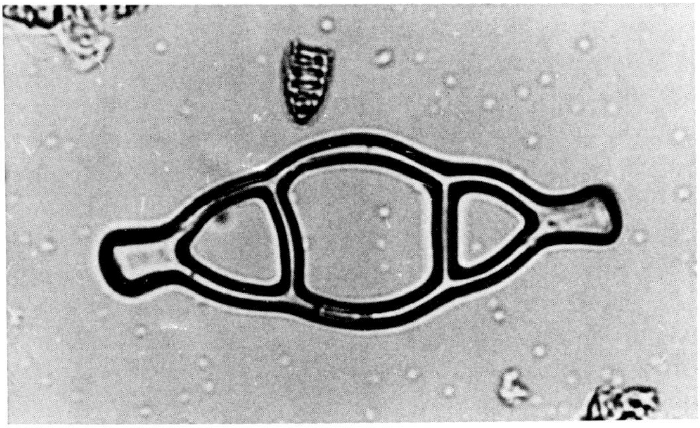

FIG. 4. *Deflandryocha cymbiformis* Jerković. Kozjak (Sveta Nedelja), Yugoslavia, lower Sarmatian, approximately ×1000.

Ocean found, that in the South Atlantic *Dictyocha* (*Dichtyocha fibula* and closely related forms) occur in low and middle latitudes with an optimal temperature ranging from 18°C to 20°C, while "*Distephanus*" (*Dichtyocha speculum*) is restricted to high latitudes, the optimal temperature being close to 0°C.

A later study of the distribution of these species and water temperature by Yanagisawa (1943) established a scheme of dependence between the ratios of the two species and temperature. These ratios, slightly modified are as follows:

D. fibula: *D. speculum*		water temp. (°C)
1:12	=	0– 5
1: 4	=	5–10
1: 2	=	10–15
1: 1.5	=	15–20
2: 1	=	20–25
21: 1	=	25–30

Mandra (1969) used the ratio of two silicoflagellate genera *Dictyocha*: *Distephanus* (forms with an apical ring) as a means of determining the water temperature of the South Atlantic during the Eocene.

Like all phototrophic organisms, silicoflagellates are dependent upon a certain amount of light. In high latitudes, where the light intensity is low, maximal numbers of silicoflagellates occur close to the surface, although an appreciable number of specimens may be found to a depth of 50 m. In equatorial regions these organisms are found mainly at a depth between 15 and 70 m. Seasonal fluctuations in silicoflagellate abundances are known, but data are sparse and further research is required.

In the modern ocean silicoflagellates are most commonly found in areas of high diatom concentrations. This relationship also appears to be true for the ancient seas. Most of the well-known fossil occurrences of silicoflagellates, some of which have already been described by Ehrenberg, are in diatomaceous sediments or in diatomites. Studies of continuous cores, recovered by JOIDES, show a conspicuous correlation between the occurrence of diatoms and silicoflagellates. Fluctuations in the abundance of diatoms are accompanied by identical changes in the abundance of silicoflagellates (Martini, 1971a).

Diatomites form in areas where upwelling caused by intensive atmospheric and oceanic circulation, brings silica into the euphotic zone. This factor together with light for photosynthesis determines areas of high primary production. Because upwelling is generally restricted to certain parts of the oceans, and may have occurred in these areas for long periods of time, diatomites and silicoflagellates are most commonly found in equatorial regions, along western sides of continents, and in high latitudes (Lipps, 1970).

Stratigraphical occurrence and silicoflagellate zonation

The oldest known occurrence of silicoflagellates has been reported from the Lower Cretaceous (Albian) of Saxonia, Germany, and the Thermopolis Shale of Wyoming, U.S.A. Several occurrences of Upper Cretaceous assemblages which contain the genera *Corbisema*, *Dictyocha*, *Cornua*, *Lyramula* and *Vallacerta*, have been described from Denmark, Germany, Russia and the United States. Of these genera *Cornua*, *Lyramula* and *Vallacerta* appear to be restricted to the Upper Cretaceous. Silicoflagellates are quite abundant in some Tertiary marine sediments, particularly in the well-known Lompoc (California) and Oamaru (New Zealand) diatomites. According to Loeblich *et al.* (1968) there are 17 Tertiary genera, including some forms better classed as *incertae sedis* or even not related to silicoflagellates (Lipps, 1970). Present knowledge indicates a restriction of the genera *Hannaites* and *Phyllodictyocha* to the Eocene, and *Pseudorocella* from the Eocene to the Miocene. The genus *Naviculopsis* seems to be restricted from the Palaeocene to early Miocene, the genus *Cannopilus* occurs only in the Oligocene and Miocene, whereas the genus *Mesocena* ranges from the Palaeocene to the Pleistocene. A reduction in the number of genera and species occurred at the end of the Tertiary. The remaining stock in the modern ocean belongs to the genus *Dictyocha*, if one considers *Distephanus* as a synonym of *Dictyocha*. Reported occurrences of *Mesocena* in the Recent are possibly aberrant forms of *Dictyocha*, and need verification.

The stratigraphical value of silicoflagellates has been demonstrated by Stradner (1961), Glezer (1966) and Mandra (1968), but these investigations were based mainly on isolated samples and a zonation was not attempted. During the first phase of the JOIDES, two continuous sequences from the Miocene to the Quaternary, which contained silicoflagellates together with abundant diatoms and radiolaria, have been recovered at sites 65 and 66 in the equatorial Pacific. Based on the silicoflagellates found in these sequences and of those in Swedish Deep Sea Core 76 and the Experimental Mohole Cores a preliminary zonation for the interval Middle Miocene/Quaternary was established (Martini, 1971a), and some long distance correlations indicated. The *Dictyocha octacantha* horizon occurs, according to correlations based on other fossil groups, at stratigraphically similar positions in the Experimental Mohole cores near Guadalupe and the JOIDES hole 66–1 in the western equatorial Pacific. The *Dictyocha pseudofibula*/*Dictyocha pseudocrux* horizon is present in California, the Experimental Mohole cores and in Japan, and the last occurrence of *Dictyocha* (*Corbisema*) *triacantha* is recorded from stratigraphically similar positions in California, the Experimental Mohole cores, the western equatorial Pacific, Japan and in Europe.

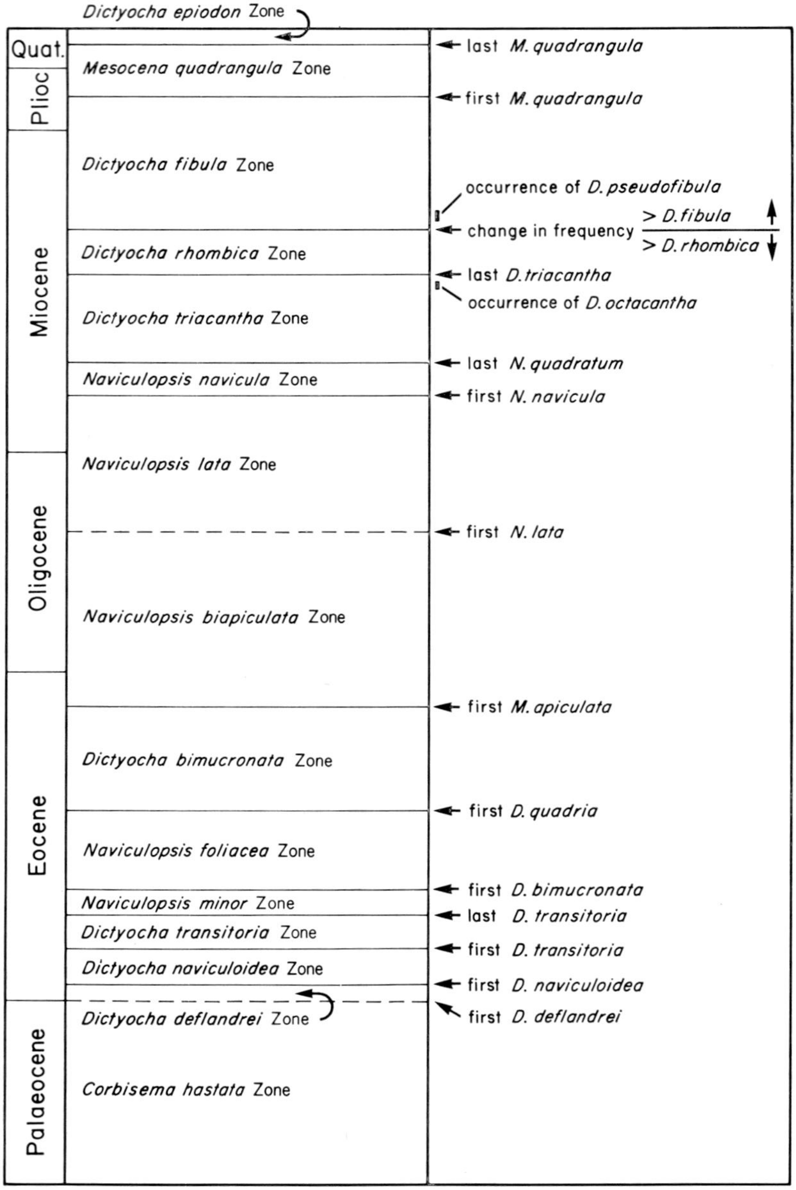

FIG. 5. Preliminary silicoflagellate zonation in the Tertiary and Quaternary, with indication of silicoflagellate events (right column).

Martini (1971b) made the first attempt to correlate the silicoflagellate zones with zonation schemes established for other fossil groups. The author (1972 and 1973) extended the silicoflagellate zonation into the Palaeogene. The present stage in the development of silicoflagellate zones for the Tertiary is shown on Fig. 5. Many of the species listed on this figure are illustrated on Plates 1 and 2. It is hoped that the stratigraphical position of the silicoflagellate "events" listed on Fig. 5, some of which are as yet not widely recorded, will be confirmed at other localities.

Preparation Technique

Smear-slides of soaked material mounted in caedax or canadabalsam provide a means of rapid examination.

In order to obtain cleaned assemblages for photographic or other purposes the following procedure is used: the sample is treated with concentrated hydrochloric acid in order to dissolve any calcareous material. After washing and treatment with concentrated sulphuric acid to destroy the organic components, chilesalpetre is added to oxidize the carbon, and the sample again is washed. Part or most of the treated sample may be passed through a 0.03 mm-sieve to exclude the very small particles. Then slides mounted in Caedax or Canadabalsam are prepared from the cleaned material for inspection with the lightmicroscope. If work with the transmission-electronmicroscope or scanning-electronmicroscope is intended, special shadowing and treatment is necessary, described elsewhere.

References

For detailed bibliography see Loeblich et al., 1968

Bachmann, A. (1970). Flagellata (Silicoflagellata). *Catalogus fossilium Austriae* **1b,** 1–28.

Bachmann, A. and Keck, A. (1969). Die Oberflächenstruktur der Silicoflagellaten. *Mikrokosmos*, 58(7), 204–207, 1 text-fig., pls. 1–2.

Bukry, D. and Foster, J. H. (1973). Silicoflagellate and diatom stratigraphy, Leg 16, Deep Sea Drilling Project. *Initial Rep. Deep Sea Drilling Project* **16,** 815–871, 1 text-fig., 12 tabs., pls. 1–17.

Deflandre, G. (1950). Contribution à l'étude des Silicoflagellidés actuels et fossiles. *Microscopie* **2**, 72–153, 243 text-figs.

Dumitrica, P. (1973). Paleocene, late Oligocene and post-Oligocene silicoflagellates in Southwestern Pacific sediments cored on DSDP Leg 21. *Initial Rep. Deep Sea Drilling Project* **21,** 837–883, 6 text-figs., pls. 1–13. U.S. Govt. Printing Office, Washington.

Ehrenberg, C. G. (1837). Eine briefliche Nachricht des Hrn. Agassiz in Neuchatel über den ebenfalls aus mikroskopischen Kiesel-Organismen bebildeten Polirschiefer von Oran in Afrika. *Ber. Verh. Kgl. Preuss. Acad. Wiss.* **1837,** 59–61.

Frenguelli, J. (1940). Consideraciones sobre los silicoflagellados fósiles. *Rev. Museo de la Plata* n. ser., **2,** 37–112, 38 text-figs., pls. 1–4.

Gemeinhardt, K. (1930). Silicoflagellatae. *In* (Rabenhorst, L.) *Kryptogamen-Flora von Deutschland, Österreich und der Schweiz* **10,** 1–87, 68 text-figs., pl. 1.

Glezer, Z. I. (1966). Silicoflagellatophyceae. *Cryptogamic plants of the U.S.S.R.* **7,** 363 pp., 28 text-figs., 58 tabs., pls. 1–33 (English translation Jerusalem, 1970).

Jerković, L. (1969). Les silicoflagellides fossiles des environs de Zagreb, de Bosanska Kostajnica et de Derventa (Yougoslavie). *Godišnjaka Biol. Inst. Univ. Sarajevu* **22**, 21–127, 239 text-figs., 4 maps, pls. 1–12.

Lemmermann, E. (1901). Silicoflagellatae. *Ber. dtsch. bot. Ges.* **19,** 247–271, pls. 10–11.

Lipps, J. H. (1970). Ecology and Evolution of Silicoflagellates. *Proc. North Amer. Paleont. Convent.* **1969,** part G, 965–993, 11 text-figs.

Loeblich, A. R. III, Loeblich, L. A., Tappan, H. and Loeblich, A. R. Jr. (1968). Annotated index of fossil and recent silicoflagellates and ebridians with descriptions and illustrations of validly proposed taxa. *Geol. Soc. Amer. Mem.* **106,** 317 pp., 12 text-figs., pls. 1–53.

Locker, S. (1974). Revision der Silicoflagellaten aus der mikrogeologischen Sammlung von C. G. Ehrenberg. *Eclog. geol. Helv.* **67** (3), 631–646, pls. 1–4.

Mandra, Y. T. (1968). Silicoflagellates from the Cretaceous, Eocene, and Miocene of California, U.S.A. *Proc. Calif. Acad. Sci.* 4. ser. **36**(9), 231–277, 83 text-figs., 7 tabs.

Mandra, Y. T. (1969). Silicoflagellates: A new tool for the study of Antarctic Tertiary climates. *Antarctic J. U.S.* **4**(5), 172–174, 4 text-figs.

Mandra, Y. T. (1969). A new genus of silicoflagellata from an Eocene South Atlantic deep-sea core (Protozoa: Mastigophora). *Occ. pap. Calif. Acad. Sci.* no. **77,** 1–7, 7 text-figs.

Martini, E. (1971a). Neogene silicoflagellates from the equatorial Pacific. *Initial Rep. Deep Sea Drilling Project* **7**(2), 1695–1708, 3 text-figs., 4 tabs., pl. 1. U.S. Govt. Printing Office, Washington.

Martini, E. (1971b). Standard Tertiary and Quaternary calcareous nannoplankton Zonation. *In* (Farinacci, A., ed.) *Proc. II. Plankt. Conf.* **2,** 739–785, 6 tabs., pls. 1–4. Tecnoscienza, Roma.

Martini, E. (1972). Silicoflagellate zones in the late Oligocene and early Miocene of Europe. *Senckenbergiana lethaea*, 53(1/2), 119–122, 4 text figs.

Martini, E. (1974). Silicoflagellate zones in the Eocene and early Oligocene. *Senckenbergiana lethaea*, **54**(5/6), 527–532, 1 text-fig.

Perch-Nielsen, K. (1975). Late Cretaceous to Pleistocene silicoflagellates from the southern Southwest Pacific, DSDP, Leg 29. *Initial Rep. Deep Sea Drilling Project* **29,** 677–721, 3 text-figs., 9 tabs., pls. 1–15. U.S. Govt. Printing Office, Washington.

Schulz, P. (1928). Beiträge zur Kenntnis fossiler und rezenter Silicoflagellaten. *Bot. Arch.* **21**(2), 225–292, 83 text-figs.

Stradner, H. (1961). Über fossile Silicoflagelliden und die Möglichkeit ihrer Verwendung in der Erdölstratigraphie. *Erdöl u. Kohle, Erdgas, Petrochemie* **14,** 87–92, 1 text-fig., pls. 1–3.

Yanagisawa, T. (1943). Klishitsu-benmochu nitsuite (Silicoflagellatae). *Umi to Sora* **23,** 11–29 (454–469), 6 text-figs., 7 tabs.

PLATES

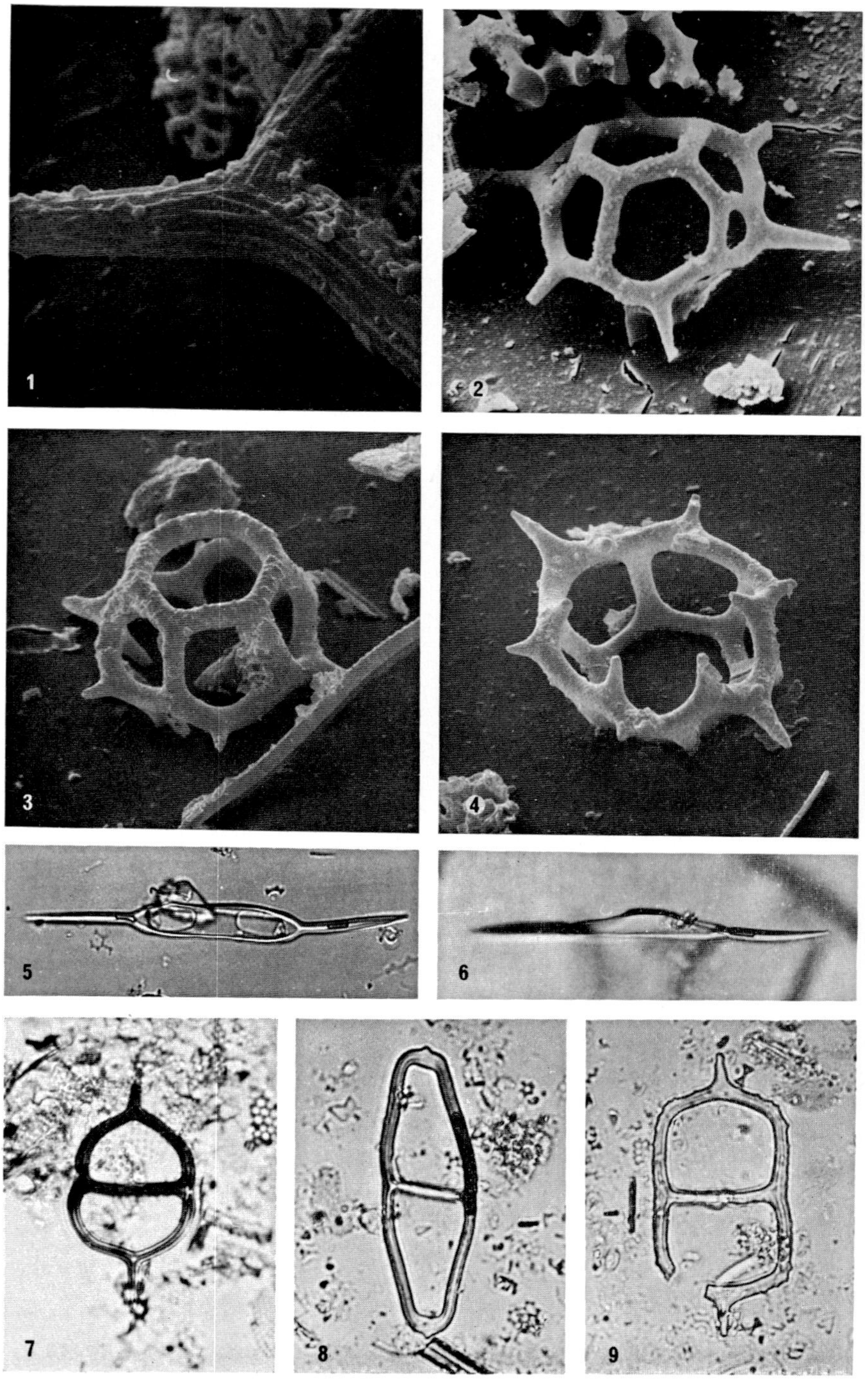
1
2
3
4
5
6
7
8
9

Plate 1

Common or important silicoflagellates

(All specimens except figs. 1–4 approximately ×500)

fig. 1. *Dictyocha fibula* Ehrenberg
Ornamentation on spine and part of the basal ring, SEM (×3480).
Sepulveda 1, Los Angeles (Miocene, Upper Mohnian, *Dictyocha fibula* zone).

fig. 2. *Dictyocha pseudofibula* (Schulz)
Apical side, SEM (×1100).
Sepulveda 1, Los Angeles (Miocene, Upper Mohnian, *Dictyocha fibula* zone).

figs. 3, 4. *Dictyocha speculum* Ehrenberg
3, apical side, SEM (×1200); 4, basal view, SEM (×1320).
Sepulveda 1, Los Angeles (Miocene, Upper Mohnian, *Dictyocha fibula* zone).

figs. 5, 6. *Naviculopsis foliacea* Deflandre
5, Basal side; 6, side view.
DWBG 23 B Catcher, equatorial Pacific (Middle Eocene, *Naviculopsis foliacea* zone).

fig. 7. *Naviculopsis lata* (Deflandre)
WRE 67/101, 20 m NE of cemetery at Casatico, Italy (early Miocene, *Naviculopsis lata* zone).

fig. 8. *Naviculopsis navicula* (Ehrenberg)
WRE 67/77, 400 m WSW of the church of Monte Calvo near Bologna, Italy (early Miocene, *Naviculopsis navicula* zone).

fig. 9. *Naviculopsis quadratum* (Ehrenberg)
WRE 67/76, 200 m WSW of the church of Monte Calvo near Bologna, Italy (early Miocene, *Naviculopsis navicula* zone).

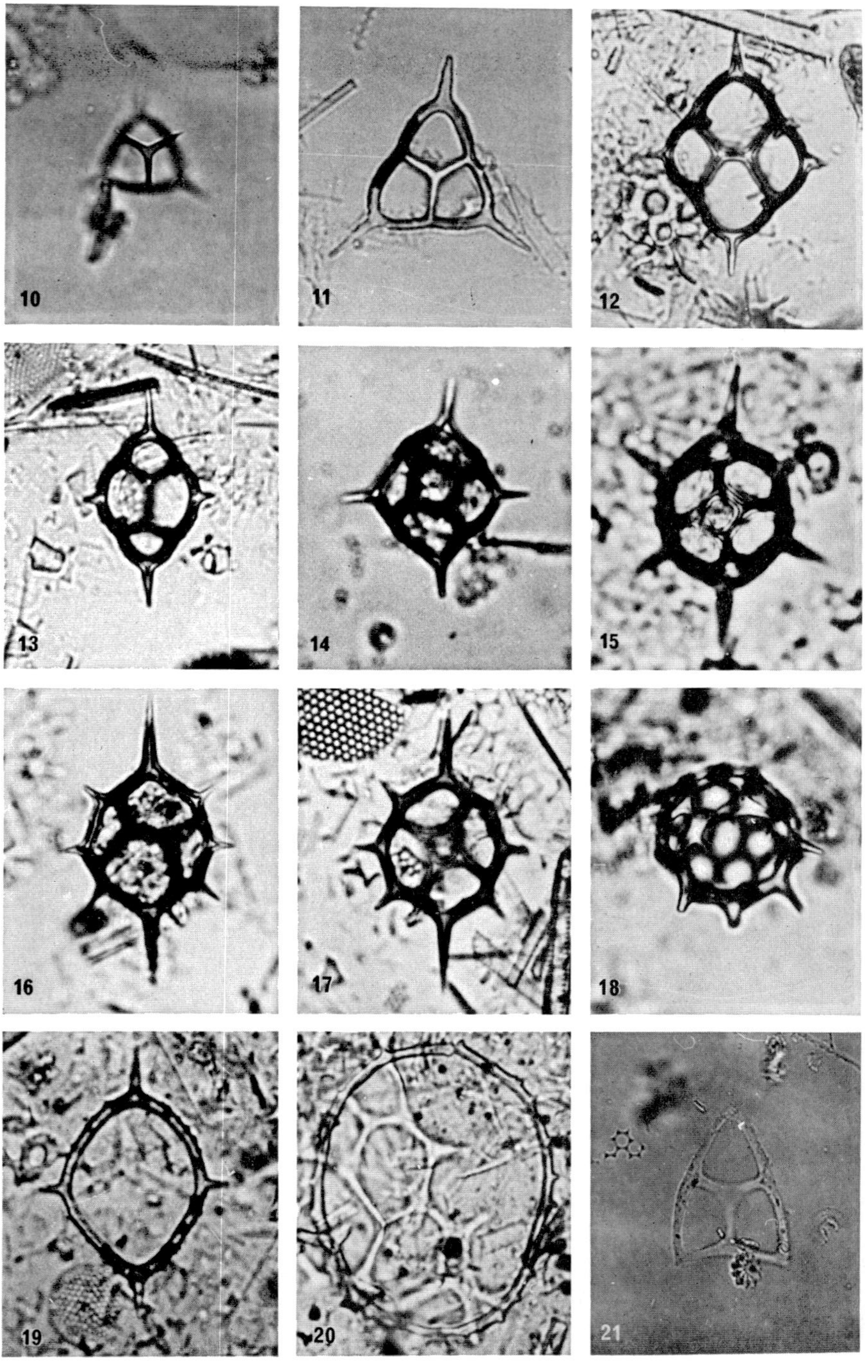
10
11
12
13
14
15
16
17
18
19
20
21

Plate 2

Common or important silicoflagellates

(All specimens approximately ×500)

fig. 10. *Dictyocha spinosa* (Deflandre)
Apical side, high focus.
DWBG 23 B Catcher, equatorial Pacific (Middle Eocene, *Naviculopsis foliacea* zone).

fig. 11. *Dictyocha triacantha* Ehrenberg
DSDP 66–1–7–5:75–75 cm, equatorial Pacific (Middle Miocene, *Dictyocha triacantha* zone).

fig. 12. *Dictyocha rhombica* (Schulz)
DSDP 66–1–5–1:75–76 cm, equatorial Pacific (Middle Miocene, *Dictyocha rhombica* zone).

fig. 13. *Dictyocha fibula* Ehrenberg
DSDP 66–1–4–1:108–109 cm, equatorial Pacific (Upper Miocene, *Dictyocha fibula* zone).

fig. 14. *Dictyocha crux* Ehrenberg
Newport Lagoon 12, California (Miocene, Lower Mohnian, *Dictyocha triacantha* zone).

fig. 15. *Dictyocha pseudocrux* (Schulz)
Experimental Mohole core EM 8–9:150–152 cm, near Guadalupe, eastern Pacific (Upper Miocene, *Dictyocha fibula* zone).

fig. 16. *Dictyocha octagona* Tsumura
Experimental Mohole core EM 7–1:36–38 cm, near Guadalupe, eastern Pacific (Middle Miocene, *Dictyocha triacantha* zone).

fig. 17. *Dictyocha octacantha* (Desikachary and Maheshwari)
DSDP 66–1–7–6:75–76 cm, equatorial Pacific (Middle Miocene, *Dictyocha triacantha* zone).

fig. 18. *Cannopilus sphaericus* Gemeinhardt
Basal side. S of Plum Point, Maryland (Miocene, Clavert Formation, Plum Point Marl, *Dictyocha triacantha* zone).

fig. 19. *Mesocena quadrangula* Ehrenberg ex Haeckel
DSDP 66–1–1–2:top, equatorial Pacific (Upper Pliocene, *Mesocena quadrangula* zone).

fig. 20. *Mesocena circulus* (Ehrenberg)
DSDP 65–0–2–CC, equatorial Pacific (Lower Pliocene, *Dictyocha fibula* zone).

fig. 21. *Phyllodictyocha recta* Deflandre
DWBG 23 B Catcher, equatorial Pacific (Middle Eocene, *Naviculopsis foliacea* zone).

17: O^{18} as a Tool for Micropalaeontologists

JAN VAN DONK*

Lamont–Doherty Geological Observatory of Columbia University, Palisades, New York 10964, U.S.A.

To indicate the remaining analytical problems inherent in the oxygen isotope method, a short description of analytical techniques used for carbonate, water, and biogenic silica samples is given. Special emphasis is placed on the fact that O^{18} variations are the combined results of ice volume changes and biological effects, as well as the "well-known" temperature effect. The application and accuracy of isotopic results as temperature and/or Pleistocene glacial ice volume estimates is discussed in detail. The conclusions are:

1. It is necessary to establish an empirical relationship between the O^{18} value and growth temperature for each species of microorganism.
2. A significant part of the O^{18} variation seen in Pleistocene microfossils is due to a glacial ice volume effect, which could account for as much as 1.2‰ of the O^{18} variation.
3. High sedimentation rate cores give a more accurate picture of the maximum amplitude of O^{18} changes than low sedimentation rate cores, due to bioturbation and sampling problems in the latter cores.

Not all microorganisms deposit their calcareous shells in isotopic equilibrium. The degree of disequilibrium may serve as an indicator to which degree matabolic CO_2 is used in carbonate secretion by the various species of microorganisms.

A general order of depth habitats for the various species of planktonic foraminifera, based on oxygen isotope results has been established, and agrees well with the general order of depth rankings based on tow data.

Conflicting evidence concerning the reliability of the isotope analyses of biogenic silica exists. Further studies are required, to reveal the usefulness of the oxygen isotope method for biogenic silica, as a climatological and palaeontological tool.

Outline

Introduction

The study of oxygen isotope variations in natural materials as a tool in palaeoclimatic studies was first suggested by Urey (1947). It was further developed

* Present address: University of Botswana, Lesotho and Swaziland, Private Bag 0022, Gaborone, Botswana.

by Urey (1948), McCrea (1950), and Epstein *et al.* (1953). Since then the isotope method has been applied, with varying degrees of success, to a variety of calcareous and non-calcareous hard parts of organisms from nearly all geological periods since the Devonian. Oxygen isotope measurements on Mesozoic belemnites have been made by Urey *et al.* (1951), Lowenstam and Epstein (1954), Epstein (1959), Bowen (1961), Naidin *et al.* (1966), and Spaeth *et al.* (1971). The method has been applied to molluscs by Dorman and Gill (1959), Lloyd (1964), Keith and Parker (1965), Dorman (1966), Devereux (1967a), Stuiver (1968), and Mook (1971); to corals by Keith and Weber (1965), and Weber and Woodhead (1972); to echinoderms by Weber and Raup (1966, 1968); and to fish otoliths by Devereux (1967b). Furthermore, important isotopic results have been obtained on the phosphate of the hard parts of belemnites, pelecypods, gastropods, brachiopods, and fish teeth (Longinelli, 1965, 1966; Longinelli and Nuti, 1968, 1972). The phosphate temperature scale seems to be nearly parallel to the carbonate temperature scale. Cortecci and Longinelli (1971) have studied the oxygen isotope variation in sulfates from marine molluscs, but have not found any temperature relationship.

This chapter, however, will be concerned only with the application of oxygen isotopic studies to microfossils such as foraminifera, pteropods, and radiolaria.

After a short discussion on the theoretical basis for the method, and a brief description of the analytical techniques used for the various materials, each group of microfossils will be discussed separately, with emphasis on foraminifera. An attempt will be made to indicate the questions the method can address itself to, as well as the weakness of the method.

Theoretical considerations

The oxygen isotope method basically utilises the temperature dependence of the fractionation of the two stable isotopes of oxygen (O^{16} and O^{18}), between calcite and water, or any other pairs, such as silica–water; phosphate–water; and sulphate–water. In nature, the overall abundance ratio of O^{18}/O^{16} is close to 1:500. The isotopic equilibrium reaction for the deposition of $CaCO_3$ in equilibrium with the environmental water can be written as follows:

$$\tfrac{1}{3}\,CaCO_3{}^{16} + H_2O^{18} \leftrightharpoons \tfrac{1}{3}\,CaCO_3{}^{18} + H_2O^{16}.$$

The isotopic fractionation factor for this reaction is defined as:

$$\alpha\frac{CaCO_3}{H_2O} = \frac{(O^{18}/O^{16})CaCO_3}{(O^{18}/O^{16})H_2O}$$

The value of α only depends on the temperature of equilibrium between the two phases. The lower the temperature the larger the value for α. For example, at 0°C, α for the above reaction is 1.034 while at 25°C it is 1.028. The temperature dependence of this fractionation factor α for this reaction in the range of 0–25°C is about 0.00024°C. This temperature dependence of the enrichment in O^{18} in the carbonate, relative to the water from which it precipitated, is the basis for the isotopic palaeotemperature method, first proposed by Urey (1947). It is evident from the foregoing discussion that the oxygen isotopic composition of the carbonate precipitated in equilibrium from ocean water depends on both the temperature and the oxygen isotopic composition of the ocean water.

McCrea (1950) was the first worker to relate the isotopic composition of the carbonate to the equilibrium temperature and the isotopic composition of the water from which he inorganically precipitated the carbonate. Epstein *et al.* (1953) determined empirically a new isotopic temperature scale, which was based solely on organically precipitated $CaCO_3$. These investigators cultured molluscs with various shell morphologies, such as, abalones, limpets, snails, and bivalves, at known temperatures in the laboratory. Because Epstein *et al.* (1953) did not allow for any systematic errors in the measurements on their mass-spectrometer, Craig (1965) applied correction factors to their data. He formulated a temperature equation which is the basis for all present-day temperature calculations from O^{18} variations in carbonates formed at equilibrium with the surrounding waters. It is:

$$t(°C) = 16.9 - 4.2(\delta_C - \delta_W) + 0.13(\delta_C - \delta_W)^2$$

where $\delta_C = \delta O^{18}$ for CO_2 obtained from the *carbonate* by the acid extraction technique at 25°C with respect to mass spectrometer working standard gas, and $\delta_W = \delta O^{18}$ for CO_2 equilibrated isotopically at 25°C with the *water* from which the carbonate was precipitated and measured against the same mass spectrometer working standard gas.*

Analytical techniques

FORAMINIFERA AND PTEROPODA

The shells of these organisms can be obtained from either plankton tows or from sediment samples.

In the case of plankton tows, the shelled organisms are concentrated in

* The variations in the O^{18}/O^{16} ratio are recorded as per mil (‰) deviations from a standard, δ, which is defined as:

$$\delta = [O^{18}/O^{16}{}_{\text{sample}}/O^{18}/O^{16}{}_{\text{std.}} - 1] \times 1000$$

several ways–manually under a binocular microscope; use of high density salt solutions (Bé, 1959; McGowan and Fraundorf, 1964) and the low temperature oxidation method (Vilks, 1970).

The high-density salt solution method is based on the principle that shelled organisms will settle to the bottom of a cylinder containing such a solution because of their greater specific gravity, while the non-shelled invertebrate plankters will float to the top. The shelled organisms are recovered from the bottom of the cylinder through a drain.

The low-temperature oxidation method makes use of a low-temperature asher (Tracerlab LTA-600) in which the organic matter in a plankton sample is exposed to a flow of oxygen excited by a radio-frequency electromagnetic field. The oxidation takes place in a partial vacuum at temperatures less than 150°C, and leaves behind a clean concentration of skeletal material. At present, we are testing in this laboratory what effect the latter procedure has on the oxygen isotope composition of the shells.

For sediment samples the microfossils are concentrated by washing the sediment sample through a 63 micron sieve, after which the coarse fraction is left to dry.

After concentration of the shelled organisms from either plankton tows or sediment samples, the appropriate planktonic or benthonic species are sorted manually under a binocular microscope. Enough shells of each species are sorted to make up at least 5 mg of material, as this amount for most laboratories represents the minimum amount of sample needed for a reliable analysis. However, Shackleton (1965; *et al.*, 1973) reports analyses on samples as small as 0.04 mg. After separation the shells are lightly crushed and cleaned in an ultrasonic vibrator, in order to remove any foreign debris from the chambers, such as clay, coccoliths, and encrusted material. Some workers roast their samples, either in a stream of helium or in vacuum at 475°C, in order to remove any organic matter which might be contained in the shells, while others omit this step, when dealing with fossil foraminifera. For samples from live populations, however, the samples need to be roasted, since the organic material present has a different isotopic composition than the carbonate shell. Emiliani (1966) showed that for foraminifera, which were ground to a fine powder, the average difference in the oxygen isotopic composition between helium roasted and unroasted samples was about 0.5‰, which is significant. The unroasted samples were isotopically heavier. He also showed that not only using different purification procedures may give different isotopic results, but the type of sample material used affects the final results significantly. Van Donk (1970) found an average difference in isotopic composition between vacuum roasted and unroasted fossil foraminifera of 0.2‰ for δO^{18} and 0.14‰ for δC^{13}. Again, the unroasted samples were isotopically heavier. After the cleaning procedure, the samples are air dried and reacted with 100% H_3PO_4, according to the procedure established by McCrea (1950). The reaction takes place in a thermostatically controlled bath at a constant temperature (most workers choose 25°C). The reason is, that the acid reaction involves an isotopic fractionation, because two-thirds of the carbonate oxygen goes into CO_2 and one-third into H_2O. This fractionation is constant for any given temperature, and is independent of the original isotopic composition of the sample. After the reaction, the CO_2 gas is extracted on a vacuum line by freezing all the gases in a pyrex U-tube surrounded by liquid nitrogen. The CO_2 is separated from the H_2O by freezing the H_2O in a pyrex U-tube surrounded by dry-ice. At this temperature the water will freeze but not the CO_2. Then the CO_2 is transferred to, and analysed in a dual-collector Nuclide-analysis isotope ratio mass spectrometer.

RADIOLARIA

The sediment from which the radiolaria are to be extracted is ultrasonically dispersed in water and the detrital minerals are allowed to settle out. The liquid with suspended clays and radiolaria is then washed through a 50 micron sieve. The coarse fraction, consisting mainly of radiolaria, is then ground and ultrasonically cleaned in order to remove any trapped clays or diatoms. Carbonate is eliminated by treatment with dilute HCl. Then the samples, weighing about 20 mg, are put in a dry box for several days in order to remove any loosely bound water in the opaline structure (Mopper and Garlick, 1971). However, Labeyrie (1972) does not think this is sufficient, and he heated the samples under vacuum to 1000°C in order to drive off any water of hydration. He also treated the samples with NaOCl in order to eliminate organic material. After these treatments the samples are reacted with BrF_5. The released oxygen is converted to CO_2 by reaction with a red-hot carbon rod (Clayton and Mayeda, 1963). The carbon dioxide gas is then analysed in the mass spectrometer.

WATER SAMPLES

The preparation of water samples is essentially the method described by Epstein and Mayeda (1953). Briefly, 25 ml of water is put into a 200 cc round-bottom flask with 2 drops of 0.1 N H_2SO_4 to bring the pH down to 6 or less. The flask is evacuated by keeping the water frozen, and commercial CO_2 is introduced at slightly less than atmospheric pressure. Approximately 175 standard cc of CO_2 gas is allowed to equilibrate with the 25 ml of water for several days in a constant temperature bath at 25°C. The flask is shaken frequently to insure equilibration. The fractionation factor at 25°C between CO_2 and H_2O has been determined by several workers, but there still is no agreement on the precise value. Compston and Epstein (1958) and O'Neil and Epstein (1966) determined a value of 40.7‰, with the CO_2 being enriched in O^{18}. Bottinga and Craig (1969) report a fractionation factor of 40.9‰, while Horibe *et al.* (1973) report a value of 41.4‰, all at 25°C. Even though it is still unclear what the precise value is, the fractionation is constant for a particular temperature, which is the reason for preparing all samples at the same constant temperature. It is clear that the final isotopic composition of the equilibrated CO_2 depends on the O^{18} content of the water. An aliquot of the CO_2 gas is withdrawn from the flask, and after purification it is analysed on the mass spectrometer.

A more direct method, described by O'Neil and Epstein (1966), involves the fluorination of milligram quantities of water with BrF_5, liberating oxygen, which is converted to CO_2 by reacting it with a red-hot carbon rod. However, there are problems with samples containing polyvalent ions, such as seawater samples. It appears that these polyvalent ions surround themselves with some water molecules, which then do not get frozen into the finger in which the reaction with BrF_5 takes place. Therefore, not all water from the original sample gets converted to oxygen, causing non-systematic errors. The problem is alleviated by adding NaF crystals to the sample (Friedman and Manheim (personal communication through J. Lawrence)).

Application and accuracy of δO^{18} as a temperature estimate of marine environments

This section will draw heavily upon a paper by Imbrie *et al.* (1973). In order for the oxygen isotopic composition of the carbonate to be a precise thermometer, the following three conditions must be fulfilled:

1. The carbonate must be deposited in equilibrium, or out of equilibrium by a constant amount which can be determined.
2. The isotopic composition of the water must be known, or can be inferred.
3. No change in the isotopic composition has occurred since the time of deposition.

Whether all of these conditions are met is still a much debated issue in the recent literature, and is still being researched. Each of these points will be discussed separately in greater detail below.

CONDITION 1

Epstein *et al.* (1953) have shown that some macro-organisms do deposit their carbonate shell in near equilibrium with the surrounding water. They drilled holes in shells of live snails, abalones, oysters, and limpets, and placed the organisms in a tank with water whose temperature could be controlled. As soon as the organism had repaired the hole this carbonate was removed and analysed for δO^{18}. They replated the isotopic composition of the carbonate to the water temperature and the isotopic composition of the water itself and found a temperature relationship which was very much similar to that established for inorganically precipitated carbonate (McCrea, 1950).

Smith and Emiliani (1968) showed that isotopic palaeotemperatures, calculated for recent benthonic foraminifera off the coast of Central America, agreed closely with *in situ* temperatures. Therefore, it seemed to have been demonstrated that the first condition was fulfilled.

However, van Donk and Mathieu (1969) showed that benthonic foraminifera from an Arctic core top gave too high an isotopic temperature, using the accepted temperature equation given before. Similarly, van Donk (1970) demonstrated that many planktonic foraminifera also gave too high isotopic temperatures, and concluded that many foraminiferal species do precipitate their carbonate out of isotopic equilibrium (Fig. 1). This was attributed to what Urey (1947) called a "vital" effect. Duplessy *et al.* (1970) analysed several species of benthonic foraminifera from the same level in a short Atlantic core off the French coast, and found differences between species up to a maximum of 1‰. Most species gave too high isotopic temperatures. The same thing has again been confirmed by Shackleton *et al.* (1973) for several species of planktonic foraminifera.

Not only are different species from the same environment giving different isotopic values, but individuals within a species have significantly different isotopic values. Vinot-Bertouille and Duplessy (1973) found differences in the δC^{13} and δO^{18} values among individuals of the same species of benthonic foraminifera as large as several per ml. They concluded that characteristic isotope values for each species can be obtained by analysing large populations, rather than basing it upon a few individuals.

The amount of disequilibrium is probably related to the amount of CO_2 of metabolic origin that is used in construction of the carbonate shell. All indications are that many planktonic and benthonic foraminifera do grow out of equilibrium, but by a constant amount, which is species dependent. Once the amount of disequilibrium for the different species has been determined, they would still be very useful as an isotopic temperature record.

CONDITION 2

The second condition of knowing the isotopic composition of the water back in time is still a controversial issue, as no water samples are available from the geologic past. The oxygen isotope composition of water can change under the influence of several processes, such as influx of meltwater from glacial ice, runoff from rivers, precipitation, and evaporation. These factors are directly related to the process of enrichment in O^{18} of the remaining water during the

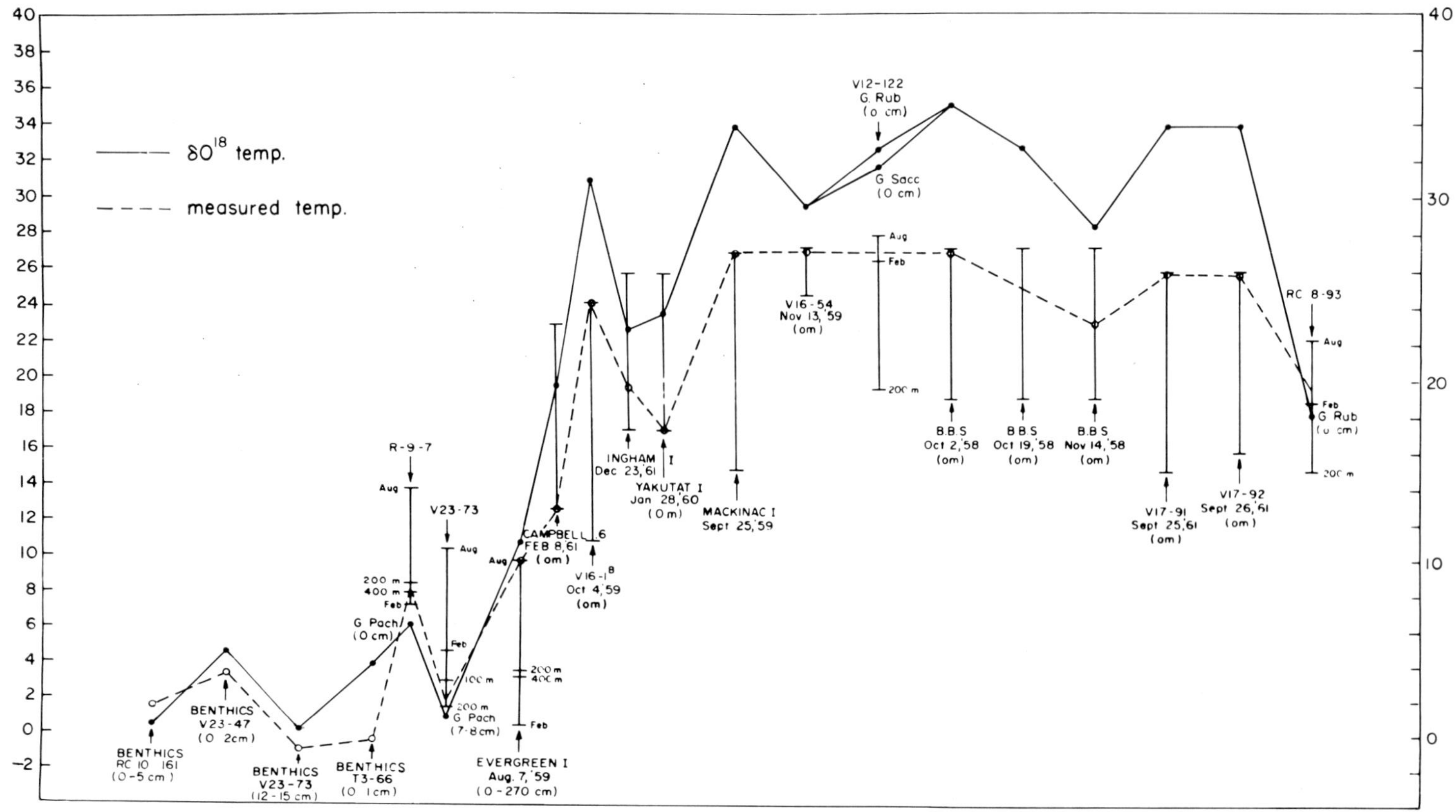

FIG. 1. Comparison of oxygen isotope temperatures for foraminifera from plankton tows and core tops, with the *in situ* temperatures. (The bars indicate the range between the summer and winter temperature of the surface waters.)

process of evaporation, or conversely, the depletion in O^{18} of the water vapour. As a result of the progressive loss of the heavy oxygen atom during each stage of precipitation, water vapour that reaches high-latitude regions, is isotopically very light. For example, snow falling over the Arctic and Antarctic regions are depleted in O^{18} by as much as 40‰, compared to average ocean water.

As the oceanic system constitutes a relatively well-mixed, large reservoir, it is buffered against very large fluctuations. At present, the largest observed difference between various parts of the major oceans is 1.4‰, although surface water in the Arctic Ocean is as light as −4‰ relative to mean ocean water (van Donk and Mathieu, 1969).

Despite the fact that the oceans are buffered against major fluctuations, during glacial periods when large amounts of isotropically light ice are tied up in continental glaciers, an observable change in the oxygen isotopic composition of the ocean water will occur. Estimates of the maximum difference in the isotopic composition of sea water between glacial and interglacial periods vary from a low of 0.5‰ to a high of 1.7‰ (Emiliani, 1955a, 1966; Craig, 1965; Olausson, 1965; Shackleton, 1967; Dansagaard and Tauber, 1969).

A study by Imbrie *et al.* (1973), in which temperature variations with time were obtained independently from the isotope method, through quantitative faunal analysis, indicates that the best estimate is perhaps close to 1.3‰ in agreement with Shackleton (1967).

CONDITION 3

The last condition of preservation of the original isotopic composition seems to be met for foraminifera as far back as the Oligocene. Emiliani (1954a, 1956) has demonstrated this by the fact that benthonic and planktonic foraminifera in the Pacific from the Oligocene through the Miocene, give much higher isotopic temperatures than the present-day bottom-water temperature. Although the shells have been exposed to colder bottom waters for many millions of years, they still indicate 8°C warmer temperatures. Saito and van Donk (1971, 1973) have found that foraminifera have not changed their isotopic composition to any large extent as far back as the Cretaceous. They found that Cretaceous planktonic and benthonic foraminiferal species still show significant differences in isotopic compositions. If they had exchanged their isotopic composition completely, then they would all have the same isotopic composition.

Isotopic results in foraminifera

Since these microorganisms occur in relatively large abundances in many parts of the major world oceans, they were the first group of calcareous organisms from oceanic sediment samples to be studied for isotopic variations.

Most oxygen isotope studies on foraminifera cover the time span from the Oligocene to the Recent, with emphasis on the Late Pleistocene period. Emiliani (1954, 1954b, 1955a, 1956, 1958, 1964, 1966, 1971a, 1971b) pioneered the study of oxygen isotope variations in foraminifera from deep sea cores. The first work on planktonic foraminifera (Emiliani, 1954b) was an investigation of the depth habitats of different species of recent planktonic foraminifera. This was based on the assumption that, if different species from the same level in a core gave different isotopic results, this meant that they lived at different levels in the water-column where the temperatures were different. It should be emphasised that in all cases the measured isotopic compositions are the values for the total carbonate. Therefore, when depth habitats are inferred from calculated isotopic temperatures, these depth habitats only represent the average depths at which the bulk of the carbonate of the shells of the various species was secreted. It does not preclude the possibility that some carbonate for the shell construction was secreted at a shallower depth and some at a deeper depth than the average depth. Emiliani (1954b) found that *Globigerinoides ruber*, d'Orbigny, and *G. sacculifer* (Brady) live very close to the ocean surface while the various species of *Globorotalia* generally occupy the deepest habitats down to about 220 m. These conclusions, based on isotopic results, are in good agreement with field observations by Jones (1967), using opening and closing plankton nets.

Lidz *et al.* (1968) extended the work by Emiliani (1954b) through different glacial and interglacial intervals, as shown by a Caribbean deep sea core. They ranked fourteen different pelagic species of foraminifera in order of increasing depth (decreasing temperature) based on their recorded isotopic temperatures for the postglacial period, and at several warm and cold stages of interglacial and glacial periods, respectively. They concluded that the sequence is in general the same for all these different periods in time. Four species and subspecies of *Globigerinoides* occupy the shallowest habitat (0–40 m), while six species and subspecies of *Globorotalia* occupy the deeper habitats (120–200 m). *Orbulina universa*, d'Orbigny, *Pulleniatina obliquiloculata* (Parker and Jones), *Hastigerina pelagica* (d'Orbigny), and *Sphaeroidinella dehiscens* (Parker and Jones) were found to prefer the intermediate depths (80–110 m).

Oba (1969) did a similar study by comparing the δO^{18} values of 19 concomitant pelagic species from one level in an Indian Ocean core and found essentially the same order of depth habitats as Lidz *et al.* (1968).

The depth habitats of the various planktonic foraminifera species as reported by eight workers, and deduced from either oxygen–isotope results or from field observations with opening and closing nets, are summarised in Tables 1 and 2.

In light of the recent findings of non-equilibrium isotopic fractionation between sea water and foraminiferal tests of several species (van Donk and Mathieu, 1969; van Donk, 1970; Duplessy *et al.*, 1970; Vinot, 1971; Glaçon *et al.*, 1971; Shackleton *et al.*, 1973; Vinot-Bertouille and Duplessy, 1973), it is not certain that all the observed isotopic differences between concomitant species in earlier studies represent differences in depth habitat. Some of the differences might actually represent the differences in the amount of metabolic CO_2 incorporated in the shells, while in other cases they do indeed represent differences in depth habitat.

It should be emphasised that when samples from deep sea core tops are compared with those from plankton tows, that core top samples generally represent thicker shelled populations than those from plankton tows. This is probably due to the fact that several species thicken their shells as they gradually descend to greater depth habitats (Bé and Ericson, 1963; Bé and Lott, 1964; Bé, 1965; Bé and Hemleben, 1970). This will be discussed in more detail in a later section.

Despite these complicating factors, it does appear that in general terms the isotopic depth classification of the different species of pelagic foraminifera into three categories (shallow, intermediate, and deep habitats) agree well with the depth classification based on plankton tow data with opening and closing nets (Tables 1 and 2).

However, without further detailed knowledge of the amount of isotopic disequilibrium for each species and whether or not this is constant in time and place, it is not warranted to use the isotopic differences between concomitant species as a precise measure of their depth habitat. At the same time, it is clear that the oxygen isotope method might open up new doors to a better understanding of the biochemical process of shell formation, through a study of the degree of disequilibrium exhibited by the different species of foraminifera from samples of living populations. Such work is now in progress by several workers (van Donk, Kahn, Shackleton, Duplessy, and Glaçon).

Another interesting palaeontological study was made by Longinelli and Tongiorgi (1964). They measured the oxygen isotope differences in concomitant right- and left-coiled populations of six species of foraminifera from living and fossil faunas from different habitats. The species were, *Rotalia beccarii* (Linné), *Cibicides lobatulus* (Walker and Jacob), *Cassidulina crassa*, d'Orbigny, *Discorbis isabelleana* (d'Orbigny), *Globigerina bulloides*, d'Orbigny, and *G. inflata*, (d'Orbigny). They observed that both planktonic and benthonic species consistently showed lower δO^{18} values for right-coiled populations relative to left-coiled populations, with one exception (*Cassidulina crassa*). They interpreted this in terms of average carbonate secretion by the organisms taking place during different seasons. Although, in light of the

TABLE 1. Summary of depth rankings of planktonic foraminifera[1] based on the oxygen isotope method.

Emiliani (1954b)	Depth (m)	Lidz *et al.* (1968)	Depth (m)	Oba (1969)	Isotopic Temp. (°C)	Hecht and Savin (1972)	Depth range (m)	Average depth (m)
				"SHALLOW"[2]				
Globigerinoides ruber	0–45	*Globigerinoides ruber*	20	*Globigerina bulloides*	27.1	*Globigerinoides ruber* (normal form)	0–85	20
G. sacculifer	0–25	*G. sacculifer* (with sac.)	20	*Globigerinita glutinata*	26.9	*G. ruber* (kummerform)	0–120	66
G. conglobatus	80	*G. sacculifer* (without sac.)	8–40	*Globigerinoides ruber*	23.4	*Globigerinella aequilateralis*	—	45
		G. conglobatus	16–30	*G. sacculifer*	23.0	*Globigerinoides sacculifer* (with sac.)	0–150	60
				G. conglobatus	21.5	*G. sacculifer* (without sac.)	0–150	60
						G. conglobatus (bulla)	55–135	63
						G. conglobatus (non-bulla)	70–130	71
				"INTERMEDIATE"[2]				
Globoquadrina dutertrei	33–100	*Orbulina universa*	81	*Orbulina universa*	21.4	*Globoquadrina dutertrei*	60–120	80
Pulleniatina obliquiloculata	15–110	*Pulleniatina obliquiloculata*	85	*Globoquadrina dutertrei*	21.2	*Pulleniatina obliquiloculata*	60–105	80

Globorotalia menardii	43–130	*Hastigerina pelagica*	88	*Pulleniatina obliquiloculata*	20.7	*Orbulina universa*	50–220	—
Orbulina universa	130	*Sphaeroidinella dehiscens*	106	*Sphaeroidinella dehiscens*	20.1	*Globorotalia menardii*	65–135	90
		Globoquadrina dutertrei	121	*Globorotalia crassaformis* (right)	20.1			
				Globigerinella aequilateralis	19.8			
				"DEEP"[2]				
Globorotalia tumida	55–140	*Globorotalia tumida*	132	*Globoquadrina conglomerata*	19.3	*G. tumida*	60–170	102
G. truncatulinoides	220	*G. menardii*	139	*Globorotalia menardii*	18.6	*G. truncatulinoides*	200–255	230
		G. menardii flexuosa	166	*G. crassaformis* (left)	17.8	*G. crassaformis*	—	330
		G. truncatulinoides	181	*Globigerina calida*	16.2			
				Globorotalia tumida	16.0			
				G. hirsuta	12.7			
				G. truncatulinoides	11.8			
				Globoquadrina hexagona	11.2			
				Globorotalia scitula	7.6			

[1] The author has taken the liberty to standardise the species names, used by the various authors, to conform to the classification of Bé (1967).

[2] The grouping into "Shallow", "Intermediate" and "Deep" has been made to roughly coincide with the equivalent grouping in Table 2, in order to facilitate comparison.

TABLE 2. Summary of depth rankings of planktonic foraminifera[1] based on abundance data from plankton tows, with opening and closing nets.

Jones (1967)	Berger (1969) after Bradshaw (1957)	Bé and Tolderlund (1971)[3]	Williams (1971)[4]	Depth range of 75–100% of population (m)
		"SHALLOW"[2]		
Globigerinoides ruber	*Globigerinoides ruber*	*Globigerinoides ruber*	*Globigerinoides ruber*	0–40
G. sacculifer	*G. sacculifer*	*G. sacculifer*	*G. sacculifer*	0–60
	G. conglobatus	*G. conglobatus*		
	Globigerinita uvula	*Globigerina rubescens*		
	G. glutinata	*G. quinqueloba*		
	Globigerina bulloides			
	Globorotalia menardii			
	G. hirsuta			
		"INTERMEDIATE"[2]		
Globoquadrina dutertrei	*Globigerina quinqueloba*	*Globigerinella aequilateralis*	*Hastigerina pelagica*	0–90
Pulleniatina obliquiloculata	*Orbulina universa*	*Globigerina bulloides*	*Globorotalia menardii*	0–90
Globorotalia menardii	*Globoquadrina dutertrei*	*Globigerinita glutinata*	*G. tumida*	20–80
	G. hexagona	*Pulleniatina obliquiloculata*	*Globoquadrina dutertrei*	0–160
	Globigerinella aequilateralis	*Orbulina universa*	*Orbulina universa*	30–120
	Globigerina calida	*Candeina nitida*	*Globigerinella aequilateralis*	0–180
	Globorotalia inflata			

		"DEEP"[2]		
Globorotalia crassaformis	*Pulleniatina obliquiloculata*	*Globoquadrina dutertrei*	*Pulleniatina obliquiloculata*	60–160
	Hastigerina pelagica	*Globorotalia inflata*	*Globorotalia truncatulinoides*	30–320
	Sphaeroidinella dehiscens	*G. menardii*	*G. crassaformis*	90–550
	Candeina nitida	*G. crassaformis*		
	Globoquadrina conglomerata	*Sphaeroidinella dehiscens*		
	Globigerina pachyderma	*Hastigerina pelagica*		
	Globorotalia tumida	*Hastigerinella digitata*		
	G. truncatulinoides	*Globigerina pachyderma*		
	Globigerinella adamsi	*Globorotalia tumida*		
		G. hirsuta		
		G. truncatulinoides		
		G. scitula		

[1] The author has taken the liberty to standardise the species names, used by the various authors, to conform to the classification of Bé (1967).

[2] "Shallow" refers mainly to the mixed layer, "Intermediate" to the water immediately below the mixed layer and "Deep" to waters below the upper thermocline.

[3] "Shallow" is surface species (max. abundance in 0–10 m tows); "Intermediate" is equally abundant in 0–10 m and 0–300 m tows; "Deep" is sub-surface species (more abundant in 0–300 m than in 0–10 m tows).

[4] Grouping into "Shallow", "Intermediate" and "Deep" made to roughly coincide with other three columns for ease of comparison.

proven disequilibrium isotopic fractionation by several species, this difference between right- and left-coiled populations could be due to a "vital" effect, it does not appear likely, because the two populations belong to the same species. The explanation of deposition of carbonate in different seasons by these two populations is probably the more likely reason.

Hecht and Savin (1970, 1972) used oxygen isotopes as a tool in the study of various phenotypes of Recent species of planktonic foraminifera. They found significant differences in the δO^{18} values of populations of single species of Recent planktonic foraminifera with normal final chambers and those with diminutive final chambers (kummerform growth). They concluded that the latter populations develop as a response to environmental stress. For example, the kummerform phenotypes of the *Globigerinoides ruber* species record isotopic temperatures that are from 1–4.5°C cooler than those recorded by the normal form. This is interpreted in terms of a deeper habitat, rather than growth in a cooler season at the same depth, as Berger (1970) presented evidence that in plankton tows the percentage of kummerform individuals increases with depth. Furthermore, Hecht and Savin (1970, 1972) analysed *G. sacculifer* with a sac-like final chamber and those without this final chamber. Both gave the same isotopic results, and their conclusion was that the two phenotypes secrete their tests at essentially the same temperatures. Emiliani (1954b, 1971b) and Hecht and Savin (1972) concluded from their isotopic studies on depth habitats that water density seems to be the controlling factor as to where a species lives and not depth *per se*. For example, Emiliani (1954b) observed, on the basis of oxygen isotope measurements, that *Globorotalia tumida* (Brady) on the average lives at depth of 140 m in the Gulf of Mexico and at 55 m in the equatorial Atlantic.

Oba (1969) and Emiliani (1971b) did another interesting study in which they separated several size groups of the same species of pelagic foraminifera in order to check if there is any difference in the isotopic composition. Oba (1969) found that in the Indian Ocean the smaller size group of *G. menardii* (d'Orbigny) has slightly lighter δO^{18} values than the large size group. This would confirm Bé's contention (1960, 1966), that *G. menardii* shows selective thickening of its shell when it grows into adulthood, and this thickening appears to take place in deeper (colder) waters. Emiliani (1971b) found that the species *G. ruber* and *G. sacculifer* show no difference in the isotopic composition of the large and small size groups, confirming that these species deposit their shell material at constant depths. On the other hand, *G. menardii* and *G. truncatulinoides* (d'Orbigny) show an increase of 0.8‰ in the isotopic values for the large size groups as compared to the small size groups in the Caribbean, thus favouring the explanation of migration with depth during their life spans. Observations from plankton tows by Bé and Lott (1964), and

Bé *et al.* (1966) are consistent with this model of continual growth of these shells, while they are sinking slowly through the water column. In contrast to the Caribbean, *G. menardii* does not appear to change its depth habitat appreciably in the Pacific and Indian Oceans, based on isotopic data (Oba, 1969; Emiliani, 1971b).

Palaeoclimatic interpretations

ISOTOPIC ANALYSES OF FORAMINIFERA

The most well-known and widely used application of the isotope method has been in terms of reconstructing palaeoclimatic curves form deep sea cores based on isotopic analyses on single species of foraminifera. Many cores in different parts of the world oceans have been studied. Oba (1969) studied Indian Ocean curves and found a maximum amplitude of 1.6‰ in δO^{18} fluctuations for *G. menardii.*

In the Mediterranean, Emiliani (1955b) found about 5 isotopic cycles with a maximum amplitude of more than 2.8‰ for the species *G. ruber*. Vergnaud-Grazzini and Herman-Rosenberg (1969) found a maximum amplitude of close to 4‰ in O^{18} fluctuations for the same species in a Mediterranean core.

Emiliani (1955a, 1966) published data on several Caribbean and equatorial Atlantic cores. For the Caribbean cores he observed maximum variations in δO^{18} of close to 2‰ between glacial and interglacial extremes. Two of his cores and a generalised temperature curve are shown in Fig. 2. Broecker and van Donk (1970) found similar variations in the Caribbean (Fig. 3).

However, in a short Arctic core, van Donk and Mathieu (1969) found no variation at all in the species *Globigerina pachyderma* (Ehrenberg) for the last 25 000 years, which they interpreted in terms of the Arctic Ocean having been ice-covered for that length of time.

Van Donk (1970) measured a maximum difference of 0.8‰ between extremes in a complete Pleistocene core in the South Pacific. However, the core had such a slow sedimentation rate, that this 0.8‰ difference might not be representative of the true maximum difference, either due to slight mixing or due to a lack of very close sampling. Shackleton and Opdyke (1973) determined a maximum amplitude of about 1.2‰ in a long equatorial Pacific core, while van Donk (in preparation) finds a similar maximum amplitude for a complete Pleistocene core in the equatorial Atlantic. Barash and Gromova (1969) reported a maximum of 2.9‰ variation in the δO^{18} values (1969) reported a maximum of 2.9‰ variation in the δO^{18} values of the upper metre of two North Atlantic cores northwest of Ireland on the southern slope of the Rockall Bank. The analyses were performed on mixtures of all the planktonic foraminiferal species present. Compared to all other Atlantic cores the difference seems to be excessive.

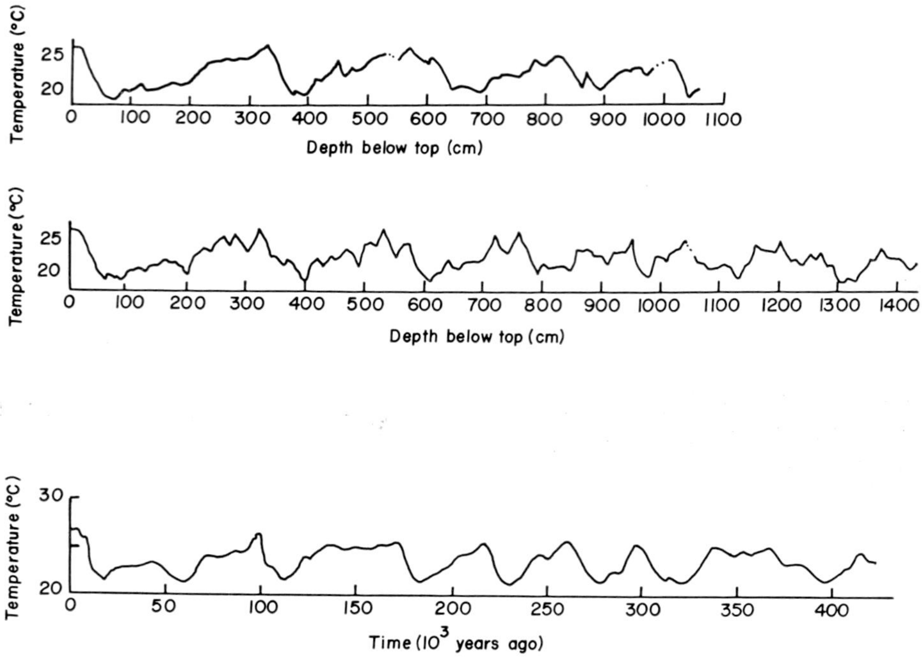

FIG. 2. Isotopic temperatures for two Caribbean cores as a function of depth and a generalised temperature curve for Caribbean surface water as a function of time (from Emiliani, 1966).

As mentioned before, there is still some controversy about how much of the variation can be ascribed to temperature fluctuations and how much to an ice-volume effect. Based on analyses of benthonic foraminifera from glacial and interglacial extremes Shackleton (1965) believes that about 1.2‰ is ice-volume effect. Van Donk (1970) summarised all the available data on benthonic foraminifera up to 1970, and found that the maximum amplitude in δO^{18} for benthonic foraminifera in the Pleistocene was about 1.2‰. Assuming no temperature change of the bottom water, and ascribing all of the variation in δO^{18} to the ice-volume effect, it would be about 1.2‰. Recently Imbrie *et al.* (1973) have evaluated a quantitative faunal analysis technique for palaeotemperature determinations, and compared it to the isotope record of the same core. They concluded that about 1.2‰ of the isotopic variations should be ascribed to the ice-volume effect. Whatever the interpretation of the observed variations in δO^{18}, there is no doubt that these variations are directly linked with climatic changes.

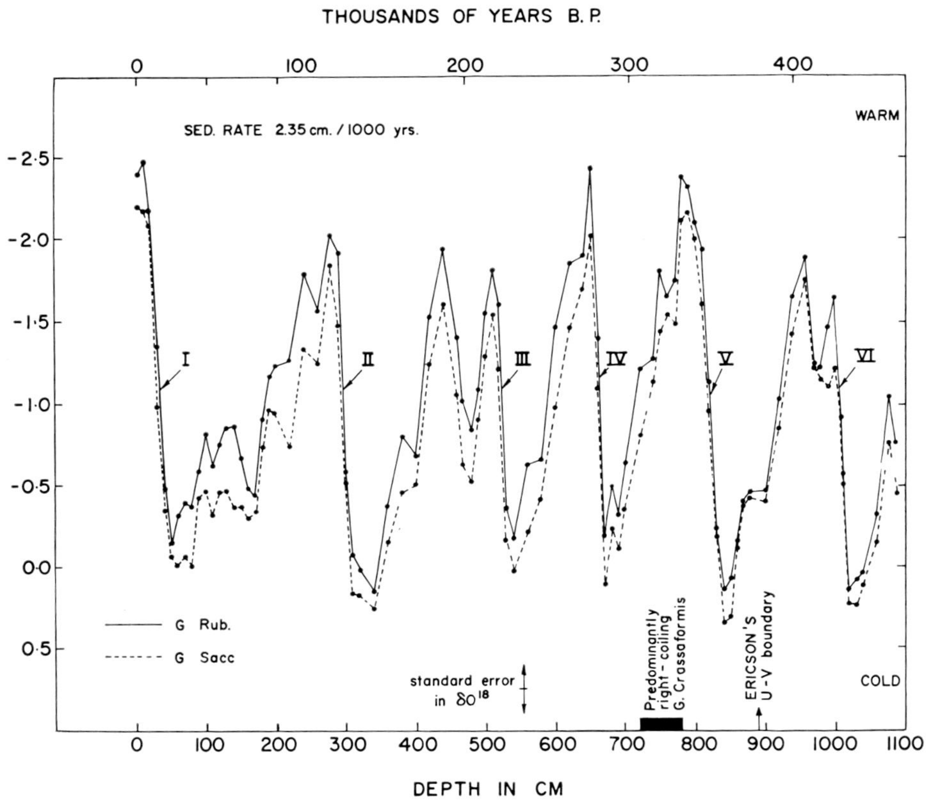

FIG. 3. Plots of δO^{18} versus depth and time for two species of planktonic foraminifera (*G. ruber* and *G. sacculifer*) from a Caribbean core (from Broecker and van Donk, 1970). Roman numerals designate terminations of Broecker and van Donk, *op. cit.*, and represent times of sudden O^{18} decrease and presumed rapid deglaciation.

ISOTOPIC ANALYSES OF PTEROPODA

The few analyses on pteropoda that have been made were performed on samples from the central Red Sea and the Persian Gulf. Deuser and Degens (1969), observed that foraminifera and pteropoda from the same horizon in the core, differed only slightly in their carbon and oxygen isotope contents. The δO^{18} values for pteropoda are in general somewhat higher, and the δC^{13} values slightly lower than those values for foraminifera. This suggests a deeper depth habitat for the pteropoda. However, there is a remarkable parallelism between the O^{18} curves for the two groups of microfossils, and it appears that pteropoda are equally useful for oxygen isotope studies as foraminifera. In the Red Sea study by Deuser and Degens (1969), they observed at least 7 drastic changes in the O^{18} and C^{13} records as recorded by pteropoda during the last 80 000 to 100 000 years. The maximum change in δO^{18} was as large as 5.5‰. They interpreted their isotopic results in terms of repeated occurrences of periods of evaporation in the Red Sea during this time, resulting in very high salinities and concurrently higher O^{18} content of the water. These periods most likely coincided with periods of lowered sea level in response to continental glaciations, which severely restricted the exchange of water between the Red Sea and the Indian Ocean. The reduction in the exchange between these two bodies of water could also have been caused by tectonic activity near the Strait of Bab el Mandeb.

Hoefs and Sarnthein (1971) investigated the pteropod species *Cavolinia longirostris* (Lesueur) from the Persian Gulf and concluded, on the basis of the oxygen isotope temperatures, that it builds its skeleton within the Gulf and does not drift in from the Indian Ocean. This is in contrast with the planktonic foraminifera, which are believed to drift into the Persian Gulf from the Indian Ocean (Sarnthein, 1971). The isotopic temperatures were close to the average *in situ* temperatures of the sea water column at each sample site in the Persian Gulf, although it is not known at which depth or season the skeleton is constructed. Furthermore, they analysed transparent and white opaque shell material from the same samples and found differences in isotopic composition as large as 0.5‰. The white opaque shells were usually depleted, although both the transparent and opaque shells consisted of pure aragonite. No correlation existed between wall structure and shell type. Therefore, the authors were not able to give an explanation for the difference.

ISOTOPIC ANALYSES OF BIOGENIC SILICA

Mopper and Garlick (1971) tested the suitability of siliceous radiolaria skeletons in deep sea sediments for oxygen isotope investigations. The oxygen isotope compositions of radiolarian skeletons from equatorial and antarctic deep-sea cores averaged around +38‰ ± 0.7‰ relative to SMOW (Standard Mean Ocean Water). Only radiolaria from tops of equatorial cores were 1.5‰ lighter than those from lower portions in the same cores. The indication is that equatorial radiolaria isotopically equilibrate with cold bottom water within several thousand years. Thus the δO^{18} values of radiolaria may only serve as a tool for distinguishing Recent sediments from Pleistocene sediments, and does not appear useful in palaeoclimatic studies. The poor reproducibility is attributed to fluctations in the amount of loosely bound water within the opaline silica. This brings out the importance of dehydrating the samples prior to oxygen extraction.

Labeyrie (1971, 1972) observed an oxygen isotope fractionation between biogenic silica and water of 38.7‰ at 10°C. The variation in this fractionation with temperature is about 0.3‰/1°C. Fresh water diatom samples from Lake Pavin, and sponge spicules from around the Kerguelen Islands were used in his study. He concluded that the biogenic silica is deposited in equilibrium with the environment and that such samples will be very useful in palaeoclimatic studies. Labeyrie also stressed the importance of the preparation procedure used. The samples should be treated for 3 days with a dilute NaOCl solution in order to eliminate organic material, and should be heated up under vacuum to 1000°C in order to eliminate all water of hydration.

In areas where both calcareous and siliceous organisms occur together, one would have a unique opportunity, by measuring the oxygen isotope composition of both types of shell material, to establish a unique solution for the temperature as well as the isotopic composition of the water in which these organisms grew together. This would eliminate the uncertainty of the ice-volume effect on the oxygen isotope temperature record based on carbonate only. Therefore, future isotopic palaeoclimatic research should be directed towards this end, and a precise relationship between δO^{18} of biogenic silica and temperature over the range from 0–30°C should be established.

ISOTOPIC ANALYSES OF COCCOLITHOPHORIDAE

Lastly, it should be mentioned that a study is underway by the author, to check the possibility of using the preserved isotopic record in the calcareous platelets of Coccolithophoridae as a tool in palaeoclimatic studies. To this end, Coccolithophoridae are cultured in the laboratory under controlled conditions, in order to establish the relationship between growth-temperature and the isotopic composition.

Furthermore, fossil samples extracted from deep-sea sediment cores, will be analysed to establish preservation of the original composition. Since these organisms are plants, they are primarily confined to the photic zone of the water column. As a result they are more indicative of surface conditions than the other micro-organisms such as foraminifera or radiolaria.

Acknowledgements

The author wishes to thank Drs A. Bé and J. Lawrence, who critically read the manuscript and made many helpful suggestions. Furthermore, the assistance of Mrs Zickl in typing the manuscript is greatly appreciated.

The writing of this paper was made possible through the support of the National Science Foundation under grant GA 35381-X.

References

Barash, M. S. and Gromova, T. S. (1969). Palaeotemperature analysis of cores by planktonic Foraminifera. *In: Prinicipal Problems of Micropaleontology and of Organic Sedimentation in the Oceans and Seas*, 153–163. Pub. Nauke, Moscow (In Russian).

Bé, A. W. H. (1959). A method for rapid sorting of Foraminifera from marine plankton samples. *J. Paleont.* **33,** 846–848.

Bé, A. W. H. (1960). Ecology of Recent planktonic foraminifera. Part 2, Bathymetric and seasonal distributions in the Sargasso Sea off Bermuda. *Micropaleont.* **6,** 373–392.

Bé, A. W. H. (1965). The influence of depth on shell growth in *Globigerinoides sacculifer* (Brady). *Micropaleont.* **11,** 81–97.

Bé, A. W. H. (1967). Foraminifera, Families: Globigerinidae and Globorotaliidae, Sheet 108. *In* (Fraser, J. H., ed.) *Fiches d'identification du zooplancton*, 10 pp. Conseil International pour l'Exploration de la Mer, Charlottenlund, Denmark.

Bé, A. W. H. and Ericson, D. B. (1963). Aspects of calcification in planktonic Foraminifera (Sarcodina). *Ann. N.Y. Acad. Sci.* **109,** 65–81.

Bé, A. W. H. and Lott, L. (1964). Shell growth and structure of planktonic foraminifera. *Science* **145,** 823–824.

Bé, A. W. H., McIntyre, A. and Breger, D. L. (1966). Shell microstructure of a planktonic foraminifer, *Globorotalia menardii* (d'Orbigny). *Eclog. geol. Helv.* **59,** 885–896.

Bé, A. W. H. and Hemleben, C. (1970). Calcification in a living planktonic foraminifer, *Globigerinoides sacculifer* (Brady). *N. Jb. Geol. Paläont. Abh.* **134,** 221–234.

Bé, A. W. H. and Tolderlund, D. S. (1971). Distribution and ecology of living planktonic foraminifera in surface waters of the Atlantic and Indian Oceans. *In* (Funnell, B. M. and Riedel, W. R., eds) *Micropaleontology of Oceans*, 105–149. Cambridge University Press, London.

Berger, W. H. (1969). Ecologic patterns of living planktonic foraminifera. *Deep-Sea Res.* **16,** 1–24.

Berger, W. H. (1970). Planktonic foraminifera: Differential production and expatriation off Baja California. *Limn. Oceanogr.* **15,** 183–204.

Bottinga, Y. and Craig, H. (1969). Oxygen isotope fractionation between CO_2 and H_2O, and the isotopic composition of marine atmospheric CO_2. *Earth Planet. Sci. Lett.* **5,** 285–295.

Bowen, R. (1961). Paleotemperature analyses of Belemnoidea and Jurassic paleoclimatology. *J. Geol.* **69,** 309–320.

Bradshaw, J. S. (1957). Ecology of living planktonic foraminifera in the North and Equatorial Pacific Ocean, 256 pp. Ph.D. Thesis, University of California, Los Angeles.

Broecker, W. S. and van Donk, J. (1970). Insolation changes, ice volumes, and the O^{11} record in deep-sea cores. *Rev. Geophys. and Space Phys.* **8,** 169–198.

Clayton, R. N. and Mayeda, T. K. (1963). The use of bromine pentafluoride in the extraction of oxygen from oxides and silicates for isotopic analysis. *Geochim. Cosmochim. Acta* **27,** 43–52.

Compston, W. and Epstein, S. (1958). A method for the preparation of CO_2 from water vapor for oxygen isotope analysis. *Trans. Am. Geophys. Union* (Abstract) **39,** 511.

Cortecci, G. and Longinelli, A. (1971). O^{18}/O^{16} ratios in sulfate from living marine organisms. *Earth Planet. Sci. Lett.* **11,** 273–276.

Craig, H. (1965). The measurement of oxygen isotope paleotemperatures. *In: Proc. Spoleto Conf. on Stable Isotopes in Oceanographic Studies and Paleotemperatures,* **3,** 161–182. Spoleto, Italy.

Dansgaard, W. and Tauber, H. (1969). Glacier O^{18} content and Pleistocene ocean temperatures. *Science* **166,** 499–502.

Deuser, W. G. and Degens, E. T. (1969). O^{18}/O^{16} and C^{13}/C^{12} ratios of fossils from the Hot-Brine Deep area of the Central Red Sea. *In* (Degens, E. T. and Ross, D. A., eds) *Hot Brines and Recent Heavy Metal Deposits in the Red Sea*, 336–347. Springer-Verlag, New York.

Devereux, I. (1967a). Oxygen isotope paleotemperature measurements on New Zealand Tertiary fossils. *N.Z. J. Science* **10,** 988–1011.

Devereux, I. (1967b). Temperature measurements from oxygen isotope ratios of fish otoliths. *Science* **155,** 1684–1685.

Dorman, F. H. (1966). Australian Tertiary paleotemperatures. *J. Geol.* **74,** 49–61.

Dorman, F. H. and Gill, E. D. (1959). Oxygen isotope paleotemperature measurements on Australian fossils. *Proc. Roy. Soc. Victoria* **71,** 73–98.

Duplessy, J. G., Lalou, C. and Vinot, A. C. (1970). Differential isotopic fractionation in benthic foraminifera and palaeotemperatures reassessed. *Science* **168,** 250–251. Emiliani, C. (1954a). Temperatures of Pacific bottom waters and polar superficial waters during the Tertiary. *Science* **119,** 853–855.

Emiliani, G. (1954b). Depth habitats of some species of pelagic foraminifera as indicated by oxygen isotope ratios. *Am. J. Sci.* **252,** 149–158.

Emiliani, C. (1955a). Pleistocene temperatures. *J. Geol.* **63,** 538–578.

Emiliani, C. (1955b). Pleistocene temperature variations in the Mediterranean. *Quaternaria* **2,** 87–98.

Emiliani, C. (1956). Oligocene and Miocene temperatures of the equatorial and subtropical Atlantic Ocean. *J. Geol.* **64,** 281–288.

Emiliani, C. (1958). Paleotemperature analysis of core 280 and Pleistocene correlations. *J. Geol.* **66,** 364–275.

Emiliani, C. (1964). Paleotemperature analysis of the Caribbean cores A 254-BR-C and CP-28. *Geol. Soc. Am. Bull.* **75,** 129–144.

Emiliani, C. (1966). Paleotemperature analysis of Caribbean cores P6304–8 and P6304–9 and a generalized temperature curve for the past 425,000 years. *J. Geol.* **74,** 109–126.

Emiliani, C. (1971a). The amplitude of Pleistocene climatic cycles at low latitudes and the isotopic composition of glacial ice. *In* (Turekian, K., ed.) *The Late Cenozoic Glacial Ages*, 183–197. Yale University Press, New Haven.

Emiliani, C. (1971b). Depth habitats of growth stages of pelagic Foraminifera. *Science* **173,** 1122–1124.

Epstein, S. (1959). The variations of the O^{18}/O^{16} ratio in nature and some geloogic applications. *In* (Abelson, P. H., ed.) *Researches in Geochemistry*, 217–240. John Wiley and Sons, Inc., New York.

Epstein, S. and Mayeda, T. K. (1953). Variation of O^{18} content of waters from natural sources. *Geochim. Cosmochim. Acta* **4,** 213–224.

Epstein, S., Buchsbaum, R., Lowenstam, H. A. and Urey, H. C. (1953). Revised carbonate–water isotopic temperature scale. *Geol. Soc. Am. Bull.* **64,** 1315–1325.

Glaçon, G., Vergnaud-Grazzini, C. and Sigal, M. J. (1971). Premiers résultats d'une série d'observations saissonnières des Foraminifères du plancton Méditerranéen. *In* (Farinacci, A., ed.) *Proc. II Plankt. Conf.*, 555–574. Tecnoscienza, Roma.

Hecht, A. D. and Savin, S. M. (1970). Oxygen-18 studies of Recent planktonic Foraminifera: Comparisons of phenotypes and of test parts. *Science* **170,** 69–71.

Hecht, A. D. and Savin, S. M. (1972). Phenotypic variation and oxygen isotope ratios in Recent planktonic foraminifera. *J. Foram. Res.* **2,** 55–67.

Hoefs, J. and Sarnthein, M. (1971). O^{18}/O^{16} ratios and related temperatures of Recent pteropod shells from the Persian Gulf. *Mar. Geol.* **10,** M.11–M.16.

Horibe, Y., Shigehara, K. and Takakuwa, Y. (1973). Isotope separation factor of carbon dioxide–water system and isotopic composition of atmospheric oxygen. *J. Geophys. Res.* **78,** 2625–2629.

Imbrie, J., van Donk, J. and Kipp, N. G. (1973). Paleoclimatic investigation of a Late Pleistocene Caribbean deep-sea core: Comparison of isotopic and faunal methods. *Quat. Res.* **3,** 10–38.

Jones, J. I. (1967). Significance of distribution of planktonic foraminifera in the equatorial Atlantic undercurrent. *Micropaleont.* **13,** 489–501.

Keith, M. L. and Parker, R. H. (1965). Local variation of C^{13} and O^{18} content of mollusk shells and the relatively minor temperature effect in marginal marine environments. *Mar. Geol.* **3,** 115–129.

Keith, M. L. and Weber, J. N. (1965). Systematic relationships between carbon and oxygen isotopes in carbonates deposited by modern corals and algae. *Science* **150,** 498–501.

Labeyrie, M. L. (1971). Problèmes poses par la mesure de la composition isotopique des silicates: application a l'étude des diatomees. 62 pp. Ph.D. Thesis, Faculty of Science, University of Paris.

Labeyrie, M. L. (1972). Composition isotopique de l'oxygène de la silice biogénique. *C. R. Acad. Sc. Paris* **274,** 1605–1608.

Lidz, B., Kehm, A. and Miller, H. (1968). Depth habitats of pelagic foraminifera during the Pleistocene. *Nature* **217,** 245–247.

Lloyd, R. M. (1964). Variations in the oxygen and carbon isotope ratios of Florida Bay mollusks and their environmental significance. *J. Geol.* **72,** 84–111.

Longinelli, A. (1965). Oxygen isotopic composition of orthophosphate from shells of living marine organisms. *Nature* **207,** 716–719.

Longinelli, A. (1966). Ratios of oxygen-18:oxygen-16 in phosphate and carbonate from living and fossil marine organisms. *Nature* **211,** 923–927.

Longinelli, A. and Tongiorgi, E. (1964). Oxygen isotopic composition of some right- and left-coiled foraminifera. *Science* **144,** 1004–1005.

Longinelli, A. and Nuti, S. (1968). Oxygen–isotope ratios in phosphate from fossil marine organisms. *Science* **160,** 879–882.

Longinelli, A. and Nuti, S. (1972). Revised and extended phosphate–water isotopic temperature scale. *Trans. Am. Geophys. Union* (Abstract) **53,** 555.

Lowenstam, H. A. and Epstein, S. (1954). Paleotemperatures of the post-Aptian Cretaceous as determined by the oxygen isotope method. *J. Geol.* **62,** 207–248.

McCrea, J. M. (1950). On the isotopic chemistry of carbonates and a paleotemperature scale. *J. Chem. Phys.* **18,** 849–857.

McGowan, J. A. and Fraundorf, V. J. (1964). A modified heavy fraction zooplankton sorter. *Limnol. Oceanogr.* **9,** 152–154.

Mook, W. G. (1971). Paleotemperatures and chlorinates from stable carbon and oxygen isotopes in shell carbonate. *Palaeont. Palaeoecol. Palaeochim.* **9,** 245–263.
Mopper, K. and Garlick, G. D. (1971). Oxygen isotope fractionation between biogenic silica and ocean water. *Geochim. Cosmochim. Acta* **35,** 1185–1187.
Naidin, D. P., Teis, R. V. and Zadorozhny, I. K. (1966). Isotopic paleotemperatures of the Upper Cretaceous in the Russian Platform and other parts of the U.S.S.R. (In Russian). *Geokhimiya* **11,** 1286–1299 (translation published in *Geochemistry International* **3,** (6) 1038–1051).
Oba, T. (1969). Biostratigraphy and isotopic paleotemperature of some deep-sea cores from the Indian Ocean. *Tohoku Univ. Sci. Rep. 2nd ser.* (Geol.) **41,** 129–201.
Olausson, E. (1965). Evidence of climatic changes in North Atlantic deep-sea cores, with remarks on isotopic paleotemperature analysis. *Prog. Oceanography* **3,** 221–252.
O'Neil, J. M. and Epstein, S. (1966). A method for oxygen isotope analysis of milligram quantities water and some of its applications. *J. Geophys. Res.* **71,** 4955–5961.
Saito, T. and van Donk, J. (1971). Oxygen isotopic measurements of Late Cretaceous sediments. *Geol. Soc. Am. Ann. Meeting* (Abstract) **3,** 692–693.
Saito, T. and van Donk, J. (1973). Oxygen isotopic measurements of Late Cretaceous and Early Tertiary Foraminifera. (In preparation.) To be published in *Micropaleont.*
Sarnthein, M. (1971). Oberflächen sediment im Persischen Golf und Golf von Oman, II—Quantitative Komponentenanalyse der Großfraktion. *In* (Seibold, E. and Closs, H., eds) "*Meteor*" *Forschungsergebnisse, Reihe C*(5), Pub. Borntraeger, Stuttgart.
Shackleton, N. J. (1965). The high-precision isotopic analysis of oxygen and carbon in carbon dioxide. *J. Sci. Instrum.* **42,** 689–692.
Shackleton, N. J. (1967). Oxygen isotope analyses and Pleistocene temperatures re-assessed. *Nature* **215,** 15–17.
Shackleton, N. J. and Opdyke, N. D. (1973). Oxygen isotope and paleomagnetic stratigraphy of Equatorial Pacific core V28-238: Oxygen isotope temperatures and ice volumes on a 10^5 year and 10^6 year scale. *Quat. Res.* **3,** 39–55.
Shackleton, N. J., Wiseman, J. D. H. and Buckley, H. A. (1973). Non-equilibrium isotopic fractionation between seawater and planktonic Foraminiferal tests. *Nature* **242,** 177–179.
Smith, P. B. and Emiliani, C. (1968). Oxygen isotope analysis of recent tropical Pacific benthonic Foraminifera. *Science* **160,** 1335–1336.
Spaeth, C., Hoefs, J. and Vetter, U. (1971). Some aspects of isotopic composition of Belemnites and related paleotemperatures. *Geol. Soc. Am. Bull.* **82,** 3139–3150.
Stuiver, M. (1968). O^{18} content of atmospheric precipitation during last 11,000 years in the Great Lakes region. *Science* **162,** 994–997.
Urey, H. C. (1947). The thermodynamic properties of isotopic substances. *J. Chem. Soc.* **1947,** 562–581.
Urey, H. C. (1948). Oxygen isotopes in nature and in the laboratory. *Science* **108,** 489–496.
Urey, H. C., Lowenstam, H. A., Epstein, S. and McKinney, C. R. (1951). Measurements of paleotemperatures and temperatures of the Upper Cretaceous of England, Denmark, and the south-eastern United States. *Geol. Soc. Am. Bull.* **62,** 399–416.

van Donk, J. (1970). The oxygen isotope record in deep-sea sediments, 228 pp. Ph.D. Thesis, Columbia University, New York.

van Donk, J. (1973). A complete Pleistocene oxygen isotope record for the Equatorial Atlantic. (In preparation.) To be published in *Geol. Soc. Am. Bull.*

van Donk, J. and Mathieu, G. (1969). Oxygen isotope compositions of Foraminifera and water samples from the Arctic Ocean. *J. Geophys. Res.* **74,** 3396–3407.

Vergnaud-Grazzini, C. and Herman-Rosenberg, Y. (1969). Étude paléoclimatique d'une carotte de Méditerranée orientale. *Rev. de Géogr. Phys. et de Géol. Dyn.* (2), **11,** 279–292.

Vilks, G. (1970). New method for separation of skeletal material from a plankton sample. *Maritime Sediments* **6,** 72–73.

Vinot, A. C. (1971). Le fractionnement isotopique chez les Foraminifères benthiques; consequences pour l'interpretation paleoclimatique, 66 pp. Ph.D. Thesis, University of Paris.

Vinot-Bertouille, A. C. and Duplessy, J. C. (1973). Individual isotopic fractionation of carbon and oxygen in benthic Foraminifera. *Earth Planet. Sci. Lett.* **18,** 247–252.

Weber, J. N. and Raup, D. M. (1966). Fractionation of the stable isotopes of carbon and oxygen in marine calcareous organisms—the Echinoidea. *Geochim. Cosmochim. Acta* **30,** 681–736.

Weber, J. N. and Raup, D. M. (1968). Comparison of C^{13}/C^{12} and O^{18}/O^{16} in the skeletal calcite of recent and fossil Echinoids. *J. Paleont.* **42,** 37–50.

Weber, J. N. and Woodhead, P. M. J. (1972). Temperature dependence of oxygen-18 concentrations in reef coral carbonates. *J. Geophys. Res.* **77,** 463–473.

Williams, L. K. (1971). Selected planktonic Foraminifera as biological indicators of hydrological conditions in the eastern Gulf of Mexico, 153 pp. M.S. Thesis, Florida State University, Tallahassee.

18: Sedimentological Clues to Palaeo-oceanography

A. T. S. RAMSAY

Department of Geology,
University College of Swansea,
Singleton Park,
Swansea SA2 8PP

This chapter presents an attempt to elucidate elements in the palaeo-oceanography of the ocean since the Late Eocene. The following inter-related parameters have been examined: (*a*) the history of the calcite compensation depth, (*b*) the history of the 10% sand sized foraminiferal boundary, (*c*) the occurrence and geographical distribution of hiatuses.

The history of the calcite compensation depth (CCD) in the Atlantic Ocean, Caribbean Sea, Indian and Pacific Oceans is reconstructed for the Pliocene to Quaternary interval (1–3 m.y. B.P.) and for intervals in the Late Miocene, Middle Miocene, Late Oligocene and Late Eocene. The history of this boundary is compared both within and between ocean basins and is also compared with the history of an arbitrarily defined level, the 10% sand sized foraminiferal boundary which in this report delineates foraminiferal ooze from nannofossil ooze. The depths of these boundaries are interpreted within a depth—latitude framework (Pliocene–Quaternary interval), and within a palaeodepth–latitude/palaeolatitude frameworks for the Late Eocene to Late Miocene. The histories of these boundaries in the North Atlantic and Caribbean Sea are compared in a depth–longitude framework (Pliocene–Quaternary) and in palaeodepth–longitude frameworks for the Late Eocene to Late Miocene.

Changes in the depth of the CCD between glacial and interglacial cycles in the equatorial Pacific, on the East Pacific Rise and in the North Atlantic Basin are examined and used as the basis of a model for glacial and interglacial carbonate accumulation.

The history of the CCD and 10% foraminiferal boundary since the Late Eocene is considered in terms of possible changes in the pattern of deep-ocean circulation associated with climatic changes in Antarctica, the separation of Antarctica and Australia, the opening of the Drake Passage and the tectonic isolation of the Caribbean basin. Parallel trends in the history of the CCD and 10% foraminiferal boundary between ocean basins are attributed to variations in the global patterns of shelf-ocean basin fractionation of carbonate and to the latitudinal fractionation of carbonate since the Middle Miocene. Variations in the histories of these boundaries within and between ocean basins are speculatively attributed to differences in the patterns of abyssal circulation and mixing within the ocean.

The origin of regional hiatuses in deep-sea sediments is considered and their stratigraphical occurrence in the South Atlantic, northwest Atlantic, Indian Ocean and southwestern Pacific and eastern North Pacific is analysed. The geographical distribution of regional hiatuses and changes in their stratigraphical abundance are considered within the framework of a model of the modern abyssal circulation pattern, in terms of the history of the CCD, and in the light of climatic changes in Antarctica.

Outline

Introduction

Most contributions to this book are concerned with biological and palaeobiological investigations of modern and fossil pelagic organisms, with the view of elucidating information concerning their taxonomy, geographical and biostratigraphical distribution. Exceptions are the chapter by Schneidermann on factors which influence the sedimentation and preservation of coccoliths, and the chapter by Van Donk on the isotopic composition of biogenic carbonate and silica.

This account is concerned with the oceanic palaeoenvironments since the late Eocene, based primarily on sedimentological data obtained from the Deep Sea Drilling Project sites shown in Fig. 1. This kind of study relies heavily upon palaeontological dates assigned to sediments from different ocean basins. The study is concerned with changes in three parameters. These are the Calcite Compensation Depth, hereafter referred to as the CCD, a boundary based on the 10% value of sand sized foraminiferal tests in pelagic sediments, and the abundance and geographical distribution of hiatuses.

On the basis of these data conclusions are drawn concerning past patterns of oceanic circulation and mixing. The results of this study do not present a comprehensive review of palaeooceanography since the late Eocene but are a first attempt at viewing the Ocean, insofar as the available data permit, as a complete system for various intervals since this time.

The nature and distribution of Recent deep-sea sediments

A generalised map of the distribution of Recent sediments on the ocean floor is shown in Fig. 2. The sediment distribution pattern which has formed on those regions of the deep-sea floor which are isolated from continental or volcanic sedimentary sources is a product of the rate of plankton production within the photic zone of the ocean, of the inhibiting influence of low temperatures on the abundance of the calcareous plankton in high latitudes, and of the

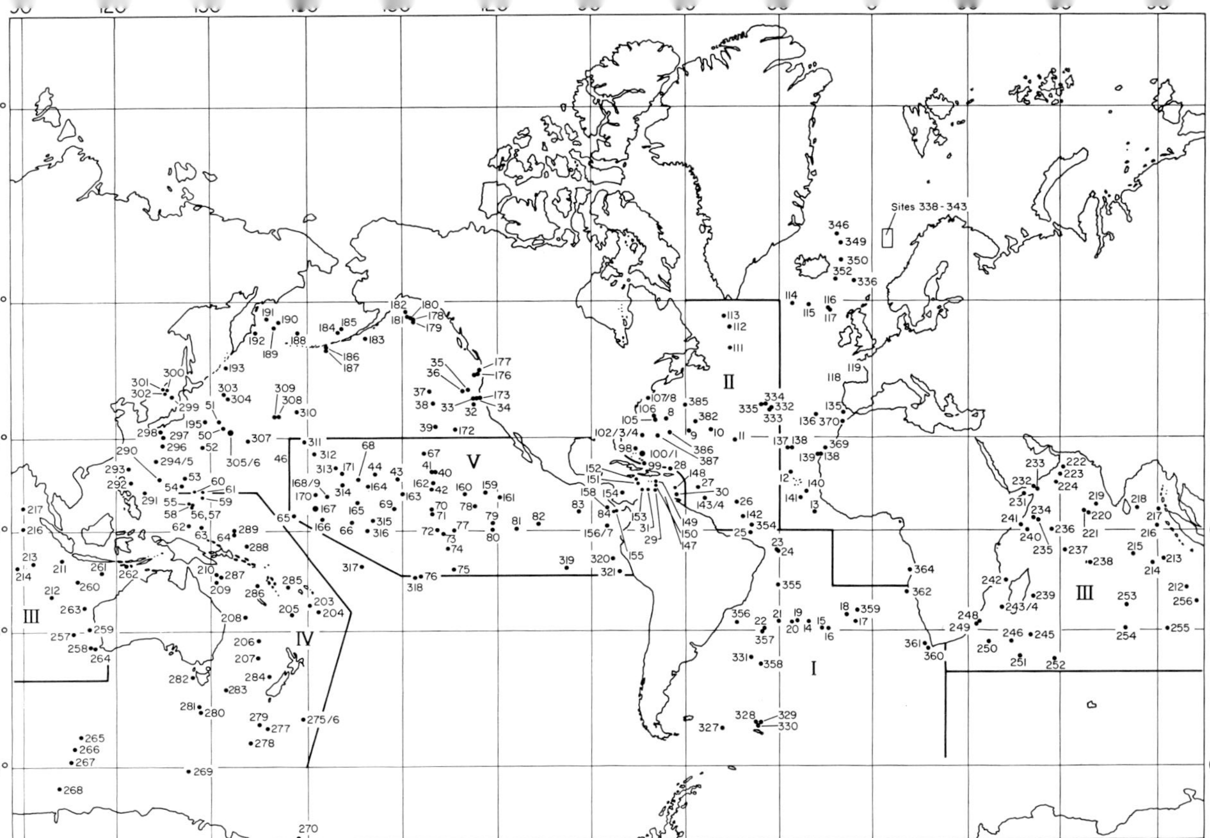

FIG. 1. The distribution of Deep Sea Drilling Project Sites used in this chapter. Regions designated I, II, III, IV and V indicate the areas for which hiatus abundances have been determined.

dissolution of elements in the calcareous plankton within the progressively undersaturated deep water of the ocean. Thus the CCD forms the most obvious and significant facies boundary in the deep-sea deposits shown on the map. Areas of the ocean floor which are elevated above the calcite compensation depth are characterised and defined by calcareous ooze. Deep basins are characterised by the accumulation of residual deposits of two kinds. A red clay facies, an essentially inorganic sediment whose constituents are derived from marine volcanic and continental sources (Griffin *et al.*, 1968; Heath, 1969b; Berger, 1974), which accumulates beneath less fertile water masses. Biogenic siliceous ooze which accumulates beneath the polar front regions, the equatorial belts of the Pacific and Indian Oceans, and the ocean margins; all are areas of high fertility.

Factors which influence the distribution of carbonate in the ocean

Most of the calcium carbonate supplied to the ocean via continental run off is precipitated organically within the saturated to supersaturated photic zone. Here organisms utilise various forms of this compound, aragonite, low and high magnesium calcite, to construct hard parts. Within the tropics the activities of various carbonate secreting organisms has led to the accumulation of rigid framed reefs and biogenic sediments on continental shelves, atols and around oceanic islands. Inorganically precipitated calcium carbonate is rare and is possibly restricted to processes such as ooid growth, the formation of intertidal high magnesian calcite and aragonite encrustations, and to intertidal and subtidal high magnesian calcite and aragonite cements.

THE SEDIMENTATION AND DISSOLUTION OF CALCIUM CARBONATE IN THE DEEP SEA

In the pelagic realm of the ocean virtually all precipitation occurs within the shell building pteropods, which utilises aragonite, and the planktonic foraminifera and coccolithophores which utilise low magnesium calcite. The accumulation and preservation of these shells as pelagic ooze occurs only on more elevated areas of the ocean floor. In deep basins they are dissolved in seawater which becomes progressively undersaturated with respect to calcium carbonate as pressure increases and temperature decreases. The aragonitic shells of Pteropods are dissolved at relatively shallow oceanic depths, less than 300 m in the Pacific, and 2000 to 2500 m in the Atlantic (Li *et al.*, 1969; Berger, 1970a; Chen, 1971). The dissolution of calcite tests within the calcite under-saturated zone is selective and, with increasing depths, results in the progressive removal of the less solution-resistant species of the calcareous nanoplankton (McIntyre and McIntyre, 1971; Schneidermann, this volume) and

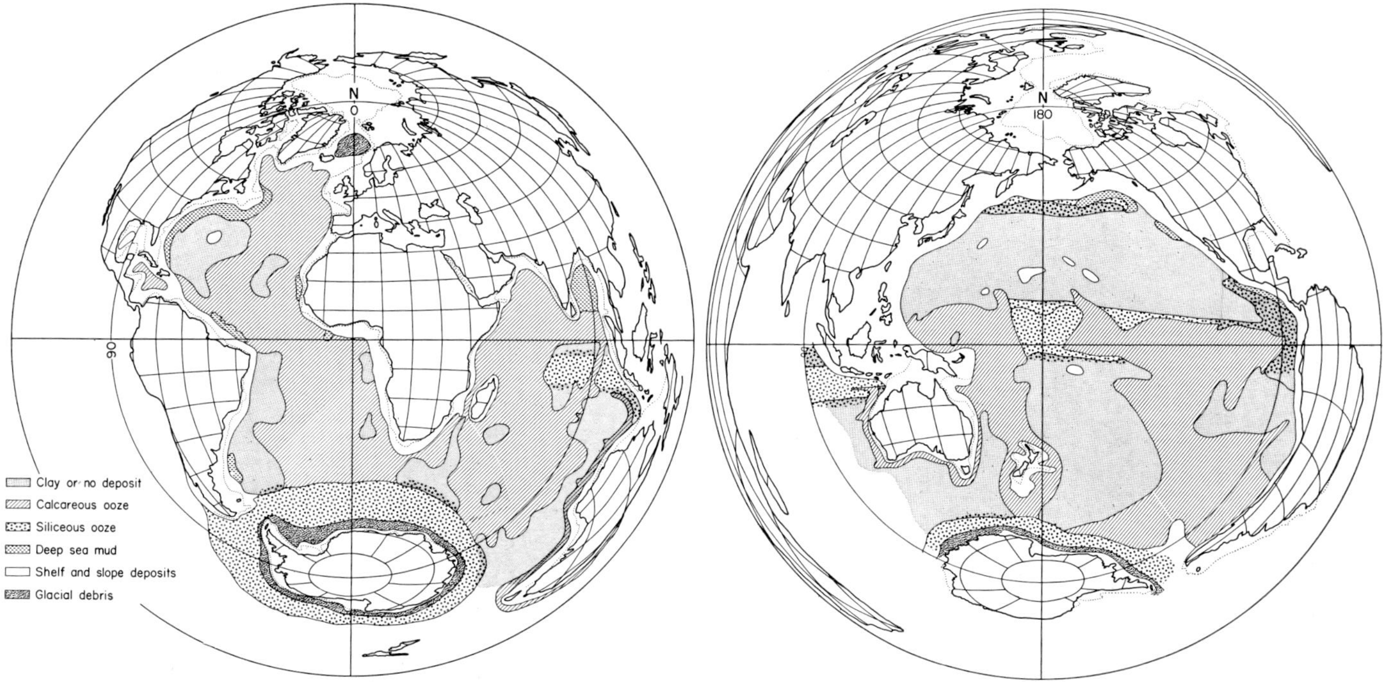

FIG. 2. Sediment distribution on the deep ocean floor (after Berger, 1974).

planktonic foraminifera (Berger, 1968, 1970a, 1971; Bé, this volume). Most solution probably occurs while the particles rest on the bottom (Bramlette, 1961; Adelseck and Berger, 1974). Ultimately coincidence between the sedimentation and dissolution rates results in the total destruction of carbonate at the CCD.

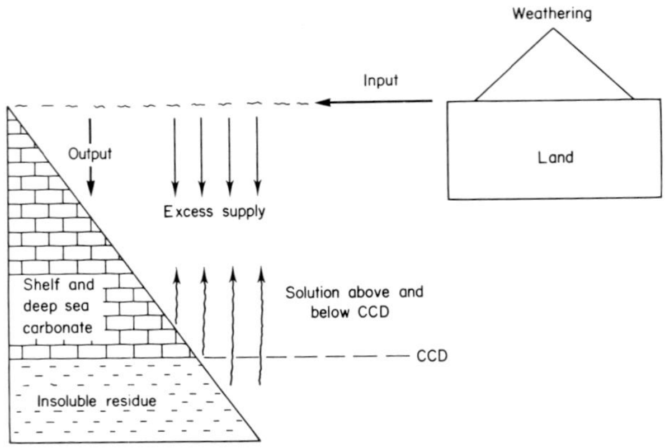

FIG. 3. The relationship between the supply of calcium carbonate to the ocean, and its sedimentation through oceanic precipitation and solution.

Sea water becomes undersaturated because organisms utilise and supply calcium carbonate to the bottom at a rate which exceeds its rate of input into the ocean via continental run off and thus deplete the ocean of calcium carbonate. The dissolution of the excess supply in the undersaturated deep water (see Fig. 3) probably maintains a dynamic steady state between rates of input and sedimentation (Pytkowicz, 1968; Li *et al.*, 1969; Berger, 1970b). The undersaturated state of sea water is caused by the production of excess CO_2 in ocean deep water as a result of the oxidation of organic matter which accounts for approximately 80%, and the dissolution of calcite which accounts for approximately 20% (Li *et al.*, 1969). The variation in the depths of cross-over from saturation to undersaturation within an ocean basin and between ocean basins (see Fig. 4) is probably related to the influence of the deep oceanic circulation pattern on the residence time of bottom waters. Thus values for dissolved CO_2 are almost twice as high for the relatively poorly mixed Pacific as the values obtained for the comparatively well flushed North Atlantic.

Two levels of calcite dissolution are recognised within the undersaturated zone, the Lysocline and the calcite compensation depth. The Lysocline which lies above the CCD marks a maximum change (i.e. a 50% loss) in the composition of foraminiferal assemblages through dissolution (Berger, 1970a, 1971). This level is complex, it is depressed beneath the less fertile regions of the ocean, and elevated in the more fertile regions, e.g. the equatorial Pacific and near continents. In less fertile areas of the Atlantic and apparently

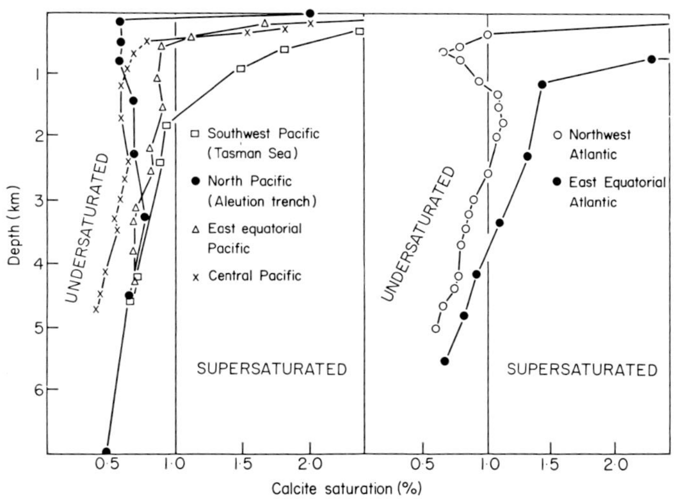

FIG. 4. Calcite saturation curves for the Pacific and Atlantic Oceans (after Berner, 1974).

the Pacific the Lysocline may correspond with the top of the Antarctic Bottom Water. Calcite tests are aggressively dissolved wherever this current is present either because it achieves "critical undersaturation" (Morse and Berner, 1972) or because its high rates of flow transport and keep tests in suspension. In fertile regions no such correlation is observed, and the elevation of the Lysocline is possibly associated with an increased production of CO_2 in interstitial waters as a result of the high rate of supply of organic matter.

The CCD is the level where, at any one region of the ocean, the sedimentation and dissolution rates of calcite are equal. The CCD results from an approximately linear increase in the rate of calcite dissolution with depth below the Lysocline (Heath and Culberson, 1970). The characteristics of the CCD and some of the factors which are considered to control its depth in the ocean are shown in Fig. 5. The CCD is depressed beneath the fertile equatorial belt, where a high rate of input of calcite tests (Arrhenius, 1952) and possibly low interstitial CO_2 values, caused by the efficient removal of organic detritus by feeding organisms (Berger, 1974) ensure a suitable environment for high rates of carbonate accumulation. In contrast the calcite compensation depth shallows towards the continental margins despite high rates of plankton production. This phenomenon may be caused in part by the dilution of carbonate with terrigenous sediments and in part by the enhanced solution of calcite in contact with pore waters enriched in CO_2 through the intense biological activity which occurs in well aerated sediments with a high content of organic matter (Berger and Soutar, 1970).

The CCD is elevated beneath the less fertile central water masses and at high latitudes where the abundance of calcareous plankton is restricted by low temperatures.

The effect of the deep ocean circulation pattern on the residence time of bottom waters in different regions of the ocean may also contribute to variations in the level of the CCD both within an ocean and between ocean basins. The CCD is shallower in the Caribbean Sea than in the North Atlantic basin, and deeper in the well-flushed CO_2 poor North Atlantic than in the poorly mixed CO_2 rich North Pacific.

THE INFLUENCE OF TOPOGRAPHY ON DEEP-SEA SEDIMENTATION

Ocean floor topography is to a large extent generated by the spreading and subsidence of new ocean crust (Hess, 1962; Menard, 1969) about mid-ocean ridges which are elevated as a result of thermal expansion and phase changes in the mantle. The topography produced by sea floor spreading has a significant influence on the distribution of pelagic facies. It is however, unnecessary to consider the effects of the vertical component of motion in the spreading process on Quaternary to Recent sediments which are deposited on crust older than 10 million years. The depth changes involved are insignificant and decrease to less than 20 m per million years on crust older than 50 million years (Berger, 1973; see also Fig. 9). The horizontal component in sea-floor spreading is also insignificant over such a short interval of time. Both components of motion are, however, important in the interpretation of sequences of Tertiary and Mesozoic pelagic sediments.

THE EFFECT OF HORIZONTAL MOVEMENTS OF OCEAN CRUST

Horizontal movements of the ocean floor coupled with the younging of oceanic crust towards mid-ocean ridge crests results in the formation of chronostratigraphic units which thin towards ridge crests (Ewing and Ewing, 1967) and which overstep onto the ridges in the manner of transgressive sequences (Hess, 1962; Berger and Winterer, 1973).

Horizontal movements also result in significant latitudinal displacements

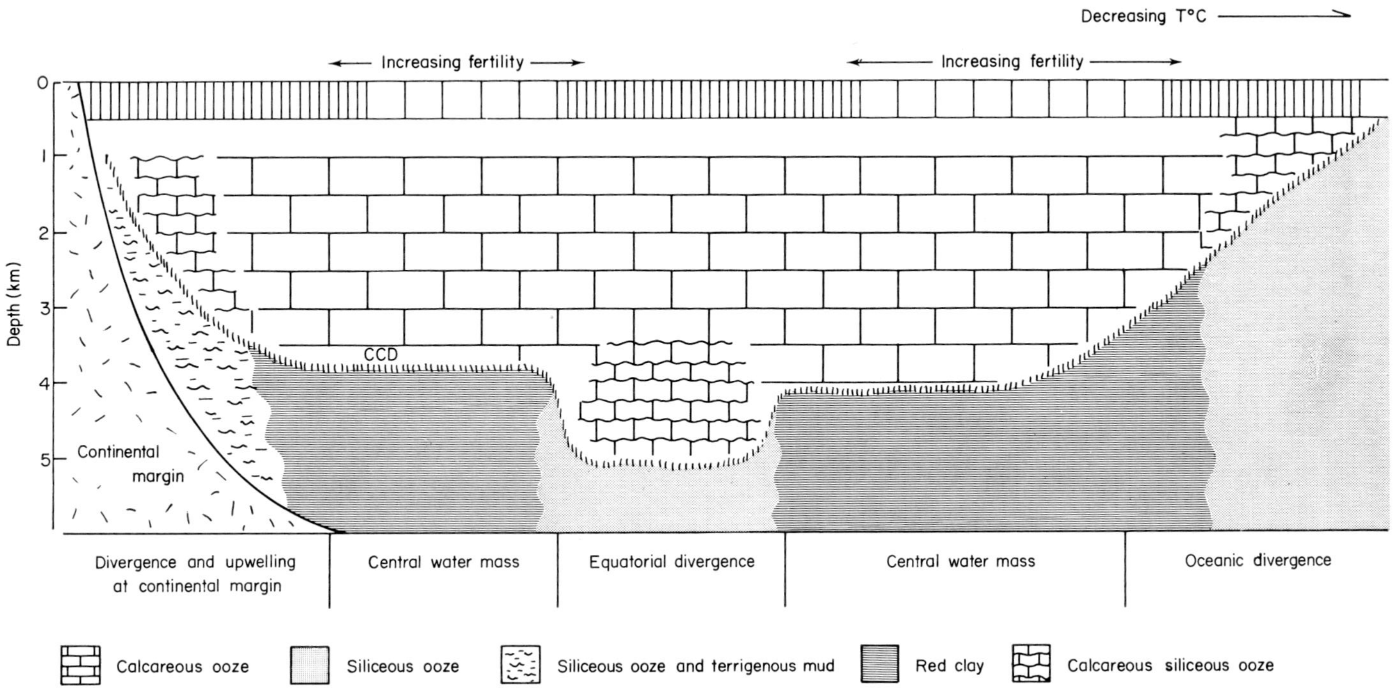

FIG. 5. Distribution of deep-sea sedimentary facies and characteristics of the CCD relative to major patterns of oceanic circulation and near surface fertility (schematic diagrams based on the Pacific Ocean).

of large areas of the ocean crust and can result in the development of sedimentary sequences which contain a record of their passage beneath fertile and non-fertile regions of the ocean. Examples of this effect are shown in Figs 6 and 7. Figure 6 records the progressive displacement of a zone of thick

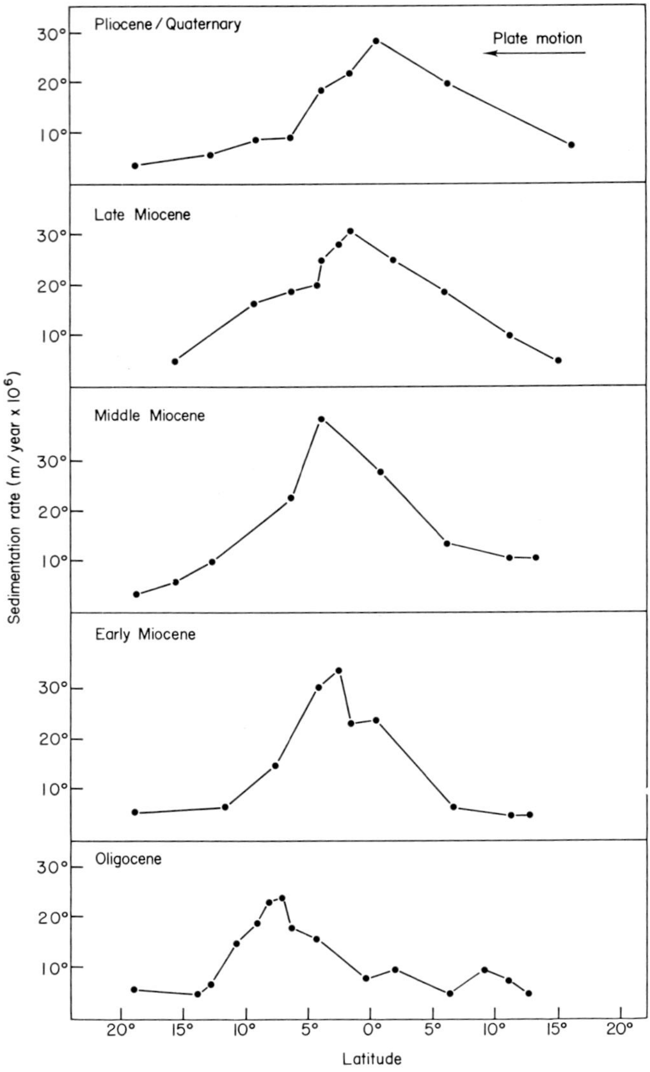

FIG. 6. A comparison between sedimentation rates (not corrected for compaction) and for biogenic calcareous ooze, latitude and time in the central Pacific. (See Appendix 2 for sources of data.)

sediments which accumulates at present beneath the equatorial Pacific (Ewing *et al.*, 1968). This displacement is associated with the northward motion of the Pacific Plate (Francheteau *et al.*, 1970; Sclater and Jarrard, 1973; Claque and Jarrard, 1973). Winterer (1973) has determined a northward component of plate motion of 0.23°/m.y. on the basis of the displaced sediment, thickness patterns.

Figure 7 shows evidence of the northward displacement of sediments which accumulated on the Indian Ocean crust during its passage beneath the Antarctic Circumpolar Current. This current is a part of the highly fertile southern zone of the modern ocean (Lisitzin, 1971a; 1971b). Details of the development of the Indian Ocean and the northward migration of Indian Ocean Crust are given in papers by McKenzie and Sclater (1971) and Sclater and Fisher (1974).

THE EFFECT OF VERTICAL MOVEMENTS IN SEDIMENTARY FACIES

The sequence of pelagic facies which develop on the flanks of spreading ocean ridges is determined in part by the vertical component of motion of the oceanic crust and in part by the vertical stability of the CCD and solution rates above this level. Given a steady state CCD and fixed solution rates, the sediment facies pattern would be simple and sites at increasing distances from the ridge crests would contain a sequence of sediments which have undergone increasing degrees of dissolution until red clays or siliceous oozes are developed (see Fig. 8a).

The fact that many of the DSDP sites contain non-calcareous clay (e.g. sites 10, 15, 19, 146/149) or non-calcareous siliceous ooze (e.g. sites 42 and 162) interbedded with calcareous ooze implies the occurrence of temporal vertical fluctuations in the CCD. This situation is shown in Fig. 8b and, for comparison, a model of sedimentation on a non-spreading ridge with a steady state CCD is shown in Fig. 8c.

Methods used to investigate sedimentary facies, palaeodepths and hiatuses

THE RECOGNITION OF SEDIMENTARY FACIES

In this study vertical changes in the distribution of calcareous and non-calcareous sediments for selected time intervals since the Late Eocene have been used to reconstruct the position of the CCD and of major sedimentary facies for these intervals of time.

The data used were, whenever possible, extracted from the comprehensive accounts contained in the Initial Reports of the Deep Sea Drilling Project of which Volumes 1 to 21 and 26 and 27 were available to the author. For

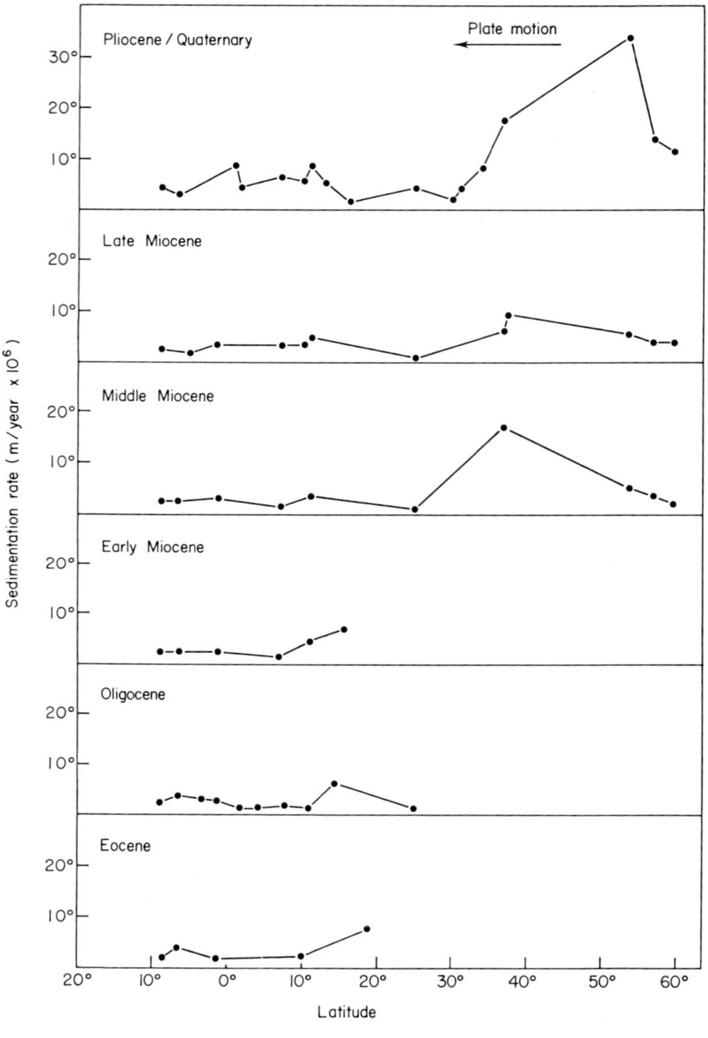

FIG. 7. A comparison between sedimentation rates (not corrected for compaction) for biogenic and siliceous ooze, latitude and time in the Indian Ocean. (See Appendix 2 for sources of data.)

all other legs the preliminary reports published in Geotimes were used and the data from these reports, i.e. characteristics of the sediment, age of carbonate/non-carbonate boundaries and basement ages are tentative.

Where quantitative values for $CaCO_3$ in the sediment were available the boundary between calcareous and non-calcareous sediments was set at 10%.

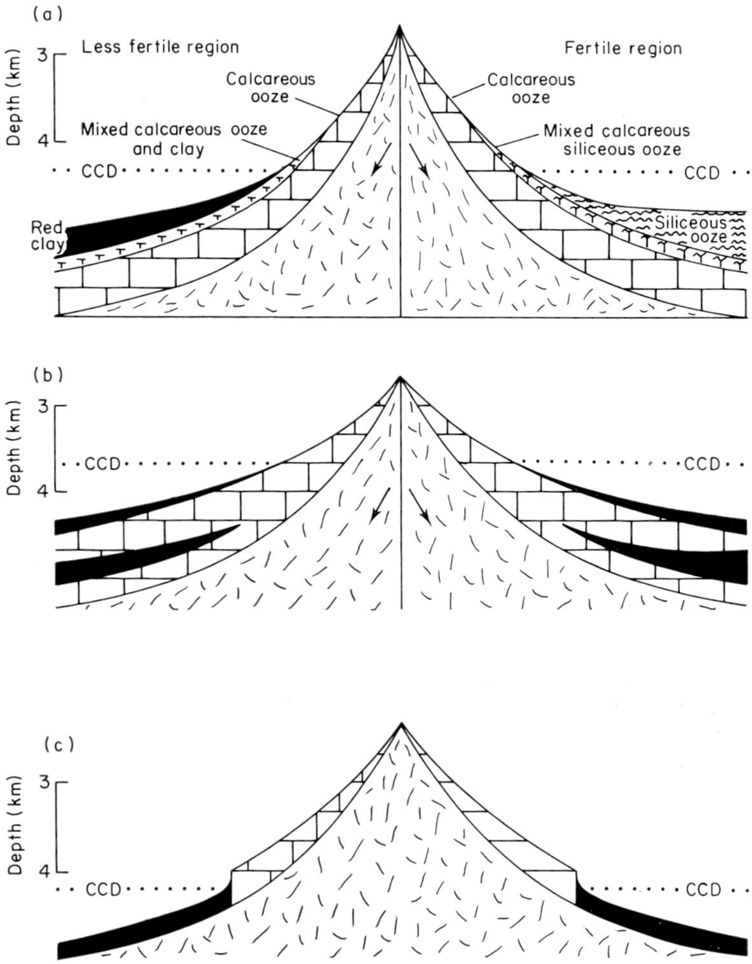

FIG. 8. The sedimentation of calcareous ooze about a spreading ridge with subsiding flanks for (a) steady state and (b) fluctuating CCD levels. Fig. 8(c) shows the sedimentation of calcareous ooze on a non-spreading ridge with a steady state CCD. Note that the red clay is not underlain by calcareous ooze.

Qualitative lithological descriptions and smear slide descriptions were used when C_aCO_3 determinations were unavailable.

The CCD provides a geologically useful facies boundary. Unfortunately the palaeo-Lysocline could not be defined in this general synthesis. Indeed, the recognition of this boundary over wide regions of the Tertiary ocean will only be possible when extensive site coverage at a variety of depths in several regions is achieved. Detailed studies and comparisons of the taxonomic composition and states of preservation of nannofossil and planktonic foraminiferal assemblages at various intervals of time within and between these sites will have to be made in order to define this boundary. This task, though immense, would provide useful data concerning the history of carbonate dissolution and the vertical structure of ancient water masses.

In this account a distinction is made between foraminiferal and nannofossil ooze. A boundary between these facies is arbitrarily defined at the 10% value for sand-sized planktonic foraminiferal tests. Like the CCD the depth of this boundary is probably a function of the supply of foraminifers versus their rate of dissolution. Thus this boundary should be deep beneath regions where planktonic foraminifers are abundant in the near surface waters and shallow beneath regions where they are less common. Since the tests of planktonic foraminifers become fragmented and reduced in size through dissolution (Berger, 1967; Cita and Gartner, 1968) the 10% foraminiferal boundary provides an indication of the degree of preservation of foraminiferal tests in some regions. This boundary was chosen in order to compare the history of another, albeit arbitrarily chosen, facies boundary with the history of the CCD.

THE DETERMINATION OF PALAEODEPTHS

The depths of pelagic facies boundaries at their time of origin, or of sediments at their time of deposition (i.e. palaeodepths) can be ascertained by determining the substance history of a site. This approach was used by Ramsay to investigate the distribution of discoasters in North Atlantic and Caribbean sediments and of facies distributions in Atlantic, Caribbean and Pacific cores (Ramsay, 1972, 1974). The technique used was based on using the spreading rate history of a ridge to relocate the positions of sites on the flanks of that ridge assuming that its slopes had remained constant. Berger (1973a) developed a more refined and simpler technique based on a hypothesis of age/depth constancy derived from the investigations of Menard (1969) and Sclater and Francheteau (1970). This technique has been applied with considerable success by various workers (Berger and von Rad, 1972; Berger, 1973; Berger and Winterer, 1974; Luyendyk and Davies, 1974) and is also used in this investigation.

Starting with the basement age of a site a curve is plotted parallel to an average basement age/basement depth curve (see Fig. 9). Palaeodepths are obtained by correcting the curve for isostatic adjustment of the basement to the sediment load by the addition of one half of the depth in the core of the boundary or sediment in question. An example of the estimation of

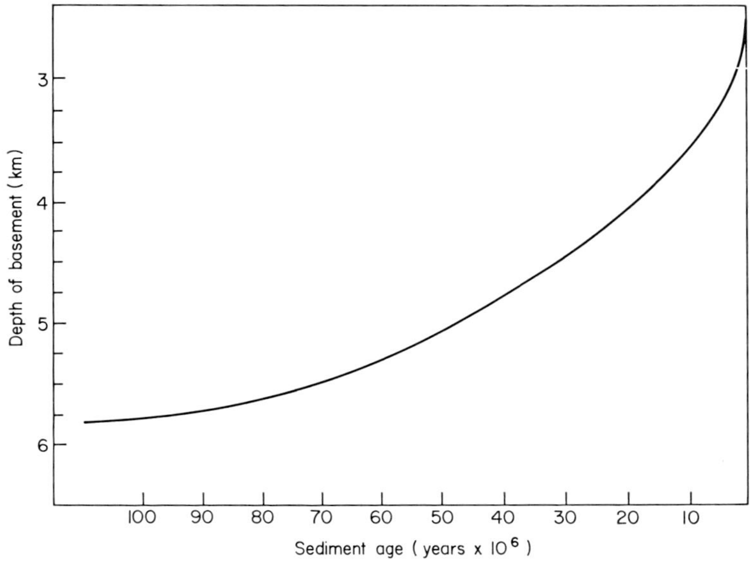

FIG. 9. Schematic age/depth relationship for spreading oceanic crust after Berger and von Rad (1972, Fig. 55).

palaeodepths is shown in Fig. 10. In this study all sedimentary ages and Tertiary basement age are interpreted within the framework of Berggren's 1969 time scale illustrated in Maxwell *et al.* (1970, p. 20; see also Fig. 11).

This technique has been applied to all sites where basement ages are recorded or where they can be assumed from the cruise reports, irrespective of whether the site occurs in a marginal ocean basin, a major ocean basin, or on the anomalous crust of the Magellan and Hess Rises, the Ontong Java Plateau and Caribbean Sea. At shallow sites where basement ages are not known the depths of sediments in the holes were adjusted for isostatic loading and these depths were used.

Errors in palaeodepths determined by means of this technique are probably introduced by uncertainties concerning the true age of the basement and possible regional divergence of actual subsidence histories from the idealised history. Continued refinement of the technique in the light of future data may help to reduce the degree of error. The results for each site are documented in a series of palaeodepth–latitude/palaeolatitude matrices (see Figs 14, 17, 19). This allows the reader to observe and judge the magnitude of the depth measurements which have been made and the efficacy and reliability of this approach to sites situated in tectonically distinct areas of the ocean.

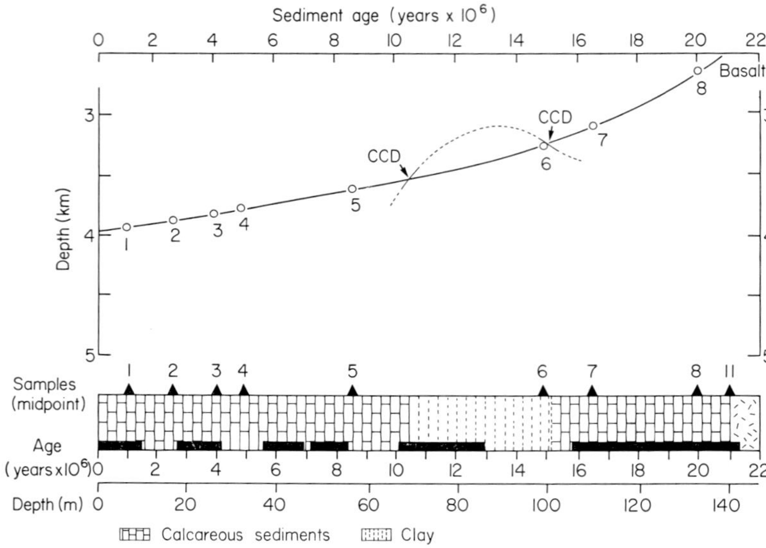

FIG. 10. An example of the establishment of palaeodepths by means of a backtracking procedure (Berger, 1972) for Site 15, Leg 3 (Maxwell *et al.*, 1970) after Berger and Winterer (1974, Fig. 22).

THE INVESTIGATION OF HIATUSES

The data concerning hiatuses were extracted from the Initial Reports of the Deep Sea Drilling Project and from articles in Geotimes. No attempt has been made to re-interpret the raw data other than to fit them into the framework of the time scale used in this report. Only proven hiatuses were used to establish the hiatus stratigraphy for regions I–V shown in Fig. 1.

The history of the CCD

THE QUATERNARY RECORD OF CCD FLUCTUATIONS

Carbonate cycles which coincide with climatic events have been recorded in Quaternary sediments from the equatorial Pacific (Arrhenius, 1952; Hays *et al.*, 1969), the south-east Pacific (Broecker and Broecker, 1974) and the North Atlantic basin (McIntyre *et al.*, 1972). In the equatorial Pacific between depths of approximately 4400 m and 5000 m (the depth of the modern calcite compensation depth, Berger and Winterer, 1974) carbonate ooze which accumulated during glacial periods is interbedded with siliceous or siliceous carbonate oozes which were deposited during interglacial interludes.

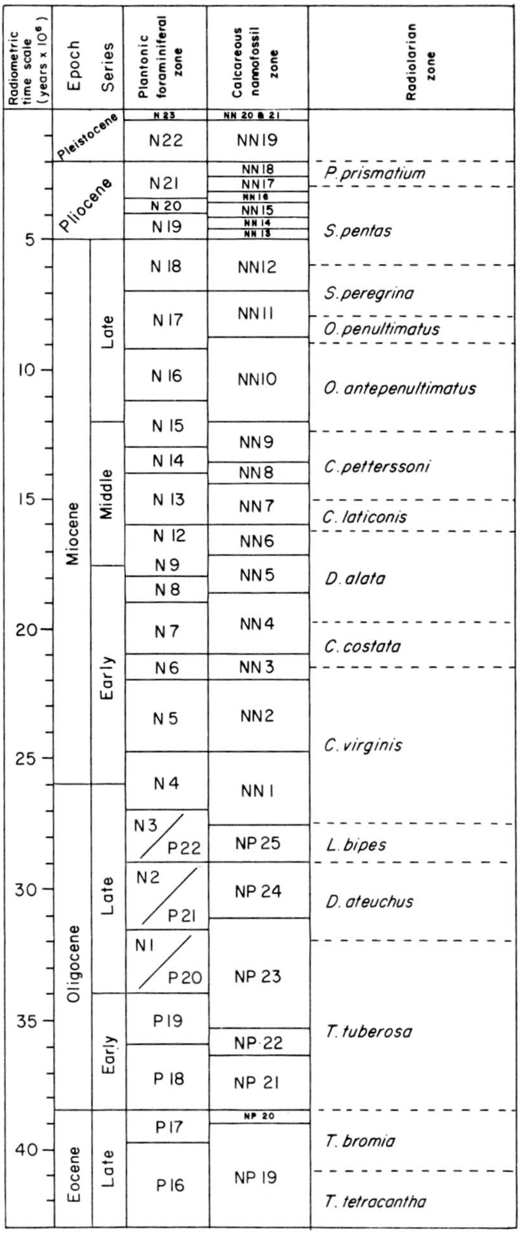

FIG. 11. Planktonic foraminiferal zones, calcareous nannofossil zones and radiolarian zones plotted against Berggren's 1969 time scale (in Maxwell *et al.*, 1970).

Arrhenius (1952) and Olauson (1971) interpreted these cycles in terms of increased plankton production during glacial periods. Berger (1973b) re-examined Arrhenius' data and concluded that the variations in the ratios of planktonic to benthic foraminifers, planktonic foraminifers to radiolarians and of resistance to easily dissolved planktonic foraminifers are more indicative of changes in dissolution rates than of changes in plankton production. Sediments which accumulated during glacial periods are less dissolved than interglacial deposits.

The carbonate cycles of the equatorial Pacific probably coincided with vertical fluctuations in the CCD. In Fig. 12a the position of the modern CCD, which may be close to glacial positions, is compared with the CCD for the Brunhes 3 interglacial cycle of Hays *et al.* (1969) which has been calculated for a value of 10% $CaCO_3$ assuming a linear decrease in $CaCO_3$ percentages at depths below 4 km (see Fig. 12b). According to this calculation the CCD was approximately 100 m shallower during the Brunhes 3 interglacial cycle.

Broecker and Broecker (1974) have described carbonate cycles in piston cores recovered from the East Pacific Rise between 15° and 18°S. Their data for the percentage of $CaCO_3$ at depths of 10 cm and 30 cm in the cores have been used to construct Fig. 13a in which the positions of the interglacial and glacial calcite compensation depths are calculated for a value of 10% $CaCO_3$ again assuming a linear decrease in the percentage of $CaCO_3$ at depths below 4 km (see Fig. 13b). The sedimentation patterns shown in Fig. 13a indicate a vertical fluctuation of 150 m in the calcite compensation depth for the cycles examined.

It is not possible using the published data to correlate the Brunhes 3 equatorial cycle at about 180 000 years BP (Hays *et al.*, 1969) with the cycles described by Broecker and Broecker, 1974). Indeed, it is likely that the 30 and 10 cm depth intervals in the South East Pacific cores are not time synchronous in all cores. The data do, however, provide an indication of the possible depth changes in the position of the CCD during the Quaternary. In this respect it is interesting to note that the vertical fluctuations in the CCD is greater beneath the less fertile south east Pacific region than beneath the highly productive equatorial region.

Outside the Pacific evidence for enhanced solution in mid and low latitudes during interglacial cycles is recorded from the Atlantic Ocean (Berger, 1968), Caribbean Sea and the eastern equatorial Indian Ocean (Broecker, 1971; Ramsay, 1974). In North Atlantic Quaternary sequences (north of approximately 45°) the record of pelagic carbonate deposition is the reverse of that described for the equatorial Pacific, and carbonate cycles appear to be related to climatically controlled fluctuations in calcareous plankton production. In this region of the ocean McIntyre *et al.* (1972) demonstrate an association

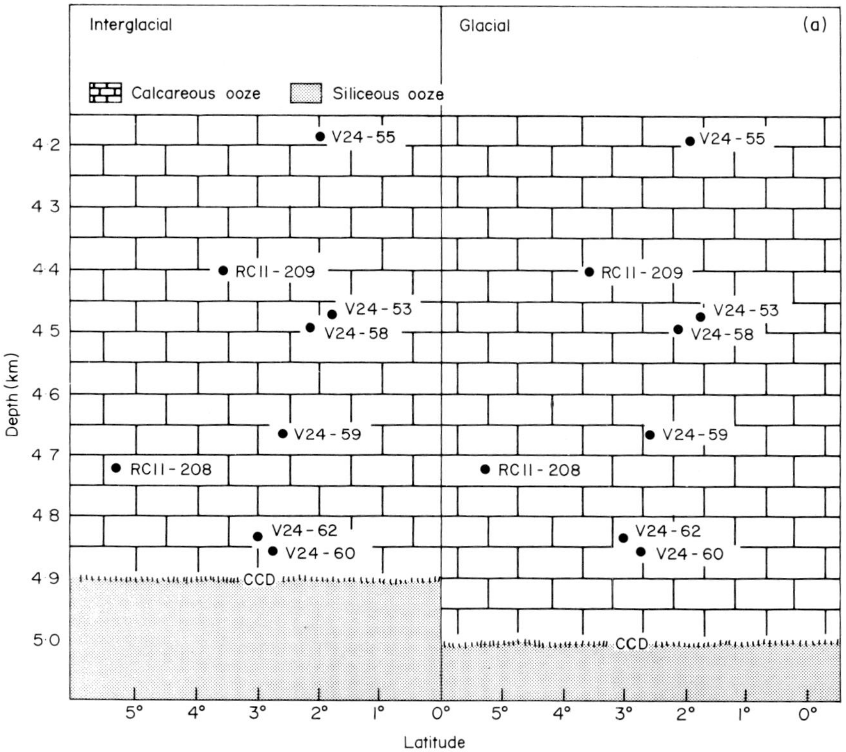

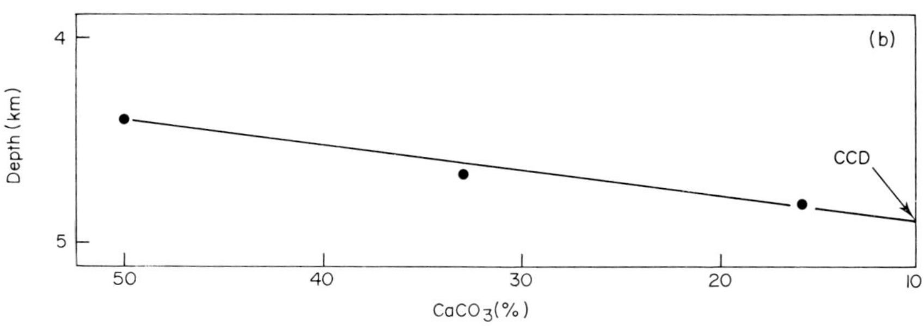

FIG. 12. (a) A comparison between the depth of the modern CCD glacial and estimated depth of the CCD during the Brunhes 3 interglacial cycle for the equatorial Pacific. Data from Hays *et al.* (1969). (b) Depth/$CaCO_3$ relationship from which the level of the interglacial CCD in Fig. 12(a) was calculated. Data from Hays *et al.* (1969).

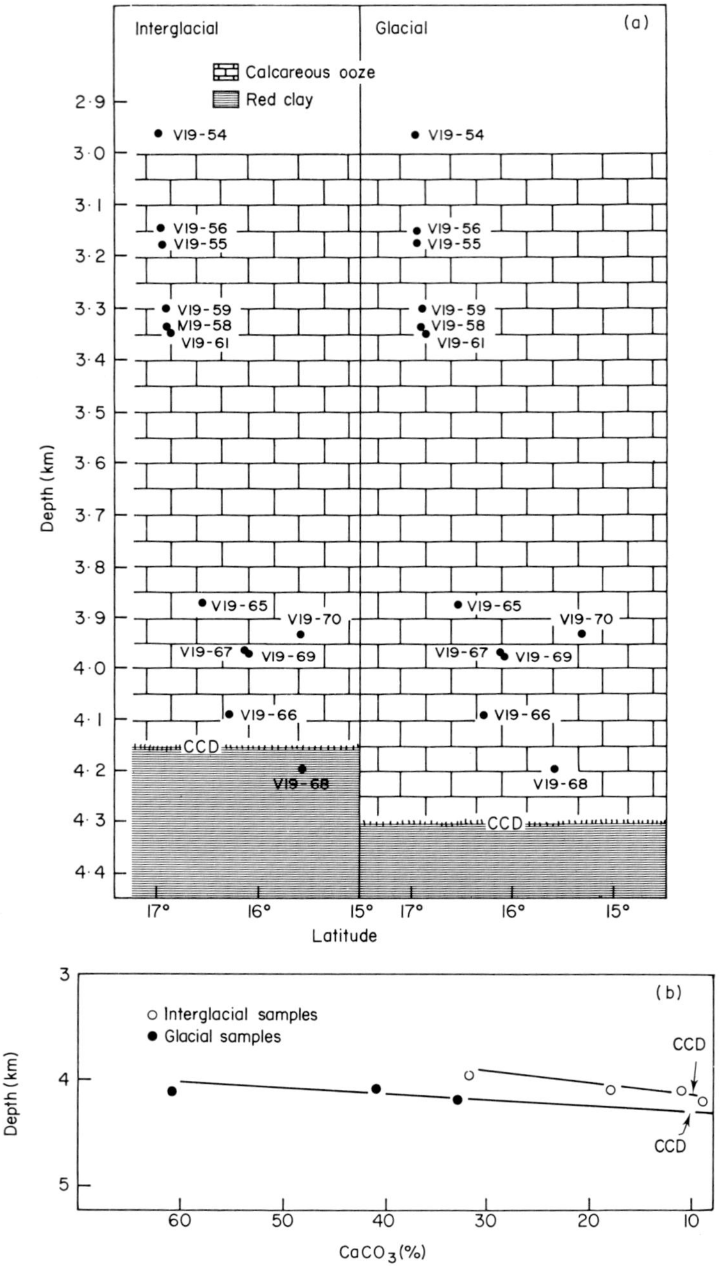

FIG. 13. (a) A comparison between the depths of the interglacial and glacial CCD on the east Pacific Rise at 17°S. Data from Broeker and Broeker (1974). (b) Depth $CaCO_3$ relationship for an interglacial and glacial cycle from which the levels of the CCD shown in Fig. 13(b) were calculated.

between glacial advance, the southward penetration of cold polar water, low production of calcareous plankton and carbonate minima. Carbonate maxima coincide with interglacial polar water retreat and increased production of the calcareous plankton.

FLUCTUATIONS IN THE CCD SINCE THE LATE EOCENE IN THE ATLANTIC OCEAN, CARIBBEAN SEA, INDIAN AND PACIFIC OCEANS

The occurrence of non-calcareous clays or siliceous oozes interbedded with carbonate ooze in cores from the deep sea has been interpreted as evidence for temporal fluctuations in the CCD (Hay, 1970; Berger, 1972; Berger and von Rad, 1972; Berger, 1973a; Berger and Winterer, 1974; Ramsay *et al.*, 1973; Ramsay, 1974; van Andel and Moore, 1974). In the following account depth corrected data of this nature have been used to reconstruct the history of the CCD and of the 10% foraminiferal boundary at selected time intervals, starting with the Late Eocene, for the Atlantic Ocean, Caribbean Sea, Indian and Pacific Oceans.

The Atlantic Ocean

The data used to reconstruct the histories of the CCD and 10% foraminiferal boundary for this ocean are illustrated in Figs 14a–f. The CCD curves derived for the Late Eocene, Late Oligocene, Early, Middle and Late Miocene, and Pliocene/Quaternary intervals are compared in Figs 15a–f.

The palaeolatitudes shown in these figures are derived from palaeolatitude reconstructions for the Atlantic by Smith and Drewry (see Figs 34–38) which indicate a northward migration of the Atlantic crust by approximately 0.4° per m.y.

The data indicate a latitudinally influenced pattern of CCD fluctuations imposed on a CCD curve which has maintained a similar shape since the Late Eocene. The most marked change in the depth of the CCD occurs between the Late Eocene, when it was shallow, and the Late Oligocene when the CCD was depressed at all latitudes. Subsequent to the Late Oligocene the following trends in the history of the CCD are recorded:

1. The gradual depression of the CCD in the North Atlantic basin north of palaeolatitude 30°.
2. The initiation of a shallowing of the CCD in the South Atlantic basin during the Early Miocene. This phenomenon had extended into the North Atlantic basin by the Middle Miocene and persisted until the Late Miocene.
3. In the Pliocene to Recent interval the CCD is depressed at all latitudes north of 30°S but becomes even more elevated south of this latitude.

Unfortunately there are too few available data to reconstruct a history of the 10% foraminiferal boundary for the whole Atlantic basin at each of the selected time intervals. The most complete records are for the Pliocene/Quaternary and Late Miocene intervals. These records show that the shape of the curve obtained for this boundary is not a mirror image of the CCD curve. It is strongly depressed in a zone extending from the equator to 25°N and is elevated north of 25°N despite the depressed nature of the CCD. These differences probably reflect differences in the rates of the near surface production of planktonic foraminiferal tests, i.e. the boundary is depressed below areas of high shell production and elevated beneath areas of low shell production.

Though palaeodepths for this boundary have only been reconstructed for limited areas of the Atlantic basin, the data obtained suggest that this level fluctuated in sympathy with the vertical fluctuations in the CCD.

The Caribbean Sea versus the North Atlantic

The history of the CCD in the Caribbean Sea and the data from which it was formulated is shown in Figs 16a–f. These are palaeodepth–longitude matrices and were chosen in order to facilitate a comparison between the histories of the CCD for the Caribbean and North Atlantic basin.

The depths of the Caribbean sites are based on isostatically adjusted modern depths. Unfortunately the precise age for the Caribbean crust is unknown, but if it is older than Late Cretaceous (Edgar *et al.*, 1973) the influence of crustal subsidence would have had a negligible effect on palaeodepth since the Late Eocene. The possibility of depth changes as a result of vertical tectonics in the Caribbean Sea (Edgar *et al.*, 1973) presents a more serious problem in palaeodepth interpretation. Depth changes of this nature should, however, be revealed by inconsistencies in depths obtained for the CCD. Since no inconsistencies are observed the palaeodepth interpretations are regarded as reliable.

Figures 16a–f show similar patterns of CCD fluctuations for the Caribbean and North Atlantic basins. The CCD was shallower in the Caribbean Sea subsequent to the Late Miocene. Prior to this time the depth of the CCD could have been equal in these basins.

The history of the 10% foraminiferal boundary can be inferred as far back as the Oligocene. The data show that this boundary oscillated in sympathy with the CCD.

FIG. 14. Pliocene to Quaternary bathymetric facies distribution in the Atlantic Ocean in a north–south vs. depth of water projection (a) and palaeobathymetric facies distributions for the Late Miocene (b), Middle Miocene (c), Early Miocene (d), Late Oligocene (e) and Late Eocene (f), compared with latitude and palaeolatitude.

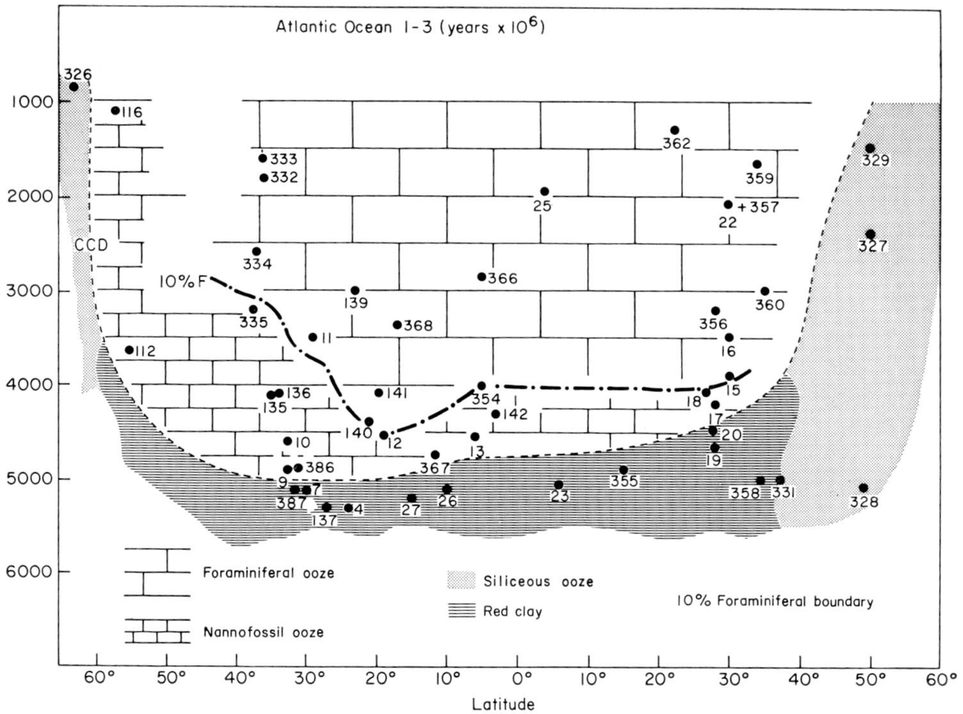
Atlantic Ocean 1-3 (years x 10^6)
CCD
10% F
Foraminiferal ooze
Nannofossil ooze
Siliceous ooze
Red clay
10% Foraminiferal boundary
Latitude

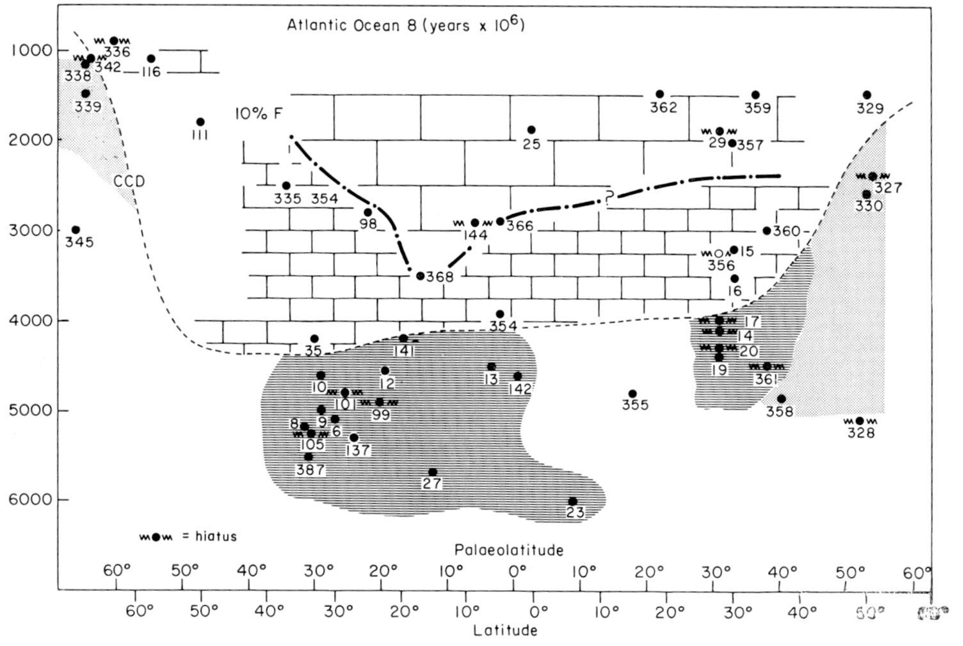
Atlantic Ocean 8 (years x 10^6)
10% F
CCD
= hiatus
Palaeolatitude
Latitude

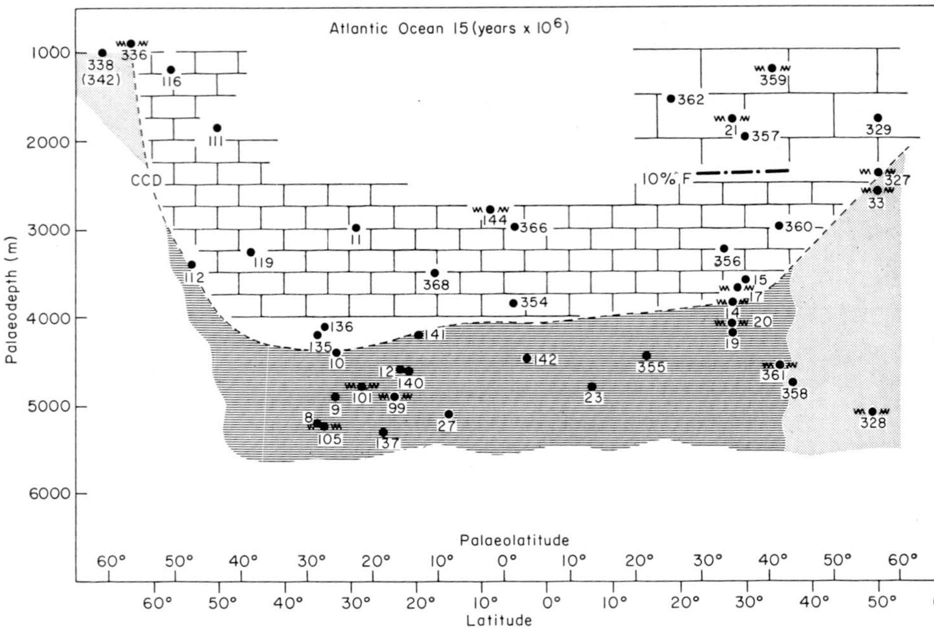

Atlantic Ocean 15 (years x 10⁶)
Palaeodepth (m)
1000
2000
3000
4000
5000
6000
CCD
10% F
Palaeolatitude
Latitude

(c)

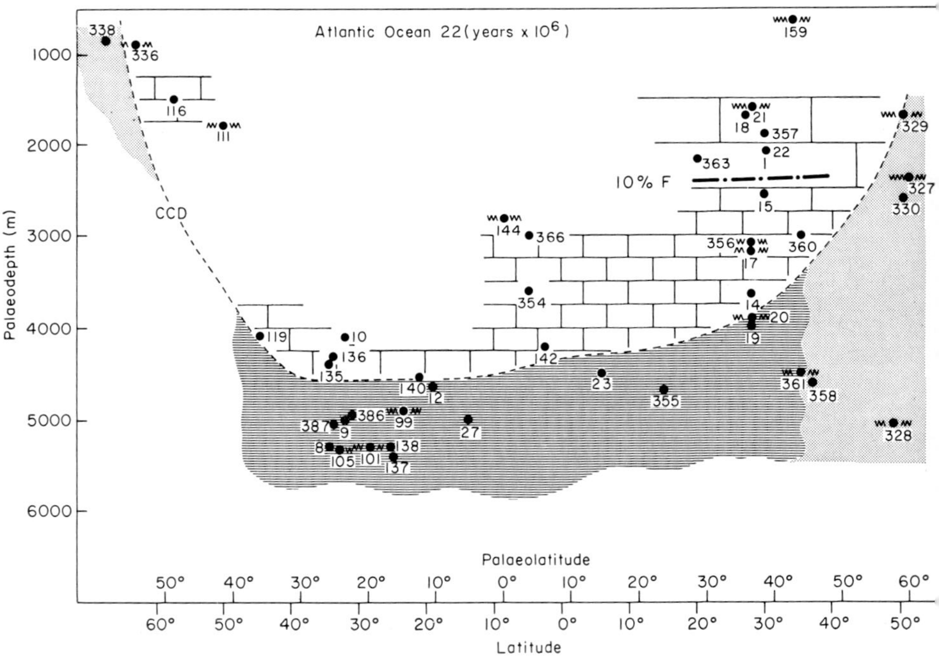

Atlantic Ocean 22 (years x 10⁶)
Palaeodepth (m)
1000
2000
3000
4000
5000
6000
CCD
10% F
Palaeolatitude
Latitude

(d)

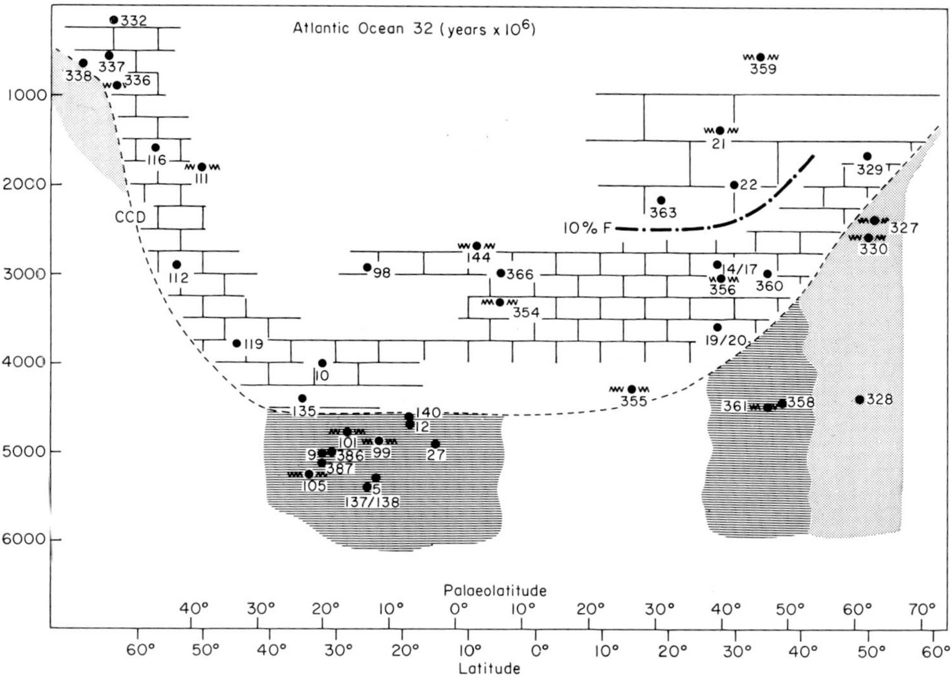
Atlantic Ocean 32 (years x 10⁶)
1000
2000
3000
4000
5000
6000
CCD
10% F
332
337
338
336
116
111
112
119
10
135
140
12
101
99
9
386
387
105
5
137/138
27
98
144
366
354
355
359
21
363
22
329
327
330
14/17
356
360
19/20
361
358
328
Palaeolatitude
40° 30° 20° 10° 0° 10° 20° 30° 40° 50° 60° 70°
60° 50° 40° 30° 20° 10° 0° 10° 20° 30° 40° 50° 60°
Latitude

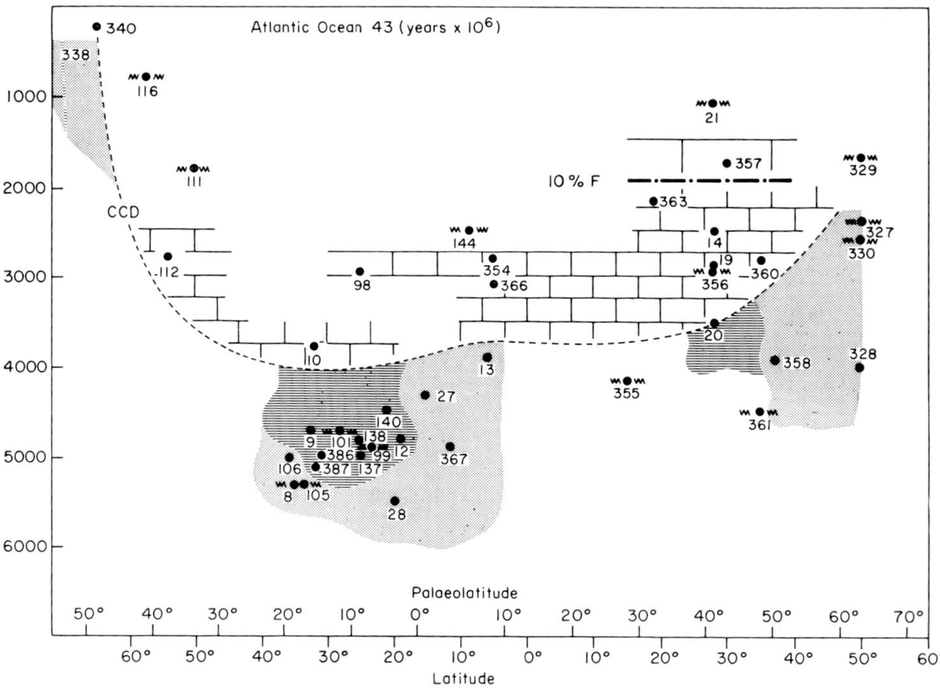
Atlantic Ocean 43 (years x 10⁶)
1000
2000
3000
4000
5000
6000
CCD
10% F
340
338
116
111
112
98
144
354
366
110
13
27
140
9
101
138
12
386
99
137
387
106
8
105
28
367
355
21
357
363
14
19
356
360
20
358
328
361
329
327
330
Palaeolatitude
50° 40° 30° 20° 10° 0° 10° 20° 30° 40° 50° 60° 70°
60° 50° 40° 30° 20° 10° 0° 10° 20° 30° 40° 50° 60°
Latitude

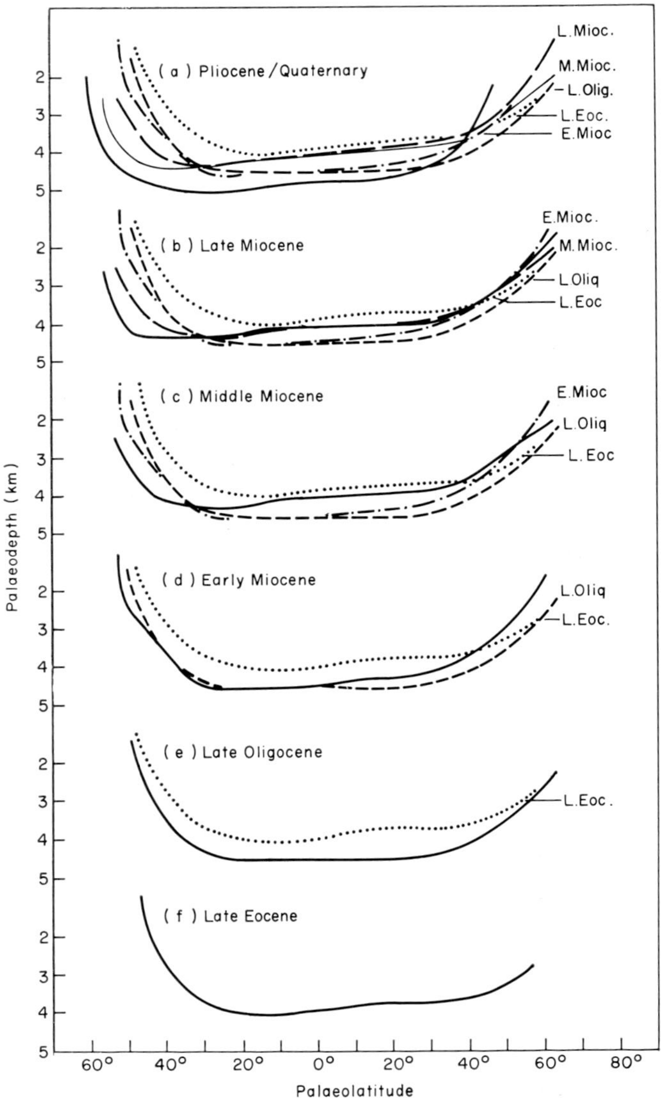

FIG. 15. Comparisons between the Quaternary to Pliocene (a), Late Miocene (b), Middle Miocene (c),Early Miocene (d), Late Oligocene (e) and Late Eocene (f). CCD curves derived for the Atlantic Ocean.

FIG. 16. A comparison between the history of the CCD in the Atlantic and Caribbean basins for (a) Pliocene Quaternary interval, (b) the Late Miocene, (c) the Middle Miocene, (d) the Early Miocene, (e) the Late Oligocene and (f) the Late Eocene.

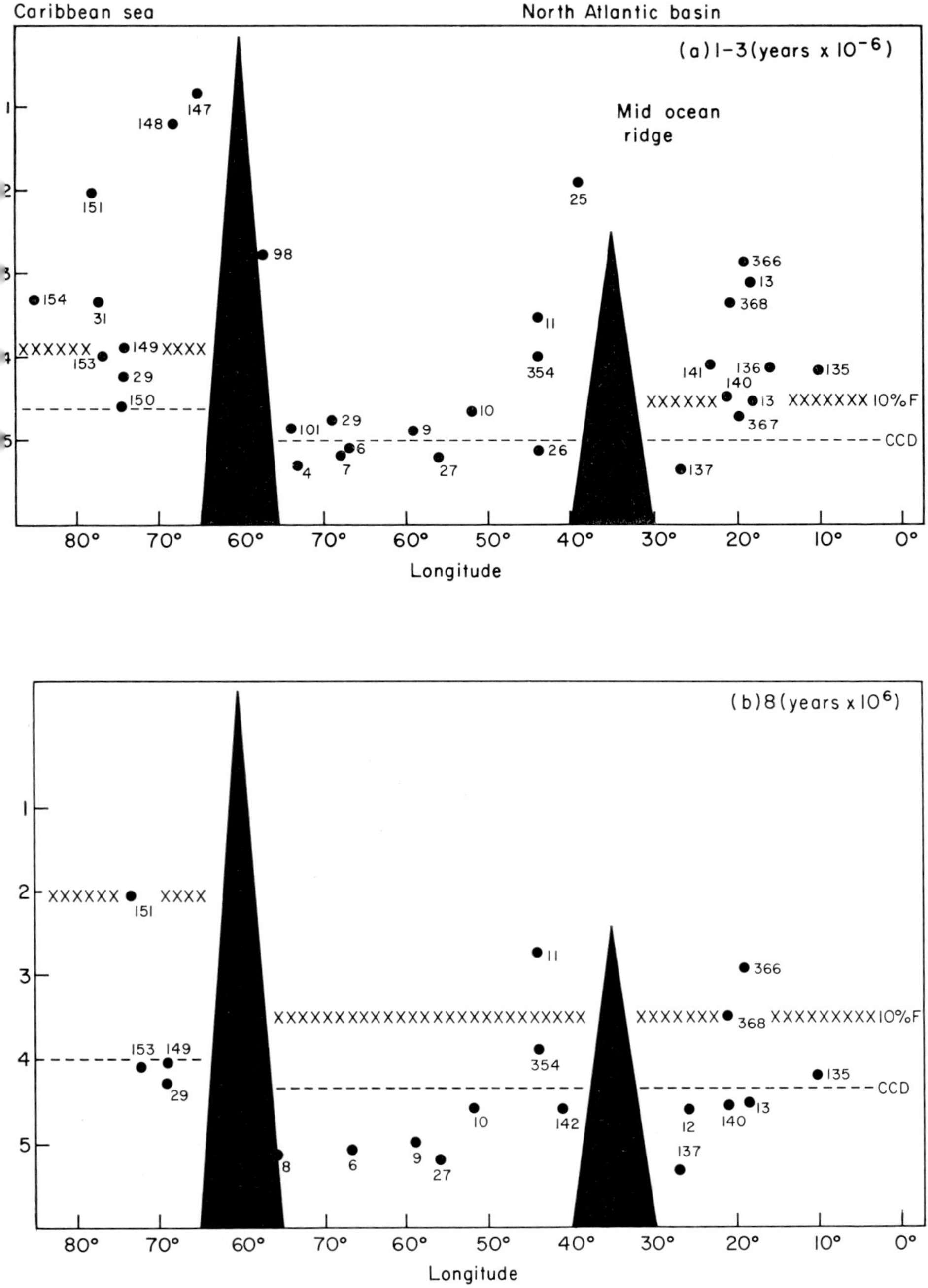

Caribbean sea
North Atlantic basin
(a) 1–3 (years x 10⁻⁶)
Mid ocean ridge
147
148
151
98
25
366
13
368
154
31
11
149
153
354
29
141
136
135
140
150
13
10%F
10
29
101
9
367
CCD
6
26
4
7
27
137
80°
70°
60°
70°
60°
50°
40°
30°
20°
10°
0°
Longitude
(b) 8 (years x 10⁶)
151
11
366
368
10%F
153
149
354
29
135
CCD
10
142
12
140
13
8
6
9
27
137
80°
70°
60°
70°
60°
50°
40°
30°
20°
10°
0°
Longitude

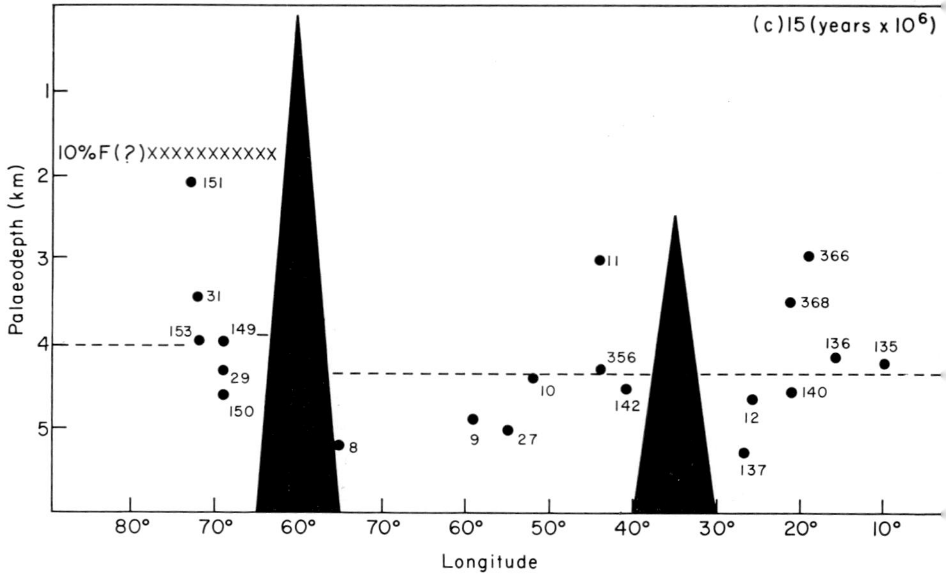
(c) 15 (years x 10⁶)
10%F(?)XXXXXXXXXXX
Palaeodepth (km)
1
2
3
4
5
151
31
153
149
29
150
8
9
27
10
11
356
142
12
137
366
368
136
135
140
80°
70°
60°
70°
60°
50°
40°
30°
20°
10°
Longitude

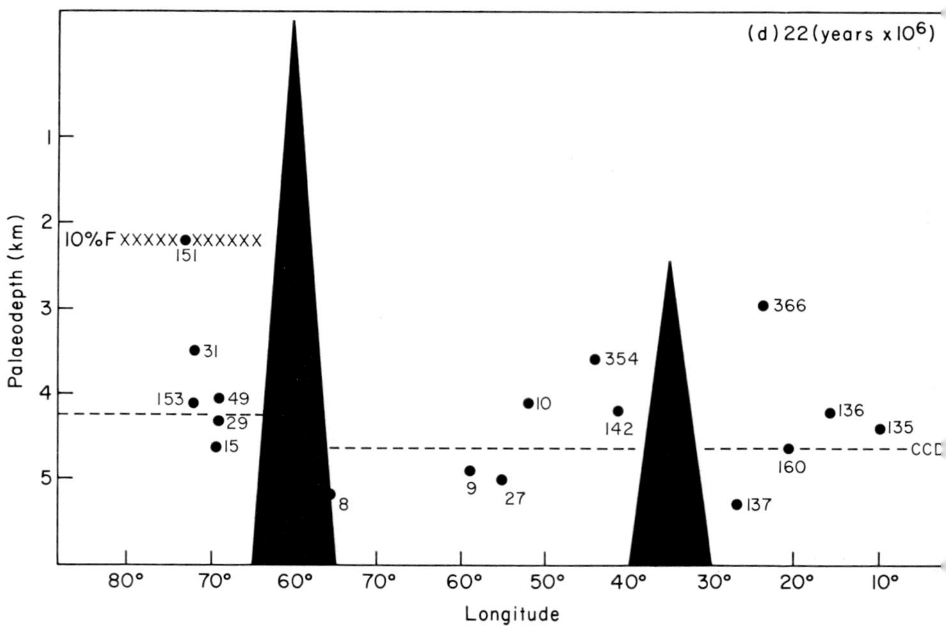
(d) 22 (years x 10⁶)
10%F XXXXXXXXXXXX
Palaeodepth (km)
1
2
3
4
5
151
31
153
49
29
15
8
9
27
10
354
142
366
136
135
160
137
CCD
80°
70°
60°
70°
60°
50°
40°
30°
20°
10°
Longitude

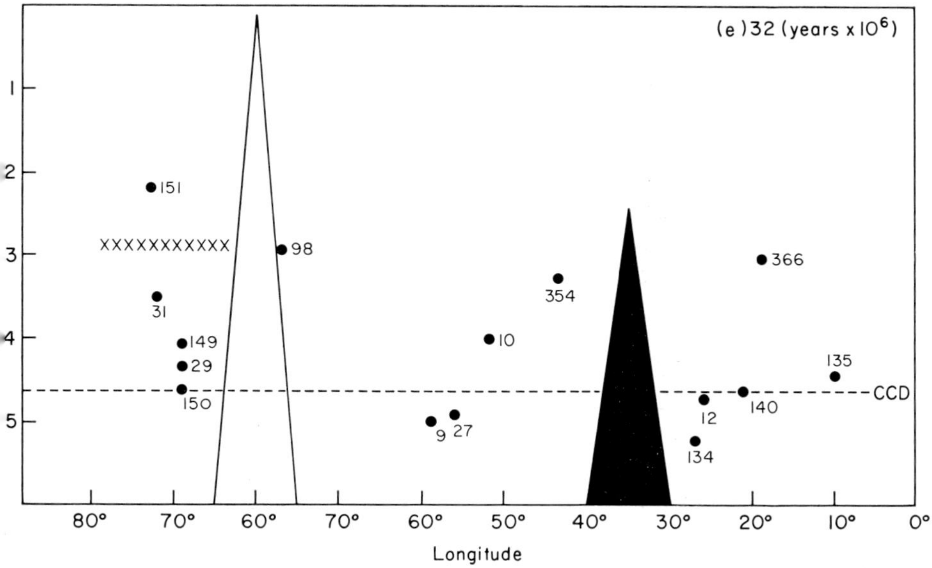

(e) 32 (years x 10^6)
1
2
3
4
5
151
98
31
149
29
150
354
10
9
27
366
135
CCD
12
140
134
80°
70°
60°
70°
60°
50°
40°
30°
20°
10°
0°
Longitude

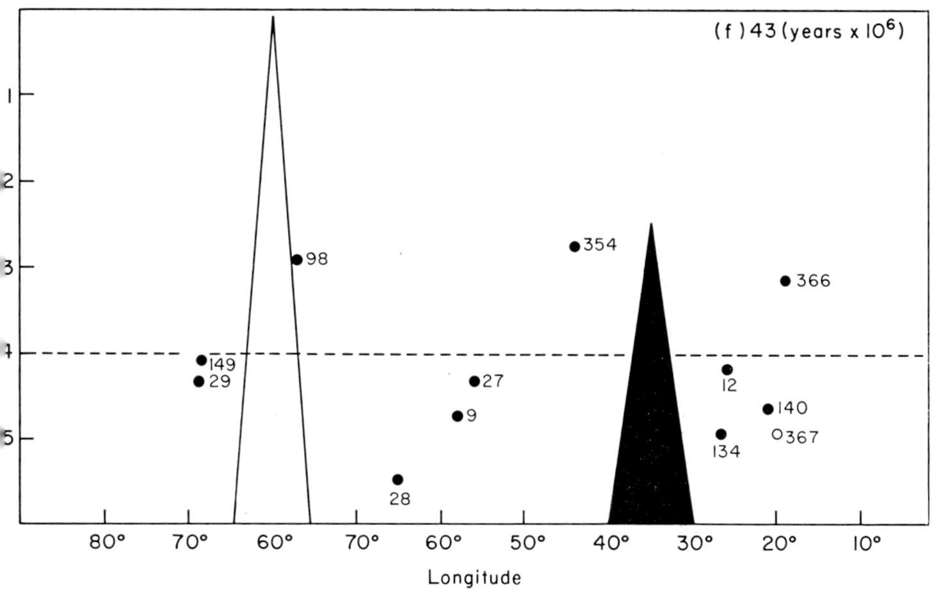

(f) 43 (years x 10^6)
1
2
3
5
98
354
366
149
29
27
9
28
12
140
367
134
80°
70°
60°
70°
60°
50°
40°
30°
20°
10°
Longitude

The Indian Ocean

The information used to reconstruct the histories of the CCD and 10% foraminiferal boundary since the Late Eocene is shown in Figs 17a–f. The CCD curves derived for the intervals shown in these figures are compared in Figs 18a–f. The palaeolatitudes used in these figures are derived from palaeogeographical reconstructions of the Indian Ocean by McKenzie and Sclater (1971) and Sclater and Fisher (1974) which indicate a 0.62° northward composed of motion for the Indian Ocean crust from the Oligocene.

Figures 17 and 18 show two trends in the history of the CCD. These are:

1. The gradual deepening of the CCD in mid and low latitudes from a shallow Late Eocene depth to a Pliocene/Quaternary maximum.

2. The shoaling of the CCD in high latitudes from middle miocene to a shallow depth in the Pliocene/Recent interval.

The 10% foraminiferal boundary was not recognised in the Late Eocene sediments. Subsequent to the Oligocene this boundary gradually deepens in sympathy with the downward migration of the CCD in mid and low latitudes.

The Pacific Ocean

The data which have been used to reconstruct the history of the Pacific CCD and 10% foraminiferal boundary are shown in Figs 19a–f. The CCD curves derived for the intervals shown in these figures are compared in Figs 20a–f. The palaeolatitudes used in these figures are derived from the estimated northward migration of the Pacific plate by 0.23° per m.y. (Winterer, 1973) and an estimated northward shift of the eastern sector of the Indian plate by 0.38° per m.y. derived from the palaeogeographical reconstructions of McKenzie and Sclater (1971).

The data indicate that the distinctive shape of the CCD curve, i.e. depressed beneath the equatorial zone and shallowing towards higher latitudes has been maintained since the Late Eocene. They also suggest that the phenomenon of a shallow North Pacific and deeper South Pacific CCD was already established by the Late Eocene, The most striking change in the depth of the Pacific CCD occurred between the Late Eocene and Late Oligocene, when the CCD migrated from a shallow position to greater depths at all latitudes. Subsequent to the Oligocene the following trends in the history of the CCD are recorded.

1. A decrease in the width of the equatorial carbonate belt from its Oligocene maximum. Also the distortion of its Middle and Late Miocene bilateral

---

FIG. 17. Pliocene to Quaternary bathymetric facies distribution in the Indian Ocean on a north–south vs. depth of water projection (a) and palaeobathymetric facies distributions for the Late Miocene (b), Middle Miocene (c), Early Miocene (d), Late Oligocene (e) and Late Eocene (f), compared with latitude and palaeolatitude.

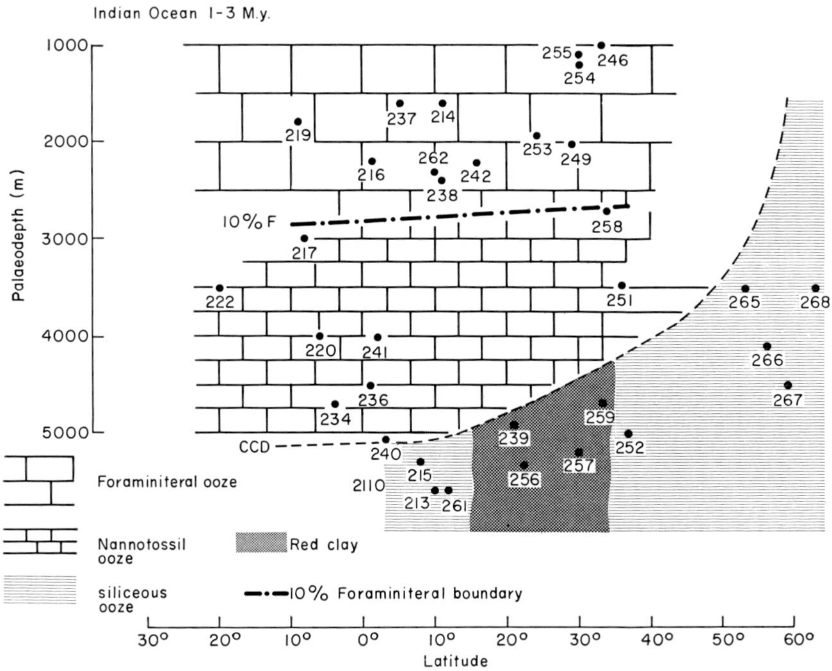

Indian Ocean 1-3 M.y.
Palaeodepth (m)
1000
2000
3000
4000
5000
255
246
254
237
214
219
253
249
262
216
242
238
10% F
258
217
222
251
265
268
220
241
266
236
267
234
259
CCD
239
252
240
257
256
215
2110
213
261
Foraminiteral ooze
Nannotossil ooze
Red clay
siliceous ooze
10% Foraminiteral boundary
30°
20°
10°
0°
10°
20°
30°
40°
50°
60°
Latitude

(a)

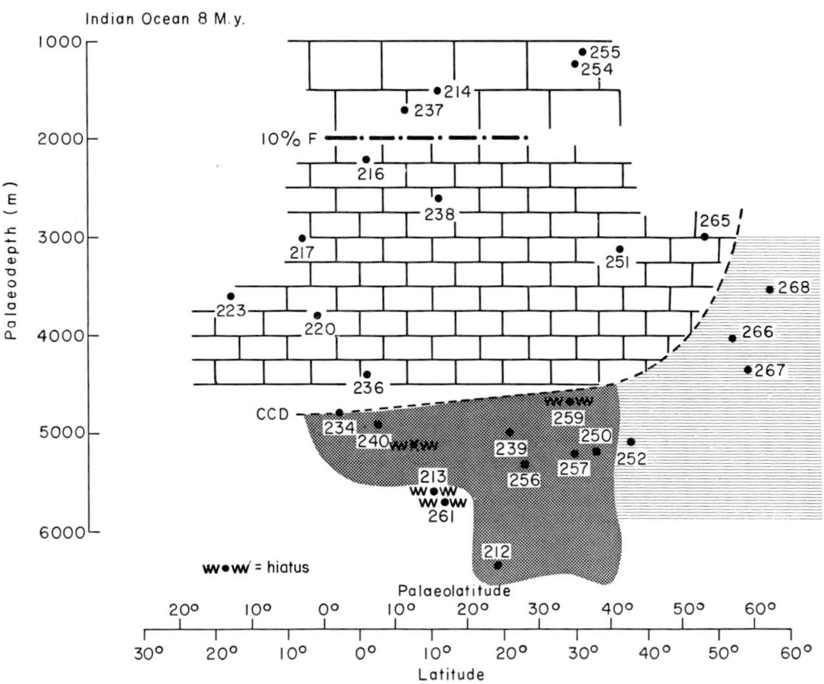

Indian Ocean 8 M.y.
Palaeodepth (m)
1000
2000
3000
4000
5000
6000
255
254
214
237
10% F
216
238
265
217
251
268
223
220
266
267
236
259
CCD
234
240
250
239
252
257
256
213
261
212
w•w = hiatus
Palaeolatitude
20°
10°
0°
10°
20°
30°
40°
50°
60°
30°
20°
10°
0°
10°
20°
30°
40°
50°
60°
Latitude

(b)

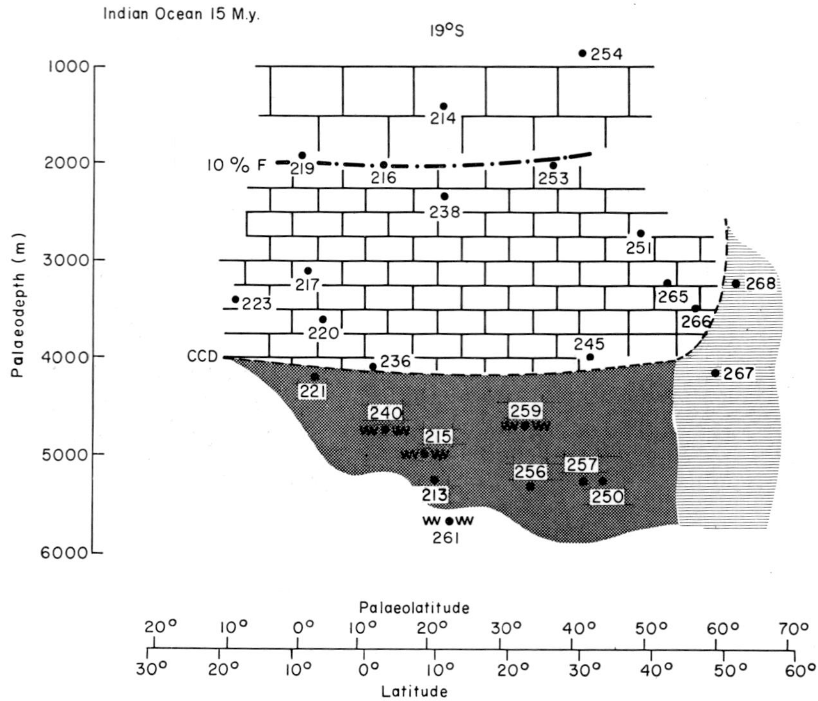
Indian Ocean 15 M.y.
19°S
Palaeodepth (m)
1000
2000
3000
4000
5000
6000
10 % F
CCD
254
214
219
216
253
238
251
217
223
220
265
266
268
245
236
267
221
240
259
215
256
257
213
250
261
Palaeolatitude
20° 10° 0° 10° 20° 30° 40° 50° 60° 70°
30° 20° 10° 0° 10° 20° 30° 40° 50° 60°
Latitude

(c)

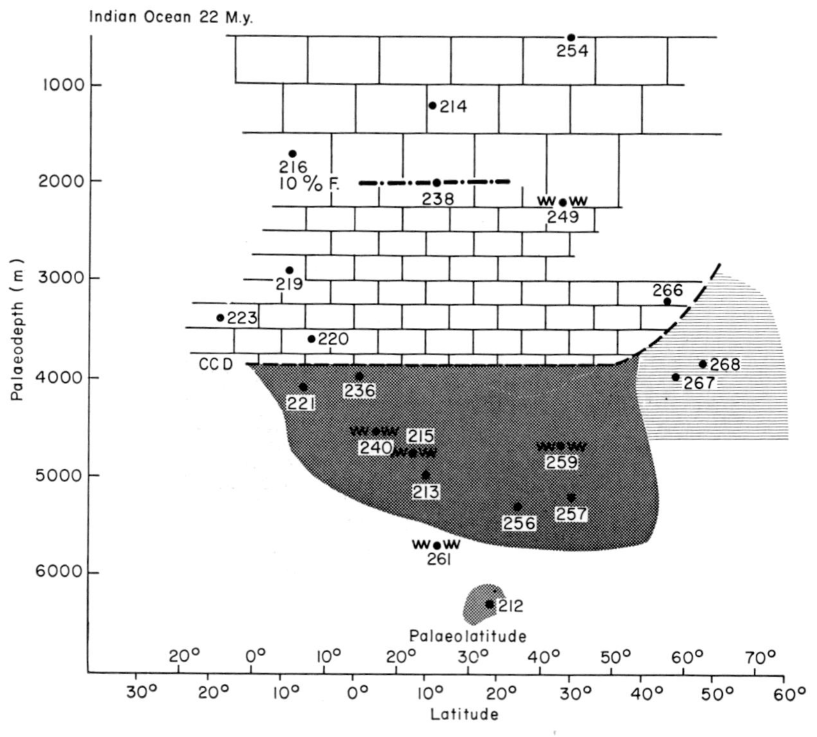
Indian Ocean 22 M.y.
Palaeodepth (m)
1000
2000
3000
4000
5000
6000
254
214
216
10 % F.
238
249
219
266
223
220
CCD
268
267
221
236
215
240
259
213
256
257
261
212
Palaeolatitude
20° 0° 10° 20° 30° 40° 50° 60° 70°
30° 20° 10° 0° 10° 20° 30° 40° 50° 60°
Latitude

(d)

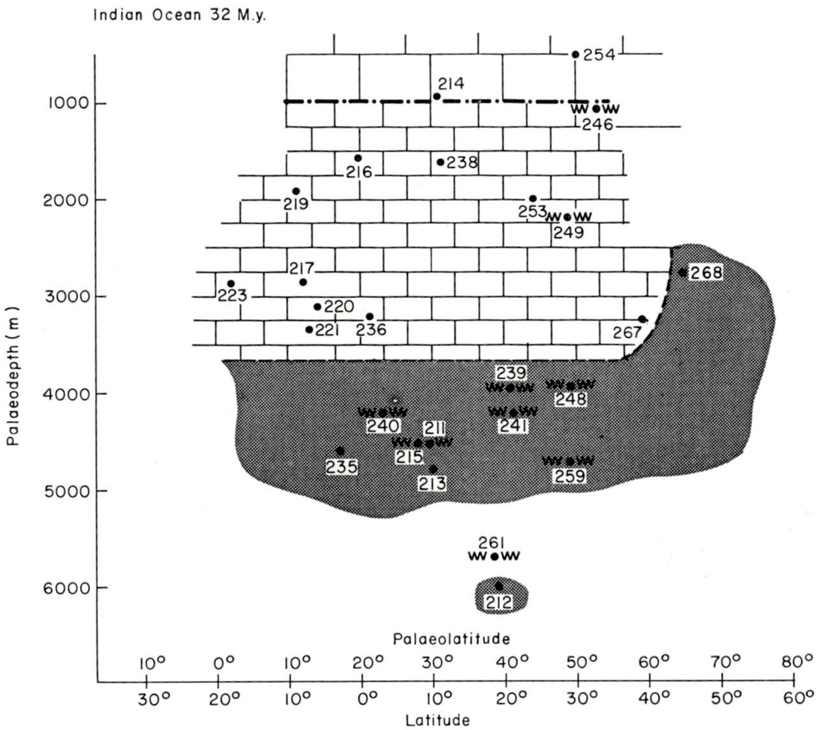
Indian Ocean 32 M.y.
Palaeodepth (m)
1000
2000
3000
4000
5000
6000
254
214
246
238
216
219
253
249
268
217
223
220
221
236
267
239
248
240
211
241
215
235
213
259
261
212
Palaeolatitude
10°
0°
10°
20°
30°
40°
50°
60°
70°
80°
30°
20°
10°
0°
10°
20°
30°
40°
50°
60°
Latitude

(e)

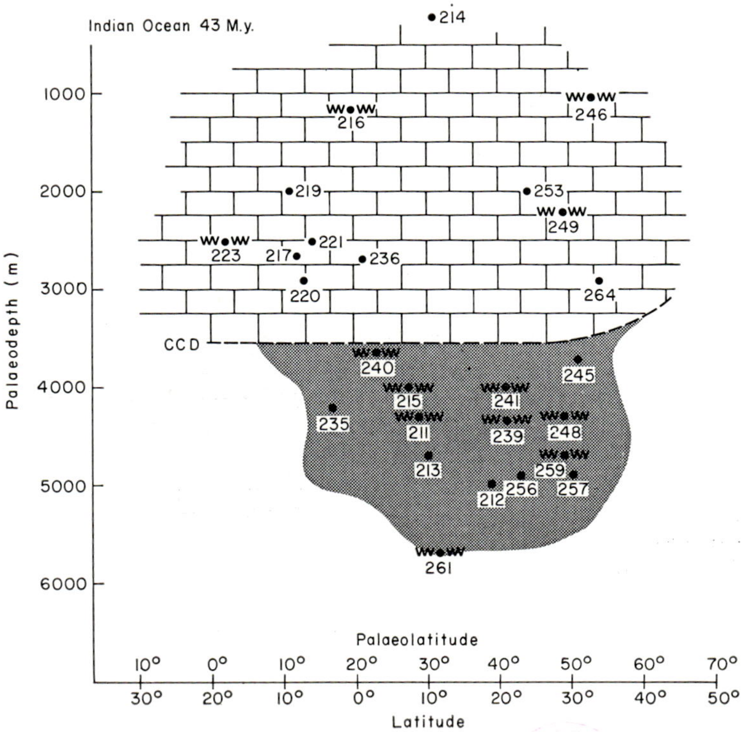
Indian Ocean 43 M.y.
Palaeodepth (m)
1000
2000
3000
4000
5000
6000
214
216
246
219
253
249
223
221
217
236
220
264
CCD
240
245
215
241
235
211
239
248
213
259
212
256
257
261
Palaeolatitude
10°
0°
10°
20°
30°
40°
50°
60°
70°
30°
20°
10°
0°
10°
20°
30°
40°
50°
Latitude

(f)

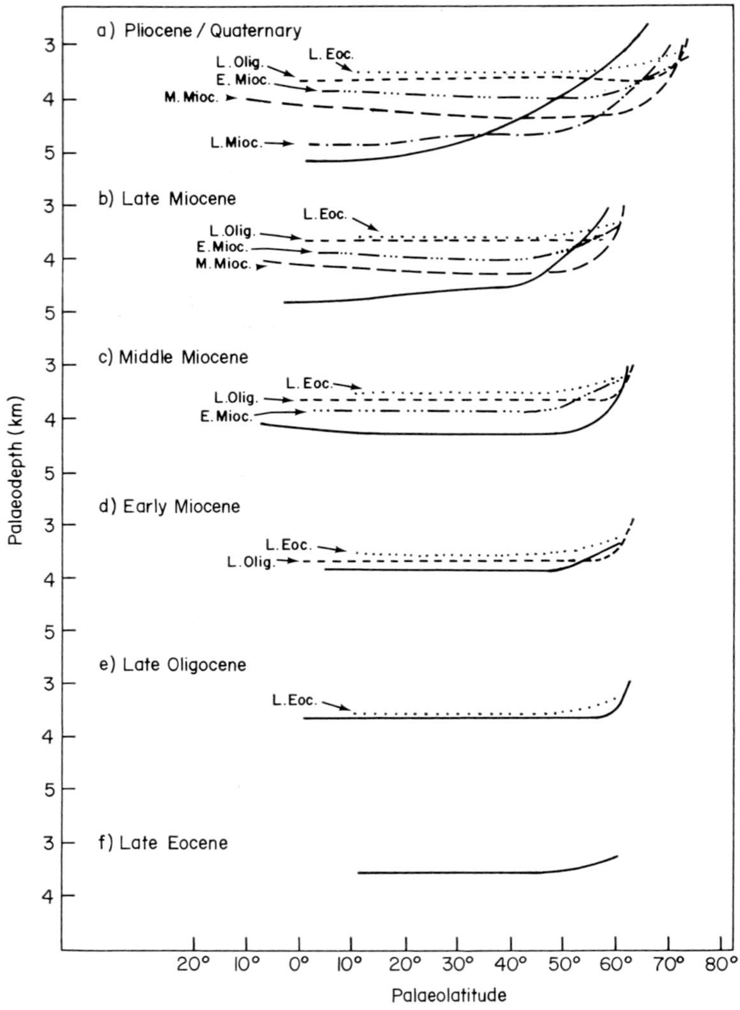

FIG. 18. Comparisons between the Pliocene to Quaternary (a), Late Miocene (b), Middle Miocene (c), Early Miocene (d), Late Oligocene (e) and Late Eocene (f), CCD curves derived for the Indian Ocean.

FIG. 19. Pliocene to Quaternary bathymetric facies distribution for the Pacific Ocean in a north–south vs. depth of water projection (a) and palaeobathymetric facies distributions for the Late Miocene (b), Middle Miocene (c), Early Miocene (d), Late Oligocene (e) and Late Eocene (f), compared with latitude and palaeolatitude.

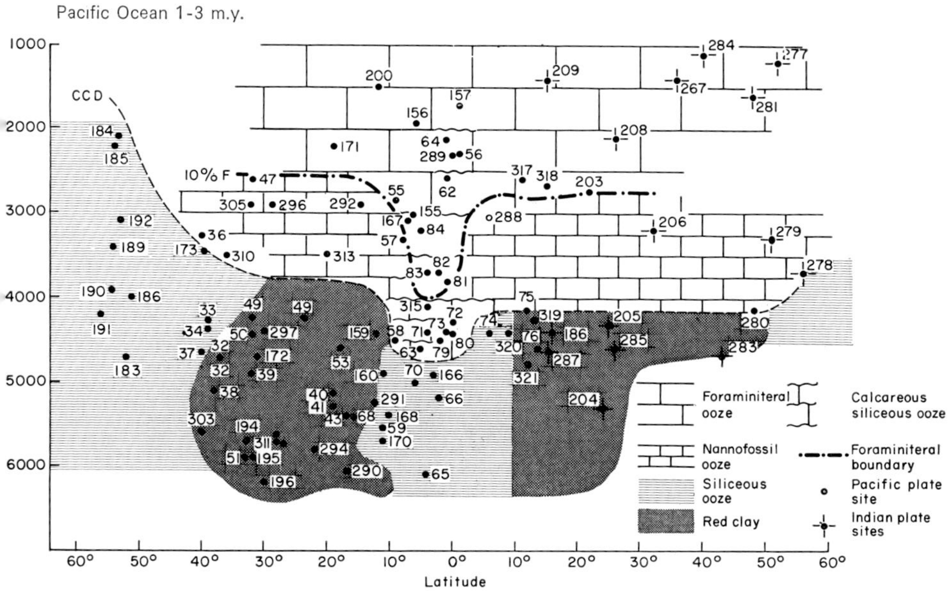

Pacific Ocean 1-3 m.y.
1000
2000
3000
4000
5000
6000
CCD
10% F
60°
50°
40°
30°
20°
10°
0°
10°
20°
30°
40°
50°
60°
Latitude
Foraminiferal ooze
Calcareous siliceous ooze
Nannofossil ooze
Foraminiferal boundary
Siliceous ooze
Pacific plate site
Red clay
Indian plate sites

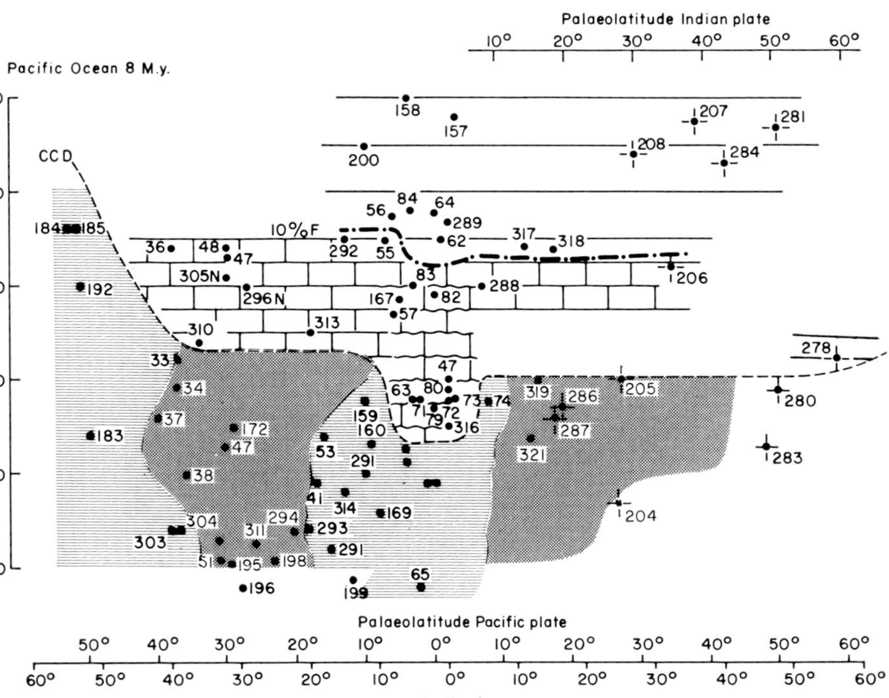

Pacific Ocean 8 M.y.
Palaeolatitude Indian plate
10°
20°
30°
40°
50°
60°
1000
2000
3000
4000
5000
6000
CCD
10% F
Palaeolatitude Pacific plate
50°
40°
30°
20°
10°
0°
10°
20°
30°
40°
50°
60°
Latitude

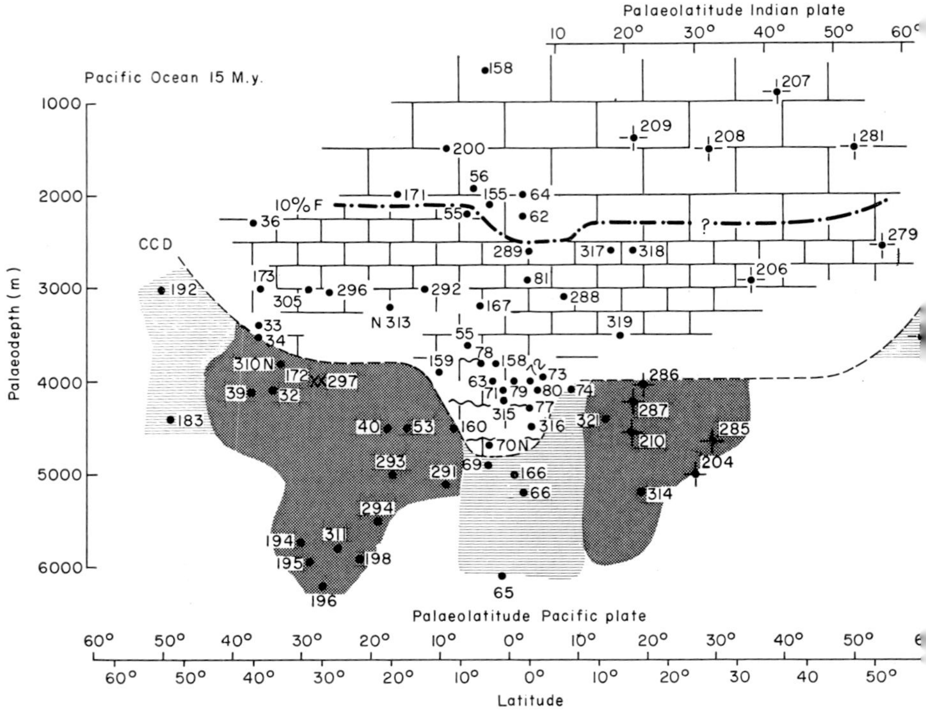

Palaeolatitude Indian plate
10
20°
30°
40°
50°
Pacific Ocean 15 M.y.
1000
2000
3000
4000
5000
6000
Palaeodepth (m)
10% F
CCD
Palaeolatitude Pacific plate
60°
50°
40°
30°
20°
10°
0°
10°
20°
30°
40°
50°
Latitude

(c)

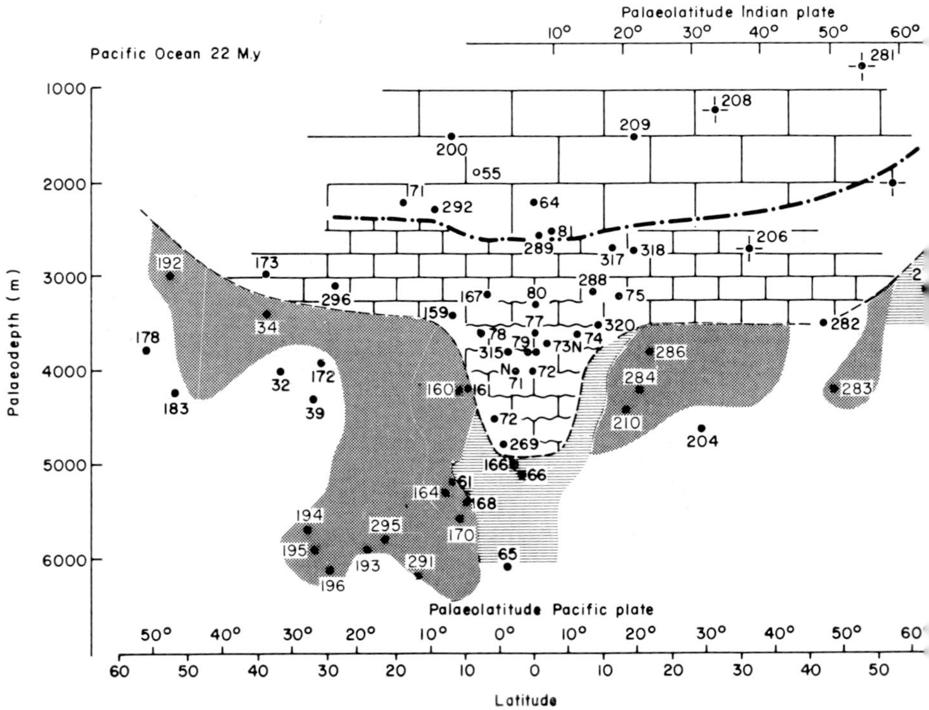

Palaeolatitude Indian plate
10°
20°
30°
40°
50°
60°
Pacific Ocean 22 M.y
1000
2000
3000
4000
5000
6000
Palaeodepth (m)
Palaeolatitude Pacific plate
50°
40°
30°
20°
10°
0°
10°
20°
30°
40°
50°
60
Latitude
60
50
40
30
20
10
0
10
20
30
40
50

(d)

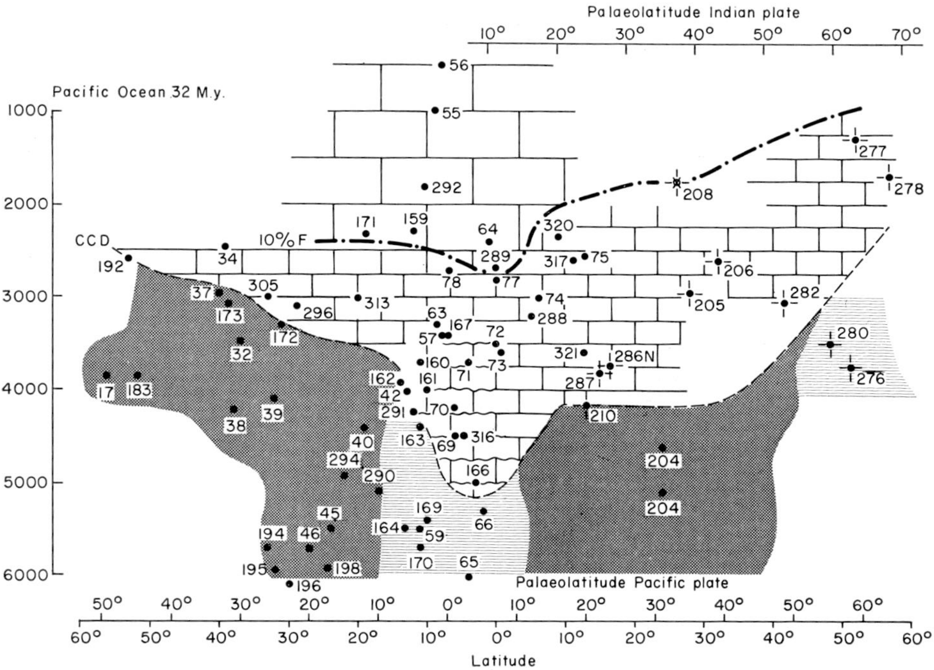

Pacific Ocean 32 M.y.
Palaeolatitude Indian plate
Palaeolatitude Pacific plate
Latitude
CCD
10% F

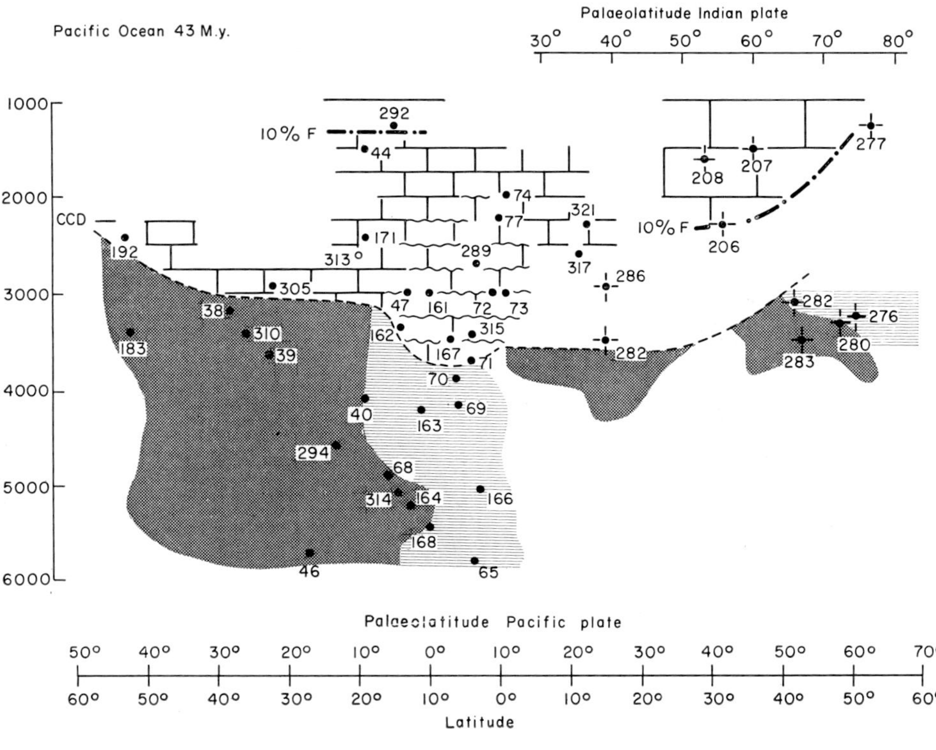

Pacific Ocean 43 M.y.
Palaeolatitude Indian plate
Palaeolatitude Pacific plate
Latitude
CCD
10% F

symmetry about the equator prior to and subsequent to these intervals. These changes may be associated with changes in the width of the equatorial current system coupled with possibly variations in the topography of the East Pacific Rise.

2. The continued depression of the CCD in the North Pacific north of palaeolatitude 10°N.

3. The Early Miocene shallowing of the CCD between palaeolatitudes 10°S and 60°S. This phenomenon continued in the Equatorial zone until the Late Miocene and terminated at the depression of the CCD between this interval and the Pliocene/Quaternary interval.

4. The depression of the CCD approximately south of palaeolatitude 10°S subsequent to the early Miocene.

The 10% foraminiferal boundary fluctuates in sympathy with the CCD in the North and Equatorial Pacific. In the South Pacific this boundary migrated upwards from its Late Eocene level to a shallower position during the Late Oligocene despite the considerable deepening of the CCD between these intervals. Subsequent to the Late Oligocene the downward migration of this boundary coincides with a deepening CCD.

THE HISTORY OF THE CCD IN THE ATLANTIC, INDIAN OCEAN AND PACIFIC BASINS AS A POSSIBLE INDICATOR OF DEEP-WATER CIRCULATION

Parallel trends between Ocean basins in the history of the CCD and 10% foraminiferal boundary are probably caused by the response of the ocean to global events. Examples of parallel trends are provided by the following:

1. The deepening of the CCD between the Late Eocene and Oligocene in all ocean basins.

2. The similarity between the pattern of fluctuations in the depth of the CCD and 10% foraminiferal boundary for the Caribbean Sea and the Atlantic and Pacific Oceans subsequent to the Oligocene.

3. The increase at low latitudes in the depths of the CCD and 10% foraminiferal boundary between the Late Miocene and Pliocene/Quaternary intervals in all ocean basins.

4. The downward migration of the CCD in the North Atlantic, Indian Ocean and North Pacific basins subsequent to the Oligocene, and in the South Pacific subsequent to the Early Miocene.

All of the above changes could have been induced by oceanwide changes in plankton production, carbonate dissolution rates, oceanic mixing rates, or combinations of all these factors.

Variations in the depth and history of the CCD between Ocean basins and within Ocean basins may have been caused by differences and changes in the

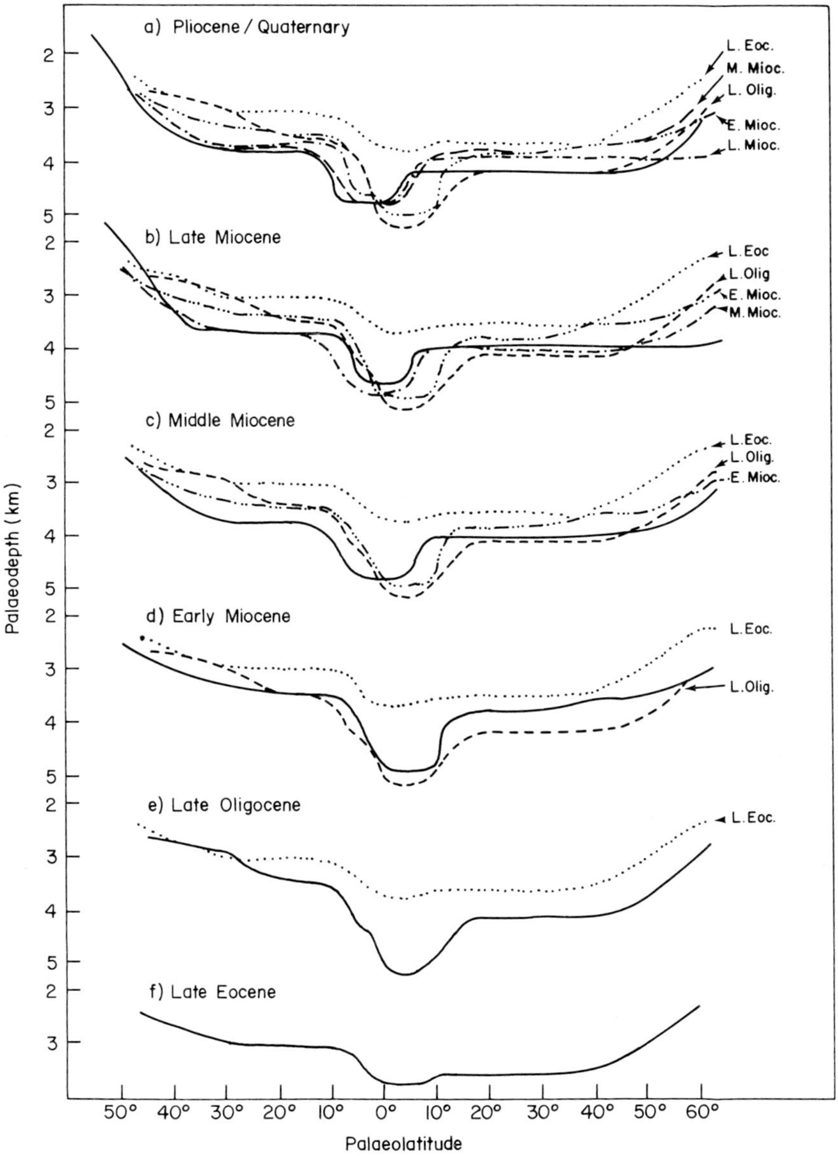

FIG. 20. Comparisons between the Pliocene to Quaternary (a), Late Miocene (b), Middle Miocene (c), Early Miocene (d), Late Oligocene (e) and Late Eocene (f), CCD curves derived for the Pacific Oean.

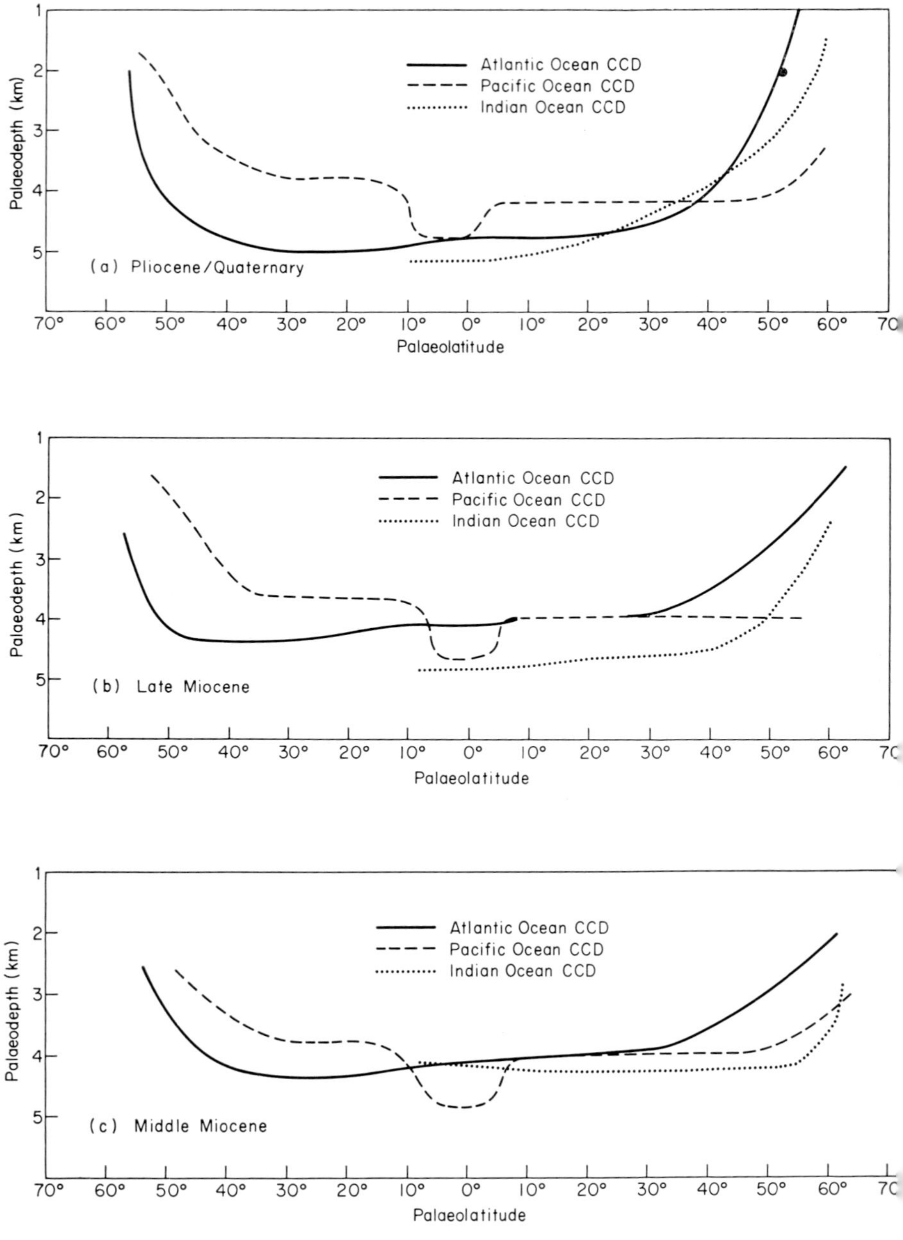

FIG. 21. A comparison between the CCD curves derived for the major ocean basins for (a) Pliocene to Quaternary interval, (b) Late Miocene, (c) Middle Miocene, (d) Early Miocene, (e) Late Oligocene and (f) Late Eocene.

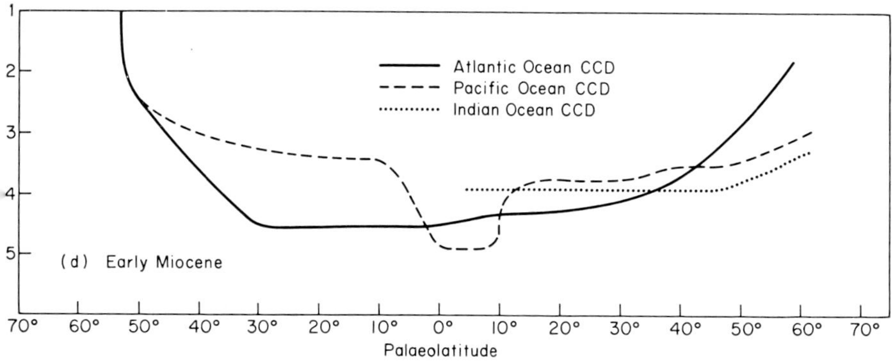
Atlantic Ocean CCD
Pacific Ocean CCD
Indian Ocean CCD
(d) Early Miocene
1
2
3
4
5
70° 60° 50° 40° 30° 20° 10° 0° 10° 20° 30° 40° 50° 60° 70°
Palaeolatitude

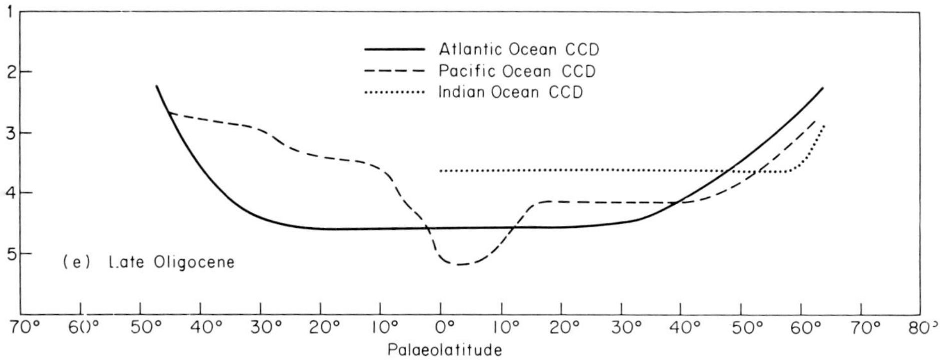
Atlantic Ocean CCD
Pacific Ocean CCD
Indian Ocean CCD
(e) Late Oligocene
1
2
3
4
5
70° 60° 50° 40° 30° 20° 10° 0° 10° 20° 30° 40° 50° 60° 70° 80°
Palaeolatitude

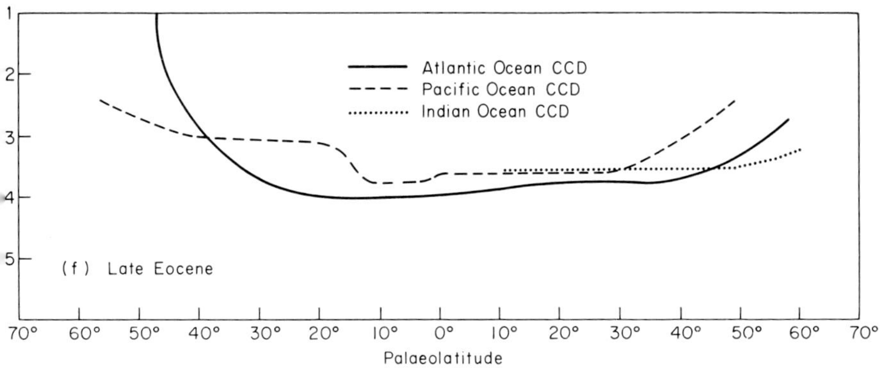
Atlantic Ocean CCD
Pacific Ocean CCD
Indian Ocean CCD
(f) Late Eocene
1
2
3
4
5
70° 60° 50° 40° 30° 20° 10° 0° 10° 20° 30° 40° 50° 60° 70°
Palaeolatitude

patterns of deep and surface water exchange between and within the ocean basins. Figures 21a–f show a comparison between the CCD curves for the major ocean basins from the Late Eocene. These figures show that a shallow Pacific and deep Atlantic CCD have existed since the Late Eocene. This difference which is more pronounced for the northern basins of each ocean indicates that relative to the Pacific the Atlantic already had "lagoonal" characteristics, i.e. an outflow of oxygen-rich deep water, low rates of calcite dissolution and carbonate deposition at depth (Berger, 1970b) during the Late Eocene.

The variation in the depths of the CCD between the North Atlantic and Caribbean basins since the Early Miocene (see Fig. 16a–f) may also have been caused by differences between their deep water circulation patterns which produce a less efficiently flushed Caribbean basin. Prior to the early Miocene the depth of the CCD may have been equal in both basins. The separation of these basins possibly through the elevation of an island arc barrier or the closure of a tectonic valve, e.g. a fault valley during the Late Oligocene or Early Miocene, could have led to the establishment of separate CCD levels in these basins.

The downward migration of the CCD between the Late Eocene and Late Oligocene is less pronounced in the Indian Ocean than in the other ocean basins, and probably reflects the nature of the deep and surface water circulation patterns of this basin during this interval. Kennett *et al.* (1972) and Kennett *et al.* (1974) have suggested that prior to the complete separation of Antarctica and Australia, Antarctic Bottom Water and southerly derived surface waters were deflected northward into the Indian Ocean basin by Australia. The occurrence of extensive Late Eocene and Oligocene regional hiatuses in the Indian Ocean (Davies *et al.*, 1975) is indicative of bottom water circulation and mixing. The absence of extensive biogenic siliceous deposits of Late Eocene to Late Oligocene age however suggests that this basin was not typically estuarine, with inflowing oxygen poor nutrient rich deep water, high rates of calcite dissolution shallow carbonate accumulation and the deposition at depth of lime poor muds and siliceous ooze (Berger, 1970). It is possible that inflowing deep water, derived from the closing Tethys Ocean, to the north, which may have been warm saline, oxygen rich and depleted in nutrients contributed both to the ventilation and nature of the Indian Ocean during the Late Eocene to Late Oligocene and possibly Early Miocene. The shallow CCD possibly reflects low near surface fertility in response to cool southerly derived surface water masses. The gradual downward migration to the CCD and 10% foraminiferal boundary since the Late Eocene may record a gradual increase in near surface plankton production in response to the waning influence of Tethys and the diversion of cool surface currents away from the Indian Ocean as a result of the increasing separation of Australia and Antarctica.

The Miocene pattern of CCD shoaling in the Atlantic basin, i.e. shoaling was initiated in the South Atlantic during the Early Miocene, had extended into the North Atlantic by the Middle Miocene (see Figs 15c and d) and persisted until the Late Miocene, suggests the increasing influence of a southerly derived water mass. This may have been provided by the generation of increasingly stable northward flowing supply of Antarctic Deep and Bottom Water in response to changing climatic conditions in the southern hemisphere since the Oligocene (Shackleton and Kennett, 1975; Savin *et al.*, 1975).

It is possible that the downward migration of the CCD in the North and South Pacific were also associated with changes in bottom mixing which were caused in part by the generation of a stable flow of Antarctic Bottom Water. The history of the CCD in the South Pacific in particular was probably influenced by the changes in bottom mixing and surface circulation associated with the breakup and separation of Australia and Antarctica, and with the opening of the Drake Passage between Antarctica and South America. The development of this channel, which was apparently wide enough to permit the passage of Antarctic Bottom Water from the southeast Pacific basin into the Atlantic Indian basin by the Middle Miocene (Barker *et al.*, 1974), could have had a significant influence on the residence time of deep water in the southern Pacific.

Fluctuations in the CCD and global events

THE QUATERNARY RECORD

The Quaternary contains a record of carbonate cycles which are of two origins. The high latitude cycles which are produced primarily by changes in calcareous plankton production in response to climatically induced fluctuations in surface temperature (McIntyre *et al.*, 1972). The geographically widespread mid- and low-latitude cycles which are produced by vertical changes in the CCD and in solution rates above the CCD. These changes coincide with climatic cycles and could have occurred in response to variations in any of the following potentially climatically controlled phenomena: plankton production, deep ocean mixing, variations in the extent of carbonate sedimentation on the continental shelves and in high latitudes. Investigations by Berger (1973b) and Broecker and Broecker (1974) indicate that glacial to interglacial changes have a negligible effect on calcareous planktonic producton both in the highly fertile equatorial Pacific and beneath the less fertile central water mass of the south-east Pacific. According to Broecker and Broecker (1974) the rates and patterns of ocean mixing may not be significantly affected by glacial to interglacial changes. Quaternary climatic events were accompanied by sea-level changes of approximately 130–150 m which resulted mainly from the growth and recession of the northern continental ice sheets (Flint, 1971). Evidence for an oscillating sea level and of associated marine transgressions and regressions is recorded in the erosional and depositional features around many of the world's coastlines. In tropical shelf areas of carbonate sedimentation, e.g. the Florida Shelf, Caribbean Islands, the East African and Red Sea Coasts and Western Australia exposed sequences of Quaternary limestones provide evidence of more extensive carbonate deposition during higher interglacial stands in sea level. In terms of the ocean's $CaCO_3$ budget more carbonate was lost to the continental shelves during interglacial periods than during glacial periods. Maximum shelf carbonate sedimentation also coincided with the maximum extent of high latitude carbonate sedimentation in the deep sea.

Climatically controlled changes in the patterns of high versus low latitude fractionation of carbonate and shallow shelf versus deep ocean basin carbonate fractionation appear in the light of present knowledge to provide the most promising explanation for mid- and low-latitude carbonate cycles.

Figures 22a and 22b present a model of oceanic carbonate deposition for the Quaternary. The model is based on the influence of changing climate on shelf versus basin and high- versus low-latitude fractionation of carbonate. During glacial intervals high latitudes are characterised by low rates of carbonate accumulation and a shallow CCD as a result of reduced calcareous plankton production. The extent of the continental shelves and of shelf carbonate sedimentation is reduced by marine regression, and subaerial weathering of exposed carbonates probably increases the supply of dissolved $CaCO_3$ to the ocean. The CCD in mid and low latitudes is depressed during interglacial periods, more equable climates result in increased calcareous plankton production in high latitudes, increased sedimentation rates and a depressed CCD. The continental shelves and shelf carbonate sedimentation are extended by marine transgression and there is an increase in the ratio of littoral to deep-sea carbonate accumulation rates. In mid and low latitudes an elevated CCD and increased solution rates compensate for the excess sedimentation of carbonate on the shelves and in high latitudes.

THE TERTIARY RECORD

The parallelism between oscillation in the CCD for the Atlantic Ocean, Caribbean Sea and the Pacific Ocean and, to a lesser extent, the Indian Ocean reflects the response of the CCD to global rather than local events. Models of carbonate deposition which are based on climatically controlled changes in

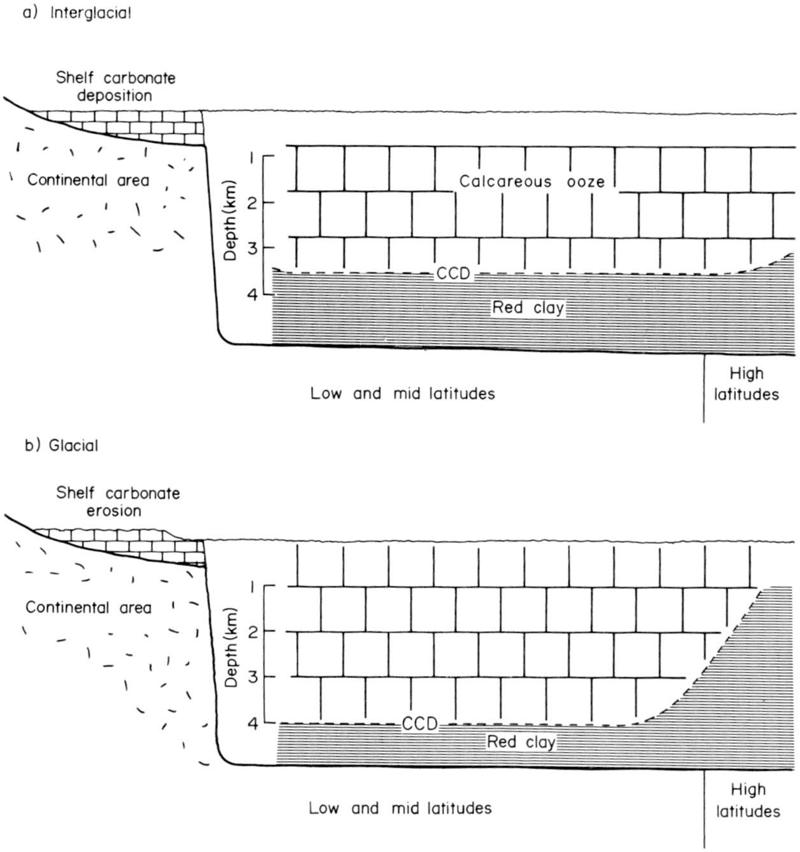

FIG. 22. Schematic model of deep-sea carbonate accumulation during the Quaternary, (a) for an interglacial cycle and (b) a glacial cycle.

the patterns of high versus low latitude fractionation of carbonate and shallow shelf versus ocean basin fractionation of carbonate, may also be applicable to fluctuations in the depth of the CCD since the Late Eocene.

Evidence for basin shelf fractionation of carbonate

At present the evidence for the influence of this process is based on observations of the similarity between patterns of CCD fluctuations and curves of worldwide sea level changes during the past 150 million years (van Andel, 1975). The fact that fluctuations in the CCD show parallel fluctuation with palaeotemperature, and that warm climates which may coincide with extensive marine transgressions are unfavourable for carbonate deposition in mid and low latitudes, has been used as indirect evidence for the removal of carbonates to the continental shelves (Berger, 1973a; Berger, 1974; Berger and Winterer, 1974). No attempt has been made to date to determine the influence of changes in shelf carbonate accumulation on the global carbonate budget.

The data produced in this report also show the coincidence between a shallow mid-latitude CCD in the Late Eocene and Middle Miocene and warm climates (see Fig. 23). The marked change from a shallow Late Eocene to a deep Oligocene CCD may have been associated with a marine regression which accompanied more extensive Oligocene glaciation of Antarctica. This regression may have been responsible for a change in the Tethyan and Indo-Pacific region from tropical shelf carbonate sedimentation in the Late Eocene to terrigenous sedimentation in the Oligocene (based on data in Kummel, 1970). A change of this nature could have made more carbonate available to the deep-sea basins. The temperature rise in high southern latitudes during the Middle Miocene (Shackleton and Kennett, 1975) may have contributed to a marine transgression, extended shelf carbonate accumulation and an elevated CCD.

The predominance of modern narrow shelf seas is high and mid-latitudes, which, because of their size and location, are inefficient carbonate traps, probably contributes to the deep nature of the modern CCD (van Andel, 1975).

Evidence of latitudinal fractionation of calcite in the deep sea

Ramsay (Ramsay *et al.*, 1973; Ramsay, 1974) suggested a possible connection between climatic changes and the latitudinal fractionation of carbonate during the Tertiary. Figure 23 shows a comparison between palaeotemperatures deduced for the tropical Pacific from the oxygen isotope composition of foraminiferal tests (Savin *et al.*, 1975) and the palaeodepths of the CCD at the equator in the Atlantic, Pacific and Indian Ocean. This figure demonstrates a negative relationship between palaeotemperature and CCD depths in the Atlantic and Pacific Oceans.

The Indian Ocean CCD does not show as clear a relationship, instead the CCD deepens from the Late Eocene to the Quaternary. This apparent independence between palaeotemperature changes and changes in the depth of the CCD may be related to the nature of the deep and surface water circulation pattern of this ocean from the Late Eocene to the Early Miocene. During this interval Antarctic bottom and surface currents were probably diverted into the Indian Ocean by Australia. This flow pattern was gradually modified by the separation of Australia and Antarctica (Kennett *et al.*, 1972; Kennett *et al.*, 1974). During the Late Eocene and Oligocene, and possible the Lower Miocene, the "estuarine" Indian Ocean may have been cool and infertile relative to the other ocean basins. Low plankton production and reduced rates of carbonate accumulation probably resulted in the shallow Late Eocene to Early Miocene CCD in this basin and probably contributed to the deep CCD in the other ocean basins.

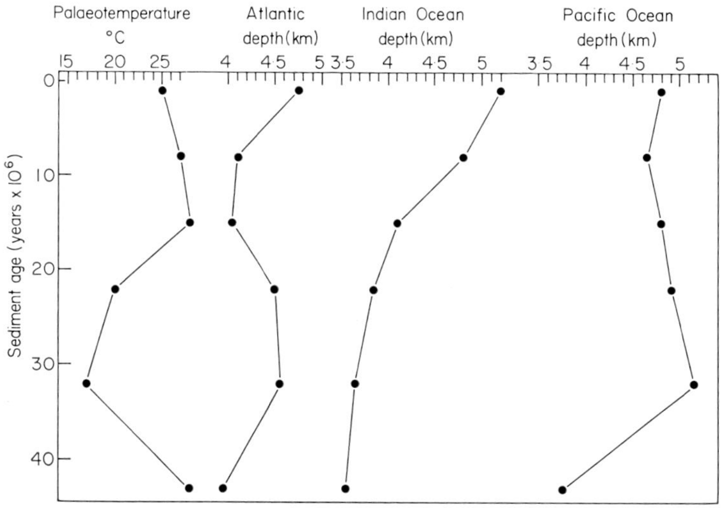

FIG. 23. A comparison between the inferred depth of the CCD at the equator (Pliocene/Quaternary) and palaeoequator for the Atlantic, Indian and Pacific Oceans and palaeotemperature since the Late Eocene.

The relationship between palaeotemperatures inferred from subantarctic cores by Shackleton and Kennett (1975) and the palaeodepths of the CCD at palaeolatitude 60°S is presented in Fig. 24. This figure shows that, as in low latitudes, a negative relationship existed between palaeotemperature and the depth of the CCD prior to the Middle Miocene. Subsequent to the Middle Miocene changes in the palaeodepths of the CCD show a positive correlation with changes in palaeotemperature. These data suggest that prior to the Middle Miocene the climatically controlled latitudinal fractionation of carbonate may have been unimportant or certainly a less important factor in the carbonate budget of the ocean.

The divergence between the response of the CCD in high and low latitudes to post Middle Miocene changes in surface temperature was probably associated with the change in the mixing patterns of water masses in the

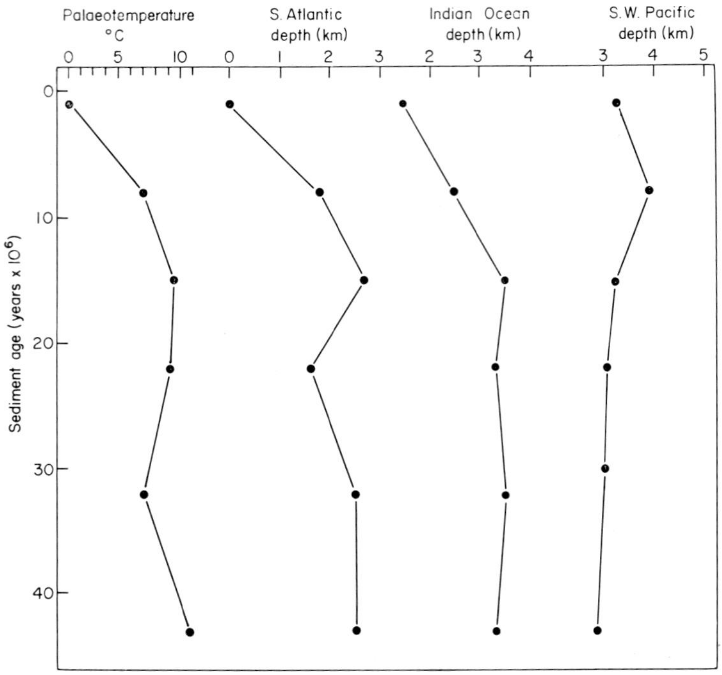

FIG. 24. A comparison between the inferred depth of the CCD at 60°S (Pliocene/Quaternary) and palaeolatitude 60° for the Atlantic, Pacific and Indian Oceans and palaeotemperature since the late Eocene.

southern hemisphere which accompanied the development of circum-Antarctic circulation similar to the Recent. This was possibly achieved by the opening or more probably widening of the Drake Passage to permit a substantial flow of deep water flow into the southern Atlantic by the Middle Miocene (Barker *et al.*, 1974). Savin *et al.* (1975) record a sudden divergence of high- and low-latitude palaeotemperatures in the Middle Miocene. Prior to this time parallel trends are recorded in high and low latitude patterns of temperature variation. Savin *et al.* attribute these changes to the development of a circum-Antarctic circulation and the near absence of mixing water masses across the circum-Antarctic current. It is probable that these conditions contributed to the development of a stable Antarctic ice sheet which is recorded by Shackleton and Kennett (1975). The establishment of the circum-Antarctic current and Antarctic convergence was probably a major factor in increased plankton production in the high southern latitudes during the Early to Middle Miocene. This is indicated by the increasing significance of biogenic silica in southeastern Pacific pelagic sediments during this interval (Hollister *et al.*, 1974) and the increased accumulation rates of pelagic sediments in the southwestern Atlantic (site 329) and southern Indian Ocean (see Fig. 7) during the Middle Miocene.

DISSOLUTION VERSUS TERRIGENOUS DILUTION AND DISSOLUTION VERSUS PLANKTON PRODUCTION

Figures 25 and 26 show a comparison between the percentages of calcium carbonate and solution resistant discoasters in sediments from Deep Sea Drilling Project sites in the Caribbean Sea and south Atlantic. The strongly negative correlation between these parameters indicates that the reduced carbonate content in the sediments is probably a product of increased dissolution of carbonate rather than mechanical dilution by an increasing supply of insoluble mineral fractions. These data do not, however, show whether migrations in the depth of the CCD are related to changes in rates of dissolution, to changes in the rates of the near surface production of calcite tests or to concomitant changes in both parameters. Changes in plankton production would cause changes in the calcite versus organic matter ratios of the bottom sediment and affect the production of interstitial CO_2 and probably the CO_2 levels in the deep water of the ocean. Thus high rates of fertility would result automatically in increased solution through the increased production of interstitial CO_2 and increased CO_2 levels in oceanic deep water. On the other hand reduced plankton production would result in a decrease in sedimentation rates and an increase in the resistance time of shells at the sediment water interface. If extended residence times occurred within the calcite undersaturated zone of the ocean then solution which may have been inhibited by higher rates of sedimentation would occur.

Regional hiatuses in the ocean

The recognition of regional hiatuses in deep sea sediments (Kennett *et al.*, 1972; Kennett *et al.*, 1974; Pimm and Hayes, 1972; Davies *et al.*, 1975) provides evidence of past advection and the production deep-ocean currents which were capable of transport and erosion. Together with the occurrence of bioturbated deposits they also indicate that the late Eocene to Recent ocean basins were not stagnant, but were mixed and aerated by deep and bottom water currents. The effects of mixing patterns on the levels of CO_2 and rates of solution in the ocean have probably contributed to the different depth patterns of the CCD in the major basins of the Late Eocene to Recent Ocean.

A suggested model for the origin of regional hiatuses in deep sea sediments is shown in Fig. 27. In this model it is assumed that three factors contribute to the formation of a deep-sea hiatus; these are:

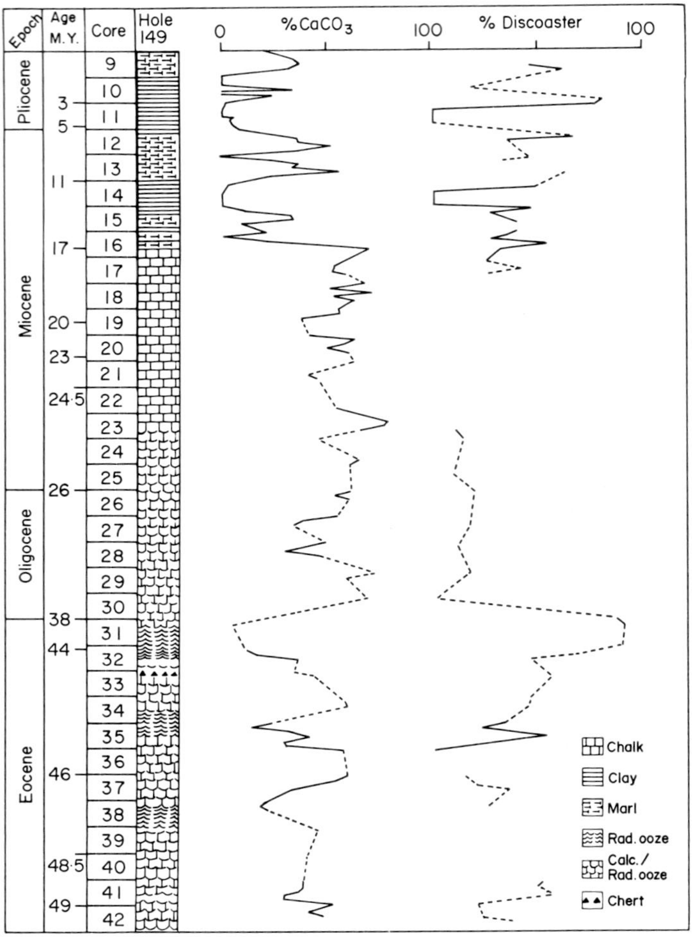

FIG. 25. A comparison between lithology and percentage of $CaCO_3$ and discoaster in sediments from the Deep Sea Drilling Project, Site 149 (after Ramsay *et al.*, 1973). Time scale after Berggren, 1969 (in Maxwell *et al.*, 1970).

1. The presence of a deep or bottom water current capable of transport and erosion.

2. Differential rates of plankton production which give rise to regional variations in sedimentation rates.

3. The coincidence of regions of low rates of plankton production and sedimentation with regions where high rates of dissolution at the sediment water interface are produced either by the presence of a cold corrosive current or by the build-up of CO_2 due to slow mixing and ponding of oceanic deep-water.

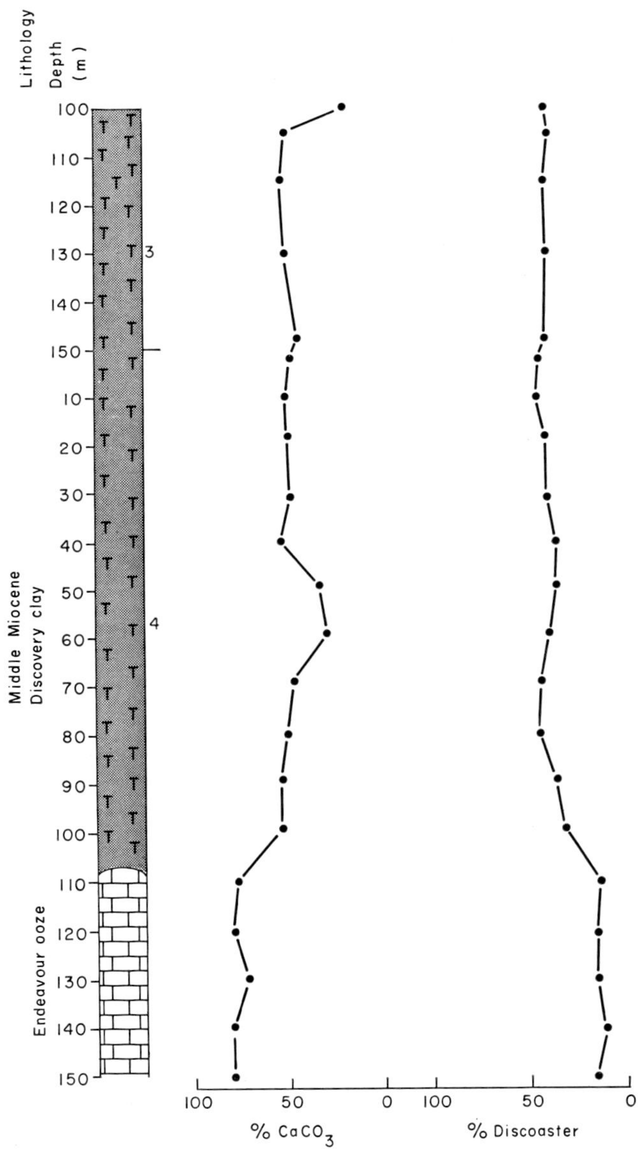

FIG. 26. A comparison between lithology and the percentages of $CaCO_3$ and discoasters in cores 15.6.3 and 15.6.4 of the Deep Sea Drilling Project, Site 15. (The $CaCO_3$ analyses were made by Mr Abdul el Hameed.)

Changes in any of these factors would influence the balance between deposition and erosion.

The separation of continents or the formation of transform fault valleys would provide new tectonic conduits for deep ocean currents and could alter regional patterns of erosion. The closure of such channels as a result of plate movements would have a similar effect. The displacement of ocean crust through sea-floor spreading could remove an area from a region of high

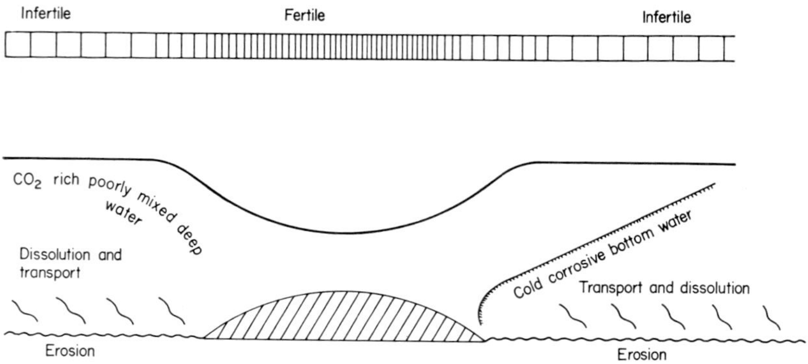

FIG. 27. A diagram of deep-sea erosion and sedimentation in an ocean fertility-deep water circulation frame.

fertility and sedimentation to one of low fertility and erosion. Regional changes in surface productivity or in rates of dissolution could also influence patterns of erosion.

THE STRATIGRAPHY OF HIATUS ABUNDANCE SINCE THE LATE EOCENE

The interpretation of the duration of hiatuses is not simple. Since erosion is involved the time of initiation of a hiatus cannot be precisely determined. The time of cessation of an erosional event could be obliterated through the removal of deposited sediments by subsequent erosional events. Regional variations in the velocity of deep-water currents and/or changes in near-surface plankton production could lead to erosional events whose duration may differ regionally. The synchroneity of renewed deposition does, however, enable one to date the termination of an erosional event in pelagic sequences on a regional scale.

The stratigraphical distribution of hiatuses in the North and South Atlantic, Indian Ocean, southwestern Pacific and east central Pacific basins is represented graphically in Figs 28 to 32. In these figures the distribution of hiatuses is expressed as a percentage, at million year intervals, of Deep Sea Drilling Project Sites where these time intervals were cored. These figures demonstrate the temporal nature of the occurrences of hiatus abundances in the ocean basins since the Late Eocene.

The occurrence of hiatuses in the Atlantic Ocean

Hiatus time graphs for the North and South Atlantic basins are shown in Figs 28 and 29. In the South Atlantic (see Fig. 28) a short interval characterised by fewer hiatuses (23 m.y. B.P. to 26 m.y. B.P.) separates the Late

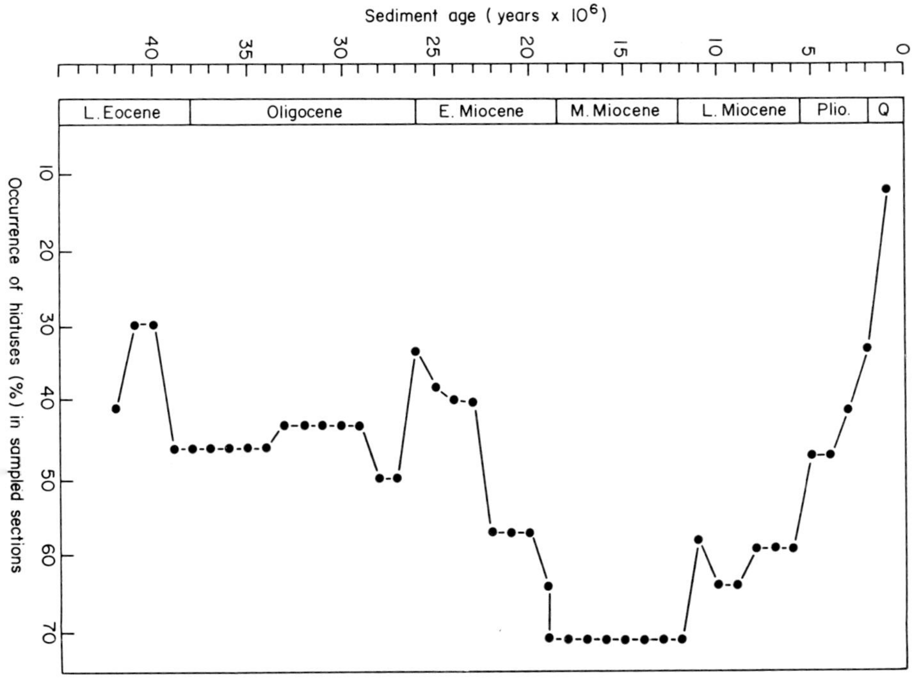

FIG. 28. The occurrence of hiatuses in the South Atlantic.

Eocene to Late Oligocene peak of abundant hiatuses from an Early Miocene to Pliocene peak of even more abundant hiatuses. It is possible, however, that the broad peaks which define the hiatus abundant intervals contain an erosion induced amalgamation of a number of hiatuses of short duration.

The hiatus time graph for the North Atlantic (see Fig. 29) indicates that the hiatuses were more common from the Late Eocene to Early Miocene and

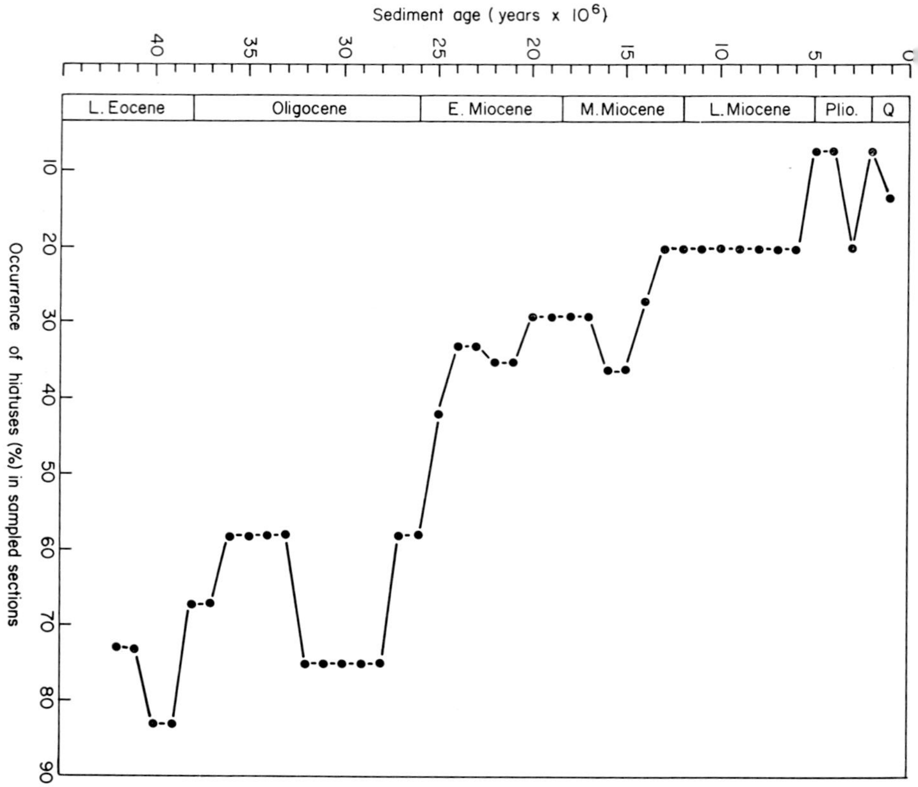

FIG. 29. The occurrence of hiatuses in the eastern North Atlantic.

that with the exception of minor abundance peaks at 15–16 m.y. B.P. and 3 m.y. B.P. the abundance of hiatuses in this basin declined from approximately 25 m.y. B.P. In contrast to the South Atlantic the Late Eocene to Early Miocene interval contains a major abundance peak in the Late Eocene and another in the Late Oligocene.

The temporal nature of hiatus abundances in the basins is probably related to changes in deep water circulation. The fact that the abundance peaks in the South Atlantic are broader probably reflects either a greater abundance of hiatuses or more intense erosion in this basin.

The Late Eocene to Early Miocene phase of erosion in the South Atlantic was possibly associated with the initiation of Antarctic Bottom Water or with an increased flow of deep-water from the Antarctic which accompanied deteriorating climatic conditions in the southern hemisphere. In the North

Atlantic erosive action by two currents may have contributed to the formation of hiatuses during this interval: the southerly flowing North Atlantic Deep Water (Western Boundary Undercurrent) and the southerly derived Antarctic Bottom Water. It is at present impossible to differentiate between the relative importance of these currents in the North Atlantic Basin during the Late Eocene.

In the South Atlantic the increased abundance of hiatuses in the Early Miocene to Pliocene intervals suggests that erosion intensified during the Early Miocene, perhaps in conjunction with the establishment of a more stable supply of Antarctic Bottom Water. The decline in the abundance of hiatuses in the North Atlantic Basin subsequent to 25 m.y. B.P. however appears to indicate a decreasing competence of the southerly derived Antarctic Bottom Water to erode in this basin.

The occurrence of hiatuses in the Indian Ocean

The hiatus time graph for the Indian Ocean (see Fig. 30) shows similarities with the hiatus time curves for the North and South Atlantic basins. Like the South Atlantic maximum hiatus abundance is represented by broad peaks which span the Late Eocene to Late Oligocene and Early to Late Miocene. These are separated by an Early Miocene interval of fewer hiatuses which is of longer duration than a similar interval in the South Atlantic. A trend towards the decline in the abundance of hiatuses subsequent to 27 m.y. B.P. is similar to the North Atlantic trend. It is possible that this trend reflects a waning of the Antarctic Bottom Water current in both basins. The influence of this current on sedimentation in the Indian Ocean probably decreased as increasing proportions of this water mass were channelled away from the Indian Ocean as a consequence of the separation of Australia and Antarctica.

The occurrence of hiatuses in the Pacific Ocean

The stratigraphy of hiatus abundance in the southwestern Pacific and eastern central Pacific areas is shown in Figs 31 and 32. The central Pacific curve (Fig. 31) shows a minor peak of hiatus abundance in the Late Eocene. Subsequent to 33 m.y. B.P. the abundance of hiatuses increases gradually to 20 m.y. B.P. From this time a major interval of abundant hiatuses persisted until the Late Pliocene (2 m.y. B.P.). This trend is opposed to the trends recorded for the North Atlantic and Indian Ocean basins, but crudely follows the trend in hiatus abundances recorded for the South Atlantic. The gradual increase in hiatus abundance in this region may record the influx of an increasingly stable supply of Antarctic Bottom Water into this region subsequent to the Late Oligocene. Prior to this time a substantial proportion

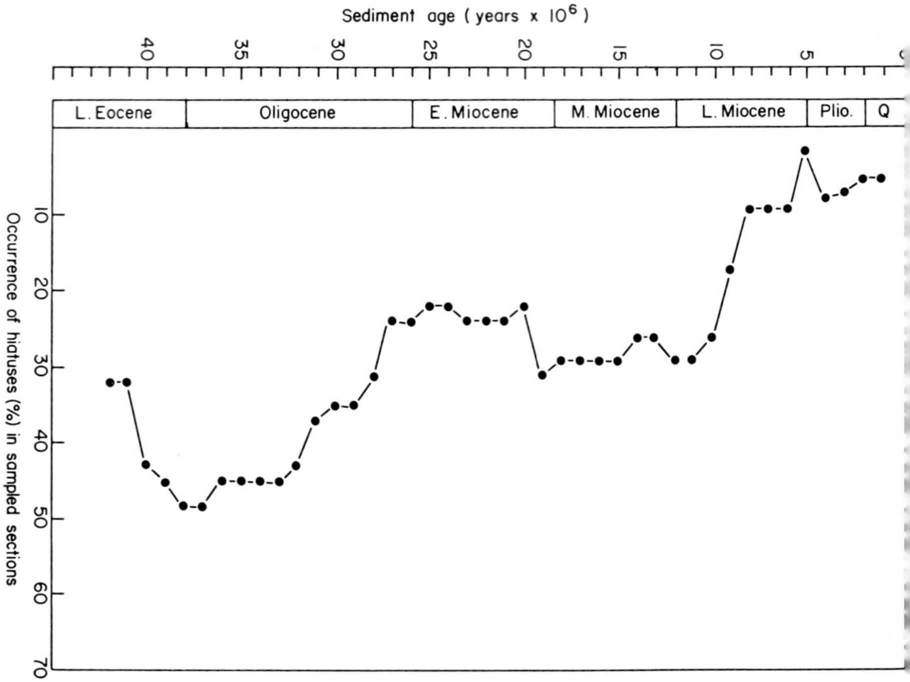

FIG. 30. The occurrence of hiatuses in the Indian Ocean.

of the Antarctic Bottom Water may have been diverted southwards through the Tasman Sea and away from the Central Pacific (Kennett *et al.*, 1972).

The hiatus time curve for the southwestern Pacific (see Fig. 32) shows that hiatuses were more common in this region than in the central Pacific. A broad Late Eocene–Oligocene peak is followed by a decrease in hiatuses through the Oligocene. Subsequently two major peaks of hiatus abundance are developed. A broad Late Oligocene to Middle Miocene peak and a narrower Late Miocene to Pliocene peak. The hiatus stratigraphy of this region is similar to that of the South Atlantic. The fact that both regions were characterised by a consistantly greater abundance of hiatuses from the Late Eocene to Pliocene possibly reflects the continuous influence of Antarctic Bottom Water flow. Differences between these regions probably express variations in their tectonic histories.

THE GEOGRAPHICAL DISTRIBUTION OF HIATUSES

The maximum extent of hiatuses for the Late Eocene, Oligocene, Early and Middle Miocene, Late Miocene and Pliocene to Recent intervals is shown in

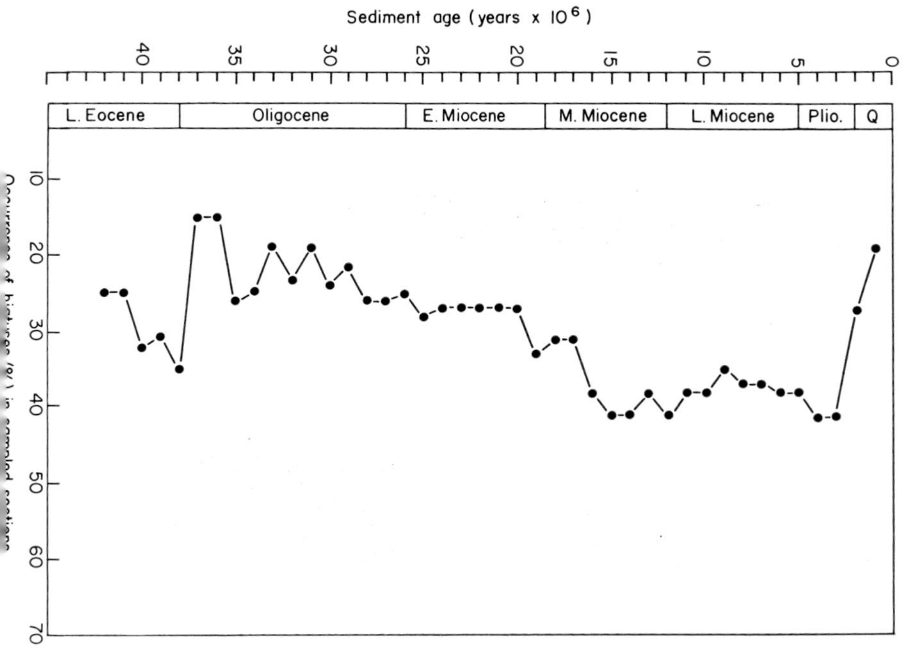

FIG. 31. The occurrence of hiatuses in the eastern central Pacific.

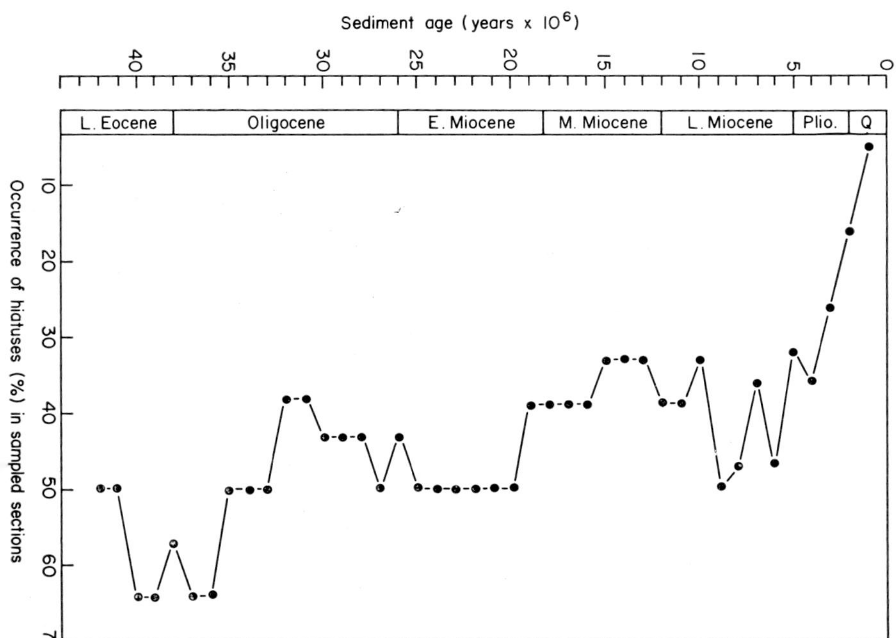

FIG. 32. The occurrence of hiatuses in the southwestern Pacific.

Figs 34 to 38. A model of the pattern of Recent abyssal circulation compiled by Heezen and Hollister (1971) shown in Figs 33a and b is used as a basis for interpreting the changing patterns of hiatuses shown in Figs 34–38. In this model the influence of two currents is indicated. Firstly, the Antarctic Bottom Water Current which, from its primary sources in the Ross and Weddell seas, forms a circum-Antarctic current and flows northward into the major ocean basins through transform fault valleys and other tectonic channels (Heezen and Hollister, 1971). Secondly, the Arctic Deep Water Current which flows southward through the Norwegian sea into the North Atlantic basin where it mixes with the deep water which is advected in the North Atlantic (North Atlantic Deep Water) and flows southward as a western boundary current along the western margin of the Atlantic into the South Atlantic basin. Here it overrides and mixes with the denser northward-flowing Antarctic Bottom Water.

The Pliocene to Quaternary Interval

This interval defines a period of minimal erosion in the Western Atlantic, Indian Ocean and the Coral and Tasman seas (see Figs 34a, b). With the exception of the Norwegian Sea the regions in which hiatuses are recorded are situated in areas which at present lie in the path of Antarctic Bottom Water Flow. The well-defined region of hiatuses in the Norwegian Sea possibly reflects erosion by the southerly flow of Arctic Bottom Water.

The Late Miocene Interval

In contrast to the subsequent interval the Late Miocene is characterised by the more extensive development of hiatuses in the Western basins of the North and South Atlantic, off the Western margin of Australia, in the Tasman and Coral Seas and in the North Pacific. Hiatuses are also recorded in the Indian Ocean and Norwegian Sea (see Figs 35a, b). The pattern of hiatuses shown in Figs 35a and b could have been produced by the present abyssal flow regime. The more extensive development of erosion surfaces may indicate changes in the intensity of abyssal flow.

The Early to Middle Miocene Interval

The regional pattern of erosion during the Early and Middle Miocene (see Figs 36a and b) could have been produced by a pattern of abyssal circulation similar to the Recent.

In contrast to the preceding and succeeding intervals the regional extent of hiatuses is reduced in the western Atlantic, Tasman and Coral Seas and in the North Pacific, this possibly indicates a change in the intensity of bottom

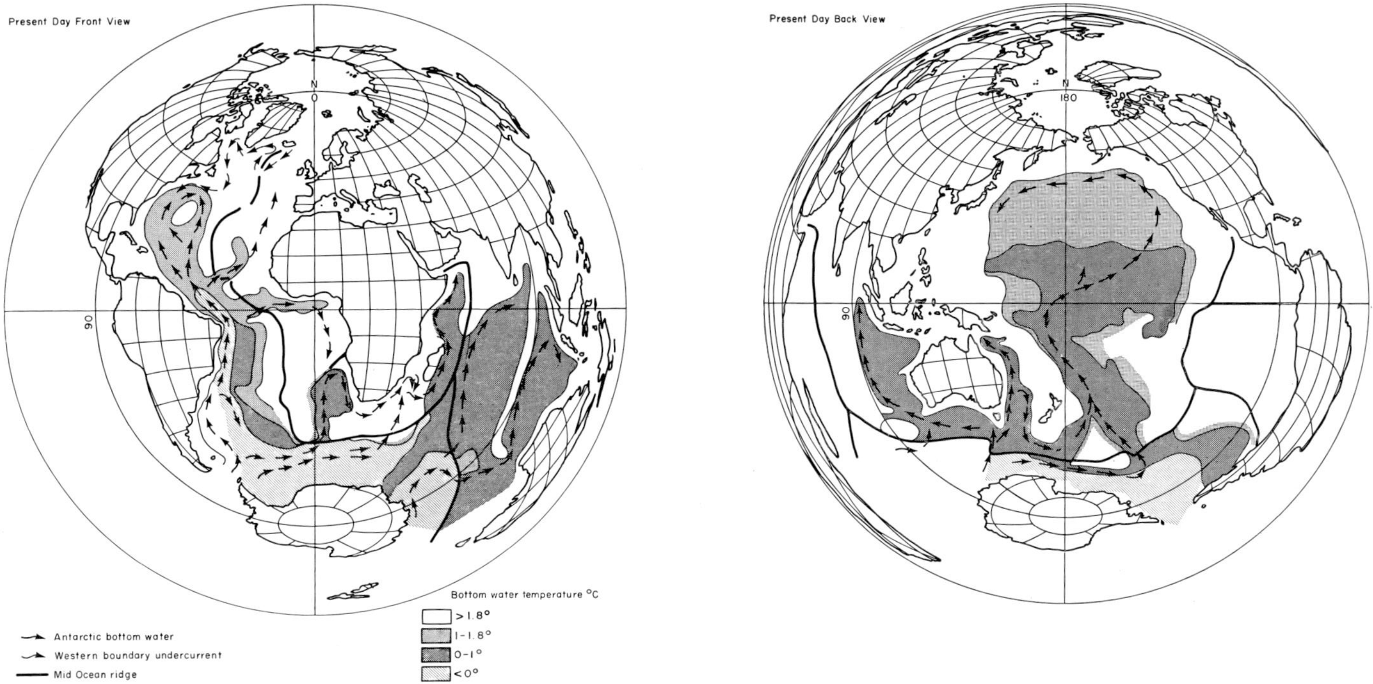

FIG. 33. Oceanic abyssal circulation (after Heezen and Hollister, 1971).

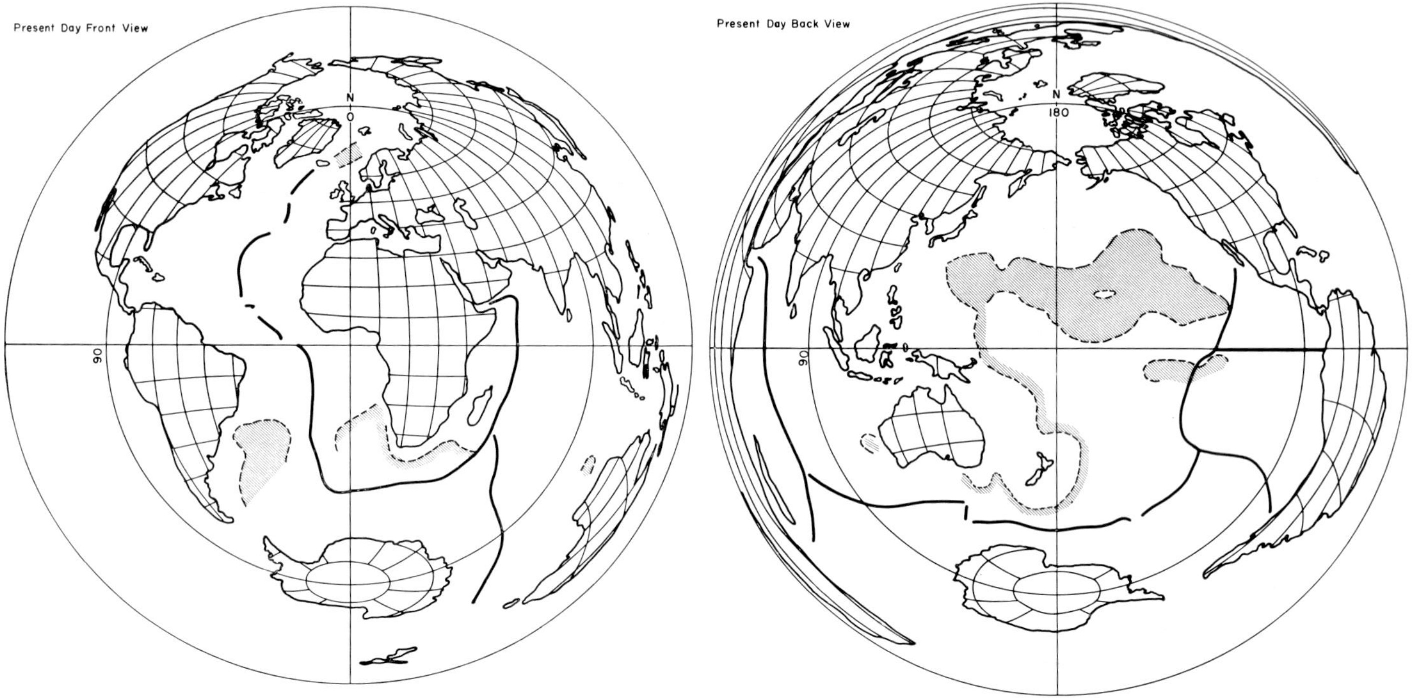

FIG. 34. The geographical distribution of hiatuses during the Pliocene/Quaternary interval.

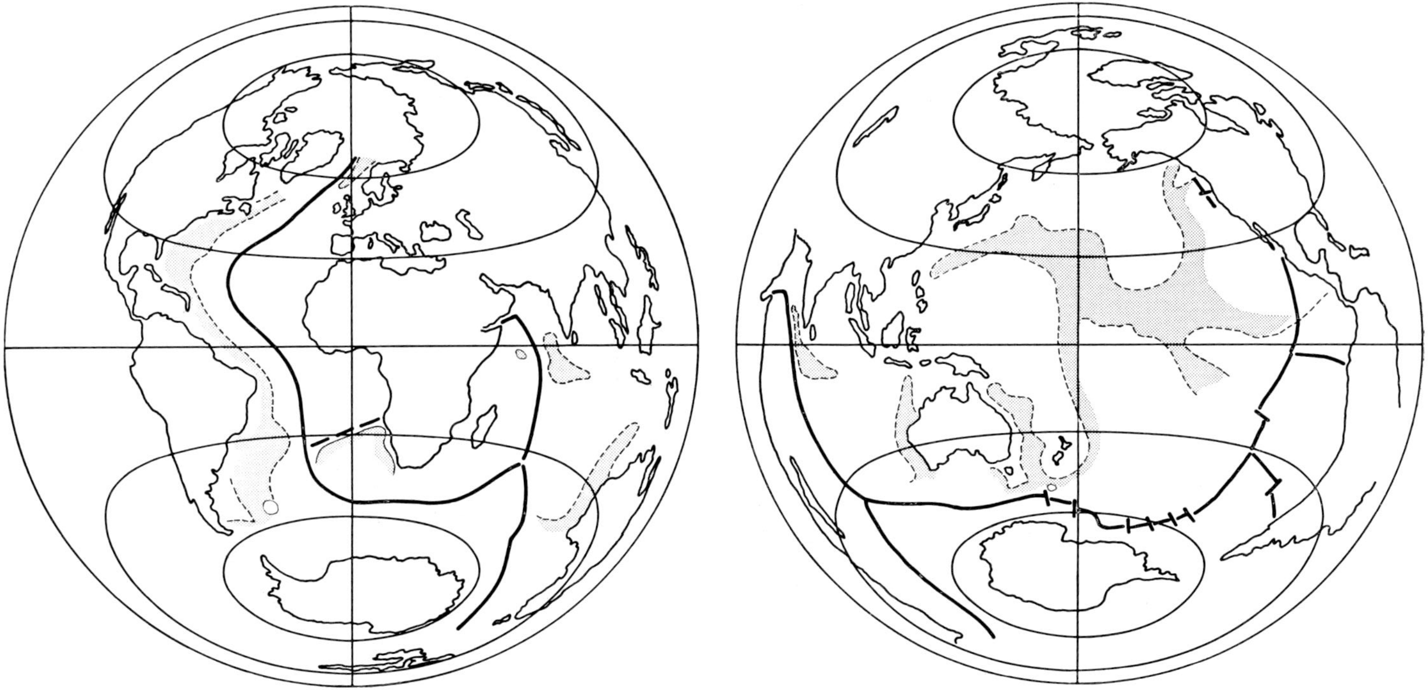

FIG. 35. The geographical distribution of hiatuses during the Late Miocene interval. Lambert Equal Area Projection for 10 m.y. B.P.

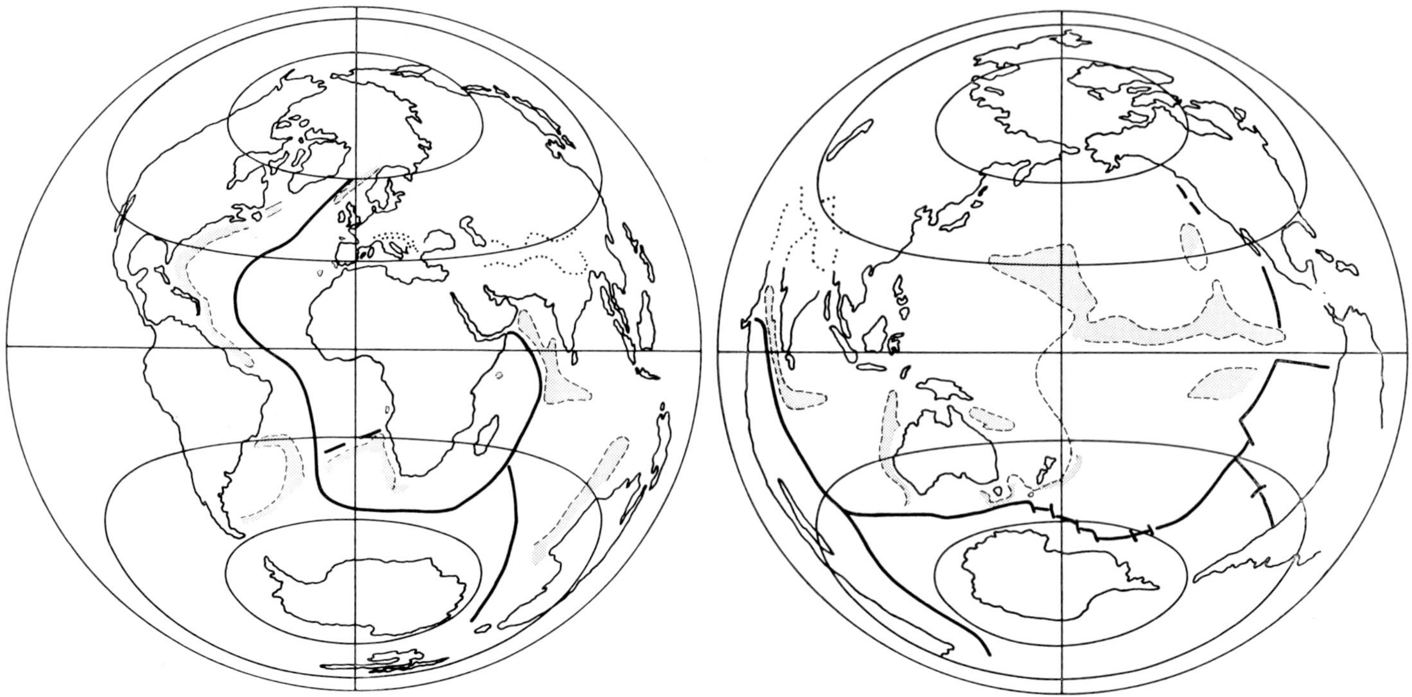

FIG. 36. The geographical distribution of hiatuses during the Early Miocene to Middle Miocene. Lambert Equal Area Projection for 20 m.y. B.P.

circulation or possibly to the inhibiting influence of increased sedimentation rates on erosion.

The Oligocene Interval

The regional pattern of hiatuses during the Oligocene is shown in Figs 37a and b. In contrast to the succeeding interval erosion was more extensive in the western Atlantic where it continued into the Venezuelan Basin (Edgar *et al.*, 1973) in the Indian Ocean, and Coral and Tasman Seas.

The pattern of erosion indicated for the Oligocene could have been produced by tectonic modifications of the present abyssal flow pattern. The absence of a sill of a fault controlled conduit between the Caribbean and North Atlantic basins would have allowed the inflow of Antarctic Bottom Water into the Venezuelan basin. The deflection of Antarctic Bottom Water into the Indian Ocean by Australia may account for its erosion pattern during the Oligocene (Davies *et al.*, 1975). The gradual diversion of this current away from the Indian Ocean as Australia and Antarctica separated probably accounts for a decrease in the extent of erosion by the Early Miocene (Davies *et al.*, 1975).

The presence of a well-defined region of erosion in the Norwegian Sea is possibly associated with the development of a southerly flow of Arctic Bottom Water into the North Atlantic Basin associated with Late Oligocene subsidence of the Iceland Faroe Ridge.

The Eocene Interval

The pattern of hiatuses for the Norwegian Sea, Tasman and Coral Seas and the Pacific basin differs from the erosion patterns derived for the Oligocene (see Figs 38a and b).

In the Norwegian Sea the absence of a well-defined region of erosion probably reflects the inhibiting influence of an elevated Iceland Faroe Ridge on the southerly flow of Arctic Bottom Water into the North Atlantic basin (Talwani *et al.*, 1975).

In the southwestern Pacific region the occurrence of hiatuses suggests that the flow of Antarctic Bottom Water which was diverted into the Indian Ocean may have entered the Pacific North of the Australian continental block and flowed northwards into the North Pacific and southwards through the Coral and Tasman Seas. From the Tasman Sea this current was deflected northwards into the South Pacific.

This flow pattern may have terminated by the Oligocene as a result of tectonism in the western Pacific (Moberly, 1972) and along the northern margin of Australia (Veevers, 1969) which was probably associated with the northward movement of Australia.

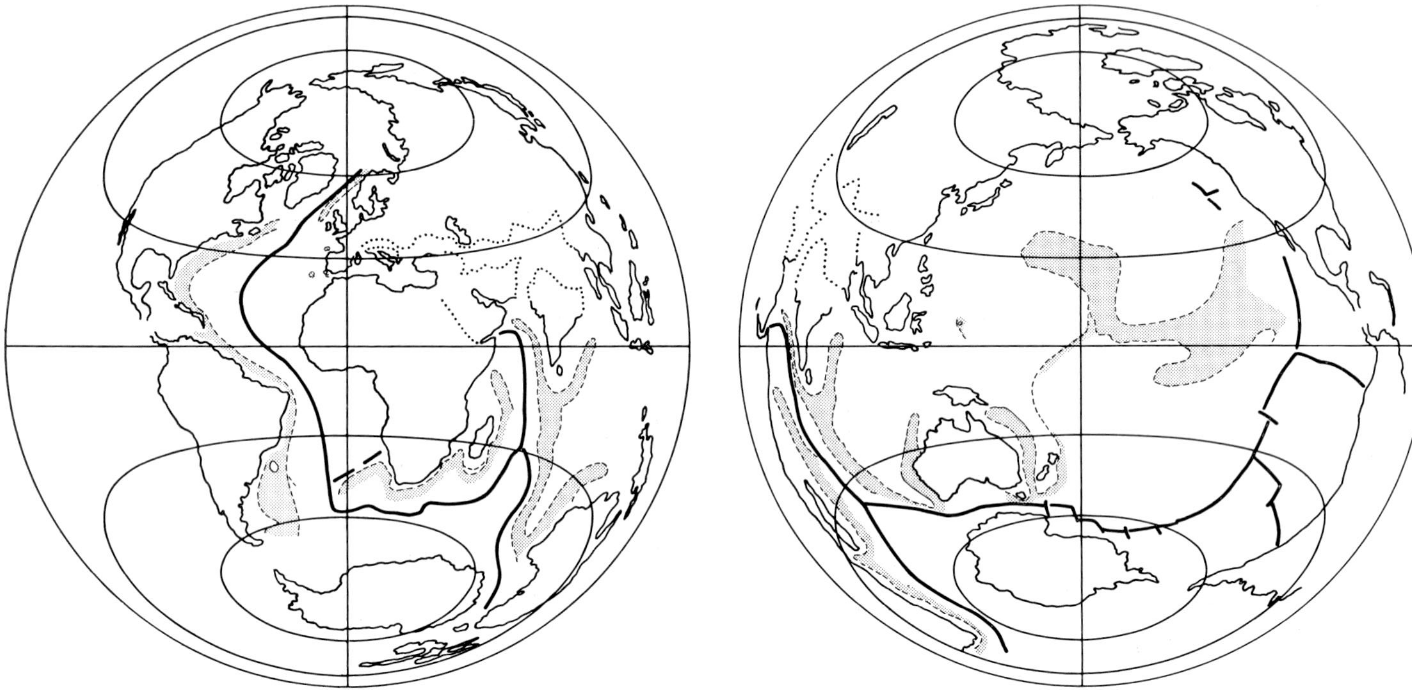

FIG. 37. The geographical distribution of hiatuses during the Oligocene. Lambert Equal Area Projection for 40 m.y. B.P.

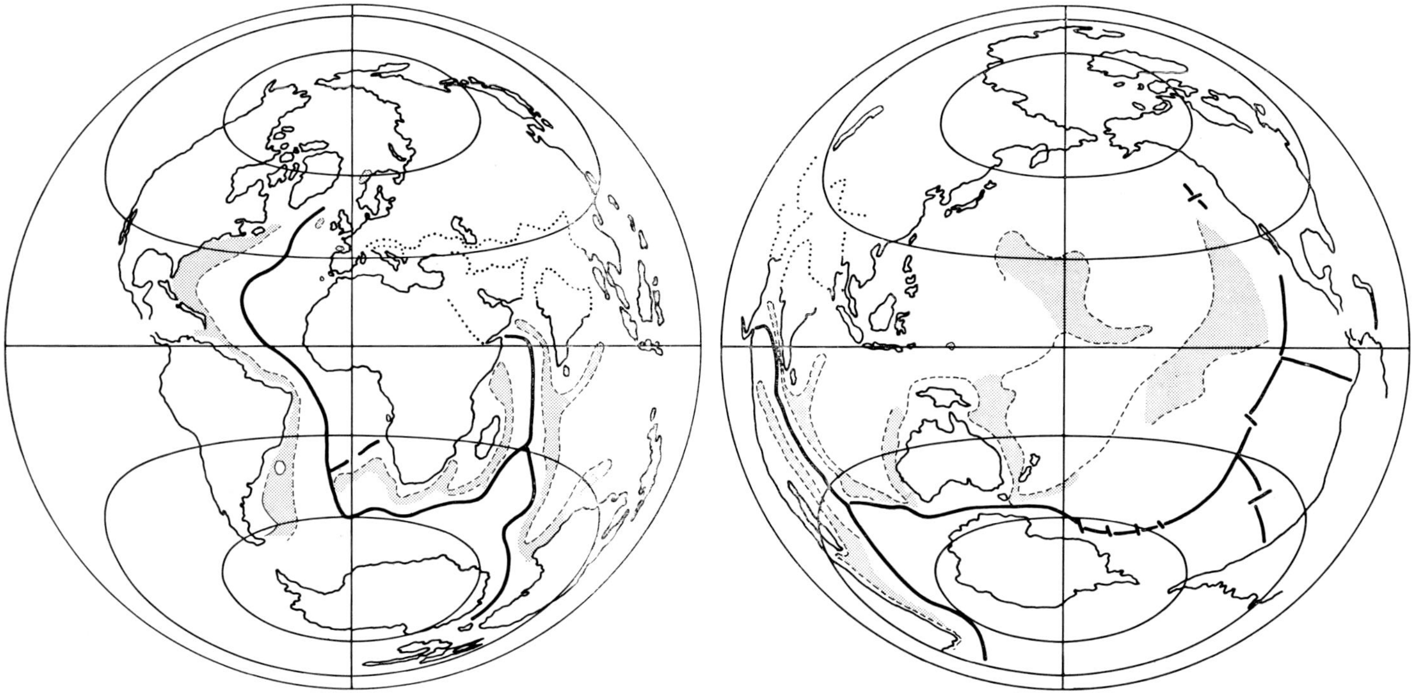

FIG. 38. The geographical distribution of hiatuses during the Late Eocene. Lambert Equal Area Projection for 40 m.y. B.P.

Hiatus abundance, the history of the CCD and ocean circulation: a summary

The history of hiatuses on the ocean floor like the history of the CCD is related to the ocean's response to global and regional events, and it is thus not surprising to find coincidences between the histories of these parameters in the ocean basins. From the foregoing discussions two events appear to have had a significant influence on the Ocean since the Late Eocene. One event, the climatic deterioration in the southern hemisphere caused the formation of Antarctic sea ice and the growth of Antarctic glaciers to sea level by the Late Eocene to Early Oligocene, and resulted in the production of Antarctic Bottom Water at a temperature close to freezing (Shackleton and Kennett, 1975). The other event was concerned with the diversion of this current into circum-polar flow as a result of the gradual separation of Australia and Antarctica during the Late Eocene and Oligocene (Kennett *et al.*, 1972; Kennett *et al.*, 1974) and the Late Oligocene opening of the Drake Passage (D. G. Jenkins, personal communication). This had widened sufficiently to permit substantial circumpolar flow possibly by the Middle Miocene (Barker *et al.*, 1974). The formation of the circumpolar current probably contributed to climatic deterioration in the southern hemisphere and to the establishment of a stable Antarctic ice cap by the Late to Middle Miocene.

The response of the ocean to these changes is reflected in the changing patterns of erosional and sedimentary events in the ocean basins.

THE LATE EOCENE INTERVAL

The Late Eocene Ocean was characterised by abundant and geographically widespread hiatuses and by an elevated CCD and shallow carbonate accumulation. It is possible that a low supply of carbonate to the ocean floor as a result of low plankton production, removal of carbonate on extended continental shelves, high rates of solution or a combination of all these factors may have been the dominant factor in this imbalance between sedimentation and erosion.

There are few data concerning the bottom and surface circulation of the Late Eocene Ocean. Data published by Shackleton and Kennett (1975) suggest that warm bottom water (approx. 9°C) formed near the Antarctic during the Late Eocene. The geographical pattern of hiatus distribution (see Figs 38a and b) suggests that this water flowed northward into the Atlantic, and Indian Oceans, and that some of this flow may have been directed north of Australia and entered the central Pacific, via an equatorial channel, and the southern Pacific via the Coral and Tasman Seas.

The extension of hiatuses from the North Atlantic into the Venezuelan Basin and the probable coincidence between the depth of the CCD in these basins during the Late Eocene suggests that their deep and bottom waters were equally well mixed, and that Antarctic Bottom Water may have entered the Caribbean Sea.

The deep nature of the Atlantic CCD relative to the other major basins suggests that this basin was lagoonal relative to the Indian and Pacific Ocean and was, as now, characterised by an outflow of deep water probably generated by advection in the north Atlantic. It is possible that this southward flowing current eroded and became well mixed with the warm Antarctic Bottom Water and also contributed to the development of hiatuses in the North Atlantic Basin.

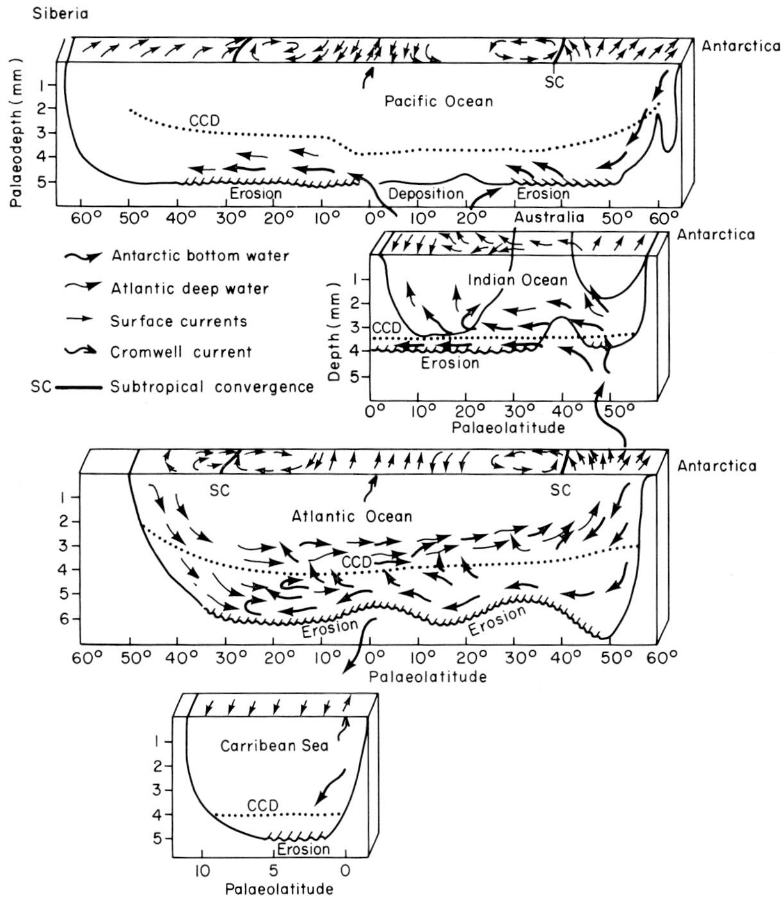

FIG. 39. Schematic block diagram illustrating presumed major elements of Late Eocene abyssal circulation and the influence of tectonically controlled assymetry on this circulation pattern.

Figure 39 shows, schematically, the proposed pattern of Late Eocene deep-water circulation and deep ocean mixing discussed above, and includes a suggested scheme for surface water circulation. In this pattern the absence of a well-defined Antarctic convergence in the southern Ocean is based on the fact that the CCD at the palaeoequator and at palaeolatitude 60°S responded in a similar manner to temperature changes from the Late Eocene to at least 22 m.y. B.P. The proposed pattern of equatorial surface flow is also different from the Recent. It is suggested that there was some flow of equatorial water out of the Indian Ocean into the Pacific, and that Pacific equatorial water flowed across the Isthmus of Panama into the Caribbean and Atlantic basins via an east-flowing equatorial surface countercurrent and subsurface Cromwell current. The latter conclusion is based on the occurrence of Eocene biogenic siliceous deposits in the Caribbean and Atlantic basins (Ramsay, 1971).

THE OLIGOCENE INTERVAL

This interval was characterised by abundant geographically widespread hiatuses (see Figs 37a, b) and by the deepening of the CCD which was more pronounced in the Atlantic and Pacific basins than in the Indian Ocean.

The persistence of a geographically extensive imbalance between erosion and sedimentation despite deep carbonate accumulation suggests that the intensity of bottom water circulation and its competence to erode may have increased during the Oligocene. Shackleton and Kennett (1975) record a 5°C decline in bottom water temperature close to the Eocene–Oligocene boundary which coincides with a deteriorating Antarctic climate and may reflect the production of Antarctic Bottom Water at temperatures close to freezing. A major change in the species assemblage structure of deep-sea benthic ostracodes about the Eocene–Oligocene boundary (Benson, 1975) also indicates an important palaeoenvironmental event in the ocean at this time.

The rapid deepening of the CCD about the Eocene–Oligocene boundary in the Atlantic and Pacific Oceans possibly occurred in response to the increased rates of ocean mixing, and reduction in the extent of shallow shelf seas and shelf carbonate sedimentation which may have accompanied climatic deterioration and ice accumulation in Antarctica. In the Indian Ocean a low input of calcite shells and a shallow CCD contributed to the deep CCD of the other ocean basins and to the extensive nature of hiatuses in that basin.

The geographical distribution of hiatuses suggests that the Late Eocene pattern of oceanic deep-water circulation was modified during the Oligocene. The flow of Arctic Bottom Water into the North Atlantic basin as a result of the subsidence of the Iceland Faroe Ridge probably contributed to the efficiency of the flushing mechanism of the Atlantic Ocean, and to the development of hiatuses in the western North Atlantic. The separation of Australia and Antarctica and the closure of the equatorial passage between the Indian and Pacific Oceans diverted a large proportion of the eastward flowing Antarctic Bottom Water into the South Pacific. From the Late Oligocene Antarctic Bottom Water probably entered the Central and North Pacific via the southwestern Pacific.

THE MIOCENE INTERVAL

Miocene trends in hiatus abundance and distribution and in the history of the CCD probably record changes in oceanic circulation and mixing during this interval.

In the North Atlantic basin a decline in hiatus abundance and the trend towards the downward migration of the CCD may reflect increasing production of oxygen-rich North Atlantic Deep Water through the Miocene. This would have resulted in a more efficiently flushed lagoonal basin in which the deep accumulation of carbonate and the erosion of Antarctic Bottom Water by southerly flowing North Atlantic Deep Water altered the balance between erosion and deposition in favour of sediment accumulation.

A minor interlude of hiatus abundance during the Middle Miocene which coincides with a period of maximum CCD elevation may be associated with an increase in the intensity and supply of cold corrosive Antarctic Bottom Water.

In the South Atlantic the trend in the abundance of hiatuses from an Early Miocene low to a Middle Miocene peak and a Late Miocene–Pliocene decline coincides with the Early to Middle Miocene elevation of the CCD and 10% foraminiferal boundary and the downward migration of these levels from the Late Miocene to Pliocene. Both phenomena may be related to changes in the supply and intensity of Antarctic Bottom Water flow which possibly occurred in response to the widening of the Drake passage and the growth of a stable Antarctic ice cap.

The absence of hiatuses in the Venezuelan basin since the Early Miocene apparently corresponds with a divergence between the CCD levels in the Caribbean and North Atlantic basins. This probably reflects a change in the deep mixing pattern of the Caribbean basin in response to its tectonic isolation at depth from the Atlantic.

In the Indian Ocean a reduction in the geographical extent of hiatuses was accompanied by the depression of the CCD and 10% foraminiferal boundary. These changes probably reflect modifications in the deep and surface circulation patterns of this ocean which occurred in response to the separation of Australia and Antarctica, the northward drift of India, and the establishment of circum-Antarctic flow by the Middle Miocene. The changes in circulation which probably resulted in increased near surface plankton production, improved bottom mixing and deep carbonate accumulation changed the nature of the Indian Ocean in favour of deposition by the Early Miocene. The establishment of essentially modern oceanographic conditions during the Middle–Late Miocene is indicated by the increased accumulation rates of biogenic siliceous sediments in the southern Indian Ocean during the Middle Miocene, and the occurrence of equatorial siliceous deposits in the eastern Indian Ocean by the Late Miocene.

The Pacific trends in hiatus abundance from an Early Miocene low to Middle and Late Miocene peaks and a Late Miocene to Quaternary decline coincide with a downward migration of the CCD, from its elevated Lower Miocene position, in the northern and southern basins of this ocean. In the equatorial Pacific the CCD shallowed during the Early to Late Miocene and deepened subsequent to 8 m.y. B.P. Post Middle Miocene oscillations in the CCD are recorded south of 50° south.

It is possible that the downward migration of the CCD in the northern and southern basins and increasing hiatus abundance reflect an increase in the intensity of Antarctic Bottom Water flow and in the efficiency of deep-water mixing of the Pacific basin. The changes in the CCD depth at the equator and south of 50°S possibly record a response to the changing oceanography of the Pacific, and to external global phenomena, i.e. climatic changes in the southern hemisphere, and variations in the shelf versus deep sea fractionation of calcium carbonate.

The Early–Middle Miocene pattern of bottom water circulation was probably similar to the Late Oligocene but presumably increasing proportions of Antarctic Bottom Water and surface water were diverted into the widening passage between Antarctica and the northward drifting Australia. The most significant event in the evolution of oceanic circulation during the Miocene was the widening of the Drake Passage and development of the present circum-Antarctic current system by the Middle Miocene. The establishment of circumpolar flow and the thermal isolation of surface water in a circum-Antarctic ring probably contributed to the development of the temperature regime in the southern hemisphere and to the establishment of a stable Antarctic ice cap. The development of the Antarctic convergence during the Early to Middle Miocene which probably occurred in response to these changes in climate and circulation resulted in increased plankton production in the southern hemisphere and established latitudinal fractionation as an important factor in the carbonate budget of the ocean.

Figure 40 presents a schematic model of the circulation and bottom mixing pattern of the Ocean which was established by the Late to Middle Miocene and has persisted to the Recent.

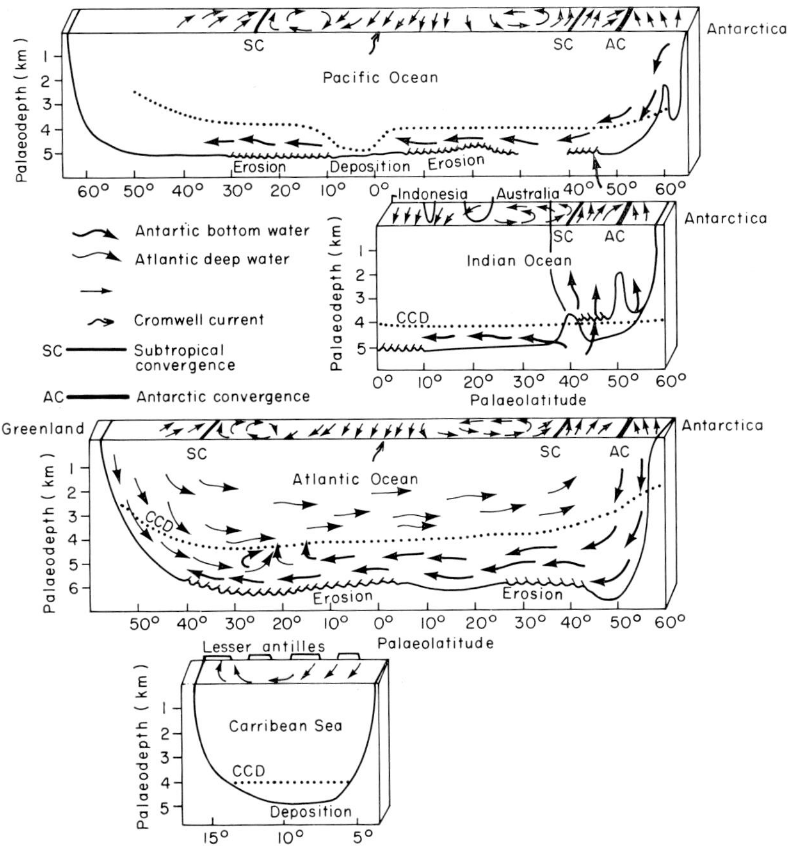

FIG. 40. Schematic block diagram of presumed main elements of Late–Middle Miocene abyssal circulation and the influence of tectonically controlled assymetry on this circulation pattern.

THE PLIOCENE TO QUATERNARY INTERVAL

During this interval a decrease in hiatus abundance in all ocean basins is accompanied by a marked downward migration of the CCD and 10% foraminiferal boundary. These phenomena may reflect an alteration in the erosion versus sedimentation relationship in favour of sedimentation as carbonate accumulation increased in mid and low latitudes. The downward migration of the CCD could have occurred in response to more efficient ocean mixing, to decreased carbonate accumulation in the cold Arctic and Antarctic regions, or to a shift in carbonate deposition away from the continental shelves to the deep sea.

Acknowledgements

I am grateful to Dr. A. G. Smith and Mrs. G. Drewry for allowing me to use their unpublished Lambert Equal Area Projections for 10 m.y. B.P., 20 m.y. B.P. and 40 m.y. B.P. (Figs 35–38), and to Mr. Abdul El-Hameed for the $CaCO_3$ analyses used in Fig. 26. The sediment samples quoted in Figs. 25 and 26 were provided by the Deep Sea Drilling Project.

APPENDIX 1. Localities of cores used to construct CCD history for the Quarternary.

Core	Latitude (°N)	Longitude (°W)	Depth (m)
	EQUATORIAL PACIFIC		
RC11–208	05°21′	139°58′	4720
RC11–209	03°39′	140°04′	4400
V24–53	01°51′	129°01′	4473
V24–55	02°03′	134°38′	4189
V24–58	02°16′	141°40′	4490
V24–59	02°34′	145°32′	4662
V24–60	02°48′	149°00′	4859
V24–62	03°04′	153°35′	4834
	EAST PACIFIC RISE		
V19–54	17°02′	113°54′	2964
V19–55	17°00′	114°11′	3177
V19–56	17°00′	114°32′	3147
V19–58	16°58′	115°12′	3334
V19–59	16°58′	115°33′	3309
V19–61	16°57′	116°18′	3422
V19–65	16°39′	124°23′	3867
V19–66	16°24′	127°38′	4096
V19–67	16°10′	129°52′	3952
V19–68	15°37′	131°58′	4199
V19–69	15°55′	134°17′	3970
V19–70	15°39′	138°35′	3939

APPENDIX 2. Localities and estimated basement ages of DSDP sites.

Site	Latitude	Longitude	Depth (m)	Basement Age
		ATLANTIC OCEAN		
4	24°28.68′N	73°47.52′W	5319	>Tithonian
5	24°43.59′N	73°38.46′W	5361	>Tithonian
6	30°50.39′N	67°38.86′W	5125	†135 m.y.
7	30°08.04′N	68°17.80′W	5185	†135 m.y.
8	35°23.00′N	76°33.20′W	5169	†140 m.y.
9	32°46.40′N	59°11.70′W	4965	†100 m.y.
10	32°51.73′N	52°12.92′W	4697	75 m.y.
11	29°56.58′N	44°44.80′W	3556	21 m.y.
12	19°41.73′N	26°00.03′W	4557	†155 m.y.
13	06°02.40′N	18°13.71′W	4585	†135 m.y.
14	28°19.89′S	20°56.46′W	4343	40 m.y.
15	30°53.38′S	17°58.99′W	3927	24 m.y.
16	30°20.15′S	15°42.79′W	3527	11 m.y.
17	28°02.74′S	06°36.10′W	4265	33 m.y.
18	27°58.72′S	08°00.70′W	4018	26 m.y.
19	28°32.08′S	23°40.63′W	4677	49 m.y.
20	28°31.47′S	26°50.73′W	2113	67 m.y.
22	30°00.31′S	35°15.00′W	2134	†80 m.y.
23	06°08.75′S	31°02.60′W	5079	†70 m.y.
26	10°53.55′N	44°02.57′W	5169	¶
27	15°51.39′N	56°52.76′W	5251	†80 m.y.
98	25°22.95′N	77°18.68′W	2796	†180 m.y.
105	34°53.72′N	69°10.40′W	5251	†180 m.y.
106	36°26.01′N	69°27.69′W	4500	†180 m.y.
111	50°25.57′N	46°22.05′W	1797	¶
112	54°01.00′N	46°36.24′W	3657	65 m.y.
116	57°29.76′N	15°23.97′W	1038	§
135	35°20.80′N	10°25.46′W	4152	150 m.y.
136	34°10.13′N	16°18.19′W	4169	121 m.y.
137	25°55.53′N	27°03.64′W	5361	105 m.y.
139	23°31.14′N	18°42.26′W	3047	164 m.y.
140	21°44.97′N	21°47.52′W	4483	151 m.y.
141	19°25.16′N	23°59.91′W	4148	¶
142	03°22.15′N	42°23.49′W	4372	¶
327	50°52.28′S	46°47.02′W	3812	¶
328	49°48.67′S	36°39.53′W	5013	¶
329	50°55.19′S	46°53.00′W	2626	¶
330	50°55.19′S	46°53.00′W	2626	¶
332	36°52.72′N	33°38.46′W	1818	3 m.y.
333	36°50.45′N	33°40.05′W	1665	3 m.y.
334	37°02.13′N	34°24.87′W	2632	10 m.y.
335	37°17.74′N	35°11.92′W	3198	11 m.y.

APPENDIX 2 (*contd.*)

Site	Latitude	Longitude	Depth (m)	Basement Age
		ATLANTIC OCEAN		
336	63°21.06′N	07°47.27′W	811	45 m.y.
337	64°52.30′N	05°20.15′W	2631	†38 m.y.
338	67°47.11′N	05°23.26′E	1297	†60 m.y.
341	67°20.10′N	06°06.64′E	1439	†60 m.y.
342	67°57.04′N	05°45.73′E	1303	†60 m.y.
354	05°53.95′N	44°11.78′W	4052	†70 m.y.
355	15°42.59′S	30°36.03′W	4896	†80 m.y.
356	28°17.22′S	41°05.28′W	3203	†105 m.y.
357	37°39.31′S	35°33.59′W	2109	—
359	34°59.10′S	04°29.83′W	1658	†45 m.y.
360	35°50.75′S	18°05.79′E	2949	†105 m.y.
361	35°03.97′S	15°26.91′E	4549	†105 m.y.
362	19°45.45′S	10°31.95′E	1325	—
363	11°34.32′S	11°58.30′E	2248	†105 m.y.
		CARIBBEAN SEA		
29	14°47.11′N	69°19.36′W	4247	>Turonian
31	14°56.60′N	72°01.63′W	3369	>Turonian
147	10°42.68′N	65°10.45′N	892	>Turonian
148	13°25.12′N	63°43.25′W	1232	>Turonian
149	15°06.25′N	69°21.85′W	3972	>Turonian
150	14°30.69′N	69°21.35′W	4545	>Turonian
151	15°01.02′N	73°24.58′W	2029	>Turonian
153	13°58.33′N	72°26.08′W	3932	>Turonian
		INDIAN OCEAN		
211	09°46.53′S	102°41.95′E	5528	¶
*212	19°11.34′S	99°17.84′E	6240	100 m.y.
213	10°12.71′S	93°53.77′E	5609	†63 m.y.
*214	11°20.21′S	88°43.08′E	1671	†60 m.y.
215	08°07.30′S	86°47.50′E	5321	60 m.y.
*216	01°27.73′N	90°12.48′E	2262	65 m.y.
217	08°55.57′N	90°32.33′E	3020	100 m·y.
*219	09°01.75′N	72°52.67′E	1764	¶
*220	06°30.97′N	70°59.02′E	4036	†57 m.y.
222	20°05.49′N	61°30.56′E	3546	†57 m.y.
234	04°28.96′N	51°13.48′E	4738	¶
*236	01°40.62′S	57°38.85′E	4504	†57 m.y.
*237	07°04.99′S	58°07.48′E	1640	¶
*238	11°09.21S	70°31.56′E	2844	32 m.y.
239	21°17.67′S	51°40.73′E	4971	60 m.y.
240	03°29.24′S	50°03.22′E	5082	¶
241	02°22.24′S	44°40.77′E	4054	‡90 m.y.

APPENDIX 2 (*contd.*)

Site	Latitude	Longitude	Depth (m)	Basement Age
		INDIAN OCEAN		
*242	15°50.65′S	41°49.23′E	2275	¶
245	31°32.02′S	52°18.11′E	3879	63 m.y.
246	33°37.21′S	45°09.60′E	1030	¶
248	29°31.78′S	37°28.48′E	4994	¶
249	29°56.99′S	36°04.62′E	2088	¶
250	33°27.74′S	39°22.15′E	5119	80 m.y.
*251	36°30.26′S	49°29.08′E	3489	26 m.y.
252	37°02.44′S	59°14.33′E	5032	¶
*253	24°52.65′S	87°21.97′E	1962	¶
254	30°58.15′S	87°53.72′E	1253	38 m.y.
255	31°07.87′S	93°43.72′E	1144	¶
256	23°27.35′S	100°46.46′E	5361	105 m.y.
257	30°59.16′S	108°29.99′E	5278	110 m.y.
258	33°47.69′S	112°28.42′E	2793	¶
259	29°00.37′S	112°00.42′E	5709	¶
261	12°00.57′S	117°00.54′E	5687	¶
262	10°00.52′S	123°00.51′E	2315	¶
*265	53°32.45′S	112°02.68′E	3582	†20 m.y.
*266	56°24.13′S	109°56.74′E	4173	†28 m.y.
*267	59°15.74′S	104°29.30′E	4564	†38 m.y.
268	63°56.99′S	105°09.34′E	3544	†50 m.y.
		PACIFIC OCEAN		
32	37°07.63′N	127°33.38′W	4758	38 m.y.
33	39°28.48′N	127°33.81′W	4284	32 m.y.
34	39°28.21′N	127°16.54′W	4322	31 m.y.
36	40°59.08′N	130°06.58′W	3273	15 m.y.
37	40°58.74′N	140°43.11′W	4682	32 m.y.
38	38°42.12′N	140°21.27′W	5137	56 m.y.
39	32°48.28′N	139°34.29′W	4929	60 m.y.
42	13°50.56′N	140°11.31′W	4848	51 m.y.
44	19°18.50′N	169°00.00′W	1478	¶
47	32°26.90′N	157°42.70′E	2689	¶
49	32°24.10′N	156°36.00′E	4282	¶
51	33°28.50′N	153°24.30′E	5981	¶
53	18°02.00′N	141°11.50′E	4629	¶
*55	09°18.10′N	142°32.90′E	2850	30 m.y.
*56	08°22.40′N	143°33.60′E	2508	30 m.y.
57	08°40.90′N	143°32.00′E	3300	30 m.y.
59	11°46.80′N	147°34.90′E	5554	¶
*62	01°52.20′N	141°56.30′E	2602	35 m.y.

APPENDIX 2 (*contd.*)

Site	Latitude	Longitude	Depth (m)	Basement Age
		PACIFIC OCEAN		
63	00°50.13′N	147°53.39′E	4486	38 m.y.
64	01°44.53′S	158°36.58′E	2060	¶
65	04°21.21′N	176°59.16′E	6142	100 m.y.
66	02°22.61′N	166°07.31′W	5310	†95 m.y.
69	06°00.00′N	152°51.93′W	5326	87 m.y.
*70	06°20.08′N	140°21.72′W	5059	†60 m.y.
*71	04°28.28′N	140°18.91′W	4419	†60 m.y.
72	00°26.49′N	138°52.02′W	4326	†57 m.y.
73	01°54.58′N	137°28.12′W	4387	54 m.y.
*74	06°14.20′S	136°05.80′W	4431	†53 m.y.
*75	12°31.00′S	134°16.00′W	4181	†39 m.y.
76	14°05.90′S	145°39.64′W	4597	¶
77	00°28.90′N	133°13.70′W	4290	38 m.y.
78	08°57.00′N	127°21.35′W	4377	34 m.y.
*79	02°33.02′N	121°34.00′W	4573	26 m.y.
80	00°57.72′S	121°33.22′W	4411	26 m.y.
81	01°26.49′N	113°48.54′W	3865	18 m.y.
*82	02°35.48′N	106°56.52′W	3706	9 m.y.
*83	04°02.08′N	95°44.25′W	3645	10 m.y.
84	05°44.92′N	82°53.29′W	3096	8 m.y.
155	06°07.38′N	81°02.62′W	2752	19 m.y.
157	01°45.70′S	85°54.17′W	2591	9 m.y.
*158	06°37.36′N	85°14.16′W	1953	17 m.y.
159	12°19.92′N	122°17.27′W	4484	29 m.y.
*160	11°42.27′N	130°52.81′W	4940	36 m.y.
161	10°50.25′N	139°57.a1′W	4939	45 m.y.
162	14°52.19′N	140°02.61′W	4854	50 m.y.
163	11°14.66′N	150°17.52′W	5320	73 m.y.
164	13°12.14′N	161°30.98′W	5499	115 m.y.
165	08°10.07′N	164°51.06′W	5040	85 m.y.
166	03°45.07′N	175°04.08′W	4950	120 m.y.
*167	07°04.01′N	176°49.05′W	3176	135 m.y.
168	01°42.02′N	173°35.09′E	5420	¶
170	11°48.00′N	177°37.00′E	5774	¶
*171	19°07.09′N	169°27.06′W	2290	120 m.y.
172	31°32.23′N	133°22.36′W	4767	35 m.y.
173	39°57.71′N	125°27.12′W	2927	26 m.y.
183	52°34.30′N	161°12.33′W	4708	53 m.y.
184	53°42.64′W	170°55.39′W	1910	¶
185	54°25.73′N	169°14.59′W	2110	¶
186	51°07.81′N	174°00.34′W	5422	¶
189	54°02.14′N	170°13.38′E	3437	¶
190	55°33.55′N	171°38.42′E	3875	¶
191	56°56.70′N	168°10.72′E	3860	¶
192	53°00.37′N	164°42.01′E	3014	70 m.y.

APPENDIX 2 (*contd.*)

Site	Latitude	Longitude	Depth(m)	Basement Age
		PACIFIC OCEAN		
*200	12°50.20′N	156°46.96′E	1479	¶
203	22°09.22′S	177°32.77′W	2720	¶
204	24°57.27′S	174°06.69′W	5354	¶
205	25°30.99′S	177°53.95′E	4320	†30 m.y.
206	32°00.75′S	165°27.15′E	3169	†70 m.y.
207	36°57.75′S	165°26.06′E	1389	†70 m.y.
208	26°06.61′S	161°13.27′E	1545	†70 m.y.
209	15°56.19′S	152°11.27′E	1428	¶
210	13°45.99′S	152°53.78′E	4643	*70 m.y.
277	15°13.43′S	166°11.48′E	1222	¶
278	56°33.42′S	160°04.29′E	3698	†30 m.y.
279	51°20.14′S	162°38.10′E	3371	†20 m.y.
280	48°57.44′S	147°14.08′E	4181	†53 m.y.
281	47°59.84′S	147°45.85′E	1591	†40 m.y.
283	43°54.60′S	154°16.96′E	4756	†65 m.y.
284	40°30.48′S	167°40.81′E	1068	¶
285	26°49.16′S	175°48.24′E	4658	¶
286	16°31.92′S	166°22.18′E	4465	†49 m.y.
287	13°54.67′S	153°15.93′E	4632	†52 m.y.
*288	05°58.35′S	161°49.53′E	3000	†110 m.y.
*289	00°29.92′S	158°30.69′E	2206	†110 m.y.
*291	12°48.43′N	127°49.85′E	5217	†44 m.y.
292	15°49.11′N	124°39.05′E	2943	†40 m.y.
294	22°34.74′N	131°23.13′E	5784	¶
296	29°20.41′N	133°31.52′E	2920	¶
297	30°52.36′N	134°09.89′E	4458	¶
303	40°48.50′N	154°27.07′E	5609	¶
304	39°20.27′N	155°04.19′E	5630	¶
*305	32°00.13′N	157°51.00′E	2903	*130 m.y.
307	28°35.26′N	161°00.28′E	5696	¶
310	36°52.11′N	176°54.09′E	3516	110 m.y.
311	28°07.46′N	179°44.25′E	5775	¶
*313	20°10.52′N	170°57.15′W	5484	†75 m.y.
*315	04°10.26′N	158°31.54′W	4164	†80 m.y.
*316	00°05.44′N	157°07.71′W	4464	†120 m.y.
*317	11°00.09′S	162°15.78′W	2625	†110 m.y.
*318	14°49.63′S	146°51.51′W	2659	†110 m.y.
*319	13°01.04′S	101°31.46′W	4296	†20 m.y.
*320	09°00.50′S	83°31.80′W	4487	†31 m.y.
321	12°01.29′S	81°54.24′W	4827	†42 m.y.

* Sites for which rates of sedimentation shown in Figs 6 and 7 were estimated.
† Author's assessment of basement age.
§ The palaeodepths assigned in the Cruise Report have been used.
¶ No adjustment has been made to the depth.

References

The reference list contains references cited in the text and references to the data (*Initial Reports of the Deep Sea Drilling Project* and *Geotimes* on which this investigation was based.

Adelseck, C. G., Jr. and Berger, W. H. (1974). On the dissolution of planktonic foraminifera and associated microfossils during settling on the sea floor. *Cushman Foundation for foraminiferal research, Spec. Publ.* **13,** 70–81.

Andrews, J. E. *et al.* (1973). Leg 30. Deep Sea Drilling Project Southwest Pacific Structures. *Geotimes* **18,** 18–21.

Arrhenius, G. (1952). Sediment cores from the east Pacific. *Rept. Swedish Deep Sea Exped.* (1947–1948) **5,** pts. 1–4, 1–228.

Bader, R. G. *et al.* (1969). *Initial Reports of the Deep Sea Drilling Project* **4,** U.S. Government Printing Office, Washington, 753 p.

Barker, P. F. *et al.* (1974). Southwestern Atlantic, Leg. 36. *Geotimes* **19,** 16–18.

Benson, R. H. (1973). The origin of the psycrosphere as recorded in changes of deep-sea ostracode assemblages. *Lethaia* **8,** 69–83, 69–83.

Berger, W. H. (1967). Foraminiferal ooze: Solution at depths. *Science* **156,** 383–385.

Berger, W. H. (1968). Planktonic foraminifera: selective solution and paleoclimatic interpretation. *Deep-Sea Res.* **15,** 31–43.

Berger, W. H. (1970a). Planktonic foraminifera: selective solution and the lysocline. *Marine Geol.* **8,** 111–138.

Berger, W. H. (1970b). Biogenous deep-sea sediments: fractionation by deep-sea circulation. *Bull. geol. Soc. Am.* **81,** 1385–1402.

Berger, W. W. (1971). Sedimentation of planktonic foraminifera. *Marine Geol.* **11,** 325–358.

Berger, W. H. (1972). Deep-sea carbonates: dissolution facies and age-depth constancy. *Nature* **236,** 392–395.

Berger, W. H. (1973a). Cenozoic sedimentation in the eastern tropical Pacific. *Bull. geol. Soc. Am.* **84,** 1941–1954.

Berger, W. H. (1973b). Deep-sea carbonates: Pleistocene dissolution cycles. *J. Foraminiferal Research* **3,** 187–195.

Berger, W. H. (1974). Deep-sea sedimentation. *In* (Burk, C. A. and Drake, C. L., eds) *The Geology of Continental Margins*, 213–235. Springer-Verlag, New York, Heidelberg, Berlin.

Berger, W. H. and Soutar, A. (1970). Preservation of plankton shells in an anaerobic basin off California. *Bull. geol. Soc. Am.* **81,** 275–282.

Berger, W. H. and Winterer, E. L. (1974). Plate Stratigraphy and the fluctuating carbonate line. *In* (Hsu, K. J. and Jenkyns, H. C. eds) *Pelagic sediments on land and under the sea. Spec. Publs. int. Ass. Sediment* **1,** 11–48.

Berger, W. H. and von Rad, U. (1972). Cretaceous and Cenozoic sediments from the Atlantic Ocean. *In* (Hayes, D. E., *et al.*) *Initial Reports of the Deep Sea Drilling Project* **14,** 787–984. U.S. Government Printing Office, Washington.

Berner, R. A. (1974). Physical chemistry of carbonates in the oceans. *In* (Hay, W. W., ed.) *Studies in Paleo-oceanography. Soc. Econ., Paleont. Mineral Spec. Publ.* **20,** 37–43.

Bolli, H. M. *et al.* (1975). Basins and Margins of the eastern South Atlantic, Leg 40. *Geotimes* **20,** 22–24.

Bramlette, M. N. (1961). Pelagic sediments. *In* (Sears, M., ed.) *Oceanography. Am. Assoc. Adv. Sci. Publ.* **67,** 345–366.

Broecker, W. S. (1971). Calcite accumulation rates and glacial to interglacial changes in oceanic mixing. *In* (Turekian, K. K., ed.) *The Late Cenozoic glacial ages*, 239–265. Yale University Press.

Broecker, W. S. and Broecker, S. (1974). Carbonate dissolution on the western flank of the East Pacific Rise. *In* (Hay, W. W., ed.) *Studies in Paleo-oceanography. Soc. Econ. Paleont. Mineral Spec. Publ.* **20,** 44–57.

Burns, R. E. *et al.* (1972). Glomar Challenger Down Under. Deep Sea Drilling Project Leg 21. *Geotimes* **17,** 14–16.

Cita, M. B. and Gartner, S. (1968). Biostratigraphy. *In* (Peterson *et al.*) *Initial Reports of the Deep Sea Drilling Projects* **2,** 391–411. U.S. Government Printing Office, Washington.

Chen, C. (1968). Occurrence of pteropods in pelagic sediments. *In* (Funnell, B. M. and Riedel, W. R., eds) *The Micropalaeontology of Oceans*, 351. Cambridge University Press.

Clague, D. A. and Harrard, R. D. (1973). Tertiary Pacific plate motion deduced from the Hawaiian–Emperor chain. *Bull. geol. Soc. Am.* **84,** 1135–1154.

Creager, J. S. *et al.* (1973). *Initial Reports of the Deep Sea Drilling Project* **19,** 913 pp. U.S. Government Printing Office, Washington.

Davies, T. A. *et al.* (1974). *Initial Reports of the Deep Sea Drilling Project* **26,** 1129 p. U.S. Government Printing Office, Washington.

Davies, T. A., Weser, O. E., Luyendyck, B. P. and Kidd, R. B. (1975). Unconformities in the sediments of the Indian Ocean. *Nature* **253,** 15–19.

Edgar, N. T., Holcomb, T., Ewing, J. L. and Johnson, W. (1973). Sedimentary hiatuses in the Venezuelan basin. *In* (Edgar, N. T. *et al.*) *Initial Reports of the Deep Sea Drilling Project* **15,** 1051–1062. U.S. Government Printing Office, Washington.

Ewing, J. and Ewing, M. (1967). Sediment distribution on the mid-ocean ridges with respect to spreading of the sea floor. *Science* **156,** 1591–1592.

Ewing, M. *et al.* (1968). *Initial Reports of the Deep Sea Drilling Project* **1,** 672 pp. U.S. Government Printing Office, Washington.

Ewing, J., Ewing, M., Aitken, T. and Ludwig, W. J. (1968). North Pacific sediment layers measured by seismic profiling. *Geophys. Monogr.* **12,** 147–173.

Fisher, A. G. *et al.* (1971). *Initial Reports of the Deep Sea Drilling Project* **6,** 1329 pp. U.S. Government Printing Office, Washington.

Fisher, R. L. *et al.* (1972). Deep Sea Drilling Project in dodo land. *Geotimes* **17,** 17–21.

Flint, R. F. (1971). *Glacial and Quaternary geology*, 892 pp. John Wiley, New York.

Francheteau, J., Harrison, C. G. A., Sclater, J. G. and Richards, M. L. (1970). Magnetization of Pacific seamounts: a preliminary polar curve for the northeastern Pacific. *J. geophys. Res.* **75,** 2035–2061.

Griffin, J. J., Windom, H. and Goldberg, E. D. (1968). The distribution of clay minerals in the world ocean. *Deap-Sea Res.* **17,** 245–266.

Hart, S. R. *et al.* (1974). Leg 34. Oceanic basalt and the Nazca Plate. *Geotimes* **19,** 20–24.

Hay, W. W. (1970). Calcium carbonate compensation. *In* (Bader, R. G., *et al.*) *Initial Reports of the Deep Sea Drilling Project* **4,** 669, 672. U.S. Government Printing Office, Washington.

Hays, J. D., Saito, T., Opdyke, N. and Burckle, L. H. (1969). Pliocene–Pleistocene sediments of the equatorial Pacific: their paleomagnetic, biostratigraphic and climatic record. *Bull. geol. Soc. Am.* **80,** 1481–1514.
Hays, J. D. *et al.* (1972). *Initial Reports of the Deep Sea Drilling Project* **9,** 1205 pp. U.S. Government Printing Office, Washington.
Hayes, D. E. *et al.* (1972). *Initial Reports of the Deep Sea Drilling Project* **14,** 975 pp. U.S. Government Printing Office, Washington.
Hayes, D. E. *et al.* (1973). Leg 28. Deep sea drilling in the southern ocean. *Geotimes* **18,** 19–24.
Heath, G. R. (1969). Mineralogy of Cenozoic deep-sea sediments from the Equatorial Pacific Ocean. *Bull. geol. Soc. Am.* **80,** 689–694.
Heath, G. R. and Culberson, C. (1970). Calcite: degree of saturation, rate of dissolution, and the compensation depth in the deep oceans. *Bull. geol. Soc. Am.* **81,** 3157–3160.
Heezen, B. and Hollister, C. D. (1971). *The Face of the Deep*, 659 pp. Oxford University Press.
Heezen, B. C. *et al.* (1973). *Initial Reports of the Deep Sea Drilling Project* **20,** 958 pp. U.S. Government Printing Office, Washington.
Hess, H. H. (1962). History of ocean basins. *In* (Engel, A. E. J., James, H. L. and Leonard, B. L., eds) *Petrologic studies: a volume in honour of A. F. Buddington*, 599–620. *Geol. Soc. America.*
Hollister, C. D. *et al.* (1970). *Initial Reports of the Deep Sea Drilling Project* **11,** 1077 pp. U.S. Government Printing Office, Washington.
Hollister, C. D. *et al.* (1974). Deep Drilling in the southeast Pacific basin. *Geotimes* **19,** 16–19.
Kennett, J. P. *et al.* (1972). Australian–Antarctic continental drift, palaeocirculation changes and Oligocene deep-sea erosion. *Nature* **239,** 51–55.
Kennett, J. P. *et al.* (1973). Deep sea Drilling in the roaring 40s. *Geotimes* **18,** 14–17.
Kennett, J. P. *et al.* (1974). Development of the circum-Antarctic current. *Science* **186,** 144–147.
Kulm, L. D. *et al.* (1973). *Initial Reports of the Deep Sea Drilling Project* **18,** 1077 pp. U.S. Government Printing Office, Washington.
Kummel, B. (1970). *History of the Earth.* 2nd edn, 707 pp. Freeman, San Francisco.
Larson, R. L. *et al.* (1973). Leg 32. Deep Sea Drilling Project. *Geotimes* **18,** 14–17.
Laughton, A. S. *et al.* (1972). *Initial Reports of the Deep Sea Drilling Project* **12.** 1243. U.S. Government Printing Office, Washington.
Li, Y. H., Takahashi, T. and Broecker, W. S. (1969). Degree of saturation of $CaCO_3$ in the oceans. *J. geophys. Res.* **74,** 5507–5525.
Lisitzin, A. P. (1971a). Distribution of siliceous microfossils in suspension and in bottom sediments. *In* (Funnell, B. M. and Riedel, W. R., eds) *The Micropalaeontology of Oceans*, 172–195. Cambridge University Press.
Luyendyck, B. P. and Davies, T. A. (1974). Results of DSDP Leg 26 and the geological history of the southern Indian Ocean. *In* (Davies, T. A. *et al.*) *Initial Reports of the Deep Sea Drilling Project* **26,** 909–943. U.S. Government Printing Office, Washington.
Maxwell, A. E. *et al.* (1970). *Initial Reports of the Deep Sea Drilling Project* **3,** 806 pp. U.S. Government Printing Office, Washington.
McIntyre, A. and McIntyre, R. (1971). Coccolith concentrations and differential

solution in oceanic sediments. *In* (Funnell, B. M. and Riedel, W. R., eds) *The Micropalaeontology of Oceans*, 253–261. Cambridge University Press.

McIntyre, A., Ruddiman, W. F. and Jantzen, R. (1972). Southward penetrations of the North Atlantic Polar Front: faunal and floral evidence of large-scale surface water mass movements over the last 250,000 years. *Deep-Sea Res.* **19,** 61–77.

McKenzie, D. P. and Sclater, J. G. (1971). The evolution of the Indian Ocean since the late Cretaceous. *Geophys. J. R. astr. Soc.* **25,** 437–528.

Lisitzin, A. P. (1971b). Distribution of carbonate microfossils in suspension and in bottom sediments. *In* (Funnell, B. M. and Riedel, W. R., eds.) *The Micropalaeontology of Oceans*, 197–218. Cambridge University Press.

McManus, D. A. *et al.* (1970). *Initial Reports of the Deep Sea Drilling Project* **5,** 827 pp. U.S. Government Printing Office, Washington.

Melson, W. G. *et al.* (1974). Deep sea Drilling Project, Leg. 37—the volcanic layer. *Geotimes* **19,** 16–18.

Menard, H. W. (1969). Elevation and subsidence of oceanic crust. *Earth Planet Sci. Lett.* **6,** 275–284.

Moberly, R. (1972). Origin of the lithosphere behind island arcs with reference to the western Pacific. *In* (Shagam, R., ed.) *Studies in earth and space sciences. Mem. geol. Soc. Am.* **132,** 288–291.

Morse, J. W. and Berner, R. A. (1972). Dissolution kinetics of calcium carbonate in sea water: II. A kinetic origin for the Lysocline. *Am. J. Sci.* **272,** 840–851.

Olausson, E. (1971). Quaternary correlations and the geochemistry of oozes. *In* (Funnell, B. M. and Riedel, W. R., eds.) *The Micropalaeontology of Oceans* 375–398. Cambridge University Press.

Perch-Nielsen, K. *et al.* (1975). Leg 39 examines facies changes in South Atlantic. *Geotimes* **20,** 26–28.

Peterson, M. N. *et al.* (1968). *Initial Reports of the Deep Sea Drilling Project* **2,** 501 pp. U.S. Government Printing Office, Washington.

Pimm, A. C. and Hayes, D. E. (1972). General Synthesis. *In* (Hayes, D. E. *et al.*) *Initial Reports of the Deep Sea Drilling Project* **14,** 955–975.

Pytkowicz, R. M. (1968). The carbon dioxide–carbonate system at high pressure in the oceans. *Ann. Rev. Oceanogr. Marine Biol.* **6,** 83–135.

Ramsay, A. T. S. (1972). Aspects of the distribution of fossil species of calcareous nannoplankton in North Atlantic and Caribbean sediments. *Nature* **236,** 67–70.

Ramsay, A. T. S. (1974). The distribution of calcium carbonate in deep sea sediments. *In* (Hay, W. W., ed.) *Studies in Paleo-oceanography. Soc. Econ., Paleont. Mineral Spec. Publ.* **20,** 58–76.

Ramsay, A. T. S., Schneidermann, N. and Finch, J. W. (1973). Fluctuations in the past rates of carbonate solution at Site 149: a comparison with other ocean basins and an interpretation of their significance. *In* (Edgar, N. T., *et al.*) *Initial Reports of the Deep Sea Drilling Project* **15,** 805–811.

Ross, D. A. *et al.* (1972). Deep sea Drilling Project in the Arabian Sea. *Geotimes* **17,** 22–216.

Savin, S. M., Douglas, R. C. and Stehli, F. G. (1975). Tertiary marine paleotemperatures. *Bull. geol. Soc. Am.* **86,** 1499–1510.

Schlanger, S. O. *et al.* (1974). Leg 33, Deep sea Drilling Project testing a hot-spot theory. *Geotimes* **19,** 16–20.

Sclater, J. G. and Jarrard, R. D. (1971). Preliminary paleomagnetic results. *In*

(Winterer, E. L. *et al.*) *Initial Reports of the Deep Sea Drilling Project* **7,** 1227–1234. U.S. Government Printing Office, Washington.
Sclater, J. G. and Fisher, R. L. (1974). The evolution of the east central Indian Ocean. *Bull. geol. Soc. Am.* **85,** 683–702.
Sclater, J. G. and Francheteau, J. (1970). The implications of terrestrial heat flow observations on current tectonic and geochemical models of the crust and upper mantle of the earth. *Geophys. J. R. astr. Soc.* **20,** 509–542.
Shackleton, N. J. and Kennett, J. P. (1975). Palaeotemperature history of the Cenozoic and the initiation of Antarctic glaciation: Oxygen and carbon isotope analyses in DSDP sites, 277, 279 and 281. *In* (Kennett, J. P. *et al.*) *Initial Reports of the Deep Sea Drilling Project* **29,** 743–755. U.S. Government Printing Office, Washington.
Talwani, M., *et al.* (1975). Leg 38. *Geotimes* **20,** 24–26.
Tracey, J. I. *et al.* (1971). *Initial Reports of the Deep Sea Drilling Project* **8,** 1037 pp. U.S. Government Printing Office, Washington.
Tucholke, B. *et al.* (1975). Glomar Challenger Drills in the North Atlantic. *Geotimes* **20,** 18–21.
Van Andel, T. H. and Moore, T. C. Jr. (1974). Cenozoic calcium carbonate distribution and calcite compensation depth in the central equatorial Pacific. *Geology* **2,** 87–92.
Van Andel, T. H. (1975). Mesozoic/Cenozoic calcite compensation depth and the global distribution of calcareous sediments. *Earth Planet Sci. Lett.* **26,** 187–194.
Van Andel, T. H. *et al.* (1973). *Initial Reports of the Deep Sea Drilling Project* **16,** 949 pp. U.S. Government Printing Office, Washington.
Veevers, J. (1969). Palaeogeography of the Timor Sea region. *Palaeogeog. Palaeoclim. Palaeoecol.* **6,** 125–140.
Veevers, J. *et al.* (1974). *Initial Reports of the Deep Sea Drilling Project* **27,** 1060 pp. U.S. Government Printing Office, Washington.
Von der Borch, C. C. *et al.* (1974). *Initial Reports of the Deep Sea Drilling Project* **22,** 890 pp. U.S. Government Printing Office, Washington.
Winterer, E. L. *et al.* (1971). *Initial Reports of the Deep Sea Drilling Project* **7,** 2 vols, 1740 pp. U.S. Government Printing Office, Washington.
Winterer, E. L. (1973a). Sedimentary facies and plate tectonics of the equatorial Pacific. *Bull. Am. Ass. Petrol. Geol.* **57,** 265–282.
Winterer, E. L. *et al.* (1973b). *Initial Reports of the Deep Sea Drilling Project* **17,** 930 pp. U.S. Government Printing Office, Washington.

General Index

Taxonomic Index